ANALYSIS AND DESIGN OF MARINE STRUCTURES

T0227542

PROCEEDINGS OF MARSTRUCT 2009, THE 2nd INTERNATIONAL CONFERENCE ON MARINE STRUCTURES, LISBON, PORTUGAL, 16–18 MARCH 2009

Analysis and Design of Marine Structures

Editors

C. Guedes Soares
Instituto Superior Técnico, Technical University of Lisbon, Portugal

P.K. Das
Universities of Glasgow and Strathclyde, UK

CRC Press
Taylor & Francis Group
Boca Raton   London   New York   Leiden

CRC Press is an imprint of the
Taylor & Francis Group, an **informa** business

A BALKEMA BOOK

Cover photograph: Bulk carrier INA from Portline, Portugal

MARSTRUCT Book Series

1. Advancements in Marine Structures (2007)
Edited by Carlos Guedes Soares & P.K. Das
ISBN: 978-0-415-43725-7 (hb)

2. Analysis and Design of Marine Structures (2009)
Edited by Carlos Guedes Soares & P.K. Das
ISBN: 978-0-415-54934-9 (hb)
ISBN: 978-0-203-87498-1 (e-book)

First issued in paperback 2017

CRC Press/Balkema is an imprint of the Taylor & Francis Group, an informa business

©2009 Taylor & Francis Group, London, UK

'The importance of welding quality in ship construction' by: Philippa Moore
© 2009 TWI Ltd.

Typeset by Vikatan Publishing Solutions (P) Ltd., Chennai, India

Published by: CRC Press/Balkema
P.O. Box 447, 2300 AK Leiden, The Netherlands
e-mail: Pub.NL@taylorandfrancis.com
www.crcpress.com – www.taylorandfrancis.co.uk – www.balkema.nl

ISBN 13: 978-1-138-11647-4 (pbk)
ISBN 13: 978-0-415-54934-9 (hbk)

Table of Contents

Preface   IX

Organisation   XI

Methods and tools for loads and load effects

A study on the effect of heavy weather avoidance on the wave pressure distribution along the
midship transverse section of a VLCC and a bulk carrier   3
Zhi Shu & Torgeir Moan

Comparison of experimental and numerical sloshing loads in partially filled tanks   13
S. Brizzolara, L. Savio, M. Viviani, Y. Chen, P. Temarel, N. Couty, S. Hoflack, L. Diebold,
N. Moirod & A. Souto Iglesias

Experiments on a damaged ship section   27
T.W.P. Smith, K.R. Drake & P. Wrobel

Estimation of parametric rolling of ships—comparison of different probabilistic methods   37
Jelena Vidic-Perunovic & Jørgen Juncher Jensen

Local hydro-structure interactions due to slamming   45
Š. Malenica, F.X. Sireta, S. Tomašević, J.T. Tuitman & I. Schipperen

Methods and tools for strength assessment

Finite element analysis

Methods for hull structure strength analysis and ships service life evaluation, for a large LNG carrier   53
Leonard Domnisoru, Ionel Chirica & Alexandru Ioan

Parametric investigation on stress concentrations of bulk carrier hatch corners   67
Dario Boote & Francesco Cecchini

A study on structural characteristics of the ring-stiffened circular toroidal shells   77
Qing-Hai Du, Zheng-Quan Wan & Wei-Cheng Cui

Application developments of mixed finite element method for fluid-structure interaction
analysis in maritime engineering   83
Jing Tang Xing, Ye Ping Xiong & Mingyi Tan

Efficient calculation of the effect of water on ship vibration   93
Marc Wilken, G. Of, C. Cabos & O. Steinback

Finite element simulations of ship collisions: A coupled approach to external dynamics
and inner mechanics   103
Ingmar Pill & Kristjan Tabri

Ultimate strength

Discussion of plastic capacity of plating subject to patch loads   113
Claude Daley & Apurv Bansal

Ultimate strength characteristics of aluminium plates for high speed vessels   121
S. Benson, J. Downes & R.S. Dow

Improving the shear properties of web-core sandwich structures using filling material   133
Jani Romanoff, Aleksi Laakso & Petri Varsta

Stability of flat bar stiffeners under lateral patch loads   139
Jacob Abraham & Claude Daley

Ultimate strength of stiffened plates with local damage on the stiffener   145
M. Witkowska & C. Guedes Soares

Approximate method for evaluation of stress-strain relationship for stiffened
panel subject to tension, compression and shear employing the finite element approach   155
Maciej Taczala

Residual strength of damaged stiffened panel on double bottom ship   163
Zhenhui Liu & Jørgen Amdahl

Assessment of the hull girder ultimate strength of a bulk carrier using nonlinear
finite element analysis   173
Zhi Shu & Torgeir Moan

Ultimate strength performance of Suezmax tanker structures: Pre-CSR versus CSR designs   181
J.K. Paik, D.K. Kim & M.S. Kim

Coatings and corrosion

Large scale corrosion tests   193
Pawel Domzalicki, Igor Skalski, C. Guedes Soares & Yordan Garbatov

Anticorrosion protection systems—improvements and continued problems   199
Anders Ulfvarson & Klas Vikgren

Prospects of application of plasma electrolytic oxidation coatings for shipbuilding   207
Alexander N. Minaev, Natalie A. Gladkova, Sergey V. Gnedenkov & Vladimir V. Goriaynov

Corrosion wastage statistics and maintenance planning of corroded hull structures of bulk carriers   215
Yordan Garbatov & C. Guedes Soares

Numerical simulation of strength and deformability of steel plates with surface pits
and replicated corrosion-surface   223
Md. Mobesher Ahmmad & Yoichi Sumi

Effect of pitting corrosion on the collapse strength of rectangular plates under axial compression   231
S. Saad-Eldeen & C. Guedes Soares

Fatigue and fracture

Fracture mechanics procedures for assessing fatigue life of window and door corners in ship structures   239
Mika Bäckström & Seppo Kivimaa

Experimental and numerical fatigue analysis of partial-load and full-load carrying fillet welds
at doubler plates and lap joints   247
O. Feltz & W. Fricke

Global strength analysis of ships with special focus on fatigue of hatch corners   255
Hubertus von Selle, Olaf Doerk & Manfred Scharrer

Structural integrity monitoring index for ship and offshore structures   261
Bart de Leeuw & Feargal P. Brennan

Effect of uncertain weld shape on the structural hot-spot stress distribution   267
B. Gaspar, Y. Garbatov & C. Guedes Soares

A study on a method for maintenance of ship structures considering remaining life benefit   279
Yasumi Kawamura, Yoichi Sumi & Masanobu Nishimoto

Impact strength

Impact behaviour of GRP, aluminium and steel plates   293
L.S. Sutherland & C. Guedes Soares

Impact damage of MARK III type LNG carrier cargo containment system due
to dropped objects: An experimental study   301
J.K. Paik, B.J. Kim, T.H. Kim, M.K. Ha, Y.S. Suh & S.E. Chun

Simulation of the response of double bottoms under grounding actions using finite elements   305
I. Zilakos, M. Toulios, M. Samuelides, T.-H. Nguyen & J. Amdahl

Fire and explosion

CFD simulations on gas explosion and fire actions   315
J.K. Paik, B.J. Kim, J.S. Jeong, S.H. Kim, Y.S. Jang, G.S. Kim, J.H. Woo, Y.S. Kim,
M.J. Chun, Y.S. Shin & J. Czujko

The effects of reliability-based vulnerability requirements on blast-loaded ship panels   323
S.J. Pahos & P.K. Das

Structural monitoring

Structural monitoring of mast and rigging of sail ships   333
Giovanni Carrera, Cesare Mario Rizzo & Matteo Paci

Assessment of ice-induced loads on ship hulls based on continuous response monitoring   345
B.J. Leira, Lars Børsheim, Øivind Espeland & J. Amdahl

Materials and fabrication of structures

Welded structures

The importance of welding quality in ship construction   357
Philippa L. Moore

A data mining analysis to evaluate the additional workloads caused by welding distortions   365
Nicolas Losseau, Jean David Caprace, Philippe Rigo & Fernandez Francisco Aracil

3D numerical model of austenitic stainless steel 316L multipass butt welding and comparison
with experimental results   371
A.P. Kyriakongonas & V.J. Papazoglou

Adhesive joints

Fabrication, testing and analysis of steel/composite DLS adhesive joints   379
S. Hashim, J. Nisar, N. Tsouvalis, K. Anyfantis, P. Moore, Ionel Chirica, C. Berggreen, A. Orsolini,
A. Quispitupa, D. McGeorge, B. Hayman, S. Boyd, K. Misirlis, J. Downes, R. Dow & E. Juin

The effect of surface preparation on the behaviour of double strap adhesive joints with
thick steel adherents   387
K.N. Anyfantis & N.G. Tsouvalis

Pultrusion characterisation for adhesive joints   393
J.A. Nisar, S.A. Hashim & P.K. Das

Buckling of composite plates

Studies of the buckling of composite plates in compression   403
B. Hayman, C. Berggreen, C. Lundsgaard-Larsen, A. Delarche, H.L. Toftegaard, R.S. Dow,
J. Downes, K. Misirlis, N. Tsouvalis & C. Douka

Buckling strength parametric study of composite laminated plates with delaminations   413
N.G. Tsouvalis & G.S. Garganidis

Buckling behaviour of the ship deck composite plates with cut-outs   423
Ionel Chirica, Elena-Felicia Beznea & Raluca Chirica

Buckling behaviour of plates with central elliptical delamination   429
Elena-Felicia Beznea, Ionel Chirica & Raluca Chirica

Methods and tools for structural design and optimization

Structural design of a medium size passenger vessel with low wake wash   437
Dario Boote & Donatella Mascia

Multi-objective optimization of ship structures: Using guided search vs.
conventional concurrent optimization   447
Jasmin Jelovica & Alan Klanac

Digital prototyping of hull structures in basic design   457
José M. Varela, Manuel Ventura & C. Guedes Soares

Structural reliability safety and environmental protection

Still water loads

Probabilistic presentation of the total bending moments of FPSO's   469
Lyuben D. Ivanov, Albert Ku, Beiqing Huang & Viviane C.S. Krzonkala

Stochastic model of the still water bending moment of oil tankers   483
L. Garrè & Enrico Rizzuto

Statistics of still water bending moments on double hull tankers   495
Joško Parunov, Maro Ćorak & C. Guedes Soares

Ship structural reliability

Structural reliability of the ultimate hull girder strength of a PANAMAX container ship   503
Jörg Peschmann, Clemens Schiff & Viktor Wolf

Sensitivity analysis of the ultimate limit state variables for a tanker and a bulk carrier   513
A.W. Hussein & C. Guedes Soares

Ultimate strength and reliability assessment of laminated composite plates under axial compression   523
N. Yang, P.K. Das & Xiong Liang Yao

Environmental impact

Modelling of environmental impacts of ship dismantling   533
I.S. Carvalho, P. Antão & C. Guedes Soares

Fuel consumption and exhaust emissions reduction by dynamic propeller pitch control   543
Massimo Figari & C. Guedes Soares

Author index   551

Preface

This book collects the papers presented at the second International Conference on Marine Structures, MARSTRUCT 2009, which was held in Lisbon 16 to 18 March. This Conference follows up from the initial one that was held in Glasgow, Scotland two years before and aims at bringing together researchers and industrial participants specially concerned with structural analysis and design. Despite the availability of several conferences, it was felt that there was still no conference series specially dedicated to marine structures, which would be the niche for these conferences.

The initial impetus and support has been given by the Network of Excellence on Marine Structures (MARSTRUCT), which is now in its 6th year of funding by the European Union and brings together 33 European research groups from Universities, research institutions, classification societies and industrial companies that are dedicated to research in the area of marine structures. However this Conference is not meant to be restricted to European attendees and a serious effort has been made to involve in the planning of the Conference participants from other continents that could ensure a wider participation, which is slowly happening.

The conference reflects the work conducted in the analysis and design of marine structures, in order to explore the full range of methods and modelling procedures for the structural assessment of marine structures. Various assessment methods are incorporated in the methods used to analyze and design efficient ship structures, as well as in the methods of structural reliability to be used to ensure the safety and environmental behaviour of the ships. This book deals also with some aspects of fabrication of ship structures.

The 60 papers are categorised into the following themes:

- Methods and tools for establishing loads and load effects
- Methods and tools for strength assessment
- Materials and fabrication of structures
- Methods and tools for structural design and optimisation
- Structural reliability, safety and environmental protection.

The papers were accepted after a review process, based on the full text of the papers. Thanks are due to the Technical Programme Committee and to the Advisory Committee who had most of the responsibility for reviewing the papers and to the additional anonymous reviewers who helped the authors deliver better papers by providing them with constructive comments. We hope that this process contributed to a consistently good level of the papers included in the book.

Carlos Guedes Soares
Purnendu Das

Analysis and Design of Marine Structures – Guedes Soares & Das (eds)
© 2009 Taylor & Francis Group, London, ISBN 978-0-415-54934-9

Organisation

Conference Chairmen

Prof. Carlos Guedes Soares, IST, Technical University of Lisbon, Portugal
Prof. Purnendu K. Das, Universities of Glasgow & Strathclyde, UK

Technical Programme Committee

Prof. N. Barltrop, University of Glasgow & Strathclyde, UK
Dr. N. Besnard, Principia Marine, France
Dr. M. Codda, CETENA, Italy
Prof. L. Domnisoru, UGAL, Romania
Prof. R.S. Dow, University of Newcastle upon Tyne, UK
Prof. W. Fricke, TUHH, Germany
Prof. Y. Garbatov, IST, Technical University of Lisbon, Portugal
Prof. J.M. Gordo, IST, Technical University of Lisbon, Portugal
Dr. B. Hayman, DNV, Norway
Prof. A. Incecik, University of Newcastle upon Tyne, UK
Prof. T. Jastrzebski, TUS, Poland
Prof. J.J. Jensen, DTU, Denmark
Prof. B.J. Leira, NTNU, Norway
Prof. V. Papazoglou, NTUA, Greece
Prof. P. Rigo, University of Liège, Belgium
Prof. E. Rizzuto, University of Genova, Italy
Prof. R.A. Shenoi, University of Southampton, UK
Prof. P. Temarel, University of Southampton, UK
Prof. A. Ulfvarson, Chalmers University of Tech., Sweden
Prof. P. Varsta, Helsinki University of Technology, Finland
Dr. A. Vredeveldt, TNO, The Netherlands

Advisory Committee

Dr. R.I. Basu, ABS, USA
Prof. F. Brennan, Cranfield University, UK
Dr. F. Cheng, Lloyd's Register, UK
Prof. Y.S. Choo, Nat. Univ. Singapore, Singapore
Prof. W.C. Cui, CSSRC, China
Prof. C. Daley, Memorial University, Canada
Prof. A. Ergin, ITU, Turkey
Prof. S. Estefen, COPPE, Brazil
Prof. M. Fujikubo, Osaka University, Japan
Prof. D. Karr, University of Michigan, USA
Prof. H.W. Leheta, Alexandria University, Egypt
Dr. O. Valle Molina, Mexican Inst. of Petroleum, Mexico
Prof. J.K. Paik, Pusan National University, Korea
Prof. P.T. Pedersen, DTU, Denmark
Dr. N.G. Pegg, DND, Canada
Prof. M. Salas, University Austral of Chile, Chile

Prof. Y. Sumi, Yokohama National University, Japan
Prof. V. Zanic, University of Zagreb, Croatia

Conference Secretariat

Ana Rosa Fragoso, IST, Technical University of Lisbon, Portugal
Maria de Fátima Pina, IST, Technical University of Lisbon, Portugal
Sandra Ponce, IST, Technical University of Lisbon, Portugal

Methods and tools for loads and load effects

Analysis and Design of Marine Structures – Guedes Soares & Das (eds)
© 2009 Taylor & Francis Group, London, ISBN 978-0-415-54934-9

A study on the effect of heavy weather avoidance on the wave pressure distribution along the midship transverse section of a VLCC and a bulk carrier

Z. Shu & T. Moan
CeSOS and Department of Marine Technology
Norwegian University of Science and Technology, Trondheim, Norway

ABSTRACT: This paper is concerned with the effect of heavy weather avoidance on the long-term wave-induced pressure along the midship transverse section of a VLCC and a bulk carrier. A practical model is proposed to consider the effect of the heavy weather avoidance on the wave pressure along the midship transverse section. This model is based on the modification of the wave scatter diagrams according to the operational limiting criteria. The wave pressure distributions of a VLCC and a Bulk carrier are investigated in the case study using the computer code WASIM with linear analysis option. Two different wave scatter diagrams from IACS and OCEANOR were considered for the long-term predictions of wave pressures. The results show that the influence of heavy weather avoidance on the extreme values of wave pressure along the midship transverse section is dependent on how the heavy weather avoidance is accounted for.

1 INSTRUCTIONS

Wave induced loads are very important for the safety of seagoing vessels. From the design point of view, the wave induced loads can generally be categorized into two classes, (1) global loads, i.e. hull girder bending moments and shear forces; (2) local loads, i.e. wave pressures.

The main external loads on the seagoing vessels are the wave induced hydrodynamic pressures. With the hydrodynamic pressures, the motions and global hull girder loads in waves can be directly obtained by integrating the hydrodynamic pressure and inertia forces over the ship hulls. In order to evaluate the structural strength of the vessels under the wave induced loads, a finite element analysis of the midship region or the entire ship model is generally required in the direct calculation. The accuracy of structural response analysis is highly depends on prediction of extreme values of the wave induced hydrodynamic pressure distributions, acceleration and the hull girder bending moments.

The extreme values of wave induced loads can be obtained either from the ship rules or direct calculations. The ship rules express the loads by simplified formulas obtained based on experience with different kinds of vessels. However, these formulas are only functions of the ship's length, breath, and block coefficients without explicit consideration of the effect of heavy weather avoidance such as the route, scatter diagram, operational profiles.

In reality, ship masters are expected to operate the ship to avoid heavy sea states by adopting speed reduction, course change and even fully stop in order to decrease the hull damage and ease the motions. This means that the probability of the severe wave climate encountered by the ships will be reduced to some extent compared with the planned route. Hence the wave induced loads are expected to be reduced as a direct consequence of the vessel operation. Since sea states with large significant wave height contribute most to the extreme loading, it is particularly important to consider the effect of avoiding those sea states on the extreme loading. Some researchers have investigated the effect of heavy weather avoidance on the global loads such as wave induced vertical bending moment. Guedes Soares (1990) performed a study on the effect of heavy weather maneuvering on the wave-induced vertical bending moments by assuming non-uniform distribution heading under different sea states. Shu & Moan (2008) proposed a practical model to consider the effect of the heavy weather avoidance on the wave induced hull girder loads. This model is based on the modification of the wave scatter diagrams according to the operational limiting criteria. Various wave scatter diagrams exist for the long-term prediction of extreme wave induced loads. Two possibilities are wave scatter diagrams from IACS (2000) and OCEANOR (2005). The IACS wave scatter diagram describes the wave data of the North Atlantic, covering the area as defined in the Global Wave Statistics (GWS) (1980) with more realistic considerations of the wave steepness. Moreover, the IACS data have been smoothed by fitting an analytical probability density function to the raw data. OCEANOR data are hindcast data of the North Atlantic in the last 12 years from 1994 to 2005 based on satellite data and calibrations to wave buoy measurements. Compared with the wave scatter diagram from IACS, it should be noted that OCEANOR wave data were not obtained by the on-board observations and measurements. Hence for sure, OCEANOR data do not reflect any influence of heavy weather avoidance. Therefore these data have more probability in the tail of the marginal distribution of the significant wave height.

In this paper, this model is adopted to evaluate the effect of heavy weather avoidance on the wave pressure along the midship transverse section. The wave pressure distributions along the midship transverse sections of a VLCC and a Bulk carrier are investigated in the case study. The long term predictions of wave pressure distributions along the midship transverse section will change depending on the heavy sea states that are avoided by modifying the wave scatter diagram. The characteristic values corresponding to exceedance probability of 10^{-8} from direct calculation are compared with those from rule simplified formulae.

2 HEAVY WEATHER AVOIDANCE

When sailing on a seaway, the shipmasters will in general try to avoid severe sea states by adopting actions such as reducing speed, changing course or both of the formers according to certain limiting operational criteria relating to the safety and comfort of passengers and crew, to the safety and capacity of the vessel or to operational considerations. Operational restrictions for high speed and light craft (HSLC) on coastal voyages of short duration with short distance to shelter can be fulfilled based on relevant sea state forecasts. For voyages of long duration without the opportunity to seek shelter, fulfillment of operational restrictions depends upon additional information about on-board monitoring of responses and sea state forecast as well as time to carry out the necessary change of operation: speed reduction and change of heading or generally a combination of both. These actions will generally be based on the observations and measurements of the waves and vessels responses. On the other hand with the development of forecast technology of wave climate, most sea-going vessels have installed facilities to receive weather data onboard. The shipmaster will have an advanced knowledge of what kind of sea states are ahead on the planned route and what kind of actions should be adopted to avoid running directly into the severe sea states according to the corresponding operational restrictions. The success of these actions depends on how fast the sea

state changes and the time available. The operational criteria can be expressed in terms of allowable significant wave height Hs. However when and how these actions should be taken is also greatly dependent on the shipmasters experiences.

It has been reported by Sternsson (2002) that avoidance of heavy weather can significantly reduce the probability of encountering large waves according to on-board measurements. However so far, this kind of on-board measured data usually have only covered a very short period such as 1–2 years or even less. For this reason it can not be directly used to perform the long term prediction to investigate the effect of heavy weather avoidance on wave-induced loads.

Systematic studies of bulk carriers, tankers and containerships have been performed by ISSC Committee V.1 (Moan et al. 2006) and it is found that the operational envelopes can be transformed into operational criteria on significant wave height Hs vs ship length. Therefore, a simplified estimate of the effect of heavy weather avoidance can be obtained by identifying the operational limiting boundaries which for simplicity may be expressed by Hs $\leq$ H$_{slim}$. The limiting significant wave height is approximated in this study by a constant H$_{slim}$, independent of Tp. If severe sea states are avoided by ships, the occurrences of actual sea states encountered by ships during its service life must be different from those given by the scatter diagrams for the geographical area in which the ship is operating. This especially must be true for high sea states. It is also known that high sea states usually contribute most to the long term prediction values. Moe et al. (2005) had compared the wave observations from a bulk carrier (Lpp = 285 m) with meteorological data and found that the probability of encountering waves with Hs > 6 m was less for the ship observations which indicates the effect of avoidance of heavy weather on the actual encountered sea states. Despite the uncertainty in these observations and comparisons, they support the assumption that the effect of operational restrictions can be modelled by modifying the original wave scatter diagram according to the limiting significant wave height.

The operability limiting boundaries are obtained from short term statistics combined with seakeeping criteria. The limiting curves can be presented as the envelope curves which are functions of the limiting significant wave height H$_{slim}$ and corresponding peak period Tp.

Shu & Moan (2008) proposed a practical model to consider the effect of the heavy weather avoidance on the wave induced hull girder loads. This model is based on the modification of the wave scatter diagrams according to the limiting significant wave height determined by operational limiting criteria. Three simplified methods as shown in Table 1 are used to carry out this kind of modification of wave scatter diagram.

Table 1.	Methods about modification of wave scatter diagram.
	Descriptions of the method
Method 1	truncate the wave scatter diagram at H_{slim} and add the probability of the truncated area directly to the areas just below H_{slim}.
Method 2	truncate the wave scatter diagram at H_{slim} and add the probability of the truncated area uniformly to all the areas below H_{slim}.
Method 3	use a reduction factor, α less than 1.0, for the tail above H_{slim} and an inflation factor β greater than 1.0 for the part below H_{slim}.

Table 2. Main particulars of vessels.

	VLCC	Bulk carrier
Displacement (ton)	342400	245000
Length (m)	320	294
Breadth (m)	58	53
Draught (m)	22.4	18.62

Method 1 can be comparable to a situation where the shipmasters avoid the heavy weather by maneuvering the vessel into a just calmer sea state. If the shipmasters have some knowledge of the wave climate ahead under the help of wave climate forecast, he will be sure to take some actions in advance to avoid running directly into those severe sea states. Then an alternative way is to add the probability of truncated area uniformly to the all the areas below the limiting significant wave height H_{slim}. This approach is denoted by Method 2. However, due to the uncertainty of the wave climate forecasts and the decision of the shipmaster, the severe sea sates with significant wave height greater than H_{slim} can not be perfectly avoided in reality. This is supported by the wave data measured onboard (Olsen et al. 2006). Method 3 accounts for this situation in a simplified manner by using a reduction factor, α less than 1.0, for the tail above H_{slim} and an inflation factor β greater than 1.0 for the part below the limiting significant wave height. After these modifications, the total probability should still be unit.

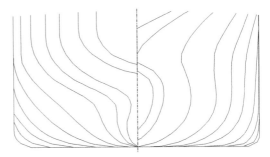

Figure 1. The body plan of a VLCC.

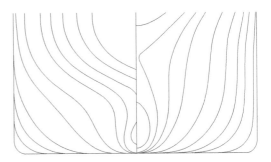

Figure 2. The body plan of a Capesize bulk carrier.

3 CASE STUDY

3.1 *Vessel particulars*

In this case study, two representative vessels, including a VLCC and a bulk carrier are considered to study the effect of avoidance of heavy weather on the wave induced pressure along the midship transverse section. The main dimensions of the relevant vessels are listed in Table 2 The body plans are shown in Figure 1 and Figure 2.

3.2 *Hydrodynamic analysis*

The linear version of WASIM, DNV (2006) is used to calculate wave induce pressures with zero speed.

WASIM is a 3D time-domain hydrodynamic code base on Rankine panel method. The theory core of WASIM is based on the method developed by Kring et al. (1998). The wave induced pressures are calculated at 24 locations as shown in Figure 3 from p13 to p36 along the midship transverse sections of both the VLCC and the bulk carrier. Although it is usually accepted that the heave and pitch motions can be reproduced very well by potential theory, the roll motion predicted by potential theory considering only wave making damping is questionable. The wave pressure distributions along the midship transverse section are consequently subjected to great uncertainty due to the questionable prediction of roll motion in beam sea. A sensitivity analysis was carried out to assess the influence of the roll damping on the wave induced pressures. Four damping levels in roll as a percentage of the critical damping are considered for the hydrodynamic calculation by WASIM, namely 2.5%, 5%, 7.5% and 10%, respectively (Ferrari & Ferreira, 2002).

The damping coefficient in roll motion is crucial especially for the roll motion around the resonance frequency.

Figure 3. The pressure locations along the midship cross section.

3.3 Long term prediction considering heavy weather avoidance

The characteristic extreme value of the wave pressure can be conveniently obtained by a linear long-term response analysis. The long-term predictions is performed according to the procedure recommended by IACS (2000). The procedure is generally based on the North Atlantic wave condition, uniform distribution of mean wave heading, Pierson-Moskowitz wave spectrum, and short-crested waves with spreading function $\cos^2 \theta$. The speed is assumed to be zero since the ship masters are expected to just keep a very slow speed for steering or even fully stop in order to decrease the hull damage and ease the motions in severe sea states.

Two wave scatter diagrams for the North Atlantic are considered for the long term prediction in this study. They are those provided by IACS and OCEANOR. Wave scatter diagram from IACS is issued by classification society and recommended for evaluation of extreme loads of hull girder. OCEANOR data are hindcast data of the last 12 year from 1994–2005 and have more probability in the upper tail with Hs > 14 m compared to the classification society scatter diagrams. It is noted that the IACS procedure does not describe explicitly that heavy weather avoidance should be accounted for. However, since the IACS scatter diagram is mainly based on onboard observations and measurements, they could have implicitly included the effect of heavy weather avoidance compared with the OCEANOR data which describe the real wave condition on sea.

In the long term analysis of the wave pressures, these wave scatter diagrams are modified at four limiting significant wave heights, i.e., 8 m, 10 m, 12 m and 14 m, Shu & Moan (2008), to account for heavy weather avoidance.

4 RESULTS AND DISCUSSIONS

4.1 Comparison of wave pressure between direct calculation and Common Structural Rule

The comparison of the wave pressure distribution envelop along the midship transverse sections obtained by simplified rule formulas and that obtained by long

term prediction with different roll damping for the full load condition of a VLCC and a bulk carrier at exceedance probability level of 10^{-8} are shown in Figure 4 and Figure 5, respectively. The calculations of wave pressure are different between Common Structural Rules (CSR) for tankers (IACS 2006a) and Common Structural Rules for bulk carriers (IACS 2006b). In the CSR, simplified formulas are given to calculate the dynamic wave pressure. According to CSR for tankers, the wave pressure is taken as the greater one resulting from simplified formulae p_1 and p_2 denoted as CSR p_1 and CSR p_2 in Figure 4. p_2 is dominant in the midship region. In CSR for bulk carriers, the wave pressure is assigned the maximum value from the simplified formulas for load case **F**, **H**, **R** and **P** where **F** represents load case with equivalent design wave in following sea, **H** head sea, **R** beam sea with maximum roll motion and **P** beam sea with maximum wave pressure at the waterline on the weather side, respectively. The detailed simplified formulas for dynamic wave pressure can be found in the CSR. The nonlinear effects in large waves are not considered in the long termprediction based on linear hydrodynamic analysis. Therefore, the nonlinear correction factors

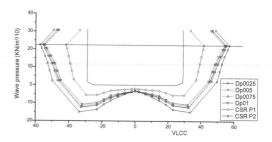

Figure 4. Wave pressure distributions of a VLCC in the full load condition (IACS wave data). Dp0025, Dp005, Dp0075 and Dp01 represent 2.5%, 5%, 7.5% and 10% of critical damping, respectively.

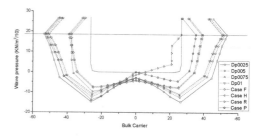

Figure 5. Wave pressure distributions of a bulk carrier in the full load condition (IACS wave data). Dp0025, Dp005, Dp0075 and Dp01 represent 2.5%, 5%, 7.5% and 10% of critical damping, respectively.

Table 3. Comparison of wave pressure distributions between those obtained by long term prediction and simplified formulas (IACS 2006a) with different roll damping for typical locations along the midship transverse section for a VLCC with IACS wave data.

Location along the cross section	Wave pressure			
	$\dfrac{P_{2.5\%*}}{P_{rule}}$	$\dfrac{P_{5\%*}}{P_{rule}}$	$\dfrac{P_{7.5\%*}}{P_{rule}}$	$\dfrac{P_{10\%*}}{P_{rule}}$
P13 (P36)	1.07	0.99	0.97	0.95
P15 (P34)	1.11	0.99	0.94	0.91
P18 (P31)	1.20	1.01	0.94	0.89
P21 (P28)	1.50	1.29	1.21	1.15
P24 (P25)	1.06	1.04	1.03	1.02

* Percentage of critical damping.

Table 4. Comparison of wave pressure distributions between those obtained by long term prediction and simplified formulas (IACS 2006b) with different roll damping for typical locations along the midship transverse section for a bulk carrier with IACS wave data.

Location along the cross section	Wave pressure			
	$\dfrac{P_{2.5\%*}}{P_{rule}}$	$\dfrac{P_{5\%*}}{P_{rule}}$	$\dfrac{P_{7.5\%*}}{P_{rule}}$	$\dfrac{P_{10\%*}}{P_{rule}}$
P13 (P36)	1.04	0.97	0.93	0.91
P15 (P34)	1.08	0.97	0.91	0.88
P18 (P31)	1.07	0.90	0.84	0.79
P21 (P28)	1.01	0.87	0.82	0.78
P24 (P25)	0.74	0.73	0.73	0.72

* Percentage of critical damping.

used in the rule simplified formulas are not included. The envelop of the dynamic wave pressures represents the maximum dynamic wave pressure distribution. The wave pressure is proportional to the distance from the envelop to the corresponding point on the hull.

From Figure 4, it is observed that the long term predictions with roll damping of 5% agree well with the values from p_2 according to CSR-tanker for this VLCC. For the bulk carrier, it is seen from Figure 5 that load case P is dominant for the wave pressure at the midship transverse section except the area near the centre bottom which is dominant by load case H according to CSR-bulk carrier and the long term predictions with roll damping of 5% seems to agree well with those obtained by simplified formulas. The comparisons of the wave pressure between the simplified formulae and those by long term predictions at 5 typical locations with different roll dampings are summarized in Table 3 and Table 4 corresponding to a VLCC and a bulk carrier, respectively. From Table 3 and Table 4, it can be seen that the roll damping has a significant influence on the wave pressures especially at area around the bilge keel while the wave pressure

at the centre bottom is almost not affected by the roll damping.

4.2 Extreme wave pressure considering the effect of heavy weather avoidance

The heavy weather avoidance is considered by modifying the wave scatter diagram according to the limiting significant wave height. It has been seen that the roll damping is of importance for the wave pressures along the midship transverse section and the determination of the accurate roll damping is subjected to a large uncertainty. For illustrative purpose, the long term predictions of wave pressure with consideration of heavy weather avoidance are based on the hydrodynamic analysis with 5% of linear critical roll damping for both the VLCC and the bulk carrier in the present study. Figure 6 and Figure 7 show the wave pressures considering heavy weather at typical locations along the midship transverse sections of the VLCC and the bulk carrier, respectively. In the legend of Figures 6 and 7, **original** represents no consideration of heavy weather avoidance. **add**, **cut** and **alfa010** represent consideration of heavy weather avoidance by method 1, method 2 and method 3, respectively. From Figure 6, it is noted that if the heavy weather avoidance is not considered, the direct calculations using the 3D program with IACS wave scatter diagram agree very well with the rule values for the VLCC at various typical locations except P21 along the midship transverse sections. P21 is about in the middle of the bilge keel and centre bottom. At this location, the wave pressure obtained by long term prediction is 29% larger than that by simplified formula. For bulk carrier without consideration of heavy weather avoidance, it is observed that the direct calculations using the 3D program with IACS wave scatter diagram generally give lower extreme values of wave pressure than rule simplified formulae as shown in Figure 7.

The underestimation increases from the water line to the centre bottom. It is noted that wave pressure at the centre bottom obtained by long term prediction with IACS wave data is 27% lower than that obtained by simplified formulas.

The IACS wave scatter diagram is conditioned on the presence of the vessel to some extent and has been smoothed from 10^3 to 10^5 by fitting an analytical distribution to the raw data. It may be argued that the observations on board would imply non-conservative predictions of severe sea states, because they may have been avoided by the vessel. On the other hand, fitting of an analytical probability density function to the raw data represents a smoothing and possible extrapolation and may increase the probability content in the distribution tail. The OCEANOR data are believed to be more severe than the IACS wave scatter diagrams.

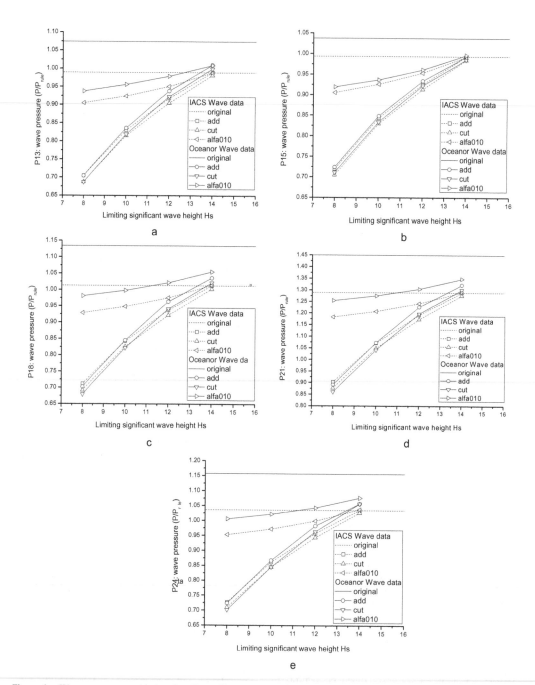

Figure 6.   Wave pressure considering the effect of heavy weather avoidance for a VLCC in the full load condition. **a** P13, **b** P15, **c** P18, **d** P21, **e** P24. **original**: with full scatter diagram; **add**: method 1; **cut**: method 2; **alfa010**: method 3.

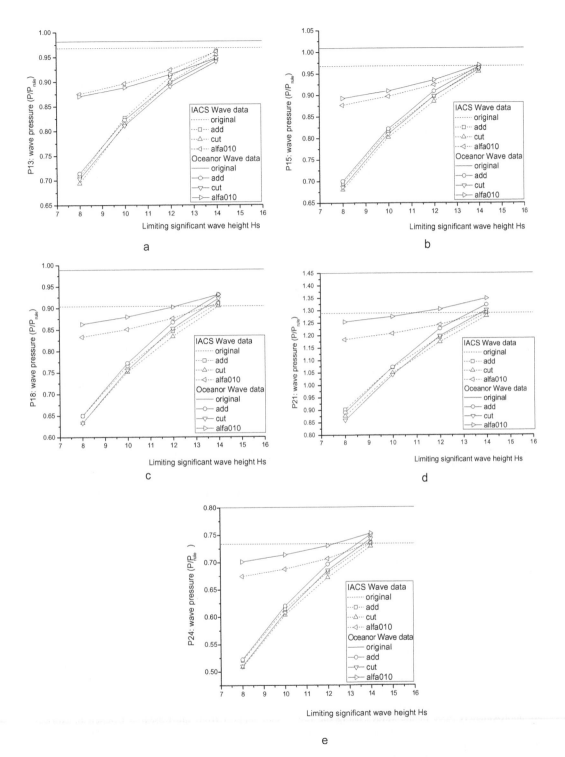

Figure 7. Wave pressure considering the effect of heavy weather avoidance for a bulk carrier in the full load condition. **a** P13, **b** P15, **c** P18, **d** P21, **e** P24. **original**: with full scatter diagram; **add**: method 1; **cut**: method 2; **alfa010**: method 3.

The OCEANOR data is found to yield wave pressures at various locations along the midship transverse section which are about 5–18% higher than those obtained from the IACS wave scatter diagram issued by classification society by using the full scatter diagram for this VLCC while the overestimations are 3–9% for bulk carrier. This may indicate that the IACS scatter diagram has inherently included the effect of heavy weather avoidance compared with OCEANOR scatter diagram. However this fact does not lead to the conclusion that the rule formulae give non-conservative results and should be increased to have enough safety margins since the effect of heavy weather avoidance is expected to reduce the characteristic extreme value.

In the present work, the heavy weather avoidance has been modeled by modifying the original wave scatter diagram according to operational restrictions expressed as limiting significant wave height. As expected it is found that Methods 1 and 2 (see Table 1), in general result in a larger reduction in wave pressure than Method 3. Method 1 and Method 2 imply that the operational restrictions are perfectly satisfied by proper action adopted by the shipmaster and the wave climate forecast available is absolutely correct. It is also found that the long term predictions at exceedance probability of 10^{-8} based on Method 1 were very close to Method 2 independent of the wave scatter diagram applied. Moreover it is found that the characteristic extreme values of wave pressure can be reduced significantly when the limiting significant wave height is less than Hs = 12 m for the VLCC and bulk carrier compared with those using the full scatter diagram. In case of Hs = 14 m, the extreme values from Method 1 and Method 2 agree very well with those extreme values without consideration of heavy weather avoidance.

However due to the uncertainty in the forecast and shipmaster's decision, the ship can not avoid those severe sea states with Hs > H$_{slim}$ absolutely. The sea states experienced by passing ship are expected to have a smaller probability in the upper tail as compared with the scatter diagrams issued by the classification societies and meteorological data. Method 3 is established to account for this fact with a reduction factor of $\alpha = 0.1$ to reduce the tail above the limiting significant wave height. Compared to Method 1 and Method 2, Method 3 yields much higher extreme values at limiting significant wave height of 8 m and 10 m. If the sea states with significant wave height above 8 m are avoided, the 10^{-8} predictions of wave pressure at various locations along the midship transverse section are approximately 90% of the values using the full scatter diagram for both vessels independent of wave scatter diagrams adopted. The factor α in Method 3 is expected to have an important influence on the extreme values.

5 CONCLUSIONS

The effect of the avoidance of heavy weather on the wave pressure along the midship transverse section has been assessed in this paper. The hydrodynamic analysis of the wave pressure with various roll damping has been run by WASIM with linear analysis option. The heavy weather avoidance is accounted for by modifying the wave scatter diagram according to the limiting significant wave height.

It is found that the roll damping is of significant importance for the long term prediction values of wave pressures along the midship transverse section of both VLCC and bulk carrier, especially for the area between the bilge and the centre bottom. The extreme value of wave pressure at the centre bottom is not affected by the roll damping.

Compared with OCEANOR wave data, which are believed to describe the real wave condition on sea, the IACS wave data usually yields lower extreme values of wave pressures along the midship transverse section, which indicate to some extent that the IACS wave data has implicitly included the effect of heavy weather avoidance.

The influence of heavy weather avoidance on the extreme values of wave pressure along the midship transverse section is dependent on how the heavy weather avoidance is accounted for. For both Method 1 and Method 2, the sea states above the H$_{slim}$ are assumed to be absolutely avoided. As expected, Method 1 and Method 2 can effectively reduce the response when H$_{slim}$ is less than 12 m. Method 3 accounts for that the vessel can not completely avoid the heavy weather. In this case, if the sea states with significant wave height above 8 m are avoided, the 10^{-8} predictions of wave pressures along the midship transverse section are approximately 90% of the values using the full scatter diagram. The reduction factor α in Method 3 is expected to have an important influence on the characteristic extreme values.

ACKNOWLEDGMENTS

The work has been performed in the scope of the project MARSTRUCT, Network of Excellence on Marine Structures, (www.mar.ist.utl/marstruct/), which has been financed by the EU through the GROWTH Programme under contract TNE3-CT-20030-506141. The authors wish to give thanks to the support from this project. The authors acknowledge the financial support from the Research Council of Norway (RCN) through the Centre for Ship and Ocean Structures (CeSOS). The authors also wish to thank Fugro OCEANOR for providing the valuable wave climate hindcast data.

REFERENCES

Barstow, S. 2005. OCEANOR wave data, Private Communication.

DNV. 2006. Wasim-user's manual, Det Norske Veritas, Hovik, Norway.

Ferrari, J.A. & Ferreira, M.D. 2002. Assessment of the effectiveness of the bilge keel as an anti-roll device in VLCC-sized FPSOS, *Proceedings of 12th ISOPE*, Kitkyushu, Japan.

Guedes Soares, C. 1990. Effect of heavy weather maneuvering on the wave induced vertical bending moments in ship structures, *Journal of Ship Research*, Vol. 34: 60–68.

Hogben, N., Dacunha, N.M.C. & Oliver, G.F. 1986. Global wave statistics, British Maritime Technology Ltd., Feltham.

IACS. 2000. Recommendation, No. 34, Standard Wave Data, International Association of Classification Societies, http://www.iacs.org.uk/publications.

IACS. 2006a. Common Structural Rules for double hull oil tankers, International Association of Classification Societies, http://www.iacs.org.uk/publications.

IACS. 2006b. Common Structural Rules for bulk carriers, International Association of Classification Societies, http://www.iacs.org.uk/publications.

Kring, C.G., Huang, Y.F., Sclavounos, P., Vada, T. & Braathen, A. 1996. Nonlinear ship motions and wave-induced loads by a rankine method, *Proceedings of 21st Symposium on Naval Hydrodynamics*, Trondheim, Norway.

Moan, T., Das, P.K., Gu, X.K., Friis-Hansen, P., Hovem, L., Parmentier, G., Rizzuto, E., Shigemi, T. & Spencer, J. 2006. Report of ISSC special task committee VI.1. Reliability based structural design and code development, *Proceedings of 16th ISSC*, University of Southampton.

Moe, E., Holtsmark, G. & Storhaug, G. 2005. Full scale measurements of the wave induced hull girder vibrations of an ore carrier trading in the North Atlantic, *International Conference on Design and Operation of Bulk Carriers*. RINA.

Olsen, A.S., Schrøter, C. & Jensen, J.J. 2006. Wave height distribution observed by ships in the North Atlantic, *Journal of Ships and Offshore Structures*, Vol. 1: 1–12.

Sternsson, M. & Bjorkenstam, U. 2002 Influence of weather routing on encountered wave height, *Int. Shipbuild. Progr.*, Vol. 49: 85–94.

Shu, Z. & Moan, T. 2008. Effects of avoidance of heavy weather on the wave induce load on ships, *Journal of Offshore Mechanics and Arctic Engineering*, Vol. 130: 021002-1-8.

Analysis and Design of Marine Structures – Guedes Soares & Das (eds)
© 2009 Taylor & Francis Group, London, ISBN 978-0-415-54934-9

Comparison of experimental and numerical sloshing loads in partially filled tanks

S. Brizzolara, L. Savio & M. Viviani
Department of Naval Architecture and Marine Engineering, University of Genoa, Italy

Y. Chen & P. Temarel
Ship Science, School of Engineering Sciences, University of Southampton, UK

N. Couty & S. Hoflack
Principia, France

L. Diebold & N. Moirod
Bureau Veritas, France

A. Souto Iglesias
Naval Architecture Department (ETSIN), Technical University of Madrid (UPM), Spain

ABSTRACT: Sloshing phenomenon consists in the movement of liquids inside partially filled tanks, which generates dynamic loads on the tank structure. Resulting impact pressures are of great importance in assessing structural strength, and their correct evaluation still represents a challenge for the designer due to the high nonlinearities involved, with complex free surface deformations, violent impact phenomena and influence of air trapping. In the present paper a set of two-dimensional cases for which experimental results are available are considered to assess merits and shortcomings of different numerical methods for sloshing evaluation, namely two commercial RANS solvers (FLOW-3D and LS-DYNA), and two own developed methods (Smoothed Particle Hydrodynamics and RANS). Impact pressures at different critical locations and global moment induced by water motion for a partially filled tank with rectangular section having a rolling motion have been evaluated and results are compared with experiments.

1 INTRODUCTION

The sloshing phenomenon is a highly nonlinear movement of liquids inside partially filled tanks with oscillatory motions. This liquid movement generates dynamic loads on the tank structure and thus becomes a problem of relative importance in the design of marine structures in general and an especially important problem in some particular cases (Tveitnes et al. 2004).

In some cases, this water movement is used for dampening ship motions (passive anti-roll tanks), especially for vessels with low service speed (fishing vessels, supply vessels, oceanographic and research ships, etc.) and for which active fin stabilizers would not produce a significant effect (Lloyd 1989).

The sloshing problem has been to a great extent investigated in the last 50 years, with increasing levels of accuracy and computational efforts.

First attempts were based on mechanical models of the phenomenon by adjusting terms in the harmonic equation of motion (Graham & Rodriguez 1952, Lewison 1976). These types of techniques are used when time-efficient and not very accurate results are needed (Aliabadi et al. 2003).

The second series of investigations solves a potential flow problem with a very sophisticated treatment of the free-surface boundary conditions (Faltinsen et al. 2005) that extends the classical linear wave theory by performing a multimodal analysis of the free-surface behavior. This approach is very time efficient and accurate for specific applications but it does not allow to model overturning waves and may present problems when generic geometries and/or baffled tanks are considered.

The third group of methods solves the nonlinear shallow water equations (Stoker 1957) with the use of different techniques (Lee et al. 2002, Verhagen & Van Wijngaarden 1965).

The fourth group of techniques used to deal with highly nonlinear free-surface problems is aimed at solving numerically the incompressible Navier—Stokes equations. Frandsen (2004) solves the nonlinear potential flow problem with a finite difference method in a 2-D tank that is subjected to horizontal and vertical motion, with very good results, but this approach suffers from similar shortcomings to the multimodal method. Celebi and Akyildiz (2002) solve the complete problem by using a finite difference scheme and a VOF formulation for tracking the free-surface. Sames et al. (2002) present results carried out with a commercial finite volume VOF method applied to both rectangular and cylindrical tanks. Schellin et al. (2007) present coupled ship/sloshing motions with very promising results.

In general, numerical techniques present significant problems when considering highly nonlinear waves and/or overturning waves, the effect of air cushions and fluid-structure interactions. Considering the first problem, Smoothed Particle Hydrodynamics (SPH) meshless method appears as a promising alternative to standard grid based techniques because of their intrinsic capability to capture surface deformations. Literature about SPH applications to typical marine problems is not very abundant; in Colagrossi et al. 2003 one of the first applications is shown. In successive years, a certain number of applications devoted to the assessment of slamming phenomenon (Oger et al. 2006, Viviani et al. 2007a, b 2008) are found. Sloshing phenomenon is considered by Souto Iglesias et al. 2006 and Delorme et al. 2008b, in which a comprehensive series of calculations is performed, focusing attention on resulting global moment and its dependence with tank oscillating frequency, but problems related to the evaluation of local impact pressures are still considerable, with presence of significantly oscillating results.

The activity described in the present paper, which was carried out in the framework of the EU funded MARSTRUCT Network of Excellence, covered three two-dimensional (or infinite length) cases, focusing on impact pressures and global moments into a partially filled tank with rectangular section, which has an oscillatory rolling motion with different periods and different water levels. For these tests, experimental measurements were carried out by the Model Basin Research Group (CEHINAV) of the Naval Architecutre Department (ETSIN) of the Technical University of Madrid (UPM), in the context of a comprehensive analysis of sloshing phenomenon, as reported by Delorme et al. (2007 and 2008a).

A series of numerical techniques has been applied by various participants to assess their merits and shortcomings, and in particular:

– a RANS code own-developed by UoS (University of Southampton)

– two available commercial software for the solution of RANS equations, namely FLOW-3D applied by BV (Bureau Veritas) and LS-DYNA applied by PRI (Principia)
– a SPH code own developed by DINAV (University of Genoa).

2 EXPERIMENTAL SET-UP

The experimental tests which are used for benchmarking the various numerical techniques were performed by CEHINAV-ETSIN-UPM, as reported in Delorme et al. 2007 and 2008a.

In particular, a rectangular tank having dimensions (in centimeters) reported in Fig. 1 was considered. The tank is cylindrical, and the dimension perpendicular to those reported in Fig. 1 is 62 mm; a sinusoidal rolling motion has been imposed during experiments, with a rolling axis located 18.4 cm over the bottom line, an amplitude of 4° in all cases and different periods.

The tank was fitted for a series of sensors in different locations, as indicated in Fig. 1. During experiments, pressures in correspondence to the two most critical positions were recorded. The sensors are BTE6000—Flush Mount, with a 500 mbar range.

In parallel to pressure measurements, global torque measurement were conducted. In particular, torque time history was measured during experiments with water inside the tank and with empty tank, then the first harmonic of the moment response of the liquid with respect to the tank rotating centre for every case was obtained postprocessing data.

In table 1, the two water levels considered in present analysis are reported, together with the correspondent natural period of oscillation.

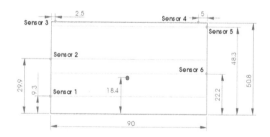

Figure 1.   Tank geometry and position of the sensors.

Table 1.   Water levels considered.

Level symbol	Level [cm]	Level/ tank height	Natural Period T_0 [s]
A	9.3	18.3%	1.91
B	22.2	43.7%	1.32

14

Table 2. Test Cases description.

Case	Level/tank height	Period [s]	Period/T_0
1	A—18.3%	1.91	1.0
2	B—43.7%	1.19	0.9
3	B—43.7%	1.32	1.0

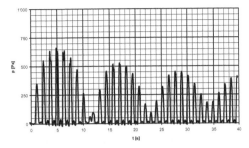

Figure 3. Test Case 2—Level B—$T = 0.9\,T_0$—Pressure at Sensor 6.

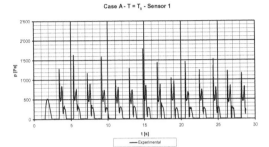

Figure 2. Test Case 1—Level A—$T = T_0$—Pressure at Sensor 1.

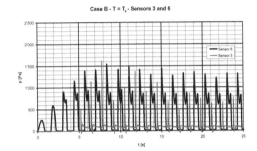

Figure 4. Test Case 3—Level B—$T = T_0$—Pressure at Sensors 3–6.

Table 3. Oscillating moment first harmonic values.

Case	M_0[Nm]	M_0[Nm/m]	ϕ [deg]
1	4.52	72.87	84.4
2	1.07	17.22	173.6
3	6.70	108.03	112.0

For both water levels, experiments carried out in correspondence to the resonance period have been considered (case 1 and 3 for level A and B respectively), to have large free surface deformation and analyse codes' capability to capture it. Moreover, for case B a lower oscillating period (90% of the resonance one) has been analysed (case 2), in which a marked beating phenomenon has been observed.

In table 2, main characteristics of cases analysed are briefly summarized for a better understanding.

In correspondence to case 1, pressure sensors are located at positions 1 and 6, and impacts were recorded on sides in correspondence to lower sensor, as presented in following Fig. 2.

In correspondence to cases 2 and 3, pressure sensors were located at positions 3 and 6; as anticipated, case 2 is interesting for the presence of beating type kinematics and pressures, with peak values oscillating, as presented in following Fig. 3 for sensor 6. Regarding case 3, pressure peaks were recorded in correspondence to both locations as reported in Fig. 4, with impact events at the tank top (sensor 3) and pressure rises due to the incoming wave and to water fall after impact at sensor 6.

In addition to pressure measurement, in following table 3 the resulting first harmonic of the liquid moment is reported.

In particular, M_0 is the first harmonic amplitude and ϕ is the phase lag with respect to oscillating motion, resulting in equation (1) for time history, considering the motion period T:

$$M = M_0 \sin\left(\frac{2\pi}{T}t - \phi\right) \tag{1}$$

First harmonic M_0 value is reported both for the real experimental case (tank with 62 mm third dimension) and for the equivalent two-dimensional configuration, for which moment per unit length is given.

Rolling motion during experiments was a pure sinusoidal motion apart at the very beginning of the experiment to avoid infinite accelerations.

Finally, a proper uncertainty analysis has not been conducted yet since it is hard to find a consistent approach to it in this case. This is due to the strong chaotic character of the pressure peaks, as discussed in Delorme et al. 2008, where the initial steps to such analysis were given.

3 DESCRIPTION OF METHODS

Methods used for modeling the sloshing phenomenon are summarized in Tables 4 and 5 for test case 1 and

Table 4. Details of method and idealization—Test case 1.

Participant	Method	Idealisation details
UoS	RANS	55 × 50 grid
DINAV	SPH	37138 real particles
		1876 virtual particles
PRI	LS-DYNA	90 × 51 grid
BV	FLOW 3D	5 × 220 × 122 grid

Table 5. Details of method and idealization—Test case 2–3.

Participant	Method	Idealisation details
UoS	RANS	90 × 50 grid
DINAV	SPH	88652 real particles
		1876 virtual particles
PRI	LS-DYNA	90 × 51 grid
BV	FLOW 3D	5 × 90 × 62 grid

Table 6. Outline of sloshing cases.

Case	ε	Time step [s]
Case 1	$1.5 * \Delta z$	1.0×10^{-3}
Case 2	$1.5 * \Delta z$	1.0×10^{-3}
Case 3	$1.5 * \Delta z$	1.0×10^{-3}

for test cases 2 and 3, respectively; in the same tables, some details of the idealization are also included.

In SPH real particles are used to represent water, while virtual ones are used to represent the moving tank. For both of them a diameter of 1.5 mm has been adopted. BV calculations are 3-dimensional, and the experimental setup was effectively reproduced, while in all other cases a two-dimensional problem was analyzed. Only UoS considered air (and pressure variations inside it), while in all other cases only water was considered (in VoF methods, air is schematized as void). Finally, effective time histories were applied both by BV (case 1 only) and PRI, while pure harmonic oscillations were applied by BV (case 2 and 3), UoS and DINAV; it resulted from calculations that this has not a significant impact once phenomena are stabilized.

3.1 RANS approach adopted by UoS

The level set formulation in a generalized curvilinear coordinates system has been developed to simulate the free surface waves generated by moving bodies or sloshing of fluid in a container. The Reynolds-averaged Navier-Stokes (RANS) equations are modified to account for variable density and viscosity for the two-fluid flows (i.e. water-air). By computing the flow fields in both the water and air regions, the location and transport of free surface inside a tank is automatically captured. A detailed description of the numerical method can be found in Chen et al. 2008.

In the simulation of sloshing, appropriate boundary conditions need to be imposed to calculate the impact loads. Traditionally, in a single phase flow solver, a thin artificial buffer zone is adopted near the tank ceiling and a linear combination of free surface and rigid

wall conditions imposed inside the buffer zone. The magnitude of the impact pressure is affected by several factors such as the choice of time step and thickness of the buffer zone, as discussed in Chen et al. 2008. In the present investigation there is no special treatment required for the free surface as a two-fluid approach is used to solve the RANS equations in both water and air regions in a unified manner and the interface is only treated as a shift in fluid properties. The solid wall boundary condition is imposed by vanishing the normal velocity to the wall or setting the components of velocity on the wall to zero. The wall pressure is obtained by projecting the momentum equation along the normal to the wall.

In addition to data already provided in previous paragraph, other computational conditions are listed in Table 6, indicating the values of half the finite thickness of the interface in which density and viscosity change and time step increment examined in this study (Δz represents the vertical distance between two grid nodes and ε the interface half thickness).

Regarding torque calculations, for this method they are provided for case 1 only since some problems have been encountered for other cases, which are still under investigation.

3.2 LS-DYNA approach adopted by PRI

The LS-DYNA software is used. This is an explicit finite element code solving the mechanics equations (here corresponding more precisely to the RANSE with Reynolds stresses neglected). The fluid domain is modelled using a multi-materials Eulerian formulation. For the advection of the variables at the integration points, a Van Leer scheme (second order precision) is used and a Half Index Shift (HIS) method for the nodal variables. Each material is characterized by a volume fraction inside each element. It interacts in the calculation of the "composite" pressure for the partially filled cells within a Volume Of Fluid (VOF) method. The dispersion of the volume fraction for these particular cells is limited with an interface reconstruction method (boundary between materials modelled as a plane). Air is modeled as void.

The problem is addressed in two dimensions, although under-integrated 8-noded elements (one layer in the direction normal to the plane) are used

16

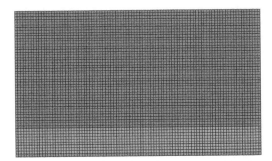

Figure 5. Eulerian mesh for case 1 (water in red, void in blue).

to model the fluid domain in practice. The size of the fluid cells is of 10 mm edge length, chosen close to the sensor size. The mesh, using 90×51 solid elements, can be seen in Fig. 5 for cases 1. Cases 2 and 3 are schematized in the same way, except that level of water is modified in accordance to test setup.

The behaviour of water is modelled with a polynomial equation of state, and no viscosity is defined.

The fluid domain is surrounded with one layer of rigid elements (in grey on Fig. 5), modeling the wall and preventing normal flow. Besides, the movement (as given by the roll time histories) is imposed to this rigid part, then the rigid movement of the Eulerian grid (following the walls) is achieved by forcing the grid to follow the movement of three nodes of the rigid part (Aquelet et al. 2003).

The time scheme is based on the Finite Difference method, which is conditionally stable. Thus, the time step is linked to the shortest duration for an acoustic wave to cross any element of the model (fluid or solid), close to 6 μs here.

Pressure histories are directly post-treated in solid elements where sensors are located.

3.3 Smoothed particle hydrodynamics method, adopted by DINAV

SPH is a Lagrangian meshless CFD method, initially developed for compressible fluids (Gingold and Monaghan 1977) and successively adapted and corrected for hydrodynamic problems (Liu and Liu, 2003). The continuum is discretized in a number of particles, each representing a certain finite volume of fluid, which are followed in a Lagrangian way during their motions induced by internal forces between nearby interacting particles and external mass forces or boundary forces. Internal forces derive from the usual Navier-Stokes and continuity equations, made discrete in space by means of a kernel formulation. In general, if A is a field variable and W is the kernel function, the

following equations (already in their discretised form) are adopted:

$$\langle A(r)\rangle = \sum_{j=1}^{N} \frac{m_j}{\rho_j} A(r_j) W(r - r_j, h) \qquad (2)$$

where m_j and ρ_j are mass and density of the jth particle, r and r_j are position vectors in space and for the jth particle, respectively. The term h represents the smoothing length, which determines the extent to which a certain particle has influence on the others. Different kernel functions may be utilized (Liu and Liu 2003); in particular, in the present work, Gaussian kernel was adopted. The resulting formulae, in the particles approximation, for the continuity equation and the momentum equation are:

$$\partial \rho_i / \partial t = \sum m_j (v_i - v_j) \cdot \nabla W(r - r_j, h),$$

$$\partial v_i / \partial t = -\sum m_j (p_i / \rho_i^2 + p_j / \rho_j^2) \nabla W(r - r_j, h) + g \qquad (3)$$

Moreover, in order to close the problem, an equation of state (Monaghan 1994) which relates density to pressure for each particle, considering the fluid as "weakly compressible", is adopted:

$$p = \frac{c_0^2 \rho_0}{\gamma} \left[\left(\frac{\rho}{\rho_0} \right)^{\gamma} - 1 \right] \qquad (4)$$

where ρ_0 is reference density (1000 kg/m³) and ρ is density of each particle. Constant γ is set to 7 in accordance to Batchelor 1967, providing satisfactory results for various SPH applications. The value of sound speed c_0 cannot, in general, be set to its effective value for practical reasons (i.e. time steps become too small); thus, it is usually set in order to limit the Mach number to a value below 0.1 (Monaghan 1994) and, consequently, density variations in the incompressible flow to acceptable values. In particular, in accordance to previous works (Viviani et al. 2008), sound speed was set in order to limit density variations to values below 1%; for all cases considered, a value of 50 m/s was sufficient to achieve this condition; according to this, a time step of 10^{-5} s was adopted, which is about half the one required by the simple application of Courant condition (according to previous experience).

Since SPH can be affected by a lack of stability, various authors developed different methods which can help in reducing this problem, such as Artificial Viscosity and XSPH (Monaghan 1992); in present calculations, on the basis of previous experiences (Viviani et al. 2007a, b), it was decided to avoid XSPH, and to consider an artificial viscosity term as reported in equations (5) where c, ρ, and h are sound speed, density, smoothing length and kernel function respectively and the subscript 'ij' represents a mean value.

$$\Pi_{ij} = \begin{cases} \dfrac{-\alpha c_{ij}\phi_{ij} + \beta\phi_{ij}^2}{\bar{\rho}_{ij}} & \vec{v}_{ij} \cdot \vec{x}_{ij} < 0 \\ 0 & \vec{v}_{ij} \cdot \vec{x}_{ij} \geq 0 \end{cases} \quad (5)$$

$$\phi_{ij} = \dfrac{h_{ij}\vec{v}_{ij} \cdot \vec{x}_{ij}}{\left|\vec{x}_{ij}\right|^2 + \left(\varepsilon h_{ij}\right)^2}$$

A comprehensive analysis was carried out to obtain the best setup of parameters α, β and ε (results cannot be included for space limitations), and after it they were all set to the low value of 0.01, thus almost neglecting also artificial viscosity.

With reference to boundary treatment, repulsive forces are used as suggested by Monaghan (1994), adopting a force which is dependent on the inverse of the distance between fluid and boundary particles according to the formulation of Lennard-Jones for molecular force, as described by Equation (6), where $p_1 = 12$ and $p_2 = 6$, r_0 is the cutting-off distance, approximately equal to the smoothing length, and r is distance between real and boundary particles:

$$f(r) = D\frac{r}{r^2}\left(\left(\frac{r_0}{r}\right)^{p1} - \left(\frac{r_0}{r}\right)^{p2}\right) \quad \text{when r} < r_0$$

$$f(r) = 0, \quad \text{elsewhere} \quad (6)$$

Finally, in order to evaluate pressure an approach similar to the one presented by Oger et al. (2006) was adopted, albeit simplified since pressure is evaluated as a mean of values computed for real particles in proximity of the boundary particle, i.e. in a rectangular region with width parallel to boundary surface and height perpendicular to boundary surface which are multiples of the smoothing length (in particular, a 4×10 region was considered).

3.4 Flow-3D approach, adopted by BV

FLOW-3D solves the transient Navier-Stokes equations by a finite volume/finite differences method in a fixed Eulerian rectangular grid. One of its distinctive features is the Fractional Area Volume Obstacle Representation (FAVOR) technique, which allows for the definition of solid boundaries within the Eulerian grid. Using such a technique definition of boundaries and obstacles is carried out independent of the grid generation.

In particular, the 1-phase flow option was used for this calculation; regarding time, the real time roll series was used as input motion for Case A, while for case B (both $T = 0.9T_0$ and $T = T_0$), harmonic excitations were adopted for the sake of simplicity, since it was

found that differences were not important. Finally the x-direction was also considered, with a "depth" equal to 6.2 cm, as utilized also in the experiments.

4 RESULTS AND DISCUSSION

4.1 Test case 1—Water Level A—$T = T_0$

In general, this is the case for which the most comprehensive analysis was carried out. In particular, a rather large number of computations was performed with SPH to analyse the effect of some parameters, and in particular of time step and of artificial viscosity; results of this analysis are not included in present paper for space limitations, and only final setup results are presented (time step equal to 10^{-5} s, XSPH and artificial viscosity parameters set to 0.01). It has to be mentioned, however, that influence of artificial viscosity is very pronounced, and in particular higher values result in an overdamping of all phenomena (with impact pressures almost vanishing in correspondence to values of α and β in artificial viscosity equal to 0.3), without overturning waves, and with lower pressure peaks.

In Figs. 6 and 7, pressure at Sensor 1 results obtained with all methods are summarized (Fig. 6 represents results obtained in the first part of the simulation/experiment, while Fig. 7 represents a larger

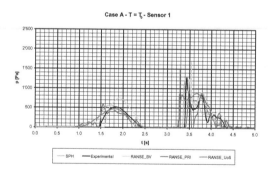

Figure 6. Test Case 1—Level A—$T = T_0$—Sensor 1 pressure time history—first two oscillations (upper) and up to $t = 10$s (lower).

18

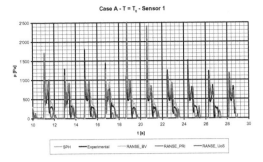

Figure 7. Test Case 1—Level A—T = T_0—Sensor 1 pressure time history for t = 10–30 s.

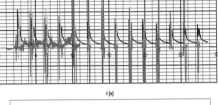

Figure 8. Test Case 1—Level A—T = T_0—Sensor 6 pressure time history up to t = 30 s.

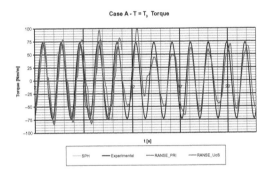

Figure 9. Test Case 1—Level A—T = T_0—Global torque time history up to t = 25 s.

scale); in general, pressure peak order of magnitude is captured by all methods applied.

For all of them, first peak is captured correctly (considering mean values in the whole time history), and the second peak is also captured, even if its amplitude accuracy is not as good as the one of the first peak. SPH calculations provide a higher peak at fourth oscillation, similar to some values obtained by PRI at longer time, while UoS results provide a peak which has the correct order of magnitude but a narrower form.

Looking at Fig. 7 for longer times (only BV and PRI calculations have been extended to that), both methods still provide accurate results, even if BV calculations are probably more robust (with lower oscillations) and similar to experimental ones. Calculations with SPH were not extended to these higher times because of long calculation times, and attention was paid to calibration of various parameters as explained before. Once a satisfactory setup was obtained, this was applied to other cases to better analyze its general nature.

In following Fig. 8 results obtained for sensor 6 are reported. In this case, pressure values are lower by an order of magnitude with respect to those encountered for previous Sensor 1, representing more a "sensor wetting" (thus a less critical point) rather than a real impact phenomenon.

Keeping this in mind, only BV and UoS methods are able to reproduce the experimental pressure peaks, while SPH overestimates them and provides in some cases anomalous negative values (which are present also for UoS); in all cases a time history different from the experimental one is produced, with faster pressure decay than recorded.

PRI method seems only to be capable of signaling the "senor wetting" occurrences after the first 10 seconds of calculations, with the presence of a repetitive phenomenon with the same period of the experimental one, however pressure peaks seem to be most of time overestimated, and anomalous negative values result.

In following Fig. 9, results in terms of global torque, as evaluated by DINAV, UoS and PRI are reported.

Only results after the first period are reported, in order to look at the "stabilized values" cutting the initial transient phase; this allows a better comparison with experimental results which, as already underlined, are referred only to the first harmonic of oscillating torque time history.

Regarding SPH calculations, results reported are referred as before to the best setup from preliminary analysis. It has to be noticed, however, that once global torque is considered instead of local pressure, differences between various setups are reduced, with the presence of similar results for a wide range of settings of artificial viscosity terms. This result is due to the fact that pressure integration needed to obtain the resultant torque acts like a low-pass filter, smoothing pressure oscillations and cutting the effect of localized pressure peaks. Differences can still be found in correspondence to torque peaks, with values obtained in correspondence to the lowest α and β values in artificial viscosity which are about 10–15% higher than the correspondent ones obtained with high α and β.

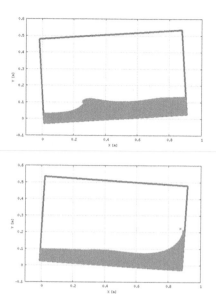

Figure 10–11. Kinematics capturing with SPH—overturning wave (upper) and flow rise after impact (lower).

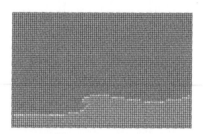

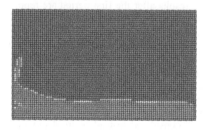

Figure 14–15. Kinematics during experiments—overturning wave at two successive periods.

Finally, in following figures some examples of kinematic capturing are presented (10–11 for SPH, 12–13 for LS-DYNA, 14–15 for experiments), showing a good qualitative capturing of sloshing impact phenomena and of overturning waves.

Figure 12–13. Kinematics capturing with LS-DYNA—overturning wave (upper) and flow rise after impact (lower).

Comparing results to experiments, SPH tends to underestimate by about 10% the effective oscillating torque amplitude. PRI results are more consistent with experiments, even if in the central part of the simulation they present an anomalous behavior, with a sort of beating phenomenon. UoS results are probably the best with a lower mean error and a higher stability, even if they are limited to the first 10 seconds.

4.2 Test case 2—Water Level B—T = 0.9 T₀

In Figs. 16–17, results obtained with all methods for Sensor 6 are summarized; as already mentioned, this case is considerably different from the others, since strong impacts are not expected, but just flow oscillations; interest in this case is due to the presence of a beating type phenomena, with pressure peaks of oscillating amplitude and not periodic with the same value for each period.

For all methods applied the beating phenomenon is captured qualitatively, even if most of the numerical codes do not allow to simulate the almost complete quiescence of fluid with pressures near to zero. Only PRI calculations allow obtaining the low pressures in correspondence to the beating, even if they seem

20

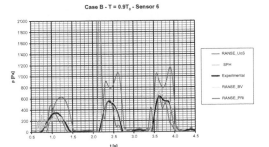

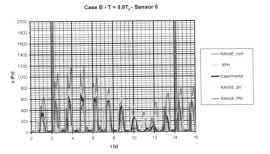

Figure 16. Test Case 2—Level B—T = 0.9 T$_0$—Sensor 6 pressure time history—first three oscillations (upper) and up to t = 16 s (lower).

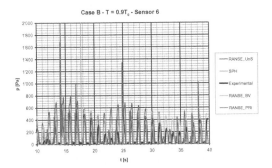

Figure 17. Test Case 2—Level B—T = 0.9 T$_0$—Sensor 6 pressure time history—t = 10–40 s.

to present higher instability than obtained with other methods.

Regarding pressure peaks, UoS calculations overestimate them in the whole time history, while other methods seem to capture them with a higher accuracy; in particular, SPH and PRI capture correctly maximum pressures of the first beating, while BV overestimate them. SPH and BV methods capture pressure curve shapes in the best way, with a sort of secondary peak during "excitation phase" and a more "bell-shaped" curve in the "damping phase"; two anomalous pressure peaks were encountered during calculations by

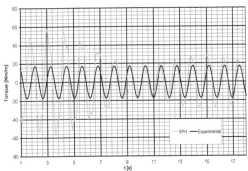

Figure 18. Test Case 2—Level B—T = 0.9 T$_0$—Global torque time history up to t = 20 s.

means of SPH, showing some instability, while BV calculations seem more robust.

Looking at longer times, PRI calculations continue to capture the beating amplitude (even if with higher peaks in the "excitation phase" and lower peaks in the "damping phase"), while other methods tend to obtain a smoother phenomenon, with lower oscillations due to a possible overdamping and, in general, with overestimated pressures; timing of the beating phenomenon is not correctly captured by PRI and UoS, which show an apparent shift with increasing calculation time, while BV methods captures it correctly. SPH calculations have not been extended to the highest time to avoid too high computational times.

In following Fig. 18, results in terms of global torque, as evaluated with SPH are reported.

In this case the comparison with experimental results is more difficult than in previous paragraph; this is due to the fact that experimental data are provided in terms of first harmonic of torque oscillations, which in this case of the beating phenomenon does not describe satisfactorily the complete time history. As already mentioned regarding pressure time histories, the method is able to capture the beating phenomenon. Torque values are qualitatively reasonable for the first beating even if, as already noted for pressure results, torque does not vanish in correspondence to the quiescent part. For the second beating results seem to be more damped, consistently with pressure results. Finally, In following Figs. 19–20 and 21–22 some examples of kinematics capturing with SPH and RANS method by PRI, respectively, are reported; in particular, maximum and minimum fluid oscillation during the pseudo-period are shown.

Maximum oscillations are similar for SPH and PRI, while lowest oscillations are not completely captured by SPH, while PRI captures an almost horizontal surface, in accordance with pressure results.

21

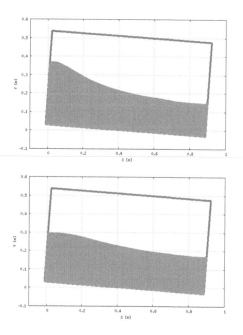

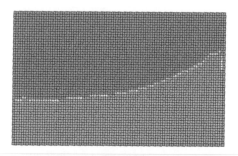

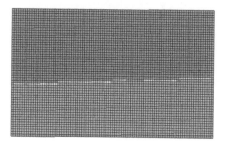

Figure 21–22. Kinematics capturing with LS-DYNA-max peak (upper) and min peak (lower).

Figure 19–20. Kinematics capturing with SPH—max peak (upper) and min peak (lower).

4.3 Test case 3—Water Level B—T = T₀

While in previous test case very violent impact phenomena were not recorded during experiments, in this case impacts were recorded in correspondence to the pressure gage on the tank top (sensor 3), and pressure peaks were recorded also in correspondence to the pressure gage placed on the tank side in correspondence to the calm water level (sensor 6), despite smoother flow motions are found.

In Figs. 23–24 and 25, results obtained with all methods for Sensor 6 and Sensor 3 respectively are summarized.

In general, pressure peaks order of magnitude for sensor 6 have been captured by all methods applied, while sensor 3 presents higher problems, similarly to what was remarked for sensor 6 in case 1.

In particular, if Sensor 6 results are considered, mean experimental pressure peak value is about 1300 Pa and all methods tend to overestimate it, in particular BV, SPH and UoS results are about 25–30% higher, while PRI has a more oscillating behavior, with maximum values about 80% higher and minimum values 30% lower than the experimental ones, even if the trend is captured satisfactorily. Regarding pressure time history, SPH and BV results seem to be the ones which reproduce better the secondary pressure peak, while UoS presents a larger second peak, and PRI has an intermediate behavior.

Also in this case PRI and BV calculations are those which were run for a longer time, showing again a good stability, considering also the high nonlinearities and surface deformations involved in this case.

For what regards Sensor 3, different considerations have to be made. All methods are capable of capturing the sloshing impact on the tank top, however considering pressures, ranging the experimental value between 700 and 1000 Pa apart two initial higher peaks, all numerical methods present a rather variable behavior, with BV presenting oscillating values with higher peaks up to 1500 Pa and lower values with almost no impact, PRI and SPH with a similar trend and even higher peaks in some cases (2000 and 2500 Pa, respectively) and UoS with less oscillating values with a decreasing tendency.

Moreover, both PRI and BV find a "quiet period" between t = 15 s and t = 20 s in which pressure peaks are very low; comparing it with the experiments, pressure peaks are effectively lower in that time range, even if not as low as obtained by calculations. SPH results are not available in correspondence to those times, and UoS seems to capture the pressure decrease even if data are not available to evaluate the possible pressure rises after t = 17 s.

It is worth underlining that experimental results in this case should be considered only as a signal of impact, while the absolute values can be affected by some problem because of very high frequency of impacts.

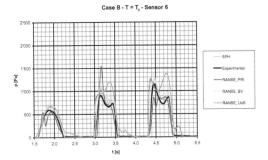

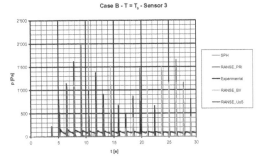

Figure 25. Test Case 3—Level B—T $= T_0$—Sensor 3 pressure time history up to 30 s.

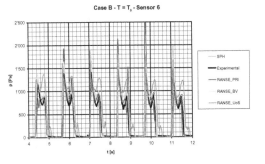

Figure 23. Test Case 3—Level B—T $= T_0$—Sensor 6 pressure time history—first oscillations (upper) and t $= 4$–12 s (lower).

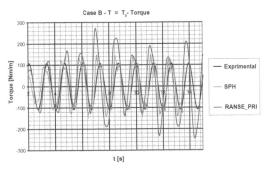

Figure 26. Test Case 2—Level B—T $= T_0$—Global torque time history up to t $= 15$ s.

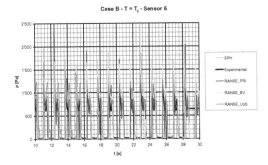

Figure 24. Test Case 3—Level B—T $= T_0$—Sensor 6 pressure time history for t $= 10$–30 s.

In following Fig. 26, results in terms of global torque, as evaluated with SPH and LS-DYNA are reported.

Global torque calculation with SPH in this case seem to be in a good agreement with experimental results, with a similar harmonic behavior and higher superimposed peaks (about 35%), while RANS method by PRI shows a general overestimation of peaks, together with higher oscillations.

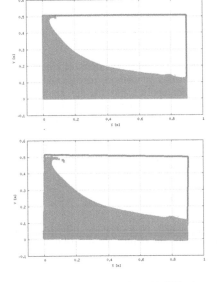

Figure 27–28. Kinematics capturing with SPH—impact on tank top.

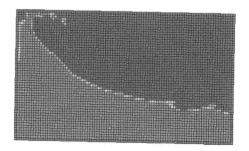

Figure 29.   Kinematics capturing with LS-DYNA—impact on tank top.

Figure 30–31.   Kinematics during experiments—sloshing impacts at the two sides.

In Figs. 27–31, some examples of kinematics capturing with different methods adopted and experimental results are reported, showing the capability of each code to capture the impact on the tank top.

Calculations with SPH and LS-DYNA are in rather good agreement with each other from this point of view, and are able to capture the most significant phenomenon of sloshing impact on the tank top.

5   CONCLUSIONS

In present paper, an analysis of different numerical techniques for the evaluation of sloshing was carried out, comparing simulations with experimental results provided by CEHINAV-ETSIN-UPM for three significantly different test cases. In particular, attention was concentrated on pressures predicted by means of SPH, an own developed RANS code and commercial software LS-DYNA and FLOW-3D; moreover, comparisons were also made to assess global torque calculation and kinematics capturing capability of some of the codes adopted.

From the analysis of results obtained, it can be concluded that:

– pressures predicted by the various numerical methods are rather satisfactory in general, with a sufficient correspondence with experiments, even if there are differences between each other (with pressures overestimation in some cases), with different capacity of different codes to capture correctly the sloshing phenomenon;
– the most problematic cases proved to be, as expected, the two involving large free surface deformations and violent impact phenomena, i.e. case 1 and 3 with roll period set equal to natural oscillation period of fluid; for those cases, most difficult calculations are those carried for sensors which are out of water for most the time;
– also for the less challenging case 2, some methods tend to overestimate peak pressures;
– application of commercial CFD codes, such as Flow-3D and LS-DYNA, is the most successful; codes provide satisfactory results and present an intrinsic robustness (especially the one adopted by BV, while PRI results show anomalous higher oscillations) which own developed codes still have to reach, together with a comparably lower computational and/or setup time;
– SPH technique seems promising, and a satisfactory setting of parameters (which is one of the main shortcomings of the method) was achieved and utilized for all tests; moreover, pressure values are captured correctly without presence of strongly oscillating time histories, which is another important shortcoming of SPH; with these two achievements, results are comparable or even better in some cases than those obtained with commercial codes; however, the long computational time (about 10 and 40 hours for each computed second for case A and B respectively with respect to about 1.5–2 h with UoS method for case B and about 1.4 h for LS-DYNA, all on a conventional CPU), still prevents a systematic use of this technique, and requires further research efforts to consider it as applicable as the more usual RANS codes;
– also own developed RANS code adopted by UoS seems promising, even if it still has to be further analyzed and calibrated, especially for what regards free surface treatment (and resulting pressures in its proximity) and time step/grid density adopted, with the aim of improving results already obtained, especially for test cases 2 and 3.

– regarding global torque, results obtained with SPH and PRI RANS methods are in rather good agreement with each other and with the first harmonics values provided by UPM group, even if PRI values show higher oscillations and a tendency to overestimate peaks, while SPH produces a better correspondence.

Possible future research issues will be a further development of the own-made software, with the aim of getting a better insight into some problems still existing, like the long computational time with SPH (analyzing possible particles reductions without loosing accuracy and, in general, other different strategies to accelerate calculations) and the free surface treatment with the RANS code.

Moreover, possible future work may be related to the analysis of codes capability of capturing influence of air, fluid-structure interactions and possible different tank shapes (e.g. with baffles introduction).

ACKNOWLEDGEMENTS

The work was performed in the scope of the project MARSTRUCT, Network of Excellence on Marine Structures (<http://webmail.mar.ist.utl.pt/exchweb/bin/redir.asp?URL=http://www.mar.ist.utl.pt/marstruct/>http://www.mar.ist.utl.pt/marstruct/), which has been financed by the EU through the GROWTH Programme under contract TNE3-CT-2003-506141.

Experimental data for the test cases were obtained in the framework of the project STRUCT-LNG (file number CIT-370300-2005-16) leaded by the Technical University of Madrid (UPM) and funded by the Spanish Ministerio de Educacion y Ciencia in the program PROFIT 2005.

Messrs Chen and Temarel would like to acknowledge the support from Lloyd's Register Educational Trust, University of Southampton, University Technology Centre in Hydrodynamics, Hydroelasticity and Mechanics of Composites.

REFERENCES

Aliabadi, S., Johnson, A. & Abedi, J., 2003. Comparison of finite element and pendulum models for simulation of sloshing. Computational Fluids 32, 535–545.
Aquelet, N., Souli, M., Gabrys, J. & Olovson, L., 2003. A new ALE formulation for sloshing analysis. Structural Engineering and Mechanics 16 (4), 423–440.
Batchelor, G.K., 1967. An introduction to fluid mechanics. Cambridge Press.
Celebi, M.S. & Akyildiz, H., 2002. Nonlinear modeling of liquid sloshing in a moving rectangular tanks. Ocean Engineering 29 (12), 1527–1553.
Chen, Y.G., Djidjeli, K., & Price W.G., 2008. Numerical Simulation of Liquid Sloshing Phenomena in Partially Filled Containers. Computers & Fluids, in press, available online.
Colagrossi, A. & Landrini, M., 2003. Numerical simulation of interfacial flows by smoothed particle hydrodynamics, Journal of Computational Physics 191: 448–475.
Delorme, L. & Souto Iglesias, A., 2007. Impact Pressure Test Case Description, ETSIN Report.
Delorme, L. & Souto Iglesias, A., 2008. Impact Pressure Test Case Description, updated version with global moment calculations, ETSIN Report.
Delorme, L., Colagrossi, A., Souto-Iglesias, A., Zamora-Rodriguez, R. & Botia-Vera, E., 2008. A set of canonical problems in sloshing, Part I: Pressure field in forced roll—comparison between experimental results and SPH, Ocean Engineering, In Press, Corrected Proof, Available online 17 October 2008, DOI: 10.1016/j.oceaneng.2008.09.014.
Faltinsen, O.M., Rognebakke, O.F. & Timokha, A.N., 2005. Resonant three-dimensional nonlinear sloshing in a square-base basin. Part 2. Effect of higher modes. Journal on Fluid Mechanics 523, 199–218.
Frandsen, J.B., 2004. Sloshing motions in excited tanks. Journal of Computational Physics 196 (1), 53–87.
Gingold, R.A. & Monaghan, J.J., 1977. Smoothed particle hydrodynamics: theory and application to non-spherical stars, Royal Astronomical Society 181: 375–389.
Graham, E.W. & Rodriguez, A.M., 1952. The characteristics of fuel motion which affects airplane dynamics. Journal of Applied Mechanics 19, 381–388.
Lee, T., Zhou, Z. & Cao, Y., 2002. Numerical simulations of hydraulic jumps in water sloshing and water impacting. Journal of Fluid Engineering 124, 215–226.
Lewison, G.R.G., 1976. Optimum design of passive roll stabiliser tanks. RINA Transactions and Annual Report, pp. 31–45.
Liu, G.R. & Liu, M.B., 2003. Smoothed Particle Hydrodynamics—A Meshfree Particle Method, World Scientific Publishing.
Lloyd, A.R.J.M., 1989. Sea Keeping—Ship Behavior in Rough Weather, Ellis Horwood Limited, Chichester.
Monaghan, J.J., 1992. Smoothed Particle Hydrodynamics, Annual Rev. Astron. Astrophysics 30: 543–574.
Monaghan, J.J., 1994. Simulating free surface flows with SPH, Journal of Computational Physics 110: 399–406.
Oger, G., Doring, M., Alessandrini, B. & Ferrant P., 2006. Two-dimensional SPH simulations of wedge water entries" Journal of Computational Physics 213: 803–822.
Sames, P.C., Marcouly, D. & Schellin, T.E., 2002. Sloshing in rectangular and cylindrical tanks. Journal on Ship Research 46 (3), 186–200.
Souto-Iglesias, A., Delorme, L., Perez-Rojas, L. & Abril-Perez, S., 2006. Liquid moment amplitude assessment in sloshing type problems with smooth particle hydrodynamics, Ocean Engineering Volume 33.
Stoker, J.J., 1957. Water waves. Pure and Applied Mathematics, vol. IV. Interscience, New York.
Schellin, T.E., Peric, M., El Moctar, O., Kim, Y.S. & Zorn, T., 2007. Simulation of Sloshing in LNG Tanks, OMAE 2007, 26th International Conference on Offshore Mechanics and Arctic Engineering, San Diego, USA, 10–15 June 2007.

Temarel, P. (editor)., 2005. Slamming and green water scenarios and a comparison of theoretical slamming and green water impact models with model test results, MARS-TRUCT Report MAR-D1-3-UoS-01 (2).

Temarel, P. & Xu, L. (editors), 2007. Slamming scenarios and a comparison of theoretical slamming impact models with model test results, MARSTRUCT Report MAR-D1-3-UoS-02 (1).

Tveitnes, T., Ostvold, T.K., Pastoor, L.W. & Sele, H.O., 2004. A sloshing design load procedure for membrane LNG tankers. Proceedings of the Ninth Symposium on Practical Design of Ships and Other Floating Structures, Luebeck-Travemuende, Germany.

Verhagen, J.H.G. & Van Wijngaarden, L., 1965. Non-linear oscillations of fluid in a container. Journal on Fluid Mechanics 22 (4), 737–751.

Viviani, M., Savio, L., & Brizzolara, S., 2007. Evaluation of Slamming Loads on Ship Bow Section Adopting SPH and RANSE Method, 32nd Congress of IAHR, International Association of Hydraulic Engineering and Research, Venice.

Viviani, M., Brizzolara, S., & Savio, L. 2007. Evaluation of Slamming Loads on a Wedge-Shaped Section at Different Heel Angles adopting SPH and RANSE Methods, Conference of the International Maritime Association of the Mediterranean (IMAM 2007), Varna, ISBN 978-0-415-45521-3.

Viviani, M., Brizzolara, S. & Savio, L., 2008. Evaluation of Slamming Loads using SPH and RANS Methods, Journal of Engineering for the Maritime Environment, in press.

Analysis and Design of Marine Structures – Guedes Soares & Das (eds)
© 2009 Taylor & Francis Group, London, ISBN 978-0-415-54934-9

Experiments on a damaged ship section

T.W.P. Smith, K.R. Drake & P. Wrobel
Department of Mechanical Engineering, University College London, London, UK

ABSTRACT: This paper details a series of experiments undertaken on a damaged thin-walled prismatic hull form. The damage consisted of a longitudinal orifice the full length of the keel. The model length spanned the width of the tank so that the fluid behaviour could be approximated as two-dimensional. Various widths of orifice were investigated, with the damaged hull form in two configurations, open and airtight (to investigate the effect of aerostatic stiffness). The hull was force oscillated for a range of amplitudes and frequencies and measurements of hydrodynamic force and internal free surface elevation are obtained and discussed.

1 INTRODUCTION

1.1 Background

A need exists for analysis methods capable of predicting the motions and structural loads on a damaged ship. Where the damaged ship experiences flooding, the intact ship methods used to assess the structural loads on a damaged ship will not be able to account for floodwater behaviour and could provide misleading quantifications on structural integrity for the applications:

– decision support, guidance on the structural integrity of a damaged ship;
– analysis of the damage tolerance of new designs, for example with reference to the ALS (Accident Limit State).

A damage event that causes flooding can be described by the phases listed below.

1. An initial transient response, characterised by large motions and violent floodwater motion.
2. A period of progressive flooding when water ingress slowed by internal obstructions finds an approximate equilibrium.
3. Steady state floodwater behaviour, the oscillation of floodwater about an approximate equilibrium water level.

The third case is of most interest to the application of decision support analysis tools and is the phase considered in this paper, whereas an understanding of all three phases is relevant to analysis of new vessel designs.

A scenario that provides context for the analysis is that of a ship that has experienced damage and consequent flooding in a low sea-state. The ship has reached a new equilibrium trim and draught and the flooding has been contained. A forecast of an imminent worsening of sea-state is obtained and the ship must make an assessment of structural integrity in order to guide their decisions. This example is similar to the real-life experience of HMS Nottingham, as described by Groom (2003).

Smith et al. (2007) presented an investigation into how the magnitude of global structural loads, vertical bending moment and shear force, could vary depending on damage size and position. Changing damage size and position was shown to modify the peak global loading (experienced in wavelengths approximately equal to ship length) by up to 30%. The structural loads were obtained using a modified strip theory coupled to a flooding algorithm that calculated the instantaneous rate of ingress and egress of floodwater using Torricelli's formula (Fox (1974)):

$$Q = C_d A_0 \sqrt{2gh}. \tag{1}$$

Where C_d is a discharge coefficient, A_0 is the area of the orifice, g is acceleration due to gravity and h is a hydrostatic head. Torricelli derived the relationship through experimentation, however without the discharge coefficient it becomes a specific case of the steady Bernoulli equation.

This initial investigation suggested that ingress and egress effects could not be ignored, but the assumptions used in the computational model were substantial and lacked validation. This justified a simplified experiment to assess the limits of the assumption's application.

1.2 Experiment design

An assumption used in the analysis of intact structural loads is that the component of flow velocity along the ship's length makes a negligible contribution to overall structural loading when compared to velocity components perpendicular to the ship's length. This

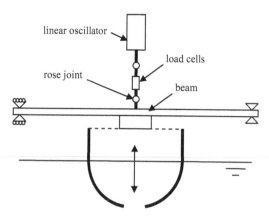

linear oscillator

load cells

rose joint

beam

Figure 1. Experiment configuration for the forced oscillation of a damaged body.

assumption enables the fluid domain around a ship to be divided into a sequence of two-dimensional fluid domains around prismatic strips formed from the dimensions of a ship's local cross section (Salvesen et al. (1970)). For an intact ship, the dominant structural loads are vertical bending moment and shear force, for which only the force acting on each strip in a single degree of freedom, heave, is required. Whilst the structural integrity of a damaged ship might not be as dominated by failure modes associated with vertical bending as an intact ship, it will still be an important load case.

Vugts (1968) described an experiment for an intact prismatic hull form. The hull spanned the wave tank so that the waves generated, as the body was forced in oscillation, were approximately two-dimensional. The force on the body was measured and compared favourably with theoretical computations. A similar experiment was designed for the forced oscillation of a damaged prismatic hull form in a wave tank at the Department of Mechanical Engineering at University College London (UCL), with a depth of 0.7 m and a length of 20 m (At low frequency, this limits the number of oscillations that will be unaffected by wave reflections).

A schematic of the arrangement that was used is shown in Figure 1. The linear motion system comprised a precision ball screw assembly driven by an AC servo motor and digital drive controller. The system was configured to move in accordance with prescribed time histories of vertical displacement, a periodic sinusoid varied in amplitude and frequency. The model was constructed in plywood above the waterline and an 8 mm thick PVC half-pipe was used for the underwater form (semi-circular cross section). The principal dimensions are:

– Length = 1200 mm
– Beam = 315 mm

– Draught = 157.5 mm

Two 25 kgf load cells were fitted between the cone and the lower end of the connecting rod. These were isolated from any bending moments experienced in the forcing assembly by two rose joints. To constrain the model in vertical plane motion a beam, sized to have high stiffness in lateral and torsional bending but low stiffness in vertical bending, acted as the interface between the driving assembly and the model, with one end clamped on a fulcrum and the other simply supported. The beam, without the model attached, was force oscillated at every amplitude and frequency to provide a dataset that could be subtracted from measurements made with the model to provide just the fluid force acting on the body and its D'Alembert inertial force.

Three twin wire wave probes were fixed on the model to measure time histories of the free surface elevation relative to the body, two inside the model (on and off the centre line) and one outside. Six analogue signals were sampled using a data acquisition unit. The sampling frequency was set at 100 Hz on each channel for all of the tests.

1.3 Data analysis

The raw time histories of measured force contained response components that were at frequencies well above the input frequency. The observed behaviour was attributed to control aspects of the linear motion system and, to a lesser extent, elasticity effects in the support arrangement. Because of this, low-pass digital filtering was applied during post-processing of the experimental data.

The motion of an intact body undergoing forced sinusoidal oscillation F in heave displacement w can be described by Equation 2:

$$(m + a)\ddot{w} + b\dot{w} + cw = F, \qquad (2)$$

where m is its dry mass, a is its hydrodynamic added mass coefficient, b is its hydrodynamic damping coefficient and c its hydrostatic restoring force coefficient. We can use the same equation for a damaged body. For that case, if m includes the mass of the equilibrium quantity of floodwater and the hydrostatic restoring force is considered to be the same as that of an equivalent intact body, then the dynamic floodwater behaviour will be represented by a modification to the hydrodynamic coefficients a and b. The experiment measured the force experienced by the body for a prescribed (and measured) displacement. By differentiating the displacement signal twice, to give an acceleration time history, both the D'Alembert inertial force and hydrostatic restoring force time histories can be calculated. Application of these time histories to the

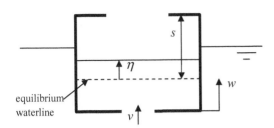

Figure 2. Instantaneous internal free surface elevation η, relative to its equilibrium level, for a body undergoing a forced displacement w.

measured force was used to produce a measured hydrodynamic force time history according to Equation 3.

$$a\ddot{w} + b\dot{w} = F - m\ddot{w} - cw \tag{3}$$

This can be further decomposed using Fourier analysis to find the in-phase and quadrature force components and hence the magnitude of the added mass and damping coefficients a and b.

1.4 Formulation of a mathematical model

A body in still water experiencing a single degree of freedom displacement w (heave) is shown in Figure 2.

If we assume that the flow through an orifice can be approximated using steady Bernoulli and that $v \gg \dot{w}$, then an expression for the spatial average velocity v (across the orifice) is

$$v = C_d\sqrt{2g(w + \eta)}. \tag{4}$$

Applying mass continuity provides a relationship between v and the rate of change of free surface elevation:

$$\dot{\eta} = \alpha C_d\left(\sqrt{2g(w + \eta)}\right), \tag{5}$$

where α is the ratio of orifice area to internal free surface area. This can be rearranged, taking into account the directionality of the flow, to

$$|\dot{\eta}|\,\dot{\eta} = -(\alpha C_d)^2 2g(w + \eta). \tag{6}$$

This can be rearranged to become the differential Equation 8 with coefficient β,

$$\beta = \alpha^2 C_d^2 2g; \tag{7}$$

$$|\dot{\eta}|\,\dot{\eta} + \beta\eta = -\beta w. \tag{8}$$

Equation 8 can be solved to find the free surface elevation η resulting from a periodic displacement w.

For a linearised mathematical model (Patel (1989)), both the body and the free surface will undergo a periodic sinusoidal oscillation, which can be described as:

$$w = Re\left[\bar{w}e^{j\omega t}\right] \tag{9}$$

$$\eta = Re\left[\bar{\eta}e^{j\omega t}\right]. \tag{10}$$

Where $\bar{w}$ and $\bar{\eta}$ are complex numbers describing the magnitude of a function periodic in time t and frequency ω, then we can express Equation 8 as

$$\frac{8j\omega^2}{3\pi\beta}|\bar{\eta}|\,\bar{\eta} + \bar{\eta} = -\bar{w}. \tag{11}$$

This can be solved iteratively to find $\bar{\eta}$, for a prescribed amplitude of oscillation $\bar{w}$.

1.5 Model including aerostatic stiffness

The application of steady Bernoulli used to formulate Equation 4 made the assumption that the pressure above the floodwater's free surface was the same as the pressure above the external free surface. If the damaged compartment is sealed, or there is significant restriction to the airflow trying to equilibrate any pressure differential, then this assumption is no longer valid. For small changes in volume, the change in pressure Δp can be calculated as

$$\Delta p = \frac{\gamma \eta p_a}{s}, \tag{12}$$

where γ is the ratio of the specific heats, p_a is atmospheric pressure and s is the distance between the still water equilibrium waterline and the airtight lid (see Figure 2). Applying steady Bernoulli to find the velocity through the orifice, taking into account the differential between internal and external air pressure, Equation 4 becomes

$$v = C_d\left(\sqrt{\left(2g\left(w + \eta\left(1 + \frac{\gamma p_a}{\rho g s}\right)\right)\right)}\right), \tag{13}$$

where ρ is the density of air. Rewriting in terms of the rate of change of the free surface elevation,

$$|\dot{\eta}|\,\dot{\eta} + \beta\eta\left(1 + \frac{\gamma p_a}{\rho g s}\right) = -\beta w. \tag{14}$$

This can be solved in a linearised form similar to Equation 11.

2 RESULTS

2.1 *Intact model*

To ensure that the experimental setup was producing valid data, an initial set of experiments were undertaken for an intact model using an identical underwater form to one of those tested in Vugts (1968). Figures 3 and 4 compare data obtained using the UCL apparatus with the classical results at two amplitudes. The hydrodynamic coefficients a and b correspond to a unit length hull and have been non-dimensionalised with the model's underwater cross-sectional area A and beam B.

The data shows predominantly good agreement between the two experiments and a high degree of amplitude linearity. Discrepancies occur between the measured coefficients and data obtained by Vugts for

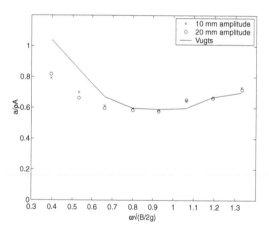

Figure 3. Measurement of added mass coefficient at two amplitudes of oscillation.

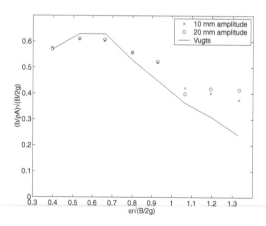

Figure 4. Measurement of damping coefficient at two amplitudes of oscillation.

values of low frequency added mass and high frequency damping. The magnitude of the differences at the extremes of frequency are of the order of 20% for added mass and 30% for damping. Because of the scale of the UCL facilities a number of compromises were made. These were

- a draught-to-depth ratio (model to tank) of 4.4 compared with a value of 13 used by Vugts; and
- a length-to-beam ratio for the model of 7.6 compared with a value of 28 used by Vugts.

At either ends of the model, a gap of 2–3 mm separated the model from the tank wall. This gap was filled with water and it was expected that there would be a small hydrodynamic force measured associated with shearing of this fluid as the model was oscillated. This artificiality is a greater proportion of the measured force for the UCL apparatus as it has a smaller length to beam ratio than the apparatus used by Vugts. The finite depth of the tank prevents the fluid particle kinematics from approximating the behaviour of infinite depth, and this artificiality is greater for the UCL apparatus.

2.2 *Damage hydrodynamic force results*

Because the experiment was constrained to a single degree of freedom, only damage cases symmetrical about the centreline (similar to a grounding damage event) of the model were investigated. Data were collected for four different damage sizes, all applied along the full length of the model and with varying widths. The damage sizes are categorised according to the percentage of the model's internal free surface that they represent. In the limit, as the damage size reduces, the results will tend towards those for an intact body (0% damage) filled to its draught with water.

Data were collected for each damage size at four different amplitudes, both with and without an airtight lid on the model. For brevity only a few key results are shown here. All data are plotted alongside data for the intact body, in order to provide a benchmark.

Figures 5 and 6 show the added mass and damping coefficients respectively for all four damage cases for the same amplitude of oscillation, 10 mm. In these figures, the damaged compartment was open to the atmosphere, so there is no effect of an aerostatic spring.

The results show that as the damage size increases, there is a significant effect on the magnitude of the coefficients. At low frequency, added mass increases with damage size, an effect associated with reduction in hydrostatic restoring force as floodwater flows more freely into and out of the model. At high frequency a reduction in added mass is observed, relative to an intact body, with the greater reduction occurring for the larger damage size. Throughout the frequency range, the ranking of damage size is maintained, with

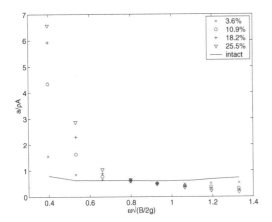

Figure 5. Non-dimensional added mass coefficient for 10 mm amplitude oscillation, all damage sizes.

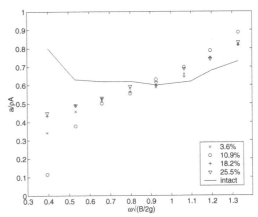

Figure 7. Non-dimensional added mass coefficient for 10 mm amplitude oscillation, all damage sizes, with airtight lid.

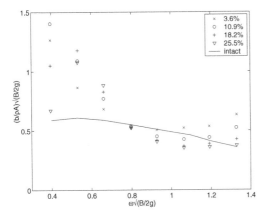

Figure 6. Non-dimensional damping coefficient for 10 mm amplitude oscillation, all damage sizes.

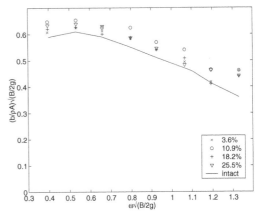

Figure 8. Non-dimensional damping coefficient for 10 mm amplitude oscillation, all damage sizes, with airtight lid.

the smallest damage size having results most similar to those of the intact body, in keeping with expectations.

The results for damping are also consistent in ranking, exhibiting a low frequency turning point at a frequency that is damage size dependent. At high frequency, the largest damage size has results most like the intact model, but at low frequency, the results change ranking, with the lowest frequency data point for the 25.5% damage case appearing to be off-trend. This apparent outlier is consistent with data measured for all amplitudes, and the reduction in damping at low frequency can be observed from theoretical simulation to be associated with the amplitude of the internal free surface oscillation tending to the amplitude of the model's oscillation as frequency reduces.

Figures 7 and 8 show the same data and damaged cases measured for the model with the airtight lid in place. The air in the compartment was at atmospheric

pressure when the model was at static equilibrium. The air pressure significantly reduces the amount of flow of floodwater into and out of the model, and so even the largest damage case experiences little modification to damping, relative to when the model had no airtight lid. There is, however, a significant reduction in the added mass at low frequency. This is thought to be associated with the equilibration of pressure between exterior and interior surfaces, noting that minimal flow is required to achieve such an effect.

For an intact body, hydrodynamic coefficients are approximately linear with respect to amplitude of oscillation. Figures 9 and 10 show the added mass and damping coefficients for a single damage case, 10.9%, at four different amplitudes of oscillation. There is a pronounced non-linearity with amplitude for this damage case, in particular for the results for

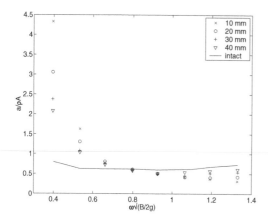

Figure 9. Non-dimensional added mass coefficient for 10.9% damage case, all amplitudes of oscillation.

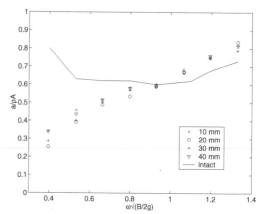

Figure 11. Non-dimensional damping coefficient for 10.9% damage case, all amplitudes of oscillation.

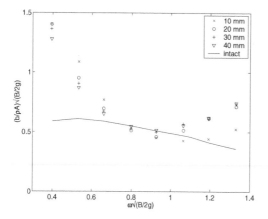

Figure 10. Non-dimensional damping coefficient for 10.9% damage case, all amplitudes of oscillation.

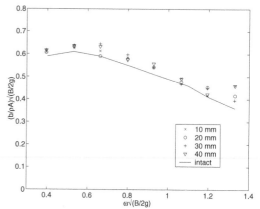

Figure 12. Non-dimensional damping coefficient for 10.9% damage case, with airtight lid, all amplitudes of oscillation.

added mass, which is consistent with the non-linearity inherent in the rate of flow of floodwater as shown in Equation 4. As damage size increases, the height of the internal free surface will approach the amplitude of the body's oscillation for a given frequency of oscillation and so any effect associated with ingress and egress will be amplified. For this reason, the same trend in results for added mass is seen for variations in amplitude as for variations in damage size. The ranking with amplitude is less clear for the damping coefficient, for which the data for the three largest amplitudes are similar and the smallest amplitude, particularly at high frequency, is significantly closer to the intact body data.

Figures 11 and 12 show the hydrodynamic coefficients measured for the case of 10.9% damage with the airtight lid in place. As with the variations for damage size, there is little modification or variation

in the damping coefficient relative to the intact measurement. At low frequency, the added mass tends to zero, and throughout the frequency range shows limited amplitude non-linearity.

2.3 *Measurements of mean force*

Forces acting on an intact body are approximated as oscillations about a static (or still water) equilibrium condition. For a damaged body, there will still be a static equilibrium condition, but any oscillation about this condition will have a non-zero mean value if the mean floodwater height is not the same as the equilibrium floodwater height. This can be the case if the rates of ingress and egress are different, which is shown in Equation 4 when the value of C_d is flow direction dependent.

Figures 13 to 16 show the non-dimensionalised mean force for a variety of amplitudes, frequencies and damage sizes. This is obtained by taking the mean value of the measured periodic force and dividing by the amplitude of the measured periodic force. A positive value for the mean force is equivalent to an increase in the mean weight of the model during the period of oscillation.

Figure 13 shows the results for all damage cases, at an amplitude of 10 mm. For the largest damage cases there is only small modification to the mean force, most markedly a reduction in the mean force at low frequency of approximately 5% of the force amplitude, whereas the smaller damage cases see an increase in mean force at high amplitude of between 10% and 15% of the force amplitude.

The data for variations in amplitude of oscillation and the 10.9% damage case, Figure 14, show a clear trend to approximately zero mean at low frequency and an increase of between 10% and 15% at high frequency, clearly dividing the dynamic behaviour of the floodwater into two regimes.

Figures 15 and 16 show the same data for the experiment performed with the airtight lid in place. Even with a smaller amplitude internal free surface oscillation, a significant mean force was measured, varying between −5% and 5% of the measured hydrodynamic force amplitude, depending on frequency, amplitude and damage case.

2.4 Measurement of discharge coefficient

The mathematical model for the damaged model requires a discharge coefficient that represents viscous and rotational losses associated with flow through an orifice. Whilst there are measurements of this coefficient in literature (Ruponen et al. 2007), the variability of these measurements justified an experiment to calculate the coefficient for the specific body and damage geometries investigated in this study.

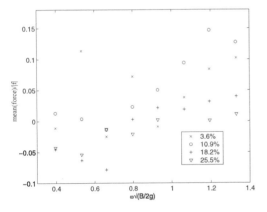

Figure 13. Non-dimensional mean force for 10 mm amplitude, all damage sizes.

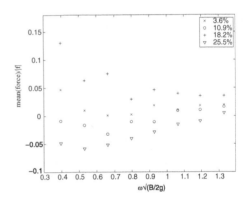

Figure 15. Non-dimensional mean force for 10 mm amplitude, all damage sizes, with airtight lid.

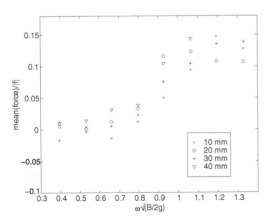

Figure 14. Non-dimensional mean force for 10.9% damage case, all amplitudes of oscillation.

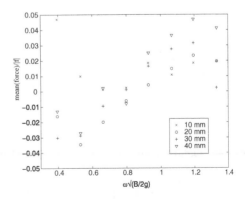

Figure 16. Non-dimensional mean force for 10.9% damage case, with airtight lid, all amplitudes of oscillation.

Table 1. Values of discharge coefficient C_d measured for each of the four damage cases.

Damage %	C_{din}	C_{dout}
3.6	0.85	0.75
10.9	0.6	0.6
18.2	0.6	0.6
25.5	0.6	0.6

The model was displaced at constant velocity and high amplitude through the equilibrium waterline, in order to approximate steady flow through the orifice. Measurements were taken of the free surface height inside and outside the body, and used in conjunction with Equation 5 to calculate a value for the discharge coefficient. Values were calculated for both ingress and egress of floodwater (C_{din} and C_{dout} respectively) and are listed in Table 1.

As damage size increases, the accuracy of the measurement drops and so the larger damage results are rounded to 1 decimal place. There is a practical limit on the velocity of the body above which the flow behaviour is no longer approximately steady, which prevents a higher velocity being used for the experiment.

Apart from for the smallest damage, the size of the orifice does not appear to affect the value of C_d, nor is any asymmetry observed between ingress and egress. The smallest damage has a C_d closer in magnitude to the values obtained by Ruponen et al. There is also an asymmetry in the flow behaviour for this damage case, which is significant in the context of the results in Figures 13 and 15; if the flow into the model is less impeded than out of the model, then the mean interior free surface level will be higher than the static equilibrium water level and will manifest itself as an increase in the mean hydrodynamic force measured for the model.

2.5 Comparison of linearised theory and experiment

Section 1.4 describes a mathematical model for the calculation of the free surface elevation inside the damaged compartment. To first order, the floodwater will have its most significant effect on the restoring force F_{res} experienced by the damaged section. This can be expressed as

$$F_{rest} = -\rho g A_w (w + \eta), \qquad (15)$$

where A_w is the waterplane area of the internal free surface. The solution to Equation 14 provides the magnitude and phase of η relative to w, and this information can be used to find the components of

force that are in phase with the body's acceleration and velocity to modify the added mass and damping coefficients respectively. Adding these 'floodwater' components of the hydrodynamic coefficients to the intact coefficients provides a linearised estimate of the damaged body's hydrodynamic coefficients, which can be directly compared with the measured data decomposed according to Equation 3.

Figures 17 and 18 show the results for the experiment compared with the mathematical model at all amplitudes. The damage case is 3.6% damage and the compartment is open to atmospheric pressure. The mathematical model was solved using a value of $C_d = 0.8$, an average of the ingress and egress coefficients for this damage size in Table 1. Only the results for the smallest damage case are shown here, for brevity. Discrepancies between experiment and theory increase with damage size, as the magnitude of ingress and egress increases.

As the frequency approaches zero, the linearised model picks up the diverging magnitudes of the

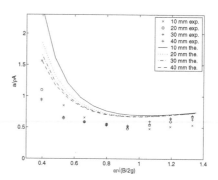

Figure 17. Comparison of added mass coefficients obtained from the experiment and the linearised model for the 3.6% damage case.

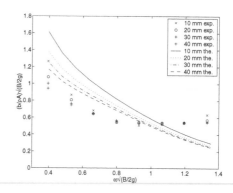

Figure 18. Comparison of damping coefficients obtained from the experiment and the linearised model for the 3.6% damage case.

coefficients of the measured data with amplitude. The ranking of the added mass and damping at low frequency is consistent between theory and experiment.

Above a non-dimensional frequency of 0.8, there is less agreement between theory and practice. The damping coefficient measurements show a trend to increase in magnitude with frequency, which opposes the theoretical prediction. This leads, at the highest frequency, to a measurement of damping coefficient more than 100% greater than the value predicted using linearised theory.

The measured added mass coefficients for different amplitudes diverge, whereas the theoretical predictions converge. Throughout the frequency range the measured values for added mass coefficient are significantly lower than those predicted using linear theory.

3 DISCUSSION

For the model without the air tight lid, the variation in the magnitude of the measured data across the frequency range suggests three regimes of fluid flow through the orifice (damage).

1. At low frequency, the ingress and egress of fluid is approximately quasi-steady and the trends are represented by a linearised simplified mathematical model (albeit not quantitively correctly).
2. At a non-dimensional frequency of approximately 0.8, there appears to be a 'dead band' in orifice flow. Qualitatively, the amplitude of internal freesurface oscillation is low. The measured data show little amplitude non-linearity and results agree very closely with data measured for the intact body.
3. At high frequency, depending on damage case and amplitude of oscillation, large variations to the hydrodynamic coefficients occur which are not consistent with linearised theory.

For a UK frigate, the non-dimensional frequency that corresponds to the worst case loading event (wave length ~ ship length) is approximately 0.6. This occurs in a frequency range where the behaviour is approximately quasi-steady, which adds some justification to the use of a quasi-steady flooding algorithm in Smith et al. (2007). However, even for the smallest damage case investigated here, the magnitudes of the measured hydrodynamic coefficients differ from the linearised theory prediction by up to 50%, depending on amplitude of oscillation.

3.1 Scale effects

In practice, a ship damaged by grounding is likely to have some restriction to the flow of air into and out of the damaged compartment. In the limit of low air flow, the compartment would be approximately airtight. The experiment indicates that if this is the case, the modification to a damaged vessel's hydrodynamic coefficients could be small compared to the intact condition. However, there is a significant scaling effect for the air in the damaged compartment. This can be seen in Equation 12. At model scale, s is 0.1m, and so

$$\frac{\gamma p_a}{\rho g s} = 140. \tag{16}$$

At full scale, s will be at least an order of magnitude bigger, if not more, and so the significance of aerostatic effects will be reduced (for a large volume of trapped air the hydrodynamic coefficients will tend to the results for the open model).

4 CONCLUDING REMARKS

Measurements have been obtained for a variety of damage cases and amplitudes. The data show trend similarity between experiments and consistency with theory in ranking and asymptotic behaviour at low frequency.

The experiment necessarily simplifies the damaged ship to a prismatic model with thin-walled structure and no internal layout oscillated in a single degree of freedom. Whilst this is only a subset of the 6 degree of freedom behaviour of a ship in waves, the experiment provides a dataset for the validation of theoretical analysis tools for damaged ship motion and load prediction.

For low frequencies of oscillation, this dataset raises questions over the 'calibration' of a quasi-steady flooding algorithm. At high frequency, the dataset raises questions on the validity of the assumption of quasi-steady fluid flow.

The importance of the observed behaviour to global structural loads and the efficient calculation of the structural integrity of a damaged ship now needs to be considered.

ACKNOWLEDGMENTS

The authors are indebted to the staff of the UCL Mechanical Engineering Workshop for their hard work preparing the model and the apparatus. We are also grateful for funding provided by Lloyds Register and the UK Ministry of Defence.

REFERENCES

Fox, J.A. 1974. *An Introduction to Engineering Fluid Mechanics*, London: The Macmillan Press.

Groom, I.S. 2003. HMS Nottingham—The view from HQI. *Journal of Marine Design and Operations*, B3:33–40.

Patel, M.H. 1989. *Dynamics of Offshore Structures*, 291–292, London: Butterworths.

Ruponen, P., Sundell, T. & Larmela, M. Validation of a simulation method for progressive flooding. *International Shipbuilding Progress*, 54:305–321.

Salvessen, N., Tuck, E.O. & Faltinsen, O. 1970. Ship motions and sea loads. *Transactions of the Society of Naval Architects and Marine Engineers,* 78: 250–287.

Smith, T.W.P., Drake, K.R. & Rusling, S. 2007. Investigation of the variation of loads experienced by a damaged ship in waves. *Advancements in Marine Structures,* Leiden: Balkema.

Vugts, Ir.J.H. 1968. The hydrodynamic coefficients for swaying, heaving and rolling cylinders in a free surface. *Report 112S*. Netherlands Ship Research Centre.

Analysis and Design of Marine Structures – Guedes Soares & Das (eds)
© 2009 Taylor & Francis Group, London, ISBN 978-0-415-54934-9

Estimation of parametric rolling of ships—comparison of different probabilistic methods

Jelena Vidic-Perunovic & Jørgen Juncher Jensen
Technical University of Denmark, Department of Mechanical Engineering, Coastal,
Maritime and Structural Engineering, Lyngby, Denmark

ABSTRACT: Probability of extreme roll angles due the parametric roll of the ship has been evaluated by use of different probabilistic methods. A containership in head sea has been considered. It is free to roll and surge, whereas the vertical motion only has been statically included in the model. Probability of the roll event in irregular waves has been estimated using the first order reliability method (FORM) and the results have been compared to direct Monte Carlo simulation. The calculations are made for different sea states and operational conditions and for the relevant maximum pre-conditioned roll angles.

1 INTRODUCTION

Different statistical methods can be used for estimation of the probability level of an event in structural mechanics. For the evaluation of the safety of an intact ship it is important to find the probability of maximum prescribed roll angle exceedence. Most general method of direct response simulation can be very time consuming especially when dealing with elaborate non-linear structural models exposed to a complex excitation (as for the parametric roll in ships analyzed in the paper). The first order reliability method FORM is much more efficient method but the question is how accurate predictions it gives compared to direct simulation.

In this paper the applicability of FORM to the prediction of the probability of parametric roll has been investigated.

A panamax container ship has been taken for the analysis. The routine for roll prediction in long-crested irregular sea includes dynamically coupled roll and surge degree of freedom of the ship, calculated at each time instant, Jensen et al. (2007; Vidic-Perunovic et al. (2008). The hydrodynamic routine has been linked to a probabilistic tool in order to find the probability of the extreme roll event. Approximate reliability methods are generally preferred over the direct simulation methods for their efficiency and simplicity. The First Order Reliability Method (FORM) implies linearization of the limit state surface around the design point. In the Second Order Reliability Method (SORM) solution is approximated by second order surface. Probability estimation by use of Monte Carlo simulations gives almost an "exact" solution, taking care that the number of simulations is large enough and that the necessary number of up-crossings the threshold level is satisfied. Various structural reliability methods have been described in e.g. Ditlevsen & Madsen (1996).

Results have been generated for a sea state relevant for parametric roll and in head sea condition, for the range of ship velocities and preconditioned roll angles of in the range of 0.3–0.5 rad. Both cases with the constant and instantaneous ship speed due to the surge velocity have been examined. Predictions by use of the FORM have been compared to the results by Monte Carlo method. Both methods have been implemented in the tool for probabilistic calculations PROBAN (DNV).

If it comes out that FORM is able to estimate probabilities of the roll event accurately, especially aiming on the case where the surge degree of freedom of the ship is taken to account, this may be very beneficial for the use in on-board decision support systems where the response is to be simulated in real time.

2 STRUCTURAL MODEL

When dealing with large amplitude ship motions one can always refer to accurate numerical codes such as WASIM, Vada (1994) or LAMP, France et al. (2003); Shin et al. (2004). They can quite accurately deal with the parametric roll dynamics but they are usually restricted to regular waves due to long realization time. In the present study an efficient methodology is presented using 2-DOF hydrodynamic routine for extreme roll prediction, based on Jensen (2006) and modified in Jensen et al. (2007). Part of it will be repeated here for the completeness. Vertical motions are determined as the frequency dependent transfer functions by closed form expressions (Jensen, 2004) and surge is calculated from the longitudinal Froude-Krylov force, Jensen et al. (2007). The damping is modeleld by a standard combination of a linear, a quadratic and a cubic variation in roll velocity. Only the head sea condition is studied in following. The surge interference with roll has been previously studied by Spyrou (2000) in following sea condition. The procedure is somewhat simplified but similar to ROLLS, Krüger et al. (2004).

Equilibrium equation for roll reads:

$$\ddot{\phi} = -2\beta_1\omega_\phi\dot{\phi} - \beta_2\dot{\phi}\,|\dot{\phi}| - \frac{\beta_3\dot{\phi}^3}{\omega_\phi} - \frac{(g - \ddot{w})\,GZ(\phi)}{r_x^2} \tag{1}$$

where r_x is the roll radius of gyration and g the acceleration of gravity. The roll frequency ω_ϕ is given by the metacentric height GM_{sw} in still water:

$$\omega_\phi = \frac{\sqrt{gGM_{sw}}}{r_x} \tag{2}$$

The surge motion u is determined from the equilibrium equation

$$\ddot{u} = \frac{1}{1.05M}F_x + sg\left(\frac{\dot{u}}{V}\right)^3 \tag{3}$$

where the surge force F_x has been calculated according to the linear theory

$$F_x = \int_L \left\{ \int_{-T}^0 B(x,z)\frac{\partial p(X,z,t)}{\partial x} \right\} dzdx \tag{4}$$

In the stochastic sea the following approximation for the righting arm has been applied:

$$GZ(\phi,t) = GZ_{SW}(\phi) + \frac{h(t)}{0.05L}(GZ(\phi,x_c(t)) - GZ_{SW}(\phi)) \tag{5}$$

The added mass of water in surge is taken to be 5 per cent of ship displacement and the pressure p is the incident linear wave pressure i.e. all radiation and diffraction effects are ignored. The vertical integration is carried out up to the mean water level and over the ship length in horizontal direction. The second term on the right hand side, taken as cubic power in surge velocity variation, is an attempt to model additional thrust i.e. the action to maintain the speed V in waves, where acceleration of gravity g = 9.81 m/s² and coefficient s is taken to be $s = 15$ in order to be in the good agreement with predictions by Gerritsma-Beukelman added resistance, Vidic-Perunovic et al. (2008).

A panamax containership with the same main particulars as given in Jensen and Pedersen (2006) as Ship #1 is considered ($Lpp = 284.72$ m, $B = 32.22$ m,

$T = 10.5$ m). The roll damping coefficients are taken $\beta_1 = 0.012$, $\beta_2 = 0.40$, $\beta_3 = 0.42$, Bulian (2005).

The restoring moment arm has been fitted for the wave length equal to the ships length and the wave height equal to 5% of the ships length. The analytical solution using fifth order polynomial has been fitted to the exact calculation as given in Jensen & Pedersen (2006).

The instantaneous wave height $h(t)$ and the position of the crest x_c are determined by an equivalent wave procedure:

$$a(t) = \frac{2}{L_e}\int_0^{L_e} H\left(X(x,t),t\right)\cos\left(\frac{2\pi x}{L_e}\right)dx$$

$$b(t) = \frac{2}{L_e}\int_0^{L_e} H\left(X(x,t),t\right)\sin\left(\frac{2\pi x}{L_e}\right)dx$$

$$X(x,t) = (x + (V + \dot{u})t)\cos\psi \tag{6}$$

$$h(t) = 2\sqrt{a^2(t) + b^2(t)}$$

$$x_c(t) = \begin{cases} \dfrac{L_e}{2\pi}\arccos\left(\dfrac{2a(t)}{h(t)}\right) & \text{if } b(t) > 0 \\[2ex] L_e - \dfrac{L_e}{2\pi}\arccos\left(\dfrac{2a(t)}{h(t)}\right) & \text{if } b(t) < 0 \end{cases}$$

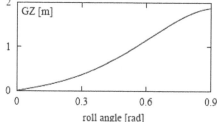

Figure 1. GZ curve in still water.

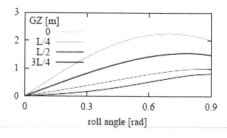

Figure 2. GZ curves in regular waves with wave length equal to the ship length L and wave height of 5% of L. Wave crest positions x = 0, 0.25 L, 0.5 L, 0.75 L, L.

The GZ curve in still water is given in Fig. 1 and for regular wave with different crest positions along the hull in Fig. 2. The length of the wave equals the ship length (critical wave for parametric roll onset) and the wave height 5% of the ship length. GZ (ϕ, t) is linearly scaled on different wave amplitudes, as given in Eq. (5).

3 RELIABILITY METHOD

In the present paper dealing with the roll motion of a ship, linear, long-crested waves are assumed and the normal distributed wave elevation $H(X,t)$. Possible corrections due to severity and non-linearity of the sea can be included, Jensen & Capul (2006). Hence, the wave elevation for moderate sea states as a function of space X and time t can be written

$$H(X,t) = \sum_{i=1}^{n} (u_i c_i(X,t) + \bar{u}_i \bar{c}_i(X,t)) \qquad (7)$$

where the variables $u_i, \bar{u}_i$ are uncorrelated, standard normal distributed variables to be determined by the stochastic procedure and with the deterministic coefficients given by

$$c_i(x,t) = \sigma_i \cos(\omega_i t - k_i X)$$
$$\bar{c}_i(x,t) = -\sigma_i \sin(\omega_i t - k_i X) \qquad (8)$$
$$\sigma_i^2 = S(\omega_i) d\omega_i$$

where ω_i, $k_i = \omega_i^2/g$ are the n discrete frequencies and wave numbers applied. Furthermore, $S(\omega)$ is the wave spectrum and $d\omega_i$ the increment between the discrete frequencies.

From the wave elevation, Eqs. (7)–(8), and the associated wave kinematics, any non-linear wave-induced response $\phi(t)$ of a marine structure can in principle be determined by a time domain analysis using a proper hydrodynamic model:

$$\phi = \phi(t | u_1, \bar{u}_1, u_2, \bar{u}_2, \ldots, u_n, \bar{u}_n, \text{initial conditions}) \qquad (9)$$

Each of these realisations represents the response for a possible wave scenario. The realisation which exceeds a given threshold ϕ_0 at time $t = t_0$ with the highest probability is sought. This problem can be formulated as a limit state problem, well-known within time-invariant reliability theory, Der (2000):

$$g(u_1, \bar{u}_1, u_2, \bar{u}_2, \ldots, u_n, \bar{u}_n) \equiv \phi_0 - \phi(t_0 | u_1, \bar{u}_1, u_2,$$
$$\bar{u}_2, \ldots, u_n, \bar{u}_n) = 0 \quad (10)$$

The integration in Eq. (10) must cover a sufficient time period $\{0, t_0\}$ to avoid any influence on $\phi(t_0)$ of the initial conditions at $t = 0$, i.e. to be longer than the memory in the system.Proper values of t_0 would usually be 1–3 minutes, depending on the damping in the system. Hence, to avoid repetition in the wave system and for accurate representation of typical wave spectra $n = 15$–50 would be needed.

An approximate solution can be obtained by use of the First-Order Reliability Method (FORM). The limit state surface g is given in terms of the uncorrelated standard normal distributed variables $\{u_i, \bar{u}_i\}$, and hence determination of the design point $\{u_i^*, \bar{u}_i^*\}$, defined as the point on the failure surface $g = 0$ with the shortest distance to the origin, is rather straightforward, see e.g. Jensen (2007). A linearization of the limit state surface around this point is then performed with a hyperplane in 2n space. The distance β_{FORM}

$$\beta_{\text{FORM}} = \min \sqrt{\sum_{i=1}^{n} \left(u_i^2 + \bar{u}_i^2 \right)} \qquad (11)$$

from the hyperplane to the origin is denoted the (FORM) reliability index. The calculation of the design point $\{u_i^*, \bar{u}_i^*\}$ and the associated value of β_{FORM} can be performed by standard reliability codes (e.g. Det Norske Veritas (2003).

The deterministic wave profile

$$H^*(X,t) = \sum_{i=1}^{n} (u_i^* c_i(X,t) + \bar{u}_i^* \bar{c}_i(X,t)) \qquad (12)$$

Can be considered as a design wave or a critical wave episode since it has the highest probability of occurrence leading to the exceedance of the conditioned response level ϕ_0.

Probabilistic software PROBAN, Det Norske Veritas (2003) has been used for the present calculations, with its already existing Rackwitz-Fiessler search algorithms in approximate methods, Rackwitz & Fiessler (1978).

The FORM is significantly faster than direct Monte Carlo simulations for low probabilities of exceedance, but most often very accurate. In Jensen & Pedersen (2006) dealing with parametric rolling of ships in head sea the FORM approach was found to be two orders of magnitude faster than direct simulation for realistic exceedance levels i.e. for small failure probabilities.

Second Order Reliability Method (SORM), implemented in the same standard probabilistic tool is able to deal with the higher non-linearities in the wave-induced response $\phi(t)$ and consequently in the limit state surface $g = 0$, since the area around the design point is approximated by a second order surface.

Monte Carlo method is one of the direct simulation methods available in PROBAN, Det Norske Veritas (2003). By random simulation of the response, points are obtained within the safe domain or in the failure domain. It is not an efficient method especially for large complexity structural models or low probabilities of failure. Here the method will be used to check the accuracy of the approximate reliability method.

4 NUMERICAL RESULTS

In the following results are presented in order to demonstrate that the model of the ship is able to predict parametric roll and after that, safety indices have been calculated and compared using different structural reliability methods for large roll angle events.

Using the hydrodynamic model, the response has been simulated from $t = 0$ till $t = 180$s, for number of frequencies $n = 25$. In Fig. 3 the results of the simulation for the design point corresponding to the maximum roll angle 0.5 rad are shown. The sea state is taken to be Hs = 12 m and Tz = 11.7 s, and the constant ship speed as V = 10 m/s. The zero-crossing period is chosen such that parametric roll can be expected due to occurrence of encounter frequencies in the range of twice the roll frequency. The influence of the surge

velocity on the ship speed has been presented in Fig. 3 (lower to the right). The surge acceleration is given in the upper right figure. The critical wave episode gets the non-linear shape due to the time variation in the velocity (upper left). From the ship it looks as if the ship spends more time on the crest than on the through of the wave. The associated most probable roll response is given in Fig. 3 (lower left). Approximately around the time $t = 175$ s the wave crest is located amidships.

At this point the ship passes through the vertical upright position i.e. the roll angle is close to zero. The short period speed variation is zero and the acceleration is at its maximum when the ship has climbed the wave crest.

The calculation of the safety index β gas been conducted for the sea state with Hs = 12 m and the two zero up-crossing periods Tz = 11 s and Tz = 11.7 s. Different ship speeds are taken in calculation and roll angles between 0.3–0.5 rad have been considered as they may be relevant in on-board decision support systems. In Fig. 4 results for the roll angle 0.4 rad are shown in case of constant and variable ship speed. The calculation for the other maximum roll angles is given in Fig. 5 and Fig. 6.

In principle, the following two situations have been experienced during utilization of FORM analysis implemented in PROBAN:

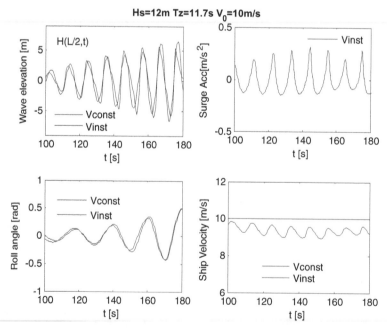

Figure 3. Time domain variation of exciting critical wave episode shown at amidships (upper left figure) conditioned on a roll angle of 0.5 rad at $t_0 = 180$ s, surge acceleration (upper right figure), roll angle (lower left figure), ship velocity including surge effect (lower right figure).

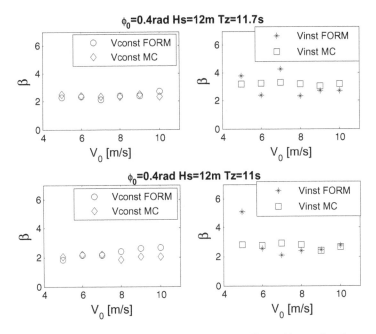

Figure 4. Reliability index for maximum roll angle 0.4 rad, calculated at different ship speeds and sea states, using FORM and Monte Carlo procedure in case of constant (upper and lower left) and non-constant ship speed (upper and lower right hand side).

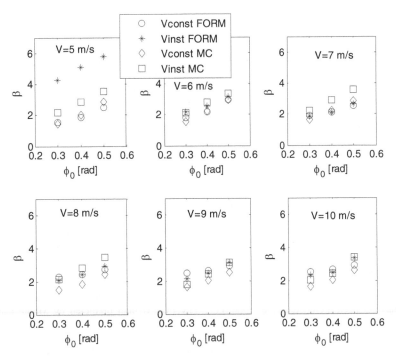

Figure 5. Reliability index by use of FORM and Monte Carlo method, for sea state Hs = 12 m Tz = 11 s and roll angles 0.3–0.5.

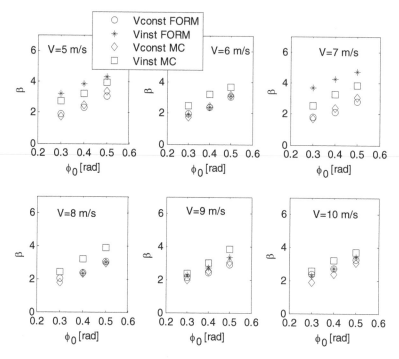

Figure 6. Reliability index by use of FORM and Monte Carlo method, for sea state Hs = 12 m Tz = 11.7 s and roll angles 0.3–0.5.

- The solution is 'wrong' i.e. safety index β is found much larger than the one predicted by Monte Carlo method.
- Safety index by FORM is generally lower than the prediction by Monte Carlo method. The probability of the event is still of the correct order of magnitude. This behavior is more pronounced with the increase of non-linearity in the response and hence, in the limit state surface.

The latter situation is present in the largest number of cases meaning that the accuracy of FORM generally is satisfactory. Similar findings about FORM conservatism have been recently addressed in the work by Yang et al., 2006. An attempt to improve the searching algorithm has been suggested in Wang and Grandhi, 1996.

The probability level by FORM and Monte Carlo with 90% confidence interval is given in Table 1 and Table 2, for the range of ship speeds, for two sea states and three different maximum roll angles where the surge effect has not been included in the model i.e. the ship speed is kept constant at all times. The analogous results have been given in Table 3 and Table 4 for the case of instantaneous ship speed, i.e. when the surge has been accounted for in the hydrodynamic model. Tables are given in Appendix. The comparison

shows generally good agreement in order of magnitude, especially in case of constant ship speed, with some conservatism, as already mentioned.

Second order reliability method (SORM) is in general more accurate to use since largely curved failure surfaces can be better approximated. The comparison with the SORM analysis for one speed V = 6 m/s and the sea state Hs = 12 m, Tz = 11.7s (Fig. 7) shows very good agreement with the other methods in case of constant ship speed and, as expected, better agreement between the SORM and the predictions by the direct simulation method.

5 CONCLUSIONS

A reasonably good agreement has been found between the statistical predictions by FORM and Monte Carlo probabilistic methods for parametric rolling in the case of constant ship speed. In the case of instantaneous ship speed due to the influence of surge velocity, generally lower value of reliability index is found by FORM than predicted by Monte Carlo method. Though FORM gives conservative results, still the probabilities of the event are of the correct order of magnitude. The exceptions are cases when the searching algorithm implemented in FORM fails to find a solution with the minimum distance to the limit state surface. This problem should be investigated more.

The deviations from Monte Carlo predictions increase with the level of non-linearity in the system.

FORM requires by far the shortest computational time (order of magnitude of several minutes per calculation) and can therefore be very beneficial for application to on-board decision support systems.

ACKNOWLEDGEMENT

The first author greatly acknowledges the financial support from the European Commission through the Network of Excellence on Marine Structures project MARSTRUCT, Contract No. FP6-PLT-506141.

REFERENCES

Bulian, G. Nonlinear parametric rolling in regular waves—a general procedure for the analytical approximation of the GZ curve and its use in time simulations. *Ocean Engineering*, 2005, Vol. 32, pp. 309–330.

Der Kiureghian, A. The geometry of Random Vibrations and Solutions by FORM and SORM. *Probabilistic Engineering Mechanics*, 2000, Vol. 15, pp. 81–90.

Det Norske Veritas. Proban, General Purpose Probabilistic Analysis Program, Version 4.4, 2003.

Ditlevsen, O. and Madsen, H.O. Structural Reliability Methods. 1996, Wiley, Chichester.

France, W.N., Levadou, M., Treakle, T.W., Paulling, J.R., Michel, R.K. and Moore, C. An Investigation of Head-Sea Parametric Rolling and Its Influence on Container Lashing Systems. *Marine Technology*, 2003, Vol. 40, No. 1, pp. 1–19.

Rackwitz, R. and Fiessler, B. Structural Reliability Under Combined Random Load Sequence. *J. Computers and Structures*, 1978, Vol. 9, pp. 489–494.

Jensen, J.J., Mansour, A.E. and Olsen, A.S. Estimation of Ship Motions using Closed-Form Expressions. *Ocean Engineering*, 2004, Vol. 31, pp. 61–85.

Jensen, J.J. and Pedersen, P.T. Critical Wave Episodes for Assessment of Parametric Roll. *Proc. IMDC'06*, 2006, Ann Arbor, May, pp. 399–411.

Jensen, J.J. and Capul, J. Extreme Response Predictions for Jack-up Units in Second Order Stochastic Waves by FORM. *Probabilistic Engineering Mechanics*, 2006, Vol. 21, No. 4, pp. 330–337.

Jensen, J.J. Efficient Estimation of Extreme Non-linear Roll Motions using the First-order Reliability Method (FORM). *J. Marine Science and Technology*, 2006.

Jensen, J.J., Pedersen, P.T. and Vidic-Perunovic, J. Estimation of Parametric Roll in a Stochastic Seaway. IUTAM Symposium on Fluid-Structure Interaction in Ocean Engineering, 2007, Hamburg, Germany.

Krüger, S., Hinrichs, R. and Cramer, H. Performance Based Approaches for the Evaluation of Intact Stability Problems. *Proc. PRADS'2004*, 2004, Travemünde, September, Germany.

Shin, Y.S., Belenky, V.L., Paulling, J.R., Weems, K.M. and Lin, W.M. Criteria for Parametric Roll of Large Containerships in Head Seas. *Transactions of SNAME*, 2004, Vol. 112, pp. 14–47.

Spyrou, K.J. On the Parametric Rolling of Ships in a Following Sea Under Simultaneous Nonlinear Periodic Surging. *Phil. Trans. R. Soc. London*, 2000, No. 358, pp. 1813–1834.

Vada, T. SWAN-2 Theory and Numerical Methods. Technical Report No 94–2030, Det Norske Veritas, 1994, Høvik.

Vidic-Perunovic, J., Rognebakke, O., Jensen, J.J. and Pedersen, P.T. Influence of Surge Motion on Extreme Roll Amplitudes. 2008, Proc. of the 27th International Conference on OffshoreMechanics and Arctic Engineering - OMAE2008. Estoril, Portugal.

Wang, L. and Grandhi, R.V. Safety Index Calculation Using Intervening Variables for Structural Reliability Analysis. *J. Computers and Structures*, 1996, Vol. 59, pp. 563–571.

Yang, D., Gang, L. and Gengdong, C. Convergence analysis of first order reliability method using chaos theory. *J. Computers and Structures*, 2006, Vol. 84, No. 6, pp. 1139–1148.

APPENDIX

Table 1. Probability of the response exceedance by FORM and Monte Carlo procedure with 90% confidence interval, sea state $H_s = 12$ m $T_z = 11.7$ s, case with constant speed V_{const}.

$H_s = 12$ m $T_z = 11.7$ s (V_{const})

Roll = 0.3 rad

V_0 [m/s]	Monte Carlo Probability 90% confidence interval	FORM Probability
5	$3.514 \cdot 10^{-2}$–$4.146 \cdot 10^{-2}$	$2.917 \cdot 10^{-2}$
6	$3.742 \cdot 10^{-2}$–$4.591 \cdot 10^{-2}$	$2.465 \cdot 10^{-2}$
7	$3.947 \cdot 10^{-2}$–$4.613 \cdot 10^{-2}$	$3.606 \cdot 10^{-2}$
8	$3.389 \cdot 10^{-2}$–$4.086 \cdot 10^{-2}$	$2.198 \cdot 10^{-2}$
9	$2.243 \cdot 10^{-2}$–$2.757 \cdot 10^{-2}$	$1.747 \cdot 10^{-2}$
10	$2.809 \cdot 10^{-2}$–$3.271 \cdot 10^{-2}$	$8.432 \cdot 10^{-3}$

Roll = 0.4 rad

5	$5.309 \cdot 10^{-3}$–$7.246 \cdot 10^{-3}$	$1.073 \cdot 10^{-2}$
6	$6.626 \cdot 10^{-3}$–$9.574 \cdot 10^{-3}$	$9.285 \cdot 10^{-3}$
7	$7.790 \cdot 10^{-3}$–$1.010 \cdot 10^{-2}$	$1.622 \cdot 10^{-2}$
8	$8.477 \cdot 10^{-3}$–$1.112 \cdot 10^{-2}$	$8.568 \cdot 10^{-3}$
9	$4.909 \cdot 10^{-3}$–$7.491 \cdot 10^{-3}$	$8.207 \cdot 10^{-3}$
10	$8.042 \cdot 10^{-3}$–$1.062 \cdot 10^{-2}$	$3.103 \cdot 10^{-3}$

Roll = 0.5 rad

5	$1.472 \cdot 10^{-4}$–$6.306 \cdot 10^{-4}$	$1.059 \cdot 10^{-3}$
6	$4.796 \cdot 10^{-4}$–$1.187 \cdot 10^{-3}$	$1.165 \cdot 10^{-3}$
7	$7.027 \cdot 10^{-4}$–$1.520 \cdot 10^{-3}$	$2.628 \cdot 10^{-3}$
8	$1.119 \cdot 10^{-3}$–$2.103 \cdot 10^{-3}$	$1.181 \cdot 10^{-3}$
9	$8.860 \cdot 10^{-4}$–$1.781 \cdot 10^{-3}$	$1.775 \cdot 10^{-3}$
10	$7.027 \cdot 10^{-4}$–$1.520 \cdot 10^{-3}$	$3.279 \cdot 10^{-4}$

Table 2. Probability of the response exceedance by FORM and Monte Carlo procedure, sea state $H_s = 12$ m $T_z = 11$ s, case with constant speed V_{const}.

$H_s = 12$ m $T_z = 11$ s (V_{const})

Roll = 0.3 rad

V_0 [m/s]	Monte Carlo Probability 90% confidence interval	FORM Probability
5	$7.551 \cdot 10^{-2} - 8.276 \cdot 10^{-2}$	$6.013 \cdot 10^{-2}$
6	$5.623 \cdot 10^{-2} - 6.257 \cdot 10^{-2}$	$3.691 \cdot 10^{-2}$
7	$4.461 \cdot 10^{-2} - 5.032 \cdot 10^{-2}$	$2.741 \cdot 10^{-2}$
8	$6.481 \cdot 10^{-2} - 7.159 \cdot 10^{-2}$	$1.089 \cdot 10^{-2}$
9	$4.532 \cdot 10^{-2} - 5.108 \cdot 10^{-2}$	$6.942 \cdot 10^{-3}$
10	$4.902 \cdot 10^{-2} - 5.498 \cdot 10^{-2}$	$6.054 \cdot 10^{-3}$

Roll = 0.4 rad

5	$1.907 \cdot 10^{-2} - 2.293 \cdot 10^{-2}$	$3.143 \cdot 10^{-2}$
6	$1.249 \cdot 10^{-2} - 1.565 \cdot 10^{-2}$	$1.556 \cdot 10^{-2}$
7	$1.198 \cdot 10^{-2} - 1.509 \cdot 10^{-2}$	$1.647 \cdot 10^{-2}$
8	$8.477 \cdot 10^{-3} - 1.112 \cdot 10^{-2}$	$7.922 \cdot 10^{-3}$
9	$1.907 \cdot 10^{-2} - 2.293 \cdot 10^{-2}$	$4.599 \cdot 10^{-3}$
10	$1.767 \cdot 10^{-2} - 2.139 \cdot 10^{-2}$	$3.954 \cdot 10^{-3}$

Roll = 0.5 rad

5	$1.514 \cdot 10^{-3} - 2.753 \cdot 10^{-3}$	$6.201 \cdot 10^{-3}$
6	$1.119 \cdot 10^{-3} - 2.214 \cdot 10^{-3}$	$1.608 \cdot 10^{-3}$
7	$1.457 \cdot 10^{-3} - 2.677 \cdot 10^{-3}$	$5.441 \cdot 10^{-3}$
8	$6.557 \cdot 10^{-3} - 8.910 \cdot 10^{-3}$	$3.474 \cdot 10^{-3}$
9	$4.719 \cdot 10^{-3} - 6.747 \cdot 10^{-3}$	$1.890 \cdot 10^{-3}$
10	$3.391 \cdot 10^{-3} - 5.142 \cdot 10^{-3}$	$1.684 \cdot 10^{-3}$

Table 3. Probability of the response exceedance by FORM and Monte Carlo procedure with 90% confidence interval, sea state $H_s = 12$ m $T_z = 11.7$ s, case with variable speed V_{inst}.

$H_s = 12$ m $T_z = 11.7$ s (V_{inst})

Roll = 0.3 rad

V_0 [m/s]	Monte Carlo Probability 90% confidence interval	FORM Probability
5	$1.839 \cdot 10^{-3} - 4.161 \cdot 10^{-3}$	$6.543 \cdot 10^{-4}$
6	$4.504 \cdot 10^{-3} - 7.829 \cdot 10^{-3}$	$2.704 \cdot 10^{-2}$
7	$4.373 \cdot 10^{-3} - 6.827 \cdot 10^{-3}$	$3.985 \cdot 10^{-5}$
8	$6.649 \cdot 10^{-3} - 8.795 \cdot 10^{-3}$	$2.745 \cdot 10^{-2}$
9	$7.422 \cdot 10^{-3} - 9.912 \cdot 10^{-3}$	$1.111 \cdot 10^{-2}$
10	$4.053 \cdot 10^{-3} - 5.947 \cdot 10^{-3}$	$1.175 \cdot 10^{-2}$

Roll = 0.4 rad

5	$3.502 \cdot 10^{-4} - 9.831 \cdot 10^{-4}$	$7.097 \cdot 10^{-5}$
6	$3.081 \cdot 10^{-4} - 9.141 \cdot 10^{-4}$	$8.857 \cdot 10^{-3}$
7	$2.259 \cdot 10^{-4} - 7.741 \cdot 10^{-4}$	$2.648 \cdot 10^{-6}$
8	$3.929 \cdot 10^{-4} - 1.059 \cdot 10^{-3}$	$9.715 \cdot 10^{-3}$
9	$7.890 \cdot 10^{-4} - 1.744 \cdot 10^{-3}$	$3.475 \cdot 10^{-3}$
10	$3.200 \cdot 10^{-4} - 1.013 \cdot 10^{-3}$	$3.289 \cdot 10^{-3}$

(continued)

Table 3. *(continued)*

Roll = 0.5 rad

5	$7.104 \cdot 10^{-6} - 7.290 \cdot 10^{-5}$	$8.532 \cdot 10^{-6}$
6	$6.302 \cdot 10^{-5} - 1.770 \cdot 10^{-4}$	$9.168 \cdot 10^{-4}$
7	$2.790 \cdot 10^{-5} - 1.054 \cdot 10^{-4}$	$2.182 \cdot 10^{-7}$
8	$2.518 \cdot 10^{-6} - 9.748 \cdot 10^{-5}$	$1.254 \cdot 10^{-3}$
9	$1.184 \cdot 10^{-5} - 1.215 \cdot 10^{-4}$	$3.899 \cdot 10^{-4}$
10	$3.285 \cdot 10^{-5} - 1.671 \cdot 10^{-4}$	$3.057 \cdot 10^{-4}$

Table 4. Probability of the response exceedance by FORM and Monte Carlo procedure, sea state $H_s = 12$ m $T_z = 11$ s, case with variable speed V_{inst}.

$H_s = 12$ m $T_z = 11$ s (V_{inst})

Roll = 0.3 rad

V_0 [m/s]	Monte Carlo Probability 90% confidence interval	FORM Probability
5	$1.375 \cdot 10^{-2} - 1.633 \cdot 10^{-2}$	$1.034 \cdot 10^{-5}$
6	$1.550 \cdot 10^{-2} - 1.850 \cdot 10^{-2}$	$1.523 \cdot 10^{-2}$
7	$1.140 \cdot 10^{-2} - 1.415 \cdot 10^{-2}$	$3.326 \cdot 10^{-2}$
8	$1.552 \cdot 10^{-2} - 1.902 \cdot 10^{-2}$	$1.857 \cdot 10^{-2}$
9	$3.350 \cdot 10^{-2} - 3.805 \cdot 10^{-2}$	$1.709 \cdot 10^{-2}$
10	$1.876 \cdot 10^{-2} - 2.258 \cdot 10^{-2}$	$9.428 \cdot 10^{-3}$

Roll = 0.4 rad

5	$1.747 \cdot 10^{-3} - 2.753 \cdot 10^{-3}$	$1.671 \cdot 10^{-7}$
6	$2.096 \cdot 10^{-3} - 3.304 \cdot 10^{-3}$	$5.442 \cdot 10^{-3}$
7	$1.309 \cdot 10^{-3} - 2.358 \cdot 10^{-3}$	$1.717 \cdot 10^{-2}$
8	$1.743 \cdot 10^{-3} - 3.057 \cdot 10^{-3}$	$8.202 \cdot 10^{-3}$
9	$7.011 \cdot 10^{-3} - 9.211 \cdot 10^{-3}$	$6.605 \cdot 10^{-3}$
10	$3.152 \cdot 10^{-3} - 4.848 \cdot 10^{-3}$	$2.991 \cdot 10^{-3}$

Roll = 0.5 rad

5	$5.510 \cdot 10^{-5} - 3.616 \cdot 10^{-4}$	$4.471 \cdot 10^{-9}$
6	$1.324 \cdot 10^{-4} - 5.676 \cdot 10^{-4}$	$8.521 \cdot 10^{-4}$
7	$8.404 \cdot 10^{-6} - 3.249 \cdot 10^{-4}$	$4.246 \cdot 10^{-3}$
8	$4.738 \cdot 10^{-5} - 4.860 \cdot 10^{-4}$	$1.833 \cdot 10^{-3}$
9	$5.235 \cdot 10^{-4} - 1.254 \cdot 10^{-3}$	$8.698 \cdot 10^{-4}$
10	$8.817 \cdot 10^{-5} - 5.785 \cdot 10^{-4}$	$3.815 \cdot 10^{-4}$

Local hydro-structure interactions due to slamming

Š. Malenica & F.X. Sireta
Burau Veritas - Research Department, Neuilly sur Seine, France

S. Tomašević
Faculty of Mechanical Engineering and Naval Architecture, Zagreb, Croatia

J.T. Tuitman & I. Schipperen
TNO, Delft, The Netherlands

ABSTRACT: It is fair to say that the general seakeeping problem of ship advancing with arbitrary forward speed in rough waves is still a challenge and no fully satisfactory numerical model exists up to now. Indeed physics involved in this problem is extremely complex and includes many different physical phenomena which can not be handled by a single complete model. The usual practice consists in subdividing the problem into simpler problems targeting the specific applications. One of the very important practical problems concerns the slamming loads and their effects on ship structural responses. This is the main focus of the present paper. Only potential flow models for fluid flow are discussed and that both for seakeeping and slamming. Seakeeping model in time domain is constructed from the frequency domain data using the method proposed by Cummins 1962 and Ogilvie 1964. Once in the time domain slamming events are identified and treated using the strip approach with Generalized Wagner Model (GWM) for each strip. Pressure, which is calculated by GWM, is finally transferred to the FE model using the interpolation scheme and the final loading cases for FEM calculations are created.

1 INTRODUCTION

It is enough to take a look on Figure 1, to understand how complex general seakeeping problem is. As briefly mentioned in the abstract, lot of different physical phenomena are involved (waves, ship speed, large ship motions, complex free surface deformations, hull flexibility, slamming, sprays, wind, ...) and it is impossible to take all them into account by a single numerical model. The usual practice is to subdivide the overall problem into simpler parts which target the specific applications. As far as the ship structural response is concerned, the rough subdivision can be made as follow:

- Global hydro-structural issues

  - quasi static linear rigid body structural response
  - quasi static weakly non linear rigid body structural response
  - linear hydro-elastic ship response
  - weakly non-linear hydroelastic ship response

- Local hydro-structural issues (quasi static and hydroelastic)

  - slamming
  - sloshing
  - green water ...

Slamming represents a very important source of structural loading both from local and global points of view. Indeed, the severe local pressures, induced by slamming, can cause local structural deformations of the ship's bow but, on the other hand, the importance of the integrated slamming action can also induce global ship vibrations called whipping. In this paper the main focus is on the local hydro-structure interactions due to slamming in the context of the potential flow models of fluid flow and quasi static approach for structural response.

Figure 1. Ship sailing in rough waves.

2 CONCEPTUAL MODEL

The overall procedure for slamming calculations was presented in Malenica et al. 2007, and here below we just briefly recall the basic principles.

As already mentioned, the potential flow model is adopted. The analysis starts with the linear seakeeping calculations in frequency domain, after what the time domain equivalent model is built using the method proposed in Cummins 1962 and Ogilvie 1964. In this way, the following well known motion equation is obtained:

$$([\mathbf{m}] + [\mathbf{A}^\infty])\{\ddot{\xi}(t)\} + ([\mathbf{k}] + [\mathbf{C}])\{\xi(t)\}$$

$$+ \int_0^t [\mathbf{K}(t - \tau)]\{\dot{\xi}(\tau)\}d\tau = \{\mathbf{F}^{DI}(t)\} + \{\mathbf{Q}(t)\} \tag{1}$$

where the over-dots denote the time derivatives. Thanks to the fact that the motion equation is written in the time domain, nonlinear effects (denoted by $\{\mathbf{Q}(t)\}$) can be included relatively easily. Among these nonlinear effects, the most well known is probably the so called non linear Froude-Krylov loading which can be calculated using the linear results (relative wave elevation) as a input. This model is known as a weakly-nonlinear seakeeping model, and in spite of the several approximations which are included in it, it seems to give reasonable approximation for ship motions (Fig. 2), so that the input parameters for calculation of slamming loading can be calculated with sufficient accuracy.

2.1 Hydrodynamic slamming model

Slamming represents probably the most complicated part of the wave loading. The present state of the art do not allow for the use of the 3D slamming models, and the so called strip approach is usually employed. Within this approach the part of the ship which is likely to experience slamming, is subdivided into several strips which are treated separately and combined together afterward. Typical example of the slamming sections is shown in Figure 3.

The main input for slamming calculations at each section are the relative geometry and the relative velocity in between the section and the wave surface. These quantities are calculated at each time step, using the above described seakeeping model. Here below we briefly describe the exact procedure for determination of the relative motion and the relative velocity for one particular section (see also Tuitman et al. 2009).

The lowest point at the section is used as reference point to calculate the relative motions. The intersection of a vertical line in the section plane which goes trough the reference point and the wave is found using a numerical iteration. This gives the relative displacement. Figure 4 illustrates the numerical approach which is used to calculate the relative velocity. The relative displacement, rd, is calculated for a time step before and after the actual time step. The wave elevation is known for every time step, but the displacements of this ship is not known for the next time step. Therefore the velocities of the ship are used to make an estimation of the displacement of

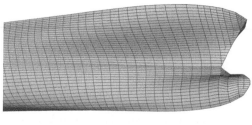

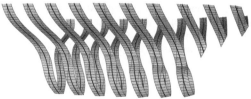

Figure 3. 3D hydrodynamic mesh and corresponding 2D sections.

Figure 2. Slamming event in experiments and in numerical simulations.

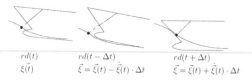

Figure 4. Calculation of relative velocity.

the ship at the two time step. The relative velocity is equal to:

$$\dot{rd} = \frac{rd(t + \Delta t) - rd(t - \Delta t)}{2\Delta t} \qquad (2)$$

The velocity calculated by equation 2 is used for the decision to start a slamming calculation. During the slamming calculation the relative velocity is calculated using a higher order differentiating scheme to obtain a smoother velocity input which is important for the slamming calculation.

Once the relative velocity is know, a separate module is used for calculation of the slamming pressures and forces. The method which is used here, is based on the so called Generalized Wagner Model (GWM) first presented in Zhao et al. 1996. GWM represents an improvement of the original Wagner model in the sense that the body boundary condition is satisfied on the exact body surface (Figure 5), the condition on the free surface ($\phi = 0$) is the same as in the original Wagner model but it is imposed on the horizontal lines at the splash-up height. The splash-up height is unknown in advance and is determined with the help of the linearized kinematic condition and the "Wagner condition", which states that the elevation of the free surface is equal to the vertical coordinate of the body surface at the contact point (in symmetric case).

The numerical method which is used for solving the boundary value problem for velocity potential is based on the classical Boundary Integral Equation Technique, in which the body geometry is subdivided into a finite number of segments over which a constant strength singularity distribution is assumed. Once the velocity potential evaluated, the pressures are calculated by the help of the nonlinear Bernoulli equation. Due to some theoretical inconsistencies in the GWM approach, some special treatment of the pressure near the contact point is necessary.

Once the slamming pressure calculated, they need to be integrated over the wetted part of the slamming section and can be added to the nonlinear loading term $\{\mathbf{Q}(t)\}$ in the motion equation (2). Due to the different time integration schemes in the seakeeping and slamming models, and very different time steps (much shorter for slamming), the inclusion of the slamming loading into the seakeeping motion

equation should be performed very carefully (eg see Tuitman et al. 2009). Another option consists in neglecting the influence of slamming on ship motions. Indeed, experience shows that, in many cases, the slamming loading is relatively small compared to the overall ship inertia and global wave loading, so that the overall motions are not much affected. In this way the procedure is much simpler since the seakeeping and slamming calculations are performed separately.

Anyhow, whatever the coupling method which is chosen, the technical difficulties associated with the pressure transfer from 2D hydrodynamic slamming model(s) to 3D FEM structural model remains the sames.

2.2 Pressure transfer

The slamming impact remains a 3D problem even if the pressures and forces are calculated with a 2D

Figure 6. Typical distribution of the boundary elements at each section, in GWM approach.

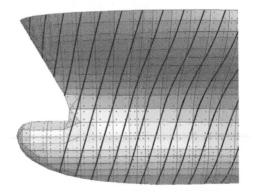

Figure 7. Gauss integration points at the FEM model.

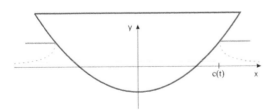

Figure 5. Generalized Wagner model.

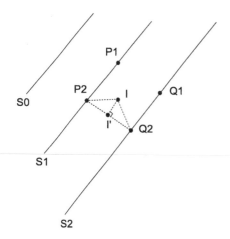

Figure 8. Space interpolation of pressure from pressure points to integration point.

at the Gauss points and divided the found force by the element area. The interpolation of the slamming pressure, onto each Gauss or integration point (I) of the FE model, is done within 6 main steps (Figure 8):

1. Find two slamming sections which are closest to the integration point [sections S1 & S2].
2. Find two points on each section which are closest to the integration point [P1, P2, Q1 & Q2].
3. In between the 4 points found in the previous step, chose 2 closest ones [P2 & Q2].
4. Project the real integration point onto the line defined by the 2 closest points [I goes to I'].
5. Found the pressure at the 2 points from the previous step, by linear interpolation of the associated time histories.
6. Linear space interpolation at I' using the pressures at two points from previous step. This is the pressure which is attributed to the integration point I.

Let us also note that some more complex time and space interpolation schemes were also tested but no significant improvement of the results was found.

The final output of the above procedure is the corresponding loading case for 3DFEM structural code, for required time instant. Indeed, as mentioned at the beginning, the structural calculations are performed in quasi static manner so that each time step can be treated independently.

approach. Figure 5 illustrates the slamming calculation using the GWM during the seakeeping calculation. The 2D slamming calculations starts when the section hits the water with large enough relative velocity. This happens at a different time for the different sections. Also the GWM uses an iterative time step for solving the slamming problem. This implies that the time steps which are used for each particular slamming calculation will be different and an interpolation in time is needed to transfer the slamming pressure to the FE-model for a fixed time step.

In order to calculate the slamming pressure as accurate as possible, the GWM distributes a fixed number of boundary elements over the wetted section at each time step. In particular, there are more elements at the intersection with the free surface of fluid, because of the larger pressure gradient. The typical distribution of the boundary elements can be seen in Figure 6. In order to transfer the GWM pressure to the FE-mesh, it is possible to use the original GWM pressure distribution over these boundary elements, but this would require the spatial interpolation procedure at each time step. Within the present approach, this is avoided by re-calculating the pressure inside the GWM program at fixed points. Using these fixed points makes a single spatial interpolation valid for all time steps!

Several types of interpolation procedures can be employed and here we present probably the simplest possible option. The first step is to define integration points at the FE-mesh. Typical example of integration points is shown in Figure 7. So called Gauss points are used for an accurate integration. The accuracy of the integration and total interpolation can be changed by using a different number of Gauss points per FE-element. The total pressure at a FE-element is obtained by integrating the pressure

3 EXAMPLES

Here below we present one example of calculations which were performed using the above described methodology. It is the case of 7800 TEU Container vessel. The hydrodynamic mesh is shown in Figure 10. and the structural FE model of the bow part in Figure 11. From the time domain simulations for one particular sea spectra the most severe slamming situation is chosen. The typical time histories at several points of one slamming section which are obtained using the GWM,

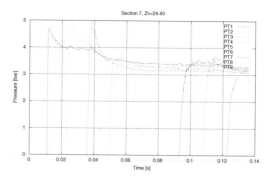

Figure 9. Time histories of the pressure at several points of one particular slamming section obtained by GWM.

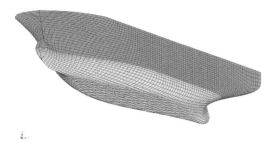

Figure 10. Hydrodynamic mesh for linear and weakly nonlinear seakeeping calculations.

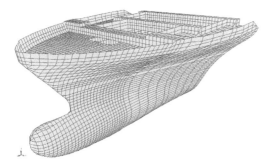

Figure 11. Finite element model of the fore part of 7800 TEU Container Ship.

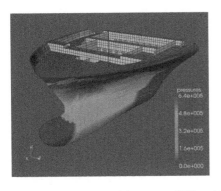

Figure 12. Pressure loading of the structural FE model at one particular time instant.

are shown in Figure 9. These pressures are transfered to the structural FE model using the above described procedure and the corresponding slamming pressure field at one particular time instant is presented in Figure 12.

4 CONCLUSIONS & FURTHER WORK

We presented here a general methodology for evaluation of the local structural response of the ship under the slamming loading. It is clear that the proposed method is not fully consistent and lot of assumptions are made at several steps of the procedure, but we believe that the proposed approach is good starting point for investigation of this complex phenomena. In spite of all the limitations, the method has a merit to keep the most important parts of the physics and to combine the state of the art models for each particular part of the analysis (seakeeping, slamming, structural FE model, ...). At the same time the method is very robust, automatized and easy to use, so that the overall calculation procedure can be performed very quickly starting from scratch.

Let us finally note that the method presented here is certainly not the only one which can be used for this type of applications and, among others, the methods based on the modern CFD techniques (VOF, SPH, ...) seem to be good candidates in view of the important advances made in this field during the last decade. The main advantages of the potential flow models lie in their simplicity, robustness and small CPU time requirements.

ACKNOWLEDGEMENTS

Important part of this work was done within the MARSTRUCT European project and the Cooperative Research Ship Program.

REFERENCES

Cummins W.E., (1962): "The impulse response function and the ship motions.", *Schiffstechnik*, Vol. 47, pp. 101–109.

Faltinsen O.M., (1990): *"Sea loads on ships and off-shore structures."*, Cambridge University Press.

Malenica Š. & Korobkin A.A. (2007): "Some aspects of slamming calculations in seakeeping", In Proc. of 9th Int. Conf. on Numerical Ship Hydrodynamics, Ann Arbor, Michigan USA.

Ogilvie T.F., (1964): "Recent progress toward the understanding and prediction of ship motions.", 5th Symposium on Naval Hydrodynamics.

Tuitman J.T. & Malenica Š, (2009): "Direct coupling between seakeeping and slamming calculation.", 5th Int. Conf. on Hydroelasticity in Marine Technology, Southampton, UK. (in preparation)

Zhao R., Faltinsen O.M. & Aarsnes J.V., (1996): "Water entry of arbitrary two dimensional sections with and without flow separation.", 21st Symp. on Naval Hydrodynamics, Trondheim, Norway.

Methods and tools for strength assessment

Finite elements analysis

Analysis and Design of Marine Structures – Guedes Soares & Das (eds)
© 2009 Taylor & Francis Group, London, ISBN 978-0-415-54934-9

Methods for hull structure strength analysis and ships service life evaluation, for a large LNG carrier

L. Domnisoru, I. Chirica & A. Ioan
University "Dunarea de Jos" of Galati, Naval Architecture Faculty, Galati, Romania

ABSTRACT: This paper is focused on the numerical methods for the ship hull structure strength and fatigue analyses, under extreme hydroelastic wave loads. The ship model is a large LNG carrier, with membrane type cargo-tanks and length of 289.3 m. Two significant load cases are considered: full cargo and ballast. The numerical analyses are divided in three-interlinked parts. The first part includes the method for the hull strength analysis, based on 3D and 1D FEM models, under equivalent quasi-static head wave loads. The second part includes the method for the ship hull dynamic response analysis, based on non-linear hydroelasticity theory, under irregular head waves, model Longuet-Higgins. The last part includes the method for the initial ship hull structure fatigue analysis, based on the cumulative damage ratio criterion, for the World Wide Trade wave significant height histogram. The numerical results outline the extreme wave loads, from slamming-whipping and springing hydroelastic dynamic responses, and the ship initial service life evaluation.

1 INTRODUCTION

Based on the methods improved in the frame of MARSTRUCT Project, we focus in this study on the link between: the ship strength analysis with the finite element method, the waves induced ship hydroelastic response analysis and the ship structure fatigue analysis. The assessment of the ship hull strength and the evaluation of the ship service life are obtained, base on the initial ship hull structure design data concept.

In this study the numerical analyses are carried out on a large LNG—Liquefied Natural Gas Carrier with double hull and membrane cargo-tanks, having a complete separate secondary insulation barrier (GL 2008). For the LNG carrier are considered the full cargo and ballast load cases, under head wave condition. The initial design data are presented in Chapter 2.

The ship structural requirements impose to develop three-dimensional (3D) models, based on the FEM-Finite Element Method (Frieze & Shenoi 2006). The accuracy of the strength analysis for the ship structure is increased by using the 3D-FEM full ship length models (Lehmann 1998, Rozbicki et al. 2001, Domnisoru 2006), instead of models extended only on several cargo-holds (Hughes 1988, Servis et al. 2003). This topic is approached in Chapter 3.

In the standard analysis of the wave induced ship dynamic response are included only the ship rigid hull oscillations (Bhattacharyya 1978, Bertram 2000). For the design of large and elastic ships, the wave induced ship hull dynamic response has to be determined in the hypotheses of the hydroelasticity theory, based

on more realistic non-linear and statistical models, including the low frequency (oscillations) and high frequency (vibrations) components (Hirdaris et al. 2003, 2005). The hydroelastic analysis includes: the linear and non-linear oscillation components, taking into account the bottom and side slamming phenomena, and also the vibration components, on the first and higher order elastic modes, the springing phenomenon, linear and non-linear steady state dynamic response, due to the ship structure-wave resonance, and the whipping phenomenon, slamming induced transitory dynamic response (Bishop & Price 1979, Guedes Soares 1999, Domnisoru 1998, Perunovic & Jensen 2005, Park & Temarel 2007). This topic is approached in Chapter 4.

Based on the stress values induced by the extreme wave loads in the ship girder (see Chapters 3–4), the ship hull structure fatigue strength analysis is carried out, using the cumulative damage ratio D method (Palmgren-Miner method), for steel standard S-N design curves (GL 2000, GL 2008, Garbatov & Guedes Soares 2005) and the World Wide Trade wave significant height histogram. Based on the fatigue analysis, an approximate initial prediction of the ship service life at the early design stage is obtained, considering a travel scenario with equal probabilities of the two loading cases. This topic is approached in Chapter 5.

In this study, the numerical analyses are based on own codes and Cosmos/M (SRAC2001) FEM program. The conclusions of this study are included in Chapter 6.

2 THE LNG CARRIER 3D AND 1D MODELS

In this study is considered a large LNG 150000 m³ carrier, with the main dimensions in Table 1 and the ship offset section lines in Figure 1. The amidships structure is presented in Figure 2, with scantlings in accordance with Germanischer Lloyd Hull Structure Rules (GL 2008).

Table 2, based on the amidships structure, presents the following section characteristics: the moment of inertia I_y [m⁴], the total transversal section area A [m²],

Table 1. LNG 150000 m³ carrier main dimensions.

LOA [m]	289.30	d_{MLD}[m]	11.80
LBP [m]	276.00	c_B	0.732
B_{MLD}[m]	43.30	$Vcargo$ [m³]	150000
D_{MLD}[m]	25.60	Dw [t]	77700

Table 2. LNG 150000 m³ carrier amidships section characteristics.

I_y [m⁴]	1034.782	e_B [m]	14.563
A [m²]	7.275	e_D [m]	17.937
A_{fz} [m²]	3.381	W_B [m³]	71.057
$K_{\tau n-n}$ [m⁻²]	0.563	W_D[m³]	57.689

Table 3. The material isotropic steel characteristics.

ρ_m[t/m³]	7.7	G [N/m²]	8.1 10¹⁰
E [N/m²]	2.1 10¹¹	β	0.001
R_{eH-A}[N/mm²]	235	$R_{eH-AH36}$ [N/mm²]	355
R_{m-A} [N/mm²]	400	R_{m-AH36} [N/mm²]	490
σ_{adm-A} [N/mm²]	175	$\sigma_{adm-AH36}$ [N/mm²]	243
τ_{adm-A}[N/mm²]	110	$\tau_{adm-AH36}$[N/mm²]	153

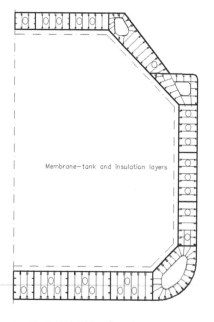

Figure 1. The LNG 150000 m³ carrier offset section lines.

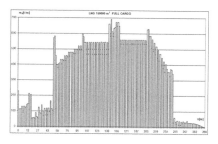

Figure 3a. The mass diagram at Full cargo load case 1.

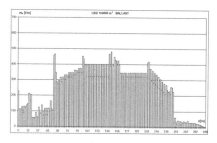

Figure 3b. The mass diagram at Ballast load case 2.

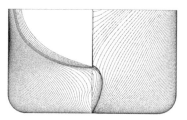

Membrane-tank and insulation layers

Figure 2. The LNG 150000 m³ carrier amidships structure.

the equivalent shearing area A_{fz} [m²], the coefficient of maximum shearing tangential stress in the neutral axis $K_{\tau n-n}$ [m⁻²], the deck and bottom bending modules

W_D, W_B[m³].

$$W_{B,D} = \frac{I_{yy}}{e_{B,D}}; \quad K_{\tau\, n-n} = \frac{S_{y\, n-n}}{I_{yy} \cdot 2 \cdot t_{n-n}};$$

$$t_{n-n} = t_S + t_{DS} \qquad (1)$$

Table 4. LNG 150000 m³ carrier ship loading cases*.

No	Load case	Δ [t]	d_m[m]	d_{aft}[m]	d_{fore}[m]	v_s[Knots]
1	Full cargo	105881.4	11.80	11.80	11.80	20
2	Ballast	73166.8	8.50	10.00	7.00	20

*where Δ is the displacement; d_m, d_{aft}, d_{fore} are the ship draughts: medium, aft peak, fore peak; v_s is the ship navigation speed.

Table 5. LNG 150000 m³ carrier natural modes frequencies f [Hz].

Modes			Oscillations		Vibrations		
No.	Load case		0	1	2	3	4
1	Full	dry	–	–	1.377	2.900	4.523
	cargo	wet	0.109	0.118	0.953	2.085	3.303
2	Ballast	dry	–	–	1.552	3.320	5.217
		wet	0.097	0.112	1.071	2.327	3.675

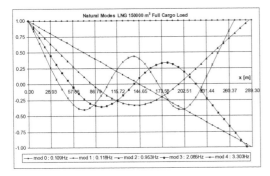

Figure 4. The natural modes at Full cargo load case.

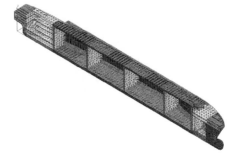

Figure 5a. The LNG 150000 m³ carrier 3D-FEM hull model.

Table 6. The LNG carrier 3D-CAD/FEM model characteristics.

Number of points	PT_{max}	19144
Number of curves	CR_{max}	103927
Number of surfaces	SF_{max}	46560
Number of nodes	ND_{max}	27623
Number of shell3T elements	EL_{max}	93435
Number of element groups	EG	583
Average/maximum EL size [m]	average/max	0.8/1.6
Support condition aft-peak node	ND_{aft}	2007
Support condition fore-peak node	ND_{fore}	26665

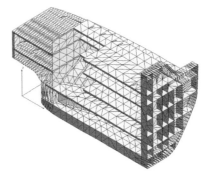

Figure 5b. The LNG 3D-FEM model, the aft peak domain.

where: S_{yn-n} is the transversal section static moment at the neutral axis; t_S, t_{DS}, are the thickness of the side and double side shells; e_D, e_B are the distances of the neutral axis to the deck and bottom shells.

Table 3 presents the material isotropic steel characteristics, the admissible stresses, according to the Germanischer Lloyd's Rules (GL 2008), and also the structural damping coefficient β, according to Johnson & Tamita (Bishop & Price 1979).

In accordance with the Ship Classification Society Rules for the LNG carrier, two main loading cases are

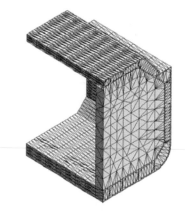

Figure 5c. The LNG 3D-FEM model, detail in the cargo-holds.

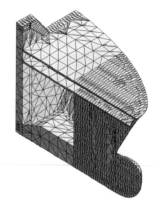

Figure 5d. The LNG 3D-FEM model, the fore peak domain.

considered, presented in Table 4. The mass diagrams are as following: in Figure 3a the full cargo load case 1 and in Figure 3b the ballast load case 2.

In Table 5, based on the rigidity and inertial characteristics of the LNG carrier, the natural hull oscillation and vibration mode frequencies are calculated (1D-FEM model), for dry hull or with hydrodynamic masses (wet hull). In Figure 4 are presented the modal form functions for the full cargo loading case (dry hull at vibration modes).

For the ship dynamic response analysis, presented in Chapter 4, based on the hydroelasticity theory, the LNG 150000 m^3 carrier hull structure is modelled according to the 1D-FEM equivalent beam ship girder model (Domnisoru & Ioan 2007), using the above basic input data. The equivalent ship girder 1D-FEM has 40 equal length elements along the ship, so that from the original ship offset lines (Fig. 1) are selected only the transversal sections disposed at the middle of

each element. The diagrams of the transversal section characteristics (Table 2) are idealised with trapezoidal distributions along the ship.

For the LNG 150000 m^3 carrier strength analysis, presented in Chapter 3, based on 3D-CAD/FEM model developed on the whole ship length (Stoicescu & Domnisoru 2007), the Cosmos/M (SRAC 2001) FEM program is used. In Table 6 the main characteristics of the 3D-CAD/FEM model of the LNG carrier are presented. The 3D-FEM model (including stiffeners) is based on auto-mesh procedure with thick triangular shell elements, which ensures meshing convergence also on surfaces with high curvature. In Figures 5a–d the 3D-FEM LNG carrier hull model is presented. The boundary conditions and the loading cases are introduced on the ship hull 3D-FEM model according to the method presented in Chapter 3. The results based on 1D and 3D models are presented in Chapters 3–5.

3 THE 3D-FEM LNG STRENGTH ANALYSIS

In this section the analysis is focused on the hull structure strength for the LNG 150000 m^3 carrier, for the initial design concept, based on the 3D-FEM model extended on the whole ship length, under the equivalent quasi-static head wave loads.

In the following are included the main steps for the ship strength analysis, based on 3D-FEM model, in detail presented in (Domnisoru 2006).

– the 3D-CAD ship hull offset lines generation;
– the 3D-CAD ship hull structure model;

Table 7. The maximum LNG 150000 m^3 vertical deflection $|w_z|$ [m].

h_w [m]	Full cargo load case		Ballast load case	
	Sagging	Hogging	Sagging	Hogging
0	0.0445	0.0445	0.0532	0.0532
5	0.1850	0.0627	0.0433	0.1426
10.715	0.3724	0.1701	0.2116	0.2428
$w_{z\,max}/w_{adm}$	0.64	0.29	0.37	0.42

Table 8. The LNG buckling factor, reference wave $h_w = 10.715$ m.

Full cargo load case		Ballast load case		
Sagging	Hogging	Sagging	Hogging	Buckling criterion
1.348	1.371	1.758	1.763	>1

Table 9. The maximum LNG 150000 m³ carrier stresses, for the full cargo load case, under sagging and hogging quasi-static equivalent head wave conditions.

h_w[m]	Sagging			Hogging		
	1D-girder	3D-FEM	3D/1D	1D-girder	3D-FEM	3D/1D
1. Deck normal stresses $|\sigma_{xD-max}|$ [N/mm²] : $\sigma_{adm}-A = 175$						
0	30.30	48.15	1.59	30.30	48.15	1.59
5	79.77	97.73	1.23	21.98	45.80	2.08
10.715	144.24	174.82	1.21	64.95	65.81	1.01
$\sigma_{max}/\sigma_{adm}$	0.82	1.00		0.37	0.38	
2. Bottom normal stresses $|\sigma_{xB-max}|$ [N/mm²]: $\sigma_{adm-AH36} = 243$						
0	24.60	39.92	1.62	24.60	39.92	1.62
5	64.77	62.44	0.96	17.84	38.39	2.15
10.715	117.10	116.30	0.99	52.73	54.55	1.03
$\sigma_{max}/\sigma_{adm}$	0.48	0.48		0.22	0.22	
3. Tangential stresses in neutral axis $|\tau_{xznn-max}|$ [N/mm²]: $\tau_{adm-A} = 110$						
0	24.22	43.75	1.81	24.22	43.75	1.81
5	36.26	67.56	1.86	21.27	23.53	1.11
10.715	57.12	96.20	1.68	30.54	23.30	0.76
τ_{max}/τ_{adm}	0.5211	0.87		0.28	0.21	
4. Deck von Mises stresses $|\sigma_{vonD-max}|$ [N/mm²]: $\sigma_{adm-A} = 175$						
0	30.30	61.40	2.03	30.30	61.40	2.03
5	79.77	95.00	1.19	21.98	57.50	2.62
10.715	144.24	170.20	1.18	64.95	63.55	0.98
$\sigma_{max}/\sigma_{adm}$	0.82	0.97		0.37	0.36	
5. Bottom von Mises stresses $|\sigma_{vonB-max}|$ [N/mm²]: $\sigma_{adm-AH36} = 243$						
0	24.60	44.94	1.83	24.60	44.94	1.83
5	64.77	61.70	0.95	17.84	44.61	2.50
10.715	117.10	104.90	0.90	52.73	50.18	0.95
$\sigma_{max}/\sigma_{adm}$	0.48	0.43		0.22	0.21	

- the 3D-FEM ship hull structure mesh model;
- the boundary conditions on the 3D-FEM model.

The boundary conditions are of two types: the symmetry condition at the nodes disposed in the centre plane of the ship (the model is developed only one side); the vertical support condition at two nodes disposed at the ship hull extremities (in the centre plane), noted ND_{aft} at aft peak and ND_{fore} at fore peak. At the vertical equilibrium conditions, for still water or equivalent quasi-static head wave cases, the reaction forces RFZ (ND_{aft}), RFZ (ND_{fore}) in the two vertical supports have to become zero.

- the loading conditions on the 3D-FEM model.

The loads acting on the ship hull structure are of three types: the gravity load from the hull structure weight and other mass components of the displacement; the cargo, the membrane and insulation layers weight; the equivalent quasi-static head wave pressure load, on external ship hull shell, for the following cases: $h_w = 0$ (still water) and $h_w \neq 0$, according the statistical values from Ship Classification Society Rules (GL 2008). An iterative procedure for the free floating and trim condition equilibrium is used, implemented with own GEO macro-commands file (Domnisoru 2006), in the Cosmos/M (SRAC 2001) FEM program, as structural solver at each iteration. According to Germanischer Lloyd (GL 2008), I-Part 1, Ch. 6, Sec. 4.2.2 & Tab. 4.3, the LNG carrier membrane tanks are non-self-supporting tanks, which consist of a thin layer supported through insulation by the adjacent hull structure. The membrane is designed in such a way that thermal and other expansion or contraction is compensated, so that, in the 3D-FEM hull

Table 10. The maximum LNG 150000 m^3 carrier stresses, for the ballast load case, under sagging and hogging quasi-static equivalent head wave conditions.

	Sagging			Hogging		
	1D-girder	3D-FEM	3D/1D	1D-girder	3D-FEM	3D/1D
1. Deck normal stresses $|\sigma_{xD-max}|$ [N/mm^2]: $\sigma_{adm-A} = 175$						
0	12.05	33.93	2.82	12.05	33.93	2.82
5	41.24	44.14	1.07	44.31	52.34	1.18
10.715	99.60	115.40	1.16	82.37	100.00	1.21
$\sigma_{max}/\sigma_{adm}$	0.57	0.66		0.47	0.57	
2. Bottom normal stresses $|\sigma_{xB-max}|$ [N/mm2]: $\sigma_{adm-AH36} = 243$						
0	10.36	39.78	3.84	10.36	39.78	3.84
5	33.48	47.87	1.43	35.97	38.24	1.06
10.715	80.87	80.46	0.99	66.88	70.70	1.06
$\sigma_{max}/\sigma_{adm}$	0.33	0.33		0.28	0.29	
3. Tangential stresses in neutral axis $|\tau_{xznn-max}|$ [N/mm^2]: $\tau_{adm-A} = 110$						
0	21.03	28.48	1.35	21.03	28.48	1.35
5	20.46	38.00	1.86	21.61	24.28	1.12
10.715	41.88	55.36	1.32	37.81	37.20	0.98
τ_{max}/τ_{adm}	0.38	0.50		0.34	0.34	
4. Deck von Mises stresses $|\sigma_{vonD-max}|$ [N/mm2]: $\sigma_{adm-A} = 175$						
0	12.05	54.55	4.53	12.05	54.55	4.53
5	41.24	54.05	1.31	44.31	50.55	1.14
10.715	99.60	110.70	1.11	82.37	96.34	1.17
$\sigma_{max}/\sigma_{adm}$	0.57	0.66		0.47	0.55	
5. Bottom von Mises stresses $|\sigma_{vonB-max}|$ [N/mm2]: $\sigma_{adm-AH36} = 243$						
0	10.36	34.68	3.35	10.36	34.68	3.35
5	33.48	44.08	1.32	35.97	34.54	0.96
10.715	80.87	72.87	0.90	66.88	65.35	0.98
$\sigma_{max}/\sigma_{adm}$	0.33	0.30		0.28	0.27	

model no thermal loads are included. The cargo, the membrane and insulation layers weight is modelled as pressure load over the cargo-tanks inner hull shell.

The iterative procedure includes two main parts:

a. the free floating condition, having as objective numerical function the sum of vertical reaction forces at the two nodes from the ship extremities:

$$RFZ_{eq} = RFZ(ND_{aft}) + RFZ(ND_{fore}) \longrightarrow 0 \quad (2a)$$

b. the free trim and floating condition, having as objective numerical functions the vertical reaction forces at each two nodes from the ship extremities:

$$RFZ(ND_{aft}) \longrightarrow 0 \qquad RFZ(ND_{fore}) \longrightarrow 0 \quad (2b)$$

– the numerical results evaluation.

Based on 3D-FEM models with own GEO macro-commands file (Domnisoru 2006), the following numerical results are obtained: the free floating and trim equilibrium parameters, the deformations of the ship hull, the stress distributions over the whole ship hull length, and also the prediction of the higher risk structural domains. For the equivalent von Mises stress values in deck and bottom structures are calculated the influence coefficients:

$$k_{\sigma vonM} \mid_{D,B} = \sigma_{vonM\_3DFEM} \mid_{D,B} / M_{3DFEM} \quad (3)$$

Based on the classical ship equivalent 1D-girder method, using own program codes (Domnisoru 2006), the results for comparing to the 3D-FEM-model approach are obtained (Tables 9–10).

The numerical data of the 3D-FEM and 1D-girder models of the LNG carrier are presented in Chapter 2.

The numerical LNG 150000 m^3 hull strength analyses are carried out for the two loading cases described in Table 4 and Figures 3a,b.

The external quasi-static head wave pressure, with height $h_w = 0$–12 m, step $\delta h_w = 1$ m is applied on the 3D-FEM hull model, using the iterative procedure for the vertical in plane equilibrium condition. The sagging and hogging head wave conditions are considered. Based on the Germanischer Lloyd's Rules (GL 2008), the Ship Classification Society equivalent quasi-static statistical wave height for the LNG 150000 m^3 carrier is $h_w = 10.715$ m.

The numerical results for the ship strength analyses are synthesized in Tables 7–10 and Figures 6–9, for both loading cases (Table 4), (see Conclusions 1–4).

In the following tables are included the next numerical results:

– the maximum ship girder deflection in vertical direction w_z [m], based on 3D-FEM model, with the

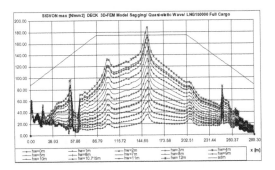

Figure 6a. Maximum von Mises deck stress [N/mm^2], sagging, full.

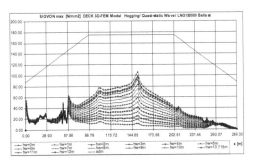

Figure 7b. Maximum von Mises deck stress [N/mm^2], hogging, ballast.

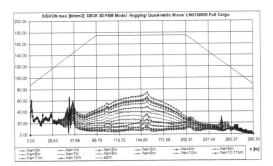

Figure 6b. Maximum von Mises deck stress [N/mm^2], hogging, full.

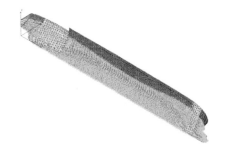

Figure 8a. Wave pressure at $h_w = 10.715$ m, sagging, full cargo.

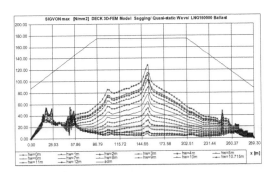

Figure 7a. Maximum von Mises deck stress [N/mm^2], sagging, ballast.

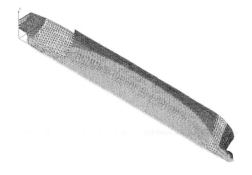

Figure 8b. Wave pressure at $h_w = 10.715$ m, hogging, full cargo.

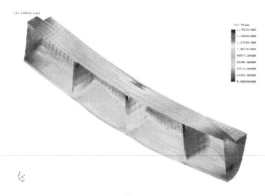

Figure 9a. σ_{vonM} [KN/m²], h_w = 10.715 m, x = 53.6–246.4 m, sagging, full.

Figure 9b. σ_{vonM} [KN/m²], h_w = 10.715 m, x = 53.6–246.4 m, hogging, full.

admissible hull deflection value w_{adm} = $L/500 = 0.579$ m, for both loading cases (Table 7);
– the ship hull girder buckling factor values, based on 3D-FEM model, for quasi-static wave height $h_w = 10.715$ m, for both loading cases (Table 8);
– the maximum deck shell normal stresses $\sigma_{xD\,max}$ [N/mm²], based on 3D-FEM and 1D-girder models, inclusive their ratio 3D/1D (Tables 9–10(1));
– the maximum bottom shell normal stresses $\sigma_{xB\,max}$ [N/mm²], based on 3D-FEM and 1D-girder models, inclusive their ratio 3D/1D (Tables 9–10(2));
– the maximum tangential stresses in the neutral axis $\tau_{xzn-n\,max}$ [N/mm²], based on 3D-FEM and 1D-girder models, inclusive their ratio 3D/1D (Tables 9–10(3));
– the maximum deck shell equivalent von Mises stresses $\sigma_{vonD\,max}$ [N/mm²], based on 3D-FEM and 1D-girder models, inclusive their ratio 3D/1D (Tables 9–10(4));
– the maximum bottom shell equivalent von Mises stresses $\sigma_{vonB\,max}$ [N/mm²], based on 3D-FEM and 1D-girder models, inclusive their ratio 3D/1D (Tables 9–10(5)).

In the next figures are included the following:
– the maximum deck shell equivalent von Mises stress $\sigma_{vonD\,max}$ diagrams, based on 3D-FEM model, for head wave height $h_w = 0$–12 m, full cargo and ballast load cases, sagging and hogging quasi-static equivalent wave conditions (Figs. 6–7a, b);
– the external wave pressure distribution, for head wave height $h_w = 10.715$ m, acting on the bottom and side shell, based on 3D-FEM model, applied with own iterative GEO macro-commands file, full cargo load case, sagging and hogging quasi-static equivalent wave conditions (Figs. 8a, b);
– the equivalent von Mises stress distribution, for head wave height $h_w = 10.715$ m, in the ship central part, $x = 53.6$–246.4 m, based on the 3D-FEM model, full cargo load case, sagging & hogging (Figs. 9a, b).

4 THE LNG DYNAMIC ANALYSIS, BASED ON THE HYDROELASTICITY THEORY

In this section the analysis is focused on the linear and non-linear LNG 150000 m³ carrier dynamic response in irregular head waves, based on the hydro-elasticity theory, including: ship oscillation and vibration, the bottom and side slamming phenomena, the springing and whipping phenomena. The theoretical model is based on the following hypotheses:

– the ship hull is modelled with 1D-FEM finite element method (1D-FEM), using Timoshenko elastic beam finite element (Hughes 1988);
– the ship offset lines are modelled with the conformal multi-parametric transformation method (Bishop & Price 1979);
– the hydrodynamic excitation forces are modelled according to the hydroelasticity and strip theory, with non-linear and slamming terms, based on a generalized Gerristma and Beukelman model (Domnisoru 1998);
– the hydrodynamic coefficients are calculated based on the Porter & Vugts 2D potential fluid flow method (Domnisoru 2006);
– the ship dynamic response is decomposed, according to the modal analysis technique, on ship oscillation (low frequency, rigid hull) and vibration (high frequency, dry elastic hull) modes (Bishop & Price 1979);
– the excitation is the external head wave, model Longuet-Higgins, with second order interference components (Perunovic & Jensen 2005, Domnisoru & Ioan 2007).

Based on the theoretical model, in detail presented in (Domnisoru 1998), the numerical analyses are carried out with own program DYN, which includes the linear STABY and the non-linear TRANZY solvers.

The STABY module is developed for the steady state ship dynamic response, including linear oscillation and springing components. It includes the following main steps:

- input ship data, natural modes, first order wave spectra ITTC (Price & Bishop 1974);
- the calculation of Longuet-Higgins time domain wave elevation, with random components phases;

$$\eta_w(x,t) = \sum_{(\omega_e)} \left[\eta_w^c(x,\omega_e) \cos \omega_e t - \eta_w^s(x,\omega_e) \sin \omega_e t \right]$$

(4)

where ω_e is the encountering ship-wave circular frequency of a wave component; $\eta_w^{c,s}(x,\omega_e)$ are the ω_e wave frequency domain components;

- the calculation of motion equations system terms: structural, hydrodynamic and wave excitation forces;
- the solution in frequency domain of the ship linear dynamic response on each wave component ω_e;
- the calculation of linear ship dynamic response, based on spectral composition in time domain.

The TRANZY module is developed for the non-linear and transitory ship dynamic response, including non-linear oscillations and springing, slamming (bottom and side) and whipping components. It includes the following main steps:

- the ship data input and the solution of STABY module;
- the calculation of the non-linear motion equations system terms.

Because the excitation force includes the unknown non-linear dynamic response, it is necessary to use an iterative algorithm for the time domain solution of non-linear motion equations (Domnisoru 1998).

- the solution of the differential non-linear motion equations system, using a time domain integration procedure, β-Newmark, at each iteration, with simulation time $T_s = 80$ s and time step $\delta t = 0.01$ s;
- the ship non-linear dynamic response;
- the spectral analysis of the total ship dynamic response with the Fast Fourier Transformation (FFT), short-term statistical parameters.

The dynamic analyses are carried out for the head waves first order spectra ITTC (Price & Bishop 1974) with the significant wave height $h_{1/3} = 0$–12 m, step $\delta h_{1/3} = 0.5$ m, according to the Beaufort scale $B_{level} = 0$–11.

For the numerical analysis the 1D-girder model of the LNG carrier is presented in Chapter 2, with the loading cases from Table 4.

The numerical results for the hydroelastic response are synthesized in Tables 11–13 and Figures 10–11, for both loading cases (Table 4), (see Conclusions 5–8).

In the following tables are included the next numerical results:

- the ratio for the significant deformation of the fundamental natural vibration mode and the

Table 11a. The ratios between the significant displacements and deformations $\%w_{1/3vib}/w_{1/3osc}$, full cargo (reference $h_{1/3} = 10.715$ m).

No.	x/L	% vib/osc linear	% vib/osc non-linear	bottom slamming	side slamming
C00	0.00	2.97	3.25	>3	yes
C20	0.50	3.13	3.36	–	–
C40	1.00	3.26	3.45	no	yes
Average		3.12	3.35	$h_{1/3}$ [m] limit	

Table 11b. The ratios between the significant displacements and deformations $\%w_{1/3vib}/w_{1/3osc}$, ballast (reference $h_{1/3} = 10.715$ m).

No.	x/L	%vib/osc linear	%vib/osc non-linear	bottom slamming	side slamming
C00	0.00	4.21	4.67	≥ 0	yes
C20	0.50	4.40	4.52	–	–
C40	1.00	4.18	4.69	>9	yes
Average		4.26	4.63	$h_{1/3}$ [m] limit	

Table 12a. The maximum ratios for the significant bending moments & shearing forces max ($\%M_{1/3vib}/M_{1/3osc}$, $\%T_{1/3vib}/T_{1/3osc}$), full cargo load case ($h_{1/3} = 10.715$ m).

No.	x/L	%vib/osc linear	%vib/osc non-linear	springing	whipping
C10	0.25	4.06	55.13	linear:	
C20	0.50	3.63	57.04	small	high
C30	0.75	3.33	50.14	non-linear:	
Average		3.67	54.10	medium	

Table 12b. The maximum ratios for the significant bending moments & shearing forces max ($\%M_{1/3vib}/M_{1/3osc}$, $\%T_{1/3vib}/T_{1/3osc}$), ballast case ($h_{1/3} = 10.715$ m).

No.	x/L	%vib/osc linear	%vib/osc non-linear	springing	whipping
C10	0.25	3.61	78.46	linear:	
C20	0.50	3.90	78.38	small	higher
C30	0.75	4.06	74.83	non-linear:	
Average		3.86	77.22	medium	

61

Table 13a. The maximum significant stresses (dynamic response), added to still water values, full cargo (deck and side $\sigma_{adm-A} = 175$ N/mm^2 & $\tau_{adm-A} = 110$ N/mm^2 / bottom $\sigma_{adm-AH36} = 243$ N/mm^2).

Maximum stress [N/mm^2] (taking as reference $h_{1/3} = 12$ m)		Strength criterion	$h_{1/3}$[m] limit	Beaufort limit
$\sigma_{max\_LIN} + |sw|$ deck	103.69	0.593 < 1	12	11
$\sigma_{max\_LIN} + |sw|$ bottom	84.18	0.346 < 1	12	11
$\sigma_{max\_NL} + |sw|$ deck	153.93	0.880 < 1	12	11
$\sigma_{max\_NL} + |sw|$ bottom	124.97	0.514 < 1	12	11
$\tau_{max\_LIN} + |sw|$ n-n	49.29	0.448 < 1	12	11
$\tau_{max\_NL} + |sw|$ n-n	75.24	0.684 < 1	12	11

Table 13b. The maximum significant stresses (dynamic response), added to still water values, ballast (deck and side $\sigma_{adm-A} = 175$ N/mm^2 & $\tau_{adm-A} = 110$ N/mm^2/bottom $\sigma_{adm-AH36} = 243$ N/mm^2).

Maximum stress [N/mm^2] (taking as reference $h_{1/3} = 12$ m)		Strength criterion	$h_{1/3}$[m] limit	Beaufort limit
$\sigma_{max\_LIN} + |sw|$ deck	73.99	0.423 < 1	12	11
$\sigma_{max\_LIN} + |sw|$ bottom	60.07	0.247 < 1	12	11
$\sigma_{max\_NL} + |sw|$ deck	137.29	0.784 < 1	12	11
$\sigma_{max\_NL} + |sw|$ bottom	111.46	0.459 < 1	12	11
$\tau_{max\_LIN} + |sw|$ n-n	37.46	0.341 < 1	12	11
$\tau_{max\_NL} + |sw|$ n-n	66.12	0.601 < 1	12	11

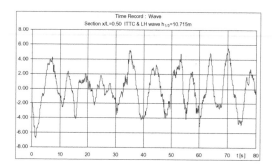

Figure 10a. The wave L-H time record ($h_{1/3} = 10.715$ m, $x/L = 0.5$).

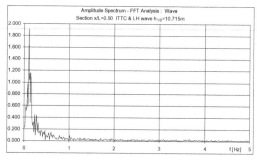

Figure 10b. The wave L-H amplitude spectrum ($h_{1/3} = 10.715$ m, $x/L = 0.5$).

significant vertical displacement of the ship rigid hull oscillations $\%w_{1/3vib}/w_{1/3osc}$ (Tables 11a, b);
- the maximum ratios for the significant bending moments and shearing forces, of fundamental natural vibration mode and the ship rigid hull oscillations, max $(\%M_{1/3vib}/M_{1/3osc}, \%T_{1/3vib}/T_{1/3osc})$ (Tables 12a, b);
- the maximum significant normal deck and bottom, tangential neutral axis stresses (Equation 5), added to still water (σ_{sw}, τ_{sw}) stresses (Tables 13a, b).

$$\sigma_{max\_D,B} = M_{1/3\,max} / W_{D,B}$$

$$\tau_{max\_n-n} = T_{1/3\,max} \, K_{\tau\,n-n}$$

$$\frac{\sigma_{max\_D,B} + |\sigma_{sw}|}{\sigma_{adm}} \leq 1 \qquad \frac{\tau_{max\_D,B} + |\tau_{sw}|}{\tau_{adm}} \leq 1$$

$$(5)$$

In the following figures are included the next numerical analysis results:

- in Figures 10a, b are presented the time record and the amplitude spectrum for the Longuet-Higgins

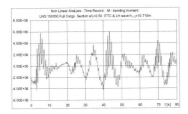

Figure 10c. Bending moment time record, non-linear analysis, wave $h_{1/3} = 10.715$ m, $v_s = 20$ Knots, $x/L = 0.5$, full cargo load case.

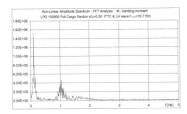

Figure 10d. Bending moment amplitude spectrum FFT, non-linear analysis, $h_{1/3} = 10.715$ m, $v_s = 20$ Knots, $x/L = 0.5$, full cargo load case.

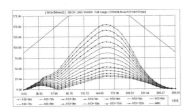

Figure 11a. Maximum significant normal deck stress [N/mm^2], non-linear analysis+still water, $h_{1/3} = 0 - 12$ m, $v_s = 20$ Knots, full cargo case.

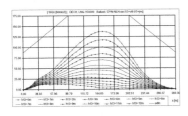

Figure 11b. Maximum significant normal deck stress [N/mm^2], non-linear analysis+still water, $h_{1/3} = 0$–12 m, $v_s = 20$ Knots, ballast case.

(L-H) wave, with first order wave spectrum ITTC $h_{1/3} = 10.715$ m, at amidships section $x/L = 0.5$;
– in Figures 10c, d are presented the time record and the amplitude spectrum FFT, for the non-linear bending moment, head wave $h_{1/3} = 10.715$ m, at amidships section $x/L = 0.5$, full cargo load case;

– in Figures 11a, b are presented the maximum significant deck normal stresses, for both loading cases.

5   THE LNG FATIGUE ANALYSIS AND THE INITIAL SHIP SERVICE LIFE EVALUATION

In this section the analysis is focused on the initial service life evaluation for the LNG 150000 m^3 carrier, based on the fatigue strength assessment of the ship hull structure, using the maximum stresses for extreme wave loads obtained in the deck shell (Chapters 3–4).

In order to evaluate the ship fatigue strength criterion, with the Ship Classification Society Methodology (GL 2000, GL 2008), the cumulative damage ratio D method, based on Palmgren-Miner method and steel standard design S-N curves, is applied.

From the short-term prediction analysis of the ship dynamic response, the significant stresses $\sigma_{1/3}$ of oscillation and vibration components, associated with head waves significant height $h_{1/3}$, are obtained.

In this case, for a reference time of $R = 20$ years, the cumulative damage ratio D has the expression:

$$D = D_{osc} + D_{vib}; \quad n_{i\_osc,vib} = p_i \cdot n_{max\_osc,vib} \quad (6)$$

$$D_{osc,vib} = \sum_{i=1}^{m} \frac{n_{i\_osc,vib}}{N_{i\_osc,vib}}; \quad N_{i\_osc,vib} = f_{SN}\left(\Delta\sigma_{i\_osc,vib}\right) \quad (7)$$

$$n_{max\_osc,vib} = 3.1536 \cdot 10^7 R \cdot f_{osc,vib};$$

$$\Delta\sigma_{i\_osc,vib} = 2\sigma_{1/3i\_osc,vib} \cdot f_c \quad (8)$$

where: $f_{osc,vib}$ the natural ship frequencies for oscillation and vibration modes (Table 5); $n_{max\_osc,vib}$ the maximum number of cycles; $p_i(h_{1/3i})$, $i = 1, m$ the probabilities of World Wide Trade (WWT) wave significant height $h_{1/3}$ histogram (Fig. 12) (Price & Bishop 1974); $n_{i\_osc,vib}$ the number of stress cycles

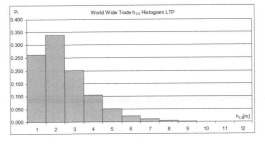

Figure 12. The World Wide Trade wave height $h_{1/3}$ histogram.

Table 14. Fatigue criterion, based on D ratio and design S-N curves.

WWT Histogram		Full	Ballast	Combined	
Analyses	Welding	D_{SN_full}	D_{SN_ball}	D_{SN}	L[years]
1D-beam model, $\sigma_{x1/3}$ deck maximum normal stress					
Linear	standard	0.37	0.11	0.24	> 20
Nonlinear	standard	0.51	0.27	0.39	> 20
Nonlinear	very good	0.24	0.12	0.18	> 20
3D & 1D model, $\sigma_{vonM1/3}$ deck maximum equivalent von Mises stress					
Linear	standard	0.74	0.30	0.52	> 20
Nonlinear	standard	1.22	0.94	1.08	18.5
Nonlinear	very good	0.53	0.39	0.46	> 20

for $h_{1/3i}$; $N_{iosc,vib}$ the number of endured stress cycles from the steel standard design S-N curves for a stress range $\Delta\sigma_{i\_osc,vib}$.

For the LNG carrier the full cargo and ballast load conditions are considered with the same occurrence probability. The cumulative damage ratio D has the following expression:

$$D = 0.5 \cdot D_{full} + 0.5 \cdot D_{ballast} \leq 1; \quad L = 20/D \quad (9)$$

where L [years] is the estimated ship service life.

Based on the dynamic response bending moments M_{1Dbeam} (Chapter 4) and the stress influence coefficients $k_{\sigma vonM}$ (Equation 3), for the 3D & 1D combined modelling, the deck equivalent von Mises stress can be approximated with the following expression:

$$\sigma_{vonM\_Deck} \cong k_{\sigma vonM\_Deck} \cdot M_{1D beam} \quad (10)$$

For the initial design not all of the structural details are defined, so that the numerical initial fatigue criterion check is carried out only in the case of transverse butt weld joints of the ship hull deck shell, which according the Germanischer Lloyds Rules (GL 2008) (Part 1, Ch. 1, Sec. 20, Tab. 20.3) is type 1 element, with 125 N/mm² the fatigue strength reference value of steel S-N curve for $2 \cdot 10^6$ cycles.

Table 14 presents a synthesize of the fatigue criterion check for the deck shell, based on the Palmgren-Miner cumulative damage ratio D and design S-N curves (Equations 6–7). The significant normal deck shell stresses $\sigma_{x1/3}$ are obtained directly from the dynamic response analyses, based on 1D-beam models (Chapter 4). The significant equivalent von Mises deck shell stresses $\sigma_{vonM1/3}$ are evaluated based on 3D & 1D combined models (Equation 8).

The influence of the butt weld joints welding quality (GL 2000, GL 2008), standard or very good welding, is also taken into account for the initial ship service life evaluation (see Conclusion 9).

6 CONCLUSIONS

Based on thenumerical results from Chapters 3–5, for the LNG 150000 m³ carrier hull structure (Chapter 2), it results the following conclusions:

1. The vertical deflection and the buckling factor satisfy the admissible limits, based on the 3D-FEM LNG hull model (Tables 7, 8).
2. From the ship strength analyses, with quasi-static head wave loads, the average stress differences between the 3D-FEM and 1D-beam models, for both loading cases (Tables 9, 10), taking as reference the wave height $h_w = 10.715$ m, are: normal stress deck +14.75% and bottom +1.75%, tangential stress in neutral axis +18.50%, von Mises stress deck +11.00% and bottom −6.75%.
3. From Figures 6a, b-9a, b, it results that the stress differences between 3D-FEM and 1D-girder models, are significant at the transversal bulkheads domain (not included in the 1D-girder model), and small in the other domains at the ship central part (between bulkheads), where the mass distribution (Figs. 3a, b), the ship lines (Fig. 1) and the ship hull structure (Fig. 2) are almost uniform.
4. The maximum stress values obtained at ship strength analyses with 3D-FEM model, under quasi-static head wave loads, for both loading cases, are smaller than the admissible values, with ratio σ, τ_{max}/σ, $\tau_{adm} = 0.21 \div 1.00 \leq 1$ (Tables 9, 10). The maximum stress values are encountered in the deck shell (Figs. 6–7a,b & Figs. 9a,b).
5. The ship dynamic response analyses, based on the hydroelasticity theory, indicate that the elastic ship girder deformations are small comparing to the ship oscillation displacements $2.97 \div 4.69\%$ (Tables 11a,b), so that the ship motion parameters can be evaluated with standard seakeeping analyses.

64

6. Bottom and side slamming have high probability to occur at ship extremities (Tables 11a, b), inducing whipping with high intensity (Tables 12a, b). The linear springing has small intensity. The non-linear springing has medium intensity, due to the ship high motion amplitudes, inducing geometrical and hydrodynamic non-linearity sources, having also a low first natural vibration frequency ($\approx$1 Hz, Table 5).

7. From the hydroelastic ship response, taking as reference the oscillation significant values for bending moments and shear forces, at statistical short-term prediction, as average of the first order vibration component (Figs. 10c, d) represent 3.86% at linear analyses, at non-linear analyses 54.10% full cargo case and 77.22% ballast load case (Tables 12a, b). The extreme wave loads are induced in the ship girder mainly due to the slamming and whipping phenomena occurrence with high intensity.

8. The numerical results from the dynamic hydroelastic analyses (Tables 13a, b), indicate that the maximum stresses are recorded in the deck shell (Figs. 11a, b).

9. The fatigue criterion based on the cumulative damage ratio method at extreme wave loads, for World Wide Trade wave significant height histogram (Fig. 12), having as reference the significant normal deck shell stresses (Table 14), for 1D-beam model, is satisfied $D_{SN} = 0.18$–$0.39 < 1$. Analogue, considering as reference the significant von Mises deck shell stresses, 3D & 1D combined models (Equation 8, Table 14), it results at non-linear analyse $D_{SN} = 1.08 > 1$ and $L = 18.5 < 20$ years. For the non-linear analyses, considering a very good welding quality condition (Table 14), it results a service life evaluation over L > 20 years ($D_{SN} < 1$).

10. Based on the 3D-FEM-hull model in combination with the hydroelastic ship dynamic response (1D-beam models), at initial design stage, the domains with higher risk on the ship hull structure, under the extreme wave loads, are predicted.

ACKNOWLEDGMENTS

The work has been performed in the scope of the project MARSTRUCT, Network of Excellence on Marine Structures, (www.mar.ist.utl.pt/marstruct/), which has been financed by the EU through the GROWTH Programme under contract TNE3-CT-2003-506141.

REFERENCES

Bertram, V. 2000. *Practical ship hydrodynamics*. Oxford: Butterworth Heinemann

Bhattacharyya, R. 1978. *Dynamics of marine vehicles*. New York: John Wiley & Sons Publication

Bishop, R.E.D. & Price, W.G. 1979. *Hydroelasticity of ships*. Cambridge: University Press Cambridge

Domnisoru, L. & Domnisoru, D. 1998. The unified analysis of springing and whipping phenomena. *Transactions of the Royal Institution of Naval Architects London* 140 (A): 19–36

Domnisoru, L. 2006. *Structural analysis and hydroelasticity of ships*. Galati: University "Lower Danube" Press

Domnisoru, L. & Ioan, A. 2007. Non-linear hydroelastic response analysis in head waves, for a large bulk carrier ship hull. *Advancements in Marine Structures, Taylor & Francis Group, London*:147–158

Frieze, P.A. & Shenoi, R.A. (editors) 2006. *Proceeding of the 16th international ship and offshore structures congress (ISSC). Volumes 1 & 2*. University of Southampton

GL. 2000. *Guidelines for fatigue strength analyses of ship structures*. Hamburg: Germanischer Lloyd

GL. 2008. Germanischer Lloyd's Rules. Hamburg

Garbatov, Y. & Guedes Soares, C. 2005. Fatigue damage assessment of a newly built FPSO hull. *Maritime Transportation and Exploitation of Ocean and Coastal Resources*, C. Guedes Soares, Y. Garbatov, N. Fonseca (Eds), Taylor & Francis, U.K.: 423–428

Guedes Soares, C. 1999. Special issue on loads on marine structures. *Marine Structures* 12(3): 129–209

Hirdaris, S.E., Price, W.G. & Temarel, P. 2003. Two and three-dimensional hydroelastic modelling of a bulk carrier in regular waves. *Marine Structures* 16: 627–658

Hirdaris, S.E. & Chunhua, G. 2005. *Review and introduction to hydroelasticity of ships*. Report 8. London: Lloyd's Register

Hughes, O.F. 1988. *Ship structural design. A rationally-based, computer-aided optimization approach*. New Jersey: The Society of Naval Architects and Marine Engineering

Lehmann, E. 1998. *Guidelines for strength analyses of ship structures with the finite element method*. Hamburg: Germanischer Lloyd

Park, J.H. & Temarel, P. 2007. The influence of nonlinearities on wave-induced motions and loads predicted by two-dimensional hydroelasticity analysis. *ABS-PRADS 1–5 Oct. 2007, Houston* (1): 27–34

Perunovic, J.V. & Jensen, J.J. 2005. Non-linear springing excitation due to a bidirectional wave field. *Marine Structures* 18: 332–358

Price, W.G. & Bishop, R.E.D. 1974. *Probabilistic theory of ship dynamics*. London: Chapman and Hall

Rozbicki, M., Das, K. & Crow, A. 2001. The preliminary finite element modelling of a full ship. *International Shipbuilding Progress Delft* 48 (2):213–225

Servis, D., Voudouris, G., Samuelides, M. & Papanikolaou, A. 2003. Finite element modelling and strength analysis of hold no.1 of bulk carriers. *Marine Structures* 16:601–626

SRAC. 2001. Cosmos/M FEM program user guide. Structural Research & Analysis Corporation (www.cosmosm.com).

Stoicescu, L. & Domnisoru, L. 2007. Global strength analysis in head waves, for a tanker with longitudinal uniform structure. *Advancements in Marine Structures*, C. Guedes Soares and P.K. Das (Eds), Taylor & Francis Group, London:283–294

Analysis and Design of Marine Structures – Guedes Soares & Das (eds)
© 2009 Taylor & Francis Group, London, ISBN 978-0-415-54934-9

Parametric investigation on stress concentrations of bulk carrier hatch corners

Dario Boote
Dipartimento di Ingegneria Navale e Tecnologie Marine, University of Genova, Italy

Francesco Cecchini
Registro Italiano Navale S.p.A—Genova, Italy

ABSTRACT: Bulk carriers are the subject of an intense research activity by Classification Societies and research centers because of their hull structure complexity and severe load conditions. Many problems affecting this kind of vessel can lead to very severe damages, up the hull global collapse.

One of the typical failure causes for bulk carriers is represented by high stress concentrations occurring in deck plating close to hatch corners in way of coaming stay. The aim of this work, performed in cooperation by the Registro Italiano Navale (RINA) and the Department of Naval Architecture and Ship Construction of the University of Genova, is to investigate the stress distribution close to hatch corners in a systematic way, in order to determine and quantify the influence of the selected parameters on this phenomenon.

1 INTRODUCTION

Since the 90's a great number of bulk carriers experienced serious structural damages caused, in most cases, by apparently insignificant failures derived by severe sea state conditions, loading and unloading operations, corrosion and fatigue phenomena. For what structure components are concerned, the most critical areas are hold frames, corrugated transverse bulkheads, cross deck structures and hatch openings. Moreover, the humidity rising from the cargo vaporization and the solar irradiation increase the corrosion process speed and the rising of failures and buckling phenomena. Frystock & Spencer (1996) presented a comprehensive assessment of structural problems of bulk carriers.

As a matter of fact the structure lay out of bulk carriers is the same since the 1960's, with a very stiff double bottom, a single deck with large hatch openings, hopper and wing tanks and a relatively weak transversely framed side. This complex structure, which could be adequate for the first 20,000 tonnes bulk carriers, begins to show its limits with the present 150,000 (and more) deadweight ships; the great difference of stiffness among bottom, side and deck areas can be the responsible of dangerous stress concentrations.

A big effort has been devoted by Classification Societies on this matter, resulting in the drafting of IACS "Common Structural Rules", (2006), where many problems seems to be solved for new ships.

In this paper the attention has been focused on failures developing from stress concentrations in correspondence of hatchways corners. Owing to the presence of hatch openings the main deck resistance is reduced to two, relatively narrow, lateral strips which must withstand hull girder longitudinal loads. In order to keep the stress level to admissible values, this deck strip is characterized by high thickness plating.

The deck zone between two consecutive hatches and between the lateral strips, called "cross deck", in the past has not been considered participating to the longitudinal strength and no particular care has been devoted to its lay out. As a proof of this the thickness of cross deck shell is always lower than that of resistant lateral strips. This thickness difference, the hatch coamings and their local strengthening reinforcements determine high stiffness variation in this area and, consequently, the occurrence of dangerous stress concentrations which can cause crack initiation and serious structural damages, up the deck collapse.

The approach followed by Classification Societies in these years consisted in deep and sistematic studies of the various typologies in order to provide designers and shipyards with a wide range of structural solutions (such as transversal stiffening for cross deck strip, elliptical hatch corner etc.) and specific formulas specifyingh the extra thickness of the hatch corner compulsory inserts.

The same approach is followed by IACS "Common Structural Rules" for bulk carriers, in force since 2006 to harmonize rules of IACS members and to improve the common safety regulations. Also in this case the problem is faced by providing both dedicated criteria and formulas; when a comprehensive understanding of the stress distribution is needed specific FEM analyses are required.

The attention of Classification Societies on bulk carriers continues to be very intense to increase even more their structural reliability and safety. With this aim in mind an investigation on the critical deck area close to the hatch corners has been carried out in a systematic way, in order to identify most important parameters and to quantify their influence on the stress distribution. This might allow to develop a more sophisticate procedure to predict the stress concentration in a preliminary design stage.

The investigation started from the assessment of six different bulk carriers built after 1990, chosen on the base of their different geometric and structural characteristics. They have been modelled and analysed by a FEM structural analysis code developed by RINA.

The results, in terms of stresses and of a stress concentration factor "k", have been related with different geometric ratios defining the geometry of hatch openings. A mathematical regression has been found as well for design purposes.

A second analysis has been carried out on secondary reinforcements such as local stiffeners and brackets. The results are presented for the second type of structural component and a regression has been extracted for design purposes.

2 PROBLEM DESCRIPTION AND IN FORCE RULES

Most common damages found in proximity of hatch corners of many bulk carriers are represented by cracking of the cross deck plates and of the corner brackets. A list of these damages have been collected by IACS (1994). They are represented in figure 1:

– cracks in correspondence of hatch corner (fig. 1a);
– cracks in correspondence of the welding between plates of different thickness (fig. 1b);
– cracks on the deck plates close to the hatch coaming bracket (fig.1c);
– cracks on the web bracket of coaming (fig. 1d).

The cause of failures type (a) and (b) is mainly the difference in thickness between the cross deck shell and the shell of the deck lateral strip, when transition plating are not present. Failures type (c) and (d) are due to the positioning and spacing of the coaming brackets and to their shapes and reinforcements.

Other factors which can have somehow an influence on the stress level at hatch corners are the alignment of the wing tank lateral bulkhead with the hatch coaming and some characteristic ratios between the geometric main dimensions of hatch openings; the most commonly used by designers and Classification Societies rules are the following (see fig. 2):

b/wa/w

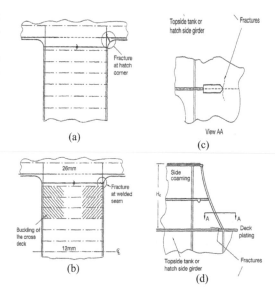

(a)

(b)

(c)

(d)

Figure 1. Failure modes at hatch corners (taken IACS 2006): (a) cracks in correspondence of the hatch corner; (b) cracks in correspondence of the welding between plates of different thickness; (c) cracks on the deck plates close the hatch coaming bracket; (d) cracks on the web bracket of the coaming.

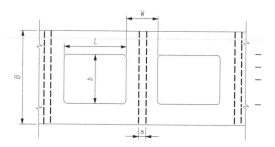

Figure 2. Geometric parameters related to hatch corner stress concentration.

where:

b = hatch breadth;
a = stool breadth of the transverse bulkhead;
w = cross deck length.

For the above mentioned ratios the rules contained in IACS 1994 suggest limit values for ships with length over 200 m and breadth over 40 m:

b/w > 2.2 a/w < 0.2

RINA (2008) and IACS (2006) have a common philosophy on the assumed criteria to limit the stress concentrations in proximity of hatch corners. For what the deck shell is concerned the Rules enforces to assume the following shrewdness:

– the hatch corners should be rounded by circular or elliptical fillets;
– the cross deck must have a transverse stiffening;
– in correspondence of the hatch corner a plating of increased thickness t_{ins} must be inserted:

$$t_{ins} = (0.8 + 0.4 \cdot \frac{l}{b}) \cdot t \qquad (1)$$

where:

  t = deck thickness;
  t_{ins} = plating insert thickness (should be $t < t_{ins} < 1.6\,t$);
  l = cross deck breadth (w in fig. 2);
  b = hatch breadth.

CSR Rules do not oblige designers to utilise transition plating with variable thickness between cross deck and lateral deck strips. This approach can be justified by the fact that, in the design phase, the hull stresses are considered supported only by lateral deck strips. This can be true if the hatch openings are very extensive in length and the cross deck is limited. On the contrary, if the hatch openings are relatively short, stresses can spread on the cross deck as well, generating undesired stress concentrations owing to the very thin shell and to the presence of hatch coaming corners. Moreover, the reduced thickness of plates in this area, can generate buckling phenomena (fig. 1b) which further weaken the cross deck structures.

Other approaches exist providing formulas and procedures derived from the analysis of plating with isolated holes. The GL Rules (2007), as an example, contain a procedure for the evaluation of notch stress σ_k in proximity of deck openings which should fullfill the following relationship:

$$\sigma_k = f\, R_{eH} \qquad (2)$$

where f is a coefficient depending on the yielding stress of the material:

  $f = 1.1$ for normal steel;
  $f = 0.9$ for HR steel with $R_{eH} = 315\ \mathrm{N/mm^2}$;
  $f = 0.8$ for HR steel with $R_{eH} = 355\ \mathrm{N/mm^2}$;
  $f = 0.73$ for HR steel with $R_{eH} = 390\ \mathrm{N/mm^2}$.

In case of proper corner rounding a further reduction of 20% of σ_k is allowed.

Lloyd's Register Rules (2007) gives some general design criteria for all type of bulk carrier and pays special attention to those ships having particular characteristics. Concerning general criteria, LR suggests longitudinal framing for lateral strip of deck, transverse stiffening for cross deck and it introduces a transition plate when the difference between

thickness of plating inside and outside the hatches line exceeds 12 mm. It also defines the thickness for this transition plate equal to the mean of adjacent plates.

More restrictions on hatch corners are imposed for bulk carriers with vertically corrugated bulkheads having at least one of the following characteristics:

– $B \geq 40$ m;
– $b/w \geq 2.2$;
– hatch side coaming and deck opening arranged inboard of the wing tank;
– bulk carriers carrying heavy cargo or with empty holds allowed.

In all these cases the corner is to be rounded with a radius of 1/20 B but not less than 1000 mm and reinforced with an insert plate having thickness 1.25 times the thickness of lateral strip with a minimum increase of 5 mm. LR provides also the extension of this insert plate which is to be extended transversely into the cross deck for a minimum distance equal to 0.75 b.

For extreme corners of end hatchways of the cargo area far from amidship the thickness of the insert plate is increased to 1.6 times the thickness of lateral strip of deck.

The Korean Classification Society (2007) mainly focuses the attention on the position of increasing of thickness and on the detail realization, leaving to the designer the task of evaluating the value and consistence of stress concentrations. Minimum radius at hatch corners (200 or 250 mm) are imposed as a function of the opening position and dimensions.

3 ANALYSIS METHODOLOGY

The aim of the investigation is to understand and to quantify the influence of b/B and w/b parameters, identified in a preliminary phase, on the stress concentration at hatch corners. A FEM analysis has been carried out on numerical models of six bulk carriers chosen as reference ships. The choice has been performed with following objectives in mind:

– ships should be different in dimensions and hatch parameters (hatch opening breadth, cross deck width) in such a way to cover a dimensional range as wide as possible;
– ships should have different local reinforcements (hatch coaming configuration, bracket shapes, bracket reinforcements).

The main characteristics of the considered reference bulk carriers are resumed in the following table 1. The analysis has been performed making reference to two parameters, usually considered by designers

Table 1. Main characteristics of reference ships.

Characteristics	Typical main section	Characteristics	Typical main section
Ship 1 Lpp = 126.0 m B = 20.0 m D = 10.5 m C_B = 0.800 b = 13 w = 9.0		**Ship 2** Lpp = 200.0 m B = 32.3 m D = 19.6 m C_B = 0.860 b = 15 w = 10.3	
Ship 3 Lpp = 230.0 m B = 32.3 m D = 19.8 m C_B = 0.810 b = 16.6 w = 8.1		**Ship 4** Lpp = 215.0 m B = 32.3 m D = 19.0 m C_B = 0.840 b = 14.4 w = 8.5	
Ship 5 Lpp = 180.0 m B = 26.0 m D = 16.0 m C_B = 0.820 b = 14.5 w = 8.9		**Ship 6** Lpp = 217.0 m B = 32.2 m D = 20.5 m C_B = 0.860 b = 15	

Table 2. Values of b/B and w/b ratios for reference ships.

	Ship 1	Ship 2	Ship 3	Ship 4	Ship 5	Ship 6
b/B	0.65	0.46	0.50	0.43	0.56	0.47
w/b	0.69	0.69	0.49	0.59	0.61	0.69

and Classification Societies when dealing with stress concentrations at hatch corners:

- w/b = the ratio of the cross deck breadth versus hatch opening breadth;
- b/B = the ratio of hatch opening breadth versus ship breadth.

While w/b ratio is already considered in many Rules, the b/B ratio, after a preliminary analysis, has been found to have a significant influence on the phenomenon herein investigated, so it has taken into consideration in this analysis. The values of the two ratios for the six reference ships are reported in table 2. It should be underlined that the bulk carrier n.6 is quite different from the others as she has a double side and a longitudinal bulkhead into the wing tanks. For this reason she has not been considered in the first part of the analysis.

For each ship a numerical model of a portion of the hull extended to three cargo holds has been created with a refinement degree suitable to identify the stress concentrations around the hatch coamings (fig. 3a). The model has been schematized entirely to allow non symmetric loading cases to be applied.

The boundary conditions of numerical models have been selected making reference to RINA and CSR rules. In the calculations performed according to RINA rules one end section has been clamped while, in the calculation performed according to CSR rules, the same section has been simply supported. In both cases the remaining end section has been fitted with "rigid link" boundary conditions; which force the points on the section to have the same rotation with respect to the neutral axis (fig. 3b).

A number of different loading conditions have been analysed, ranging from homogeneous full load, alternate full load, heavy and light ballast, in order to enlighten the most severe one with regard to stress concentrations at hatch corners. Basically, all cargo loads have been applied to the models by pressure distribution on bottom and side shells.

For each case the stress distribution has been evaluated, with particular attention to the deck area. Also the bottom has been verified in order to reject those cases for which the highest stresses overcome the maximum admissible ones of the construction material. When this happened, very seldom however, loads have been reviewed in order to lower the stress level. As an example, in fig. 3c and 3d the plots of longitudinal σ_x and transverse σ_y stresses are presented. The model in fig. 3 are relative to the reference ship n.2.

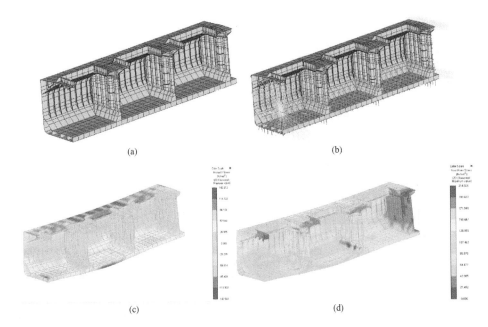

(a) (b) (c) (d)

Figure 3. Numerical model of reference ship n.2; (a) three cargo holds half model, (b) boundary and loading conditions, (c) longitudinal stress distribution σ_x, (d) transverse stress distribution σ_y.

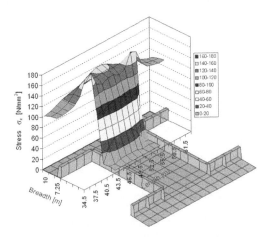

Figure 4. Longitudinal stress distribution σ_x in the deck area close to the hatch openings.

A stress concentration factor "k" has then been introduced, defined as the ratio of stress concentration σ_{1x} at hatch corner versus the hull girder longitudinal stress σ_x for each analysed loading condition:

$$k = \frac{\sigma_{1x}}{M/W} \quad \text{i.e. } \sigma_{1x} = k\frac{M}{W} = k\sigma_x \quad (3)$$

All calculations have been carried out by the FEM code Leonardo Hull 3D Ver. 2.4.1, RINA (2008b); the

software has been developed by RINA and it has a powerful preprocessing module designed specifically to model large ship portions in relatively small time. Leonardo Hull uses the linear solution algorithm of NASTRAN FEM software, NASTRAN (2003).

For each model and for each loading condition the distribution of longitudinal stress σ_x on the lateral deck strip and on the cross deck structures have been evaluated in order to determine the stress concentration and the "k" coefficient. In fig. 4, as an example, the 3D diagram of the stress distribution σ_x for a particular load case of ship n.2 is shown.

4 ANALYSIS OF B/B INFLUENCE

In this phase of the study, all the ships should have been made homogeneous with regard to w/b ratio. This implied to slightly modify the hatch opening configuration by reducing the hatch width or by widening the lateral deck strip. The latter solution has been chosen because it doesn't influence the b/B ratio, which is the main parameter of the investigation.

The FEM calculation have been run for all reference ships and loading conditions. From the stress distribution over the deck, maximum values have been identified and the results, in terms of stress concentrations σ_{1x} in N/mm², have been plotted on diagrams as a function of loading cases (coloured bars) and RINA rules wave load scenarios (A, B, C e D) (fig. 5a). The

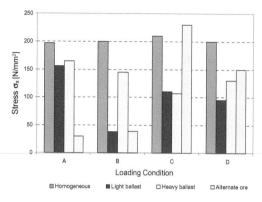

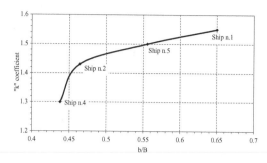

Figure 6. Maximum stress concentration coefficient "k" versus b/B ratio.

Figure 5a. Diagram of stress concentrations σ_{1x} as a function of loading condition and RINA wave load.

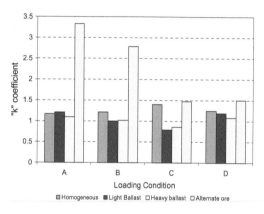

Figure 5b. Diagram of stress concentrations coefficient "k" as a function of loading condition and RINA wave load.

same presentation, but in terms of stress concentration coefficient "k", is presented in fig. 5b. The data in fig. 5 are relative to the reference ship n.2.

In fig. 5b it is possible to see that the highest value of "k" coefficient (about 3.2) takes place for the scantling draft, alternate load case for all RINA ship conditions. In remaining cases the values of "k" are quite similar each other. Nevertheless the only stress concentration coefficient is not sufficient to identify the most severe load case; in many cases to the highest "k" values it corresponds a stress concentration of about 40 N/mm², while in case of very high stress concentrations (over 200 N/mm²) "k" assumes values around 1 (that means no stress concentration).

To gather more reliable information the two diagrams of figs. 5a and 5b should be considered together. By this approach it comes out that the worst load case for stress concentrations at hatch corners is the load case C (scantling draft, alternate load, inclined ship with torque effects) with a maximum stress concen-

tration of about 230 N/mm² and a stress concentration coefficient "k" equal to 1.5.

For all five reference ships considered in this phase of the study the highest "k" values have been reported on a diagram as a function of b/B ratio (fig. 6). The curve show a high steepness for lower values of b/B and an asymptotic trend towards k = 1.6, which can be considered the highest possible stress concentration ratio.

At this point, in order to perform the same analyses on a wider range of b/B ratio, simpler models have been set up. They represent just the deck area around the hatch with increasing refinement levels: four types of different solutions have been modelled as described in the following.

a) The first model consists in a simple plate with a rectangular opening and two longitudinal box structures simulating the wing tanks. All the parts of the model have a constant thickness (fig. 7a).

b) A second series of models has then been created, similar to the first one, adding a coaming around the hatch opening. As in the first case the thickness has been maintained constant (fig. 7b).

c) The third series is similar to the first one but the thicknesses of the lateral deck strip and of the cross deck have different values, according to the real case (see fig. 7c).

d) The fourth series is similar to the second one. Again deck plates have different thicknesses (fig. 7d).

For the four series of models the b/B and w/b ratios have been parametrically varied by changing the hatch breadth and the cross deck length of the base model.

All the obtained models have been fixed at one end and on the other end a tension load, corresponding to the longitudinal stress derived by the design bending load, has been applied.

The obtained results have been plotted on the diagram of fig. 9 in terms of stress concentrations σ_x as a function of the transverse distance from the symmetry plane towards the side. The diagram shows that the results of the first two models are very close each

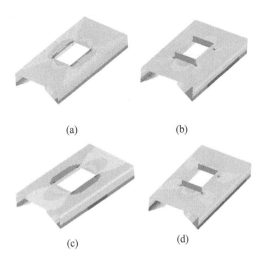

(a)   (b)

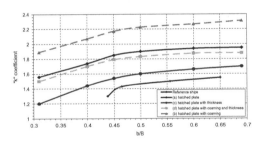

(c)   (d)

Figure 7.   Simplified models of the deck area around hatch opening: (a) simple plate with hatch opening, (b) with hatch coaming and constant thickness, (c) without hatch coaming and differentiated thickness, (d) with hatch coaming and differentiated thickness.

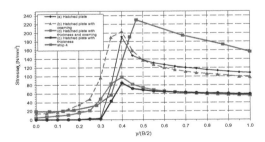

Figure 8.   Stress concentrations at hatch corners calculated by simplified models.

Figure 9.   Stress concentration factor "k" from simplified models as a function of b/B.

other; the same trend comes out for the results of the third and fourth models. From these observations it seems that the hatch coaming has not a great influence on stress concentrations. As predictable the increase

of plate thickness on the lateral deck strip dramatically lowers the stress values at hatch corners.

Then the same results have been transformed in stress concentration factor "k" and plotted on the diagram of fig. 9, together with the "k" values obtained from the larger models of reference ships. The following consideration can be done from the examination of diagrams of fig. 8 and 9.

a) The curves of fig. 8 are almost parallel each other and, as predictable, the values of stress concentrations found for reference ships are very close to those of the model (d), the geometry of which is the closer one to the real structure.

b) The ratio between higher stresses of curve (b) versus the lower ones (curve c) is quite constant and all curves show a similar slope.

c) The curves of fig. 9 have the same slope as well. Also in this case, the closer one to reference ship data is the curve relative to the fourth model (d). In this case, the homogeneity of the results would allow to gather interesting information by performing fast calculations on very simple models.

Referring to the base model, consisting in a simple plate with hatch opening (no hatch coaming, no wing tanks), it comes out the importance of the wing tank relatively to the hatch stress concentrations; they decrease with the dimensions and stiffness of wing tanks up to negligible values. As it will be assessed in the forthcoming paragraphs, this trend is very evident for the reference ship n.6 (with double side and longitudinal bulkhead in the wing tank).

5   ANALYSIS OF W/B INFLUENCE

In the second phase of the analysis, the effects of w/b ratio has been assessed. Again, reference ships n. 1, 2, 3, 4 and 5 have been considered, leaving out ship n.6 for her different structure characteristics. To maintain an homogeneous b/B ratios reference ships have been slightly modified in a way that to all of them the same b/B value of ship n.2 (0.46) has been assigned. In this way the w/b ratios has been reduced to about 0.5 and the numerical models of real ships cover a range for this ratio from 0.5 up to 0.7 approximately.

FEM calculations have been run under the same load and boundary conditions of the previous case; the results in terms of stress concentration factor "k" are shown in the diagram of fig. 10. In the same diagram the result curves relative to FEM calculation on the simplified model (a) and those coming from RINA rules calculation are shown as well (see paragraph n.2).

While the results of FEM models are very close each other, CSR rules provide "k" values lower than the others (about 75%). This is ascribable to the fact that the CSR approach takes into account other local geometric parameters of hatch openings not presented

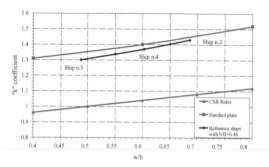

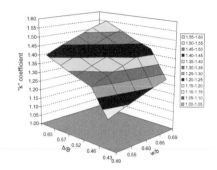

Figure 10. Stress concentration factor "k" verus w/b ratio.

Figure 11. Surface of stress concentration factor "k" as a function of b/B and w/b ratios.

in this analysis such as curvature radius and local reinforcements. These two last aspects are currently under investigation in order to evaluate their weight on hatch corner stress.

6   COMBINATION OF RESULTS

The results coming from the two analyses carried out to investigate the separate effects of b/B and w/B have been joined together in a tridimensional diagram (fig. 11). The minimum "k" value, equal to 1.16, takes place for ships with narrow hatch opening and short cross decks. The higher value (about 1.56) corresponds to wide and long hatch openings.

From all obtained results a mathematical regression has been formulated to allow its utilization for a quick determination of stress concentrations around hatch opening of bulk carriers. In order to maintain a simple formulation the surface of fig. 11 has been approximated by two planes divided each other by the intersection line of the surface versus the plane b/B = 0.46. The regression has then been splitted into two parts: one for b/B values below 0.46 and the other one for higher values:

$$k = 5 \cdot \frac{b}{B} + 0.7 \cdot \frac{w}{b} - 1.35 \quad \text{per} \quad \frac{b}{B} \leq 0.46 \qquad (4)$$

$$k = 0.57 \cdot \frac{b}{B} + 0.7 \cdot \frac{w}{b} + 0.69 \quad \text{per} \quad \frac{b}{B} > 0.46 \qquad (5)$$

Of course the regression is valid only for bulk carriers with the same main section lay out of those herein considered, with single side and wing tanks, without longitudinal bulkheads. The influence of other geometric parameters (like local reinforcements, as an example) could be assessed as well, and this is the objective of another investigation, at present under course.

The same calculations performed on the five reference ships have been repeated for the ship n.6. Her

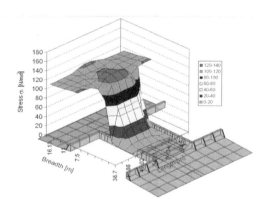

Figure 12. Numerical model of the bulk carier n.6.

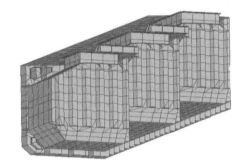

Figure 13. Longitudinal stresses σ_x at hatch corners for reference ship n.6.

structure, owing to the longitudinal bulkheads into the wing tank and to the double side, is characterised by higher strength and stiffness. The effects of double side on bulk carriers hull structure is widely assessed in Coll (1996).

For the ship n.6 the same numerical model has been set up (fig. 12) and the same loads and boundary condition applied. The results in terms of longitudinal stresses σ_x are shown in fig. 13 where it is evident the difference with respect to fig. 4. In this case the stress level on the lateral deck strip does not increase at hatch corners and falls down in the cross deck area.

7 LOCAL ANALYSIS

After having defined the most severe global loading conditions for stress concentration in correspondence of hatch corner, the investigation has been concentrated on the effects of local reinforcements such as brackets and ordinary stiffeners. All meshes have been refined and new additional details hve been modelled. Hybrid beams have been introduced in numerical models to properly model connection between 1 dimensional (ordinary stiffeners) and 2 dimensional (plating, brackets) elements in order to avoid the typical unrealistic distribution of moments of eccentric beams.

The boundary conditions to be applied to these refined models have been derived directly from global model deformations for all the considered loading conditions.

From a first, preliminary study, it could be found that the most critical structure in way of hatch opening is the coaming corner bracket in the longitudinal orientation. Other brackets just behave like hatch coaming local stiffeners in order to maintain a sufficient rigidity under the wave load. To better understand the effects of these local structures on the hatch corner stress distribution the study was performed modifying the bracket shape and scantling.

The cross deck area of reference ship n.2 was considered modelling three different arrangements:

1. Bracket extended to 3rd transv. stiffener (fig. 14a);
2. Bracket extended to 2nd transv. stiffener (fig. 14b);
3. Bracket extended to 1st transv. stiffener (fig. 14c).

The considered type of results are the deck stresses in longitudinal direction at the basis of bracket. A "k_1" coefficient was defined as the ratio between the longitudinal stress at the bracket basis σ_{2x} and the concentrated stress σ_{1x}.

"k_1" values, calculated for the three different bracket arrangements, are shown in fig. 16. On the x-axis, the longitudinal distance from transverse coaming is reported. The diagram analysis shows two important characteristics:

− a first stress concentration at bracket basis in correspondence of hatch coaming—deck welding is evident;
− a second stress concentration in correspondence of the tip of bracket takes place as well.

The first stress concentration listed above, seems not to be significantly influenced by the bracket arrangement. In all reported cases the maximum value of

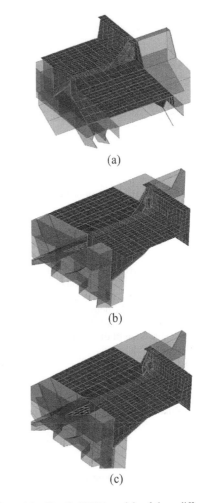

(a)

(b)

(c)

Figure 14. Detailed FEM models of three different hatch coaming brackets

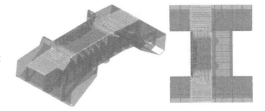

Figure 15. Von Mises stress distribution for one of the considered type of brackets.

k_1 coefficient is close to 1.25 and the curve seems to follow the same decreasing trend to reach nominal longitudinal stress.

The second concentration is clearly influenced by bracket arrangement, both for its position and its

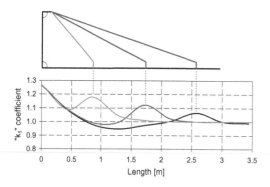

Figure 16. Longitudinal stress distribution for three types of brackets considered.

magnitude. In particular, the concentration decreases for longer brackets.

The value of k_1 coefficients in correspondence of this second peak ranges between 1.06 to 1.18 in worst case. It could be useful to calculate k_1 directly from k, thus avoiding heavy FEM calculation; for this reason, a fitting curve enveloping all peaks is defined. The mathematical expression of the fitting curve is:

$$k_1 = 1.26 \times 0.93^l \qquad (6)$$

where:

l = x coordinate of bracket starting from coaming.

8   CONCLUSIONS

In this paper an investigation on stress concentrations at hatchway corners of bulk carriers is presented. The aim of the study was to identify and to quantify the effect of some geometric ratios for design purposes. The most important parameters have been identified in the ratios b/B and w/b.

By a numerical approach two kinds of FEM models have been created:

– global numerical models of six existing bulk carriers ships; each model is extended to a hull portion of three cargo holds;

– simplified models with four different refinement degrees; the models range from a simple plate with a rectangular hole up to a deck portion with hatch coaming, wing tank and different thickness values.

The results of the analysis have been translated in terms of the stress concentration factor "k", defined as the ratio between the stress concentration and the nominal hull girder longitudinal stress. The obtained values of "k" factors for all the considered conditions have been combined together and a mathematical regression formulated which allows to quickly achieve stress concentration values from the nominal hull girder stresses.

Further investigations have been carried out on the effects on stress concentrations of local reinforcements like hatch coaming brackets. Other details are under consideration at this moment.

REFERENCES

Coll G., 1996, "Safety of Bulk Carriers—Are Two Skins Better than One?", Transactions RINA, pp. 117–129.

Frystock K. & Spencer J., 1996, "Bulk Carrier Safety", Marine Technology, The Society of Naval Architects and Marine Engineers, Vol. 33, N.4, pp. 309–318.

Germanischer Lloyd AG, 2007, "Guidance Relating to the Rules for the Classification of Steel Ships", Pt 3, Ch 5, Hamburg.

IACS, 1994, "Bulk Carriers: Guidelines for Survey, Assessment and Repair of Hull Structure", International Association of Classification Societies.

IACS, 2006, "Common structural rules for bulk carrier", International Association of Classification Societies.

Korean Register, 2007, "Guidance Relating to the Rules for the Classification of Steel Ships", Part 3, Chapter 5, Daejeon.

Lloyd Register, 2007, "Rules and Regulations for Classification of Ships", Pt 4, Ch 7, London.

NASTRAN, 2003, "User's manual", MSC.

RINA, 2008a, "Rules for the Classification of Ships", Part B, Ch 4, Sec 6.

RINA, 2008b, "Leonardo Hull 3D user's manual", Genova.

Analysis and Design of Marine Structures – Guedes Soares & Das (eds)
© 2009 Taylor & Francis Group, London, ISBN 978-0-415-54934-9

A study on structural characteristics of the ring-stiffened circular toroidal shells

Qing-Hai Du, Zheng-Quan Wan & Wei-Cheng Cui
China Ship Scientific Research Center (CSSRC), Wuxi, Jiangsu, China

ABSTRACT: The pure complete toroidal shell and the compartment of toroidal shell with ring stiffener, with circular meridian cross-section, have been investigated respectively by the numerical analysis method in this paper because of the difficulty of the theoretically solution. The thin shell element was adopted to obtain elastic analysis of a closed toroidal shell due to external pressure, which confirms the conclusion of membrane theory. Based on the nonlinear finite element method (FEM), the structural characteristics of the ring-stiffened toroidal shell have been carried out. In presented analyzing, adopting the elastic-plasticity stress-strain relations, the influences between material nonlinearity and geometry nonlinearity on the stability of structure hull are also considered. Comparing with the traditional cylindrical pressure hull, the ring-stiffened toroidal pressure hull is superior in resisting influences of the initial deflection to a certain extent and having larger reserve buoyancy. It means that the ring-stiffened toroidal shell could be used to realize better performance in the general structural form with its specific shape in underwater engineering.

1 INTRODUCTION

For the pressure hull in underwater engineering, the general shape is mostly in the form of a ring-stiffened circular cylinder blocked by end caps, or a spherical hull. Because of the advantages and disadvantages of using a circular cylindrical shell as pressure hull, the main pressure hull in a toroidal form was suggested by Ross (2005, 2006), for the design of an underwater missile launcher and the underwater space station. With its special structural shape, the main pressure hull of a toroidal form can be constructed to obtain kinds of different function.

The toroidal shell is rarely used in engineering compared to the cylindrical, spherical and conical shell, for the difficulty of theoretical solution and manufacture. Therefore it is only used as joint components or accessory and not as a main pressure hull in underwater engineering. Many researchers in the region of engineering and mechanics have spent great efforts to solve the problem in order to obtain the analysis and design method on the theoretical solution. Zhang (1944), Clark (1950) and Novozhilov (1951) had obtained an elastic asymptotic solution of the toroidal shells under axisymmetric loading. Moreover Xia & Zhang (1984) had even gained its elastic general solution due to arbitrary loading. Maching (1963) and Sobel & Flügge (1965) had explored the buckling and stability of toroidal shells. Bushnel (1967), Jordan (1973) and Panagiotopoulos (1985) had approached the stability analysis of the toroidal shells by finite difference method (FDM) or FEM. The buckling of the segments

of toroidal shells had been solved by Stein & Mcelman (1965), Hutchinson (1967) and Galletly (1995, 1996) had got the stability of closed toroidal shells with circular or non-circular cross-section. Wang & Zhang (1989) had obtained the analysis of geometric nonlinear buckling and post-buckling of toroidal shells by asymptotic approach. However all these works had been to solve the pure complete toroidal shells or segments in elastic or nonlinear analyzing, but its critical pressure could not reach higher for no ring-shaped ribs.

In this paper the complete toroidal shell, with circular meridian cross-section, is solved by the linear-elastic FEM. And to consider the influences between material nonlinearity and geometry nonlinearity on the stability of the structure hull, the nonlinear FEM has been adopted. Then under external pressure and with circular cross-section, the structural characteristics of ring-stiffened toroidal shells are presented, comparing with the circular cylindrical shells and showing the feasibility of its application as underwater structures.

2 THE ELASTIC FEM ANALYSIS OF THE COMPLETE TOROIDAL SHELL

The toroidal shell is a special type of shells of revolution for the existence of horizontal tangents at $\varphi = 0$ and π, which are called the turning points. It is that the Gaussian curvature changes its sign from positive to negative (see Figure 1, where R represents the distance of the center of the meridian circles to the axis of

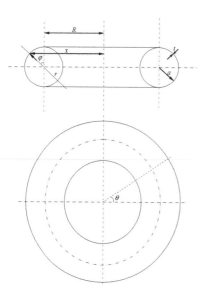

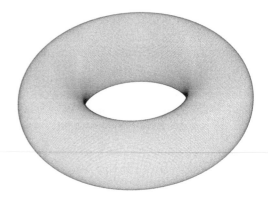

Figure 2. The finite element mesh of FEM model.

Figure 1. The toroidal shell.

rotation, φ is the tangential angle of shells, r is the curvature radius of the parallel circle on the toroidal shell to revolving direction, t is the wall thickness of shells, a is the radius of the meridian circle, θ is the rotational angle of the meridian circle).

According to the non-moment theory of shell, the components of the internal force of a closed pure toroidal shell subjected to pressure (internal or external) are:

$$N_\varphi = \frac{pa}{a\sin\varphi + R}\left(\frac{a}{2}\sin\varphi + R\right) \qquad (1)$$

$$N_\theta = \frac{pa}{2} \qquad (2)$$

(a) The stress σ_θ contour distribution.

(b) The stress σ_φ contour distribution.

Figure 3. The stress contour distribution of a closed toroidal shell ($a/t = 100$, $R/a = 2.5$, $p = 1.0$ MPa).

As was stated by Flügge (1960), however, the displacement at the turning points is discontinuous. For example, the meridian cross-section tangential displacement v and the normal displacement w are respectively

$$v = \left[\frac{paR}{2Et}\left(\frac{2a^2}{R\sqrt{R^2 - a^2}}\arctan\frac{a + Rtg\frac{\varphi}{2}}{\sqrt{R^2 - a^2}}\right.\right.$$
$$\left.\left. -\frac{a}{R}\ln tg\frac{\varphi}{2} - \cot\varphi\right) + C\right]\sin\varphi \qquad (3)$$

$$w = r\varepsilon_\theta - v\cot\varphi \qquad (4)$$

Here C is an unknown constant, p is pressure, E is Young's modulus, and ε_θ is the strain of θ direction. Wherever w is infinite at $\varphi = 0$ or π. Therefore there exist always bending moments at the turning points.

To confirm the conclusion of the membrane theory, the software ANSYS is applied to calculate the stress and deformation fields of the complete pure toroidal shell. The 8-node elastic shell elements are selected to the modeling. The finite element mesh of the model is shown as Figure 2.

With the parameters $a/t = 100$, $R/a = 2.5$ and $p = 1.0$ MPa, a closed toroidal shell model has been built to calculate. Figure 3 shows the components of stress in θ and φ direction. Figure 4 gives the deformed shape of the cross-section circle comparing with that initial shape. The numerical results verify those of membrane theoretical solution, and the bending moments exist exactly at turning points.

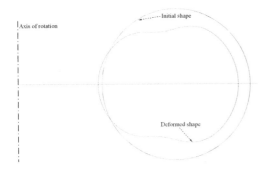

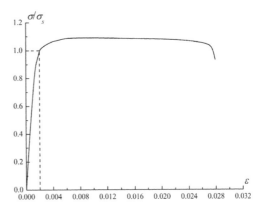

Figure 4. Deformed (and initial) shape of the cross-section circle ($a/t = 100$, $R/a = 2.5$, $p = 1.0$ MPa).

Figure 5. The curve of the stress-strain relations.

3 THE NONLINEAR FINITE ELEMENT METHOD FOR THE ANALYSIS

As a pressure hull in underwater engineering, the structure is chiefly subjected to hydrostatic external pressure. The stability of pressure hull is certainly more important than its stress intensity. Then the calculation of hull's critical pressure is significant work to the designers.

The buckling analysis includes not only large deformation and nonlinear strain-displacement relations, but also material nonlinearity, because that the material of somewhere in the pressure hull has become into plastic stage when the buckling approached. And in the paper the author adopted the stress-strain relations shown as Figure 5, to simulate the true material.

The nonlinear buckling analysis had been detailedly deduced by Wang (2003). And in the paper Von-Mises yield criterion was adopted to judge whether the local buckling has happened. So that is

$$F\,(J_2) = \sigma_{eq} - \sigma_s = \sqrt{\frac{3S_{ij}S_{ij}}{2}} - \sigma_s = 0 \qquad (5)$$

Generally the equivalent stress strength could be expressed as follows:

$$\sigma_{eq} = \sqrt{3J_2} \qquad (6)$$

$$J_2 = \frac{1}{2}S_{ij}S_{ij} = \frac{1}{6}\left[(\sigma_{XX} - \sigma_{YY})^2 + (\sigma_{YY} - \sigma_{ZZ})^2 \right.$$
$$+ (\sigma_{ZZ} - \sigma_{XX})^2\right] + \sigma_{XY}^2$$
$$+ \sigma_{YZ}^2 + \sigma_{ZX}^2 \qquad (7)$$

Here S_{ij} is deviator tensor of stress; σ_s is yield stress; σ_{XX}, σ_{YY}, σ_{ZZ} are the normal stresses of the X, Y and Z direction; σ_{XY}, σ_{YZ}, σ_{ZX} are the shear stresses of the XY, YZ and ZX plane coordinate.

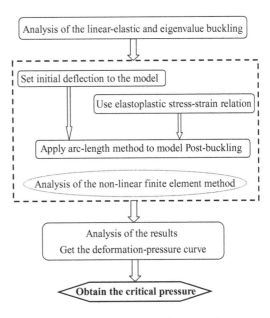

Figure 6. The nonlinear FEM analysis process chart.

All the processes of nonlinear finite element analyses (FEA) have been executed by the software ANSYS and the analysis process chart is shown in Figure 6.

4 THE CHARACTERISTICS OF THE RING-STIFFENED TOROIDAL SHELLS

Just as cylindrical and spherical form of the pressure hull, the toroidal shell would be set series of ring stiffener to achieve sufficient stability. Meanwhile the ring stiffener must be set adequately to provide not only sufficient rigidity but also effective cubage and large dimension while having less weight.

However taking account of the difficulty to build it and for the convenience of optimizing design, a quarter of a closed ring-stiffened toroidal shell is determined as a compartment for analyzing in this paper. It means that

$$L = \frac{1}{2}\pi R \qquad (8)$$

And the axis length (L) of the center of the meridian circles is given as follows:

$$1 < \frac{L}{2a} = \frac{\pi R}{4a} < 2 \qquad (9)$$

To explore the characteristics of the ring-stiffened toroidal shell, its equivalent ring-stiffened cylindrical shell is defined here by following

a. The axis length (L) of cross-section is equivalent.
b. The radius (a) of the cross-section circle is equivalent.
c. For a better performance to the structure, $1 < L/2a < 2$.

According to above definition, with the corresponding to the original ring-stiffened toroidal shell shown in Figure 7, the equivalent ring-stiffened cylindrical shell is shown in Figure 8.

Thinking about the optimization method and design experience of traditional cylindrical pressure hull, here the following parameters have been assumed to the ring-stiffened toroidal shell for analysis.

$$\frac{L}{2a} = 1.75, \quad \frac{R}{a} = 2.25, \quad \frac{L}{R} = \frac{\pi}{2} \qquad (10)$$

With above parameters, the pressure hull was set least ribs in order to appear buckling of the shells but not the ribs.

Then under the definitive primary dimensions as conditions (10) and the material property as given in

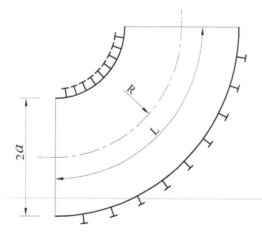

Figure 7.   The ring-stiffened toroidal shell.

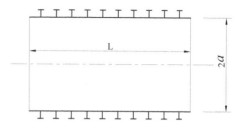

Figure 8.   The equivalent ring-stiffened cylindrical shell.

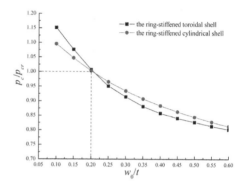

Figure 9.   The influence curves of the initial deflection on critical pressure.

Figure 5, the influences of initial deflection on critical pressure have been studied by nonlinear FEM of the previous chapter. And here the form of initial deflection of model was set as same as the first elastic buckling modal. Figure 9 gives out the influences of initial deflection on critical pressure of the ring-stiffened toroidal shell and its equivalent cylindrical shell (Here, w corresponds to the radial displacement of the meridian cross-section, t is the thickness of pressure hull, p_c is the calculated critical pressure and p_{cr} is the designed critical pressure).

It is confirmed from Figure 9 that if the maximal initial deflection is not more than $0.2\,t$, the ring-stiffened toroidal shell would be superior to the cylindrical shell. Hence to obtain the optimization design and keep better performance, the maximal initial deflection w_0 on the pressure hull should be set as

$$w_0 \le 0.2\,t \qquad (11)$$

Here t is the thickness of pressure hull. And Figure 10 gives that the normal deformation of one point on the toroidal shell, whose deflection was maximal on the hull when buckling appeared, varies with the external pressure.

It is tried to determine whether can construct the toroidal pressure hull for the depth about of 1.5 km with high strength steel here, because this form of

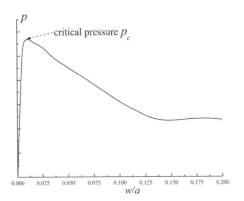

Figure 10. The pressure-deflection post-buckling track.

Table 1. The optimizational design of the two shape of pressure hull.

	The ring-stiffened toroidal hull	The equivalent cylindrical hull
a/t	52.9	58.1
w_0/t	0.2	0.2
a/w_0	264.5	290.5
p_c/p_{cr}	1.01	1.005
C_v	0.345	0.367

pressure hull is only built by a composite material for the larger depth of about 5 km (Ross, 2005, 2006).

Here the metallic material is assumed as a high strength steel with yield strength about 800 MPa according to Ross (2005, 2006). The primary parameters were set alike as conditions (10). And presented design goal was the least weight with larger cubage and enough ring stiffener for appearing the shell buckling but also convenient to construct them.

Table 1 shows the calculated results of a compartment, which include both the ring-stiffened toroidal hull and its equivalent cylindrical hull. The calculation demonstrates that the toroidal hull could have larger reserve buoyancy than the cylindrical hull. But its a/t is smaller than that of cylinder, which means the thickness of the toroidal shell would be larger than cylindrical shell with same circular cross-section. So it is seen from eq. (11) that in such condition the maximal tolerance initial deflection on the hull of toroidal shell could be larger than that of cylinder. And thus it provides some convenience for manufacturing it.

Here, C_v is the ratio of the weight of a hull compartment to its cubage.

5 CONCLUSION

By the finite element method in this paper, the closed pure toroidal shell has been discussed, revealing the conclusion of the membrane theory and the difficulty of theoretical solution. The material nonlinear and geometry nonlinear FEM is presented to solve stability of the pressure hull under external pressure. An equivalent ring-stiffened cylindrical shell for the ring-stiffened toroidal shell is defined in order to expose the influences of initial deflection and the mass-cubage ratio on the two type of pressure hull. The analytic results eventually indicate that such form of ring-stiffened circular toroidal shell could be used to a main pressure hull as the traditional ring-stiffened circular cylindrical shell, which could obtain kinds of performance in underwater engineering.

REFERENCES

Bushnell, D. 1967. Symmetric and Nonsymmetric buckling of finitely deformed eccentrically stiffened shells of revolution. *AIAA J.* 5:1455–1462.
Clark, R.A. 1950. On the Theory of Thin-Walled Toroidal Shells. *J. Math. and Phys.* 29:146.
Flügge, W. 1960. *Stresses in shells.* Berlin: Springer.
Flügge, W. & Sobel, L.H. 1965. Stability of shells of revolution: General theory and application to the torus. *Lockheed Missiles and Space Company Report* 3:6-75-65-12.
Galletly, G.D. & Blachut, J. 1995. Stability of complete circular and non-circular toroidal shells. *Proc Inst Mech. Engrs.* 209:245–255.
Galletly, G.D. & Galletly, D.A. 1996. Buckling of complex toroidal shell structures. *Thin-walled Struct.* 26:195–212.
Hutchinson, J.W. 1967. Initial post buckling behavior of toroidal shell segment. *Int. J. Solids Struct.* 3:97–115.
Jordan, P.f. 1973. Buckling of toroidal shells under hydrostatic pressure. *AIAA J.* 11:1439–1441.
Maching, O. 1963. Uber die Stabilität von Torusförmigen Schalen. *Techn. Mitt. Krupp.* 21:105–112.
Novozhilov, V.V. 1951. *Theory of thin shells.* Leningrad: National Union Press of shipbuilding industry.
Panagiotopoulos, G.D. 1985. Stress and stability analysis of toroidal shells. *Int. J. Press. Ves. & Piping* 20:47–100.
Ross, Carl, T.F. 2005. A conceptual design of an underwater missile launcher. *Ocean Engineering* 32:85–99.
Ross, Carl, T.F. 2006. A conceptual design of an underwater vehicle. *Ocean Engineering* 33:2087–2104.
Sobel, L.H. & Flügge, W. 1967. Stability of toroidal shells under uniform external pressure. *AIAA J.* 5:425–431.
Stein, M. & Mcelman, J. A. 1965. Buckling of segments of toroidal shells. *AIAA J.* 3:1704–1709.
Wang, Anwen. 1989. *Asymptotic analysis of geometric nonlinearity buckling and post-buckling of toroidal shells.* Beijing: Dissertation, Tsinghua University.
Wang, Xucheng. 2003. *Finite element method.* Beijing: Tsinghua University Press.
Xia, Zihui. 1984. *The general solution of toroidal shells due to arbitrary loads.* Beijing: Dissertation, Tsinghua University.
Zhang, Wei. 1944. Der spannungszustand in kreisringschale und ähnlichen Schalen mit Scheitelkreisringen unter drehsymmetrischer Belastung. Berlin: Arbeit zur Erlangung des Grades eines Doctor-Ingenieurs der Technichen Hochschule.

Application developments of mixed finite element method for fluid-structure interaction analysis in maritime engineering

J.T. Xing, Y.P. Xiong & M.Y. Tan
School of Engineering Sciences, University of Southampton, UK

ABSTRACT: The application developments of a mixed finite element method for transient dynamic analysis of fluid-structure interaction systems (FSIS) in maritime engineering are summarised in this paper. Following the mathematical equations governing generalised FSIS, three problems involving maritime engineering, developed in Marstruct Project, are presented to illustrate the applications of the developed numerical method. Problem 1 investigates the sloshing frequencies of a liquid storage tank and its dynamic response excited by the El Centro earthquake data. Problem 2 analyses the dynamic response of a LNG tank-water interaction system subject to an explosion pressure wave on a water boundary. Problem 3 considers the dynamic response of a structure-acoustic volume interaction system subject to human footfall impacts, which may be used to simulate ship deck vibration caused by human footfall loads to obtain a comfortable living environment of passengers. The numerical results are discussed and analysed to provide the related guidelines for the designs of maritime products involving fluid-structure interactions.

1 INTRODUCTION

Dynamic analysis of fluid-structure interaction systems subject to various dynamic loads, such as earthquakes, explosions under water, waves or impacts in maritime environments necessitates inter-disciplinary studies relating to fluids, rigid or flexible structures and their physical coupling mechanisms. For almost all problems involving fluid-structure interactions, analytical solutions are not available and recourse to numerical solutions or experimental studies are the only way forward. Based on linear approximations, fluid-solid interaction problems occurring in science and engineering have been well formulated and solved (Bishop & Price 1979, Bishop et al 1986, Liu & Ma 1982, Xing 1984, Xing 1986a, b, Liu & Uras 1988, Kock & Olson 1991, Morand & Ohayon 1995, Xing & Price 1991, Xing et al 1996, Xing et al 1997, Deruieux 2003). Among these publications, the mixed finite element approach was successfully demonstrated by analysing a wide selection of fluid-structure interaction problems of practical interest (Xing & Price 1991, Xing et al 1996, Xing et al 1997) using substructure-subdomain methods (Xing 1981, Xing & Zheng 1983, Unruh 1979). Based on this integrated generalised theory and method, the corresponding computer program FSIAP (Xing 1995a, b) has been developed and used to numerical simulations of many fluid-structure interaction problems in engineering. This paper presents a summary of the further application developments of the numerical method, which were completed at Southampton during the period of MARSTRUCT project (Xing et al 2006, Xiong et al 2006, Tan et al 2006, Toyota et al 2006, Xing et al 2007a, b, Xiong & Xing 2007, Xiong et al 2007, Xing & Xiong 2008a, b, Xiong & Xing 2008a, b). Following the mathematical formulations describing the mixed finite element substructure-subdomain method, this paper, as a report for Marstruct Project, mainly focuses the applications involving three dynamic problems in maritime engineering using the developed method. The discussions on numerical results are given as references for the designs of maritime products involving fluid-structure interactions.

2 GOVERNING EQUATIONS

As shown in Figure 1, a Cartesian coordinate system $O - x_1x_2x_3$, where the gravitational acceleration is along the negative direction of the coordinate axis $O - x_3$, is chosen as a reference frame to investigate the dynamics of fluid-structure interaction systems. The fluid-solid interaction system consists of a flexible structure of mass density ρ_s and elastic tensor E_{ijkl} within a domain Ω_s of boundary $S_T \cup S_w \cup \Sigma$ with its unit normal vector ν_i and at least one fluid (air or liquid) of sound velocity c in a domain Ω_f of boundary $\Gamma_f \cup \Gamma_w \cup \Gamma_p \cup \Sigma$ with a unit normal vector η_i. The system is excited by external dynamical forces $\hat{T}_i$, $\hat{f}_i$, $\hat{p}$ and ground acceleration $\hat{w}_i$. The Cartesian tensors (Fung 1977) with subscripts i, j, k and l (=1, 2, 3) obeying the summation convention

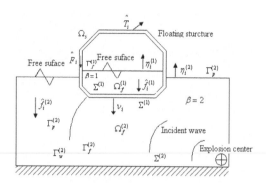

Figure 1. A fluid-structure interaction system to model a LNG tank floating on the water subject to an explosion pressure wave under the water.

are used in the paper. For example, u_i, v_i, w_i, e_{ij} and σ_{ij} represent displacement, velocity, acceleration vectors, strain and stress tensor in solid, respectively, p denotes the pressure in fluid, $p_{tt} = \partial^2 p/\partial t^2, u_{i,j} = \partial u_i/\partial x_j, v_i = \dot{u} = u_{i,t} = \partial u_i/\partial t, \quad w_i = \dot{v}_i = \ddot{u}_i = u_{i,tt} = \partial^2 u_i/\partial t^2$, etc. Let us divide this system into N_s substructures in the solid and N_f subdomains in the fluid. A typical substructure represented by $Sub^{(I)}, (I = 1, 2, 3, \ldots, N_s)$ has its domain $\Omega_s^{(I)}$ with displacement boundary $S_w^{(I)}$, traction boundary $S_T^{(I)}$, wet interface $\Sigma^{(I\alpha)} = \Sigma^{(\alpha I)}$ connecting to subdomains $Dom^{(\alpha)}, \alpha \in N^{(I\alpha)}$, where $N^{(I\alpha)}$ denotes a set of ordered numbers of the adjacent subdomains $Dom^{(\alpha)}$, and substructure interfaces $S^{(IJ)}$ connected to the N_I adjacent substructures $Sub^{(J)}, (J \in N^{(I)})$, where $N^{(I)}$ denotes a set of ordered numbers of the adjacent substructures $Sub^{(J)}$. Similarly, a typical subdomain represented by $Dom^{(\beta)}, (\beta = 1, 2, 3, \ldots, N_f)$ has its domain $\Omega_f^{(\beta)}$ with boundaries $\Gamma_f^{(\beta)} \cup \Gamma_w^{(\beta)} \cup \Gamma_p^{(\beta)}$ and f-s coupling interface $\Sigma^{(\beta K)} = \Sigma^{(K\beta)}$ connecting to substructures $Sub^{(K)}, (K \in N^{(\beta K)})$ and subdomain interfaces $\Gamma^{(\beta\gamma)}$ connected to the n_β adjacent substructures $Dom^{(\gamma)}, (\gamma \in n^{(\beta)})$ where $n^{(\beta)}$ denotes a set of ordered numbers of the adjacent subdomains $Dom^{(\gamma)}$. It is assumed that there exist N independent substructure interfaces, n subdomain interfaces and N_{sf} fluid-structure interaction interfaces in this division of the system. The governing equations describing the dynamics of the substructures and subdomains are as follows.

2.1 Solid substructure

Dynamic equation

$$\sigma_{ij,j}^{(I)} + \hat{f}_i^{(I)} = \rho_s^{(I)} w_i^{(I)}, \quad (x_i, t) \in \Omega_s^{(I)} \times (t_1, t_2). \quad (1)$$

Strain-displacement

$$e_{ij}^{(I)} = \frac{1}{2}(u_{i,j}^{(I)} + u_{j,i}^{(I)}), \quad (x_i, t) \in \Omega_s^{(I)} \times (t_1, t_2). \quad (2)$$

Constitutive equation

$$\sigma_{ij}^{(I)} = E_{ijkl}^{*(I)} e_{kl}^{(I)}, \quad (x_i, t) \in \Omega_s^{(I)} \times (t_1, t_2). \quad (3)$$

and, assuming linearity, we have

$$v_i^{(I)} = u_{i,t}^{(I)}, \quad w_i^{(I)} = v_{i,t}^{(I)},$$
$$d_{ij}^{(I)} = e_{ij,t}^{(I)} = \frac{1}{2}(v_{i,j}^{(I)} + v_{j,i}^{(I)}). \quad (4)$$

Boundary conditions

acceleration: $w_i^{(I)} = \hat{w}_i^{(I)}, \quad (x_i, t) \in S_w^{(I)} \times [t_1, t_2],$ (5-1)

traction: $\sigma_{ij}^{(I)} v_j^{(I)} = \hat{T}_i^{(I)}, \quad (x_i, t) \in S_T^{(I)} \times [t_1, t_2],$ (5-2)

Substructure interfaces

$$w_i^{(I)} = w_i^{(J)}, \quad (x_i, t) \in S^{(IJ)} \times [t_1, t_2],$$
(6-1)

$$\sigma_{ij}^{(I)} v_j^{(I)} + \sigma_{ij}^{(J)} v_j^{(J)} = 0, \quad (x_i, t) \in S^{(IJ)} \times [t_1, t_2],$$
(6-2)

2.2 Fluid subdomain

Dynamic equation

$$p_{tt}^{(\beta)} = (c^2)^{(\beta)} p_{ii}^{(\beta)}, \quad (x_i, t) \in \Omega_f^{(\beta)} \times (t_1, t_2). \quad (7)$$

Boundary conditions
free surface:

$$p_i^{(\beta)} \eta_i^{(\beta)} = -p_{tt}^{(\beta)}/g, \quad (x_i, t) \in \Gamma_f^{(\beta)} \times [t_1, t_2], \quad (8-1)$$

pressure:

$$p^{(\beta)} = \hat{p}^{(\beta)}, \quad (x_i, t) \in \Gamma_p^{(\beta)} \times [t_1, t_2], \quad (8-2)$$

acceleration:

$$p_i^{(\beta)} \eta_i^{(\beta)} = -\rho_f^{(\beta)} \hat{w}_i^{(\beta)} \eta_i^{(\beta)}, \quad (x_i, t) \in \Gamma_w^{(\beta)} \times [t_1, t_2], \tag{8-3}$$

Subdomain interfaces in a same fluid

$$p^{(\beta)} = p^{(\gamma)}, \quad (x_i, t) \in \Gamma^{(\beta\gamma)} \times [t_1, t_2], \tag{9-1}$$

$$p_i^{(\beta)} \eta_i^{(\beta)} + p_i^{(\gamma)} \eta_i^{(\gamma)} = 0, \quad (x_i, t) \in \Gamma^{(\beta\gamma)} \times [t_1, t_2], \tag{9-2}$$

2.3 Interaction interfaces

Fluid-structure interaction interfaces

$$w_i^{(I)} v_i^{(I)} = p_i^{(\alpha)} \eta_i^{(\alpha)} / \rho_f^{(\alpha)}, \quad (x_i, t) \in \sum^{(I\alpha)} \times [t_1, t_2], \tag{10-1}$$

$$\sigma_{ij}^{(I)} v_j^{(I)} = p^{(\alpha)} \eta_i^{(\alpha)}, \quad (x_i, t) \in \sum^{(I\alpha)} \times [t_1, t_2]. \tag{10-2}$$

Interaction interfaces between two different fluids

Assume that the two fluids can not mix together and there is a separate interface between them. Since the different mass densities of two fluids, the motion of the interface between the two different fluids affects the gravitational potential of the system. The kinematic and dynamic condition given in Equation 9-1, for a same fluid are not valid. The conditions describing the kinematic and dynamic relationships between two different fluids are as follows.

$$p^{(\alpha)} = p^{(\beta)}, \quad (x_i, t) \in \Gamma^{(\alpha\beta)} \times [t_1, t_2]. \tag{11-1}$$

$$\frac{1}{\rho_f^{(\alpha)}} \left(p_i^{(\alpha)} \eta_i^{(\alpha)} + \frac{p_{tt}^{(\alpha)} \eta_3^{(\alpha)}}{g} \right) + \frac{1}{\rho_f^{(\beta)}} \left(p_i^{(\beta)} \eta_i^{(\beta)} \right)$$

$$+ \frac{p_{,tt}^{(\beta)} \eta_3^{(\beta)}}{g} \right) = 0, \quad (x_i, t) \in \Gamma^{(\alpha\beta)} \times [t_1, t_2] \tag{11-2}$$

Here superscripts α and β identify two different fluids with different mass densities and $\Gamma^{(\alpha\beta)}$ denotes the interaction interface between them. The notation $\eta_3^{(\alpha)}$ or $\eta_3^{(\beta)}$ denotes the x_3-component of the unit outward normal vector on the interface of a fluid domain. Due to the gravity, the interaction interfaces between two different fluids must be in the horizontal plane in the static equilibrium condition. The heavier and lighter fluids are located under and above their inter-action interface, respectively. Therefore, under the

coordinate system shown in Figure 1, for the heavier and lighter fluids, we have that $\eta_3^{(heavy)} = 1$ and $\eta_3^{(light)} = -1$, respectively. This indicates that the effects of the two different fluids on the gravitational potential of the system can be partially cancelled. For example, if the mass densities of these two fluids are the same, Equations 11-1, 2 reduce to Equations 9-1, 2, respectively.

3 FEA EQUATION

The variational principle developed (Xing 1984, Xing & Price 1991) is extended to its substructure—subdomain form including the interactions between two different fluids as discussed above. This functional is presented as follows:

$$H_{sf}[p, w_i] = \sum_{I=1}^{N_s} \int_{t_1}^{t_2} \left\{ \int_{\Omega_s^{(I)}} \left(\frac{1}{2} \rho_s^{(I)} w_i^{(I)} w_i^{(I)} - \frac{1}{2} E_{ijkl}^{*(I)} d_{ij}^{(I)} d_{kl}^{(I)} \right. \right.$$

$$\left. - \hat{f}_i^{(I)} w_i^{(I)} \right) d\Omega_s^{(I)} - \int_{S_T^{(I)}} \hat{T}_i^{(I)} w_i^{(I)} dS^{(I)} \right\} dt$$

$$+ \sum_{\beta=1}^{N_f} \int_{t_1}^{t_2} \left\{ \int_{\Omega_f^{(\beta)}} \left[\frac{1}{2\rho_f^{(\beta)} (c^2)^{*(\beta)}} p_t^{(\beta)} p_t^{(\beta)} \right. \right.$$

$$\left. - \frac{1}{2\rho_f^{(\beta)}} p_i^{(\beta)} p_i^{(\beta)} \right] d\Omega_f^{(\beta)}$$

$$+ \int_{\Gamma_f^{(\beta)} \cup \Gamma^{(\alpha\beta)}} \frac{\eta_3^{(\beta)}}{2\rho_f^{(\beta)} g} p_t^{(\beta)} p_t^{(\beta)} d\Gamma^{(\beta)}$$

$$\left. - \int_{\Gamma_w^{(\beta)}} p^{(\beta)} \hat{w}_i^{(\beta)} \eta_i^{(\beta)} d\Gamma^{(\beta)} \right\} dt$$

$$- \sum_{L=1}^{N_{sf}} \int_{t_1}^{t_2} \int_{\Sigma^{(L)}} p^{(L)} w_i^{(L)} \eta_i^{(L)} d\Gamma^{(L)} dt. \tag{12}$$

In this functional, the acceleration in solids and the pressure in fluids are taken as the variables to be modelled by using finite element numerical method. The functional is subject to the constraints given in Equations 2–4, 5-1, 6-1, 8-2, 9-1 and 11-1 as well as the imposed variation constraints $\delta v_i = 0 = \delta p$ at the two time terminals t_1 and t_2. The stationary conditions of the functional are described in Equations 1, 5-2, 6-2, 7, 8-1, 8-3, 9-2, 10-1, 2 and 11-2.

A discretisation of the solid and fluid into finite elements expresses the displacement of solid and the pressure of fluid using FEA interpolations (Bathe 1982, Zienkiewicz & Taylor 1989, 1991) in the forms

$$\mathbf{u}^{(I)} = [u_1 \quad u_2 \quad u_3]^{(I)T} = \Phi^{(I)} \mathbf{U}^{(I)}, p^{(\beta)} = \varphi^{(\beta)} \mathbf{p}^{(\beta)}, \tag{13}$$

where $\Phi^{(I)}$ and $\varphi^{(\beta)}$ denote the interpolation function matrices and $\mathbf{U}^{(I)}$ and $\mathbf{p}^{(\beta)}$ represent the substructure/subdomain node displacement/pressure vector, respectively. The functional H_{sf} given in Equation 12 now takes the form

$$H_{sf}[\mathbf{p}, \ddot{\mathbf{U}}] = \sum_{I=1}^{N_s} \int_{t_1}^{t_2} \left(\frac{1}{2}\ddot{\mathbf{U}}^T \mathbf{M} \ddot{\mathbf{U}} - \frac{1}{2}\dot{\mathbf{U}}^T \mathbf{K} \dot{\mathbf{U}} - \ddot{\mathbf{U}}^T \hat{\mathbf{F}} \right)^{(I)} dt$$

$$- \sum_{L=1}^{N_{sf}} \int_{t_1}^{t_2} (\mathbf{p}^T \mathbf{R} \ddot{\mathbf{U}})^{(L)} dt$$

$$+ \sum_{\beta=1}^{N_f} \int_{t_1}^{t_2} \left(\frac{1}{2}\dot{\mathbf{p}}^T \mathbf{m} \dot{\mathbf{p}} - \frac{1}{2}\mathbf{p}^T \mathbf{k} \mathbf{p} - \mathbf{p}^T \hat{\mathbf{u}} \right)^{(\beta)} dt,$$

$$(14)$$

where $\mathbf{M}^{(I)}$ and $\mathbf{K}^{(I)}$ represent respectively the finite-element mass and stiffness matrices of the dry structure (I); $\mathbf{m}^\beta$ and $\mathbf{k}^{(\beta)}$ represent the finite-element matrices of the fluid domain, $\mathbf{R}^{(L)}$ denotes the fluid-structure interaction matrix.

The variational description in Equation 14 forms a basis to derive finite element equation modelling fluid-structure interaction dynamics. Taking the variation of the functional in Equation 14 and using the displacement consistent conditions on the solid substructure interfaces and the pressure equilibrium conditions on the fluid subdomain interfaces, we derive the finite element equation of the total fluid-structure interaction system in the following matrix form (Xing & Price 1991, Xing et al 1996)

$$\begin{bmatrix} \mathbf{M} & \mathbf{0} \\ \mathbf{R} & \mathbf{m} \end{bmatrix} \begin{bmatrix} \ddot{\mathbf{U}} \\ \ddot{\mathbf{p}} \end{bmatrix} + \begin{bmatrix} \mathbf{K} & -\mathbf{R}^T \\ \mathbf{0} & \mathbf{k} \end{bmatrix} \begin{bmatrix} \mathbf{U} \\ \mathbf{p} \end{bmatrix} = \begin{bmatrix} \hat{\mathbf{F}} \\ \hat{\mathbf{f}} \end{bmatrix}, \quad (15)$$

where $\mathbf{M}$, $\mathbf{K}$, $\mathbf{U}$ and $\hat{\mathbf{F}}$ represent respectively the finite-element mass, stiffness matrices, displacement vector and external force vector of the dry solid substructure; $\mathbf{m}$, $\mathbf{k}$, $\mathbf{p}$ and $\hat{\mathbf{f}}$ represent the similar finite-element matrices in a pressure form of the fluid domain and $\mathbf{R}$ denotes the fluid-structure interaction matrix. This is a non-symmetrical equation. Using the developed symmetrisation methods (Xing & Price 1991), we can obtained a symmetrical equation

$$\begin{bmatrix} \mathbf{K} & \mathbf{0} \\ \mathbf{0} & \mathbf{m} \end{bmatrix} \begin{bmatrix} \ddot{\mathbf{U}} \\ \ddot{\mathbf{p}} \end{bmatrix} + \begin{bmatrix} \mathbf{K}\mathbf{M}^{-1}\mathbf{K} & -\mathbf{K}\mathbf{M}^{-1}\mathbf{R}^T \\ -\mathbf{R}\mathbf{M}^{-1}\mathbf{K} & \mathbf{k} + \mathbf{R}\mathbf{M}^{-1}\mathbf{R}^T \end{bmatrix}$$

$$\times \begin{bmatrix} \mathbf{U} \\ \mathbf{p} \end{bmatrix} - \begin{bmatrix} \mathbf{K}\mathbf{M}^{-1}\hat{\mathbf{F}} \\ \hat{\mathbf{f}} - \mathbf{R}\mathbf{M}^{-1}\hat{\mathbf{F}} \end{bmatrix}. \quad (16)$$

If the force vectors on the right sides of Equations 15 and 16 vanish, the resultant equations describe the natural vibrations of the system. Since a constant pressure in the fluid and the corresponding static displacement configuration of the solid satisfy all the governing equations given in Section 2, there exists a natural frequency of value zero. As a result of this, the current efficient eigenvalue solver can not directly be used to fluid-structure interaction problems. To address this difficult, the frequency shift technique can be used (Xing & Price 1991, Xing & Xiong 2008b).

4 A LIQUID STORAGE TANK SUBJECT TO EL-CENTRO EARTHQUAKE EXCITATION

Figure 2a schematically illustrates a 3-dimensional air-water-tank three-phase dynamic coupled system under investigation (Xiong & Xing 2008b). The tank is a uniform spherical shell of outer diameter 300 mm and thickness 2.8 mm and is fixed at the four positions along the equator of the outer shell. The tank is filled equally by air and water. The finite element model adopts 264 four-node shell elements to describe the tank, 144 eight-node pressure elements to model the water and the same number for the air. Such a model was used to calculate the natural frequencies of the dry structure and the interaction case assuming no free surface wave to compare with experimental results (Toyota et al 2006).

The material properties of the tank structure, water and air are used as follows. For tank structure, the elastic module E = 3100 MPa, mass density $\rho_s = 0.00119$ kg/cm^3 and Poisson ratio $\nu = 0.31$. For water ($\beta = 1$) and air ($\beta = 2$), their mass

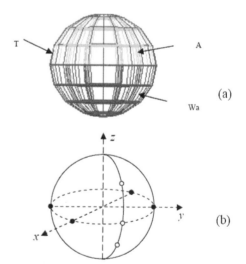

(a)

(b)

Figure 2. (a) Air-water-spherical tank interaction system, (b) coordinate system o-xyz and coordinates (cm) boundary fixed points ($\bullet$) and response point ($\bigcirc$): A (9.7078, 9.7078, -5.6867), B (10.5076, 10.5076, 0.0000), C (9.7078, 9.7078, 5.6867).

densities are $\rho_f^{(1)} = 0.1019 \times 10^{-6}$ kg/cm^3 and $\rho_f^{(2)} = 0.1249 \times 10^{-8}$ kg/cm^3, speeds of sound $c^{(1)} = 1430$ m/s and $c^{(2)} = 340$ m/s, respectively. The system is horizontally excited by El-Centro earthquake record data (Irvine 2008) shown in Figure 3.

Figure 4 shows three selected sloshing modes in which (a) is a pitch motion of 1.619 Hz, and (b) shows a bending form of 2.226 Hz while (c) behaves a convex-concave mode shape of 2.651 Hz. It is found from the simulations that the enclosed air increases the sloshing frequency of the system. This result is explained and confirmed by the following fact. Due to different mass densities and speeds of sound for the air and the water, the gravitational potentials of the water and the air caused by the motion of the air-water interface are partially canceled, as described in Equation 12. If the mass density and the speed of sound of the air are assumed to be increased to the one of the water, which represents that the shell tank is fully filled by the water. In this case, there exists no air-water interface involving the sloshing motion and the natural frequencies of the full water domain will be much higher, which is confirmed by the numerical simulations. This finding suggests an approach to control sloshing by using the pressurised air to fill sealed tanks.

Figure 5 shows the time histories of dynamic displacements (a) and pressures (b) at selected nodes in the air-water-tank system subject to El-Centro earthquake excitation. It is found that the frequency of the dominated component of the dynamic response is about 2.1 Hz which corresponds to the sloshing frequency of the air-water-tank interaction system. Both of the amplitudes of dynamic displacement and pressure in the air-water-tank system are lower than the

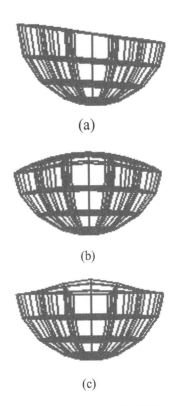

(a)

(b)

(c)

Figure 4. Three selected modes of the fluid in the tank: (a) pitch mode of 1.619 Hz, (b) bending mode of 2.226 Hz and (c) convex-concave mode of 2.651 Hz.

ones in the water-tank system in which the air effect is neglected (Xiong et al 2006).

5 A LNG-TANK-WATER INTERACTION SYSTEM SUBJECT TO A PRESSURE WAVE

Figure 6 shows the geometrical size and finite element idealisation of a 2-dimensional LNG-tank-water interaction system subject to a pressure wave excitation (Xiong & Xing 2007, Xiong et al 2007, Xiong & Xing 2008a). This pressure wave may be caused by explosions under water or earthquake or sea waves. The data used in the simulations are: external water domain: $H_s = 64$ m, $L = 120$ m; tank-internal water domain: $H = 30$ m, $h_l = 4$ m, $h_u = 9$ m, $B_l = 32$ m, $B_u = 22$ m, $b = 40$ m. The tank is treated as a uniform elastic structure of thickness 0.3 m, Young's modulus $E = 9.81 \times 10^5$ N/cm^2, Poisson ratio $\mu = 0.31$ and mass density $\rho_s = 2.4 \times 10^{-3}$ Kg/cm^3. The internal fluid is a LNG of mass density $\rho_g = 4.74 \times 10^{-4}$ Kg/cm^3 and speed of sound $C_g = 1700$ m/s. The external fluid is considered as the sea-water of mass density $\rho_w = 1.025 \times 10^{-3}$ Kg/cm^3 and speed of sound $C_w = 1430$

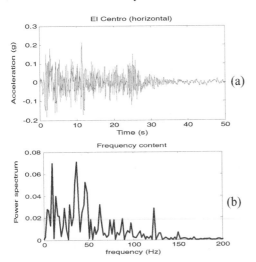

Figure 3. Ground acceleration curve (a) and power spectrum (b) of El-Centro earthquake (Irvine 2008).

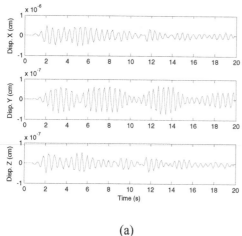

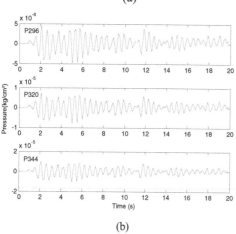

(a)

(b)

Figure 5. (a) Time histories of dynamic displacements at node C of number 169 and (b) time histories of dynamic pressures at point A of node P296, point B of node P320 and point C of P344 on the fluid-tank interface of air-water-tank system subject to El-Centro earthquake excitation.

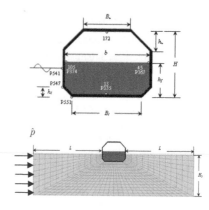

Figure 6. The geometrical size and finite element idealization of an internal liquid—LND tank—external water interaction system (Xiong & Xing 2007).

m/s. The draught is 4 m for the empty tank and 11 m for a 50% filling level $h_f = 15$ m, respectively. Figure 6 also indicates the locations of the selected displacement response points 13, 45, 105 and 172 on the tank and the pressure response points P335, P367, P374 and P541, P547 and P551 in the internal and external fluid domains, respectively. The tank is modelled using 104 four-node plane strain elements involving a total of 208 solid nodes. The internal liquid domain is divided into 336 four-node plane pressure elements with a total of 375 nodes and the external water domain is idealised by 384 four-node plane pressure elements with a total of 429 nodes.

The system is excited by a pressure wave $\hat{p}$ along the left boundary of the external water domain. This pressure wave may be caused by any explosions under

the water or an earthquake or regular sea waves. The computer program allows choosing any types of time histories, continuous or discrete time functions, to model the prescribed pressure waves (Xing 1995a, b). Here as an example, a regular sea pressure wave of unit amplitude $\hat{p} = \cos 2\pi f t$ is considered. To examine the sloshing dynamic response, we choose a lower frequency 0.1 Hz, which is near to the first sloshing frequency 0.127 Hz of the system, to conduct the numerical simulations. The mode damping coefficient of value 0.05 is used in calculation.

The simulation shows that for the internal liquid-tank-external water interaction systems like a large LNG ship floating on the water, there exist quite large numbers of internal and external sloshing frequencies due to large area of free surface on the external water. Therefore, for the design of a large LNG ship operating in sea ways, the sloshing fluid-structure interaction calculations are more important. Figure 7 shows the two selected modes of the system. In these two modes, Mode 7 has an anti-symmetric pattern of both sloshing motions of internal liquid and external water whereas and Mode 9 describes a tank rolling motion coupling with sloshing effect of internal liquid. Figure 8 shows the time histories of the vertical displacement and the dynamic fluid pressure responses at the selected points indicated in Figure 6. The tank vertical displacement response amplitude at Point 45 is about 4 m and the dynamic fluid pressure amplitude at P367 is about 5.5 MPa, which could affect the safe operation of the LNG tank.

6 A BUILDING STRUCTURE-ACOUSTIC VOLUME INTERACTION SYSTEM

Figure 9(a) shows a system of a wood chamber-air interaction system (Xing et al 2007b, Xing & Xiong

Mode 7   0.118 Hz

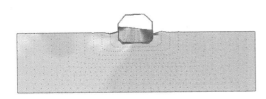

Mode 9   0.127 Hz

Figure 7.   Two natural modes of integrated interaction system.

2008a). The origin of the coordinate system $o - xyz$ is located at the centre of the bottom. The geometrical size in x, y and z directions of this chamber is $480 \times 480 \times 240$ cm. The top and walls of this chamber are made of wood plates of thickness 6 cm and gravity density $\gamma_s = 5.0 \times 10^{-4}$ Kg/cm^3. The Young's modulus and Poisson ratio of the wood are assumed as $E = 1.2 \times 10^5$ Kg/cm^2 and $\mu = 0.315$, respectively. The material properties of the air inside the chamber are used as the gravity density $\gamma_a = 1.225 \times 10^{-6}$ Kg/cm^3 and the speed of sound $C = 3.4 \times 10^4$ cm/s. In the simulation, the solid structure is modelled by 432 four node plate elements and the air volume is modelled by 864 three dimensional pressure elements. The corresponding interface elements on the air-structure interfaces between the structure and the air are used to generate the coupling matrix of the air and structure.

We investigated the dynamic response of the system excited by the footfall impacts caused by one

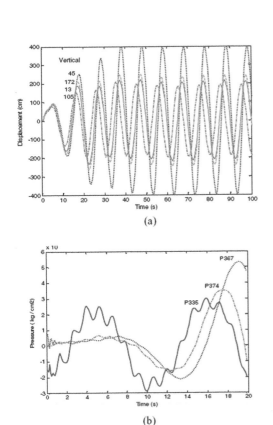

(a)

(b)

Figure 8.   The time histories of the dynamic responses of the system at chosen points of the system: (a) vertical displacement responses, (b) fluid pressure responses curves.

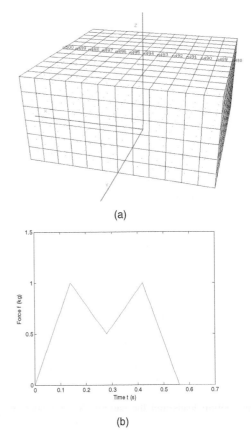

(a)

(b)

Figure 9.   (a) The finite element mesh of the wood chamber-air interaction system and (b) the time history pattern of human walking impact load (Xing, Xiong & Tan 2007b).

man (Xing et al. 2007b) and two men (Xing & Xiong 2008a) walking on the top plate of the wood chamber. It was assumed that Man 1 of weight 75 Kg was walking in x direction from the position x_0 on the intersection line of the top plate and the coordinate plane xoz with his step length $L = 80$ cm and a walking speed $v = 143$ cm/s. At the same time, Man 2 was also walking in y direction from the position y_0 on the intersection line of the top plate and the coordinate plane yoz. For a convenience, we considered that Man 2 had the same weight, step length and walking speed. The impact load caused by each right or left step was modelled in a pattern of time function shown in Figure 10(b) and Equation 17 in which the time period of walking impulse function was calculated by $\tau = L/v = 0.56$ s.

$$f(t) = \begin{cases} 4t/\tau, & 0 \le t \le \tau/4, \\ 1-2(t-\tau/4)/\tau, & \tau/4 \le t \le \tau/2, \\ 1/2 + 2(t-\tau/2)/\tau, & \tau/2 \le t \le 3\tau/4, \\ 1-4(\tau-3\tau/4)/\tau, & 3\tau/4 \le t \le \tau, \\ 0, & t \ge \tau. \end{cases}$$

(17)

Here t denotes a relative time from the starting time $t_N = 3(N-1)\tau/4$ at which the N-th step starts and the walking impulse function is added at point $x_N = x_0 + (N-1)L$. Therefore, there is a short time period about $\tau/4$ in which the walking loads are applied at two feet contacting points x_{N-1} and x_N to simulate the practical cases as measured in the experiments (Galbraith & Barton 1970, Wheeler 1982, Pavic & Reynolds 2002). The force peak-weight ratio of human footfall impacts for the walking cases is considered as 1.25 and therefore the impact load of each foot step used in the simulation equals $1.25 \times 75 f(t) = 93.75 f(t)$ Kg.

Figure 10(a) presents the time histories of sound pressures at nodes 420 (-160, 0,160), 422 (-80, 0, 160), 424 (0, 0, 160), 426 (80, 0, 160) and 428 (160, 0, 160) in the chamber (cm), and (b) gives the decibels of the sound pressures (dB) at the same nodes.

Figure 11 shows the dynamic displacement responses at points 488 and 494 on top floor of the chamber. As shown in Figure 10 the noises heard by people are mainly appeared in a time interval from 0.5 s to 2.5 s. During this period, the noise pressure shows an oscillation impulse which explains the "character" of the footstep noise claimed by people, such as "thuds", "thumps" and "booming". The values of dynamic responses depend on the footfall loads and the natural characteristics of the system. If the frequency of one harmonic component of the footfall loads is near to one of natural frequency of the air-structure interaction system, a resonance may happen, which will cause serious vibration and noise. The moving load simulation

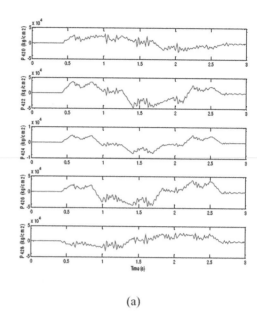

(a)

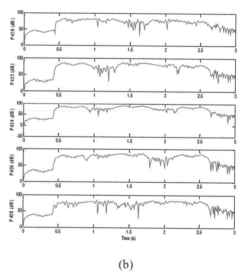

(b)

Figure 10. (a) Time histories of sound pressures at nodes 420 (-160, 0, 160), 422 (-80, 0, 160), 424 (0, 0, 160), 426 (80, 0, 160) and 428 (160, 0, 160) in chamber (cm); (b) sound pressures (dB) at nodes 420 (-160, 0, 160), 422 (-80, 0, 160), 424 (0, 0, 160), 426 (80, 0, 160) and 428 (160, 0, 160) in chamber (cm).

provides a useful approach to obtain more practical results. In this simulation, the first natural frequency of the system is about 17 Hz which is much different from the frequencies of harmonic component of the footfall load, therefore, there is no resonance

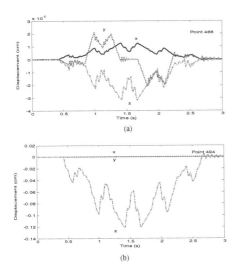

(a)

(b)

Figure 11. Dynamic displacement responses at Points 488 and 494 on top floor of chamber.

happening. However, even in this case, the level of sound pressure heard by five people sitting at the five points in the acoustic volume is higher than about 80 dB, as shown in Figure 10. This noise level is not allowed for comfort human living and working environment.

7 CONCLUSION

Following a summary of the developed mathematical model and the mixed finite element-substructure-subdomain method and its applications to dynamic analysis of fluid structure interaction systems in engineering, this paper presents the boundary conditions on the interface of two different fluids with different mass density which further confirms the theoretical foundation of the method.

Three engineering problems developed during Marstruct project period using the developed computer code are summarised. Problem 1 deals with an air-water-spherical shell system subject to El-Centro earthquake excitation. It is found from the simulations that the enclosed air increases the sloshing frequency of the system, which is explained and confirmed by theoretical formulation. This finding suggests an approach to control sloshing by using the pressurised air to fill sealed tanks. Problem 2 is a liquefied natural gas (LNG)-tank-water interaction system excited by a pressure wave. The simulation shows that for the internal liquid-tank-external water interaction systems like a large LNG ship floating on the water, there exist quite large numbers of internal and external sloshing frequencies due to large area of free surface on the external water. Therefore, for the design of a large LNG ship safely operating in sea ways, the sloshing fluid-structure interaction calculations are more important. Problem 3 is an acoustic volume-building interaction system subject to human footfall impacts. As shown in Figure 11 the noises heard by people are mainly appeared in a time interval from 0.5 s to 2.5 s. During this period, the noise pressure shows an oscillation impulse which explains the "character" of the footstep noise claimed by people, such as "thuds", "thumps" and "booming". The moving load simulation provides a useful approach to obtain more practical results. The modelling of this problem may be used to simulate ship deck vibration caused by human footfall loads to obtain a comfortable living environment for passengers.

ACKNOWLEDGEMENTS

Authors acknowledge the support in the scope of project MARSTRUCT, Network of Excellence on Marine Structure (www.mar.ist.utl.pt/marstruct) financed by EU for them to attend this conference.

REFERENCES

Bathe K.J. 1982. *Finite Element Procedures in Engineering Analysis*. Prentice-Hall.

Bishop, R.E.D. & Price, W.G. 1979. *Hydroelasticity of ships*. Cambridge University Press.

Bishop, R.E.D., Price, W.G. & Wu, Y. 1986. A general linear hydroelasticity theory of floating structures moving in a seaway. *Phil. Trans. R. Soc. Lond. A* 316: 375–426.

Deruieux, A. (Ed) 2003. *Fluid-Structure Interaction*. London: Kogan Page Limited.

Fung, Y.C. 1977. *A First Course in Continuum Mechanics*. Prentice-Hall.

Galbraith, F.W. & Barton, M.V. 1970. Ground loading from footsteps. *JASA* 48: 1288–1292.

Irvine, T. 2008. *Vibration Data*, EL Centro Earthquake Page: www.vibrationdata.com/elcentro.htm.2008.

Kock, E. & Olson, L. 1991. Fluid-solid interaction analysis by the finite element method-a variational approach. *Int. Jl. Numer. Methods Eng.* 31: 463–491.

Liu, W.K. & Ma, D.C. 1982. Computer implementation aspects for fluid-structure interaction problems. *Computer Methods in Applied Mechanics and Engineering* 31: 129–148.

Liu, W.K. & Uras, R.A. 1988. Variational approach to fluid-structure interaction with sloshing. *Nucl. Eng. Des.* 106: 69–85.

Morand, H.J.P. & Ohayon, R. 1995. *Fluid structure interaction*. Chichester: John Wiley & Sons.

Pavic, A. & Reynolds, P. 2002a. Vibration serviceability of long-span concrete building floors. Part 1: review of background information, *Shock & Vibration Digest* 34: 191–211.

Pavic, A. & Reynolds, P. 2002b. Vibration serviceability of long-span concrete building floors. Part 2: review of mathematical modelling approaches, *Shock & Vibration Digest* 34: 279–297.

Tan, M., Xiong, Y.P., Xing, J.T. & Toyoda, M. 2006. A numerical investigation of natural characteristics of a partially filled tank using a substructure method, In *Proceedings of Hydroelasticity '2006: Hydroelasticity in Marine Technology*. pp. 181–190, Beijing: National Defense Industry Press.

Toyota, M., Xing, J.T., Xiong, Y.P. & Tan, M. 2006. An experimental study on vibration characteristics of a thin spherical tank—water interaction system, In *Proceedings of Hydroelasticity '2006: Hydroelasticity in Marine Technology*. pp. 191–198, Beijing: National Defense Industry Press.

Unruh, J.F. 1979. A finite-element sub-volume technique for structure-borne interior noise prediction, *AIAA 79-585, 5th Aero. Acous. Conf. Seattle, WA 1979.*

Wheeler, J.E. 1982. Prediction and control of pedestrian induced vibration in footbridges. *JSD* 108: 2041–2065.

Xing, J.T. 1981. *Variational principles for elastodynamics and study upon the theory of mode synthesis methods.* Master thesis (In Chinese). Beijing: Dept. of Engineering Mechanics, Qinghua University, PRC.

Xing, J.T. 1984. *Some theoretical and computational aspects of finite element method and substructure-subdomain technique for dynamic analysis of the coupled fluid-solid interaction problems—variational principles for elastodynamics and linear theory of micropolar elasticity with their applications to dynamic analysis,* Ph. D. Dissertation (In Chinese). Beijing: Department of Engineering Mechanics, Qinghua University Beijing, PRC.

Xing, J.T. 1986a. A study on finite element method and substructure-subdomain technique for dynamic analysis of coupled fluid-solid interaction problems, (In Chinese). *Acta Mechanica Solida Sinica* 4: 329–337.

Xing, J.T. 1986b. Mode synthesis method with displacement compatibility for dynamic analysis of fluid-solid interaction problems. *Acta Aeronautica et Astronautica Sinica* 7: 148–156.

Xing, J.T. 1995a. *Theoretical manual of fluid-structure interaction analysis program-FSIAP*, Southampton: School of Engineering Sciences, University of Southampton.

Xing, J.T. 1995b. *User manual fluid-structure interaction analysis program-FSIAP*, Southampton: School of Engineering Sciences, University of Southampton.

Xing, J.T. & Price, W.G. 1991. A mixed finite element method for the dynamic analysis of coupled fluid-solid interaction problems. *Proc. R. Soc. Lond. A* 433: 235–255.

Xing, J.T., Price, W.G. & Du, Q.H. 1996. Mixed finite element substructure-subdomain methods for the dynamical analysis of coupled fluid-solid interaction problems. *Phil. Trans. R.S. Lond. A* 354: 259–295.

Xing, J.T., Price, W.G. & Wang, A. 1997. Transient analysis of the ship-water interaction system excited by a pressure water wave. *Marine Structures* 10 (5): 305–321.

Xing, J.T. & Xiong Y.P. 2008a. Numerical simulations of a building-acoustic volume interaction system excited by multiple human footfall impacts. In *Proceedings of 2008 ASME Pressure Vessels and Piping Division Conference,* PVP2008-61813, pp. 1–10, July 27–31, 2008, Chicago, Illinois, USA.

Xing, J.T. & Xiong, Y.P. 2008b. Mixed finite element method and applications to dynamic analysis of fluid-structure interaction systems subject to earthquake, explosion and impact loads. In *Proceedings of International Conference on Noise & Vibration Engineering-ISMA 2008*, Paper ID-562, pp. 1–15, September 14–19, 2008, Leuven, Belgium.

Xing, J.T. Xiong, Y.P., Tan, M. & Toyota, M. 2006. *Vibration problem of a spherical tank containing jet propellant: numerical simulations,* Southampton: Ship Science Report No. 141, ISSN 0140-3818, School of Engineering Sciences, University of Southampton.

Xing, J.T., Xiong, Y.P. & Tan, M. 2007a. The natural vibration characteristics of a water-shell tank interaction system. In G. Guedes Soares & P.K. Das (ed.), *Advancements in Marine Structures, Proceedings of Marstruct 2007, The 1st International Conference on Marine Structures, pp. 305–312, Glasgow, UK, 12–14 March 2007.* London: Taylor & Francis.

Xing, J.T. Xiong, Y.P. & Tan, M. 2007b. The dynamic analysis of a building structure—acoustic volume interaction system excited by human footfall impacts, In *Proceedings of Fourteenth International Congress on Sound & Vibration, Paper number 147, Cairns, Australia, 9–12 July 2007.* Cairns: IIAV.

Xing, J.T. & Zheng, Z.C. 1983. A study upon mode synthesis methods based on variational principles for elastodynamics, *Acta Mechanica Solida Sinica* 2: 250–257.

Xiong, Y.P. & Xing, J.T. 2007. Natural dynamic characteristics of an integrated liquid—LNG tank—water interaction system. In G. Guedes Soares & P.K. Das (ed.), *Advancements in Marine Structures Proceedings of Marstruct 2007, The 1st International Conference on Marine Structures, pp. 313–321, Glasgow, UK, 12–14 March 2007.* London: Taylor & Francis.

Xiong, Y.P. & Xing, J.T. 2008a. Dynamic analysis and design of LNG tanks considering fluid structure interactions, In *Proceedings of 2008 ASME 27th international conference on offshore mechanics and arctic engineering,* OMAE2008-57937, pp. 1–8, June 15–20, 2008, Estoril, Portugal.

Xiong, Y.P. & Xing, J.T. 2008b. Transient dynamic responses of an integrated air-liquid-elastic tank interaction system subject to earthquake excitations. In *2008 ASME Pressure Vessels and Piping Division Conference-PVP2008,* Paper PVP 2008–61815, pp. 1–10, July 27–31, 2008, Chicago, Illinois, USA.

Xiong, Y.P., Xing, J.T. & Price, W.G. 2006. The interactive dynamic behaviour of an air-liquid-elastic spherical tank system. In *Proceedings of 2006 ASME Pressure Vessels and Piping Division Conference, Vancouver, BC, Canada.* New York: ASME.

Xiong, Y.P., Xing, J.T. & Tan, M. 2007. Transient dynamic responses of an internal liquid-LNG tank-sea water interaction system excited by waves and earthquake loads, In *Proceedings of the 14 International Congress on Sound & Vibration, Paper number 566,* Cairns, Australia, 9–12, July 2007.

Zienkiewicz, O.C. & Taylor, R.L. 1989/1991. *The Finite Element Method.* 4th edition Vol. 1/2, McGraw-Hill.

Analysis and Design of Marine Structures – Guedes Soares & Das (eds)
© 2009 Taylor & Francis Group, London, ISBN 978-0-415-54934-9

Efficient calculation of the effect of water on ship vibration

M. Wilken
Germanischer Lloyd AG, Hamburg, Germany

G. Of
Graz University of Technology, Graz, Austria

C. Cabos
Germanischer Lloyd AG, Hamburg, Germany

O. Steinback
Graz University of Technology, Graz, Austria

ABSTRACT: Simulating global ship vibration can be split into three parts: firstly, the computation of the dry elastic vibration of the ship structure, secondly the determination of the hydrodynamic pressures caused by a given time harmonic velocity distribution on the outer shell and thirdly, the solution of the coupled vibration problem by considering the interaction of fluid and structure.

This paper describes an efficient solution to the second problem for refined models thereby forming an important element for the computation of ship vibration in higher frequencies. It is based on a fast multipole Galerkin boundary element method for the Laplace equation. The free surface boundary condition is incorporated into the method by either explicitly enforcing it on the meshed free surface or by employing a mirroring technique. Numerical examples are used to assess the accuracy and the speed of the solution procedures.

1 INTRODUCTION

Because of its high mass density, water has a significant effect on the vibration of a ship structure. Therefore, contrary to vibration in air, the interaction of fluid and structure needs to be considered in this case. This paper deals with a prerequisite for computing the interaction of the water with the vibrating ship structure, namely the computation of the hydrodynamic pressure on the ship hull caused by time harmonic deformation of this hull.

Today, typical global Finite Element models of ships have a mesh density capable of capturing the deformation due to global vibration up to approximately 30 Hz. For higher frequencies, finer meshes of the structure are required. One major problem then is, that standard Boundary Element Methods (BEM) cannot efficiently be used anymore for the computation of the hydrodynamic mass matrix. The reason is that for these methods, the necessary memory and cpu time rise quadratically or cubically, resp., with the number of wet elements. In this paper, the fast multipole method is applied to compute the effect of water on ship vibration. Through iterative computation of the hydrodynamic pressure due to harmonic ship deformation, its memory and cpu time requirements rise

significantly slower than for the standard BEM techniques. This in turn can make it possible to extend the computation of ship vibration into frequency ranges above 30 Hz.

In general, the interaction of ship and water leads to several effects as depicted in Figure 1. Below 30 Hz, the influence of compressibility on ship vibration is very small (Brunner, Junge, Wilken, Cabos, and Gaul 2009). In higher frequencies this influence can increase, but a simulation of ship vibration for the incompressible case can be regarded as a first approximation for a refined solution procedure. For this reason, the compressibility of water is ignored in this paper. Since the bouyancy forces are static forces, they also do not in uence the ship vibration.

In practical cases the structural deformation amplitudes are very small (in the range of 0.1 mm) compared to the dimensions of the structure, the structural as well as the fluid equations can be linearized with respect to the deformation amplitude. Furthermore the flow of water around the ship's hull can be assumed to be inviscid and irrotational (see e.g. (Newman 1977)). Therefore, the velocity v of the fluid particles can be expressed as the gradient of a velocity potential $v = \nabla \phi$ and mass conservation leads to the Laplace equation $\Delta \phi = 0$, which must be fulfilled in

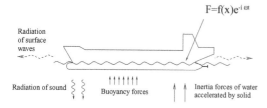

$$F = f(x)e^{-i\omega t}$$

Radiation of surface waves

Radiation of sound
Buoyancy forces
Inertia forces of water accelerated by solid

Figure 1. Physical effects of fluid-structure interaction on the ship. Only the inertia forces of the water are considered in this paper.

the fluid domain. For a more detailed account of the assumptions leading to the equations please refer to (Armand and Orsero 1979).

Assuming an inifinitely wide and deep uid domain, boundary conditions need to be specified at the wetted ship hull surface and at the free water surface.

At the wetted ship hull surface the uid velocity normal to this surface must equal the normal velocity of the structure, $\frac{\partial \phi}{\partial n} = u \cdot n$, where u denotes the structural deformation velocity and n is the normal on the hull surface pointing outwards from the ship.

The exact boundary condition at the free surface is non-linear (see e.g. (Newman 1977)) but can be linearized for the above reason to yield $\ddot{\phi} + g\frac{\partial \phi}{\partial z} = 0$ at the mean free surface position $z = 0$. Here g is the earth's gravitational acceleration. This boundary condition leads to gravity waves on the free surface. For harmonic waves, the velocity amplitude of surface waves is decreasing with distance z from the free surface with a factor of e^{-kz}, where $k = \omega^2/g$ is the surface wave number. For 10 Hz this means that the velocity amplitude is halved every 1.7 mm water depth. Therefore, surface waves will be ignored here and the infinite frequency approximation $\phi = 0$ is used as the free surface boundary condition. Since $p = -i\omega\rho f\phi$, this is also a pressure release boundary condition. As it does not lead to waves on the free surface, meshes on the free surface which are used to enforce the boundary condition $\phi = 0$ need not be refined with increasing frequency.

The main focus of this paper is the description of an efficient procedure for the solution of the above equation for the case of a known velocity field on the wetted ship hull in the case of refined meshes on the ship water interface.

2 PREVIOUS WORK ON THE COMPUTATION OF THE VIBRATION INDUCED FLOW AROUND A SHIP

Consideration of the effect of the surrounding water for the computation of global ship vibrations dates back to the first half of the 20th century. Regarding a ship

as a slender body, Lewis (Lewis 1929) showed that the inertia of the water can approximately be accounted for by analysing the two-dimensional flow around ship cross sections. In this way the total force of the water impeding the acceleration of a ship cross section leads via Newton's law to an apparent mass (called hydrodynamic mass) of the surrounding water. The hydrodynamic mass of a cylindrical cross section was generalized to more complex shapes by introducing reduction coefficients. Assuming that a ship is a slender body, Lewis succeeded in determining the hydrodynamic mass affecting vertical bending vibrations of a ship.

Since the hydrodynamic mass derived with this method depends on the particular bending mode of the ship, it is typically valid only for a specific range of frequencies around the corresponding eigenfrequency of this mode.

Wendel (Wendel 1950) and Landweber (Landweber 1957) extended Lewis' work by considering also horizontal and rotational acceleration of ship cross sections. Grim (Grim 1953), (Grim 1960) examined the reduction coefficients for higher modes.

The Lewis method is most appropriate for a Finite Element analysis, if the ship is modelled by using several beam elements. For the three dimensional analysis of ship vibrations based on a Finite Element model, the use of this method leads to the problem that only the total force onto a ship cross section (or better its total hydrodynamic mass) can be computed. The actual distribution of the hydrodynamic pressure over the contour of the section is determined with a heuristic approach. In a three dimensional vibration analysis this can lead to hydrodynamic forces having components tangential to the shell surface. Despite these approximations and shortcomings, the Lewis approach has proven to yield good results in the low frequency range. In higher frequencies, the three-dimensional effects of the flow need to be considered.

The three-dimensional effect of the vibration induced flow of water can be considered with the finite-element method. This approach has been taken e.g. by (Armand and Orsero 1979) or (Hakala 1986). It leads to considerable additional effort for building up a three-dimensional mesh of the water. Since only a finite part of the surrounding water can be modelled, the mesh is limited to some distance from the ship. At the outer boundary of the mesh appropriate boundary conditions have to be applied. Alternatively, semiinfinite elements can be used as the outmost element layer.

The latter problems can be avoided with the boundary element method. Here, only the free water surface and the ship/water interface need to be modelled with surface elements. This approach has been taken e.g. by (Kaleff 1980) and (Röhr and Möller 1997). It can be regarded as the standard approach today for the case that three-dimensional effects need to be included in the analysis, see (Cabos and Ihlenburg 2003). The drawback of the method is that—contrary to the finite element method—full system matrices arise for the water. The resulting quadratic memory requirements significantly limit its applicability for fine meshes and therefore for higher frequencies.

In the next section, it is shown how the Fast Multipole Method can overcome this problem. Moreover, using the method of image sources, the meshing of the free water surface can be avoided.

3 FAST BOUNDARY ELEMENT METHOD AND HALF-SPACE FORMULATION FOR THE FLUID DOMAIN

As described in Sect.1, the boundary value problem of the fluid domain is given by

$$- \Delta\phi(x) = 0 \quad \text{for } x \in \Omega,$$

$$t(x) := \frac{\partial\phi}{\partial n}(x) = u(x) \cdot n(x) \quad \text{for } x \in \partial S, \tag{1}$$

$$\phi(x) = 0 \quad \text{for } x \in E$$

where the velocity potential ϕ has to satisfy the radiation condition $\phi \in \mathcal{O}(\|x\|^{-1})$ for $\|x\| \to \infty$. u denotes the structural deformation velocity and n the exterior normal direction on the wetted hull surface ∂S. The free surface condition $\phi = 0$ is imposed on the water surface

$$E := \{x \in \mathbb{R}^3 : x_3 = 0, x \quad \text{outside of the ship}\},$$

where $x = (x_1, x_2, x_3)^\top$. The geometric setting is drafted in Fig. 2.

We will use a fast boundary element method to solve the boundary value problem (1). Additionally, we will consider two possibilities to handle the free water surface. We will present a boundary element method for a cut off of the free surface and a half-space formulation which utilizes the reflection principle and guarantees the free surface condition implicitly.

3.1 Boundary element method for the cut off free surface

The solution of the exterior boundary value problem (1) is given by the representation formula

$$\phi(x) = \int_\Gamma \frac{\partial}{\partial n_y} U^*(x, y)\phi(y)ds_y - \int_\Gamma U^*(x, y)t(y)ds_y \tag{2}$$

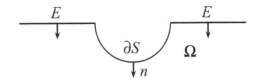

Figure 2.   Geometric setting (cross section) of the boundary value problem.

for $x \in \Omega$, the fluid domain, where $\Gamma = \overline{\partial S \cup E}$, and

$$U^*(x, y) = \frac{1}{4\pi} \frac{1}{|x - y|}$$

is the fundamental solution of the Laplacian. Thus, it is sufficient to know the values of $\phi_{|\Gamma}$ and its normal derivative $t_{|\Gamma}$ on the boundary Γ to compute the solution in Ω. In the fluid structure interaction problem these values on the boundary are even of primary interest. The unknown part of the data on the boundary can be determined by the help of two boundary integral equations

$$\phi(x) = \left(\frac{1}{2}I + K\right)\phi(x) - (Vt)(x) \tag{3}$$

and

$$t(x) = -(D\phi)(x) + \left(\frac{1}{2}I - K'\right)t(x) \tag{4}$$

for almost all $x \in \Gamma$, where V denotes the single layer potential

$$(Vt)(x) = \int_\Gamma U^*(x, y)t(y)ds_y \quad \text{for } x \in \Gamma,$$

K the double layer potential,

$$(K\phi)(x) = \int_{\Gamma\setminus\{x\}} \frac{\partial}{\partial n_y} U^*(x, y)\phi(y)ds_y \quad \text{for } x \in \Gamma,$$

K' the adjoint double layer potential,

$$(K't)(x) = \int_{\Gamma\setminus\{x\}} \frac{\partial}{\partial n_x} U^*(x, y)t(y)ds_y \quad \text{for } x \in \Gamma,$$

and D is the hypersingular operator

$$(D\phi)(x) = -\frac{\partial}{\partial n_x}\int_\Gamma \frac{\partial}{\partial n_y} U^*(x, y)\phi(y)ds_y \quad \text{for } x \in \Gamma.$$

n_x and n_y denote the exterior normal direction at point x and y, respectively. In fact, we use a combination of these two boundary integral equations. We split $t = t_0 + g$ where g is an extension of $u \cdot n$, which is defined on ∂S, and $t_0(x) = 0$ for $x \in \partial S$. Additionally, we define $\phi(x) = \phi_0(x)$ for $x \in \partial S$ with $\phi_0(x) = 0$ for $x \in \overline{E}$. Note that $\phi(x) = 0$ for $x \in E$. Now we utilize the first boundary integral equation (3) on E and the hypersingular boundary integral equation (4) on ∂S and obtain the symmetric formulation (Costabel 1987; Sirtori 1979)

$$(Vt_0)(x) - (K\phi_0)(x) = -(Vg)(x), \quad x \in E,$$

$$(K't_0)(x) + (D\phi_0)(x) = -\left(\frac{1}{2}I + K'\right)t_0(x), \quad x \in \partial S.$$
(5)

For the discretization of the symmetric formulation (5) and its equivalent variational formulation, respectively, we use a Galerkin method with conforming trial and test spaces. The potential ϕ_0 is approximated by a linear combination $\phi_{0,h}$ of piecewise linear and continuous basis functions φ_i defined with respect to a triangulation of the surface Γ. t_0 is approximated by a linear combination $t_{0,h}$ of piecewise constant basis functions ψ_k. We end up with the following system of linear equations

$$\begin{pmatrix} V_h & -K_h \\ K_h^\top & D_h \end{pmatrix} \begin{pmatrix} \underline{t}_0 \\ \underline{\phi}_0 \end{pmatrix} = \begin{pmatrix} \underline{f}_E \\ \underline{f}_{\partial S} \end{pmatrix}$$
(6)

where the matrix entries of each block are defined by

$$V_h[\ell, k] = \langle V\psi_k, \psi_\ell \rangle_E,$$

$$K_h[\ell, i] = \langle K\varphi_i, \psi_\ell \rangle_E,$$

$$D_h[j, i] = \langle D\varphi_i, \varphi_j \rangle_{\partial S}$$

for $k, \ell = 1, \ldots, N_E$ and $i, j = 1, \ldots, M_{\partial S}$. $M_{\partial S}$ is the number of nodes of the triangulation of ∂S which are not in contact with the water surface. i.e. in $\overline{E} \cdot N_E$ is the number of elements of the triangulation of the water surface $E \cdot \langle v, w \rangle_A$ is defined by

$$\langle v, w \rangle_A = \int_A v(x)w(x)ds_x.$$

For the right hand side, we use

$$f_E[\ell] = -\tilde{V}_h g_h,$$

$$f_{\partial S}[j] = -\left(\frac{1}{2}\tilde{M}_h^\top + \tilde{K}_h^\top\right) g_h$$

where

$$\tilde{M}_h[\ell, i] = \langle \varphi_i, \psi_\ell \rangle_E.$$

Here, we assume a piecewise constant approximation g_h of the piecewise linear and discontinuous $u_h \cdot n$ supplied by the structural part. This approximation is defined by an $L_2(\partial S)$ projection. For a detailed description of the boundary element method and analysis including the related error estimates see, e.g., (Steinbach 2008). The return values $\underline{r}$ for the fluid structure interaction are given by the pressure forces

$$r[j] = \frac{1}{3} \sum_{i \in \mathcal{P}(j)} n_i \Delta_i \phi_0[i]$$

for each node of the two matching grids of the structural computation and the fluid part. $\mathcal{P}(j)$ denotes the set of elements of the triangulation which share the node with index $j \cdot \Delta_i$ and n_i are the area and the normal vector of the i-th element, respectively. For the numerical realization, we cut off the water surface at a suitable distance from the ship. This defines an open surface $\partial S \cup E_d$ where

$$E_d := \{x \in E : \text{dist}\{x, \partial S\} > c_w\}$$

is the cut off water surface with a suitably chosen parameter c_w.

3.2 Half-space boundary element method

The solution of the boundary value problem (1) can be given by the modified representation formula

$$\phi(x) = \int_{\partial S} \hat{U}^*(x, y)t(y)ds_y - \int_{\partial S} \frac{\partial}{\partial n_y}\hat{U}^*(x, y)\phi(y)ds_y$$
(7)

with the half-space fundamental solution

$$\hat{U}^*(x, y) = \frac{1}{4\pi} \frac{1}{|x - y|} - \frac{1}{4\pi} \frac{1}{|\hat{x} - y|}$$

where $\hat{x} = (x_1, x_2, -x_3)^\top$ denotes the image point of $x = (x_1, x_2, x_3)^\top$ with respect to the plane E. Like the reflection principle, the half-space fundamental solution guarantees the free surface condition $\phi(x) = 0$ for $x \in E$.

In the limit case $x \to \partial S$ we obtain the boundary integral equation

$$\phi(x) = \left(\frac{1}{2}I + \hat{K}\right)\phi(x) - (\hat{V}t)(\phi)(x)$$
(8)

for almost all $x \in \partial S$ in the usual way. $\hat{V}$ and $\hat{K}$ denote the single and double layer potentials defined by the half-space fundamental solution $\hat{U}(x, y)$. As long as the point $x \in \partial S$ is not located on the plane

E, the integrals of the extra term $|\hat{x} - y|^{-1}$ are regular and no extra jump relations arise. The special case of $x \in \partial S \cap E$ is treated in (Li, Wu, and Seybert 1994; Wu and Seybert 1989; Sladek, Tanaka, and Sladek 2000) for the Helmholtz kernel. These extended jump relations are not needed for a Galerkin variational formulation which will be considered here.

Applying the normal derivative and the limit $x \to \partial S$ to the modified representation formula, the hypersingular boundary integral equation

$$t(x) = \left(\frac{1}{2}I - \hat{K}'\right) t(x) - (\hat{D}\phi)(x) \qquad (9)$$

for almost all $x \in \partial S$ is obtained. $\hat{K}'$ and $\hat{D}$ denote the adjoint double layer potential and the hypersingular operator based on the half-space fundamental solution $\hat{U}(x, y)$. Essentially, these modified boundary integral operators have the same properties except jump conditions on ∂S.

Since the free surface condition is fulfilled implicitly, the unknown Cauchy data $\phi_{|\partial S}$ can be determined by the hypersingular boundary integral equation,

$$(\hat{D}\phi)(x) = -\left(\frac{1}{2}I + \hat{K}'\right) g(x). \qquad (10)$$

The Galerkin discretization described in Sect.3.1 leads to a non-symmetric realization

$$\hat{D}_h \underline{\phi} = -\left(\frac{1}{2}M_h^\top + \hat{K}_h^\top\right) \underline{g}$$

of the symmetric continuous operator with a piecewise constant approximation g_h of the piecewise linear and discontinuous $u_h \cdot n$ supplied by the structural part. This effects the performance of the overall solver which assumes the symmetry of this operator on the discrete level, too. To fix this problem we define the same continuous operator by inserting ϕ defined by (10) into the first boundary integral equation (8),

$$\phi(x) = -\left(\hat{V} + \left(\frac{1}{2}I + \hat{K}\right)\hat{D}^{-1}\left(\frac{1}{2}I + \hat{K}'\right)\right) g(x).$$

The Galerkin discretization of this operator enables an approximation of the integral of the potential ϕ over each element τ_i of the triangulation by

$$\phi_{\text{sym}}[i] = \langle \phi(x), \varphi_i^0 \rangle_\Gamma = (T_h g)[i] \qquad (11)$$

with the symmetric matrix

$$\hat{T}_h = -\hat{V}_h - \left(\frac{1}{2}M_h + \hat{K}_h\right)\hat{D}_h^{-1}\left(\frac{1}{2}M_h^\top + \hat{K}_h^\top\right).$$

In the preprocessing, the vector $\underline{u}$ of the structural deformation velocity u, which is given in each node, is projected to a piecewise constant function g_h with the coefficients

$$g[i] = \frac{1}{3}(u[i_1] + u[i_2] + u[i_3]) \cdot n_i.$$

The return values are the pressure forces

$$r[j] = \frac{1}{3}\sum_{i \in \mathcal{P}(j)} n_i \phi_{\text{sym}}[i]$$

where $\mathcal{P}(j)$ again denotes the set of elements of the triangulation which share the node with index j. The values $\phi_{\text{sym}}[i]$ already include the weighting by the area, since they are defined as the integral of ϕ over a single element, see (11). Thus the overall operator, which we defined for the half-space fundamental solution, is symmetric.

3.3 Fast boundary element method

Boundary element methods have the advantages that only a discretization of the boundary is needed and that the dimension of the system of linear equations is reduced compared to a finite element method. But the matrices are fully populated and the memory requirements are of order $\mathcal{O}(N^2)$ with the number N of elements of the boundary triangulation. Thus, the solution of the system of linear equations by an iterative method needs $\mathcal{O}(N^2 N_{it})$ essential operations where N_{it} is the number of iteration steps.

Fast boundary element methods reduce the effort for setting up the matrices, the memory requirements and the time for a single matrix vector product to almost linear complexity. A detailed description of the application of fast boundary element methods is given by (Rajasanow and Steinbach 2007). Even for fast boundary element methods, preconditioners are recommended to bound the number of iteration steps needed in the iterative solution. Here, we use the concept of operators of opposite order (Steinbach and Wendland 1998), i.e. the single layer potential as a preconditioner of the hypersingular operator. For the preconditioning of the single layer potential, we use an algebraic multilevel preconditioner (Steinbach 2003) and an algebraic multigrid method (Langer and Pusch 2005; Of 2008), respectively, which makes use of the structure of the fast boundary element method. The generalized minimal residual method (Saad and Schultz 1986; Golub and Van Loan 1996) is used to solve the block-skew-symmetric system (6). Note that an appropriate scaling of the preconditioners of the two blocks is need to get a good performance. A conjugate gradient method(Hestenes and Stiefel 1952; Golub and Van Loan 1996) is applied to invert the matrix $\hat{D}_h$ in the application of the operator $\hat{T}_h$ of the half-space formulation.

The fast boundary element method of our choice is the fast multipole method (Greengard and Rokhlin 1987). An overview on many papers related to the use of the fast multipole method is given by (Nishimura 2002). The two main tools of the data-sparse approximation of the boundary element matrices are a partitioning of the matrix in blocks and a low rank approximation of suitable blocks. The fast multipole method provides only a fast realization of the matrix vector product.

Here, we will sketch the realization of the single layer potential V by the fast multipole method. The matrix times vector product $\underline{w} = V_h \underline{t}$ can be written component-wise by

$$w_\ell = \sum_{k=1}^{N} V_h[\ell, k] t_k = \sum_{k=1}^{N} \frac{t_k}{4\pi} \int_{\tau_\ell} \int_{\tau_k} \frac{1}{|x - y|} ds_y ds_x$$

$$(12)$$

for $\ell = 1, \ldots, N$. The low rank approximation is defined by a kernel expansion in spherical harmonics which separates the variables. An approximation of the kernel $k(x, y) = |x - y|^{-1}$ is defined by

$$k_p(x, y) = \sum_{n=0}^{p} \sum_{m=-n}^{n} \overline{S_n^m}(y) R_n^m(x) \quad \text{for } |y| > |x|, \quad (13)$$

using reformulated spherical harmonics

$$R_n^{\pm m}(x) = \frac{1}{(n+m)!} \frac{d^m}{d\widehat{x}_3^m} P_n(\widehat{x}_3)(\widehat{x}_1 \pm i \widehat{x}_2)^m |x|^n,$$

$$S_n^{\pm m}(y) = (n - m)! \frac{d^m}{d\widehat{y}_3^m} P_n(\widehat{y}_3)(\widehat{y}_1 \pm i \widehat{y}_2)^m \frac{1}{|y|^{n+1}},$$

and $\widehat{y} = y/|y| \cdot P_n$ are the Legendre polynomials

$$P_n(u) = \frac{1}{2^n n!} \frac{d^n}{du^n}(u^2 - 1)^n \quad \text{for } |u| \leq 1.$$

As the expansion (13) converges only for $|y| > |x|$, the matrix V_h has to be split into a nearfield and a farfield part. The expansion (13) can be applied in the farfield $\mathrm{FF}(\ell)$ of a boundary element τ_ℓ, while the matrix entries of the nearfield $\mathrm{NF}(\ell)$ have to be computed in the usual way. The approximation of the matrix times vector product now reads as

$$\widetilde{w}_\ell = \sum_{k \in \mathrm{NF}(\ell)} V_{L,h}[\ell, k] t_k$$

$$+ \frac{1}{4\pi} \sum_{n=0}^{p} \sum_{m=-n}^{n} M_n^m(O, \ell) \widetilde{L}_n^m(\mathrm{FF}(\ell))$$

using the coefficients

$$M_n^m(O, \ell) = \int_{\tau_\ell} R_n^m(x) ds_x, \; L_n^m(O, k) = \int_{\tau_k} \overline{S_n^m}(y) ds_y,$$

and

$$\widetilde{L}_n^m(\mathrm{FF}(\ell)) = \sum_{k \in \mathrm{FF}(\ell)} t_k L_n^m(O, k)$$

for $n = 0, \ldots, p, m = -n, \ldots, n$ with respect to a local origin O.

The coefficients $\widetilde{L}_n^m(\mathrm{FF}(\ell))$ depend on the farfield $\mathrm{FF}(\ell)$ of each boundary element and differ consequently. For their efficient computation, a hierarchical structure built upon the boundary elements is used. This cluster structure defines the partitioning of the matrix. The fast multipole method reduces the memory requirements and the time for a matrix vector product to $\mathcal{O}(N \log^2 N)$. For a detailed description we refer to (Greengard 1987), (Of, Steinbach, and Wendland 2006) and other papers cited therein.

For the fast multipole realization of $\hat{V}$, we discuss two possibilities to handle the additional kernel $|\hat{x} - y|^{-1}$. Since the distance $|\hat{x} - y|$ of the image point $\hat{x}$ from y is always larger than $|x - y|$, the partitioning of the standard part can be used for the image part, too. This approach has the benefit that it is easy to implement as each fast multipole operation of the standard part is applied to the image part in a similar manner. The overall effort is less than two times the effort for the standard kernel only. Due to the half-space formulation, the water surface is not meshed and fewer boundary elements are needed. Thus, the half-space formulation can be fast than the formulation with the cut off of the water surface. Additionally, the half-space formulation gives more accurate results. Here, we have used this approach.

The second possibility is to build up an image of the whole mesh of the ship and to apply the fast multipole method in the usual way. This method increases the memory requirements due to the image part of the mesh and the additional part of the cluster structure of the fast multipole method. Since $|\hat{x} - y|$ is always larger than $|x - y|$, this approach admits larger blocks than the first version and results in a fast realization of the matrix vector product. This second approach is examined in (Brunner, Of, Junge, Steinbach, and Gaul 2008).

An additional splitting

$$\hat{U}^*(x, y) = \frac{1}{4\pi} \frac{1}{|x - y|} - \frac{1}{8\pi} \frac{1}{|\widetilde{x} - y|} - \frac{1}{8\pi} \frac{1}{|x - \widetilde{y}|}.$$

of the kernel is needed to guaranty the symmetry of the approximation by the fast multipole method.

Thus, both the sources and the targets have to have an image in the fast multipole method. For a discussion of the symmetric approximation by the fast multipole method and for the realization of the other boundary integral operators see, e.g., (Of, Steinbach, and Wendland 2006).

4 NUMERICAL COMPARISON OF METHODS

4.1 *Subject of investigation*

A typical container vessel of 250 m length and 32 m breadth was investigated to assess the accuracy and the speed of the above described standard BE method, fast multipole method and the Lewis method. A draft of 15.58 m was used in the calculations leading to 2032 wetted elements of the outer shell (the total number of surface elements is 19970).

The different solution procedures were used to compute the pressure on the outer shell imposed from time harmonic displacements of the vibrating hull. Eight different principal modes computed from a modal analysis of the dry ship in the frequency range between 0 Hz and 2 Hz were used as typical boundary conditions.

In the lower frequency range the strucural deformation of the outer shell is of a global shape and tends to be more and more local for the higher frequencies. In case of the BE methods the free water surface was modelled by explicit meshes having a minimum distance of 10m (see Figure 3), 20 m and 40 m from the outer shell. For the fast multipole procedure the free water surface was also considered by the above described image method. In the following these free surface considerations are denoted as (10 m), (20 m), (40 m) and (image).

The standard BE solution was obtained with an implementation of the University of Rostock (Röhr and Möller 1997) used for several years at the Germanischer Lloyd AG, which is capable to retrieve the fully populated hydro mass matrix for small and medium sized models. The fast multipole BE method

was performed with an implementation developed by the Graz University of Technology and the Lewis method with an implementation of the Germanischer Lloyd AG.

In all computations, the boundary element mesh for the fuid coincides with the structural mesh of the wet part of the ship hull. The reasons for this are firstly that building up a different mesh for the fluid is considerable additional work and secondly that a specific coupling algorithm would be required for incompatible meshes.

The calculations were done on a 32 bit linux computer having two single core Intel Xeon processors with 4 GB RAM and a clock rate of 2.2 GHz.

4.2 *Pressure results*

To assess the accuracy of the different solution techniques a measure of pressure deviations is needed. We have chosen the results derived from the standard BE method (40 m) as reference pressure distribution, since this method is an established procedure regarding the inertia effects of the water of a submerged vibrating surface. The pressure retrieved with this method is compared with the pressure computed by the other methods at a wetted surface element i by

$$dev_i = \frac{|p_i - p_i^{\text{std}(40\text{ m})}|}{|p_{\max}^{\text{std}(40\text{ m})}|} \tag{14}$$

whereas $p_{\max}^{\text{std}(40\text{ m})}$ denotes the maximal element pressure computed by the standard BE method (40 m). An overall deviation norm can be constructed by computing the averaged individual element errors weighted with the element areas.

Figure 4 shows the pressure distribution on the outer shell calculated with the standard BE method (40 m) induced by a dry structural mode at 7.73 Hz. Since the mode is abitrarily scaled no absolute pressure values are given.

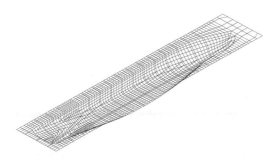

Figure 3. Wetted shell with discretized free water surface with a minimum distance of 10 m to the shell.

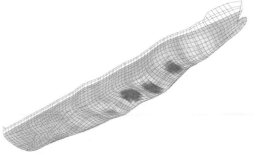

Figure 4. Pressure distribution at the outer shell caused by the deformation of the structural mode at 7.73 Hz.

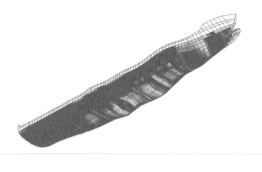

Figure 5. Deviation of the pressures computed by the multipole BE method (image) from the standard BE method (40 m) [%] (boundary condition is the same as in figure 4).

Figure 5 illustrates the individual pressure deviation for this mode according to equation 14 for each element computed with the fast multipole method (image). Only in small regions do the individual element pressures show a difference of more than 5% from the standard BE (40 m) solution (in relation to the maximum pressure from the standard BE (40 m) method). Computing the overall deviation yields a very good agreement of less than 5% between the standard and the multipole BE methods for all eight considered principal modes and free surface treatments (see Figure 7). Since the considered shell is approximately of a cylindrical shape the pressure distribution derived from the Lewis method exhibits also an acceptable but substantially greater deviation (up to 15%).

Computing the pressure with standard BE or with the fast multipole method therefore nearly yield the same results for the considered model and assumed boundary conditions.

4.3 Performance

As shown in Table 1 the fast multipole method can retrieve an accurate pressure distribution for a single given boundary condition approximately 4.5 times faster than the standard BE method. Looking more into detail concerning the time consumption of the multipole method shows that much less time for this model is needed to set up the Laplace operator but considerably more time is used to retrieve the pressure result for a single boundary condition than the standard BE method does.

The memory consumption is generally twice as high for the explicitly meshed free water surface models using the standard BE method compared to the multipole method. To validate the more efficient memory usage of the multipole BE method, the meshsize of the

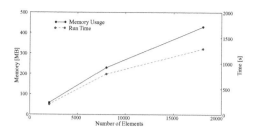

Figure 6. Memory usage and run time of multipole solution procedure for refined outer shell meshes (64-bit computer).

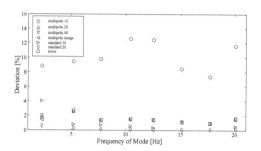

Figure 7. Comparison of computation methods and free surface considerations with standard BE (40 m).

Table 1. Memory usage, run time and deviation of different computation methods (32-bit computer).

Method	Free Surface Treatment	Setup Time [s]	Max Solution Time for Mode [s]	Memory [MB]	Max Deviation for Mode [%]
multipole BE	10m	292	4.40	75	2.70
multipole BE	20m	378	4.70	149	2.60
multipole BE	40m	445	5.28	238	2.60
multipole BE	image	482	3.00	68	2.80
standard BE	10m	1320	<1	339	4.10
standard BE	20m	1680	<1	379	0.70
standard BE	40m	2220	<1	526	0.00
Lewis	-	0.2	<1	3	13.00

wetted shell was decreased by refining the elements to one half and one third of the original element edge length. The resulting meshes of 8010 and 18052 elements as well as the original mesh were applied with artificial boundary conditions and the resulting pressure distribution (incorporating the free water surface by the image method) was computed on a 64bit linux computer having a dual core AMD Opteron 285 processor with 16 GB RAM and a clock rate of 2.6 GHz. It can be seen in Figure 6 that the memory usage tends to grow almost linearly for finer meshes instead of the quadratic memory consumption of the standard BE method. The finer meshes could even be processed nearly in core on the 32 bit machine whereas the standard BE method could not. Also, the run time of the method increases only linearly with the number of elements as compared to a quadratic or cubic run time increase for the standard BEM (see Figure 6).

5 CONCLUSIONS

Computing ship vibrations in the higher frequency range requires an efficient procedure to include the inertia effect of the surrounding water on the outer shell. Since the computation in the higher frequency range requires finer meshes, the computation of the hydrodynamic mass effect reaches its limit for standard BE methods. This is mainly due to the quadratic increase of memory usage (and cubic increase of runtime for direct methods) with the number of elements. The fast multipole method can overcome this limit and is capable of computing pressures for very large models.

The introduced image method considers the pressure release boundary condition of the free surface in a simple but exact manner. It further reduces the memory consumption as compared to the explicit free surface meshing.

The remaining step for solving the fluid-structure problem for large models is now to develop an effective coupling method that—instead of making use of the full hydrodynamic mass matrix—requires only a limited execution of the Laplacian operator to keep the run time in an acceptable range.

REFERENCES

Armand, J.-L. and P. Orsero (1979). A method for evaluating the hydrodynamic added mass in ship hull vibrations. *SNAME Transactions* 87, 99–120.

Brunner, D., M. Junge, M. Wilken, C. Cabos, and L. Gaul (2009). Vibro-Acoustic Simulations of Ships by Coupled Fast BE-FE Approaches. In *Proceedings of IMAC 2009*.

Brunner, D., G. Of, M. Junge, O. Steinbach, and L. Gaul (2008). A fast BE-FE coupling scheme for partly immersed bodies. Technical Report 2008/5, Institute of Computational Mathematics, Graz University of Technology.

Cabos, C. and F. Ihlenburg (2003). Vibrational Analysis of Ships with Coupled Finite and Boundary Elements. *Journal of Computational Acoustics* 11 (1), 91–114.

Costabel, M. (1987). Symmetric methods for the coupling of finite elements and boundary elements. In *Boundary elements IX, Vol. 1*, pp. 411–420. Southampton: Comput. Mech.

Golub, G. H. and C. F. Van Loan (1996). *Matrix computations* (Third ed.). Johns Hopkins Studies in the Mathematical Sciences. Baltimore, MD: Johns Hopkins University Press.

Greengard, L. (1987). *The Rapid Evaluation of Potential Fields in Particle Systems*. The MIT Press.

Greengard, L. and V. Rokhlin (1987). A fast algorithm for particle simulations. *J. Comput. Phys.* 73, 325–348.

Grim, O. (1953). Berechnung der durch Schwingungen eines Schiffskörpers erzeugten hydrodynamischen Kräfte. In *STG Jahrbuch.*

Grim, O. (1960). Elastische Querschwingungen des Schiffskörpers. *Schiffstechnik* 7 (35), 1–3.

Hakala, M.K. (1986). Application of the finite element method to uid-structure interaction in ship vibration. Technical report, Espoo.

Hestenes, M.R. and E. Stiefel (1952). Methods of conjugate gradients for solving linear systems. *J. Research Nat. Bur. Standards* 49, 409–436 (1953).

Kaleff, P. (1980). *Berechnung hydroelastischer Probleme mit der Singularitäten/Finite- Elemente-Methode*. Ph. D. thesis, Universität Hannover, Hamburg.

Landweber, L. (1957). Added Mass of Lewis Forms Oscillating in a Free Surface. In *Proceedings: Symposium on the Behaviour of Ships in a Seaway,* Wageningen.

Langer, U. and D. Pusch (2005). Data-sparse algebraic multigrid methods for large scale boundary element equations. *Appl. Numer. Math.* 54 (3–4), 406–424.

Lewis, F. M. (1929). The Inertia of the Water Surrounding a Vibrating Ship. *Transactions of the SNAME* 37.

Li, W. L., T. W. Wu, and A. F. Seybert (1994). A half-space boundary element method for acoustic problems with a re ecting plane of arbitrary impedance. *J. Sound Vib.* 171 (2), 173–184.

Newman, J. N. (1977). *Marine Hydrodynamics*. Cambridge, Mass.: The MIT Press.

Nishimura, N. (2002). Fast multipole accelerated boundary integral equation methods. *Appl. Mech. Rev.* 55 (4), 299–324.

Of, G. (2008). An efficient algebraic multigrid preconditioner for a fast multipole boundary element method. *Computing* 82 (2–3), 139–155.

Of, G., O. Steinbach, and W.L. Wendland (2006). The fast multipole method for the symmetric boundary integral formulation. *IMA J. Numer. Anal.* 26, 272–296.

Rjasanow, S. and O. Steinbach (2007). *The Fast Solution of Boundary Integral Equations*. Mathematical and Analytical Techniques with Applications to Engineering. Springer, New York.

Röhr, U. and P. Möller (1997). Elastische Schiffskörperschwingungen in begrenztem Fahrwasser. In *Hauptversammlung der Schiffbautechnischen Gesellschaft*, Hamburg.

Saad, Y. and M.H. Schultz (1986). GMRES: a generalized minimal residual algorithm for solving nonsymmetric linear systems. *SIAM J. Sci. Statist. Comput.* 7 (3), 856–869.

Sirtori, S. (1979). General stress analysis method by means of integral equations and boundary elements. *Meccanica* 14, 210–218.

Sladek, V., M. Tanaka, and J. Sladek (2000). Revised helmholtz integral equation for bodies sitting on an infinite plane. *Transactions of JSCES.*

Steinbach, O. (2003). Artificial multilevel boundary element preconditioners. *Proc. Appl. Math. Mech.* 3, 539–542.

Steinbach, O. (2008). *Numerical Approximation Methods for Elliptic Boundary Value Problems. Finite and Boundary Elements*. Springer, New York.

Steinbach, O. and W. L. Wendland (1998). The construction of some efficient preconditioners in the boundary element method. *Adv. Comput. Math.* 9 (1–2), 191–216.

Wendel, K. (1950). Hydrodynamische Massen und hydrodynamische Massenträgheitsmomente. In 14. *Jahrbuch der STG*, Volume 44, pp. 207–255. STG.

Wu, T.W. and A.F. Seybert (1989). Modi- fied helmholtz integral equation for bodies sitting on an infinite plane. *J. Acoust. Soc. Am.* 85 (1), 19–23.

Analysis and Design of Marine Structures – Guedes Soares & Das (eds)
© 2009 Taylor & Francis Group, London, ISBN 978-0-415-54934-9

Finite element simulations of ship collisions: A coupled approach to external dynamics and inner mechanics

I. Pill & K. Tabri

Department of Applied Mechanics, Helsinki University of Technology,
Espoo, Finland

ABSTRACT: The paper presents a method for dynamic numerical simulations of ship collision with LS-DYNA. The method can be employed both to simulate the ship motions and the structural deformations during the collision. The proposed model treats the external dynamics and the internal mechanics simultaneously and thus includes the possible interaction. The ship motions are limited to the horizontal plane of water. This limitation allows neglecting the restoring force, buoyancy and gravity. The simulation method considers precisely the most important force components—the inertia force and the contact force. The striking ship is modelled as a rigid bulb with a number of discrete mass points attached to it. The side structure of the struck ship is modelled in detail in the collision area; the rest of the struck vessel is also modelled with discrete mass elements. Validation data for the proposed method was obtained from a series of model-scale experiments, where a striking ship collided with an initially motionless struck ship. The paper describes the numerical simulation of an eccentric collision at inclined angle and compares the outcomes to the experimental measurements. The numerical simulation results agree well with the experiments implying the validity of the method.

1 INTRODUCTION

Numerical simulations are often used to assess the crashworthiness of ship structures. Such analysis is a demanding task and significant simplifications are required to achieve the outcomes in reasonable time. A usual crashworthiness analysis is displacement controlled: the struck ship is kept fixed and the striking ship moves with a constant velocity along the prescribed path. The velocity of the striking ship is often beyond the realistic value in order to shorten the simulation time. Such a method is suitable for symmetric collision scenario where the striking ship approaches under right angle to the amidships of the struck ship. In non-symmetric collisions the path of the penetration can not be defined beforehand. The contact force affects the ship motions and vice versa. The ship motions and the forces should be evaluated during the same simulation run.

The paper proposes a coupled method for the simultaneous analysis of inner mechanics and external dynamics with finite element (FE) method solver LS-DYNA. The inner mechanics deals with the investigation of the structural behaviour of the colliding ships. Contact force and energy due to the bending, crushing and tearing of the ship structures is evaluated as a function of penetration depth. The external dynamics looks into the motions of the ships and evaluates the energy involved in ship motions. In the evaluation of energy, the forces arising from the surrounding water such as restoring forces, friction forces, radiation forces due to the hydrodynamic added masses etc. are considered.

Coupling of the inner mechanics and the external dynamics means that both are included in one model and evaluated during the same calculation run. This approach enables more precise simulations of ship collisions. One obstacle for performing coupled FE collision simulations is that in most numerical codes used for structural analysis, the surrounding water cannot be modelled efficiently. Modelling the necessarily large water domain in LS-DYNA makes the numerical collision simulations very time extensive and presents a number of additional challenges. Therefore a simple and computationally fast method including all the significant phenomena is needed for the numerical modelling of the external dynamics of a ship collision.

One of the first papers concerning the dynamics of ship collisions was written by Minorsky (1959). The method evaluates the loss of kinetic energy to be absorbed by structural deformations by a simple momentum conservation equation. Minorsky's method is often used together with displacement controlled FE simulations to obtain the deformation energy. In his method, the radiation forces due to the surrounding water were modelled by constant added mass. Motora et al. (1969) showed based on model tests that the value of the added mass is dependent on the duration of the transient motion due to the contact loading; added mass coefficients given by Minorsky are valid only in case the duration of the collision is below 0.5–1 seconds. If the duration of the collision is longer, the added masses increase and their values in some cases can even be larger than the ship's own mass. This problem was solved by making a clear distinction between two components of the radiation force, a component proportional to the acceleration and the other one related to the velocity. Cummins (1962) and Ogilvie (1964) investigated the hydrodynamic effects and described the force arising from arbitrary ship motions using unit response functions. Like in the work of Motora et al. (1969), these approaches require that the frequency dependent added mass and damping coefficients of a ship are evaluated.

As already mentioned, the majority of the numerical collision simulations described in literature are displacement controlled, where the boundaries of the struck ship are fixed and the striking ship follows prescribed displacement. The analysis consists of two steps. First, inner mechanics are analysed and force-penetration and absorbed energy-penetration curves are obtained. In the second stage the external dynamics are considered and the energy absorbed in the collision is evaluated using some analytical method, e.g. Minorsky's formulas. Comparing this energy to the absorbed energy-penetration curve allows evaluating the penetration depth. This approach can precisely be used only in a symmetric collision and is therefore unsuitable in many collision cases. An example of this approach is presented in Kitamura (2002).

In Törnqvist & Simonsen (2004), the external dynamics of a symmetric collision have been taken into account by calculating the equivalent mass of the striking ship using the deformation energy measured in full-scale experiments. The velocity of the striking ship was set to be the same as in the experiments. The simulated contact force and energy histories corresponded well to the experimental measurements.

Another option for modelling the external dynamics in LS-DYNA is to use the MCOL sub-routine (Ferry, 2002). At each simulation time step, nodal forces calculated in LS-DYNA are transferred to MCOL, which calculates new coordinates of the nodes and returns the coordinates to LS-DYNA. The MCOL code has been used in collision simulations by Törnqvist (2003) for symmetric collisions and Biehl et al. (2007) for collisions between a push barge and rigid wall.

Precise modelling of the external dynamics using the FE method is difficult for many reasons. The main fluid forces that need to be taken into account are the hydrostatic restoring forces and the radiation forces due to the interaction between the ship and the surrounding water. The dynamic character of the mentioned forces presents several challenges in modelling. For example, the radiation forces depend on the direction of the movement as well as on the duration of transient motion of the colliding ships.

The buoyancy-, restoring- and radiation forces are excluded from the model limiting the motions of the ships to the horizontal plane of water, i.e. the heave, roll and pitch motions of the colliding vessels are not included. The radiation forces are simplified and modelled by constant added mass terms depending only on the direction of the motion. The velocity and time dependent part of the radiation force is neglected. This is done based on the conclusion made by Motora (1969) that if the duration of the collision is below 0.5–1 seconds, the added masses can be approximated by its value at infinite frequency. In Tabri et al. (2008a) it is shown that if the duration of the contact is around 0.5 seconds, excluding the velocity-dependent part of the radiation force results in an inaccuracy of less than 10% in the energy balance.

The method proposed in this paper differs from existing ones by not turning to additional routines such as MCOL and only uses conventional means provided by FE-simulations. Attempts to simulate the collision dynamics using MCOL were not successful. The proposed method deals with eccentric collisions, the MCOL code is validated for symmetric collisions. The external dynamics are modelled by using a robust approach in order to enable quick modelling and add as little as possible to the simulation time and obtain results with sufficient precision at the same time.

2 THEORETICAL BACKGROUND

The colliding ships experience fluid forces due to the surrounding water, gravity loading F_G and contact force F_C. Here, the background of the external forces acting on the colliding ships is discussed briefly and only the description of most important forces is given. A ship moving in water encounters frictional and residual resistance. Residual resistance is rarely included in the collision studies because it is considered small compared to other phenomena. The frictional force is proportional to the wetted surface and to the square of ships velocity. The buoyancy loading consist of buoyancy at equilibrium position and of restoring force as the ship is displaced from its equilibrium position. The gravity loading is equal and opposes the buoyancy force at equilibrium position.

The radiation forces are modelled by frequency dependent added mass and damping coefficients $a(\omega)$ and $b(\omega)$, respectively. For the representation in time domain, the radiation forces are split into two parts. One part is dependent on acceleration and is marked as F_μ the other part depends on velocity and is known as the damping part F_K. The total radiation force in time domain is calculated as a sum of these two parts:

$$F_H = F_\mu + F_K = -a_\infty \dot{u}(t) - \int_0^t K_b(\tau) u(t - \tau) d\tau \quad (1)$$

where a is the added mass at infinite frequency, u denotes the motion components of the vessel, K_b is a retardation function that takes the memory effect into account.

3 FINITE ELEMENT COLLISION MODEL

The collision model consists of the striking and the struck ship, denoted by superscripts A and B respectively. If the superscripts are omitted, the definition is universal for both vessels. Motions of the vessels are described with respect to the inertial global XY coordinate system. Body-fixed coordinate systems $x^A y^A$ and $x^B y^B$ with origins in the centres of the gravities C^A and C^B of the ships are used, mainly to present the external forces, contact force components and the penetration depth.

3.1 Structural mass and inertia; hydrodynamic added masses

The main challenge of the dynamic collision simulations it to provide the proper description of the ship masses, inertias, added masses and structural resistance. The colliding ships should be modelled with as few elements as possible in order to shorten the simulation time. Mass m and inertia about the vertical axis I_{ZZ} are the main properties of the ship when evaluating the planar motions under the external forces.

In the finite element model, the masses and mass distributions of the colliding ships are modelled by using a small number of mass points. The modelling principle of the colliding ships is presented in Figure 1.

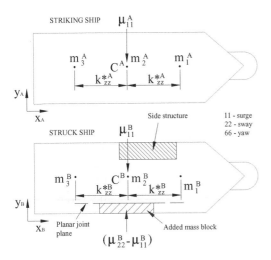

Figure 1. Modelling principle of colliding ships.

The total mass m of a ship is calculated as a summation of j mass elements as

$$m = \sum_j m_j + m_{STR} \tag{2}$$

where m_j is the mass of the j-th mass, m_{STR} is the mass of the modelled structural components. The inertia of the ship in the horizontal plane is calculated as

$$I_{ZZ} + \mu_{66} = \sum_j m_j \cdot (k_{ZZj}^*)^2 + m_{STR} \cdot k_{STR}^2, \tag{3}$$

where μ_{66} is the yaw added mass, k_{ZZj}^* is the radius of gyration of the j-th mass element, which also takes into account the yaw added mass; and k_{STR} is the radius of gyration of the side structure about the centre of gravity (COG) of the ship. Yaw added mass μ_{66} is included by calculating the suitable value for k_{ZZj}^*. Using three mass points enables to model the mass distribution and inertia of the ship with respect to the vertical axis and to control the initial location of the COG. The mass points are added to the rigid parts of the ships by creating a node set with the *SET_NODE_LIST command and then using the *CONSTRAINED_EXTRA_NODES keyword to add them to the rigid parts, the nodes can not move with respect to the rigid parts.

The accelerations of the colliding ships result in radiation force F_μ, which is direction-dependent. In the finite element model, the added mass components should also be included in a certain direction only. The surge added masses of the striking and struck ships are marked as μ_{11}^A and μ_{11}^B in Figure 1. These added masses are positioned in the centres of gravity of the ships. The approach causes the surge added mass of the ship to be included in the sway direction as well. In the case of the striking ship the sway added mass in not properly modelled as the motions are predominantly in surge direction. Neglecting the sway added mass of the striking ship results in only minor error in the simulation results. The sway added mass of the struck ship is modelled as a single block of additional mass that is located on the side opposite to the side which is hit. The mass of the sway added mass block is calculated by subtracting the surge added mass, already added to the node at the centre of gravity, from the sway added mass.

The sway added mass of the struck ship is connected to a rigid support plate on the ship using a planar joint, which restricts relative movement in one direction and allows the joined entities to move in the other directions with respect to each other. Therefore it is possible to take added masses into account when the ship is moving in one direction and disregard the effect in the other directions. In the case of surge motions, the sway added mass block remains in its initial position and is not included in the mass of the ship. The planar joint in LS-DYNA is defined using the keyword *CONSTRAINED_JOINT_PLANAR.

In order to define the planar joint, four nodes are required, two for each of the parts being joined. The nodes are necessary for indicating the position of the two parts in relation to each other and also to define the normal of the joint plane. These four nodes are presented in Figure 2. The nodes with odd numbers have to be added to one part and the nodes with even numbers to the other part.

3.2 *Structural response*

The side structure of the struck ship is one of the most important elements of the collision model. A number of aspects need to be taken into account in order to obtain reliable structural behaviour. Any type of side structure can be included in the proposed model. The side structure has to be connected to a rigid support plate, which is similar to the plate used for connecting the added mass to the ship. Instead of using a planar joint, the side structure can be connected to the support with a set of extra nodes, which can not move with respect to the support plate and are thus fixed with respect to the ship. The LS-DYNA keywords *SET_NODE_LIST and *CON-STRAINED_EXTRA_NODES can be used in the same manner as described earlier to add the side structure to the rest of the ship. Another method for adding the side structure to the rest of the model would be to merge the boundary nodes of the side structure to the nodes of the support plate. A schematic representation of connecting the side structure to the rest of the ship is seen in Figure 3, where the bold lines represent the connection between the structure and support plate.

Contact force occurs as a result of structural deformation when the striking bulb comes into contact with the side of the struck vessel. Contact between the striking bulb and side structure is defined with the *CONTACT_SURFACE_TO_SURFACE keyword.

Time history of the contact force in the LS-DYNA can be recorded as an ASCII file by using the CONTACT_FORCE_TRANSDUCER_PENALTY and *DATABASE_RCFORC keywords.

3.3 *Collision angle, location and velocity*

Desired collision angle β and eccentricity L_C are achieved by rotating the striking ship to the required angle and moving it to the proper location. The initial velocity of the striking ship u^A is defined using the LS-DYNA keyword *INITIAL_VELOCITY. The keyword enables to define separate velocity components in each direction of the inertial XY coordinate system. The complete collision setup of the numerical simulation is presented in Figure 4.

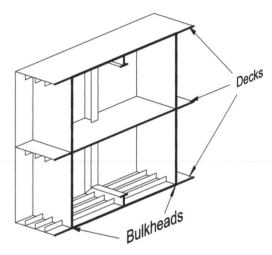

Figure 3. Principle for connecting the side structure to the struck ship.

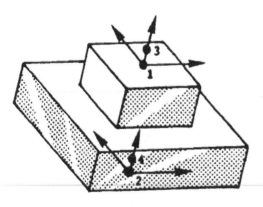

Figure 2. Planar joint in LS-DYNA, courtesy of LSTC (2003).

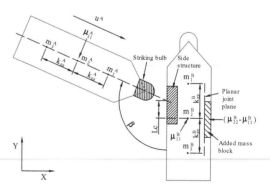

Figure 4. Collision setup for numerical simulation.

4 FE COLLISION SIMULATIONS

4.1 *Model-scale experiments*

The results of the finite element simulation are validated with data obtained from model tests. Elaborate description of the tests and the analysis is given by Määttänen (2005) and Tabri et al (2008b). During the tests, the motions of both colliding ships in six degrees of freedom as well as the contact force were measured. A force transducer between the impact bulb and the bow of the striking ship recorded the contact load in the longitudinal and transverse directions with respect to the striking vessel. The collision area of the struck model was made of polyurethane foam.

An eccentric collision under a 120° angle was selected for the validation of the proposed model because it includes strongly coupled external dynamics and internal mechanics.

4.2 *Structural response*

The impacting bulb of the striking ship was modelled with solid elements and rigid material. The exact profile of the bulb used in the model experiments was used. The profile of the bulb was calculated as a function of its radius using the relation $h(r) = 60 \ r^2$ (Määttänen, 2005). After $h = 40$ mm, the radius is constant at $r = 26$ mm. The profile of the impacting bulb is depicted in Figure 5.

The structure of the struck ship in the collision area was modelled using solid elements and foam material definition *MAT_CRUSHABLE_FOAM (MAT63). The behaviour of the material is determined by a curve, which determines the relation between stress and volumetric strain of the foam elements. Volumetric strain is the relation of the volume of a deformed element with reference to its initial volume. The yield stress of the foam material was set constant at 0.12 MPa

up to volumetric strain of 70%, after which the yield stress was increased rapidly. The stress-strain curve for MAT63 is presented in Figure 6. An elaborate study of the properties of the polyurethane foam is given by Tabri & Ranta (2007).

The crushable foam material in LS-DYNA does not have sufficient elastic recovery properties, in order to take the elastic recovery of the side structure into account in the simulations, a thin layer of *MAT_LOW_DENSITY_FOAM (MAT57) was added on the contact side of the crushable foam material. The low density foam has very good elastic recovery properties and can be used to approximate the elastic recovery of the side structure.

The material properties as required for the foam in LS-DYNA keyword definition are presented in Table 1.

4.3 *Simulation results*

An eccentric collision at an inclined angle was simulated numerically. Table 2 shows the values for the characteristics presented in Figure 4. The total mass of the colliding ships is modelled using the mass

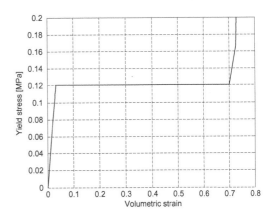

Figure 6. Stress-strain curve of crushable foam (MAT63).

Table 1. Material properties of foam.

Material parameters for MAT63	
Density [kg/m3]	27
Elasticity modulus [MPa]	3.55
Poisson's ratio	0.3
Tensile stress cutoff [MPa]	0.12
Damping coefficient	0.055
Additional parameters for MAT57	
Shape factor	5.000
Hysteretic unloading	0.100
Damping coefficient	0.000

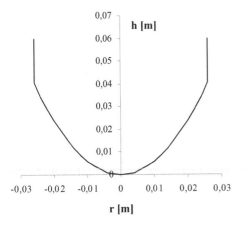

Figure 5. Profile of the impacting bulb.

Table 2. Collision data for simulated case.

Characteristic	Striking ship[A]	Struck ship[B]
u^A [m/s]	0.87	–
m_1 [kg]	14.25	10.25
m_2 [kg]	0	0
m_3 [kg]	14.25	10.25
m_{STR} [kg]	–	0.6
μ_{11} [kg]	1.425	1.025
μ_{22} [kg]	–	4.305
μ_{66} [m^4]	2.593	1.265
I_{ZZ} [m^4]	9.260	6.660
k^*_{ZZ} [m]	0.645	0.622
L_C [m]	0.37	–
β [°]	120	–

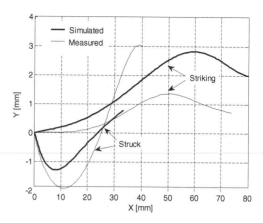

Figure 8. Displacements of colliding ships.

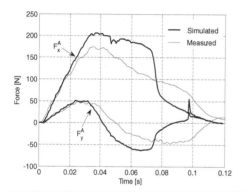

Figure 7. Contact force components.

points m_1, m_2 and m_3, the structural mass m_{STR} and added masses μ_{11}, μ_{22} and μ_{66}. The inertia of the ships is modelled to meet the value of I_{zz} by using the correct radius of gyration k^*_{ZZ}.

Simulated and measured contact force components are presented in Figure 7. The contact forces are presented in the body-fixed coordinate system of the striking ship with F^A_x being the longitudinal force component and F^A_y the transverse component.

The longitudinal contact force F^A_x is estimated at reasonable accuracy, contact duration in the simulated case is roughly the same as in the experiment. In the loading stage, the forces are almost coincident, maximum force is reached at about the same time; the simulated maximum of the longitudinal force component is about 10% higher than in the experiments.

The transverse contact force component curves with respect to the striking ship F_y are in good agreement throughout the simulation time, especially in the loading stage. Some discrepancies start to occur at about $t = 75$ ms, when the simulated force starts to decrease more rapidly than in the experiment. This is probably due to contact instabilities in LS-DYNA. The duration of the contact in the transverse direction with respect to the striking ship is also in agreement with the measured result. Maximum values of the transverse force component are predicted well with the numerical model.

Displacements of the centres of gravity in the inertial coordinate system are depicted in Figure 8. Displacement of the striking ship is predicted at good accuary, difference in the Y-directional displacement is noticeable, but due to the small magnitude of the motions, the result can be considered acceptable. The simulated displacement of the struck ship is also in agreement with the experimental data. The Y-directional displacement of both colliding vessels simulated with the proposed method is smaller than measured experimentally.

Yaw angle of the struck ship was overestimated in the simulation, see Figure 9. The figure reveals that the difference between simulated and measured yaw angles starts to differ increasingly after $t = 40$ ms, this is the time at which the maximum longitudinal contact force with respect to the striking ship is reached. As the simulated longitudinal contact force is higher, the yaw angle of the struck ship in the simulation is also larger. Furthermore, the longitudinal contact force in the simulation is higher also in the unloading stage, which in turn results further increases the yaw angle of the struck ship.

Figure 10 features the penetration curves in side structure of the struck ship, the values are given in the body-fixed coordinate system of the struck ship. The figure reveals that the proposed method estimates the penetration depth with good accuracy; the x^B-directional penetration is predicted smaller than actually measured, the y^B-directional penetration is nearly the same in the simulation and experimental result.

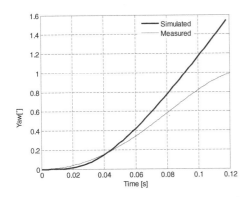

Figure 9.   Yaw angle of struck ship.

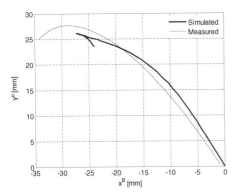

Figure 10.   Penetration of side structure.

5   SUMMARY

A robust method for coupled simulations of ship collisions has been presented. The developed model can be used for the analysis of ship motions and structural deformations as all the major interactions are included.

The mass distributions and inertias of the colliding ships in the horizontal plane are modelled by using only a few mass points, which are positioned in such a way that the centre of gravity and the moment of inertia obtain the desired value. The method of modelling added masses through planar joints is proposed, which makes it possible to take the mass into account only in one direction and disregard it in another. Any type of side structure of the struck ship can be included in the model.

An eccentric collision at an inclined angle was simulated; the results were compared to experimental measurements from the model tests (Määttänen 2005). The simulated results showed good agreement with the experimentally measured contact force components,

ship motions and penetration depth in the side structure. The yaw angle of the struck ship was overestimated due to higher longitudinal contact forces with respect to the striking ship.

Although the external dynamics of the colliding ships are taken into account using a very crude approach, good agreement of the simulation results with experimental measurements is achieved. This demonstrates that the most significant forces affecting the colliding ships are taken into account with sufficient accuracy and that the proposed method is valid for fast coupled simulations of ship collisions.

REFERENCES

Biehl, F., Kunz, K.U. & Lehmann, E., 2007. Collisions of Inland Waterway Vessels with Fixed Structures: Load—Deformation Relations and Full Scale Simulations, 4th International Conference of Collision and Grounding of Ships, pp. 71–78.

Clayton, B.R. & Bishop, R.E.D., 1982. Mechanics of Marine Vehicles, Ch. 10—Directional Stability and Control, pp. 495–515.

Cummins, W.E., 1962. The Impulse Response Function and Ship Motions, Schifftechnik 9, No 47, pp. 101–09.

Ferry, M., 2002, MCOL User's Manual (25.05.02).

Kitamura, O., 2002. FEM approach to the simulation of collision and grounding damage, Marine Structures 15, pp. 403–428.

Minorsky, V.U., 1959. An Analysis of Ship Collision with Reference to Protection of Nuclear Power Plants, J Ship Res, Vol 3, No 1, pp. 1–4.

Motora, S., Fujino, M., Suguira, M. & Sugita, M., 1971. Equivalent Added mass of ships in collision, Selected papers from J Soc. Nav. Archit. Japan 7, pp. 138–148.

Määttänen, J., 2005. Experiments on Ship Collisions in Model Scale, Master's Thesis. Helsinki University of Technology, Ship Laboratory.

Ogilvie, T.F., 1964. Recent Progress Towards the Understanding and Prediction of Ship Motions, Fifth Symposium on Naval Hydrodynamics, Bergen, Norway, pp. 3–128.

Tabri, K., Broekhuijsen, J., Matusiak, J. & Varsta, P., 2008a. Analytical modelling of ship collision based on full-scale experiments, Marine Structures doi:10.1016/j.marstruc.2008.06.002.

Tabri, K., Määttänen, J. & Ranta, J., 2008b. Model-scale experiments of symmetric ship collisions, J Mar Sci Technol 13:71–84.

Tabri, K. & Ranta, J., 2007. Study on the Properties of Polyurethane Foam for Model-Scale Ship Collisions. Helsinki University of Technology Ship Laboratory report M-297.

Törnqvist, R., 2003. Design of Crashworthy Structures, Doctoral Thesis, Technical University of Denmark, Department of Mechanical Engineering.

Törnqvist, R. & Simonsen, B.C., 2004. Safety and Structural Crashworthiness of Ship Structures; Modelling Tools and Application in Design. Proceedings of the 3rd International Conference on Collision and Grounding of Ships. pp. 285–294.

Ultimate strength

Analysis and Design of Marine Structures – Guedes Soares & Das (eds)
© 2009 Taylor & Francis Group, London, ISBN 978-0-415-54934-9

Discussion of plastic capacity of plating subject to patch loads

C. Daley
Memorial University, St. John's, Canada

A. Bansal
Indian Institute of Technology, Delhi, India

ABSTRACT: There is increasing interest of transportation and resource exploitation in the Arctic. As a result, the subject of ship shell plating under ice loads is of concern. As well, other loads such as wheel loads and collision and grounding loads can be termed as 'patch loading'. Consequently the resistance of ship shell plating under 'patch loading' is of significant interest. The recent IACS Polar Rule requirements (URI2) contain a formulation for plate design based on plastic hinge behavior (yield line theory using a rigid-plastic material model). The URI2 approach has advantages over conventional elastic design, but has also been the subject of some concern and debate. Two recent publications describing alternative formulations are reviewed. The new formulations, along with the URI2 formulation are compared to nonlinear finite element results for a range of cases. The comparison shows that one of the new proposals is an improvement on the URI2 method. However, there is still some significant differences. The present study then presents ideas for development of an elasto-plastic response formulation of ship shell plating under general patch loading. The proposed method is intended to reflect the plate behavior more realistically and with a more consistent level of permanent deformation. The suggested model consists of a folding pattern of three plastic hinges within a region of elastic deformation.

1 INTRODUCTION

The design of shell plating is a critical aspect of the design of ships and offshore structures. Design for harsh environments, such as for ice, places additional importance and challenges on the design. This paper addressed the issue of analytical design formulations for plastic design to resist patch loads. This case of needed for ice class ship design. Design equations can be developed in a variety of ways. The equations presented here are developed analytically by postulating a plastic deformation mechanism and determining the limit plate capacity prior to the mechanism. Plating will typically exhibit monotonically increasing capacity, though with changing stiffness. The IACS Unified requirements for Polar Ship Structural Design (URI2), sets the design point as the point at which plastic mechanisms begin to permit large deformations. The design deformation is not stated, but is assumed to be quite small. The challenge in formulating a design equation that consistently matches with this behavior is that plates support pressure by a combination of effects. Most analytical formulations focus on one strength mechanism and so have a limited range of validity.

2 BACKGROUND

The stiffened panel is the primary building block for ships and many other structural systems. The smallest component of the stiffened panel is the rectangular shell plate. The plate is capable of resisting a wide variety of loads. At the extreme fiber of the hull girder of a ship, plating must resist in-plane loads (see Figure 1). For loads of this type, axial yielding and elastic buckling are the primary behaviours of interest. While not the focus of this paper, it is worthwhile to review this behaviour broadly. By contrasting normal loads and axial loads, the issues discussed later in the paper should be clearer.

2.1 *Behavior of long plates under axial loads*

Assuming for the moment that the supporting stiffeners are sufficiently strong to remain elastic even while the plate fails, the general capacity of axially loaded

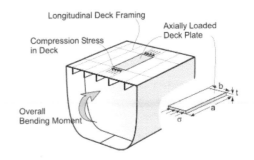

Figure 1. Axially loaded deck plates.

long plates can be described by Figure 2. Four models of plate failure are presented (see Hughes (1983), von Kármán et. al. (1932) and Caldwell (1965)). The first two models, Euler Buckling and plate yielding are based on ideal elastic stress analysis, with limits defined by instability and material yielding, respectively. The third model is an early analytical post-buckling model proposed by von Karman. In that model, stress is redistributed from the middle of the buckled plate to the edges, until the edges of the plate yield. The fourth model is a suggestion by Faulkner, based on laboratory experiments. The equation was empirically derived from data. More recent data (see Paik and Thayamballi (2003)) extends the range of issues considered, to include construction deformations and structural aging effects. The type of response can depend on the geometric and material specifics, as well as on the loading pattern. Figure 2 separates the plates into there categories; thick, intermediate and thin. Thin plate behavior is dominated by buckling and exhibits significant post buckling strength. Thick plate behavior is dominated by yielding. Intermediate plates are affected by both yielding and bucking mechanisms with the strength being potentially less than either yielding or buckling.

For the purpose of this paper, the main point of the above discussion is to show that a full understanding of behavior requires the consideration of multiple types of response, even when only one type of load is considered.

2.2 Behavior of long plates under uniform lateral loads

Now consider the case of laterally loaded plate. The plate in the side shell of a ship may be subject to lateral loads (see Figure 3). For long plates, the symmetry of

response allows for the analysis to be simplified to a 2D cross section (see Figure 4). While the complete behavior is quite complex, three simple behaviors can represent the main types of response. A comprehensive discussion of plates subject to uniform lateral loads can be found in Ratzlaff and Kennedy (1985).

The three ideal behaviors sketched in Figure 4 are elasto-plastic bending, plastic and elastic membrane. The bending equations for yield (Y), edge hinge (EH) and collapse (C) are given below. Both the pressure (P) and the total deflection (δ_C) can be found using equations 1–6.

$$P_Y = 2 \cdot \sigma_y \left(\frac{t}{b} \right)^2 \tag{1}$$

$$P_{EH} = 3 \cdot \sigma_y \left(\frac{t}{b} \right)^2 \tag{2}$$

$$P_C = 4 \cdot \sigma_y \left(\frac{t}{b} \right)^2 \tag{3}$$

$$\delta_Y = \frac{1}{384} \frac{P_Y b^4}{D} \tag{4}$$

$$\delta_{EH} = \frac{1}{384} \frac{P_{EH} b^4}{D} \tag{5}$$

$$\delta_C = \frac{2}{384} \frac{P_C b^4}{D} \tag{6}$$

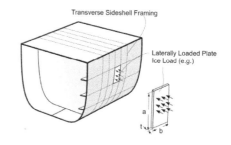

Figure 3. Ship side shell plate subject to lateral load.

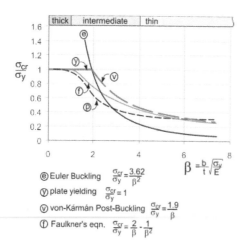

Figure 2. Strength models of axially loaded long plates.

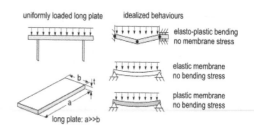

Figure 4. Three idealized behaviors of a long plate subject to a uniform lateral load.

The formulation of the plastic membrane strength, as a function of total deflection, is;

$$P_{pm} = \frac{9.24 \cdot \delta \cdot \sigma_y \cdot t}{\sqrt{b^4 + (4\delta \cdot b)^2}} \qquad (7)$$

The formulation of the elastic membrane strength, as a function of total deflection, is:

$$P_{em} = \frac{64}{3} \left(\frac{\delta}{b}\right)^3 \frac{E \cdot t}{1 - \nu^2} \frac{1}{\sqrt{b^2 + (4\delta)^2}} \qquad (8)$$

where

$$D = \frac{Et^3}{12(1 - \nu^2)} \qquad (9)$$

Figure 5 plots the three ideal strength mechanisms listed above, along with a non-linear finite element model of the behavior. It is clear from Figure 5 that the actual behavior for a thick plate does not involve practical collapse when a three-hinge condition is reached. There is considerable reserve because of membrane and strain hardening effects (note: in the ANSYS finite element models, a post yield modulus of 1 GPa was used to model strain hardening). The plate considered in Figure 5 is termed a 'thick' plate, because the membrane capacity does not exceed the bending capacity until after the central hinge has formed (which occurs at the start of the plateau of the 'bending' curve).

Figure 6 is similar to Figure 5 except that the plate is thinner and represent an 'intermediate' plate. This is because the plate forms edge hinges, but never forms a central hinge, even under large deformations. The membrane mechanism dominates the behavior after the edge hinges form. The membrane effects (elastic and then plastic) will diminish and ultimately remove any bending capacity in the plate. Clearly a model of

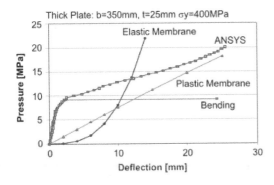

Figure 5. Plot of 3 ideal mechanisms for a long plate subject to a uniform lateral load. Also shown is the ANSYS finite element model of the complete behavior. This case represent a thick plate.

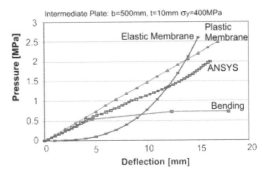

Figure 6. Plot of 3 ideal mechanisms for a long plate subject to a uniform lateral load. Also shown is the ANSYS finite element model of the complete behavior. This case represent an intermediate plate.

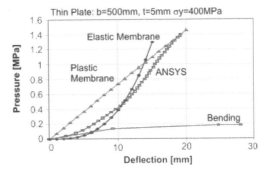

Figure 7. Plot of 3 ideal mechanisms for a long plate subject to a uniform lateral load. Also shown is the ANSYS finite element model of the complete behavior. This case represents a thin plate.

3-hinge collapse for this case would tend to underestimate the plate capacity.

Figure 7 is also similar to Figure 5 except that the plate is very thin and represents a 'thin' plate. In this case edge hinges never form, as membrane behavior dominates the strength from the beginning.

3 BEHAVIOUR OF PATCH LOADED PLATES

The above discussion of axially and uniformly loaded plate panels provides a context for considering plates subject to patch loads. Figure 8 shows the case of a plate with a rectangular patch load extending across the show dimension of a plate. This case is typical of ice loads acting on side shell plating in a transversely framed region of a ship hull.

3.1 IACS Polar Rule model for patch loads

The IACS Polar Rules (IACS 2006) contain a design equation for this case that was derived from plate folding considerations (Appolonov (2000) and Daley, Kendrick and Appolonov (2001)). The plate load capacity in the Polar rules can be expressed as;

$$P_p = P_C \cdot \left(1 + \frac{b}{2h}\right)^2 \qquad (10)$$

where PC is the nominal plastic capacity of a long plate subject to a uniform load;

$$P_C = 4\sigma_y \left(\frac{t}{b}\right)^2 \qquad (11)$$

In Appolonov (2000), it is explained that equation (10) is not the direct result of application of the theory of plate folding of the form sketched in Figure 8. The full analytical solution was simplified and adjusted to better reflect the results of non-linear finite element analysis.

There is a clear logical problem with the IACS design equation, one that the developers recognized. When the load height, h, in equation 10 tends towards zero, the plate capacity tends towards infinity. This means that, if the equation were correct, a true line load could be supported by a paper thin plate. This flaw was left in the IACS rule, because the load heights used in the Polar Rules are relatively large, where the equation works quite well. Nevertheless, this is an issue that has been investigated and should be corrected. Alternatives to equation 10 have been presented recently and will be discussed and compared here.

3.2 Double diamond model for patch loads

Hong and Amdahl (2007) discuss the issue of plastic collapse and present an alternative folding pattern as shown in Figure 9. A plate capacity equation is derived as;

$$P_p = P_C \cdot K_p \cdot K_m \cdot f_B \qquad (12)$$

where;

$$K_p = 1.9 + 1.3\frac{h}{b} + 0.18\left(\frac{h}{b}\right)^2 \qquad (13)$$

$$K_m = \sqrt{1 + \left(2z\frac{-(b/h)^2 + 8(b/h) - 4 + 48(h/b)}{-4(b/h) + 48 + 48(h/b)}\right)^2} \qquad (14)$$

$$z = \frac{\delta}{t} \qquad (15)$$

$$f_B = 1 - 0.075\sqrt{b/h} \qquad (16)$$

The above system of equations is an interesting suggestion, and does contain explicit consideration of the permanent deflection and the membrane strength. At extremely small values of load height (say <1 mm) the equations produce some odd results, but this is the result of some approximations and is of no practical concern. Of greater concern is the use of the above equations in design. While Hong and Amdahl (2007) shows good agreement with finite element results for cases of significant deformations, the agreement for cases of very small permanent deformations are not as good. This will be discussed more below.

3.3 Bound theorem model for patch loads

Nyseth and Holtsmark (2006) present an interesting method for calculation of the strength of patch loads.

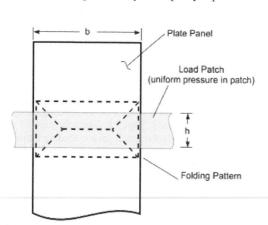

Figure 8.   Patch loaded plate, with envelope folding pattern.

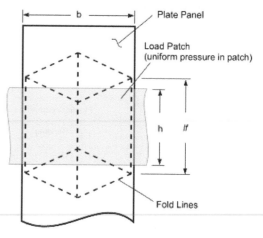

Figure 9.   Patch loaded plate, with a double diamond folding pattern.

The approach relies on plastic collapse concepts and bound theorems given in Soreide (1985).

Figure 10 illustrates the assumptions used in the collapse model. It is reasoned that the plate response to one patch is equivalent to the response to multiple patches, when the patches are sufficiently separated. This is confirmed by finite element analysis. The resulting model considers three fold lines, each of length ls. This makes the capacity of the plate relatively easy to determine. The only challenge is the calculation of ls, which is determined with the use of the bound theorems.

The capacity equation derived in Nyseth and Holtsmark (2006) is;

$$P_p = P_C \cdot \frac{ls}{h} \tag{17}$$

where the hinge length ls is;

$$ls = \frac{h}{2} \cdot \left(1 + \sqrt{1 + 4\left(\frac{b}{h}\right)^2}\right) \tag{18}$$

This can be expressed as;

$$P_p = P_C \cdot \left(\frac{1}{2} + \sqrt{\frac{1}{4} + \left(\frac{b}{h}\right)^2}\right) \tag{19}$$

Equation 19, when re-expressed in terms of required plate thickness t becomes;

$$t = 0.5\sqrt{\frac{P_p}{\sigma_y}} \cdot \sqrt{\frac{1}{\left(\frac{1}{2} + \sqrt{\frac{1}{4} + \left(\frac{b}{h}\right)^2}\right)}} \tag{20}$$

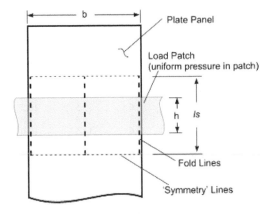

Figure 10. Patch loaded plate, 'symmetry' folding pattern.

For the case of very small load heights, a pressure is related to the 'line load' Qp as;

$$P_p = \frac{Q_p}{h} \tag{21}$$

Substituting into equation 20 and letting h go to zero gives;

$$t = 0.5\sqrt{\frac{Q_p}{b \cdot \sigma_y}} \tag{22}$$

This shows that the load height issue in equation 10 is not a problem in equation 17.

3.4 Numerical simulation of plate response to patch loads

In order to examine the above equations, a set of non-linear finite element analyses were conducted using the program ANSYS. Table 1 shows the model parameters for 27 variations of patch loads on plate panels. In all cases the plates were 3000 mm long, with a yield strength of 400 MPa, poisons ratio of 0.3 and a post

Table 1. Model parameters and results for ANSYS* runs.

run #	b mm	h mm	t mm	P_ANSYS MPa
1	400	200	20	9.3
2	400	200	30	22.4
3	400	200	40	42
4	600	200	20	5.4
5	600	200	30	12.3
6	600	200	40	22.8
7	800	200	20	3.95
8	800	200	30	8.25
9	800	200	40	15.6
10	400	400	20	5.8
11	400	400	30	14
12	400	400	40	26.1
13	600	400	20	2.85
14	600	400	30	6.9
15	600	400	40	13.15
16	800	400	20	2.4
17	800	400	30	4.3
18	800	400	40	8.35
19	400	800	20	4.15
20	400	800	30	9.3
21	400	800	40	20.4
22	600	800	20	2.0
23	600	800	30	4.45
24	600	800	40	8.85
25	800	800	20	1.45
26	800	800	30	2.7
27	800	800	40	5.1

117

yield modulus of 1 GPa. The deflection at the center of the plate under the center of the load was plotted. Figure 11 shows the plate as modeled in ANSYS. The capacity of the plate was determined as the pressure that would cause a permanent deflection equal to 0.1%

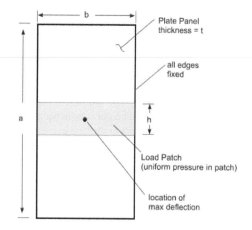

Figure 11.   Patch loaded plate for ANSYS runs.

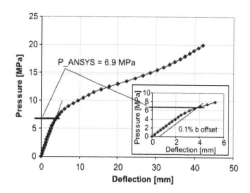

Figure 12.   Load vs. Deflection for run #14.

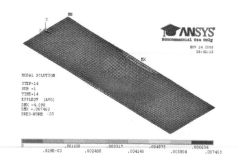

Figure 13.   ANSYS deformation for run #14 showing contours of plastic strain to highlight the plastic mechanism.

of the plate width b. Figure 12 illustrates a typical load-deflection plot from run #14, illustrating how the value of 6.9 MPa was arrived at for this case.

Figure 13 shows the ANSYS plastic deformation and contours of plastic strain for run #14.

The capacity values for all 27 ANSYS runs are tabulated in Table 1.

3.5   Comparison of models

Table 2 compares the capacity value calculated from the ANSYS run with the values form the various model presented above. To examine the various models, each will be plotted vs the ANSYS results. Figure 14 shows the agreement between the ANSYS runs and the current model used in the Polar Rules (equation 10). The agreement is reasonably good but not excellent. There as some scatter. The mean line indicates that the plates exhibit a little more capacity than the rule equation predicts, indicating a small degree of conservatism in the rule. Figure 15 shows that equation 12 tends to overestimate the capacities predicted by ANSYS, and also shows some scatter. Figure 16 show that equation 19 provides the best prediction of the ANSYS results, with the least mean error and the least scatter. Equation 19 seems to be a considerable improvement compared with the rule equation.

Table 2.   Comparison of models with ANSYS* runs.

run #	ANSYS (MPa)	PC eqn10	H&A eqn12	N&H eqn19
1	9.3	8.0	15.5	10.2
2	22.4	18.0	34.8	23.1
3	42.0	32.0	61.9	41.0
4	5.4	5.8	10.1	6.3
5	12.3	13.0	22.8	14.2
6	22.8	23.1	40.5	25.2
7	4.0	5.0	7.8	4.5
8	8.3	11.3	17.5	10.2
9	15.6	20.0	31.1	18.1
10	5.8	5.0	9.2	6.5
11	14.0	11.3	20.7	14.6
12	26.1	20.0	36.8	25.9
13	2.9	2.8	5.4	3.7
14	6.9	6.3	12.2	8.3
15	13.2	11.1	21.7	14.8
16	2.4	2.0	3.9	2.6
17	4.3	4.5	8.7	5.8
18	8.4	8.0	15.5	10.2
19	4.2	4.3	6.4	4.8
20	9.3	9.6	14.5	10.9
21	20.4	17.0	25.7	19.3
22	2.0	2.0	3.5	2.5
23	4.5	4.6	7.8	5.6
24	8.9	8.1	13.8	10.0
25	1.5	1.3	2.3	1.6
26	2.7	2.8	5.2	3.6
27	5.1	5.0	9.2	6.5

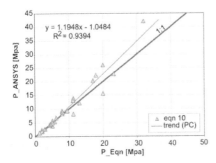

Figure 14.   ANSYS results vs. eqn 10.

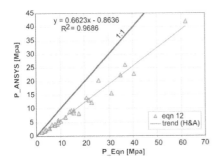

Figure 15.   ANSYS results vs. eqn 12.

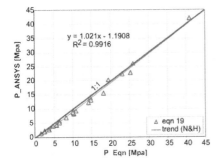

Figure 16.   ANSYS vs. eqn 19.

3.6   Discussion of models

Table 3 tabulates the % difference between the various model predictions (equations 10, 12 and 19) and the non-linear ANSYS simulations. The % was calculated by dividing the difference between the model and numerical result by the numerical result (the notionally 'true' result). A positive % means that the model over predicts the capacity (a non-conservative result). The % differences highlight issues that tend to be obscured by the plots. While the plots show that equation 19 is a great improvement, the table indicates that equation 19 can over predict the capacity by as much

Table 3.   Calculated % difference compared with ANSYS* runs.

run #	ANSYS	PC eqn10	H&A eqn12	N&H eqn19
1	0%	−14%	66%	10%
2	0%	−20%	55%	3%
3	0%	−24%	47%	−2%
4	0%	7%	87%	17%
5	0%	6%	85%	15%
6	0%	1%	78%	10%
7	0%	27%	97%	15%
8	0%	36%	112%	24%
9	0%	28%	99%	16%
10	0%	−14%	58%	12%
11	0%	−20%	48%	4%
12	0%	−23%	41%	−1%
13	0%	−3%	90%	30%
14	0%	−9%	77%	21%
15	0%	−16%	65%	13%
16	0%	−17%	61%	7%
17	0%	5%	102%	34%
18	0%	−4%	85%	23%
19	0%	2%	55%	16%
20	0%	3%	56%	17%
21	0%	−17%	26%	−5%
22	0%	1%	73%	25%
23	0%	3%	75%	26%
24	0%	−8%	56%	13%
25	0%	−14%	58%	12%
26	0%	4%	91%	35%
27	0%	−2%	80%	27%

as 34%. While equation 19 has much less scatter, its advantages are mainly in the range of higher capacity plates. For lower capacity plates, it is quite comparable to equation 10.

Equation 12 tends to generally over predict the capacities, though the approach was focused more on predicting the large deformation capacities than on the initial plastic strength (which was the focus of equations 10 and 19).

3.7   Discussion of potential ways forward

The remainder of this discussion will present to ideas for improving design models for plates subject to patch loads.

Figure 13 shows an image of the plate behavior at the design load. There are two regions of significant plastic strain along the edges. There is little plastic strain in the middle of the plate, likely because the bending and membrane stresses are canceling. (Note: there is considerable tensile plastic strain on the other side, due to the addition of membrane and bending stresses). Clearly the pattern of hinges is lot the same as supposed in equations 10, 12 or 19. Figure 17 shows a thicker plate, and shows similar edge zones of plasticity. As well in this case there is plastic strain in the

119

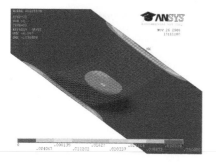

Figure 17. ANSYS deformation for a run #3 showing contours of plastic strain to highlight the plastic mechanism.

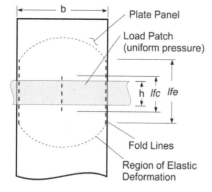

Figure 18. Suggested sketch of a plastic mechanism that may describe plate response to patch loads.

center of the plate. Even in this case, the length of the central hinge (if it can be termed as such) is not as long as the edge hinges.

Figure 18 sketches a possible model of the process. Edge hinges of length lfe, and central hinges of length lfc are assumed to form. The remainder of the plate is assumed to stay elastic. Consideration of the elastic deformations allows this pattern of deformation to be termed 'kinematically admissible', even though that actual elastic deformation is not described. It may be argued that most of the elastic work done will be recovered upon unloading. This should permit the consideration of only the plastic work when calculating plate capacity, even though the elastic deformation is needed to justify the deformation pattern.

Figure 6 indicates another issue that deserves consideration. For plates that are relatively thin, the membrane stresses will begin to support load even before central hinges form. With patch loads, the membrane action is in two directions. There is no ready analytical solution for the elastic (or plastic) membrane strength of a plate subject to a patch load that would permit the creation of a plot such as this issue numerically and this may be a fruitful avenue of research.

4 CONCLUSIONS

The above review and analysis has illustrated some of the challenges of developing a design rule for shell plating. As the focus of design shifts more towards safety and performance, and as analysis tools become increasingly complex and accurate, it is difficult to find a single simple equation that can be used for design. Nevertheless, simple equations are needed, both to permit quick and practical design and to help designers and analysts understand on a conceptual level why structures behave as they do. The analysis has shown some of the difficulties that arise when rigid-plastic (yield line) theory is applied to ship plating. It would be useful to develop new methods that allow for elastic response to part of the kinematically admissible deformation patterns, without being part of the energy balance. This may or may not be a fruitful avenue to pursue.

REFERENCES

Hughes, O.F., 1983, *Ship Structural Design*, Wiley-Interscience, Published by the Society of Naval Architects and Marine Engineers, New York, 1988.

J.B. Caldwell, "Ultimate Longitudinal Strength", *Trans RINA*, 107, p. 411, 1965; Discussion by D. Faulkner.

von Kármán, T., Sechler, E.E., Donnell, L.H., 1932, "The Strength of Thin Plates In Compression", *Trans. ASME*, vol. 54, pp. 53–57.

Paik, J.K. & Thayamballi, A.K., *Ultimate Limit State Design of Steel Plated Structures*, Wiley, 2003.

Ratzlaff, K.P. & Kennedy, D.J.L. "Analysis of Continuous Steel Plates Subjected to Uniform Transverse Loads" Can. J. Civ. Eng. vol. 12, pp. 685–699, 1985.

International Association of Classification Societies, "I2 Structural requirements for Polar Ships" March 2006.

Appolonov, E. "Background Notes to Shell Plating Thickness", Prepared for IACS Ad-Hoc Group on Polar Class Ships, March 2000.

Daley, C.G., Kendrick, A. & Appolonov, E., "Plating and framing design in the unified requirements for polar class ships" Proc. POAC01, the 16th international conference on port and ocean engineering under arctic conditions, vol. 2, Ottawa, Canada, pp. 779–791., 2001.

Nyseth, H., Holtsmark, G. "Analytical Plastic Capacity Formulation For Plates Subject To Ice Loads And Similar Types Of Patch Loadings", Proc. OMAE2006, Hamburg, Germany, 2006.

Hong, L. & Amdahl, J. "Plastic design of laterally patch loaded plates for ships", Marine Structures 20, pp 124–142, 2007.

Hayward, R.C. "Plastic Response of Ship Shell Plating Subjected to Loads of Finite Height", Master Thesis, Faculty of Engineering and Applied Science, Memorial University of Newfoundland, May 2001.

Søreide, T.H. "Ultimate Load Analysis of Marine Structures, Tapir Trondheim, 1985.

Analysis and Design of Marine Structures – Guedes Soares & Das (eds)
© 2009 Taylor & Francis Group, London, ISBN 978-0-415-54934-9

Ultimate strength characteristics of aluminium plates for high speed vessels

S. Benson, J. Downes & R.S. Dow
Newcastle University, UK

ABSTRACT: Marine grade aluminium alloy has become an established structural material for medium to high speed commercial craft and has also been used as the primary hull material for several naval vessels including a 127 m trimaran. The analysis of large high speed craft operating in deep ocean and potentially hostile environments will require rigorous methodologies to evaluate the ultimate strength of the hull girder. This paper examines the strength of plate elements with a range of geometric, imperfection and material parameters typical of a high speed vessel using a non linear finite element approach. Representative plate load shortening curves are required in simplified hull girder ultimate strength methodologies; for the case of a high speed aluminium vessel the curves need to account for the effects of parameters including alloy type, imperfection, softening in the heat affected zone, residual stresses, lateral pressure and biaxial load. A parametric study presented in this paper shows that these factors have a significant influence on the strength behaviour of aluminium plates.

1 INTRODUCTION

Marine grade aluminium alloy has become an established structural material used for medium to high speed commercial craft operating in coastal locations throughout the world. The recent development of a 127 m aluminium Littoral Combat Ship (LCS) with a range of 4500 nm has shown the potential for aluminium to be a viable material for large, high speed, ocean going vessels. The advantages of aluminium alloys over steel for construction of such vessels include a high strength to weight ratio, good resistance to corrosion and comparable ease of manufacture. Future operational requirements for large high speed craft such as LCS will likely include exposed and potentially hostile deep sea environments for significant periods with the corresponding influence on strength requirements. The strength criteria for the hull will be primarily dependent on overall girder strength. Therefore the design of such vessels requires structural prediction techniques capable of producing a light structure with high confidence in its strength and safety.

The strength and stiffness of the deck and shell of a ship is governed by the strength of the individual plate and stiffener elements which together comprise the overall hull girder structure. The vessel depends critically on the behaviour of the plates under load combinations, particularly their instability characteristics in compression. The influence of local compressive failure on the ultimate strength of a hull is encapsulated in simplified methodologies to evaluate the ultimate longitudinal bending strength of the hull girder (Smith 1977). If such methodologies are to be adapted for large high speed craft, the compressive strength characteristics of aluminium plates typical of such vessels need to be assessed.

This paper examines the strength of plate elements with a range of geometric parameters typical of the midship scantlings of a high speed vessel. A parametric study examines the influence of geometric plate imperfections, material properties and internal stress fields using a non linear finite element method. The parameters are quantified using statistical information, rules and code specifications from various sources. Results from the study are compared to various design rules including Eurocode 9.

2 METHODS TO PREDICT COLLAPSE BEHAVIOUR OF ALUMINIUM PLATES

Empirical methods to predict the collapse strength of flat, rectangular, ship type plates take into account the departure from classical plate buckling theory due to the effects of imperfections, internal stresses and the additional post buckled strength of simply supported plates. Faulkner (1975) provides a history of the development of ultimate strength formulas for steel plate. Since this publication ultimate strength methods suitable for marine plating with various boundary conditions and loading scenarios have been further progressed.

For aluminium plating much of the recent research effort is encapsulated in various design codes, simplified formulae and computer programs used to calculate the ultimate collapse strength under in-plane and lateral load. Some commonly used methodologies for long plates in uniaxial compression are compared here along with several methods to predict the strength of biaxially loaded plates. This review is not exhaustive of the available methods, but gives an indication of the values likely to be used in design.

There are limited open source experimental test results for plates in uniaxial compression. A comprehensive physical test database of 5000 and 6000 series aluminium alloy plates under uniaxial compression was established at Cambridge University in the mid 1980s by Mofflin and Dwight (Collette 2005). The results of these tests provide a good basis for comparison of empirical methods.

2.1 Uniaxial compressive load

2.1.1 Paik & duran regression formula

Paik & Duran (2004) developed a regression formula defining the ultimate strength of an unstiffened aluminium plate with a heat affected zone (HAZ). The formulation is based on finite element analysis of variously alloyed marine grade aluminium plates.

The effects of the HAZ are taken into account by defining an averaged slenderness ratio, taking into account the softened material alongside the longitudinal stiffener plate boundary:

$$\beta = \frac{b}{t}\sqrt{\frac{\sigma_{0eq}}{E}} \qquad (1)$$

σ_{0eq} is an averaged material proof stress to take into account the reduced proof strength of the HAZ material, σ_{0HAZ}, over a specified breadth of plating away from the plate edge, b_{HAZ}. In the present study, the HAZ alongside the short edge boundary was found to have a negligible effect on plate strength behaviour and is therefore excluded from the calculation of the averaged proof stress. Therefore, the equation used here is a simplified form of the formulation proposed by Paik:

$$\sigma_{0eq} = \frac{(b - 2b_{HAZ})\sigma_0 + 2b\sigma_{0HAZ}}{b} \qquad (2)$$

where σ_0 is the 0.2% proof stress of the alloy.

2.2 Faulkner formula

The Faulkner formula, as given in Equation 3, is possibly the most commonly used empirical method available in the marine industry to estimate the strength of simply supported steel plates under longitudinal compression (Guedes Soares 1988). The method has been found to correlate well with test data from various steel plate experiments (Faulkner 1975). The formula includes the effects of residual stress and distortion implicitly. Therefore the existence of a HAZ breadth should be ignored in the calculation of β.

$$\phi_x = \frac{\sigma_{xav}}{\sigma_0} = 1.0 \quad \text{when } \beta < 1$$

$$\phi_x = \frac{\sigma_{xav}}{\sigma_0} = \frac{2}{\beta} - \frac{1}{\beta^2} \quad \text{when } \beta \geq 1 \qquad (3)$$

A number of variations following the pattern of the Faulkner formula are also available, for example by the US DOD (1982) and by Wang et al. (2005). The latter reference is specific to aluminium and includes a correction to account for the softening in the HAZ.

2.2.1 Eurocode 9

Eurocode 9 (2007) is a detailed code specifying criteria in the design of structures made from wrought and cast aluminium alloys. The code includes an empirical formula for predicting ultimate plate strength in a similar form to Faulkner's method but with explicit corrections for the softening in the HAZ. The ultimate strength of a plate under uniform compression is given in the form of a limit state. For comparative purposes in this study the partial safety factors in the formulation are set to unity and, assuming a simply supported homogenous plate, the ultimate compressive force at collapse is defined in terms of an effective cross section area of the plate multiplied by the material proof stress. For the range of plates considered in this study the formulae to calculate the effective cross section can be written:

$$A_{eff} = \min \begin{cases} 2b_{HAZ}t\rho_{0HAZ} + (b - 2b_{HAZ})t\rho_c \\ bt\rho_c \end{cases}$$

$$\rho_c = \frac{C_1}{\left(\beta/\varepsilon\right)} - \frac{C_2}{\left(\beta/\varepsilon\right)^2}, \quad \rho_{0HAZ} = \frac{\sigma_{0HAZ}}{\sigma_0} \qquad (4)$$

where $\beta = b/t$, rather than as defined in Equation 1 and $\varepsilon = (250/\sigma_0)^{0.5}$. The factors C_1 and C_2 depend on the material classification; factors of 29 and 198 are specified for the marine grade alloys considered here.

2.2.2 Comparison of uniaxial strength methods

Methods to predict uniaxial compression of long plates are compared in Figure 1 alongside relevant test data from the plate tests carried out by Mofflin (Collette 2005). Calculations are made for a typical 5083-H116 alloy plate of breadth 400 mm, aspect ratio 3 and with a HAZ breadth of 25 mm.

2.3 Biaxial load

Plate strength under in-plane biaxial load is presented on interaction diagrams which compare the ultimate strength of the plate in the longitudinal and transverse

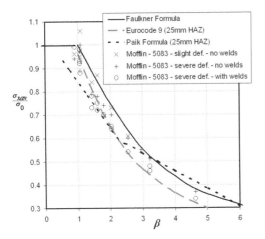

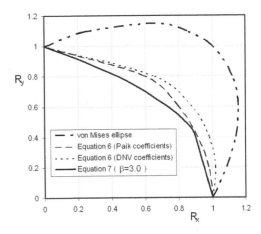

Figure 1. Comparison of methods to predict uniaxial plate strength.

Figure 2. Comparison of methods to predict biaxial plate strength.

directions. The applied longitudinal and transverse stresses are made non dimensional using the uniaxial compressive stress of the plate. Thus:

$$R_x = \frac{\sigma_{xav}(MAX)}{\sigma_{xav}(yav=0,MAX)} \quad R_y = \frac{\sigma_{yav}(MAX)}{\sigma_{yav}(xav=0,MAX)} \quad (5)$$

Various formulae are presented in literature to predict the interaction of biaxial loads. The formulas generally take a similar form to the von Mises yield criterion with coefficients replacing the power terms (Paik & Thayamballi 2003) Neglecting the effects of edge shear and lateral pressure, the expression takes the form:

$$(R_x)^{c1} + \alpha R_x R_y + (R_y)^{c2} = 1 \quad (6)$$

This equation equates to the von Mises formula by setting $c1 = c2 = 2$ and $\alpha = 1$, which is valid for plates in biaxial tension.

For biaxial compressive load, coefficients are given in a number of literature sources to predict the interaction curve (Paik & Thayamballi 2003). For example, Paik et al. (2001) suggests an adaptation of Equation 6, with $c1 = c2 = 2$ and $\alpha = 0$ when both load components are compressive. DNV HSLC rules (2001) contains an interaction formula for aluminium plates which, for the slenderness ratios considered in this study, can be approximated to Equation 6 with the coefficients $c1 = 1$, $c2 = 1.2$ and $\alpha = -0.8$.

A biaxial loading study specific to aluminium plates by Kristensen & Moan (1999), including the effects of HAZ and residual stress, found that the biaxial compressive interaction relationship is dependent on the slenderness ratio of the plate. A regression formulation defining the biaxial interaction as a function of

slenderness is presented as follows:

$$R_x = \min \begin{cases} \dfrac{1 - R_y^2}{1 - (0.213 - 0.275(\beta - 3))R_y} \\ 1 + (0.05 - 0.1\beta)R_y \end{cases} \quad (7)$$

The calculation of β should assume a zero HAZ breadth as this is accounted for implicitly in the above regression equation. If $\beta > 3$, only the first formula in Equation 7 should be used.

A more detailed formulation is given by Kristensen (2001), taking into account the aspect ratio and slenderness of the plate to calculate coefficients for the interaction formula given in Equation 6. This was found to give similar results compared to Equation 7 and therefore the latter equation is used for comparative purposes in this paper. The biaxial relationships are compared in Figure 2 for a plate with slenderness ratio of 3.

3 CHARACTERISATION OF ALUMINIUM ALLOY PLATES

A welded ship structure will inevitably have geometric imperfections and residual stresses introduced during construction and once the ship enters service. It is recognised that initial imperfections can have a significant effect on the ultimate strength of the plate (Ueda & Yao 1979). Therefore, a numerical model must be able to take into account any significant influence of geometric out of flatness, residual stresses and softening in the HAZ.

123

3.1 Material properties

The strength characteristics of aluminium differ with steel in terms of its stress-strain relationship. Unlike steel, the curved non-linear region of the stress-strain relationship of aluminium does not exhibit a clear yield point. The ultimate elongation and ultimate strength to yield strength ratio is also lower than that of steel.

The grades of aluminium alloys typically used in the marine industry are the 5000 and 6000 series. A complete list of alloys certified for use in a marine environment is given by classification authorities including DNV and ABS, as summarised by Sielski (2007). However, the present study limits itself to investigating alloys 5083-H116 and 6082-T6, which are the most common alloys currently used by the marine industry.

Due to variations in the composition of the alloys the modulus of elasticity also varies, albeit modestly. In this study it is taken as 70GPa. Representative values of proof stress as defined by DNV are used in this study. A Ramberg-Osgood approximation is used to represent the stress strain curve in the analysis:

$$\varepsilon = \frac{\sigma}{E} + 0.002 \left(\frac{\sigma}{\sigma_0} \right)^n \qquad (8)$$

There is wide variation between different codes and literature sources to define the knee factor coefficient, n, and the 0.2% proof stress of the aluminium alloy. Representative coefficients based on a review of various sources are used here. Alloy 5083-H116 is specified with a 0.2% proof stress of 215 MPa and a knee factor of 15. The 0.2% proof stress of alloy 6082-T6 is specified at 260 MPa with a knee factor of 30.

Collette (2005) highlights the possibility that the proof stress of 5000 series alloys are often higher in tension than in compression due to the strain hardening method used in their production. Quoted values in this study, based on tensile coupon tests, do not take this phenomenon into account because no established test data has been found. This effect does need to be quantified to enable a more accurate material representation in future studies.

3.2 Initial geometric imperfections

Panels are subject to a highly inhomogeneous heat treatment during welding caused by the concentrated high temperature heat input of the weld tool. The effects of welding can be broadly separated into three categories, as the process: changes the material properties in the HAZ; causes distortion of the plating due to local straining; and creates a residual stress field.

Welding induced distortion can include transverse and longitudinal shrinkage of the plate close to the weld, angular rotation of the plating around the weld bead axis, longitudinal bending of the plate-stiffener combination, rotational distortion and buckling distortion. In practice, angular change (wrap up) and longitudinal bending have been found to be the significant contributors to weld induced plate distortion (Paik & Thayamballi 2003). Misalignment and forcing together of panel elements will also result in initial out of straightness and imperfections may be further exacerbated during vessel operations.

3.2.1 Geometric imperfection amplitude

Localised distortions have been found to vary significantly for plates with different slenderness ratios, with slender plates exhibiting much larger local imperfections (Smith et al. 1988).

Numerous surveys have been carried out to measure the geometric imperfections in welded plating, with most focusing on steel plated structures. Empirical formulae to define the maximum initial imperfection amplitude, w_{opl} usually assume the imperfection magnitude is either proportional to β or to β^2 (Paik & Thayamballi 2003). Measurements of the deflection amplitude in bottom plating of a British Naval frigate in 1965 and reported by Faulkner (1975) shows closer dependence on β^2.

A well known formulation is proposed by Smith et al. (1988). Statistics of the imperfections in box girder bridges and shell plating of ships are drawn together and equations for slight (3% percentile), average and severe (97% percentile) imperfections in steel plate are derived as follows:

$$\frac{w_{opl}}{t} = \begin{cases} 0.025\beta^2 & \text{for slight level} \\ 0.1\beta^2 & \text{for average level} \\ 0.3\beta^2 & \text{for severe level} \end{cases} \qquad (9)$$

Very similar imperfection criteria are derived from an extensive study of aluminium plates recently carried out for the Ship Structure Committee (Paik et al. 2008). A more simplified alternative formulation is given by Paik & Duran (2004) based on limited data from numerous sources including Zha & Moan (2001):

$$w_{opl} = 0.009b \qquad (10)$$

Representative deflection amplitude according to Equation 10 is used in this study and compared with the amplitude as given by Equation 9.

3.2.2 Geometric Imperfection Shape

Imperfection of a rectangular steel plate is idealised as a single half wave shape along the plate length with localised distortions superimposed and a single half wave shape in the short direction.

The elastic buckling mode of a simply supported plate under uniaxial compression normally forms an approximate pattern of square sinusoidal half waves (Smith et al. 1988). However, inelastic buckling and collapse nucleates into a localised buckle with a wave length less than or equal to the aspect ratio a/b. It is well recognised that the shape of the initial deflection can significantly affect the formation of these collapse mode shapes in a uniaxially loaded plate.

Previous studies have shown that the plate collapse mode is strongly influenced by the initial deflection shape. For example, a study by Masaoka & Mansour (2004) shows snap through buckling in a uniaxially compressed plate as the deflection mode changes from the initial deflection mode to the preferred buckling mode. Certain initial imperfection shapes may even increase plate strength by stopping formation of the preferred buckling mode (Smith et al. 1988).

A Fourier series imperfection shape is typically used to define plate imperfection, where the deflection amplitude w_0 at any point on the plate surface is defined as a function of the maximum deflection, w_{opl}.

$$w_0 = w_{opl} \sum_{i=1}^{M} B_i \sin \frac{i\pi x}{a} \sin \frac{\pi y}{b} \quad (11)$$

The initial distortion coefficients, B_i, represent the influence of each mode shape contributing to the overall imperfection shape. All other terms are as shown in Figure 3.

A combination of multiple mode shapes using the Fourier series can represent a measured plate imperfection shape, but the choice of representative coefficients depends on a high number of varying factors that contribute to an assumed overall imperfection shape. A typical representation combining a single half wave and a square half wave imperfection with the ratio 0.8/0.2 is often used. A number of different Fourier representations are compared in the parametric study presented in this paper.

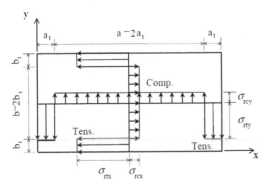

Figure 3. Idealised residual stress distribution with weld along all four edges of the plate (Paik & Thayamballi 2003).

However, a study by Dow & Smith (1984) assessing the effects of localised imperfections due to dents compared to half wave type imperfections along the entire plate length concluded that representing this form of imperfection as a Fourier series is unsatisfactory. The study found that a square wave Fourier component shape may give a non conservative representation of the imperfection if localised distortion is present. This effect is noted, although it is not considered in the present study.

3.3 Material softening in the HAZ

The welding process causes a strength reduction in the heat affected material adjacent to the weld. The formation of HAZ is due to different metallurgical processes for the 5000 and 6000 series alloys. 5000 series alloys are work hardened in the milling process. Subsequently, the high heat of fusion welding raises the temperature of material that is close to the weld above the re-crystallisation temperature, removing the work hardening and leaving the material in the weaker annealed state. In 6000 series alloy, the heat generated during welding causes the magnesium silicide precipitates, which provide extra strength in the alloy, to turn back into solution in the weld itself. Further away from the weld the precipitates grow in size, which reduces the local alloy strength.

3.3.1 HAZ material properties

Because the local transient temperature during welding decreases with increased distance from the weld, the material properties in the HAZ region also change with distance from the weld. However, for the purposes of numerical analysis, the HAZ is usually idealised to have constant reduced material properties over a defined width of plating away from the weld centre line location.

Material properties in the HAZ are determined by mechanical testing, although several inconsistencies between experiments, including the length of the test specimen and the method of forming the coupon, means that data from different sources has wide variation (Sielski 2007). In this study an idealised HAZ block with a strength knock down factor of 0.67 for 5083-H116 and 0.53 for 6082-T6 has been selected based on DNV guidance (2001).

3.3.2 HAZ breadth

The HAZ breadth can be established by testing the hardness of the area close to the boundaries of welded plates. Hardness tests are reported by Zha & Moan (2001) and Paik (2004). Both these papers propose a 25 mm HAZ breadth, originally proposed as a "1-inch rule" by Mazzolani (1995).

Zha & Moan use representative HAZ breadths of 12.5 mm and 25 mm in numerical analyses of plates.

These HAZ breadths are repeated by Collette (2005) in analysis of box sections. Paik uses a HAZ breadth of 3 times the plate thickness, consistent with DNV guidance (2001).

BS8118 (1991) and Eurocode 9 (2007) give explicit formulae to estimate the HAZ breadth. The formulas give HAZ breadth as a function of the plate/stiffener thickness. However, for the range of plate slenderness' considered in this study the codes give HAZ breadths similar to the 1 inch rule. Therefore, a representative HAZ breadth of 25 mm is used in the parametric study presented here.

3.4 Residual stress

Residual stress is the term used to describe self equilibrated internal stresses present in otherwise unstressed structural elements. They arise when some or all areas of a structural member undergoes physical or thermal induced deformation and is subsequently prevented from returning to its previous non deformed state. This creates a permanent, inhomogeneous deformation field in the structural member. It is well known that residual stress fields are generated in most metal structures and manufactured parts during construction (Stephens et al. 2001).

For welded aluminium panels the most significant cause of residual stress is the welding process itself (Sielski 2007). The residual stress field develops from the formation of a continuous weld. The liquid weld bead is resisted from contracting as it cools by the bulk of the parent material. This causes tensile residual stresses to be created within and near to the weld and corresponding balanced compressive stresses in material away from the HAZ.

The intensity of the heat input from welding is a function of a number of variables, including: the type of weld procedure used; the pass size; and the depth of penetration (Mazzolani 1995). Recent studies have found that the heat input and associated mechanical properties of the material at high temperatures is the significant contributor to residual stresses for both molten state (arc) and solid state (friction stir) welding (Peel et al. 2003; Withers 2007). This suggests that the quantity of heat introduced during welding is the primary contributor to residual stress rather than the specific levels of plastic deformation caused by the weld method.

For the purposes of numerical analysis, the welding induced residual stress field can be idealised into stress blocks, as depicted in Figure 3. A relationship between the residual stress and the width of the tensile stress field is:

$$b_t = \frac{\sigma_{rcx}}{2(\sigma_{rcx}) - \sigma_{rtx}}b \qquad (12)$$

An empirical approach to estimate the compressive residual stress of steel plates is given by Smith et al. (1988):

$$\frac{\sigma_{rcx}}{\sigma_0} = \frac{\sigma_{rcy}}{\sigma_0} = \begin{cases} -0.05 & \text{for slight level} \\ -0.15 & \text{for average level} \\ -0.3 & \text{for severe level} \end{cases} \qquad (13)$$

However, measurements in a recent Ship Structure Committee experiment programme indicate tensile residual stresses well below the parent metal strength and closer to the assumed HAZ strength (Paik et al. 2008). As an example, using representative values for b_t of 25 mm for a 400 mm plate in Equation 12 and 13 gives a predicted tensile stress well in excess of the assumed HAZ proof strength as defined previously, suggesting that the differences of aluminium compared to steel, including heat conduction rate and the HAZ softening, means that Smith's approach may not be valid for use.

Based on these findings, a tensile residual stress equal to the HAZ proof stress in a block of equal width to the HAZ breadth is used in the parametric analyses conducted in this study. Compressive residual stresses are calculated to give a self equilibrated resultant stress as given by Equation 12.

4 FINITE ELEMENT METHODOLOGY

Abaqus finite element analysis software was used to carry out the parametric study. The Abaqus solver has geometric and material non linear capabilities and has been shown to provide good comparative results to test data of plates in compressive instability (Zha & Moan 2001).

An incremental arc length analysis utilising the Riks method copes with negative slopes on the iteration curve which are typical of plate collapse analyses when the plate loses strength in the post collapse region. The incremental form of the analysis enables production of complete load shortening curves in the post processing module.

Quadrilateral shell elements with reduced integration were used for all plates modelled in the parametric study. A check on result convergence for different mesh densities was carried out and based on these results the mesh for all plates was specified so that at least one element, and more usually two, spans the HAZ region. Element type and mesh size are consistent with previous studies (Zha & Moan 2001).

Initial imperfection is introduced by applying an out of plane Fourier series displacement to each node using an external subroutine to directly edit the node input file. Residual stress is introduced using the *INITIAL CONDITIONS feature of Abaqus, which allows elements to be prescribed an initial stress field prior to any displacement of the model.

The plate is modelled to represent support on all sides by longitudinal stiffeners at the long edges and transverse frames at the short edges. The supports are assumed to be relatively stiff along the boundary, preventing out of plane deformation and constraining the edges to remain straight throughout the analysis. The plating is assumed to be continuous across several frame and stiffener spans and therefore the stiffeners are assumed to provide no rotational restraint about the edge axis, resulting in a simply supported boundary condition at all edges.

Load is applied with a displacement control on one edge perpendicular to the load direction. This ensures edges remain straight throughout the application of load. Unloaded edges are free to move in-plane but are constrained to remain straight.

In the biaxial compression analyses a fixed load is applied on the long edges as an initial load step prior to an incremental compressive displacement applied on the short edges. The long edges remain free to move in-plane, allowing the biaxial transverse force to remain constant throughout the longitudinal compressive displacement load step.

5 COLLAPSE BEHAVIOUR OF ALUMINIUM PLATES IN UNIAXIAL COMPRESSION

This section presents a parametric study which assesses the influence of geometric and material factors characterising aluminium ship plating in uniaxial compression. A series of load shortening curves, calculated using the finite element method, compare the influence of plate slenderness, aspect ratio, fabrication related distortion, material softening in the HAZ and the residual stress field.

Load shortening curves are presented in non dimensional form. The applied force on the loaded plate edge is used to calculate an average applied stress, σ_{xav}, assuming that the cross section area of the plate remains constant throughout the analysis. The applied stress is then divided by the welded plate proof stress. Similarly, the displacement of the loaded edge is divided by the unloaded plate length to give the strain and then further divided by the coupon strain, ε_0, at 0.2% proof stress to give results in non dimensional form.

5.1 *Effects of aspect ratio, slenderness and alloy*

Load shortening curves are presented in Figures 4 and 5 for various plates with slenderness ratio between 1.5 and 5.0 in uniaxial compression. Levels of imperfection and residual stress are modelled using the rules and assumptions described previously in this paper, consisting of: an imperfection magnitude of 4 mm;

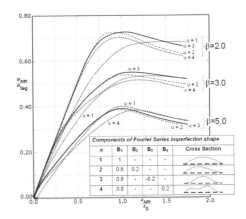

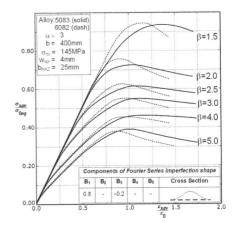

Figure 4. Load shortening curves for typical 5083 plates with average imperfections and residual stresses.

Figure 5. Load shortening curves for typical 5083 and 6082 plates with average imperfections and residual stresses.

an imperfection shape superimposing 80% single half wave and 20% square half wave distortion; a 25 mm HAZ with reduced strength material; a residual tension field equal to the softened material proof stress; and a self equilibrating residual compression field. The results for plates modelled with a residual stress field in the x-direction only are presented; initial analyses found minimal differences between equivalent plates with a biaxial residual stress field. For the same reasons results for plates with the HAZ modelled in the long edges only are given here, consistent with initial findings and in previous studies (Zunaidi 2007).

5.1.1 *Aspect ratio*

Analyses of three different thickness plates modelled with aspect ratios between 1 and 4 are compared in Figure 4. The results show similar characteristics before and after collapse, although the peak strength at collapse varies slightly. The initial imperfection shape of the square plate is equivalent to the nucleated collapse shape, causing the load shortening curve to depart immediately from a linear elastic response.

A typical longitudinal framed midship of a high speed vessel will likely be made up of long plates predominantly loaded in the longitudinal direction. Therefore the load shortening behaviour of plates with higher aspect ratios are more relevant to this study and an aspect ratio of 3 is therefore further studied here.

5.1.2 *Slenderness ratio*

In keeping with the simplified formulae based methods described previously, a progressive decrease in overall plate strength occurs as the slenderness ratio increases, as shown in Figure 5. The shape of the load shortening curves initially follows a similar pattern for all slenderness ratios and for both alloy types, following the linear elastic stress-strain relationship of the material. This indicates an initial phase of load shortening predominantly due to in-plane compression of the plate.

The load shortening curves for slender plates make a deviation from the linear elastic relationship well before collapse. This is characterised by a transition from the initially imposed imperfection to a shape equivalent to the buckling mode. The plate then exhibits stable behaviour as it continues to withstand further increases in load. Through this phase the central area of the plate increasingly deflects out of plane, escaping the applied load and causing an increasing portion to be taken by the side regions. Eventually the plate undergoes elasto-plastic deformation with a resulting loss of load carrying capacity.

The deviation point from the initial phase is less distinct for more stocky plates where elastic buckling stresses are approximately equal to or greater than the overall collapse strength. This is particularly apparent for low slenderness 6082 plates. In these cases the transition from the linear region to collapse is relatively rapid.

5.1.3 *Alloy type*

Comparing both alloys in Figure 5 shows differences in the pre and post collapse regions of the load shortening curve. These resemble the corresponding differences in the alloys stress strain curve, where 5083 has a more rounded transition away from the elastic region compared to 6082. The 6082 alloy appears to provide slightly higher ultimate strength for the lower β values and lower ultimate strength for the higher slenderness plates. In all cases the 6082 plates exhibit more rapid load shortening in the post collapse region.

5.2 *Effects of imperfection shape and amplitude*

A series of imperfection models have been analysed to demonstrate the variability of the plate behaviour in the pre and post collapse phases during the application of uniaxial compression. Figure 6 presents a sample of the results for two plates with slenderness ratio of 2.0 and 3.0. A comparison of results for both plates demonstrate significant differences in the load shortening curve in the pre collapse region, suggesting that the initial imperfection shape has significant influence on the strain response in this region. Clearly this will be of importance in a progressive collapse analysis including plate elements.

For the more stocky plate shown in Figure 6 the imperfection shape also has a significant effect on the overall collapse strength. This is less apparent for the more slender plate, despite the significant differences in the shortening curve before the collapse load is reached.

The effects of imperfection amplitude have been tested by comparing Equation 10 given by Paik with the average and severe imperfection magnitudes as given by Smith in Equation 9. For the range of plates tested here the average imperfection given by both equations are similar, particularly for the more stocky plates. This results in very similar load shortening curves in Figure 7. Use of the severe deflection amplitude equation causes a reduction in the overall collapse strength of the plate. In all cases, the load shortening curve maintains the same overall shape regardless of imperfection magnitude, with only the peak position affected.

5.3 *Effects of HAZ*

Representative HAZ breadths of 12.5 mm, 25 mm and 50 mm are compared in Figure 8 for $\beta = 2.0$ and 3.0.

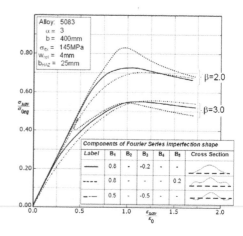

Figure 6. Load shortening curves for typical plates with different initial imperfection shapes.

128

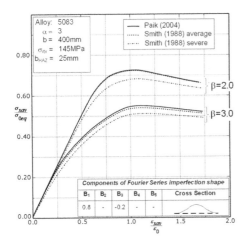

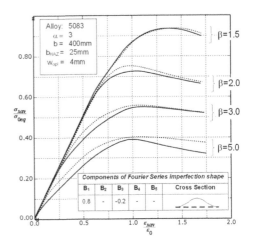

Figure 7. Load shortening curves for typical plates with different initial imperfection magnitude.

Figure 9. Load shortening curves for typical plates with residual stress field (full) and without residual stress field (dashed).

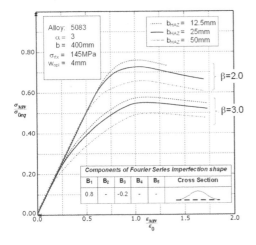

Figure 8. Load shortening curves for typical plates with different HAZ breadth and typical geometric distortions.

For each model the residual stress distribution was adjusted to match the HAZ breadth. The plots show that HAZ breadth has a significant influence on the collapse strength of the plate, primarily by shifting the transition away from the initial linear elastic phase. From this point onwards the shape of the load shortening curve remains unaffected by the size of the HAZ breadth.

5.4 Effects of residual stress

A plate with a residual stress field as proposed previously is compared to a plate with zero residual stress in Figure 9. Comparisons are made for a full range of plate slenderness. For all except the most stocky plate

analysed here, the results indicate that the residual stress has a moderate impact on the plate collapse characteristics. For intermediate slender plates the zero residual stress case diverges from the residual stress case in the pre collapse region, resulting in higher magnitude collapse strength. The very slender plate ($\beta = 5$) also shows different characteristics after the collapse load has been passed.

5.5 Effects of lateral load

Limited analyses were undertaken for plates in uniaxial compression with a fixed lateral load, P, applied uniformly across the plate surface. The load is assumed to act on the concave side of the distorted plate, exacerbating the imperfection amplitude. The load is equivalent to a hydrostatic pressure on the side or bottom shell of the vessel. Pressure loads of 25 kPa and 50 kPa, equivalent to a hydrostatic pressure depth of approximately 2.5 m and 5.0 m respectively, are compared in Figure 10. The results demonstrate an increasingly adverse effect of lateral pressure as the slenderness of the plate increases.

5.6 Comparison with empirical results

Plate collapse strength values for a limited selection of analyses from Sections 1.1 to 1.4 are compared with the Eurocode 9 formula in Figure 11. The results for the average imperfection levels, as described previously in Section 5.1, show reasonable correlation with the shape of the Eurocode 9 curve, although predicted strength here is higher in all cases, suggesting a slight conservatism in the code formula. The influence of the imperfection and HAZ parameters tested in this study

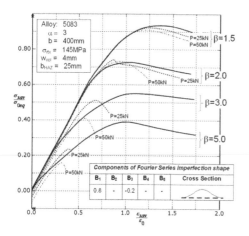

Figure 10. Comparison of typical plates with lateral load of 25 kN and 50 kN applied on the concave side of the plate.

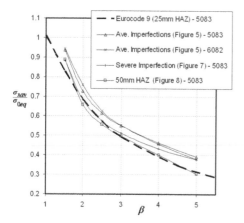

Figure 11. Ultimate strength of typical plates as a function of slenderness ratio. Comparison with Eurocode 9 formulation.

has a moderate effect on the ultimate strength value, with the most pronounced differences occurring in the intermediate slenderness region of Figure 11.

6 COLLAPSE BEHAVIOUR OF ALUMINIUM PLATES IN BIAXIAL COMPRESSION

This section presents a limited analysis set of 5083 alloy plates in biaxial compression. It is intended that the analyses presented here will be expanded in a future paper to investigate different material and residual stress representations along with the influence of further load combinations.

Interaction curves for plates with slenderness ratios between 1.5 and 4.0 are given in Figure 12. A HAZ breadth of 25 mm is assumed at all 4 edges along with a biaxial residual stress distribution as in Figure 3 with tensile stresses equal to the HAZ proof stress and balanced compressive stresses calculated using Equation 12. The results are presented as non dimensional functions of the parent material 0.2% proof stress, rather than the equivalent proof stress used in previous sections.

The curves show an initial region from the x axis where the collapse strength is relatively unaffected by the increasing biaxial load on the long edges. This region is more pronounced for the lower slenderness plates, and becomes less distinct as the slenderness increases.

The analysis results are compared with the closed form method of Kristensen & Moan (Equ. 7) and with the von Mises type interaction formula (Equ. 6) using the coefficients as given by Paik. To enable direct comparison with the analysis results the empirical formulations, which calculate non dimensional values R_x and R_y, are further multiplied by the non-dimensional uniaxial strength of the plate loaded in the corresponding direction as calculated in this study.

The results from this study show good correlation to the closed form Kristensen equation for all slenderness ratios analysed, demonstrating the effect of slenderness on the behaviour of a plate under biaxial load. The simple expression as given by the von Mises type formula cannot be adjusted to take into account the slenderness of the plate, unless specific formulations for the coefficients are specified, as is done by Kristensen (2001).

It is clear from these results that a biaxial load combination has a varying effect on the strength of aluminium panel elements depending on the slenderness

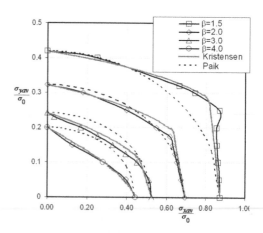

Figure 12. Ultimate strength of typical 5083 alloy plates in biaxial compression. Imperfection shape as given in Figure 5.

of the plate. The results suggest that further studies of the effects of biaxial load are required to identify the load shortening curve paths for use in ultimate strength methodologies of stiffened panels and hull girder structures under biaxial bending.

7 CONCLUSIONS

The imperfection of ship type aluminium plates are defined by a number of parameters which together describe the geometric imperfection, material properties, softening in the HAZ and the residual stress field. Code formulations such as Eurocode 9 explicitly take into account many of the parameters examined in this study and provide a good correlation to the ultimate strength of plates modelled in uniaxial compression using non linear FEA.

Representative plate load shortening curves are required in simplified hull girder ultimate strength methodologies; for the case of a high speed aluminium vessel the curves need to account for the effects of alloy type, imperfection, softening in the heat affected zone and residual stress. This study has shown that the parameters have a significant influence on the strength behaviour of aluminium plates in uniaxial compression both before and after the collapse load is reached. The effects of lateral pressure and biaxial loading also have a significant effect on the collapse strength and load shortening curve of the plates. An ultimate collapse method specific to a large high speed aluminium hull girder will need to adequately account for these parameters.

ACKNOWLEDGEMENTS

This present study was undertaken with the support of the Office of Naval Research. The first author would also like to thank Mr John Garside for his insightful discussions on the subject of this study.

REFERENCES

British Standards 1991. *Structural Use of Aluminium.* BS8118:Part 2.
British Standards 2007. *Eurocode 9: Design of aluminium structures.* BS EN 1999-1-1: 2007.
Collette, M. 2005. *The strength and reliability of aluminium stiffened panels.* PhD Thesis. School of Marine Science and Technology, Newcastle University.
DNV. 2001. *Rules for Classification of High Speed, Light Craft and Naval Surface Craft.* Det Norske Veritas.
Dow, R.S. & Smith, C.S. 1984. Effects of Localized Imperfections on Compressive Strength of Long Rectangular Plates. *Journal of Constructional Steel Research* 4: 51–76.
Faulkner, D. 1975. A review of effective plating for use in the analysis of stiffened plating in bending and compression. *Journal of Ship Research* 19(1): 1–17.
Guedes Soares, C. 1988. Design Equation for the Compressive Strength of Unstiffened Plate Elements with Initial Imperfections. *Journal of Constructional Steel Research* 9: 287–310.
Kristensen, O.H.H. & Moan, T. 1999. Ultimate Strength of Aluminium Plates under Biaxial Loading. *FAST 1999.*
Kristensen, O.H.H. 2001. *Ultimate Capacity of Aluminium Plates under Multiple Loads, Considering HAZ Properties.* Dr. Ing Thesis. Department of Marine Structures, Norwegian University of Science and Technology, Trondheim.
Masaoka, K. & Mansour, A. 2004. Ultimate Compressive Strength of Imperfect Unstiffened Plates: Simple Design Equations. *Journal of Ship Research* 48(3): 191–201.
Mazzolani, F.M. 1995. *Aluminium alloy structures.* London: E & FN Spon.
Naval Sea Systems Command. 1982. Design Data Sheet 100-4—Strength of Structural Members.
Paik, J.K. & Duran A. 2004. Ultimate Strength of Aluminum Plates and Stiffened Panels for Marine Applications. *Marine Technology* 41(3): 108–121.
Paik, J.K. & Thayamballi A.K. 2003. *Ultimate limit state design of steel-plated structures.* Wiley.
Paik, J.K., Thayamballi, A.K. et al. 2001. Advanced Ultimate Strength Formulations for Ship Plating Under Combined Biaxial Compression/Tension, Edge Shear, and Lateral Pressure Loads. *Marine Technology* 38(1).
Paik, J.K., Thayamballi, A.K. et al. 2008. *Mechanical Collapse Testing on Aluminum Stiffened Panels for Marine Applications.* Ship Structure Committee.
Peel, M., Steuwer, A. et al. 2003. Microstructure, mechanical properties and residual stresses as a function of welding speed in aluminium AA5083 friction stir welds. *Acta Materialia* 51: 4791–4801.
Sielski, R.A. 2007. *Aluminum Marine Structure Design and Fabrication Guide.* Ship Structure Committee.
Smith, C.S. 1977. Influence of local compressive failure on ultimate longitudinal strength of a ship's hull. *PRADS 1977:* 73–79.
Smith, C.S., Davidson, P.C. et al. 1988. Strength and Stiffness of Ships' Plating under In-plane Compression and Tension. *Transactions of the Royal Institution of Naval Architects.* 130: 277–296.
Stephens, R. Fatemi, A. et al. 2001. *Metal fatigue in engineering.* Wiley-Interscience.
Ueda, Y. & Yao T. 1979. Ultimate Strength of a Rectangular Plate under Thrust : with Consideration of the Effects of Initial Imperfections due to Welding. *Transactions of Japanese Welding Research Institute.* 8(2): 257–264.
Wang, X., Sun, H. et al. 2005. Buckling and Ultimate Strength of Aluminium Plates and Stiffened Panels in Marine Structures. *Fifth International Forum on Aluminium Ships.* Tokyo, Japan.
Withers, P.J. 2007. Residual stress and its role in failure. *Reports on Progress in Physics* 70: 2211–2264.
Zha, Y. & Moan T. 2001. Ultimate strength of stiffened aluminium panels with predominantly torsional failure modes. *Thin Walled Structures* 39: 631–648.
Zunaidi, A.N. 2007. *The Strength Characteristics of Aluminium Panels in Ships.* MSc Thesis. School of Marine Science and Technology, Newcastle University.

Analysis and Design of Marine Structures – Guedes Soares & Das (eds)
© 2009 Taylor & Francis Group, London, ISBN 978-0-415-54934-9

Improving the shear properties of web-core sandwich structures using filling material

J. Romanoff, A. Laakso & P. Varsta
Helsinki University of Technology, Department of Applied Mechanics, Marine Technology, Espoo, Finland

ABSTRACT: Recently the industry has shown interest on using steel sandwich panels as part of ship structure. Web-core sandwich panels have unidirectional core which causes high orthotropic to shear stiffness of these panels. The shear stiffness opposite to web plate direction can be order of magnitude lower than that in web plate direction. Often this difference leads to high shear-induced normal stresses in the face and web plates of sandwich plates. These stresses are considered to be critical when for example yield or fatigue criteria are concerned. The paper presents recent theoretical and experimental investigations on improving the shear characteristics of web-core sandwich beams using filling material. Shear stiffness, shear induced normal stresses and ultimate strength of these beams under shear are considered. The web-core sandwich beams, made from steel, are filled with H-Grade Divinycell with densities of $80 \, kg/m^3$ and $200 \, kg/m^3$, leading to a weigh increase of 6% and 15%. The theoretical investigations are carried out by Finite Element simulations. The experimental investigation is carried out on beams in four point bending. The linear-elastic response and strength of these beams are compared to those of empty beams. The shear stiffness was increased and thus the shear induced normal stresses decreased by three times and seven times with the $80 \, kg/m^3$ and $200 \, kg/m^3$ foams respectively. In ultimate strength the corresponding values are 2.5 and 3.5; thus the ultimate strength is significally lower than the elastic response would indicate. The empty beams are failing due to plastic hinge formation at laser-welded T-joint. The filled beams fail by core shear followed by the plastic hinge formation at the laser weld.

1   INTRODUCTION

Demand for lighter and safer structures has stimulated the need to study new materials and new structural configurations in ship structures. Filled web-core sandwich panels offer an option to fulfill these requirements, see Figure 1.

In this structure, the high density web plates and low density filling material separate the two face plates. The connection between the face and web plates is laser-welded and the filling material is adhesively bonded to steel. While, the web plates mainly contribute to the shear stiffness in the web plate direction, the filling material is used to stabilize the face plates and increase the shear stiffness in the other direction.

These web-core structures have been studied intensively by several authors, who have mainly focused on empty panels. Then, the shear force is carried out by the local bending of the web and face plates (Holmberg, 1950; Kolsters and Zenkert, 2002; Romanoff et al., 2007a). Hasebe & Sun (2000) and Aimmanee & Vinson (2002) assumed that, in the case of filled panels, the shear forces are carried out by the filling material alone. Kolsters and Zenkert (2002) made attempt to correct the shear stiffness of the sandwich panels by considering separately the stiffness of the steel structure and the filling material and finally summing these

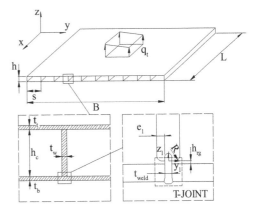

Figure 1.   Laser-welded web-core sandwich panel and the notations used.

to give the overall stiffness. However, they did not consider the stress response in their investigations.

Stress response is important when structures are designed for service. When stresses are evaluated, several papers show that the shear-induced secondary normal stresses dominate the response of empty web-core sandwich panels; see for example Knox et al. (1998), Bright and Smith (2004), Romanoff and Varsta

(2006). However, these papers do not consider influence of the filling material on the normal stress response, nor they present anything on the limiting stress values.

The purpose of present paper is to investigate the shear induced secondary normal stresses in the face and web plates of foam filled web-core sandwich structures. Finite Element analyses and experiments are carried out on beam in four point bending. Linear elastic response and ultimate strength of these beams is considered.

2 EXPERIMENTS

2.1 *Description of the investigated structures*

The steel sandwich panels were manufactured by Meyer Werft in Germany and cut to beams having length 1080 mm and breadth 50 mm; the length direction of the beam is opposite to web plate direction. The face plates were measured to have thickness of 2.52 mm. The web plates had thickness and height of 3.97 mm and 40 mm respectively. The web plate spacing was 120 mm. Thus, the beams had 9 unit cells along its length.

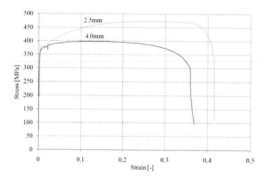

Figure 2.   Stress strain curves for the face and web plates.

The material properties have been measured in Romanoff et al. (2006); the stress strain curves are presented in Figure 2. The face plates have Young's modulus of 221 GPa, yield strength 360–368 MPa and tensile strength of 470–476. The corresponding values for the web plates are: Young's modulus 200 GPa, yield strength 360 and tensile strength 398. The Poisson's ratio is assumed to be 0.3.

The influence of different filling materials has been investigated by considering different densities of Divinycell H-grade. Calculations have been performed for all densities available by DIAB (2008). In experimental investigations H80 and H200 foams are considered. The average mechanical properties measured by the manufacturer are given in Table 1. The Poisson's ratio in all cases is assumed to be 0.33 (Gibson and Ashby, 1988).

For the experiments three specimens were manufactured for beams with H80 and H200 as filling material. In addition two empty specimens were used as reference. The filled specimens were manufactured by cutting the foam blocks to fit the voids in steel sandwich beams, see Figure 3. Then polyurethane-based Sikabond 545 adhesive was applied to foam blocks and activated by spraying water on the adhesive. While curing, the adhesive filled all gaps between the foam block and void. As a result, beams with 6% and 5% weight increase were obtained.

2.2 *Test set up and instrumentation*

Each specimen was instrumented with 12 strain gauges (5 mm) located symmetrically with respect to the mid-plane of the beam. The strain gauges were positioned at 2nd, 5th and 8th unit cells in order to guarantee that the strains were unaffected by local effects arising from application of external loads or boundary conditions, see Figure 4. Eight strain gauges were located at the constant shear region in order to capture the shear-induced secondary normal stresses. All strain gauges were glued on the top surface of the top face plate at distances 10 mm and 30 mm from the

Table 1.   Mechanical properties for the filling material (DIAB, 2008, according to ASTM).

Property	H45	H60	H80	H100	H130	H200	H250
Density [kg/m^3]	48	60	80	100	130	200	250
Compressive properties							
Strength [MPa]	0.5	0.7	1.15	1.65	2.4	4.2	5.4
Modulus [MPa]	45	60	80	115	145	200	240
Tensile properties							
Strength [MPa]	1.1	1.5	2.2	2.5	3.5	6.3	8.0
Modulus [MPa]	45	57	85	105	135	210	260
Shear properties							
Strength [MPa]	0.46	0.63	0.95	1.4	1.9	3.2	3.9
Modulus [MPa]	12	16	23	28	40	75	88
Elongation [%]	8	10	15	25	30	30	30

134

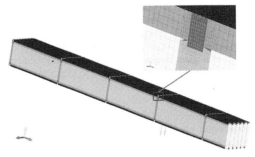

Figure 3. Test specimen before and after filling the voids of the steel sandwich beam.

Figure 5. 3D FE-model used in the calculations.

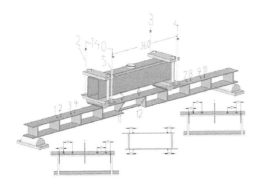

Figure 4. Instrumentation of the specimen.

nearest web plate. Four strain gauges were located at the constant bending moment region. Deflection was measured from load introduction points at $y = L/3$ and $y = 2L/3$ with displacement gauges having measuring range of ± 20 mm. The load was measured with force gauge having maximum measuring capacity of 10 kN.

2.3 Performing the test

The stiffness test was carried out first by increasing the load from 0 N to 400 N and decreasing load back to 0 N. All measured signals were recorded at 50 N intervals with increasing and decreasing load. This type of cycle was repeated three times to get the linear elastic response.

Finally, the ultimate strength of the beam was determined in similar fashion, but by increasing the load as long as the maximum range of the displacement gauge was reached at approximately 35 mm. As a rough estimate, this correspond a shear angle of approximately 10% at span $L/3$.

3 FINITE ELEMENT ANALYSES

Two types of Finite Element analyses were carried out. In the first set the linear elastic response was evaluated using 3D solid elements. The 3D finite element mesh and the load and boundary conditions are presented

in Figure 5. In the second set of analyses the ultimate strength behavior of the beams was analyzed using 2D plane strain element models. The influence of plane stress condition at the edges of the beam was taken into account by reducing the breadth of the beam from 50 mm to 48 mm, in 2D plane strain analysis. This reduction was based on the 3D solid element model on linear elastic regime and the y-direction normal stress distribution along beam breadth.

The Finite Element analyses were performed with Abaqus 6.8.1 using 20 node solid elements or 8 node plane strain elements. In both cases, symmetry condition at L/2 was used. The 3D solid element model was constructed by extruding the 2D plane strain model. The rotation stiffness of welds was modeled with equivalent weld thickness of 2 mm (see Romanoff et al. 2007b). The filling material was modeled as orthotropic since the material properties from manufacturer do not fulfill the basic relation between Young's modulus, shear modulus and Poisson's ratio; this is the case for H80 while H200 can be considered as isotropic. The Young's modulus and uni-axial strength properties were taken as the average of the compressive and tensile material properties, given in Table 1.

The non-linear analysis was carried out using the Riks—option in Abaqus. Material non-linearity for steel was modeled by transforming the engineering stresses and strains to true stress and true strain. The filling material was assumed to be bi-linear with yield point at 2% elongation.

4 RESULTS

4.1 Linear elastic response

4.1.1 Comparison of the deflection
The deflection at $y = L/3$ for different filling materials is given in Figure 6. Figure 6 clearly shows that the filling material decreases the deflection. While for H80 the deflection is about three times lower than that of empty beam, for H200 the corresponding value is

seven. In chosen location, the shear deflection dominates over that of caused by the bending moment. Thus, the filling material can be used efficiently to increase the shear stiffness of the beam opposite to web plate direction. In general the agreement between Finite Element analyses and experiments is fairly good.

4.1.2 Comparison of the normal stress

The comparison of the y-direction normal stress at the top surface of top face plate is presented in Figure 7.

Figure 7 clearly shows that the correspondence between the Finite Element analyses and experiments is very good. The shear-induced normal stresses are reduced by the same factor of three and seven as in the case of deflection. In the region of the constant bending moment the stresses are unaffected. It is seen that the shear-induced normal stresses dominate the stress response except in the case of H200 where the stresses induced by the bending moments have same order of

magnitude. It is also seen that the sudden increase of stress at location of the web plate is reduced by the filling material. This indicates that the moment introduced by the web plate is smaller with filled beams.

4.2 Stresses at the filling material

Figure 8 presents the yz-shear stress at the interface of the filling material and top face plate. Figure 9 presents the z-direction normal stress in the same location.

From Figure 8 and Figure 9 it is seen that filling material works in different ways regarding the position within unit cell. In the mid-span between two web plates, the shear stresses are highest, while the normal stress is close to zero. Thus, at the mid span of the unit cell, the filling material carries large portion of the external load by shear stresses. Near the web plates the shear stresses vanish and, the filling material gives vertical support to the face plate.

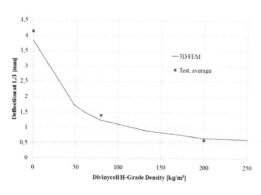

Figure 6. Comparison deflection at y = L/3 for various Divinycell H-grade densities. F = 400 N.

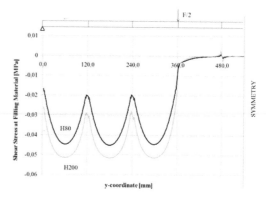

Figure 8. Comparison of yz-direction shear stress in the intersection of filling material and top face plate. F = 400 N.

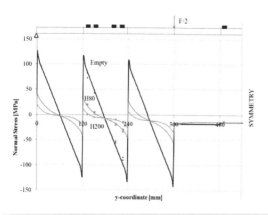

Figure 7. Comparison of y-direction surface stress in the top surface of top face plate. The experimental results are presented with dots and FE results with continuous line. F = 400 N.

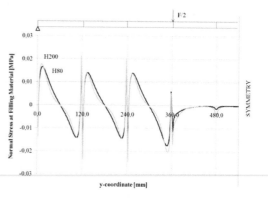

Figure 9. Comparison of z-direction normal stress in the intersection of filling material and top face plate. F = 400 N.

4.3 Bending moment induced by the web plate

Figure 10 presents the z-direction normal stress at the laser-weld, calculated with FEM. Also resulting bending moment induced by the web plate to top face plate is presented. It is seen that the stress levels and resulting bending moments are reduced by factors of three and seven when H80 and H200 are compared to empty beam. Thus, the filling material efficiently reduces the bending stresses at laser-weld and resulting web-induced bending moment.

4.4 Ultimate strength

The ultimate strength behavior of various beams is presented in Figure 11. The failure modes are presented in Figure 12.

As Figure 11 indicates the ultimate strength is increased significantly with the filling material. The ultimate strength is evaluated with relatively good accuracy with FEM. The increase in ultimate strength is not as large as the increase in stiffness or decrease in

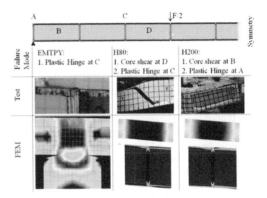

Figure 12. Failure modes for various beams.

stresses in linear elastic regime. The ultimate strength of H80 is 2.5 times higher than that of empty beam, while the corresponding value for the H200 is 3.5. Figure 12 show that, the correspondence between predicted and tested failure modes is also in good agreement. While the empty beam fails by plastic hinge formation at the laser weld, the filled beams fail by shear in the filling material followed by plastic hinge formation at the laser weld. In case of H200 the beam fails in test practically without warning. Both filled beams have higher residual strength after final failure than the empty beam.

5 CONCLUSIONS AND FUTURE WORK

This paper presented an investigation on the influence of filling material to shear characteristics of web-core sandwich structures. Beams, having span opposite to web plate direction, were tested and simulated under four point bending. The beams were standard I-core and they were filled with different grades of Divinycell H-grade foams.

Finite Element Analyses and experiments showed that that the shear deflection of the beams is significantly decreased by use of filling material. For tested H80 and H200 foams the decrease in deflection and shear-induced normal stress in the face plates was three and seven times respectively in linear elastic regime. At the same time the weight of the beams was increased only by 6% and 15%. It appears that the filling material gives vertical support to the face plates near the web plates and therefore reduces the bending stresses of the laser welds. At the mid-span between two web plates the shear stresses gets highest values, while vertical stresses are close to zero. Thus, there the external load is carried out mainly be shear stresses at the filling material.

In the ultimate strength the increase was not increased as much as the linear elastic response could

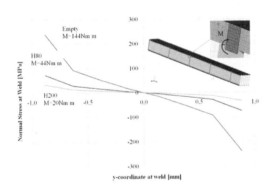

Figure 10. Comparison of z-direction normal stress in the laser-weld and resulting web-induced bending moment. F = 400 N.

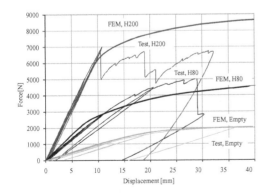

Figure 11. Comparison of the force vs. deflection at y = L/3 for empty, H80 and H200 beams.

indicate; the H80 and H200 had 2.5 and 3.5 times higher ultimate load than the empty reference beam. While the empty beams were failing by plastic hinge formation at laser-weld, the filled beams failed by shear failure of the filling material. After failure of the filling material, plastic hinge was also formed to the laser-welds in the case of filled beams. The residual strength of filled beam was in both cases higher than that of empty beam.

The present investigation has shown the potential of filling the panels with polymeric foams. However, it remains open how this potential could be used in design of ship structures. Open issues are related for example to fatigue, fire and corrosion behaviour and plate response and strength of these hybrid structures. These topics are left for future work.

ACKNOWLEDGEMENTS

This work has been performed within the project "Closed, Foam-Filled Steel Structures-SUTERA", funded by Association of Finnish Mechanical Industries. This financial help is gratefully acknowledged. I-core at Meyer Werft in Germany is thanked for producing the empty steel sandwich panels.

REFERENCES

Aimmanee, S. and Vinson, J.R., "Analysis and Optimization of Foam-Reinforced Web Core Composite Sandwich Panels Under In-Plane Compressive Loads", Journal of Sandwich Structures and Materials, Vol. 4/April 2002, pp. 115–139.

Bright S.R. and Smith J.W., "Fatigue performance of laser-welded steel bridge decks", Structural Engineering Vol. 82 (21), 2004, pp. 31–39.

DIAB, "Divinycell H Technical Manual", May 2008, p.10.

Gibson, L.J. and Ashby, M.F., "Cellular Solids—Structure and Properties", Pergamon Press, Oxford, 1988.

Hasebe, R.S. and Sun, C.T., "Performance of Sandwich Structures with Composite Reinforced Core", Journal of Sandwich Structures and Materials, Vol. 2, January 2000, pp. 75–100.

Holmberg, Å., "Shear-Weak Beams on Elastic Foundation", IABSE Publications, Vol. 10, 1950, pp. 69–85.

Kolsters, H. and Zenkert, D., "Numerical and Experimental Validation of Stiffness Model for Laser-Welded Sandwich Panels with Vertical Webs and Low Density Core", Presented in: Hans Kolsters, Licentiate Thesis–Paper B, Kunliga Tekniska Högskolan, Stockholm, 2002.

Knox, E.M., Cowling, M.J. and Winkle I.E., "Adhesively Bonded Steel Corrugated Core Sandwich Construction for Marine Applications", Marine Structures, Vol. 11, 1998, pp. 185–204.

Romanoff, J., Remes, H., Socha, G. and Jutila, M., "Stiffness and Strength Testing of Laser Stake Welds in Steel Sandwich Panels", Helsinki University of Technology, Ship Laboratory, M-291, Espoo, 2006.

Romanoff, J. and Varsta, P., "Bending Response of Web-core Sandwich Beams", Composite Structures, Vol. 73, No. 4, 2006, pp. 478–487.

Romanoff, J., Varsta, P. and Klanac, A., "Stress Analysis of Homogenized Web-Core Sandwich Beams", Composite Structures, Vol. 79, No. 3, 2007a, pp. 411–422.

Romanoff, J., Varsta, P. and Remes, H., "Laser-Welded Web-Core Sandwich Plates under Patch-Loading", Marine Structures, Vol. 20, No. 1–2, 2007b, pp. 25–48.

Analysis and Design of Marine Structures – Guedes Soares & Das (eds)
© 2009 Taylor & Francis Group, London, ISBN 978-0-415-54934-9

Stability of flat bar stiffeners under lateral patch loads

J. Abraham
Global Maritime, London, UK

C. Daley
Memorial University, St. John's, Canada

ABSTRACT: The stiffened plate is the main building block for ship structures. In most situations, stiffened panels will be sensitive to buckling under axial loads. In the case of ice reinforcement, the loads are primarily normal to the shell, with minimal axial loads. The concern for frame buckling remains, although the issue is less well understood. Some stiffeners, especially those with slender webs, show a tendency to fail by local web buckling, tripping and shear buckling, causing a sudden loss of capacity and resulting in collapse of the structure. The IACS Polar Shipping Rules (URI2) contains a requirement aimed at the prevention of web buckling by specifying a maximum web aspect ratio. While URI2 employs plastic limit states, the stability ratio is based on prevention of elastic buckling. In some cases these stability requirements have a significant impact on the design. The paper examines the buckling requirements for flat bar frames subject to local ice loads and concludes that the current rule values are very conservative and do not adequately reflect the conditions that lead to instability. The finite element method (FEM) coupled with the "design of experiments" (DOE) method is used in the study. Six factors which influence stability of a flat bar stiffener are identified—four geometric factors (span, frame spacing, thickness of shell plate and thickness of web) and two material properties (yield stress and post yield modulus). To study the effect of these six factors even at two levels (at high and low levels of each factor) requires 64 (2^6) possible combinations of factors to be considered. A significant reduction in the number of cases is achieved by employing DOE method. The paper describes the main factors affecting the plastic stability of a frame, which are quite different from the usual elastic buckling parameters. A new relationship is proposed for calculating the limiting web height and web slenderness.

1 INTRODUCTION

The total load carrying capacity of a frame is contributed by the shell plate and the stiffener. Some stiffener forms show a tendency to fail locally at large deformations, thereby resulting in a reduction in total capacity. The three major local failure mechanisms of a stiffener are local web buckling, tripping and shear buckling. These failures are critical since they can cause sudden drop in capacity thereby resulting in collapse of the structure.

Consider two frames as shown in Figure 1. The frames have two different web heights (300 mm and 600 mm respectively) but all other geometric and material properties are the same.

The response of these frames to a concentrated patch load is presented in Figures 2 and 3. The frame with the 300 mm high web shows stable load carrying characteristics even at large h/t ratios (i.e., small web thicknesses). The frame with 600 mm high web at large h/t ratios becomes unstable resulting in sudden drop in capacity.

The results show that tall and thin stiffeners tend to locally fail resulting in sudden drop in capacity

whereas short stiffeners tend to be stable even with small web thickness. This demonstrates that the slenderness ratio is not the only factor that affects stability of a stiffener. Other factors such as span, frame spacing

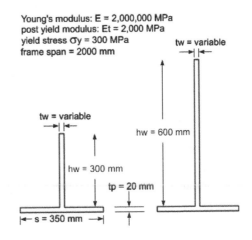

Figure 1. Two frames with different web heights.

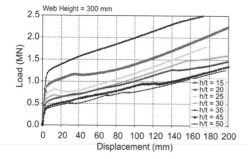

Figure 2.　300 mm web height frame (capacity vs. h/t ratio).

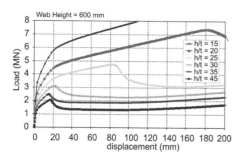

Figure 3.　600 mm web height frame (capacity vs. h/t ratio).

and thicknesses appear to affect stability in addition to the slenderness.

The current IACS UR-I2 Polar rules (and various rules of classification societies) try to prevent web instabilities by specifying a maximum slenderness ratio (hw/tw). These ratios are formulated so that the stiffener will not locally fail before yielding occurs. The current rule limiting hw/tw values depends only on the yield stress of the material. Current classification society rule limits of hw/tw for various stiffeners forms are presented in Table 1 The main factor limiting the wider use of flat bar stiffeners in ice class vessels is the h/t limit.

Non-linear analysis of the elasto-plastic response of flat bars to lateral loads led the authors to conclude that the limits given in Table 1 are conservative in the case of a flat bar stiffener for the type of loading considered. Flat bar stiffeners can exhibit stable response even when considerably more slender than allowed by the Table 1 limits.

Consequently, an investigation was carried out to check the possibility of increasing the rule h/t limits. In the absence of an analytical solution explaining the stability of a stiffener, combined use of finite element method (FEM) and design of experiment (DOE) is used in the study.

The aims of the study are:

• to identify the major factors that affects the stability of a flat bar stiffened plate

Table 1.　hw/tw limits.

	Flat Bar	Tee	Angle
IACS I2	$282/\sqrt{\sigma_y}$	$805/\sqrt{\sigma_y}$	$805/\sqrt{\sigma_y}$
ABS	$226/\sqrt{\sigma_y}$	$679/\sqrt{\sigma_y}$	

• to study the various interactions between main factors
• to find simple regression equations to predict the maximum allowable web height and web height to thickness (h/t) ratio with reasonable accuracy
• validate the regression equations for accuracy.

2 PARAMETER SPACE

Stiffened plate can occur in various combinations of geometric and material properties. The total number of factors which influence stability of a flat bar stiffener is identified as six-four geometric factors and two material properties. The ranges of each factor considered for the study are presented in Table 2 The range of each factor was chosen such that it covers the practical range of stiffener properties found in ice capable vessels.

To study these six factors even at two levels (at high and low values of each factor) results in 64 (2^6) possible combinations. Each combination requires an average of five trials to find the maximum allowable web height, making a total of 320 FEA analyses. Considering the time required for solving each FE analysis and post processing of results, it is clear that traditional methods are not efficient in studying these kinds of problems. By using the DOE method, a significant reduction in the number of FE analysis is achieved.

The number of experiments required to study the problem at a minimum run resolution V design is 22. The combinations of factors were generated using Design Expert™ software.

3 FINITE ELEMENT ANALYSIS

A single frame model is considered to be sufficient to study web instability. The study explored the stability of a flat bar stiffened plate subjected to lateral small patch loads as would arise with ice loading. Ice loads consist of localized high pressures surrounded by regions of lower pressure. For the study nodes located at the center of the frame within a small area (150 mm × 150 mm) were given a vertical prescribed displacement of 10% of the frame span. This simulates displacement of nodes due to a concentrated ice pressure during an ice-structure interaction. The loading pattern is shown in Figure 4.

The two longitudinal ends are fixed which idealizes the support provided by stringers to which the frames are welded to. The two transverse sides of the shell

Table 2. Parameter ranges for stability study.

	Factor	Low	High
1	Frame span Lf [mm]	2000	4000
2	Frame spacing S [mm]	300	600
3	Plate thickness tp [mm]	10	40
4	Yield Stress σy [MPa]	250	500
5	Post yield modulus Et [MPa]	0	2000
6	Web thickness/plate thickness	0.6	1.3

Table 3. Stability vs. web height.

Web height [mm]	Drop in capacity (%)	Remarks
300	4%	Stable
325	10%	Limiting web height
400	20%	Unstable

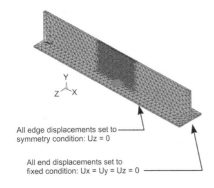

Figure 4. Loading considered for stability study.

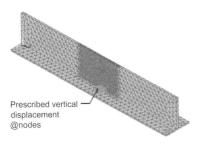

Figure 5. Boundary conditions.

plate are given symmetric boundary condition to simulate support provided by the adjacent structure. Boundary conditions considered for the analysis are presented in Figure 5.

A non-linear finite element analysis was carried out using ANSYS finite element software to find the force-deformation response of the structure.

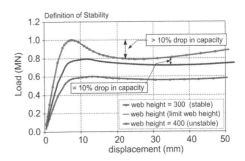

Figure 6. Definition of stability.

4 DEFINING STABILITY

The flat bar stiffener can sometimes accommodate very large deformations without sudden drop in capacity, even after the three-hinge formation. The drop in capacity due to stiffener failure is compensated by the growing membrane action of the plate and post yield modulus, thus making the frame stable. The present study is based on defining a frame as stable in situations where the capacity drop is less than or equal to 10%. Even though the stiffener at large deformations may have experienced local buckling, or various types of local folding and stretching, only when the drop in capacity is more than 10% is the frame considered to be unstable.

Figure 6 shows frame response as the web height is increased. All other geometric and material properties are kept constant. An increase in web height results in an increase in initial capacity of the frame due to addition of more material away from the neutral axis. However the behavior of the frame at large deformations becomes increasingly unstable. At some critical height the frame becomes unstable by the 10% criteria.

Based on the above results, the limiting web height for the frame is estimated as 325 mm. An increase in web height above 325 mm will result in frame becoming unstable, i.e., drop in capacity of more than 10%.

5 RESULTS OF STABILITY STUDY

5.1 Allowable web height

The current design rule limits the maximum web slenderness ratio based on material yield alone. An interesting conclusion of the study is that the maximum allowable web height of a frame is independent of material yield. The main factors affecting the allowable web height of a frame are the geometric parameters; web thickness and shell plate thickness.

A regression equation for estimating the maximum allowable web height for a given set of geometric and material properties has been generated using Design

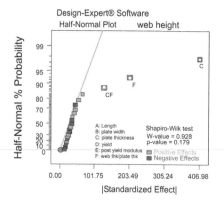

Figure 7. Half normal plot for web height.

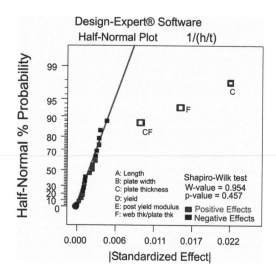

Figure 8. Half normal plot for h/t.

Expert software. The regression model has an adjusted R-squared of 0.90 and prediction R-squared of 0.86.

The maximum web height that can be achieved with a maximum allowable drop in capacity of 10% is given by:

$$\text{web height} = 81.44 + 0.033 * \text{Length} + 0.17$$
$$* platewidth + 1.59 * \text{plate thickness}$$
$$- 23.97 * \text{web thickness/plate thickness}$$
$$+ 12.38 * \text{web thickness} \quad (1)$$

5.2 *Allowable web height/web thickness ratio*

The current design rule limits the maximum allowable web height to thickness ratio based on material yield alone. An interesting result that was found from the study is that the web height to thickness ratio does not depend on yield. The main factors affecting the h/t ratio are the geometric parameters like web thickness and shell plate thickness.

A regression equation for estimating the maximum web height to thickness ratio for a given set of geometric and material properties has been found using Design Expert software. The regression model has an R-squared of 0.89 and prediction R-squared of 0.83.

The maximum web height to thickness ratio that can be achieved with a maximum allowable drop in capacity of 10% is given by:

$$1/(h/t) = 0.009 - 2.15E - 06 * \text{Length}$$
$$- 1.12E - 05 * \text{plate width}$$
$$+ 0.0007 * \text{plate thickness}$$
$$+ 0.021 * \text{web thickness/plate thickness}$$
$$(2)$$

6 VALIDATION OF REGRESSION MODELS

6.1 *Validation of web height model*

The regression model for predicting web height has been verified by independent finite element analyses. Results of validation runs are presented in Table 4.

6.2 *Validation of web slenderness model*

The regression model for predicting web height to web thickness ratio has been verified by independent finite element analyses. Results of validation runs are presented in Table 5.

The results of the validation analyses have shown that the new regression model is reasonably accurate in predicting web height. The predicted h/t ratio is much better than the current rule estimations.

7 DISCUSSION

The capacity of a frame depends on both material and geometric properties. Post yield web stability does not appear to depend on the same ratios that govern the elastic buckling mechanisms. Instead it depends on other geometric properties like span, frame spacing, plate thickness, web height and thicknesses. This is likely because post yield instabilities are the result of plastic folding mechanisms, rather than a re-direction of elastic deformation that controls elastic buckling. The current IACS I2 stability rule limits are only a function of yield stress, but the study has shown that post yield stability is independent of yield stress.

Table 4. Error percentages compared to FEA for web height estimation.

Ansys run	% error (DOE)	% error (IACS I2)
1	4%	−37%
2	−2%	−33%
3	−1%	−43%
4	−6%	−32%
5	−14%	−59%
6	−1%	−11%
7	−13%	−49%
8	1%	−11%
9	−9%	−72%
10	−10%	−32%
11	−11%	−38%
12	−12%	−43%
13	−8%	−48%
14	2%	6%
15	11%	9%
16	−6%	−24%

Table 5. Error percentages compared to FEA for web slenderness estimation.

Ansys run	% error (DOE)	% error (IACS)
1	17%	−37%
2	1%	−33%
3	12%	−43%
4	5%	−32%
5	−14%	−59%
6	2%	−11%
7	−5%	−49%
8	10%	−11%
9	−10%	−72%
10	−11%	−32%
11	−13%	−38%
12	−12%	−43%
13	−12%	−48%
14	2%	6%
15	11%	9%
16	7%	−24%

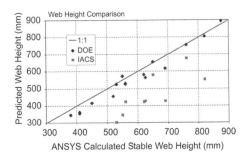

Figure 9. Web height—actual vs. predicted values.

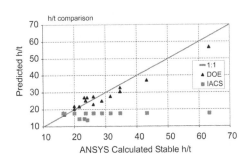

Figure 10. h/t ratio—actual vs. predicted values.

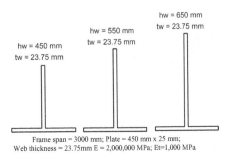

Frame span = 3000 mm; Plate = 450 mm x 25 mm; Web thickness = 23.75mm E = 2,000,000 MPa; Et=1,000 MPa

Figure 11. Three frames with different web heights.

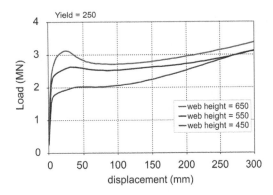

Figure 12. Capacity vs. web height (Yield = 250 MPa).

Consider three frames as shown in Figure 11. The frames have different web heights (450, 550 and 650 mm) and all other properties are kept constant.

The responses of the frames for three different yield values (250 MPa, 375 MPa and 500 MPa) are presented in Figures 12 to 14. It can be seen that web height of 450 mm is stable whereas web height of 650 mm is unstable. The frame becomes unstable at a limiting web height of around 550 mm. The stability pattern remains the same for the three different yield values. This demonstrates that yield is not a major factor affecting the stability of a frame.

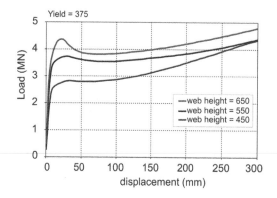

Figure 13.   Capacity vs. web height (Yield = 375 MPa).

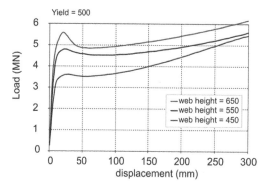

Figure 14.   Capacity vs. web height (Yield = 500 MPa).

8   CONCLUSION

A new relationship has been proposed for calculating limiting web height. The regression model depends only on geometric properties of the frame. The web height prediction model has been found to be in good agreement with the finite element results.

A new relation has also been proposed for estimating web slenderness ratio. The h/t ratio prediction model has been found to be in good agreement with the finite element results. The current IACS formulation can give very conservative limits on h/t ratio, thereby limiting the usage of flat bar stiffeners.

The DOE method has shown to be a useful tool in studying multi factored systems with manageable number of numerical simulations. The method is particularly useful in situations where a full analytical solution is difficult to achieve due to the complexities involved. The confidence level in the regression equation produced by DOE can be increased by validating using independent FEA, which has been the approach used in the present study.

ACKNOWLEDGEMENTS

Funding for the research program that enabled this analysis was provided by several sponsors, including the US Ship Structures Committee, Transport Canada, and the National Research Council of Canada (Institute for Ocean Technology). The first author would also like to gratefully acknowledge the support provided by Memorial University of Newfoundland while carrying out this work as part of his Masters studies. The problem tackled here arose from many interesting discussions within the IACS working group which developed the Polar Rules.

REFERENCES

Abraham, J., "Plastic Response of Ship Structure Subject to Ice Loading", Master Thesis, Faculty of Engineering and Applied Science, Memorial University of Newfoundland, September 2008.

Daley, C., Pavic, M., Hussein, A., Hermanski, G., Ship Frame Research Program—A Numerical Study of the Capacity of Single Frames Subject to Ice Load. OERC Report 2004-02.

Daley, C., Hermanski, G., Ship Frame/Grillage Research Program—Investigation of Finite Element Analysis Boundary Conditions. OERC Report 2005-02.

Daley, C., Kendrick, A., Direct design of Large Ice Class Ships with emphasis on the mid-body Ice Belt. OMAE2008-57846, 2008.

Daley, C., 2003, "Review of Tripping Requirements—IACS Unified Requirements for Polar Ships" Prepared for IACS Ad-hoc Group on Polar Class Ships and Transport Canada.

Daley, C.G., Kendrick, A., 2000, "Background Notes to Derivation and use of Formulations for Framing Design—IACS Unified Requirements for Polar Ships" Prepared for IACS Ad-hoc Group on Polar Class Ships and Transport Canada.

Design Expert version 7.1.3. Stat-Ease Inc.

Hughes, O.F., 1983, Ship Structural Design, Wiley-Interscience, Published by the Society of Naval Architects and Marine Engineers, New York, 1988.

International Association of Classification Societies, "I2 Structural requirements for Polar Ships" March 2006.

Paik, J.K., Thayamballi, A.K., Ultimate Limit State Design of Steel Plated Structure. Wiley, 2004.

Ultimate strength of stiffened plates with local damage on the stiffener

Malgorzata Witkowska & C. Guedes Soares
*Centre for Marine Technology and Engineering (CENTEC), Technical University of Lisbon, Instituto Superior
Técnico, Lisboa, Portugal*

ABSTRACT: The behaviour and ultimate strength of locally damaged stiffened plates is investigated. The damage is in a form of local imperfection and the affected element is the stiffener. Finite element analysis is carried out with several varying parameters like the location of the dent, its size as well as the number of them. Cases with one or two existing damages are analysed. Also several different global post-welding deflections are considered, including the initial deformation of the stiffener. It has been found that the local damage on the stiffener can change the collapse mode of the plate and decrease its ultimate strength. Reduction of strength can be significant, however it depends on the location of the dent as well as the initial global deflection. The behaviour of the plate with two damages depends mostly on only one of the locations which is more dominant.

1 INTRODUCTION

Ships are usually intended to be in use for around 20 years and various typical operations like e.g. loading and unloading the cargo can accidentally cause damages. Some of them can be severe and it is necessary to repair or even replace the damaged element. However, there can also appear minor damages which individually do not affect the overall strength but if enough number of them occurs the consequences can be significant.

For this reason it is important to understand the behaviour of locally damaged elements of the ship structure and determine the strength reduction characteristics.

The ultimate strength of plates has been investigated during decades bringing important insights. Faulkner (1975) has shown that the governing parameter of the collapse of a plate is the slenderness. However, the importance of the initial imperfections has been also widely documented. Kmiecik (1971) has shown that the strength of plates depends on the amplitude of the buckling mode component. Guedes Soares (1988) studied the effects of imperfections and generalized Faulkner's formula to account for these effects.

The random nature of the imperfections has been investigated by Guedes Soares & Faulkner (1987) who showed that it is not possible to characterize its influence on the collapse strength without capturing the variability of the results. The latter was obtained by Guedes Soares & Kmiecik (1993) by simulating the ultimate compressive strength of plates with random imperfections and analyzing the results statistically.

To predict the collapse strength of plates, analytical formulas may be used, taking properly into account the variability of the imperfections. An equation that incorporates imperfection effects and includes a proper safety margin to account for the influence of its variability has been derived by Guedes Soares (1992).

Dow & Smith (1984) compared the effects of global and local imperfections on the strength of plates. They considered different shapes and positions of the local imperfection. The consequence of combining both, global and local imperfections has also been studied, however without investigating the influence of changing the position of local imperfections while the global imperfection is already present. They concluded that the length and especially amplitude of the local imperfection, when alone, have the most impact. When added to a global one, local imperfections can significantly change the collapse strength.

Paik et al. (2003) carried out a series of non-linear finite element analyses of dented plates, varying several parameters of the plate and dent itself. It was found that the depth of the dent has little impact as long as the diameter is small. The increase of the diameter decreases the ultimate strength significantly and the depth can amplify this effect. It has also been concluded that the longitudinal position of the dent affects the strength. The plate properties, like aspect ratio and thickness, did not appear to be influential parameters to the normalized compressive strength of dented plates. The strength reduction factor has also been introduced to predict the strength of dented plates by multiplying it with the ultimate strength of intact (un-dented) plates.

Guedes Soares et al. (2005) studied the collapse strength of a single plate having two types of imperfections: global weld-induced and local damaged-induced. They generally confirmed previous conclusions that local imperfection, when added to global one, could cause severe reduction of the strength of the plate depending on its amplitude, length and position on the plate.

Using a single plate model to represent the strength of a larger panel implies a repetitive pattern of imperfections in adjacent plates. While for global post-welding imperfections such pattern can be justified, the damage-induced imperfection is localized and therefore its existence should be limited to one plate only.

Furthermore, applying simply supported boundary conditions on the single plate model simulates that imperfections are asymmetrical in the whole panel i.e. adjacent plates deflect in opposite directions. Nikolov & Andreev (2005) have studied plates having symmetrical and asymmetrical shape of imperfections and established that the highest and lowest strength is obtained for, respectively, symmetrical and asymmetrical imperfections when the amplitude of the deflection is equal in neighbouring plates. This strength-symmetry relation may however sometimes reverse, being dependant on the slenderness of the plate.

For these reasons, to analyse localized imperfection it is necessary to consider plate assemblies instead of a single plate model. Luis et al. (2006) considered a panel made of 3 plates joined transversally to the load direction showing that such model is equivalent to single rectangular plate model and allows studying isolated effects of the local damage. Luis & Guedes Soares (2006) and Luis et al. (2007) analysed assemblies made of 3 plates connected longitudinally. The model was validated and the results showed that even if the imperfection is isolated longitudinally it can affect significantly the collapse strength of the panel as well as the post-collapse behaviour.

This set of studies differ from the ones of Dow & Smith (1984) and of Paik et al (2003) because they consider panels of various plates, while in those two references only single plates were considered.

A panel made of 3 plates joined transversally and 3 plates longitudinally, has been studied by Luis et al. (2007). The damage-induced imperfection was located only in one of the nine plates of the panel. Parametric studies have been performed varying several properties of the panel as well as global and local imperfections. Different longitudinal and transversal positions of the damage have been considered. It has been found that the effect of local imperfection depends on the slenderness: increases from negligible for stocky panels, to significant for very slender ones. It may also induce steep unloading and violent collapse. It was also concluded that a local imperfection interact with the global one and the effect of the position of the local dent depends on the overall shape of the initial imperfections.

Stiffened panels with localized damage-induced imperfections on the plating were investigated by Witkowska & Guedes Soares (2008). It was shown

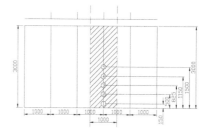

Figure 1.   Geometry of the panel.

that the stiffened panels are more resistant to the local damage than the unstiffened plates. It was also found that the presence of the local imperfection cause deformation of the stiffeners depending on the geometrical characteristics of the panel which can lead to higher reduction of the ultimate strength.

In present study the local damage-induced imperfections existing in the stiffener are investigated.

The model is made as single stiffener with associated plating being a part of a larger panel as shown by the shaded area in Figure 1. The figure shows the geometry and dimensions of the model as well as the considered locations of the damage. Longitudinal stiffeners used are flat bars with height $h_w = 0.25$ b and thickness equal to double plate thickness $t_w = 2$ t where $t = b/90$ giving slenderness of the plate $\beta = 3$.

2   MODELLING

2.1   Global post-welding initial imperfection

Several different global imperfection shapes were considered including Fourier series approximations with different number of half-waves and the elastic buckling mode shape.

The shape based on Fourier series follows the equation:

$$w = w_0 \sin \left(\frac{m\pi x}{a} \right) \sin \left(\frac{m\pi y}{b} \right) \qquad (1)$$

where m and n are the number of half-waves in, respectively, longitudinal and transverse direction, a and b are the length and breadth of the plate and w_0 is the amplitude equal to $w_0 = 0.1\beta^2 t$.

The value of n is constant for all models $n = 1$ while m changes from 1 to 4.

The stiffener imperfections are also considered. The imperfection type applied is the side-ways initial deflection of stiffeners due to angular rotation about panel-stiffener intersection line:

$$\phi_0 = \frac{a}{1000\,h_w} \sin \frac{m\pi x}{a} \qquad (2)$$

The graphical explanation of this type of stiffener imperfection is shown in Figure 2.

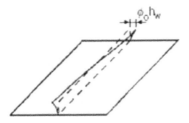

Figure 2.   Side-ways initial deflection of the stiffener.

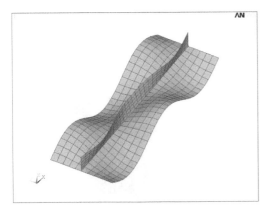

Figure 3.   Initial global imperfection—example.

The elastic buckling mode shape is derived from an eigenvalue analysis. After the mode is obtained the deflection pattern is scaled to reach the amplitude of the imperfection equal to $w_0 = 0.1\beta^2 t$ and superimposed onto the model, including the imperfections of the stiffeners. Once the initial deflection has been applied, the nonlinear analyses were carried out.

2.2   Local damaged-induced imperfection

Local imperfection, representing the stiffener damage, is expressed by:

$$w_l = w_{l0} \sin^2 \left(\frac{\pi x}{l_w} \right) \sin^2 \left(\frac{\pi z}{d} \right) \qquad (3)$$

where l_w indicates the length of damage and d the vertical extend of it $d = h_w$, Figure 4. The value of local imperfection amplitude is $w_{l0} = 0.1\,\beta^2 t$. The local imperfection is applied in a way to always increase the existing deflection of the global imperfection.

2.3   Boundary conditions

Boundary conditions are assumed simply supported. The load is applied as an imposed displacement on one of the transverse edges while the opposite one is restrained. The longitudinal edges have symmetry

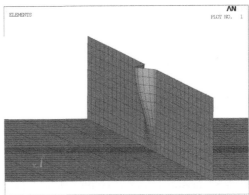

Figure 4.   Local initial imperfection of the stiffener; (a) $d = h_w$.

boundary conditions. The ultimate strength is derived from the reaction forces on the restrained edge.

3   RESULTS

3.1   One type of imperfections

The analyses were made in 3 stages. First, models having only the global imperfections were calculated. The second stage consists of models having local imperfection but no global imperfection. In the last stage the global and local imperfections were applied simultaneously.

In Figure 5 the stress-strain curves of the models with only global imperfection are shown and in Figure 6 the deflection shape at collapse for case $m = 3$. As can be seen the panels generally collapse with the same mode as the imposed imperfection.

To isolate the effects of local damage on the behaviour, the panels with only local imperfection are analysed. In Figure 7 the stress-strain curves for those models are presented. As can be seen the behaviour of all 5 cases is similar and the difference between particular locations lies only in the value of the ultimate strength. The rigidity of all models is the same and the tendency of the curves in the post-buckling region is also comparable.

In Figure 8 the progress of the deflection is shown for a panel with damage located in position number 2. As can be seen in the initial phase the deflection shape is similar to the natural buckling mode. However, with further application of load the deflection passes from having two half-waves to three half-waves at collapse showing that the presence of the damage influences the collapse mode. In the last figures of the panels after the collapse it is clearly seen that the stresses are concentrated around the local imperfection and the

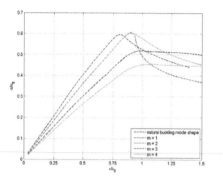

Figure 5. Stress-strain curves—models with only global imperfection.

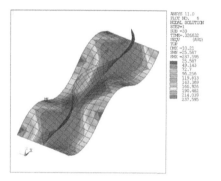

Figure 6. Von Misses stress distribution and deflection shape at collapse; $m = 3$.

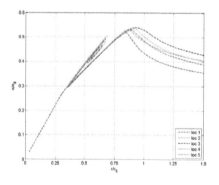

Figure 7. Stress-strain curves—models with only local imperfection.

deflection in this area is much bigger than in the rest of the panel.

3.2 Two types of imperfections

In figures below the stress-strain curves for selected imperfection cases are shown for the panels having

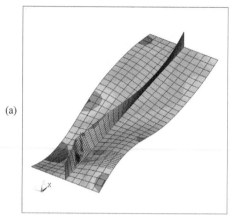

(a)

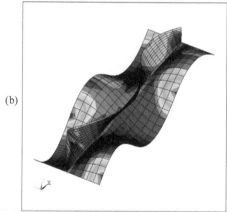

(b)

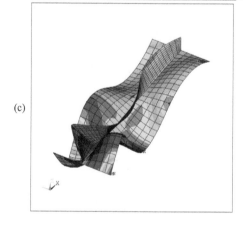

(c)

Figure 8. Progress of the deflection—models with only local imperfection; (a) before collapse, (b) at collapse, (c) after collapse.

both—global and local imperfections. In Figure 9 for the global imperfection with the shape of the natural buckling mode it can be seen that for damage locations 1 to 3 only the value of the ultimate strength is slightly reduced but the behaviour of the panel generally did

148

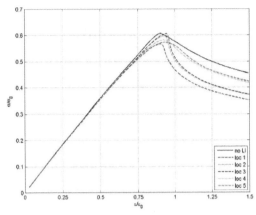

Figure 9. Stress-strain curves—both types of imperfections; natural buckling mode shape.

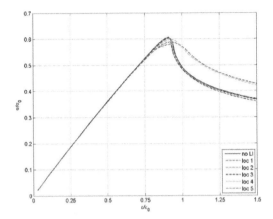

Figure 11. Stress-strain curves—both types of imperfections; $m = 2$.

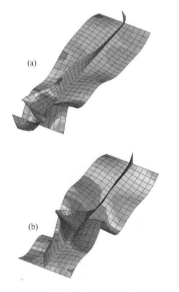

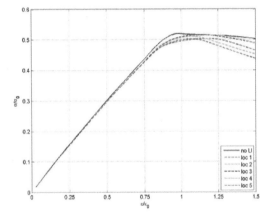

Figure 10. Von Misses stress distribution and deflection shape at post-collapse, global imperfection natural buckling mode shape; (a) no LI, (b) LI position 4.

Figure 12. Stress-strain curves—both types of imperfections; $m = 3$.

not change. However, for locations 4 and 5 the curves stand out. In Figure 10 the deflection after the collapse is presented showing the change in the collapse mode of the mentioned panel which is probably the reason of different behaviour.

Figure 11 and Figure 12 illustrate the stress-strain curves of the models with imperfection of, respectively, $m = 2$ and $m = 3$. As was shown before, the panels with no imposed global imperfection tend to start deflection in two half-waves mode but collapse with three half-waves pattern. It can be observed

that the panels initially deflected with $m = 3$ are much less affected by the presence of the damage since this is the mode that the damage is causing. For $m = 2$, depending on the location of the damage, a significant strength reduction and change in behaviour can be seen as well as some reduction of the rigidity. In

Figure 13 the deflection shape after the collapse is shown for global imperfection with $m = 2$. It can be seen again the presence of the local imperfection causes collapse mode to change what in consequence decreases the ultimate strength of the panel.

In Figure 14 the summary plot of the ultimate strength is shown for all considered cases. On the x-axis are the positions of the damage with position "0" meaning the no LI case i.e. the model with only global imperfection.

149

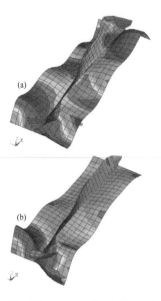

(a)

(b)

Figure 13. Von Misses stress distribution and deflection shape at post-collapse, global imperfection $m = 2$; (a) no LI, (b) LI position 1.

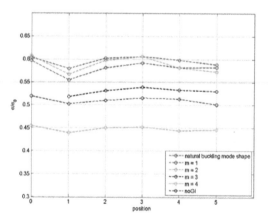

Figure 14. Summary plot of ultimate strength.

It can be seen that the lines are almost parallel with the highest strength obtained for panels with global imperfections in the shape of the natural buckling mode or $m = 2$ and the lowest results for the highest mode $m = 4$. Interesting fact is that the panels with only the local imperfection lie in between, above the $m = 3$ and below the lower modes.

The mode imposed by the presence of the damage was 3 therefore it can be said that adding the global imperfection with lower mode to the already existing local imperfection actually has a stiffening effect. However, this does not mean that the panel

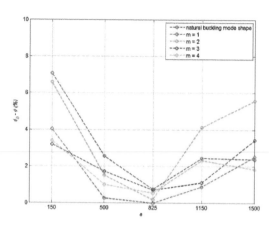

Figure 15. Strength reduction due to the presence of the damage.

with low number of half-waves are less affected by the occurrence of the damage.

In Figure 15 the strength reduction is plotted expressed as:

$$\frac{\phi_{u0} - \phi_u}{\phi_{u0}} \times 100\% \qquad (4)$$

It is seen that the highest reduction takes place for $m = 1$ and 2 reaching up to 7% while for $m = 3$ and 4 it oscillates around 2%. The highest average reduction is 3.60% for $m = 2$ and the lowest is 1.82% for $m = 4$. Though, taking into consideration the panel with imperfection shape of the natural buckling mode it can be seen it is actually least affected by the damage. The average reduction is equal to 1.57%. As it is known now that the strength reduction due to the damage depends on the initial deflection, the above plots are separated and presented with comparison to the stiffener initial deflection shape in Figure 16.

The highest reduction is observed in position 1 independently of the shape of imperfection. This might be due to the closeness of the edge. For lower modes of global imperfection the biggest reduction is at the ends of the half-wave (position 1 for $m = 1$ and positions 1 and 5 for $m = 2$) while for higher modes some increase of the reduction takes place for locations at peaks of the stiffener initial deflection i.e. position 5 for $m = 3$ and positions 4 for $m = 4$.

Some additional calculations were made with the localized imperfection amplitude increased to $w_{l0} = 0.5\beta^2 t$. In Figure 17 the stress-strain curves with increased amplitude are shown and can be compared with curves for smaller amplitude from Figure 12. It can be seen that increasing the lateral depth of the dent decreases the strength of the panel. Though it does not change the general behaviour of the plate—the

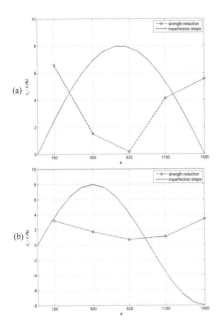

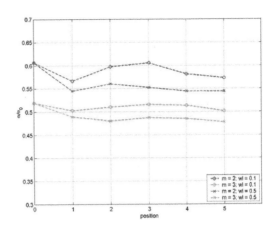

Figure 18. Summary plot of ultimate strength.

Table 1. Double damage cases.

Case number	Position of first LI	Position of second LI
1	1	2
2	1	3
3	1	4
4	1	5
5	2	3
6	2	4
7	2	5
8	3	4
9	3	5
10	4	5

Figure 16. Strength reduction; (a) $m = 2$, (b) $m = 3$.

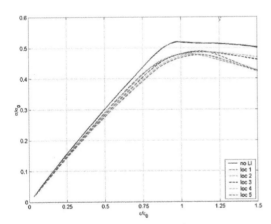

Figure 17. Stress-strain curves; global imperfection $m = 3$, $w_l = 0.5\,\beta^2 t$.

effects are only amplified in case of $w_{l0} = 0.5\beta^2 t$ in comparison to $w_{l0} = 0.1\beta^2 t$.

To analyse particular cases more carefully the summary plot of ultimate strength is presented in Figure 18. It is clear that the plates with bigger dent have lower strength with around 2% up to 5–6% difference comparing to smaller damage. However, the pattern of the lines seems to be very similar meaning that effects of other parameters like location of the damage or global

imperfection are not dependant on the amplitude of the localized imperfection.

3.3 Plates with two damages

The results above showed that the existence of a damage can significantly change the behaviour of plate and decrease its strength. Because of this it seems reasonable to verify whether adding one more damage to the one already existing can also have such an impact.

Ten cases of double damage were considered, done as an every possible combination of the 5 damage locations considered at the beginning (see Figure 1). They are gathered in Table 1.

An example of such plate is shown in Figure 19 for case number 6 with local imperfections in positions 2 and 4.

In Figure 20 a summary plot of ultimate strength of plates with two localized imperfections is presented. The case number "0" represents the plate with no damage at all and it can be seen that the strength of damaged plates is generally decreased.

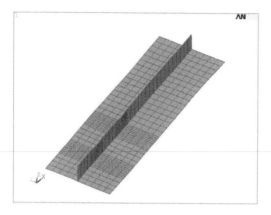

Figure 19.   Example of plate with two local imperfections.

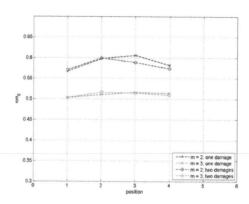

Figure 21.   Comparison of ultimate strength of single and double damage plates.

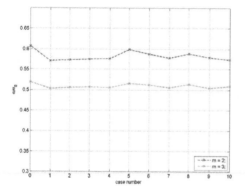

Figure 20.   Summary plot of ultimate strength of plates with two damages.

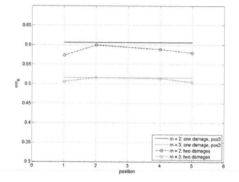

Figure 22.   Comparison of ultimate strength of single and double damage plates—position 3.

However, despite the expected that two existing damages reduce strength more that just one damage, some of the cases show the value of strength is only slightly lower then the intact plate. This seems to depend on the case considered, or in other words, on the location of at least one of the damages. For this reason the results should be analysed separately and in comparison to the plate with a single damage in particular location.

In Figure 21 the ultimate strength of plates with one and with two damages is compared. The lines marked with crosses represent the single damage plates and position 1 to 4. The lines marked with circles represent the double damage cases where the second damage was put next to the first one i.e. "1" on the x-axis signifies position 1 for single damage and position $1 + 2$ (case 1) for double damage, "2" signifies position $2 + 3$ (case 5) etc.

As can be seen, the existence of the second damage next to the first one generally decreases the resistance

of the plate, although in some cases the decrease is very small.

It was observed before that the locations that lower most the strength are positions 1 and 5 while position 3 caused the smallest strength reduction among all cases. As so, it seems worth to look at the results of plates with combination of those locations.

The comparison between effects of existence of one damage in particular position and adding second damage in various other positions is made in Figure 22 and Figure 23 for single damage, respectively, in position 3 and 4.

The continuous line represents the level of the strength of the plate with only one damage while the dashed line the plate with two damages. Each double damage case is marked with circle in a manner that the position on the x-axis signifies the position of the second damage e.g. $x = 4$ means case 8 with damages in position $3 + 4$.

Position 3 was the one least affecting the strength and Figure 22 shows that adding a second imperfection

152

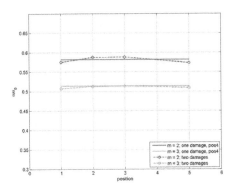

Figure 23.   Comparison of ultimate strength of single and double damage plates—position 4.

in any other place decreases the strength even further, specially adding it in the "weakest" locations i.e. 1 and 5. Similar situation occurs for plates with primary damage in position 4—cases 4 + 1 and 4 + 5 have reduced strength in comparison to single damage plate. However, adding damage in the "strongest" positions like 2 and 3 gives the results higher than just single damage in position 4.

It was shown before that the local imperfections can not only influence but also change the collapse mode of the plate depending on its location. This means that some locations are more dominant. Thus, it can be concluded that the collapse of plate follows the pattern induced by those locations and the existence of the other damage in less dominant position only decreases the value of strength.

4   CONCLUSIONS

Non-linear finite element analyses of plates with locally damaged stiffener were carried out. Different damage locations and the number of damages in one stiffener were considered as well as various global initial post-welding imperfections.

In general, it can be concluded that the existence of a local imperfection on the stiffener decreases the strength of the stiffened plate and the magnitude of this effect depends mostly on the location of the damage in relation to the global deflection. Local dent can also change the collapse mode usually increasing the number of half-waves in the initial deflection pattern and in consequence decrease the strength even further.

The reduction of strength can be significant (up to 7%) and it was determined that some locations reduce resistance of the plate regardless of the other parameters like e.g. position close to the edge of the plate. It was also shown that increasing the depth of the damage does not influence the behaviour, only amplifies

the effects caused by the occurrence of the damage itself.

Plates with two damages existing simultaneously in one stiffener were also investigated. It was established that placing the second damage next to the first one decreases the strength but does not change the general behaviour. It was also established that some of the locations can be more dominant than others. If there are two damages existing, the collapse of the plate follows the pattern induced by the damage in more dominant position and the second damage only decreases the value of strength.

ACKNOWLEDGMENTS

This paper has been prepared within the project "MARSTRUCT—Network of Excellence on Marine Structures", (www.mar.ist.utl.pt/marstruct/), which has been funded by the European Union through the Growth program under contract TNE3-CT-2003-506141.

REFERENCES

Dow, R.S. & Smith, C.S., 1984, "Effects of localized imperfections on compressive strength of long rectangular plates" *J. Construct. Steel Research*, 4: 51–76.
Faulkner, D., 1975, "A review of effective plating for use in the analysis of stiffened plating in bending and compression" *J. Ship Research*, 19: 1–17.
Guedes Soares, C., 1988, "A Code Requirement for the Compressive Strength of Plate Elements" *Marine Structures*, 1:71–80.
Guedes Soares, C., 1992, "Design Equation for Ship Plate Elements under Uniaxial Compression" *Journal of Constructional Steel Research*, 22: 99–114.
Guedes Soares, C. & Faulkner, D., 1987, "Probabilistic Modeling of the Effect of Initial Imperfections on the Compressive Strength of Rectangular Plates" *Proceedings of the 3rd International Symposium on Practical Design of Ships and Mobile Units (PRADS), Trondheim*, 783–795.
Guedes Soares, C. & Kmiecik, M., 1993, "Simulation of the ultimate compressive strength of unstiffened Plates" *Marine Structures*, 6: 553–569.
Guedes Soares, C., Teixeira, A.P., Luís, R.M., Quesnel, T., Nikolov, P. I., Steen, E., Khan, I.A., Toderan, C., Olaru, V.D., Bollero, A. & Taczala, M., 2005, "Effect of the shape of localized imperfections on the collapse strength of plates" *Maritime Transportation and Exploitation of Ocean and Coastal Resources*, C. Guedes Soares, Y. Garbatov and N. Fonseca (Eds.), Taylor & Francis Group, London, UK, 429–437.
Kmiecik, M., 1971, "Behaviour of axially loaded simply supported long rectangular plates having initial deformations" *Ship Research Institute*, Report No. 84, Trondheim.
Luís, R.M. & Guedes Soares, C., 2006, "Ultimate strength of plate assemblies with localized imperfection subjected to compressive Loads", *Proceedings of the III European Conference on Computational Mechanics Solids,*

Structures and Coupled Problems in Engineering, C.A. Mota Soares et al. (Eds.), Lisbon, Portugal.

Luís, R.M., Guedes Soares, C. & Nikolov, P.I., 2007, "Collapse strength of longitudinal plate assemblies with dimple imperfections" *Advancements in Marine Structures*, C. Guedes Soares and P.K. Das (Eds.), Taylor & Francis Group, London, UK, 207–215.

Luís, R.M., Witkowska, M. & Guedes Soares, C., 2006, "Ultimate strength of transverse plate assemblies under uniaxial loads" *Proceedings of the 25th International Conference on Offshore Mechanics and Arctic Engineering (OMAE 2006)*, ASME, Paper OMAE06-92664.

Luís, R.M., Witkowska, M. & Guedes Soares, C., 2007, "Collapse behaviour of damaged panels with a dimple imperfection" *Proceedings of the 26thInternational Conference on Offshore Mechanics and Arctic Engineering (OMAE 2007)*, ASME, Paper OMAE07-29777.

Nikolov, P.I. & Andreev, A.K., 2005, "Idealization of the plating complex initial deflections" *Maritime Transportation and Exploitation of Ocean and Coastal Resources*, C. Guedes Soares, Y. Garbatov and N. Fonseca (Eds.), Taylor & Francis Group, London, UK, 487–495.

Paik, J.K., Lee, J.M. & Lee, D.H., 2003, "Ultimate strength of dented steel plates under axial compressive loads" *International Journal of Mechanical Sciences* 45: 433–448.

Ueda, Y. & Tall, L., 1967, "Inelastic bukling of plates with residual stresses" *Publications Int. Assoc. for Bridge and Structural Engineering*

Witkowska, M. & Guedes Soares, C., 2008, "Collapse Strength of Stiffened Panels with Local Dent Damage" *Proceedings of the 27th International Conference on Offshore Mechanics and Arctic Engineering (OMAE 2008)*, ASME, Paper OMAE08-57950.

Analysis and Design of Marine Structures – Guedes Soares & Das (eds)
© *2009 Taylor & Francis Group, London, ISBN 978-0-415-54934-9*

Approximate method for evaluation of stress-strain relationship for stiffened panel subject to tension, compression and shear employing the finite element approach

M. Taczala
Szczecin University of Technology, Poland

ABSTRACT: A key aspect part of the evaluation of the ship ultimate capacity employing the approximate approach is the correct formulation of the stress-strain relationships of structural elements—longitudinally and transversally stiffened panels—subject to various loadings acting on their edges. Tensile, compressive and shear loading of the panel edge is an effect of the distribution of the stresses due to the internal forces acting in the ship hull—vertical and horizontal bending moments, vertical and horizontal shear forces, torsional moment and bimoment. Evaluation of the stress-strain relationships in terms of the finite element method calls for relatively large models which combined with the number of panels forming the hull structure in the investigated cross-section of a ship hull and necessity of the non-linear analysis require significant computational effort, as all the are subject to analysis. A model is presented in the paper for evaluation of the stress-strain relationship for stiffened panel subject to tension, compression and shear within the frame of the finite element. A model is an extension of the finite element for the analysis of stiffened plates into the non-linear range. The basic assumption of the formulation is that the stiffeners are not explicitly modeled. Their deformations are defined employing the displacement functions (shape functions in terms of the finite element method) of the plate, assuming certain dependency between displacement of plate and stiffeners. Presented are also the results of analysis for a typical ship panel.

1   INTRODUCTION

A key aspect part of the evaluation of the ship ultimate capacity employing the approximate approach is the correct formulation of the stress-strain relationships of structural elements—longitudinally and transversally stiffened panels—subject to various loadings acting on their edges. Tensile, compressive and shear loading of the panel edge is an effect of the distribution of the stresses due to the internal forces acting in the ship hull—vertical and horizontal bending moments, vertical and horizontal shear forces, torsional moment and bimoment.

Evaluation of the stress-strain relationships in terms of the finite element method calls for relatively large models which combined with the number of panels forming the hull structure in the investigated cross-section of a ship hull and necessity of the non-linear analysis require significant computational effort, as all they are subject to analysis. A solution to this problem can be a concept of approximate methods allowing for performing the analysis with sufficient accuracy at acceptable cost. One of the first methods was the one suggested by Gordo and Guedes Soares (1993). The method was developed for the axially compressed panels; it is based on the concept of the generalization of the equations defining the ultimate capacity making the involved terms dependant on the actual loading level and employing the Johnson-Ostenfeld formula for the elastic-plastic buckling. The method has been effectively applied in computation of the ship hull ultimate capacity.

A model will be presented in the paper for evaluation of the stress-strain relationship for stiffened panel subject to tension, compression and shear within the frame of the finite element. A model is an extension of the finite element for the analysis of stiffened plates into the non-linear range. The basic assumption of the formulation is that the stiffeners are not explicitly modeled. Their deformations are defined employing the displacement functions (shape functions in terms of the finite element method) of the plate, assuming certain dependency between displacement of plate and stiffeners. In the approach the position of stiffeners need not coincide with the edge of an finite element and arbitrary number of stiffeners is allowed within the element. The parameters of the problem are the degrees of freedom of the plate element therefore the element in incapable of handling the local buckling of stiffeners. A finite element is formulated being the combination of the eight-noded plane stress and four-noded nonconforming plate elements extended by the part referring to stiffeners. The stiffness matrix is derived using the principle of virtual work.

Results illustrating the efficiency of the proposed formulation will be presented during the conference.

2 APPROXIMATE FINITE ELEMENT FORMULATION

Formulating a model within the frame of the finite element method it is assumed that the stiffened panel is composed of the rectangular plates with longitudinal stiffeners and is delimited by longitudinal and transverse girders supporting plates and stiffeners (Figure 1).

It is a typical configuration of ship hull structural regions contributing to the longitudinal strength: bottom, sides and decks. The stiffener can be an arbitrary profile used in shipbuilding. The presented model comprises basic buckling modes: plate buckling, flexural buckling and flexural-torsional buckling. The panel is subject to compression in longitudinal and transverse directions and shear.

Various computational models have been used in the analysis of stiffened plates (Hughes, 1988):

- a model of orthotropic plate,
- a model, where stiffeners are idealized using eccentric beam elements,
- a hybrid model,
- a model, where stiffener webs are idealized using plate elements while flanges—beam elements,
- a model, where both webs and flanges are idealized using beam elements,
- a model composed of stiffened plate elements using displacement functions common for both plate and stiffener(s).

A concept of the stiffened plate finite element will be introduced suitable for geometrically and physically nonlinear analysis. A finite element in the isoparametric formulation for stiffened plates in plane stress was presented by Mukhopadhyay (1981). The formulation was later applied in the analysis of plates in bending employing Mindlin-Reissner theory by Mukhopadhyay and Satsangi (1984) and Mukhopadhyay et al. (1990) developing quadrilateral isoparametric plate element. The formulation was later generalized allowing for arbitrary position of stiffener(s) inside element (Satsangi, 1984, Satsangi and Ray, 1998). Sinha et al. (1992) developed an element of the stiffened

shell. Thompson et al. (1988) presented 8-noded element including arbitrarily oriented eccentric stiffeners. A model of plates subject to bending based on the Mindlin-Reissner theory, where positions of stiffeners is independent of the mesh was presented by Deb and Booton (1988). Plate and beam elements for nonlinear analysis of stiffened plates were developed by Koko and Olson (1991). Stiffened shells were analysed by Samanta and Mukhopadhyay (1998) employing the triangular Allman and plate DKT (discrete Kirchhoff triangle) elements.

The presented approach is based on the concept of the stiffened plate finite element employing Kirchhoff-Love thin plate theory (Taczala, 1998, Taczala and Banasiak, 2001). According to the essential assumption of the approach, stiffeners are not explicitly modeled, but their deformations are described employing displacement functions (shape functions in terms of the finite element method) of plate. It is necessary to establish dependence between deformations of the plate and stiffeners therefore equality of their displacements and rotations with respect to the axis being an intersection of the midsurfaces of the plate and stiffener web. The stiffeners do not need to coincide with an element edge therefore the element mesh is not related to the position of the stiffeners. Arbitrary number of stiffeners can be attached to a single element. The idea of construction of the computational model is explained in Figure 2 where a geometrical model of a panel is presented composed of the plate elements while in Figures 3 and 4 corresponding models are presented composed of the stiffened plate elements with different meshes.

Translational and rotational degrees of freedom of the plate element allow to include tension, compression and bending of a plate and tension, bending and torsion of a stiffener in the rigid web mode. Local deformations of the stiffener web are excluded therefore local buckling mode is not covered.

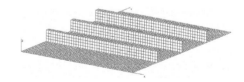

Figure 2. FE model of stiffened panel.

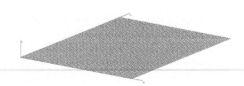

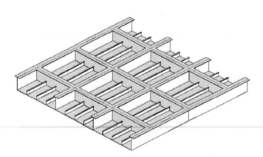

Figure 1. Typical ship structure.

Figure 3. FE model of stiffened panel—large number of elements.

Figure 4. FE model of stiffened panel—decreased number of elements.

Assumptions of the present computational model are following:

– elastic—ideally plastic material model is assumed,
– plate deformations including large deflections are governed by the von Karman plate theory,
– plane stress in the plate and uniaxial stress in the stiffener are assumed,
– torsion of stiffeners is not accounted for.

Formulation of the problem within the finite element approach yields the linear system of algebraic equations which coefficients constitute the stiffness matrix. The derived element is a combination of two elements commonly applied in FE: membrane or plane stress element and plate element. In the present analysis an element will be applied being a combination of eight-noded rectangular plane stress element and four-noded incompatible plate element (Zienkiewicz, 1977), which was effectively applied for non-linear analysis of plates and panels (Taczala, 1995). In the present formulation the element is extended to include the description of a stiffener: tension/compression and bending with respect to horizontal and vertical axes. Torsion of stiffeners is excluded due to small torsional rigidity of the open-type cross-sections therefore the strain energy related to torsion contributes to a few percent of the total strain energy and the model without torsion is considerably simpler.

Incremental form the equilibrium equation is derived beginning with the principle of virtual work. The work of the internal forces in increment ($i+1$) considering residual stresses is given by

$$\delta^{(i+1)} W_{int} = \int_{V^{(p)}} [(^{(0)}S_i^{(p)} + {}^{(i+1)}S_i^{(p)})\delta^{(i+1)}E_i^{(p)}]dV$$

$$+ \sum_{k=1}^{k=n_S} \left[\int_{V^{(w_k)}} (^{(0)}S_i^{(w_k)} + {}^{(i+1)}S_i^{(w_k)}) \right.$$

$$\times \delta^{(i+1)}E_i^{(w_k)}dV + \int_{V^{(f_k)}} \quad (1)$$

$$\left. \times (^{(0)}S_i^{(f_k)} + {}^{(i+1)}S_i^{(f_k)}) \, \delta^{(i+1)}E_i^{(f_k)}d \right]$$

where the stress vector for the plate has three components $S = \{S_{xx}\, S_{yy}\, S_{xy}\}^T$, while for the stiffener only one $S = \{S_{xy}\}$ what corresponds to uniaxial stress. Strain vectors have similar forms: for the plate $E = \{E_{xx}\, E_{yy}\, E_{xy}\}^T$ and for the stiffener $E = \{E_{xx}\}$ Formally E_{xy} is a doubled component of the strain tensor. $^{(0)}S$ is the residual stress vector, $^{(i+1)}S$—vector of the stresses induced by the loading in step ($i+1$), and $^{(i+1)}E$—corresponding strain vector. Stress tensors are second Piola-Kirchhoff tensors, and strain tensors—Green strain tensors. Equations are written in the stationary Lagrange formulation. Constitutive equations are formulated separately for the plate and stiffeners according to the assumed stress fields. For the plastic range they are derived employing the plastic flow theory.

The work of the external forces—transverse loading (pressure) p acting on the plate surface is given by

$$\delta^{(i+1)} W_{ext}^{(p)} = \int_{A^{(p)}} {}^{(i+1)}p\delta^{(i+1)}w^{(p)}dA \quad (2)$$

Displacement of the plate and stiffeners are formulated adopting the Kirchhoff hypothesis, therefore:

– for the plate

$$u^{(p)} = u - zw_{,x}$$
$$v^{(p)} = v - zw_{,y} \quad (3)$$
$$w^{(p)} = w$$

– for the stiffener web

$$u^{(w_k)} = u - zw_{,x}^{(k)} + yzw_{,xy}^{(k)}$$
$$v^{(w_k)} = -zw_{,y}^{(k)} \quad (4)$$
$$w^{(w_k)} = w^{(k)}$$

– for the stiffener flange

$$u^{(f_k)} = u - z(w_{,x}^{(k)} + yw_{,xy}^{(k)}) + (y - y_0^{(f_k)})z_0^{(f_k)}w_{,xy}^{(k)}$$
$$v^{(f_k)} = -z_0^{(f_k)}w_{,y}^{(k)} \quad (5)$$
$$w^{(f_k)} = w^{(k)} + yw_{,y}^{(k)}$$

where u, v, w are displacements with respect to x, y, z axis, respectively, y_0 and z_0 are the positions of the neutral axes in horizontal and vertical bending, respectively.

Applying the approximation of displacements typical for the finite element method, displacements u, v i

157

w can be given via shape functions applied in the plane stress and plate elements

$$u = N_j^m u_j, \quad v = N_j^m v_j, \quad w = N_j^b w_j \tag{6}$$

where $\mathbf{u}$, $\mathbf{v}$, $\mathbf{w}$ are vectors of the nodal displacements $\mathbf{u} = \mathrm{col}\{u_1 \; u_2 \; \ldots \; u_n\}$, $\mathbf{v} = \mathrm{col}\{v_1 \; v_2 \; \ldots \; v_n\}$, $\mathbf{w} = \mathrm{col}\{w_1 \; \theta_{x1} \; \theta_{y1} \; \ldots \; \theta_{yn}\}$ or in the following form

$$u = N_{1j} d_j, \quad v = N_{2j} d_j, \quad w = N_{3j} d_j \tag{7}$$

where $\mathbf{d}$ is a vector of all nodal displacements $\mathbf{d} = \mathrm{col}\{u_1 \, v_1 \, w_1 \, \theta_{x1} \, \theta_{y1} \, \ldots \, \theta_{yn}\}$ and $\mathbf{N}$ is a shape function matrix

$$\mathbf{N} = \begin{bmatrix} N_1^m & 0 & N_2^m & 0 & \ldots & 0 \\ 0 & N_1^m & 0 & N_2^m & \ldots & 0 \\ 0 & 0 & N_1^b & N_2^b & \ldots & N_{3xn}^b \end{bmatrix} \tag{8}$$

Applying the definition of displacements displacements—Eqs 3–5, displacement approximation—Eqs 6 and 7, evaluating strains as derivatives of displacements e.g. for the stiffener web

$$
\begin{aligned}
{}^{(i+1)}\Delta E_{xx}^{(w_k)} = & \Big\{ N_{1i,x} - z N_{3i,xx} + yz N_{3i,xxy} \\
& + \big[z^2 N_{3i,xy} N_{3j,xy} + N_{3i,x} N_{3j,x} \big] \big({}^{(0)}d_j + {}^{(i)} d_j \big) {}^{(i+1)}\Delta d_i \\
& + \Big[\tfrac{1}{2} N_{1i,x} N_{1j,x} + \tfrac{1}{2} z^2 N_{3i,xx} N_{3j,xx} \\
& + \tfrac{1}{2} y^2 z^2 N_{3i,xxy} N_{3j,xxy} - z N_{1i,x} N_{3j,xx} \\
& + yz N_{1i,x} N_{3j,xxy} - yz^2 N_{3i,xx} N_{3j,xxy} \\
& + \tfrac{1}{2} z^2 N_{3i,xy} N_{3j,xy} + \tfrac{1}{2} N_{3i,x} N_{3j,x} \Big]^{(i+1)} \\
& \times \Delta d_j^{(i+1)} \Delta d_i^{(i+1)} \Delta E_{xy}^{(w_k)} \\
= & N_{3i,xy} (y - y_0^{(w_l)}) {}^{(i+1)}\Delta d_i
\end{aligned}
\tag{9}
$$

to use in expression for virtual works—Eqs 1 and 2 we arrive at the equation typical for non-linear formulation in the finite element method

$$ {}^{(i)}R_p + \big({}^{(i)}K'_{pm} + {}^{(i)} K''_{pm} \big)^{(i+1)} \Delta d_m = {}^{(i+1)} P_p \tag{10}$$

where ${}^{(i)}\mathbf{K}'$ is the initial diplacement stiffness matrix, ${}^{(i)}\mathbf{K}''$—initial stress matrix, ${}^{(i)}\mathbf{R}$—internal force vector and ${}^{(i+1)}\mathbf{P}$—loading vector.

3 RESULTS OF ANALYSIS

The present approach was verified for several examples of panels subject to compression and shear. Panels having length $a = 2400$ mm, breadth b $= 3200$ mm and thickness $t_p = 10$ mm are stiffened with three equally spaced stiffeners: flat-bars 150×10, angles $200 \times 100 \times 10$ and bulbs 200×12. Young modulus was taken 210000 MPa, Poisson coefficient 0.3, and yield stress 235 MPa. Elastic-ideally plastic material model was applied.

In the numerical analysis boundary conditions of blocking translational displacements for edges of plates and stiffeners were applied. We note that such conditions correspond to clamping stiffeners since the bending of stiffeners is reproduced by translations rather than rotations. This remark applies to the elements having five degrees of freedom in corner nodes and two mid-nodes. In the case of the stiffened plate elements, for consistency of the boundary conditions, the rotational degree of freedom was also blocked for the nodes situated on the edge corresponding to the position of stiffeners. In the case of the non-linear analysis of plates it is also the boundary conditions defining restraining of mid-surfaces which is significant. In the present analysis mid-surfaces were restrained which means that the edges are not subject to in-plane displacements in the direction perpendicular to the edges. This is the case when plates or panels are situated side by side and are subject to the identical loading conditions.

The loading of the panel was realized through kinematic excitation retaining the original straight shape of the loaded edge. Initial deflections were taken In the form corresponding to the elastic buckling mode. Figure 5 with the amplitude equal to 0.3 thickness of plating.

The constant arc-length path-following method of solution, originally proposed by Crisfield (1981), was applied in the present analysis.

Results are presented in the form of diagrams presenting relative stress vs. relative strain where the term "stress" refers to the mean value of the uniform direct stress on the compressive edge, while the term "strain" refers to the strain on this edge. Relative stress is calculated dividing the actual stress σ_Y by the yield stress,

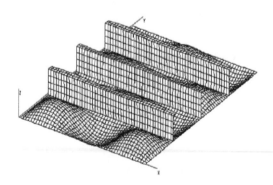

Figure 5. Initial deflections of FE model of panel.

while the relative strain is obtained dividing the actual strain by value $\varepsilon_Y = \sigma_Y/E$.

Results obtained for the FE model built of plate elements were taken as the reference level. Presented were also the results obtained using the simplified formulation proposed by Gordo and Guedes Soares (1993). They derived their approach for axially compressed panel employing the concept varying effective width with the increase of compressive loading. The method is in fact generalization of equations of the ultimate capacity by making their components dependant on actual loading level and application of the Johnson-Ostenfeld equation for elastic-plastic buckling. The following buckling modes are covered: flexural buckling of stiffeners and plating, flexural-torsional buckling of stiffeners and local buckling of stiffener webs. The method was effectively applied for the evaluation of the ship hull ultimate capacity (Béghin et al., 1998).

The results of the analysis for panels stiffened with three types of stiffeners: flat-bars 150×10, angles $200 \times 100 \times 10$ and bulbs 200×12 are presented in Figures 6–8 in the form of stress-strain curves. The results are shown by continuous lines—standard finite element analysis with the use of the plate elements, dashed lines—the present approach with the use of the stiffened plate elements and dotted lines—results obtained using the formulation derived by Gordo and Guedes Soares. In Table 1 the values of the ultimate capacities are given, and in Figures 10–13 deflections modes at collapse for the panel stiffened with angles $200 \times 100 \times 10$ are presented for standard FE model and models with the use of the stiffened plate elements. In Figure 9 the results obtained using the present approach are compared for various meshes.

Comparing to the standard finite element analysis the present formulation offers significant shortening of the time of analysis what is an effect of reduction of the computational model. The reference model

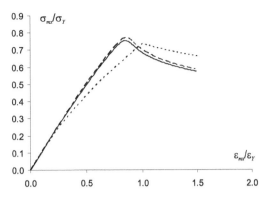

Figure 7. Stress-strain relationship for axially compressed panel stiffened with angles $200 \times 100 \times 10$ (continuous line—standard FE, dashed line—present approach, dotted line—according to Gordo and Guedes Soares).

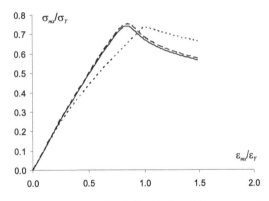

Figure 8. Stress-strain relationship for axially compressed panel stiffened with bulbs 200×12 (continuous line—standard FE, dashed line —present approach, dotted line—according to Gordo and Guedes Soares).

Table 1. Ultimate capacities for panels stiffened with various stiffeners obtained using analysed methods.

Type of stiffeners	Ultimate capacity σ_{mx}/σ_y		
	According to gordo and guedes soares	Standard FE	Present approach
flat-bars 150×10	0.662	0.715	0.725
angles $200 \times 100 \times 10$	0.735	0.752	0.769
bulbs 200×12	0.738	0.742	0.753

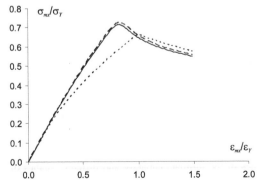

Figure 6. Stress-strain relationship for axially compressed panel stiffened with flat-bars 150×10 (continuous line—standard FE, dashed line—present approach, dotted line—according to Gordo and Guedes Soares).

composed of plate elements has 36000 degrees of freedom and halfbandwidth 1600, while the corresponding (the same mesh for plating) model composed of the stiffened plate elements—27600 DOF's and halfbandwidth 890. The agreements of results for the

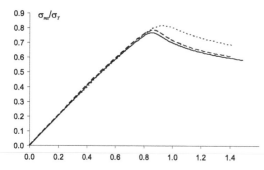

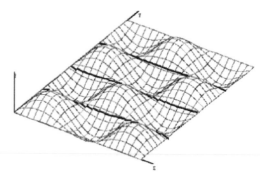

Figure 9. Stress-strain relationship for axially compressed panel stiffened with angles 200 × 100 × 10 for present approach for various meshes (continuous line—3072 elements, dashed line—540, dotted line—88).

Figure 12. Deflection mode at collapse of axially compressed panel stiffened with angles 200 × 100 × 10— present approach, 540 stiffened plate elements.

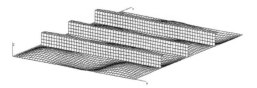

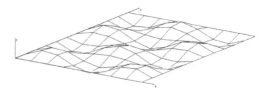

Figure 10. Deflection mode at collapse of axially compressed panel stiffened with angles 200 × 100 × 10— standard FE.

Figure 13. Deflection mode at collapse of axially compressed panel stiffened with angles 200 × 100 × 10— present approach, 88 stiffened plate elements.

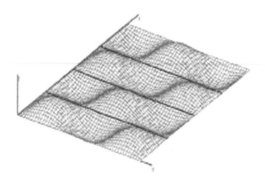

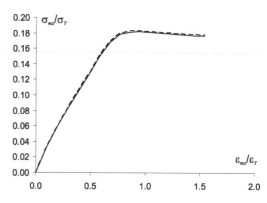

Figure 11. Deflection mode at collapse of axially compressed panel stiffened with angles 200 × 100 × 10— present approach, 3072 stiffened plate elements.

Figure 14. Stress-strain relationship for transversally compressed panel stiffened with flat-bars 150 × 10 (continuous line—standard FE, dashed line—present approach).

analyzed types of stiffeners is satisfactory. The influence of the mesh on the results in terms of ultimate capacity is, however, evident. Using mesh having 88 elements the ultimate capacity is significantly greater than for the model composed of 3072 stiffened plate elements. Ultimate capacity for the model having 540 elements is $\sigma_{xm}/\sigma_Y = 0.781$ and this value equals 0.769 for the model having 3072 elements. Deflection modes for standard and stiffened plate element models are also similar. Obviously, the stiffened plate approach is incapable of covering the deflections of stiffener webs due to local buckling. In all cases the

results obtained using the approach proposed by Gordo and Guedes Soares differ from those obtained from the finite element analysis which is due to the simplified nature of equations. Generally, they provide less values of the ultimate capacity and different shapes of stress-strain curves what can also has significance for evaluation of the hull ultimate strength.

Compression in the direction perpendicular to the stiffeners will be presented for the panel stiffened with flat-bars 150 × 10. Stress-strain relationship for this case is presented in Figure 14 for the finite element model composed of 3936 elements and model composed of 540 stiffened plate elements. Difference between the results for the two models is less for transversally than axially compressed panels. This is due to significantly less influence of stiffener type, and the collapse is governed by behaviour of the plates.

4 CONCLUSIONS

A formulation has been presented allowing for non-linear finite element analysis of stiffened panels. Instead of explicit modelling of stiffeners an approach was proposed where the stiffeners are implicitly included in the plate element as it is also the case in the linear analysis. A derivation of the finite element was presented based on the assumptions typical for analysis of such structures. The results were validated by comparison with the results of the finite element analysis using the standard plate elements and results obtained using simplified formulations for determination of stress-strain curves of stiffened panels. The present formulation can be applied for analysis of panels subject to combined loading of panels: tension/compression and shear and can be useful for the analysis of ultimate capacity of ship hulls subject to bending, shear and torsion.

REFERENCES

Béghin, D., Parmantier, G., Jastrzebski, T., Taczala, M. & Sekulski, Z. 1998. Hull girder safety and reliability of bulk carriers, Proc. 7th Int. Symp. Pract. Design of Ships and Mobile Units, Hague.

Crisfield, M.,A. 1981. A fast incremental/iterative solution procedure that handles "snap-through". Computers and Structures, 13, 55–62.

Deb, A., Booton, M. 1988. Finite element models for stiffened plates under transverse loading, Comp. Struct., 28, 361–372.

Gordo, J.,M., Guedes Soares, C., 1993. Approximate load shortening curves for stiffened plates under uniaxial compression, Proc. Conf. Integrity of Offshore Structures, Glasgow, 189–211.

Hughes, O.,F. 1988. Ship Structural Design, The Society of Naval Architects and Marine Engineers, Jersey City, New Jersey.

Koko, T.,S., Olson, M.,D. 1991. Non-linear analysis of stiffened plates using super-elements, Int. J. Num. Meth. Engng, 31, 319–343.

Mukhopadhyay, M. 1981. Stiffened plate plane stress elements for the analysis of ships' structures, Comp. Struct., 13, 563–573.

Mukhopadhyay, M., Satsangi, S.,K. 1984. Isoparametric stiffened plate bending element for the analysis of ships' structures, Trans. RINA, 126, 144–151.

Mukhopadhyay, M., Satsangi, S.,K. & Mukherjee, A. 1990. A new isoparametric plate element for the analysis of ship structures, Int. Shipbuild. Progr., 37, 79–117.

Samanta, A., Mukhopadhyay, M. 1998. Finite element static analysis of stiffened shells, Applied Mech. Engng, 3, 55–87.

Satsangi, S.,K. 1984. Structural analysis of ships' stiffened plate in plane stress, J. of Inst. of Naval Arch., 67–82.

Satsangi, S.,K., Ray, C. 1998. Structural analysis of ships' stiffened plate panels in bending, Int. Shipbuild. Progr., 45, 181–195.

Sinha G., Sheikh, A.,H. & Mukhopadhyay, M. 1992. A new finite element model for the analysis of arbitrary stiffened shells, Finite Elements in Analysis and Design, 12, 241–271.

Taczala, M. 2001. Ultimate strength of ship hull in bending and shear using simplified approach, Marine Techn. Trans., 12, 261–283.

Taczala, M. 1998. Stiffened plate element based on Kirchhoff thin plate theory, Marine Technology Transactions, 9, 199–218.

Taczala, M., Banasiak, W. 2001. Orthogonally stiffened plate element and its application for analysis of ship structures, Marine Techn. Trans, 12, 285–306.

Thompson, P.,A., Bettess, P. & Caldwell, J.,B. 1988. An isoparametric eccentrically stiffened plate bending element, Eng. Comput., 5, 110–116.

Zienkiewicz, O.,C. 1977. The Finite Element Method, McGraw-Hill, London.

Analysis and Design of Marine Structures – Guedes Soares & Das (eds)
© 2009 Taylor & Francis Group, London, ISBN 978-0-415-54934-9

Residual strength of damaged stiffened panel on double bottom ship

Zhenhui Liu & Jørgen Amdahl
Department of Marine Technology, Norwegian University of Science and Technology,
Trondheim, Norway

ABSTRACT: This paper deals with the numerical simulation of residual strength of a damaged double bottom. The damaged stiffened panel on the outer bottom is picked up as present study's research object. The typical double bottom structure is modeled from a shuttle tanker for oil transportation. A slight idealization is assumed in order to simplify the problem. Imperfections corresponding to fabrication tolerances are introduced to the whole structure. The most severe situation with the tanker in fully loaded condition is investigated. Damage is caused by a variety of indenters which are specially designed to get the desired damage extent. In total numerical calculations are carried out for 16 cases using the explicit codes LS-DYNA. The results for the intact structure are compared with predictions of PULS. The agreement is good. A single stiffener model is proposed with the aim of predicting the residual strength of the double bottom. Analytical equation is derived and reasonable results are obtained. Finally, conclusions and suggestions are presented.

1   INTRODUCTIONS

Ships are always at risk of encountering accidental loads during its service time, like collision, grounding, and dropped objects. Whenever these accidents happen, the residual strength of ship structure becomes a key issue. In this study, the residual strength of a damaged stiffened panel in the scope of the double bottom is investigated.

As to the strength of intact, stiffened panels, a lot of work has been done, e.g. by Faulkner (1975), Guedes Soares and Gordo (1997), Paik and Kim (2002). On the contrary, only a few studies are concerned with strength of damaged stiffened panels. Smith and Dow (1981) simulated the damage by applying lateral pressure causing plastic deformations and residual stresses, and then compression was applied in order to determine the compressive strength A significant strength reduction was found. Paik et al. (2003) did a non-linear numerical research on ultimate strength of dented plates. A parametric study was carried out addressing the plate aspect ratio, denting distance and the denting diameter. The research object was the plate between two neighboring stiffeners. Denting damage was simulated through geometric imperfections, no residual stresses were considered. It was found that the increasing the denting diameter decreases the ultimate strength significantly, while the depth of the denting has little influence if the diameter is small. An expression for the strength reduction due to denting was suggested by the author. Guedes Soares (2005) carried out studies on the localized imperfections by considering the possibility of having imperfections

caused by accidents. The strength was significantly reduced and so was also the post collapse strength. Paik and Kumar (2006) did nonlinear finite element analysis to get the ultimate strength of stiffened panels with cracking damage. An empirical equation was obtained through data fitting. Nikolov (2008) investigated numerically the collapse strength of damaged panel in two stages. First, forced lateral displacements in the chosen shape were applied iteratively, until the specified permanent deformation was obtained. Second, a compressive load was applied in longitudinal or transversal direction until the ultimate strength for transverse or longitudinal compression was attained. A parametric study of the plate slenderness was carried. Significant reduction of strength was found. It was claimed that GL and BV methods in the IACS Common Rules may overestimate the ultimate strength. Witkowska and Guedes Soares (2008) analyzed the collapse strength of stiffened panels with local dent damage. The dent was introduced in a form of local imperfection. It was found that the stiffened panels after being damaged in a form of local dent still demonstrates quite good performance. The strength reduction is only 1–2%, and around 5% for most slender panels, however the dent here is comparably at the same level of fabrication tolerance magnitude, say $(0.75–1)\,\beta^2 t$.

Among the studies mentioned above, no contact damage is involved. Implicit non-linear finite element codes are employed in most cases. In practice, the damage is not only caused by pressure or local imperfection, the damage extent may be even bigger, e.g. in grounding, by dropped objects. In present paper, the damage is simulated by means of a contact algorithm. Initial distortions and the residual stresses due to the damage are both considered in subsequent compression analysis. The simulations are carried out with the explicit code LS-DYNA. In order to facilitate relatively large damages, the research panel is analyzed within a double bottom structure. The panel is located from frame 65 to 68 on the outer bottom plate. By doing this the real damage and stress redistribution is hopefully obtained.

2 MODEL CHARACTERISTICS

2.1 Main dimension

The double bottom structure of a 140,000 m^3 shuttle tanker is the subject of the present study. The starboard side is modeled. A section of 3 frames from frame 65 to 68 in the drawing is modeled. This is considered to be big enough to allow for redistribution of stress/strain after damage. The target research object in present study is the stiffened panel of the outer bottom between frame 66 and 67, not including the part next to the bilge which has different type of stiffeners, refer Fig. 1 (the shaded part). The structure is slightly idealized compared to as-built, e.g. a longitudinal girder is shifted slightly in order to have uneven no. of stiffeners, the double bottom height is constant and the two frames 66, 67 in the middle are strengthened to prevent them from buckling prior to the stiffeners. The main dimensions of the FE model are as shown in Table 1.

As to the material modeling, the plastic behavior should be thoroughly considered. In this study, the true stress-strain relationship of the material in the plastic range is described by the Ludwik relation in the form of power law $\sigma = \sigma_0 + K(\varepsilon^p)^n$, where σ_0 is the yield stress and ε^p is the effective plastic strain. The following parameters are assumed based on the experiment test data. $\sigma_0 = 275$ Mpa, $K = 740$, $n = 0.24$. Rupture is not taken into account, i.e. infinite strains may develop in principle.

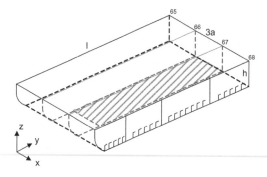

Figure 1. Research object (shaded part) within the double bottom model.

Table 1. FE model main dimension.

Dimension	Values (mm)
Outer plate	17.5
Inner plate	17
Girder	16
Frames 65,68	14
Frames 66,67	24
Length	12000
Width	21250
Height	2650

Table 2. Dimension to stiffeners.

Dimension	Value (mm)
h_w	425
t_w	11.5
b_f	120
t_f	20
b	830
b_{side}	713
β	1.72

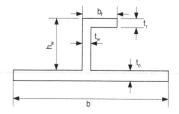

Figure 2. Properties of stiffeners.

2.2 Imperfections

Imperfections are introduced into the FE model, by means of the following formulas:

$$f(z) = \sum w_{ij} \sin\left(\frac{m\pi}{a_i}x\right) \sin\left(\frac{n\pi}{b_j}y\right) \quad (1)$$

$$f(y) = \sum w_{ij} \sin\left(\frac{m\pi}{a_i}x\right) \sin\left(\frac{n\pi}{b_j}z\right) \quad (2)$$

where x, y, z are the coordinates of the nodes, n is numbers of half-waves. (1) is for the plates and stiffeners, while (2) is for the girders. m and n are coincident set as the same as the global buckling mode, here we set $m = 5$ for the plate and girder induced longitudinal imperfection. n equals to 1, which caused asymmetrical waves transversely. The amplitudes w_{ij} are identical to those used by the PULS code, e.g. E. Steen (2001), except for the girder where the amplitude is according the rule from DNV (2004). The imperfections,

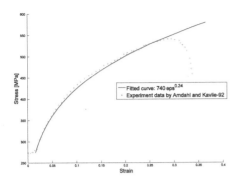

Figure 3. Material properties.

Figure 8. The whole double bottom model (scale factor = 20).

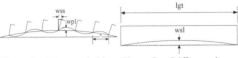

Figure 4. Plates and side ways imperfections.

Figure 5. Stiffeners imperfection.

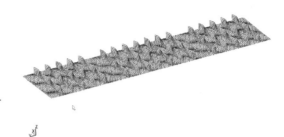

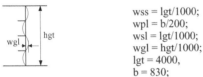

Figure 6. Girders imperfections.

wss = lgt/1000;
wpl = b/200;
wsl = lgt/1000;
wgl = hgt/1000;
lgt = 4000,
b = 830;
hgt = 2680.

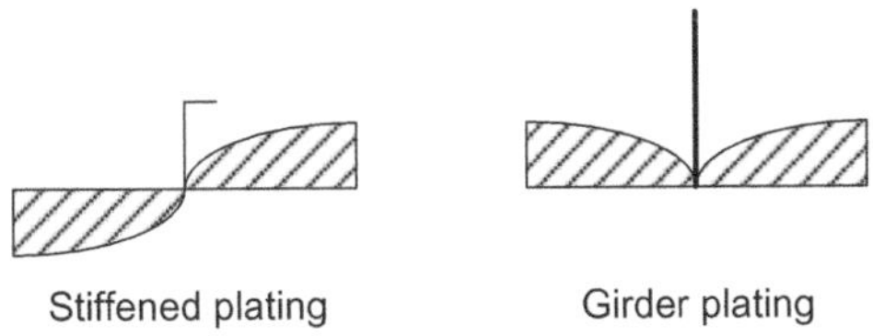

Figure 9. Present research object as part of the double bottom model (scale factor = 20).

Stiffened plating   Girder plating

Figure 7. Relation of symmetry of imperfections.

although being nominal, are considered to represent the combined effect of residual stresses and initial distortions. The final FE model, e.g. Figures 8, 9.

Both symmetric and asymmetric imperfection type to adjacent plating are employed. The former is used where the stiffeners locate, the latter is only for girders, which could provide stiff support.

2.3 Loads

Bottom damage caused by contact with the sea floor is created by rigid indenters, with a given slope in the transverse direction and the longitudinal direction. The indenter is "stretched" in the transverse direction so as to allow 1, 3, 5, 7 or 9 stiffeners (including 2 girders in the last case) to be damaged in each specific case. As seen from Figs. 10 and 11, the indenters will have the same α, β shape factor if the denting depth is the same regarding to following equations:

$$\tan \alpha = \frac{d}{l}$$

$$\tan \beta = \frac{d}{k}$$

Where $l = 800$ mm, $k = 1500$ mm, $d = 50, 100$ or 150.

Three denting depths, nominally 50 mm, 100 mm and 150 mm, are simulated. Actually, the simulation contains a loading and unloading procedure. After the indentation process is finished, the indenter is unloaded from the model. Due to elastic deformation caused by the indentation, some rebound takes place; so that the residual indentation depth differs generally form the target value, up to about 10% difference. The denting is conducted in the midsection between adjacent transverse frames 66 and 67, e.g. Figures 10–11 for the Case150b.

The indenter is placed under 10 mm lower than the base plane in order to get rid of initial penetration. The contact between the indenter and the outer shell is obtained by means of the automatic surface contact algorithm in LS-DYNA, Hallquist (2006).

Subsequent to introduction of damage the double bottom is subjected to compression created by forced inward displacement of the boundaries. The forced displacement should including global bending of the

Table 3. Case notation.

Target indentation	No. of damaged stiffeners	Notation
50	1	50a
50	3	50b
50	5	50c
50	7	50d
50	9	50e
100	1	100a
100	3	100b
100	5	100c
100	7	100d
100	9	100e
150	1	150a
150	3	150b
150	5	150c
150	7	150d
150	9	150e

Figure 10. Longitudinal View (*x*-direction).

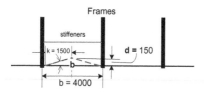

Figure 11. Transverse View (*y*-direction).

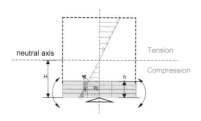

Figure 12. Moment added through angular velocity.

hull girder. In pure bending the displacement will vary over the double bottom height as illustrated in Figure 12, on the assumption that plane cross-sections remain plane. If the end shortening exhibits nonlinearities due to buckling of compressive panels or yielding in deck panels, the neutral axis will be subjected to shift.

In the linear response range, the correct inward displacement or rate of displacement for pure bending in hog is obtained by introducing an angular rate of

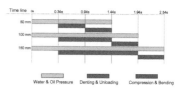

Figure 13. Loading sequences.

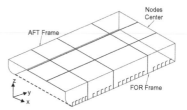

Figure 14. Boundary conditions to double bottom.

Table 4. Shortening curves at Case c.

	X	Y	Z	RX	RY	RZ
Aft. Frame	1	1	1	1	0	1
For. Frame	0	1	1	1	0	1
Nodes Center	0	1	0	0	0	1

rotation, ω in addition to the mean inward speed, V_0, given by:

$$\frac{V_0}{V_0 + \omega h/2} = \frac{H - h/2}{H} \tag{3}$$

So

$$\omega = \frac{2}{2H - h} V_0 \tag{4}$$

The tanker is assumed to be in fully loaded condition. The corresponding draft for is 15 m and the oil pressure height is 19 m. Thus the water and oil pressure was set 0.15 MPa and 0.161 MPa, respectively. The loading sequence could be easily seen from following illustration.

2.4 Boundary conditions

As shown in figure 14 the boundary conditions are divided into three groups, namely the Forward Frame, the Aft Frame and the Nodes Center line. The six-degree constraints for all three groups are found in the Table 4, in which 1 indicate fixed while 0 means free. Symmetry conditions are imposed on the Center Line nodes.

3 VALIDATION

PULS is a recognized computer code to calculate the ultimate strength of stiffened plates. It is an accepted tool for design checks of ships panels according to

Figure 15. Cross sections' view of target stiffeners.

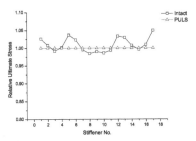

Figure 16. Comparison between double bottom model and PULS.

det Norske Veritas (DnV RP C201) The PULS code is based on series on series expansion of plate deformations combined with numerical solution strategy. In the present work a comparison is made with PULS for the intact case in order to verify that LS-DYNA is capable of providing good estimates of the ultimate strength. The study is made for the target research object. It consists of 17 L-shaped stiffeners with attached plates (Fig. 2) and 4 plates adjacent to girders (dashed lines in Fig. 14).

Figure 16 shows the ultimate strength, i.e. the maximum stress level through each individual stiffener with attached plates in the outer shell panel relatively to the PULS. The results of LS_DYNA agree quite well with those of PULS. The stiffeners midway between frames, stiffener 3, 9 and 15 have larger deflections due to water pressure. Thus they attain lower ultimate strength than those close to the longitudinal girders. It is concluded that the present FE model is capable of providing reasonable results. The intact case is used as the referenced case against which the influence of damage is measured.

4 RESIDUAL STRENGTH

The force-shortening curves for a given denting level are plotted versus number of damaged stiffeners in Figures 17–19 where the intact case is used as a reference. The origins of the shortening curves usually don't start at zero stress. This is due to the damage caused in the first stage. The significant denting distance leaves the stiffeners in the tension state. The post-collapse strength falls down quickly right after the buckling to a certain point and then the decrease slows down. It is more obvious in the intact case, but becomes hard to see when the number of damaged stiffeners increases. This specific observation is due

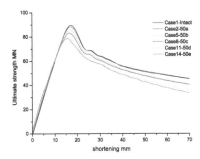

Figure 17. Shortening curves for 50 mm denting.

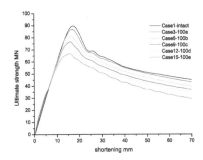

Figure 18. Shortening curves for 100 mm denting.

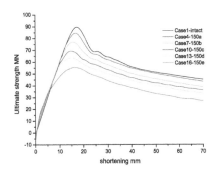

Figure 19. Shortening curves for 150 mm denting.

to the tripping failure of the stiffener profile adopted in this study. The resistance attains the peak point when the web starts to trip after which the strength decreases sharply. However, the stiffeners being supported by the transverse frames will not trip. Further, the ends are restricted against inward motion triggered by the tripping, which slows down the reduction of the strength.

As to the ultimate strength of the damaged panel, the peak loads for all cases are decreasing as the number of damaged stiffeners increases. If the denting distance is small, like 50 mm or less, the reduction is not significant, Figure 17. If the denting is larger, 100 mm or

150 mm, this effect becomes severe. In the worst case, the reduction could reach 40%.

Another observation is that the resistance in the post-collapse region is more sensitive to the number of damaged stiffeners than the level of denting, refer e.g. the values at a shortening of $3\varepsilon_y$, corresponding shortening to 45 mm.

From Figures 20–24, the force-shortening curves are presented according to the denting level where the shortening curve in the intact case again is used as a reference.

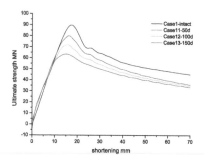

Figure 23.   Shortening curves at Case d.

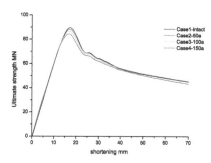

Figure 20.   Shortening curves at Case a.

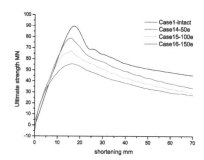

Figure 24.   Shortening curves at Case e.

As the denting distance increases the ultimate strength is increasingly reduced. It is noticed, however, that the resistance in the post-collapse range is less sensitive to the denting level.

5   SINGLE STIFFENER ANALYSIS

5.1   Single stiffener model

The question arises whether the resistance of the panel can be estimated on the basis of the behavior of individual stiffeners with associated plating. Consider the ultimate strength of the stiffener 15, which always remain undamaged during the denting phase. As shown in Figure 25 the ultimate resistance degrades slightly with the number of damaged stiffeners and denting level, but roughly speaking it is essentially constant.

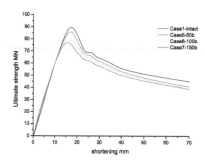

Figure 21.   Shortening curves at Case b.

Figure 26 shows the ultimate strength of stiffener 9, which is always damaged. Except for one damaged stiffener case (where the denting is not 'complete'), the ultimate resistance may be considered being fairly constant. Both observations encourages synthesizing the ultimate strength from individual stiffener behavior.

The cross section properties are found in Figure 2. A water pressure of 0.15 MPa is applied to both analysis stages, and deforms to the plate towards the stiffener.

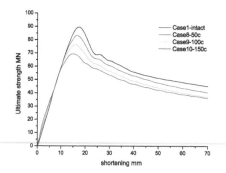

Figure 22.   Shortening curves at Case c.

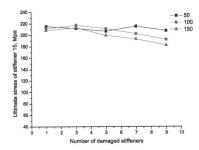

Figure 25.   Ultimate stress of stiffener 15.

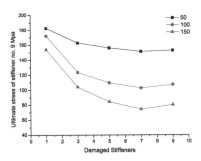

Figure 26.   Ultimate stress of stiffener 9.

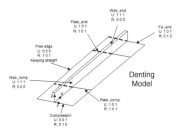

Figure 27.   Single stiffener-Denting model.

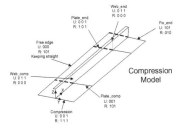

Figure 28.   Single stiffener-Compression model.

The model comprises one frame spacing plus half of the adjacent frame spacing. The boundary conditions for the denting phase and compression phase are indicated for the translational and rotational degrees of freedom (1 and 0 denote fixed and free, respectively). The stiffeners are supported by the transverse fames, where the z-displacement is fixed, e.g. Plate_comp and Plate_end in Figures 27–28. The long edges are assumed to be free against inward displacement (y direction), but the edges should remain straight. The so called *Half-symmetry* boundary condition has been applied to both ends in both models, in which z-direction displacement and y direction rotation are fixed. This setting isn't the same as the general setting for *Symmetry* boundary condition. Actually, both ends here are in the middle of a frame spacing which generally should be treated as *Symmetry* boundary condition with restriction on z-direction displacement and y-direction rotation. However in this case the *Symmetry* boundary doesn't work. The explanation to this is that the water pressure will distort the collapse mode if the boundary is free in the z-direction for we have *Free edge* here, which let the compression force can't be transferred to the middle target part. The collapse in the target area, the middle part of this single stiffener won't happen. Anyway, the primary interest lies in the middle, damaged part, the boundaries conditions used the analyses should still provide good estimates of the ultimate strength in both intact and damaged condition.

In the denting phase the stiffener is fixed against inward x-direction displacement while this restriction is removed in the compression phase.

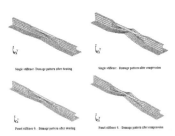

Figure 29.   Comparison of damage patterns.

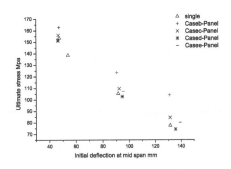

Figure 30.   Comparison of initial deflection at mid span.

169

Table 5. Calculation based on single stiffener.

Cases	m	Stress	Intact	Plates	Prediction	Actual
intact	0	210	210	11.65	88.70527	90.21
50b	3	155	210	11.65	88.05779	86.27
100b	3	95	210	11.65	87.84197	81.943
150b	3	70	210	11.65	87.62614	77.084
50c	5	155	210	11.65	85.14416	84.195
100c	5	95	210	11.65	81.25931	77.65
150c	5	70	210	11.65	79.64062	69.594
50d	7	155	210	11.65	82.77008	81.162
100d	7	95	210	11.65	76.29533	72.659
150d	7	70	210	11.65	73.59752	63.536
50e	9	155	210	11.65	80.39601	79.397
100e	9	95	210	11.65	71.33136	67.855
150e	9	70	210	11.65	67.55442	55.837

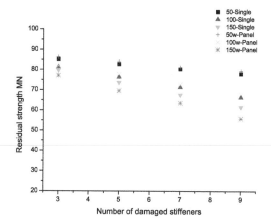

Figure 32. Comparison between prediction and double bottom.

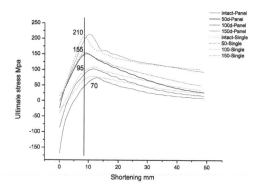

Figure 31. Shortening curves comparison at stiffener no. 9.

5.2 Simulation comparison

The single stiffener model denting distance is obtained iteratively with the aim of getting the same initial deflection comparing to the double bottom cases, i.e. Figures 29, 30. The initial deflection of the damaged panel becomes to be constant when the number of damaged stiffeners increases. It is seen that the initial deflections are both kept approximately the same. The damage pattern after compression of these two models (Figure 30) is very similar to each other, which gives very good support to the results.

In Fig. 31 the stress-end shortening curves for the single stiffener model and the damage panel in the double bottom model are plotted. Generally, they agree well, except for the 150 mm case, which is in a large initial tension mode during the compression stage, which cannot be accounted for by the single stiffener model. The non-simultaneous buckling of intact and damaged stiffeners should be noticed. Due to the initial distortion and residual stress, the damaged stiffeners will buckle at a later stage than the intact ones. So,

if the ultimate strength of the whole panel is to be assessed, the stress of the damaged stiffeners should be evaluated at the same time as the intact stiffeners reach their ultimate strengths.

6 RESIDUAL STRENGTH ASSESSMENT

Both σ_d and $\sigma_{\text{int}\,act}$ can be obtained through single stiffener analysis, see Figure 31.

It is assumed that the total force is given by

$$F_s = \sum_{i=1}^{17} \sigma_i A_i = (n\sigma_{\text{int act}} + m\sigma_d)A$$

where, n is the number of intact stiffeners, m is the number of damaged stiffeners, A is the section area.

As to the plate adjacent to girders

$$F_p = \sum_{j=1}^{6} \sigma_j A_j = 6\frac{be}{b}\sigma_y A_p$$

$$\frac{be}{b} = \frac{2}{\beta} - \frac{1}{\beta^2}$$

where b_e is the effective width, b is the plate width, here equals to 415 mm, β is the slenderness of the plate.

It could be seen that the prediction agrees quite well with the actual strength at the denting level 50 mm and 100 mm. As to the 150 mm case, the single stiffener model overestimates the resistance, which shows the limitation of the single stiffener model for large denting levels.

7 CONCLUSION

A damaged stiffened panel in a double bottom structure has been investigated. A parametric study of damage level and extent has been performed. A single stiffener model is analyzed with the aim of finding a quick way to estimate the strength of the damaged panel. The results show that both the denting distance (initial deflection) and the number of damaged stiffeners have a significant influence on the residual strength of the panel. Generally, the residual strength decreases with increasing denting distance (initial deflection) and the number of damaged stiffeners. The post-collapse resistance is less sensitive to the denting levels than the number of damaged stiffeners. A single stiffener model is capable of predicting the residual strength for moderate denting levels. The results agree well when the denting distance is 50 mm and 100 mm, while in the 150 mm case, the single stiffener model overestimates the residual strength.

ACKNOWLEDGMENTS

The work has been performed in the scope of the project MARSTRUCT, Network of Excellence on Marine Structures, (www.mar.ist.utl/marstruct/), which has been financed by the EU through the GROWTH programmed under contract TNE3-CT-2003-506141. The authors wish to give thanks to the support from this project.

REFERENCES

DNV (2004). Fabrication and Testing of Offshore Structures. *Offshore Standard DNV-OS-C401*. Oslo, DNV.

E.Steen, T.K. Ø., K.G.Wilming & E. Byklum (2001). PULS User's manual, DNV.

Faulkner, D. (1975). "A Review of Effective Plating for Use in The Analysis of Stiffened Plating in Bending and Compression." *Journal of Ship Research* 19(1).

Guedes Soares, C., et al. (2005). Effect of the shape of localized imperfections on the collapse strength of plates. *Maritime Transportation and Exploitation of Ocean and Coastal Resources*. London: 429–438.

Guedes Soares, C. & J.M. Gordo (1997). "Design methods for stiffened plates under predominantly uniaxial compression." *Marine Structures* 10(6): 465–497.

Hallquist, J.O. (2006). LS-DYNA Theory Manual, Livermore Software 2006.

Nikolov, P.I. (2008). Collapse strength of damaged plating. *Proc. of OMAE 2008*. Estori,Portugal.

Paik, J.K. & B.J. Kim (2002). "Ultimate strength formulations for stiffened panels under combined axial load, in-plane bending and lateral pressure: a benchmark study." *Thin-Walled Structures* 40(1): 45–83.

Paik, J.K. & J.M. Lee, et al. (2003). "Ultimate strength of dented steel plates under axial compressive loads." *International Journal of Mechanical Sciences* 45(3): 433–448.

Paik, J.K. & Y.V. Satish Kumar (2006). "Ultimate strength of stiffened panels with cracking damage under axial compression or tension." *Journal of Ship Research* 50(3): 231–238.

Smith, C.S. & R.S. Dow (1981). "Residual Strength Of Damaged Steel Ships And Offshore Structures" *Journal of Constructional Steel Research* 1(4): 2–15.

Witkowska, M. & Guedes Soares, C. (2008). Collapse strength of stiffened panels with local dent damage. *Proc. of OMAE 2008*, Estoril, Portugal.

Analysis and Design of Marine Structures – Guedes Soares & Das (eds)
© 2009 Taylor & Francis Group, London, ISBN 978-0-415-54934-9

Assessment of the hull girder ultimate strength of a bulk carrier using nonlinear finite element analysis

Z. Shu & T. Moan
CeSOS and Department of Marine Technology
Norwegian University of Science and Technology, Trondheim, Norway

ABSTRACT: This paper deals with evaluating the hull girder ultimate strength of a bulk carrier. The progressive collapse of the hull girder subjected to pure bending in hogging conditions is investigated using nonlinear finite element analysis with ABAQUS. Both material and geometrical nonlinearities are taken into account. The initial geometrical imperfections are considered. The ultimate bending moment obtained by the nonlinear finite element method is used as a basis to check simplified methods for hull girder strength analysis. The ultimate strength of the hull girder predicted by the simplified methods agree relatively well with obtained that by nonlinear finite element analysis. However it seems that only the nonlinear finite element method can simulate the sharp capacity reduction beyond the ultimate strength. It is shown that the nonlinear finite element analysis can provide more information about the ultimate hull girder strength and collapse mode beyond the ultimate strength than the simplified methods. The tripping or flexural-torsional buckling of the bottom longitudinal stiffeners after the plate buckling under extreme hogging condition can be clearly captured by the nonlinear finite element analysis. The results of the nonlinear finite element analysis can serve as a benchmark reference to the analyses based on simplified approaches.

1 INSTRUCTIONS

Accurate assessment of the ultimate strength and the progressive collapse of the hull girder of ships is a very important part of the safety check of the ship structures based on the ultimate limit states. During the past decades, a variety of methods have been developed by researchers for the estimation of the hull girder ultimate strength. The load shedding (strength reduction) of the individual structural member is very important for the evaluation of the hull girder ultimate strength of ships. Some early developed methods (Caldwell 1965, Mansour et al. 1990), however, can not account for the strength reduction in the individual members after they have achieved their ultimate strength locally. The so-called Progressive collapse methods can take into account the load shedding. The simplified method (Smith's method) (Smith 1977), the idealized structural unit method (ISUM) (Ueda et al. 1984) and finite element method (FEM) can be grouped into this category. Usually, simplified methods are applied for the progressive collapse analysis of a ship's hull under longitudinal bending. The applications of nonlinear finite element method are very few (Chen et al. 1983, Kutt et al. 1985, Valsgaard et al. 1991) due to the great demand on the computer resource and manpower. However the rapid development of the computer technology makes it possible to carry out the nonlinear finite element analysis for the ultimate

hull girder strength with reasonable computation cost and proper modeling.

The ISSC Committee VI.2 (Yao et al. 2000) carried out a benchmark study on the ultimate hull girder strength of five vessels including a Capesize bulk carrier by seven different approaches without results from nonlinear finite element analysis. It is found that the scatter in the ultimate strength is relatively small especially when the hull is subjected to pure hogging bending moment. But the load capacity beyond the ultimate strength is somewhat scattered. The behavior strongly depends on whether or not the capacity reduction in the elements beyond their ultimate strength is correctly accounted for. Although it requires considerably engineering and computational efforts, a proper nonlinear finite element analysis is capable of accurately predicting the ultimate strength and capturing the behavior of post buckling. Moreover, it is required to accurately account for the effect of the double bottom bending which will be considered in a separate paper.

Amlashi and Moan (2008) investigated the ultimate strength of this bulk carrier under double bottom bending using nonlinear finite element analysis by modeling of $1/2 + 1 + 1/2$ hold tanks. In this paper, based on their work, a extended model which covers 3 cargo holds and four transverse bulkhead in the midship region is set up to investigate the ultimate strength of this bulk carrier. The nonlinear finite element analysis is carried out using the finite element code ABAQUS. The ultimate strengths in hogging condition are evaluated and compared with results from simplified methods. As expected, the nonlinear finite element analysis can provide more information about the ultimate strength and collapse mode beyond the ultimate strength. The tripping or flexural-torsional buckling of the bottom longitudinal stiffeners after plate buckling under extreme hogging condition as reported by simplified methods can be clearly captured by the nonlinear FE analysis. The results of the nonlinear FE analysis can serve as a benchmark reference to the analyses based on simplified approaches.

2 FEATURE OF THE NONLINEAR FINITE ELEMENT ANALYSIS

A progressive collapse analysis based on nonlinear finite element analysis is performed in this study. The nonlinear finite element analysis is carried out using nonlinear finite element computer program ABAQUS, which can take into account both geometrical and material nonlinearities.

The ship's hull is a very complicate stiffened panel structure consisting of many components such as the deck, bottom, side shell, bulkhead, transverse frame and longitudinal. Due to the simplicity and efficiency, simplified methods like Smith's method and ISUM method are usually adopted for the progressive collapse analysis of a ship's hull under longitudinal bending with some approximation. A major assumption introduced by Smith's method is the hull girder transverse sections remain rigid plane during the linear and nonlinear behavior. The interactions between different collapse modes and individual structural components can not be accounted for in Smith's method. The Smith method can not also take into account the transverse stress during failure process involving redistribution of stress. The ISUM method will yield reliable results when a proper element is applied.

The nonlinear finite element method with proper modeling and analysis procedure is believed to be the potentially most accurate method for progressive collapse analysis of a ship's hull. Due to the significant computer resources and modeling work required, such an analysis has been seldom adopted for complicated ship structures. However with the rapid development of the computer technology, it is now possible to carry out the nonlinear finite element analysis to obtain accurate results in reasonable computation time with proper modeling of the ship's hull.

The main advantage of the nonlinear finite element method is that no assumptions adopted in simplified method are needed and the interaction between structural components for different buckling modes can be automatically taken into account. The results of the nonlinear finite element analysis will serve as a benchmark reference for the analyses based on simplified methods.

In the present study, the ABAQUS/Standard (Simulia 2007) nonlinear finite element method is adopted for the ultimate hull girder. The progressive collapse of the structure of the ship hull is preceded by a series of local failure of components due to buckling and yielding. Due to the localization of the instabilities, there will be a local transfer of strain energy from one part of the model to the neighbouring part and the well known global load control method such as 'RIKS' method in ABAQUS/Standard may not work. This class of problem can be solved with the aid of damping. This volume proportional damping works in such a way that the artificial viscous forces introduced are sufficiently large to prevent instantaneous buckling or collapse but small enough not to affect the behavior while the problem is stable. This feature is available in ABAQUS. Obtaining an optimal value for the damping factor is a manual process requiring trial and error test until a converged solution is obtained and the dissipated stabilization energy is a sufficiently small fraction of the total strain energy. It is noted that the damping factor is dependent on the mesh size and the model extent.

3 STRUCTURAL MODEL

3.1 Vessel particulars

The target vessel of the present study is a Capesize single side bulk carrier which has been benchmarked by the ISSC Committee VI.2 (Yao et al. 2000) against a number of simplified methods. The cross-section of the hull girder of the vessel is shown in Figure 1. The main particulars and the dimensions of the longitudinal stiffeners are summarized in Tables 1 and 2, respectively. The ship is longitudinally stiffened except for the side shells between the wing tank and the hopper tank which are transversely stiffened. The longitudinal stiffener spacing is 880 mm. The transverse frame spacing in the bottom is 2610 mm.

3.2 Material properties

For assessment of ultimate hull girder strength of a ship, it is important to establish the nonlinear material model in terms of stress versus strain relationship. In the present study, a bi-linear elastoplastic material model with the kinematic hardening is adopted to describe the nonlinear material properties. The material parameters are given in Table 3.

3.3 Finite element modeling of the structure

The nonlinear finite element model for the capsize bulk carrier is built up with the finite element code

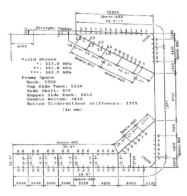

Figure 1. Cross-section of a Capesize bulk carrier.

Table 1. Main dimensions of a Capesize bulk carrier.

Length (m)	285
Breath (m)	50
Depth (m)	26.7
Design Draft (m)	19.808
Scantling Draft (m)	19.83

Table 2. Dimensions of the longitudinal stiffeners of a Capesize bulk carrier.

No.	Dimensions	Type	σY(MPa)
1	390×27	Falt-bar	392
2	$333 \times 9 + 100 \times 16$	Tee-bar	352.8
3	$283 \times 9 + 100 \times 14$	Tee-bar	352.8
4	$289 \times 9 + 100 \times 18$	Tee-bar	352.8
5	$333 \times 9 + 100 \times 17$	Tee-bar	352.8
6	$283 \times 9 + 100 \times 16$	Tee-bar	352.8
7	$180 \times 32.5 \times 9.5$	Bulb-bar	235.2
8	$283 \times 9 + 100 \times 17$	Tee-bar	352.8
9	$333 \times 9 + 100 \times 18$	Tee-bar	352.8
10	$333 \times 9 + 100 \times 19$	Tee-bar	352.8
11	$383 \times 9 + 100 \times 17$	Tee-bar	352.8
12	$383 \times 10 + 100 \times 18$	Tee-bar	352.8
13	$383 \times 10 + 100 \times 21$	Tee-bar	352.8
14	300×27	Flat-bar	392

Table 3. Material properties of a Capesize bulk carrier.

	MS	Hs32	Hs36	Hs40
Young's modulus (N/mm^2)	2.1e5	2.1e5	2.1e5	2.1e5
Poisson ratio	0.3	0.3	0.3	0.3
Strain hardening parameter (N/mm^2)	825	625	675	600
Yielding Stress (N/ mm^2)	235	313.6	352.8	392

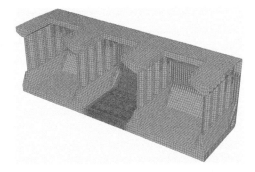

Figure 2. Three-hold nonlinear finite element model of a Capesize bulk carrier.

ABAQUS. The model used in the present study is illustrated in Figure 2. This is a 3-hold model which is located in the midship area of the bulk carrier. The longitudinal extent of FE model is to cover three cargo holds and four transverse bulkheads. The transverse bulkheads at the ends of the model extent should be included, together with their associated stools. Both ends of the model form vertical planes and include any transverse web frames on the planes if any.

The extent of the model in the longitudinal direction is required by the global strength assessment according to CSR-bulk carrier (IACS 2006). Symmetry about the centre plane is employed to keep the model size within manageable limits. The 3-hold model is suitable to simulate the simultaneous application of global loads such as vertical bending moment and local loads such as sea pressure and cargo pressure. The structural dimensions are based on the values of the midship section. The corrugated bulkheads are design according to the CSR rules which are strong enough to experience no buckling collapse.

The present focuses on the progressive collapse of the centre hold subjected to pure bending in hogging condition. A critical issue for a large nonlinear FE-model is the mesh size, the element type and the total number of elements using in the model. Fine meshes imply more accurate prediction at a greater cost of memory size and computational time. On the other hand, coarse mesh in critical area will make the model too stiff to capture the local buckling, causing unrealistically high ultimate strength. However for a ship in hogging condition, it is only the bottom part that will exhibit buckling. Hence, it is only in that part a fine mesh is required.

Before doing a large nonlinear FE analysis, a mesh convergence study (test analyses) must be conducted for the whole model or at least for the critical (typical) region. The mesh size and element type adopted in this paper are based on the work by Amlashi and Moan (2008) and Ostvold (2004). Based on the above considerations, relative coarse meshes are applied in the holds on forward and aft ends as well as their adjacent part of the centre hold which are far away from the region of interest. This coarse mesh part has 1 × 3 elements in the plate bounded by longitudinal stiffeners and transverse web frames in the double bottom. The webs of stiffeners are still modeled as shell element while the flanges are represented using beam elements. Since the elastic buckling stress of the plate of the lower part is usually much higher than the yielding stress, the materials of this coarse mesh part can be described as linear elastic. The middle part of the centre hold acting as the fully nonlinear part of interest is modeled with very fine mesh density in the lower part. The mesh density in the fine mesh region as shown in Figure 3 can be summarized as follows:

– 5 × 15 shell elements in the plate bounded by longitudinal stiffeners and transverse web frames.
– Five shell elements across the height of the longitudinal stiffeners.
– Twenty one shell elements across the height of double bottom girders and floors
– Two shell elements across the flange of the longitudinal stiffeners.

This mesh strategy is considered to be good enough to capture the proper collapse mode of the stiffened panels at practically reasonable computational time and requirement on the computer memory. However, this type of fine mesh could not be used everywhere in the model, as the problem size might grow quickly beyond the computer capacity. Therefore, only critical regions where the collapse is expected are modeled by fine mesh. This kind of fine mesh will be employed in the double bottom of the centre cargo hold, where the axial collapse of the stiffened panels is most likely to occur. The coarse mesh is used for the side shell and upper part of this fine mesh part. The mesh strategy is similar to the coarse mesh part. Nonlinear material properties are imposed on this centre part with coarse and fine mesh.

The total number of elements in model is around 186000 and the total number of degrees of freedom is about 990000. It is noted that the focus of the present study is the progressive collapse of the double bottom in hogging condition. Coarse mesh is employed for the upper deck, which indicates that this finite element model is not suitable for the ultimate limit state assessment in sagging loading.

The boundary conditions of the three-hold finite element model are adopted according to the CSR- bulk carrier rule (IACS 2006). The nodes on the longitudinal members at both end sections are rigidly linked to independent points at neutral axis on centreline as shown in Table 4 The independent points of both ends are fixed as shown in Table 5 Since only half of the width is modeled, symmetry conditions are applied to all members in the centre plane. The loading of the hull girder is considered as a prescribed rotation at the independent points. The independent points are

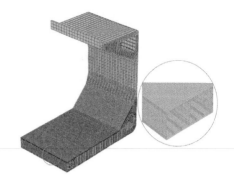

Figure 3. Finite element model of the centre hold with fine mesh and coarse mesh.

Table 4. Rigid link of both ends.

Nodes on longitudinal members	Translational			Rotational		
	Dx	Dy	Dz	Rx	Ry	Rz
All longitudinal members	RL	RL	RL	–	–	–

RL means rigidly linked to the relevant degrees of freedom of the independent point.

Table 5. Boundary conditions of independent point.

Location of independent point	Translational			Rotational		
	Dx	Dy	Dz	Rx	Ry	Rz
Aft end of model	–	fix	fix	–	–	–
Fore end of model	fix	fix	fix	fix	–	–

chosen as the intersections of the neutral axis and the centre plane.

3.4 Initial imperfections

The weld-induced initial imperfections including residual stress and geometric imperfection can significantly affect the stiffened-plate structures. It is very important to properly account for these imperfections in the ultimate strength assessment. In the present study, only the geometric imperfections are considered although the residual stress can also be included in the analysis by ABAQUS. Three types of initial imperfections are accounted for the plates and stiffeners as the following:

– Plate out-of-flatness with the amplitude of 0.005b
– Stiffener out-of-straightness with the amplitude of 0.001a
– Stiffener sideway with the amplitude of 0.001a.

where a and b are the length and width of the plate between the longitudinal stiffener and transverse frame, respectively. In the present study, the initial geometric imperfections are accounted for in the bottom and inner bottom panel of the nonlinear part with fine mesh by performing relevant finite element analysis (Amlashi & Moan 2008). No geometric imperfections are exerted to the longitudinal girders and transverse web frames in the double bottom.

4 RESULTS AND DISCUSSIONS

In this study, four cases subjected to pure bending in hogging condition have been run with respect to the effect of geometric imperfection, gross scantlings and net scantlings as shown in Table 6 Case CN.3 and CN.4 are modeled on the net scantlings according to CSR-bulk carrier rule (IACS 2006).

The calculations of the progressive collapse of the hull girder of a Capesize bulk carrier have been carried out on a HP workstation xw8200 with Intel Xeon (TM) 3.40 GHz (4 CPUs) and 8 GB RAM. A typical nonlinear run including the progressive collapse beyond the ultimate strength is about 24 hours on this workstation.

4.1 Results from nonlinear finite element analysis

The relationships between the bending moment and the end-rotation are given in Figure 4 for all the four cases by the nonlinear finite element analysis on the bulk carrier. It is observed that the ultimate strengths of the hull girder in the present study are not sensitive to the geometric imperfection. The difference of ultimate strength of the hull girder in hogging is very small with or without consideration of the geometric imperfections. However the ultimate strength based on gross scantlings are about 15% higher than those based on net scantlings. The sharply reductions of the load capacity after the ultimate strength are mainly due to a large part of the bottom collapses simultaneously as the bending moment is increased to collapse.

For case CN.2, 5 different moment levels are marked as numeral 1 through 5 as shown in Figure 4. The collapse phases at the five different moment levels are shown in Figure 5a through Figure 5g. The structural failure starts with yielding at the deck close to the hatch as shown in Figure 5a, marked as numeral 1 in Figure 4. At numeral 2, the collapse with yielding spread over all the deck, upper side plate and wing tank plate while there is no buckling in the bottom as shown in Figure 5b. Due to this, the slope of moment end-rotation curve changed obviously at this phase. From numeral 2 to 3, part of the bottom begins to buckle with the increase of the rotation as shown in Figure 5c. However the bending moment is still increasing due to the stress redistribution between bottom and inner bottom. At numeral 3, the structural

Table 6. Summary of analysis cases.

No.	Scantling	Imperfection
CN.1	gross scantling	No
CN.2	gross scantling	Yes
CN.3	net scantling	No
CN.4	net scantling	Yes

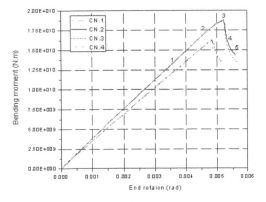

Figure 4. Relationships between moment and end-rotation.

achieved its ultimate load capacity when a large part of the bottom buckled. In this phase the stress redistribution can no longer compensate the load shedding in the bottom. From numeral 3 to 4, the collapse in bottom continues to grow to cover the full width of the region between two adjacent transverse frames and the inner bottom panels begin to collapse with buckling as shown in Figure 5d and Figure 5e. In this phase, the load capacity drops rapidly. From numeral 4 to 5, the load capacity curve continues to drop while the collapse region develops into the hopper tank plate and part of the lower side plate along a line between the two adjacent transverse frames as shown in Figure 5f and Figure 5g.

From the above, it can be seen that the progressive collapse based on the nonlinear finite element analysis on the bulk carrier is restricted to the structural components between two adjacent transverse frames. The transverse frames are strong enough to have no buckling/yielding. No collapse phenomena are observed in the remaining region of the fully nonlinear part. This leads to a similar jack-knife collapse which may lead to the sinking of M.V. Derbyshire (Paik et al. 2008).

4.2 Comparison of ultimate hull girder strength between nonlinear finite element method and simplified methods

The bulk carrier in the present study has been investigated by seven different approaches without nonlinear

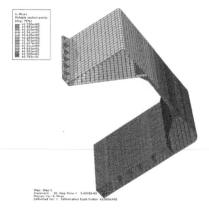

Figure 5a. Progressive collapse of a bulk carrier for case CN.2: Yielding at the deck.

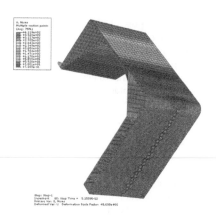

Figure 5d. Progressive collapse of a bulk carrier for case CN.2: Buckling of the total bottom.

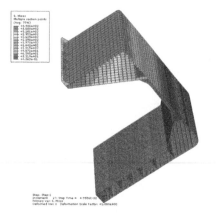

Figure 5b. Progressive collapse of a bulk carrier for case CN.2: Yielding at the deck, upper side and top wing tank.

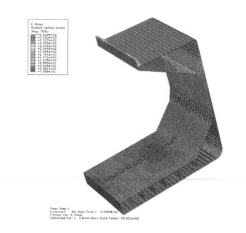

Figure 5e. Progressive collapse of a bulk carrier for case CN.2: Buckling of the inner bottom.

Figure 5c. Progressive collapse of a bulk carrier for case CN.2: Buckling of a large part of the bottom plating.

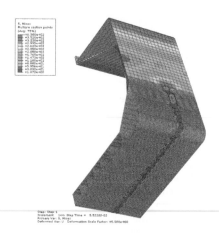

Figure 5f. Progressive collapse of a bulk carrier for case CN.2: Buckling of bottom and lower side.

178

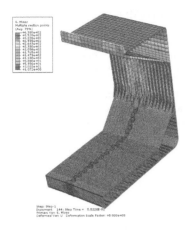

Figure 5g. Progressive collapse of a bulk carrier for case CN.2: Buckling of total inner bottom and hopper tank plate.

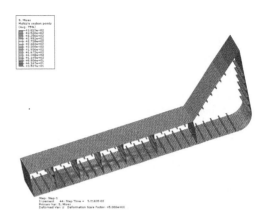

Figure 7. Bottom collapse of a Capesize bulk carrier at ultimate strength: Plate buckling.

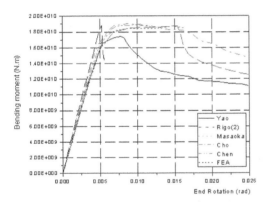

Figure 6. Relationship between moment and end-rotation of a Capesize bulk carrier in hogging by different methods.

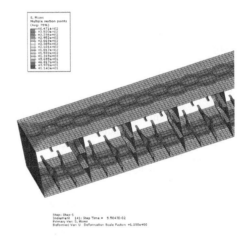

Figure 8. Structural collapse of a Capesize bulk carrier beyond the ultimate strength: Tripping and plate buckling.

finite element method in the benchmark study of ISSC Committee VI.2 (Yao et al., 2000).

In this paper, the ultimate hull girder strength obtained by the nonlinear finite element analysis has been compared with that determined by simplified method. It is noted that the bending moments by simplified methods are functions of the curvature as given in the report, while the results by nonlinear finite element analysis are moment-end rotation curves. In order to make the comparison with the same unit, we define a simple linear relation between the curvature and the end-rotation as,

$$\theta = C \cdot \frac{L}{2} \qquad (1)$$

where θ is the angle of rotation of the end-section of the finite element model, L is the length of the finite element model and C is the curvature. In Figure 6, the

results from different methods are plotted as moment-end rotation curves.

The comparison of the ultimate hull girder strength shows that the results obtained by simplified methods agree relatively well with that by nonlinear finite element analysis. This may attribute to the relatively thick bottom plate and weak interaction between different buckling modes and adjacent structural members when the structure achieves its ultimate strength. Figure 7 shows the collapse mode of the bottom at the ultimate strength. It can be seen that a large part of the bottom collapses due the plate buckling. The capacities beyond the ultimate strength obtained by the nonlinear finite element method and simplified methods show great discrepancy. Compared to the finite element results, both Smith's method (Yao, Rigo and Cho) and ISUM method (Masaoka and

Chen) can not predict the sharp decrease in the load capacity. When the ultimate hull girder strength is obtained, a large part of the bottom plate collapses simultaneously. With respect to the increase of the angle of the end-rotation beyond ultimate strength, the failure of the bottom panels involves several collapse modes such as the overall buckling of the stiffened panel of bottom and inner bottom and the tripping of the longitudinal stiffeners as shown in Figure 8. The yielding and the deformation are concentrated to the region between two adjacent transverse frames while the remaining region experiences elastic unloading. This proves the strong interactions between buckling modes and adjacent structural components which can be accurately simulated by the nonlinear finite element analysis. However due to the rigid transverse plane assumption and without considering interaction between structural components, the Smith's method can not predict the significant load shedding beyond ultimate hull girder strength. As for the ISUM method, more sophisticated elements are still needed to be developed to simulate the load shedding properly.

5 CONCLUSIONS

The ultimate hull girder strength of a Capesize bulk carrier subjected to pure bending in hogging conditions has been investigated by the nonlinear finite element method. The computer code ABAQUS is used in the analysis. Based on the present study, the following conclusions can be drawn,

1. The difference of ultimate strength of the hull girder in hogging condition is very small with or without consideration of the geometric imperfections included in the analysis. However the ultimate strength based on gross scantlings are about 15% higher than those based on net scantlings.
2. The sharp reduction of the load capacity after reaching the ultimate strength is mainly due to the fact that a large part of the bottom collapses simultaneously as the bending moment is increased to achieve the ultimate hull girder strength. The collapse of structural components is concentrated to the region between two adjacent transverse frames while the remaining region is undamaged.
3. The prediction of the ultimate hull girder strength by the simplified methods agree relatively well with that from the nonlinear finite element analysis. However the sharply reduction in the load capacity beyond the ultimate strength can not be captured by the simplified method due to those approximations assumed in the methods.
4. The results of the nonlinear finite element analysis can serve as a benchmark reference to the analyses based on the simplified methods.

ACKNOWLEDGMENTS

The work has been performed in the scope of the project MARSTRUCT, Network of Excellence on Marine Structures, (www.mar.ist.utl/marstruct/), which has been financed by the EU through the GROWTH Programme under contract TNE3-CT-2003-506141. The authors wish to give thanks to the support from this project. The authors acknowledge the financial support from the Research Council of Norway (RCN) through the Centre for Ship and Ocean Structures (CeSOS).

REFERENCES

Amlashi, H.K. & Moan, T. 2008. Ultimate strength analysis of a bulk carrier hull girder under alternate hold loading condition—A case study—Part 1: Nonlinear finite element modelling and ultimate hull girder capacity, *Marine structures*, Vol. 21: 327–352.

Caldwell, J.B. 1965. Ultimate longitudinal strength. *Trans RINA*, Vol. 107: 411–430.

Chen, Y.K., Kutt, L.M., Piaszczyk, C.M. & Bieniek, M.P. 1983. Ultimate strength of ship structures, *SNAME Trans*, Vol. 91: 149–168.

IACS, 2006. Common structural rules (CSR) for bulk carriers, International Association of Classification Societies, London.

Kutt, L.M., Piaszczyk, C.M., Chen, Y.K. & Liu, D. 1985. Evaluation of the longitudinal ultimate strength of various ship hull configurations, *SNAME Trans*, Vol. 93: 33–53.

Mansour, A.E., Yang, J.M., & Thayamballi. 1990. An experimental investigation of ship hull ultimate strength, *SNAME Trans*, Vol. 98: 411–439.

Ostvold, T.K., Steen, E. & Holtsmark, G. 2004. Nonlinear strength analyses of a bulk carrier—a case study. *Proceedings of the ninth PRADS, STG, Hamburg*.

Paik, J.K., Seo, J.K. & Kim, B.J. 2008. Ultimate limit state assessment of the M.V. Derbyshire hull structure, *J. Offshore Mechanics and Arctic Engineering*, Vol. 130.

Simulia, 2007. ABAQUS Analysis User's Manual, Providence, RI, USA,

Smith, C.S. 1977. Influence of local compressive failure on ultimate longitudinal strength of a ship's hull. *Trans PRADS 1977*: 73–79.

Ueda. Y. & Rashed S.M.H. 1984. The idealized structural unit method and its application to deep girder structures, *Comput Struct*, Vol. 18: 277–293.

Valsgaard, S., Jorgensen, L., Boe, A.A. & Thorkildsen, H. 1991. Ultimate hull girder strength margins and present class requirements, *SNAME symposium 91 on marine structural inspection, maintenance and monitoring, Arlington*: B 1–19.

Yao T, et al. 2000. Report committee VI.2: ultimate hull girder strength, *Proceedings of the 14th international ship and offshore structures congress, Nagasaki, Japan*.

Analysis and Design of Marine Structures – Guedes Soares & Das (eds)
© 2009 Taylor & Francis Group, London, ISBN 978-0-415-54934-9

Ultimate strength performance of Suezmax tanker structures: Pre-CSR versus CSR designs

J.K. Paik & D.K. Kim
Pusan National University, Busan, Korea

M.S. Kim
Lloyd's Register Asia, Busan, Korea

ABSTRACT: The objective of the present paper is to investigate the ultimate strength performance charac-teristics of Suezmax class double hull oil tanker structures designed by pre-CSR (Common Structural Rules) versus CSR methods in terms of buckling collapse of stiffened plate structures at deck and bottom part, and also of hull girder collapse. ALPS/ULSAP code for ultimate strength calculations of stiffened plate structures and ALPS/HULL code for progressive hull collapse analysis are employed for this purpose. Three types of structural scantlings, namely net, gross (as-built) and 50% corrosion margin scantlings are considered in association with ultimate strength performance. The insights and conclusions obtained from the present study are documented.

1 INTRODUCTION

It is now well recognized that the ultimate strength is much better basis for design and strength assess-ment of ship structures than the allowable working stress (Paik & Thayamballi 2003, 2007, ISO 2007, 2008). This is also true for condition assessment of aged structures (Paik& Melchers 2008).

This is because the true margin of structural safety is not determined as long as the ultimate strength remains unknown, while the allowable working stress is based on past experience which is often formulated as a fraction of material properties such as yield stress.

Since early 2006, classification societies have applied the common structural rules (CSR) for struc-tural design of tankers (IACS 2008a) and bulkers (IACS 2008b), which are more extensively utiliz-ing the method of ultimate strength ever than before (Hussein et al 2007).

It is interesting to compare the ultimate strength performance of ship structures designed by pre-CSR and CSR methods. In this regard, the research group of the authors have previously performed the ultimate strength investigations for AFRAMAX class tankers (Paik et al 2007), for 300k tankers (Paik et al 2008a), and for 170 k bulkers (Paik et al 2008b). The present study is a sequel to these previous contributions, but with the focus on Suezmax double hull oil tanker structures.

2 THE OBJECT SHIP STRUCTURES

Table 1 indicates the principal dimensions of the object tanker structures, designed by pre-CSR and CSR methods.

It is noted that the two tanker designs are not exactly the same in terms of principal dimensions. It is noted that the two tanker designs are nearly identical except deadweight, length scantlings (L_s) and block coeffi-cient (C_b). The deadweight of CSR design was a bit reduced due to net weight increase in conjunction with stringent CSR strength requirements.

Figure 1 shows a schematic of the mid-ship section design of the object ships, where the stiffened panels

Table 1. The principal dimensions of the object ships.

Parameter	Pre-CSR	CSR
Deadweight	159,000 ton	158,000 ton
LBP	264.0 m	264.0 m
Ls	260.8 m	261.0 m
Breadth	48.0 m	48.0 m
Depth	23.2 m	23.2 m
Draft	16.0 m	16.0 m
C_b(scantlings)	0.836	0.843

LBP = length between perpendiculars, C_b = block coeffi-cient, Ls = length scantlings.

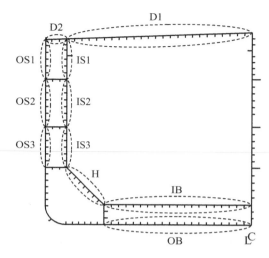

Figure 1. A schematic of the mid-ship section design of the object ship.

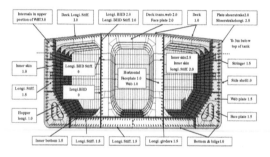

Figure 2a. Corrosion margin values of pre-CSR based tanker structures.

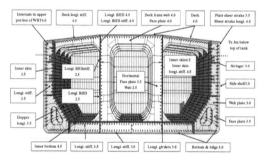

Figure 2b. Corrosion margin values of CSR based tanker structures.

at deck and bottom, surrounded by strong support members, are targeted for the comparison of ultimate strength performance.

It is noted that the structural scantlings of the two designs under consideration are those for building,

indicating that all structural design criteria are satisfied, including possible additions of backing brackets and/or structural scantlings associated with fatigue issues. In fact, the two tanker designs were finalized by fitting backing brackets for the fatigue performance, but rather than increasing stiffener scantlings.

It is important to realize that the corrosion margin values for CSR based tanker structures differ from pre-CSR, as shown in Figure 2.

3 DETERMINATION OF DEMANDS

The ultimate strength performance of stiffened panels can be evaluated as a ratio of the capacity to the demand. For stiffened panels, the ultimate strength is meant to the capacity, while the working stress under design actions is equivalent to the demand.

The working stresses of the stiffened panels can be obtained by the action effect analysis of the hull structure under design hull girder actions. Multiple sets of suspicious situations on the working stresses in terms of possible buckling collapse may need to be considered depending on various design actions including sag or hog conditions and ballast or full load conditions.

Tables 2 and 3 indicate the characteristics values of working stress components of the target stiffened panels at deck and bottom part for pre-CSR and CSR structures, respectively. The values of Table 2 were obtained by LR's ShipRight Structural Design Analysis (SDA) (LR 2008) using 3-holds model under design hull girder actions as shown in Figure 3. The results of Table 3 were obtained by the latest CSR structural strength analysis.

As per the classification societies instruction of the action effect analysis, the working stresses of pre-CSR structures indicated in Table 2 were obtained with

Table 2. Characteristics values of 'suspicious' working stresses and lateral pressure actions of the target stiffened panels of pre-CSR Suezmax tanker on gross (as-built) scantlings.

Panel	σ_x(MPa)	σ_y(MPa)	p (MPa)
D1	180	27	0.030
	190	20	0.030
	200	10	0.030
OB	140	65	0.189
	165	57	0.189
	189	44	0.189
IB	135	30	0.242
	150	10	0.242
	157	20	0.242

D = deck, IB = inner bottom, OB = outer bottom, σ_x, σ_y = longitudinal compressive stresses working in the direction parallel or normal to the ship length direction, respectively, p = lateral pressure.

Table 3. Characteristics values of 'suspicious' working stresses and lateral pressure actions of the target stiffened panels of CSR Suezmax tanker with 50% corrosion margin values.

Panel	σ_x(MPa)	σ_y(MPa)	p (MPa)
D1	218	25	0.052
	220	11	0.052
	222	23	0.052
OB	120	98	0.240
	170	77	0.240
	180	66	0.240
IB	77	62	0.262
	138	48	0.262
	150	40	0.262

D = deck, IB = inner bottom, OB = outer bottom, σ_x, σ_y = longitudinal compressive stresses working in the direction parallel or normal to the ship length direction, respectively, p = lateral pressure.

gross scantlings, while those of CSR structures indicated in Table 3 were obtained with 50% corrosion margin values, i.e., gross (as-built) scantlings minus 50% corrosion margin values.

It is interesting to note that the working stresses of CSR designs tend to be greater than those of pre-CSR designs.

On the other hand, for the ultimate strength performance analysis of hull girders, the capacity is meant to the hull girder ultimate strength obtained by the progressive hull collapse analysis, and the demand is the maximum hull girder bending moment which is a sum of still-water and wave-induced bending moment components at design conditions.

Table 4 indicates the requirements of vertical hull girder ultimate bending capacity specified by CSR. It is noted that all required values indicated in Table 4 were determined by CSR guidance formula, i.e., for both pre-CSR and CSR ships, applying the corresponding principal dimensions of each ship.

The values of required bending moments indicated Table 4 were obtained from the CSR guidance taking account of partial safety factors associated with uncertainties, namely

$$M_{req} \geq \gamma_u \ (\gamma_s M_s + \gamma_w M_w) \qquad (1)$$

where M_{req} = characteristic value of vertical ultimate bending capacity requirement specified by CSR, M_s = characteristic value of still-water bending moment, M_w = characteristic value of wave-induced bending moment, γ_s, γ_w, γ_u = partial safety factors for still-water bending moment ($\gamma_s = 1.0$), wave-induced bending moment ($\gamma_w = 1.2$), and ultimate hull girder strength ($\gamma_u = 1.1$), respectively.

It is noted that the CSR (IACS 2008a) originally specifies Equation (1) only for sagging. For a matter of convenience, however, the present study also

Case 1000610: B06D1 : Mid All Tanks Full (0.6 Tsc)
Von-Mises Stress Membrane (Mean = 92.22)

Figure 3a. A sample of three-holds model used for the action effect analysis of the ships using the ShipRight method in sagging.

Case 1000120: B01D2 : Mid Port Tank Empty (0.9 Tsc)
Von-Mises Stress Membrane (Mean = 107.4)

Figure 3b. A sample of three-holds model used for the action effect analysis of the ships using the ShipRight method in hogging.

Table 4. Requirement of vertical hull girder ultimate bending capacity specified by CSR.

	Required values (GNm)	
Design	Sag	Hog
Pre-CSR	−10.99	11.32
CSR	−10.72	12.58

applies Equation (1) for the requirements of hogging condition.

4 ULTIMATE STRENGTH PERFORMANCE OF STIFFENED PANELS

The characteristic values of the ultimate strength interaction relationships of the target stiffened panels between longitudinal and transverse compressive actions with or without lateral pressure are calculated

183

by ALPS/ULSAP (2008) in conjunction with the corresponding condition of actions.

The effects of the fabrication related initial imperfections are taken into account in the panel ultimate strength computations, namely

$$w_{op} = \frac{b}{200}, \quad w_{oc} = w_{os} = \frac{a}{1000}, \quad \sigma_{rcx} = -0.15\sigma_Y$$

(2)

where w_{op} = maximum plate initial deflection with *buckling mode*, w_{oc} = column type initial deflection of longitudinal stiffeners, w_{os} = sideways initial distortion of longitudinal stiffeners, a = length of stiffeners, b = stiffener spacing, σ_{rcx} = compressive residual stress of plate in the x (length) direction, σ_Y = material yield stress.

While ALPS/ULSAP program deals with the fabrication related initial imperfections as parameters of influence, it is noted that the magnitudes of initial imperfections defined in Equation (2) are meant to be an average or even severe level in today's shipbuilding industry.

As a result, it is fairly expected that the ultimate strength computations may be somewhat pessimistic. In this regard, the authors plan to perform the similar study but with more practical or slight level of initial imperfections, which will be presented in near future.

The panel ultimate strength calculations are undertaken with different scantlings associated with the corresponding corrosion margin values, i.e., net scantlings deducing 100% corrosion margin values and gross scantlings including 100% corrosion margin values.

One may be cautioned that the working stresses were predicted on partial scantlings with deducing 50% corrosion margins for CSR design, but on gross scantlings without deducing corrosion margin values for pre-CSR design.

4.1 Deck panels

The ultimate strength performance of the deck stiffened panel (D1) is now compared.

Figure 4 compares the deck panel ultimate strength interaction relationships for pre-CSR and CSR designs on net or gross (as-built) scantlings. The working stresses indicated in Tables 2 and 3 are also plotted in the corresponding ultimate strength relationships. The lateral pressure actions caused by sloshing and oil vaporization are also considered in the ultimate strength computations.

4.2 Outer bottom panels

Figure 5 compares the outer bottom (OB) panel ultimate strength interaction relationships for pre-CSR and CSR designs on net or gross (as-built) scantlings

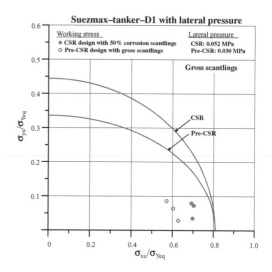

Figure 4a. Ultimate strength interaction relationships of the deck stiffened panel (D1) on gross scantlings.

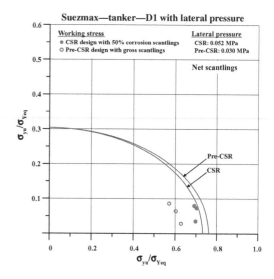

Figure 4b. Ultimate strength interaction relationships of the deck stiffened panel (D1) on net scantlings with deducing 100% corrosion margin values.

with lateral pressure actions. The working stresses indicated in Tables 2 and 3 are also plotted in the figures.

4.3 Inner bottom panels

Figure 6 compares the inner bottom (IB) panel ultimate strength interaction relationships for pre-CSR

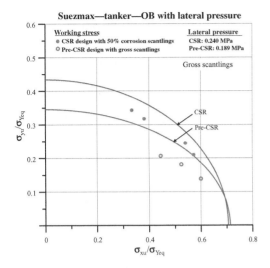

Figure 5a. Ultimate strength interaction relationships of the outer bottom stiffened panel (OB) on gross scantlings.

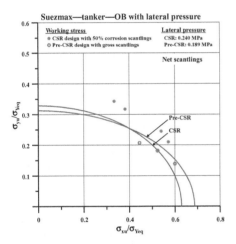

Figure 5b. Ultimate strength interaction relationships of the outer bottom stiffened panel (OB) on net scantlings.

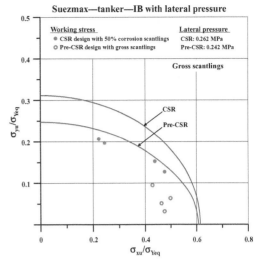

Figure 6a. Ultimate strength interaction relationships of the inner bottom stiffened panel (IB) on gross (as-built) scantlings.

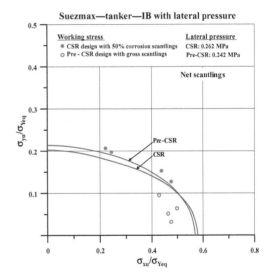

Figure 6b. Ultimate strength interaction relationships of the inner bottom stiffened panel (IB) on net scantlings.

and CSR designs on net or gross scantlings with lateral pressure actions. The working stresses indicated in Tables 2 and 3 are also plotted in the figures.

4.4 Discussions

Figure 4(a) shows that the ultimate strength performance of deck stiffened panels by CSR design is similar to that by pre-CSR in terms of gross scantlings when the longitudinal compressive actions are predominant. However, the CSR design gives much greater performance as the transverse compressive actions become dominant.

In terms of net scantlings, the ultimate strength performance of deck stiffened panels is similar between pre-CSR and CSR designs as shown in Figure 4(b), although it is surmised that the net scantlings by CSR design are slightly smaller than those by pre-CSR when the longitudinal axial compressive actions are predominant.

185

Similar observations to deck stiffened panels are relevant for outer and inner bottom stiffened panels as shown in Figures 5 and 6.

It is interesting to note that the working stresses (demand) well exceed the ultimate strength (capacity) of outer and inner bottom stiffened panels in terms of net scantlings, i.e., without corrosion margin values.

Tables 5(a) to 5(e) summarize the ultimate strength performance investigations of the target stiffened panels. In these tables, the indices associated with both working stress and ultimate strength are defined as follows

$$\sigma_D = \sqrt{\sigma_x^2 + \sigma_y^2}, \quad \sigma_C = \sqrt{\sigma_{xu}^2 + \sigma_{yu}^2} \quad (3)$$

where, σ_x, σ_y = longitudinal compressive stresses working in the direction parallel or normal to the ship

Table 5a. Summary of the ultimate strength performance investigations of the target stiffened panels with Pre-CSR Suezmax tanker on gross scantlings.

Panel	Working stress			Ultimate strength			
	σ_x	σ_y	σ_D	σ_{xu}	σ_{yu}	σ_C	σ_C/σ_D
D1	198.0	9.0	198.2	242.6	23.6	243.7	1.229
OB	189.0	44.0	194.1	201.6	45.7	206.7	1.065
IB	157.0	20.0	157.3	179.6	23.6	181.1	1.144

Note: Working stress was calculated for gross scantlings.

Table 5b. Summary of the ultimate strength performance investigations of the target stiffened panels with Pre-CSR Suezmax tanker with net scantlings.

Panel	Working stress			Ultimate strength			
	σ_x	σ_y	σ_D	σ_{xu}	σ_{yu}	σ_C	σ_C/σ_D
D1	198.0	9.0	198.2	226.8	10.0	227.0	1.145
OB	189.0	44.0	194.1	192.2	44.1	197.2	1.016
IB	157.0	20.0	158.3	170.1	22.1	171.5	1.084

Note: Working stress was calculated for gross scantlings.

Table 5c. Summary of the ultimate strength performance investigations of the target stiffened panels with CSR Suezmax tanker on gross (as-built) scantlings.

Panel	Working stress			Ultimate strength			
	σ_x	σ_y	σ_D	σ_{xu}	σ_{yu}	σ_C	σ_C/σ_D
D1	222.0	23.0	223.2	245.0	22.1	245.9	1.102
OB	170.0	77.0	186.6	179.6	81.9	197.4	1.057
IB	138.0	48.0	146.1	160.7	55.1	169.9	1.162

Note: Working stress was calculated for 50% corrosion margin scantlings.

Table 5d. Summary of the ultimate strength performance investigations of the target stiffened panels with CSR Suezmax tanker on net scantlings.

Panel	Working stress			Ultimate strength			
	σ_x	σ_y	σ_D	σ_{xu}	σ_{yu}	σ_C	σ_C/σ_D
D1	222.0	23.0	223.2	223.7	24.3	224.9	1.007
OB	170.0	77.0	186.6	152.4	70.6	167.9	0.901
IB	138.0	48.0	146.1	126.0	47.3	134.6	0.921

Note: Working stress was calculated for 50% corrosion margin scantlings.

Table 5e. Summary of the safety factor investigations for the target stiffened panels between Pre-CSR versus CSR designs.

Panel	Pre-CSR		CSR	
	Gross	Net	Gross	Net
D1	1.229	1.145	1.102	1.007
OB	1.065	1.016	1.057	0.901
IB	1.144	1.084	1.162	0.921

length direction, σ_{xu}, σ_{yu} = longitudinal ultimate compressive strengths in the direction parallel or normal to the ship length direction, respectively.

5 ULTIMATE STRENGTH PERFORMANCE OF HULL GIRDERS

The characteristic values of the hull girder ultimate strength are calculated by the progressive hull collapse analysis using ALPS/HULL (2008) in sagging and hogging conditions.

The same levels of fabrication related initial imperfections indicated in Equation (2) are taken into account. Three types of structural scantlings, namely net scantling, gross scantling and 50% corrosion margin scantlings are considered.

5.1 Pre-CSR designs

Figure 7(a) compares the vertical bending moments versus the curvature curves of pre-CSR design hulls with three types of structural scantlings. The required values of bending moments are also plotted in the figure. Figure 7(b) shows the variations of the neutral axis position of pre-CSR design hulls as the bending curvature increases. CSR designs.

5.2 CSR designs

Figure 8(a) compares the vertical bending moments versus the curvature curves of CSR design hulls with

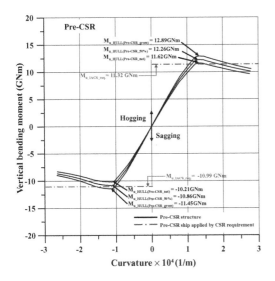

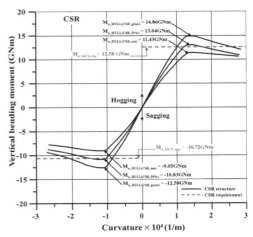

Figure 7a. Vertical bending moment versus curvature curves for pre-CSR design hulls on net, gross (as-built) and 50% corrosion margin scantlings.

Figure 8a. Vertical bending moment versus curvature curves for CSR design hulls on net, gross (as-built) and 50% corrosion margin scantlings.

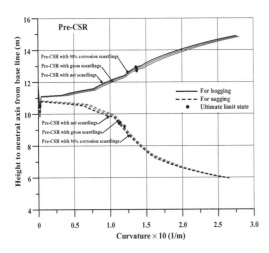

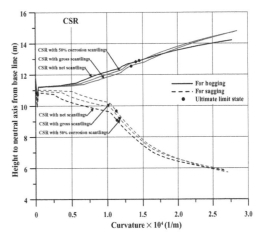

Figure 7b. Variations of the neutral axis positions for pre-CSR design hulls on net, gross (as-built) and 50% corrosion margin scantlings.

Figure 8b. Variations of the neutral axis positions for CSR design hulls on net, gross (as-built) and 50% corrosion margin scantlings.

three types of structural scantlings. The required values of bending moments are also plotted in the figure. Figure 8(b) shows the variations of the neutral axis position of CSR design hulls as the bending curvature increases.

5.3 Pre-CSR versus CSR designs

Figure 9 compares the vertical bending moments versus the curvature curves of pre-CSR versus CSR

design hulls. Figures 9(a), 9(b) and 9(c) represent the comparisons of the vertical bending moments versus the curvature curves of the hulls with net, gross (as-built) and 50% corrosion margin scantlings, respectively. Figures 10(a), 10(b) and 10(c) compare the variations of neutral axis positions of pre-CSR versus CSR design hulls on net, gross (as-built), and 50% corrosion margin scantlings, respectively.

5.4 Discussions

Table 6 summarizes the hull girder ultimate strength performance investigations of the object ships between

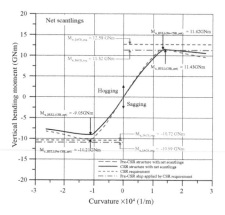

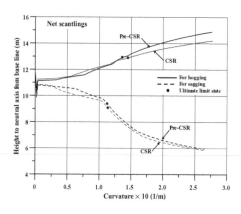

Figure 9a. Vertical bending moment versus curvature curves of pre-CSR versus CSR design hulls on net scantlings.

Figure 10a. Variations of the neutral axis positions of pre-CSR versus CSR design hulls on net scantlings.

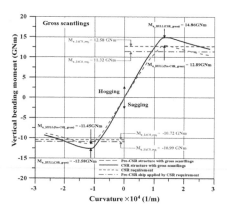

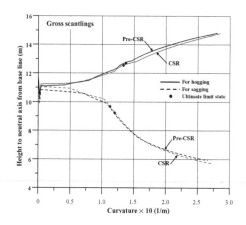

Figure 9b. Vertical bending moment versus curvature curves of pre-CSR versus CSR design hulls on gross (as-built) scantlings.

Figure 10b. Variations of the neutral axis positions of pre-CSR versus CSR design hulls on gross (as-built) scantlings.

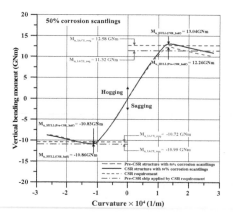

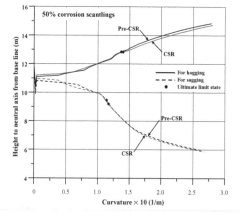

Figure 9c. Vertical bending moment versus curvature curves of pre-CSR versus CSR design hulls on 50% corrosion margin scantlings.

Figure 10c. Variations of the neutral axis positions of pre-CSR versus CSR design hulls on 50% corrosion margin scantlings.

Table 6a. Summary of the ultimate hull girder strength performance investigations of pre-CSR design in hogging.

Scantling	Demand $M_{u\_hog\_req.}$ (GNm)	Capacity $M_{u\_hog\_HULL)}$ (GNm)	S.F. $M_{u\_hog\_req.}/M_{u\_hog\_HULL}$
Gross scantlings	11.32	12.89	1.139
50% corrosion scantlings	11.32	12.26	1.083
Net scantlings	11.32	11.62	1.027

Table 6b. Summary of the ultimate hull girder strength performance investigations of pre-CSR design in sagging.

Scantling	Demand $M_{u\_sag\_req.}$ (GNm)	Capacity $M_{u\_sag\_HULL}$ (GNm)	S.F. $M_{u\_hog\_req.}/M_{u\_hog\_HULL}$
Gross scantlings	−10.99	−11.45	1.042
50% corrosion scantlings	−10.99	−10.86	0.988
Net scantlings	−10.99	−10.21	0.929

Table 6c. Summary of the ultimate hull girder strength performance investigations of CSR design in hogging.

Scantling	Demand $M_{u\_hog\_req.}$ (GNm)	Capacity $M_{u\_hog\_HULL}$ (GNm)	S.F. $M_{u\_hog\_req.}/M_{u\_hog\_HULL}$
Gross scantlings	12.58	14.86	1.181
50% corrosion scantlings	12.58	13.04	1.037
Net scantlings	12.58	11.43	0.909

Table 6d. Summary of the ultimate hull girder strength performance investigations of CSR design in sagging.

Scantling	Demand $M_{u\_sag\_req.}$ (GNm)	Capacity $M_{u\_sag\_HULL}$ (GNm)	S.F. $M_{u\_sag\_req.}/M_{u\_sag\_HULL}$
Gross scantlings	−10.72	−12.58	1.174
50% corrosion scantlings	−10.72	−10.83	1.010
Net scantlings	−10.72	−9.05	0.844

Table 6e. Summary of the ultimate hull girder safety factor investigations of pre-CSR versus CSR designs.

Scantling	Hogging Pre-CSR	CSR	Sagging Pre-CSR	CSR
Gross scantlings	1.139	1.181	1.042	1.174
50% corrosion scantlings	1.083	1.037	0.988	1.010
Net scantlings	1.027	0.909	0.929	0.844

pre-CSR and CSR designs, with three different types of structural scantlings, while the CSR originally specifies the requirements, i.e., Equation (1), only for sagging, but the present study also applies the same requirements for hogging, for a matter of convenience.

The safety factors of ships against hull girder collapse depend on the types of structural scantlings, among other factors. It is found from Table 6(e) that the safety factor of CSR design hull with net scantlings is smaller than 1.0 in both hogging and sagging. On the other hand, the safety factor of pre-CSR design hull with 50% corrosion margin scantlings or net scantlings is smaller than 1.0 in sagging.

The computed results of safety factors indicated in Table 6 were obtained without considering the effect of lateral pressure actions. The safety factors against hull girder collapse may presumably be decreased further by taking into account the effect of lateral pressure actions on plate panels.

6 CONCLUDING REMARKS

The aim of the present paper has been to investigate the ultimate strength performance characteristics of Suez-max class double hull oil tanker structures, designed in compliance with pre-CSR and CSR requirements. The safety factors for stiffened panels and hull girders were evaluated against buckling collapse.

One may be cautioned that the level of fabrication related initial imperfections considered in the ultimate strength computations is an average or even severe as defined in Equation (2) in today's shipbuilding industry practice. As a result, the present computations of ultimate strength for stiffened panels and hull girders may indicate somewhat pessimistic strength performance of ship structures.

The insights and conclusions developed in the present paper will be very useful for ultimate strength design of tanker structures in conjunction with CSR design technologies.

ACKNOWLEDGEMENTS

The present study was undertaken at the Lloyd's Register Educational Trust (LRET) Research Centre of Excellence at Pusan National University, Korea. The authors are pleased to acknowledge valuable comments from Dr. F. Cheng of Lloyd's Register London, Mr. M. Franklin of LRET, and Messrs. R. Tustin and L. Benito of Lloyd's Register Asia.

REFERENCES

ALPS/HULL. 2008. A computer program for progressive collapse analysis of ship hulls. Advanced Technology Center, DRS C3 Systems, Inc., MD, USA.

ALPS/ULSAP. 2008. A computer program for ultimate limit state analysis of stiffened plate structures. Advanced Technology Center, DRS C3 Systems, Inc., MD, USA.

Hussein, A.W., Teixeira, A.P. and Guedes Soares, C. 2007. Impact of the new common structural rules on the reliability of a bulk carrier. *Advancements in Marine Structures*, 12–14 March, Glasgow, UK.

IACS. 2008a. Common structural rules for double hull oil tankers. International Association of Classification Societies, London, UK, July.

IACS. 2008b. Common structural rules for bulk carriers. International Association of Classification Societies, London, UK, July.

ISO. 2007. International standard ISO 18072-1, Ships and marine technology—Ship structures, Part 1: General requirements for their limit state assessment, International Organization for Standardization, Geneva.

ISO. 2008. International standard ISO DIS 18072-2, Ships and marine technology—Ship structures, Part 2: Requirements for their ultimate limit state assessment, International Organization for Standardization, Geneva.

LR. 2008. User's manual for ShipRight structural design analysis (SDA). Lloyd's Register, London, UK.

Paik, J.K., Hwang, S.W., Park, J.S. and Kim, M.S. 2008a. Ultimate limit state performance of 300k double hull oil tanker structures: Pre-CSR versus CSR designs. Proceedings of ASRANet Colloquium, 25–27 June, Athens, Greece.

Paik, J.K., Kim, J.Y., Jung, J.M. and Kim, M.S. 2008b. Ultimate limit state performance of 170 k bulk carrier structures. Proceedings of ASRANet Colloquium, 25–27 June, Athens, Greece.

Paik, J.K., Kim, B.J. and Seo, J.K. 2007. Evaluation of IACS common structural rules in terms of ultimate limit state design and assessment of ship structures. Proceedings of the 26th International Conference on Offshore Mechanics and Arctic Engineering (OMAE 2007), OMAE2007-29187, 10–15 June, San Diego, California, USA.

Paik, J.K. and Melchers, R.E. 2008. Condition assessment of aged structures, CRC Press, New York, USA.

Paik, J.K. and Thayamballi, A.K. 2003. Ultimate limit state design of steel-plated structures. John Wiley & Sons, Chichester, UK.

Paik, J.K. and Thayamballi, A.K. 2007. Ship-shaped offshore installations: Design, building, and operation. Cambridge University Press, Cambridge, UK.

Coatings and corrosion

Analysis and Design of Marine Structures – Guedes Soares & Das (eds)
© 2009 Taylor & Francis Group, London, ISBN 978-0-415-54934-9

Large scale corrosion tests

Pawel Domzalicki & Igor Skalski
Ship Design and Research Centre (CTO), Gdansk, Poland

C. Guedes Soares & Y. Garbatov
Centre for Marine Technology and Engineering (CENTEC), Technical University of Lisbon,
Instituto Superior Técnico, Lisboa, Portugal

ABSTRACT: Corrosion effects of large scale specimens exposed under Baltic sea water conditions were examined in CTO floating corrosion laboratory. The tested specimens were in the form of box girders simulating the midship section and they were exposed in flowing fresh Baltic Sea water. The corrosion rate of carbon steel exposed in sea water under natural conditions is very low so it was necessary to increase the corrosion rate to observe a significant reduction of plate thickness during the experiments. The corrosion rate was increased by modification of physical and chemical parameters of the tested systems. The modification consisted of increasing the sea water temperature and its agitation. Anodic electric current supplied by an external source was used in some cases. The influence of anodic current on the corrosion phenomenon was modeled with the boundary element method using isoparametric elements. The corrosion rate distribution on the box girders' surfaces was modeled with back propagation artificial neural networks.

1 INTRODUCTION

Materials commonly used for marine structures are carbon and low alloyed steels. In seawater environment they are affected by corrosion, which is one of most important factors that influence the durability of the structures.

The controlling of corrosion degradation is usually based on periodical inspections and on installing corrosion monitoring systems (Rothwell & Tullmin 2000, Zayed et al 2007, Panayotova et al, 2008).

Because of the great importance of corrosion in the prediction of structures' lifetime and of the inconvenience of inspections and monitoring, several attempts were done to workout mathematical models of corrosion degradation processes. An example of such model applicable in seawater is given in Melchers (2006).

However to be applicable in practice, the mathematical models used for describing corrosion development are calibrated by fitting to data obtained during service (Garbatov et al., 2007) or data collected in experiments. One of the main difficulties found during designing such experiments is the dependence of corrosion on many environmental factors (Guedes Soares et al, 2008). Although several attempts have been made to identify those effects and to model them, a satisfactory situation has not yet been achieved.

On a complementary direction, there has been some work reported in predicting the ultimate strength of components and structures subjected to corrosion, (Guedes Soares and Garbatov, 1999, Teixeira and Guedes Soares, 2008), but there have been no experimental comparisons. In principle one would expect that the validations made in non-corroded structures would still apply, but there are some worries that the mechanical properties of corroded material may change also (Boon, 2007).

Therefore a programme was set up to induce corrosion in a structural test specimen, which will be later subjected to a strength test. Specimens that represent some features of a midship section and which are similar to one that has been tested in a non-corroded section (Gordo and Guedes Soares, 2008), were chosen to be subjected to corrosion under different conditions.

This paper describes large-scale corrosion experiments performed in the CTO floating corrosion laboratory (shown in Figure 1), designed for material

Figure 1. CTO floating corrosion laboratory located in Gdynia Shipyard, Poland.

and structure elements testing in natural conditions in seawater open system.

2 EXPERIMENTAL PROCEDURE

Specimens in form of box girders modelling ship's midsection were used for testing. The dimensions of the specimens were $1400 \times 800 \times 600$ mm.

The box girders were made of steel of minimum yield point equal to 235 MPa. Table 1 presents the results of chemical analysis performed using a SOPECTROLAB-type LAB 05 equipment. The box girders were welded by the FCAW method with use of OK. Tubrod 15.14 at. flux-cored wire produced by ESAB Ltd Poland.

Three box girders were fabricated and two were subjected to the corrosion programme. One of them was exposed in hot seawater and another one in cold seawater.

Corrosion tests were performed in the CTO floating corrosion laboratory. The specimens were exposed in Baltic seawater. One of the box girders was tested in cold water and the other one in hot water. The box girders were placed in large tanks and seawater was pumped into the tanks continuously.

In some cases modifications of the test conditions were applied. The temperature of seawater was increased and additionally oxygen depolarisation sub process rate was increased by agitation of seawater which resulted in corrosion rate increase.

As another method of corrosion rate acceleration anodic polarisation of the metal's surface was used. Anodic electric current supplied by an external source was applied. As a current source three phase transformer with silicon controlled rectifier was used. During experiments output voltage values of current source were recorded. The voltage drop caused by electrical connections between current source and model and cathode was measured. The value was estimated as 0.55 V. Results of three experiments are presented and compared in this paper: one test carried out in hot water with no anodic polarization and two

tests performed with application of anodic polarization in cold and hot water.

Tests durations were as follows: 30 days for experiments with application of external electrical current and 90 days for test performed without polarisation.

Corrosion wastages were estimated by thickness measurements undertaken at the same places before and after the test, with use of a DM2E instrument made by Krautkrämer-Branson. General corrosion wastage was measured with a U 361 dynamometer manufactured by BLM-USA equipped with amplifier KWS 673.D3 made by Hottinger-Germany. Physical and chemical parameters of sea water were recorded during the experiments. The models were cleaned of corrosion products once a week.

3 CALCULATIONS

Electrical fields observed during polarisation of electrochemical corrosion systems can be modelled numerically with the boundary element method (BEM) (Adey & Niku 1992).

An influence of anodic current supplied by external source on corrosion phenomenon was modelled by BEM method. The anodic current distribution on box girder surface was calculated. The calculations were based on known anodic voltage value.

To model the thickness loss distribution on the specimens' surfaces, multilayer artificial neural networks optimized with error back propagation algorithm were applied.

3.1 *BEM calculations*

Sea water resistivity value and boundary conditions on cathode and anode were needed to perform the calculations.

The influence of Baltic sea water's temperature on its resistivity was determined experimentally by heating a sample of sea water and measuring it's conductivity. Linear relationship of the values was assumed and therefore a linear approximation of the measured data by least squares method was used for obtaining the equation.

The relation of temperature and resistivity and the applicable equation are presented in Figure 2.

BEM calculations were performed for 2-D model presented in Figure 3. A cathode mounted inside the box girder is shown in Figure 4. A linear relationship was assumed between the polarisation current and the potential on both anodes (box girder) and cathode. According to the equation presented in Figure 2 the resistivity of cold seawater (for 2.1°C) was estimated as 1.4 Ohm m and of hot sea water (for 45.2°C): 0.53 Ohm.

Table 1. Chemical composition of steel used for manu-facturing of the tested models.

Element	Concentration
C	0.079
Mn	0.612
Si	0.017
P	<0.001
S	0.00133
Cr	0.0115
Ni	<0.001
Cu	0.0474
Fe	remainder

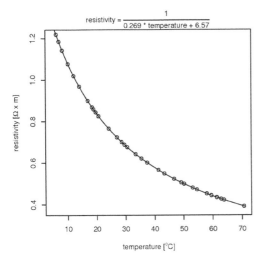

resistivity = $\dfrac{1}{0.269 \, * \, temperature + 6.57}$

Figure 2.   Baltic sea water resistivity vs temperature.

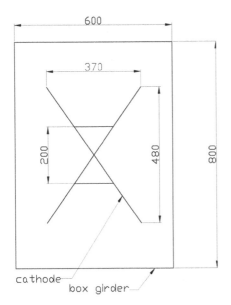

Figure 3.   Geometry of simulated model.

Figure 4.   The cathode mounted inside of the box girder.

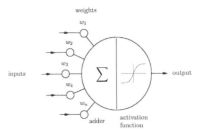

Figure 5.   Neuron—basic element of neural network.

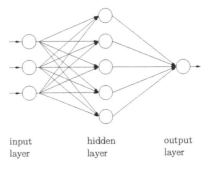

Figure 6.   Structure of an artificial neural network.

3.2   *Neural network modelling*

Artificial neural networks developed to be alike biological neural networks are applied for analysis of complex problems, difficult or impossible to be described with mathematical equations. In particular the neural networks are used for solving problems of unknown number of factors influencing the output values.

Neurons applied in neural networks—Figure 5— imitate functioning logic of biological neurons. Input elements (dendrites) supply signals to the neuron. These values, weighted with synaptic weights, are added up and after processing by the activation function are transmitted to other neurons through the output element (neurites). Learning of network consists in determination of such synaptic weights values that on assigning applicable values to the network input on its output appeared a approximated required reply. The neural network consists of neurons arranged in layers—Figure 6.

Artificial neural networks are usually constructed of three layers: input layer, hidden layer and output layer. For analysis of exceptionally complex problems four-layer networks with two hidden layers are applied. The number of neurons used in the input and output layers depend on the number of input and output data presented in sets used for learning of the network. The number of neurons used in hidden layers is determined experimentally and has a direct effect on the accuracy of the model representation.

Too low number of neurons leads to an over-generalization of the dependence and does not allow to obtain an optimum solution. It is caused by too low degrees of freedom. Too high number of neurons may result in over-learning of the network, some-times introducing dependencies not presented in the analysed case.

Learning of artificial neural networks with error backpropagation, operating in supervised learning mode, consists in repeating of writing data to the net-work input, forward propagation of this data towards output and comparing the results obtained on the out-put with expected results. Outputs' errors are passed as corrections introduced into synaptic weights in the output layer and in preceding layers.

The learning process repeated many times leads to calculation of adequate values of synaptic weights. Obtaining optimum values of weights is based on statistics of sets applied at the time of network learning (Paik et al, 1998). The learning process is continued until results with acceptable error related to expected generalization of dependences are found.

In the case presented here the coordinates of the measurement points were passed to neurons' inputs of artificial neural network. Calculated corrosion rates were passed to output neurons at the time of network learning. Changing the number of hidden neurons adequate generalisation level of relationships was achieved. After learning the networks models of tested relations were generated.

4 TESTS RESULTS

In Table 2 average values of main parameters of the tests are presented.

Figure 7. Box girder after test in hot sea water without anodic polarization.

Figure 8. Box girder after test in hot sea water with anodic polarization.

Table 2. Testing conditions and weight loss.

	Test in hot water without polarization, sea water	Test in hot water with polarization	Test in cold water with polarization
Flow [dm³/h]	308	272	3621
Temp. [°C]	48	45.2	2.1
Anodic current [A] anodic	–	51.7	50.8
Voltage [V] oxygen			
concentration	–	3.47	6.4
[mg/dm³]	5.77	2.57	12.1
pH	7.93	4.82	7.58

After the tests were completed the box girders were covered with iron corrosion products. Figure 7 shows box girder after test in hot seawater without anodic polarization.

Girders after tests with anodic polarisation and exposure in hot seawater and cold seawater are shown in Figure 8 and Figure 9 respectively.

After the tests were completed the box girders were covered with iron corrosion products. Figure 7 shows box girder after test in hot sea water without anodic polarization.

The girders are shown in Figure 8 and Figure 9 respectively after tests with anodic polarisation and exposure in hot seawater and cold sea water.

The total weight loss observed during hot sea water test without anodic polarization was 37 kg (13% of initial weight).

In cases of anodically polarized models the values were: for hot sea water test 56 kg (23% of initial weight), for cold sea water 42 kg (15% of initial weight).

Figure 9. Box girder after test in cold sea water with anodic polarization.

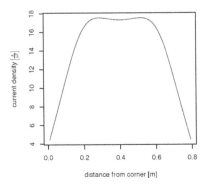

distance from corner [m]

Figure 10. BEM simulation results for left and right side of the model exposed in hot sea water.

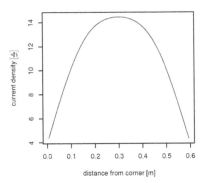

distance from corner [m]

Figure 11. BEM simulation results for top and bottom side of the model exposed in hot water.

It must be taken into consideration that the time of experiments carried out without polarisation of girders' surfaces polarisation was three times longer than in other cases so that the monthly weight loss for

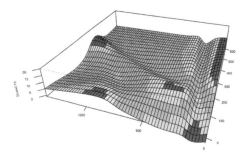

Figure 12. Box girder tested in hot sea water—corrosion rate distribution on top part of the model.

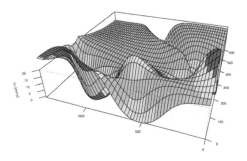

Figure 13. Box girder tested in hot sea water—Corrosion rate distribution on bottom part of the model.

model tested in hot water without polarization was 12.3 kg.

Theoretical metal weigh loss caused by anodic current influence can be expressed by Faraday's law:

$$m = k \, i \, t \qquad (1)$$

where: m—metal loss [kg], k—metal electrochemical equivalent [kg/(As)], i—electric current [A], t—time [s].

The value of k for steel is 2.9e−07 kg/(As).

Calculations based on the above equation shows the weight loss caused by current flow used in experiments was 38 kg.

The current density distributions for the model exposed in hot water were calculated by the BEM method and are presented in Figures 10 and 11. Calculated anodic voltage values required to achieve 50 A of anodic current were: 2.89 V for hot sea water and 5.86 V in case of cold sea water calculations.

Corrosion rate distributions on the box girders' surfaces were modelled with neural networks optimised with an algorithm of back propagation of errors. Measured thickness loss values were passed to the neural network's input. The results calculated for the case of test in hot seawater are presented in Figure 12 and Figure 13.

5 CONCLUSIONS

As expected increased temperature resulted in raising of the tested girders corrosion rate.

It was observed that anodic polarisation strongly influences corrosion processes of steel structures in seawater. The effect was expected indeed.

Besides the steel dissolution caused by anodic reaction some additional effects were observed. The most evident phenomenon was decreased value of pH observed in test tank as compared to the ambient sea water. The effect was highly strengthened during the tests performed in hot water. It was caused by hydrolysis of iron chloride produced in large quantities during the test.

The observed effect of this phenomenon was the evolution of hydrochloric acid and in consequence the decreasing of the pH value. Similar mechanism is usually observed in cases of local corrosion development—pitting or crevice corrosion. This is self accelerating process since hydrochloric acid increases iron dissolution rate which results in more acid production.

The phenomenon is more significant in the tests carried out in hot sea water where, besides the temperature influence, sea water flow was lower than in the test performed in cold sea water. Additional acceleration of this effect is expected during the electrolysis process because chloride ions are transported in the electrical field to the anode surface of the tested model.

The comparison of BEM simulations and experimental results shows good correlation of calculated and measured anodic voltage for both temperature values. The results of the numerical calculations show non-uniform distribution of current density on the surface. The lowest values of current density are observed in the models' corners.

The generalisation of the effect presented with an artificial neural network method indicated nonuniform distribution of corrosion rate, but it was not correlated with the calculated current density distribution. It seems that the local sea water's flow exerted higher influence on the corrosion rate than local current density variations.

Tests performed in sea water of high temperature resulted in strengthened influence of temperature on corrosion wastage: 23% of initial weight—15% wastage was observed in case of the model tested in cold water. High temperature also resulted in low oxygen concentration—it was found as more negative values of redox potential and observed in form of less oxidised (darker) corrosion deposits on model tested in hot water than on model tested in cold water.

ACKNOWLEDGEMENTS

This work has been performed within the project "MARSTRUCT—Network of Excellence on Marine Structures" (http://www.mar.ist.utl.pt/marstruct/) and has been partially funded by the European Union through the Growth programme under contract TNE3-CT-2003-506141.

REFERENCES

Adey, R.A. & Niku, S.M. (1992). Computer Modeling of Corrosion Using the Boundary Element Method. in Munn R.S., editor. Computer modelling in corrosion. ASTM STP 1154.

Boon, B. (2007). A Plea for Large Scale Testing, *Advancements in Marine Structures,* C. Guedes Soares and P.K. Das, (Editors) Taylor & Francis; London, UK, pp. 351–356.

Garbatov, Y., Guedes Soares, C. & Wang, G. (2007). "Nonlinear Time Dependent Corrosion Wastage of Deck Plates of Ballast and Cargo Tanks of Tankers", *Journal of Offshore Mechanics and Arctic Engineering*, 129(1):48–55.

Gordo, J.M. & Guedes Soares, C. (2008). Experimental Evaluation of the Behaviour of a Mild Steel Box Girder under Bending Moment, *Ships and Offshore Structures*, 3(4):347–358.

Guedes Soares, C. & Garbatov, Y. (1999). Reliability of Maintained Corrosion Protected Plate Subjected to Non-Linear Corrosion and Compressive Loads. *Marine Structures*, 12(6):425–446.

Guedes Soares, C., Garbatov, Y., Zayed, A. & Wang, G. (2008). Influence of Environmental Factors on Corrosion of Ship Structures in Marine Atmosphere. *Corrosion Science*, 50:3095–3106.

Melchers, R.E. (2006). Examples of mathematical modelling of long term general corrosion of structural steels in sea water. *Corrosion Engineering Science and Technology*, 41(1):38–44.

Panayotova, M., Garbatov, Y. & Guedes Soares, C. (2008), Corrosion Monitoring of Ship Hulls, *Maritime Industry, Ocean Engineering and Coastal Resources*, C. Guedes Soares, P. Kolev, (Editors). Taylor & Francis; London—UK: pp. 263–270.

Papik, K., Molnar, B., Schaefer, R., Dombowari, Z., Tulassay, Z. & Feher, J. (1998). Application of neural networks in medicine a review. *Medical Science Monitor, Diagnostics and Medical Technology*, 4(3):538–546.

Rothwell, N. & Tullmin, M. (2000). *The Corrosion monitoring Handbook.* Coxmoor Publishing Company.

Teixeira, A.P. & Guedes Soares, C. (2008). Ultimate Strength of Plates with Random Fields of Corrosion. *Structure and Infrastructure Engineering*, 4(5):363–370.

Zayed, A., Garbatov, Y. & Guedes Soares, C. (2007). Factors Affecting the Non-destructive Inspection of Marine Structures, *Advancements in Marine Structures*, C. Guedes Soares and P. K. Das, (Editors) Taylor & Francis; London, UK, pp. 565–576.

Analysis and Design of Marine Structures – Guedes Soares & Das (eds)
© 2009 Taylor & Francis Group, London, ISBN 978-0-415-54934-9

Anticorrosion protection systems—improvements and continued problems

A. Ulfvarson
Marine Structural Engineering, Chalmers University of Technology, Sweden

K. Vikgren
Vikgren International, Sweden

ABSTRACT: The paper is about anticorrosion protection systems for steel ships, in particular tankers and bulk carriers. The New IMO Performance Standard for Water Ballast Tank Coatings are based on the INTERTANKO recommendations for coatings lasting up to 15 years in ballast tanks. This new standard was discussed during a workshop at Chalmers University of Technology in May 2008 and was seen by participants as already being below the state of the practice of most European ship owner's requirements, not at least regarding the qualifications and experience of the coating inspectors. The new standards require one inspector with integrity and clear responsibly. It is well known that coating layers, which are too thick, may crack due to the residual stresses of the coating itself. An effort to quantify the ageing of coatings due to humidity and heat has been made in the project SAFECOAT, where test specimens after ageing were cyclically loaded. There exist field observations of cracking due to the combination of ageing of coating and local high strain due to repeated load-cycles from full load to ballast. Test results from the project SAFECOAT was discussed and related to the praxis of stripe coating around edges. The paper gives overview of the new regulations and the experience and concludes with advice for further research.

1 INTRODUCTION

In MARSTRUCT two main concepts for hull structure protection were identified, either coating protection or cathodic protection systems. The systems are used either separately or combined in both ships and off-shore structures. INTERTANKO, IACS and IMO have over the years been active promoting better corrosion protection systems. Universities and research institutes contributed with increased knowledge. The general state of the art and practice is represented by the most recent IACS documents. This field of technology is extremely experience based. There exits a lot of information between particular ship owners and manufacturers of coating that is not available for the general scientific audience.

The workshop MARSTRUCT (2008) at Chalmers University of Technology May 14 is the basis for the paper. The purpose of the workshop was to "describe state of practice" and try to answer the question "What can be done to improve corrosion protection systems from the ship owners point of view?" The purpose was also to identify research problems for universities and institutes to work with in the future and to forward any significant observations. Much attention has been focusing ballast tanks.

2 GENERAL BACKGROUND

A large percentage of the total life-cycle costs of modern ships are attributed to maintenance activities. Maintenance procedures and intervals must already be considered in the early stages of design.

Corrosion prevention is an important factor in reducing corrosion as well as fatigue damage. The most common approach to achieve satisfactory corrosion prevention is to cover all surfaces with paint. The development of improved coating systems requires an interdisciplinary effort.

Coating failures have been observed in seawater ballast tanks in locations of high stresses and strains in the underlying steel structure. These hot spots are found in bracket toes, notches and weld toes. As ships are highly stressed structures, the stresses in these hot spots occasionally exceed the yield limit of the steel. Coating failures are also found as loss of adhesion over large areas soon after painting or after some time. These losses are usually the result of poor preparation or application in wrong temperature range or on wet surfaces.

There exits efforts to develop paint that can be applied in worse weather conditions. This may be useful for extreme situations of repair but, despite efforts to develop and demonstrate corrosion protection effect, it could not be shown to be useful in SAFECOAT (2007) laboratory tests. Instead the conclusion was that the first paint system applied must be the best you can get for the money. This system is most efficient if applied in dry environment at reasonable temperatures.

Joint efforts have been invested in the corrosion protection systems by TSCF, IMO and IACS. Together these organizations have produced good procedures for long term coatings with a prospective of 15 years life. One of the few items that are not well covered is the cracking of coatings in notches, where ageing due to raised temperature has taken a toll. This is observed in this report and some efforts to bring knowledge forward through the SAFECOAT project are made.

3 BALLAST TANK CORROSION

Double hull structures has been standard design for dry cargo ships a very long time while the double hull presents a relatively recent evolution in tanker design with the purpose of reducing the risks of pollution the environment in case of grounding or collision, Rauta (2004). The space between outer hull and cargo tanks is a narrow tank which is mainly used for ballast. The surface area exposed to corrosion in the ballast tanks surrounding the cargo holds is large compared to the ballast volume. An example is that the areas of ballast tanks increased by a factor of ten, from the single skin design of abt. 35000 m^2 to the double skin design of abt. 350000 m^2 for a VLCC, Eliasson (2008). The ballast is seawater an effective electrolyte to force corrosion. In order to protect the steel one need either to insulate the surface from the electrolyte or to introduce a sacrificial anode that gives less expensive structure to the electrolyte. Coating is mandatory for ballast tanks on both bulk carriers and tankers. When we talk about coating we usually refer to a polymer coating, even if other types of coatings have recently appeared.

On the outside of the ship, offer anodes are standard practice to protect the steel, in particular around the propellers. In the old days, when there was no coating in tanks, a large steel corrosion margin was added to the plate and offer anodes were the only protection from deterioration. The system with offer anodes needs the electrolyte continuously to be present. In areas where the electrolyte is only occasionally present, as in the upper part of the tanks, coating is the only known protection system. In other parts the combination of coating and offer anodes are sometimes used. Coating systems based on alkyds or oleo resinous are unsuitable in combination with cathodic protection, while others such as epoxies and polyurethane behave well as long as they are chosen correctly Ehrman & Slätte (2006). Overprotection however, is particularly associated with Impressed Current Cathodic Protection (ICCP) systems.

The accumulation of mud generates a great threat of microbial induced corrosion. The inner shell of a double hull is subject to large and frequent temperature fluctuations and it is today understood that raised temperature will not only increase the chemical processes in favor of increased corrosion rates but also have an impact on the ageing of the polymer coating Rauta (2004). By comparison with common practice in other industries we find reasons to look at other coating systems, like galvanic coating layers, maybe in combination with a polymer coating.

4 IMO PERFORMANCE STANDARD FOR PROTECTIVE COATING

Coating is now not only a question of economy of transportation but very much considered as a safety issue. Therefore it is now regulated in SOLAS Reg. II-1/3–2—Protective Coatings. A good overview is given by Hoppe (2007). The new requirements are valid for protective coatings in dedicated seawater ballast tanks of all SOLAS-ships, i.e. most commercial ships above 500 GRT and also of double-skin spaces of bulk carriers longer than 150 m. The goal is to have a useful life of 15 years, the ship in "good" condition. The conditions are quantitatively defined so that stakeholders in principle can agree on the inspection results. Maintenance should also follow a new standard; IACS (2008) Recommendation No.87 is the basis.

The new standard enters into force for all contracts after 1st of July 2008, keels laid after 1st of January 2009 and all deliveries after 1 July 2012. IACS has already (2006–12–06) implemented the new procedures in Common Structural Rules for Tankers and Bulk Carriers ahead of the IMO time schedule.

The documents are very detailed with advice for both primary and secondary surface preparations. It is prescribing blasting quality, shop primer specification, steel conditions and surface treatments, the main coating system, environmental conditions for application and surface preparation after erection. The inspection programme is detailed with the introduction of a Coating Technical File (CTF) for continuous documentation.

There are a number of documents to comply with, namely the seven procedural documents of the IACS PR No. 34 on Coating System Approval, Assessment of Coating Inspectors Qualifications, procedures for Inspection Agreement and Verification of Application of Performance Standard for Protective Coating (PSPC), the CTF, Quality Control of Automated Shop Primer and for the review of Coating Technical Specifications. The responsibility for the different steps of PSPC is also clearly laid out, see Table 1.

The new IMO-standard which is now implemented was discussed by the workshop participants. It was

Table 1. The procedures for total system approval have pointed out the responsibility for each item.

Items of importance	Responsibility
Coating system approval	Coating manufacturer
An inspection agreement	Ship yard, owner and coating manufacturer
Coating Technical File	Ship yard
Coating inspection	Ship yard
Verification	Class societies

concluded that the new performance standard is already below the state of art and current practice of most European ship owner's requirements. In particular the procedures for qualifications and the required experience of coating inspectors were considered not safe enough. It is feared by ship owner representatives that some ship yards may suggest, that it will be sufficient standards to reach approval despite of insufficient craftsmanship and design work, which each may cause significant quality problems influencing the final result. Monitoring during coating application means "spot check". There may of course be insufficient coating thickness after spray and inspection. The new standard requires one inspector with clear responsibility. This may be good but has a potential drawback: if this particular inspector decides that the coating has passed as acceptable there may be difficulties to correct obvious deficiencies after his decision. The requirements on coating inspectors are the education according to NACE or FROSIO or equivalent and a few years of experience. The courses (NACE or FROSIO) are about one week and the candidates have no formal qualification before.

Coating "pre-qualification" must not be made by each classification society over and over again—it is a waste of efforts. Classification societies should co-operate in this. The new performance standard includes grinding and stripe coating around edges. This was considered valuable. The ship owner wants the classification societies to enforce compliance with the procedures in all steps before giving certificate. Once the certificate is given it is very difficult to have correction measures enforced.

Can 15 years of "good condition" be achieved? Can ship yards do this? One ship owner of the workshop gave examples of 25 years of life in "good condition". There are however more examples of "poor condition" in much shorter time. But the very fact that if everything is made right from the beginning the coating can last long. Intelligently implemented IMO PSPC must save money without much increase of cost. This was considered possible by the workshop participants.

It was stressed by some participants in the workshop that the Coating Technical File (CTF) must not be developed into a monster file of the numerous fitness readings. It should be cost effective and focusing the purpose of documentation.

5 ZINK OR ALUMINUM USED IN COMBINED COATINGS IN OTHER INDUSTRIES

Cars have always had difficult environment with a range of provocations. The weather conditions will vary from very cold weather in the north of Sweden to very hot weather in Dubai or Arizona; the roof may reach about 120°C in the sunshine. With a relatively thin coating cars are today well protected for the lifetime of the cars. Since about ten years ago, the new painting workshop of VOLVO has introduced a new coating system. The corrosion protection systems of VOLVO cars starts with the double sided zinc-galvanized steel plates. The first treatment after the cleaning from fat is the phosphate pretreatment.

The first real coating after phosphate pretreatment is the "electro coating", which is a process where the "body in white" is a cathode in a bath with an anode that forces the pigments to react with the cathode. The electrolyte in this "electro coating" process is water based. Then comes a primer or so called "surface", which is cured in raised temperature. Then the "base coat" with the color is applied. This is a water-based coat and is only dried out, not cured. Over that comes a clear coat and after drying the car is ready. The total thickness is then less than 110 μm, Björk (2006). Of course ships can not be painted the same way. We can learn from the use of zinc and notice that good protection can be achieved with only thin layers of coating, when composed the right way.

In the offshore industry thermally sprayed aluminum (TSA) has been used for over 20 years. Harvey (2008) presented TSA as is used for corrosion mitigation in the offshore oil and gas industry for over 20 years and pointed out that the technology has potential to resolve problems in particular areas where wear or corrosion is problematic. It was pointed out that the TSA is probably suitable for the outside of the hull rather than in the ballast tanks of tankers. Aluminum content is not allowed in ballast tank coatings due to a theoretical explosion risk. However this judgment is based only on the exothermic reaction between rusty iron and aluminum. There is no experience of explosion. If the benefit of TSA can be demonstrated an experimental test setup could give us an answer of the actual risk. Another comment is that epoxy is not a suitable top-coat for TSA as it is too tight. A suitable top coat on TSA could be silicone.

TSA is passivated with time. It means that the polarity reduces from about −1100 mV to −900 mV by the oxidation. It means that the protection property is good enough but also that the aluminum coating lasts longer time than a zinc coating would have done.

One clear benefit with the TSA is that it does not need any curing time.

Studies of compatibility with primers on steel are ongoing. The consumption of a 200 µm thickness will take about 20 years with a consumption of about 10 µm/year. The presentation included cost estimates and it was believed that LCC is less with TSA than with ordinary coatings.

6 THE AGEING OF COATING— OBSERVATIONS IN SAFECOAT PROJECT

Some results from the SAFECOAT project was presented to the workshop. For the coating it may be important to note that the fresh coatings usually have a quite large strain before failure, while aged coating may fail long before the steel fails. In tankers and bulk carriers the combination of Low-Cycle-Fatigue (LCF) and High-Cycle-fatigue (HCF) is common. Then the LCF is most damaging for the coating. The coating is very ductile when it is fresh; it loses ductility with age according to Maurer (1993). Ageing of coatings can be chemical or physical. The chemical ageing is related to thermal degradation or photo-oxidation. This results in cross-linking changes in the polymer and degradation of the coating. Physical ageing can be explained as a gradual continuation of the glass formation. In addition, the effects of changes in moisture and temperature, including exposure to sunlight, generally reduce the flexibility and toughness of organic coatings.

Experimentally, exposing samples in climate chamber can simulate the ageing of the coating. After that, tests can be done, like tension test or LCF tests.

The problem with reduced flexibility is explicitly when the substrate structure is more flexible or more ductile than the coating. In some details of ships the local strain is often larger than the strain capacity of aged coating. Then we will see cracks in the coating and, as a consequence progressive corrosion, unless early repair is Rauta (2004).

6.1 Investigation on coatings regarding residual stress effect

The drying process of coating includes curing. During this process the volume may shrink and residual stresses will develop. As the coating ages by exposure to the environment the coating may turn brittle. Small defects will initiate cracking. The crack mouth opening displacement can be a measure of the residual stresses.

There are some solvent free coatings that may not be exposed to this process. However all coatings may develop residual stresses of other reasons. If the structure is exposed for long times to high elastic strain, the coating will follow and change volume due to creep. When the elastic strain is released or replaced by strain

of opposite sign the coating will experience stresses similar to residual stresses. Another source of built in stresses in a coating layer is if the substrate is yielding. Below a simulation and experimental verification is described for a case with assumed residual stresses verified by experimental crack mouth opening.

6.2 Curing of coating

Drying and curing of paints and coatings is considered to be the combination of processes by which volatile matter evaporates from an initially liquid film to produce and end product with the desired physical coating properties.

Solvents for paints and coatings are selected so that initially they escape relatively quickly to prevent excess flow of the paint or coating, but sufficiently slowly to provide levelling and adhesion.

The evaporation of solvents will lead to a contraction of the coating. Therefore tensile stresses will appear in the film.

6.3 The residual stresses

A feasibility study by Goden (2005) was performed with the aim to simulate and demonstrate the behavior of a system coating layers with a cohesive crack. The basic idea is that the coating layers have internal tensile residual stresses after ageing, which will open the crack once developed. The opening of the crack is dependent on the distribution and magnitude of the residual stresses. The opening of the crack in an experimental set up was measured for verification.

The tests were performed at the laboratory in University of Technology of Tampere, Finland. Observations and analysis were done with a Scanning Electron Microscope (SEM).

Furthermore, simulations were made with the Finite Element Method (FEM) in 2-D linear mode. This was to investigate how different stress distributions in the system coating influence the crack opening. Only residual stresses were studied by the use of temperature gradient simulation through the coating. The opening gap at the surface was studied as a function of the depth of the crack.

State of the art, material properties, useful formulas and other knowledge were collected from a preliminary literature survey. The analysis was based on the assumption of the linear elastic behavior, which is a simplification. The linear theory in practice limits the generality of this work. However, the work is useful for the next step in the development of full understanding of coating behavior in highly strained parts of a ship.

6.4 Type of coating studied

The coatings studied in this particular experiment and in FEM simulations were of epoxy type. This type is commonly used because of their good properties:

fast cure, relatively good flexibility, and a well-known function as barrier and good adhesion to steel.

6.5 Types of cracks

The types of cracks that can be studied with the elastic FEM are cohesive failure and adhesive failures. The propagation as such was not studied but the elastic deformation of the geometry that exists in each step, i.e. each snap shot moment, of the crack propagation process. There can also be combinations of these.

A vertical crack will propagate until it reaches the substrate. Thereafter it is going to develop as a shear crack close to the substrate.

When the vertical crack approaches the substrate the substrate has a restraining effect that may cause a mixed mode of propagation close to the substrate.

6.6 Finite element modeling

The Finite Element method was used to simulate the effect of residual stresses in the coating. The commercial FE codes FEMLAB 2.3 was used for the analysis. Residual stresses were introduced with fictive temperature gradients in the material. The selected part of the steel specimen is modelled in 2D by a rectangle of 5 mm (thickness) × 6 mm (width). The coating is made of 3 layers of the same material (100 μm; 150 μm; 150 μm/from the first layer till the top layer respectively), which give a total thickness of the coating of 0.4 mm. The crack path is modelled by a starting crack's opening of 10 μm at the top layer. The modelled part has straight boundaries at some reasonable distance from the crack.

The thermal stresses are used to simulate residual stresses in a material, which we believe reproduces the best effects of the curing process. Thus, the stresses are here created without external load. No loads are added, only the effects of the temperature distribution and gradient due to the film formation are considered in this study.

FEMLAB generates automatically the mesh of the model and it uses only triangular elements. Figure 1 shows the mesh around the crack in the case of a cohesion crack down to the substrate. Not all of the modelled substrate is shown here. The three layers of coating are shown. For this model, the code produces more than 6800 nodes and 13200 elements. The mesh is finer at the surrounding of the crack and in the coating. Bigger nodes have been created for the substrate.

With different distribution of the temperature and/or different amplitude of temperature in the coating, we will obtain different results.

Results depend also on the materials used for the substrate and the coating. The results are obtained with the following data.

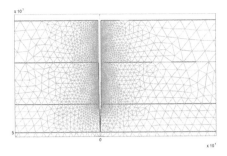

Figure 1. Mesh of the coating with the crack down to the substrate. These three layers are together 400 μm, while the model is 5 × 6 mm. Only the part of the model that is close to the crack is shown here.

Table 2. Material data used in the simulations.

	Young's modulus E	Density ρ	Expansion coeff. α	Poisson's ratio n
Substrate	200 GPa	7850	1.2E-05	0.33
Coating	4 GPa	1500	4.9E-06	0.4

6.7 Crack opening

Crack opening is shown in Figure 2. The simulations show also that the x-displacement at the gap is approximately the same for each distribution of temperature. It varies only with temperature amplitude.

After this the crack will propagate along the interface between the coating and the substrate with shear as the driving stress, Figure 3.

The material is assumed linear. The deformation of the crack opening can be estimated despite infinite stresses in the crack tip. When the residual stresses are high on the surface any irregularity may initiate a crack to occur. The evolution of the crack development was demonstrated by a series of snap shots from linear calculations. The simulations showed that the x-displacement at the gap was not sensitive to the gradient in thickness direction but varied only with temperature amplitude as seen in Table 3.

The opening indicated in the table is the total gap at the surface. Samples exposed to climate chamber environment can simulate the ageing of the coating. After that, tests were performed to measure crack opening displacements after long term exposure.

6.8 Experiments

Experiments have been made in a laboratory in a University in Finland. Different paints covered specimens. There was a 25 mm diameter hole in the middle of the specimen. They were ageing and LCF tested.

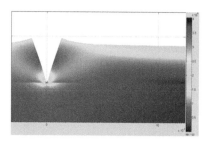

Figure 2. The crack has here reached the interface between the coating and the substrate. The crack opening can be seen.

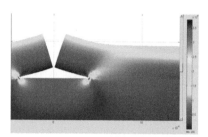

Figure 3. Further development of the crack after the tip has reached the substrate and propagated along the interface by shear stresses.

Table 3. Results of FE-simulation. The total x-displacement δ for the chosen distribution is given with surface stress on the coating.

Imposed temperature amplitude	Crack opening is 10 μm + δ	Resulting stress level at surface
−7	0.050 μm	0.225 MPa
−15	0.110 μm	0.48 MPa
−30	0.220 μm	0.96 MPa

Specimen dimensions:

- Length 200 mm,
- Width 50 mm,
- Thickness 4 mm.

Using Scanning Electron Microscope, the composition of the samples has been analyzed.

Using stereomicroscope, the opening and the length of the failures have been reported.

From this analyze, we can notice that openings and lengths are different from one lot to another since they are not using the same paint. However, among a same lot, e.g. ADO 2 or BDO 2, the ratios of the length on the opening are similar if the crack occurred in the same area (as shown in the following extract of table). We can suppose that the residual stresses present in this paint were the same. Therefore, we can make the hypothesis that the drying of the same paint will create the same internal stresses in the film. The parameters

Table 4. Relations between experiments and simulations Extract from laboratory report from Tampere referred by Goden (2005).

Specimen number and lot	Length of crack [mm]	Mouth width [Microns]
1 (ADO2)	7	15
3 (ADO2)	5	65
5 (ADO2)	4.5	15
3 (BDO2)	5.5	20
4 (BDO2)	5	20

to investigate which could influence these stresses may be the conditions under which drying occurs (temperature and humidity) and the thickness of the film.

From the simulations, we got proportional results. For a coating with materials properties similar to the data used for the simulation, an opening of 0,05 μm corresponds to 225000 Pa at the surface, thus e.g. an observed opening of 1 μm is equivalent to surface stresses away from the crack of 4.5 MPa. This result is only an indication of a promising method and the work needs to be further improved and developed into a method with high repeatability.

From the experiments we can measure an opening. If these measurements are done before tests, it becomes possible to evaluate the stress, which led to the crack.

Assuming that the crack has propagated till the substrate but not yet developed into a shear failure parallel to the substrate surface we can compare this with the simulation. We also assume a long crack in the direction not simulated by 2D. Feasibility studies were made by Ehrman & Slätte (2006), which shows that there is suitable commercial software available. What is further needed is an experimental set up which gives material data, such as critical stress intensity data.

In the workshop there was an interesting discussion on the cause of cracks. It is obvious that shrinkage is the cause of cracks in corners and close to welds, where the paint may accumulate. There are also examples of high strain in the substrate causing cracks in the coating, see figure 4. Below, which shows a transverse section of a tanker, a principal graph of pressure head in two locations and a sketch of hairline cracks along the boundaries of plates in the bottom tank ceiling according to Bengtsson (2008). As there are no cracks in the corners of the rectangular plate this indicates clearly that we have cracks induced by high strain in the substrate. The tanks with lower pressure head had no cracks of this type.

It would be interesting to know more about creep for marine's coating in order to be able to tell, with a better accuracy, under which stress a failure can be expected. We need to find out relations between stresses and strains, for polymer coatings, with respect to time.

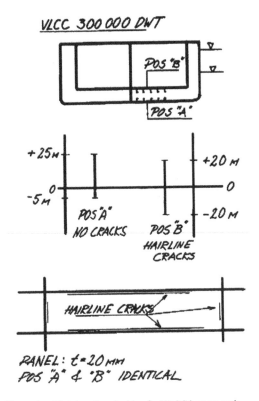

POS "A"
NO CRACKS

POS "B"
HAIRLINE
CRACKS

HAIRLINE CRACKS

PANEL: t = 20 mm
POS "A" & "B" IDENTICAL

Figure 4.   Sketches from inside of a VLCC bottom tank.

7   CONCLUSIONS AND RECOMMENDATIONS

The purpose of the workshop was to address the question "What can be done to improve corrosion protection from ship owners point of view?" The answer should of course be given relative to the very large work that is done by TSCF, INTERTANKO, IACS and IMO. It was concluded that

– Based upon experience of some excellent ships in the past it was believed by the workshop experts that 15 years of "good condition" can be achieved. In order to achieve this, every step of the procedures must be right.
– The new performance standard is already below the state of art and current practice of most European ship owner's requirements. The new standard requires one inspector with clear responsibility but the formal requirements on coating inspectors are rather too low: the education according to NACE or FROSIO or equivalent and a few years of experience.
– The Coating Technical File (CTF) must not be developed into a monster file of the numerous fitness readings.

– The use of thermally sprayed aluminium (TSA) in offshore industry can be beneficial also to ships. Further research is needed.
– Coating is aged by increased temperature and may crack before the steel due to the combination of residual stresses from the curing process and the strain of the substrate (the plate).
– The new praxis of double stripe coating around edges was considered effective by the workshop participants.
– The crack opening displacement can be simulated and measured and that is a tool for further studies of coating behavior when aged. Further research into the use of fracture mechanics to study the mechanical properties of aged coating is recommended. Both shrinkage during curing and strain in the substrate can force coating to crack.

ACKNOWLEDGEMENT

The specialists present at the workshop 2008-05-14 at Chalmers University of Technology are great fully acknowledged for their contribution to understanding of the present state of practice and descriptions of actions to be taken. These are Bo Bengtsson, Inmar, Kada Benyouce, Brostöms, Pawel Domzalicki Ship Design Research Centre, CTO S.A., Jonny Eliasson, Stolt Transportation Group B.V., Dave Harvey, The Welding Institute, Lennart Josefsson, Chalmers, Roger Karlsson, Marinvest Thorsten Lohmann, Germanischer Lloyd, Stefan Marion, Det norske Veritas, Philppa Moore The Welding Institute, Henrik Nordhammar, Stena, Jonas Ringsberg, Shipping and Marine Technology, Chalmers, Dora Tsiourva, Ship Design laboratory, National Technical University of Athens, Helge Vold, Det norske Veritas. The MARSTRUCT network is great fully acknowledged for making this co-operation possible.

REFERENCES

Bengtsson, B. 2008. Individual contribution to the MARS-TRUCT work shop, Chalmers, Gothenburg 08-06-14.
Eliasson, Stolt. 2008 Individual contribution to the MARS-TRUCT work shop, Chalmers, Gothenburg 08-06-14.
Ehrman, T. Slätte J. 2006. Coating crack propagation—A feasibility study Dept. of Shipping and Marine Technology, Chalmers Univ. of Tech., Göteborg, Sweden 2006, Report No. X-00/123.
Harvey, D. 2008. The use of thermally sprayed aluminium (TSA) for corrosion mitigation, TWI. Individual contribution to the MARSTRUCT work shop, Chalmers, Gothenburg 08-06-14.
Hoppe, H. 2007. Development of Mandatory IMO Performance Standards for Protective Coatings on Ships, IMO, Maritime Safety Division.
Goden. C. 2005. Investigation on coatings regarding residual stress effect, Dept. of Naval Architecture and Ocean

Engineering, CHALMERS Univ. of Tech., M.Sc.-thesis, Gothenburg, Sweden 2005.

IACS 2008. Requirement on Application of. the IMO Perform ance Standard, PR No. 34, <http:www.iacs.org.uk>

MARSTRUCT 2008. Workshop 2008-05-14. Full documentation is on MARSTRUCT homepage but can also be sent on request contacting the author anders@ulfvarson.se

Maurer, F. 1993. *Viscoelastic behaviour of polymers*, part 5, (Compendium), Department of Polymer Technology, Chalmers University of Technology, Gothenburg, Sweden.

Rauta, D. 2004. Double Hull and Corrosion, presentation from INTERTANKO at RINA meeting February 2004.

SAFECOAT 2007. Final reports from EU-project Project reference G3ST-CT-2002-50314.

Björk, I. 2006. E-mail communication with representative of the manufacturer February 2006.

Analysis and Design of Marine Structures – Guedes Soares & Das (eds)
© 2009 Taylor & Francis Group, London, ISBN 978-0-415-54934-9

Prospects of application of plasma electrolytic oxidation coatings for shipbuilding

A.N. Minaev & N.A. Gladkova
Far-Eastern National Technical University, Vladivostok, Russia

S.V. Gnedenkov
Institute of Chemistry, Far Eastern Branch of Russian Academy of Sciences, Vladivostok, Russia

V.V. Goriaynov
Shipyard "Zvezds", Bolshoy Kamen Sity, Russia

ABSTRACT: In this paper are introduced results of research of plasma electrolytic oxidation technologies (PEO) processing alloys. It lets to permits to form on metals and alloys surfaces the many functional coatings having the complex of practically important physical-chemical properties. PEO coatings can be: the anticorrosion coatings; antiscale coatings; wear-resistant coatings (with micro hardness up to 10000 MPa); antifriction coatings (decreasing the friction coefficient for the mechanical couple up to 0,12–0,06 usually realized with using of traditional lubricants); antifouling coatings (decreasing the intensity of biological fouling of the wares surface used in the seawater); thermally stable coatings (which do not lose their initial properties up to 800–900°C); decorative anticorrosion coatings (some colors and hues) improving the appearance and anticorrosion resistance of aluminum articles applied in the sea-climate; electro isolated coatings (R is equal to 10^{10}–10^{12} Ohm $\cdot$ cm, E—up to 10^4–10^6 V/cm, $tg\ \delta$ is equal to 0,003–0,5).

1 INTRODUCTION

PEO coatings is a very important problem for different fields of technique by Yerokhin et al (1999). During last several years a lot of research works were devoted to this problem by Yerokhin et al (2000), by Curran & Clyne (2005, 2006).

The conditions of formation on the surface of titanium alloys of anticorrosive (decreasing the contact corrosion intensity of galvanic couples: titanium-steel in the sea-water in 200 times as compared with unprotective galvanic couples), antiscale (decreasing the salt scaling intensity on the surface of heat-exchangers working under thermal flows equal to 0, wear resistant (possessing microhardness about 7000–10000 MPa), coatings were elaborated. Microarc oxidation, unlike thermal oxidation, does not change for worse corrosion-mechanical properties therefore PEO-coatings may be used in high-tension constructions working in the sea-water. For aluminium alloys the conditions of formation (electrolyte compositions, regimes of oxidation) for receiving of wear-proof, thermostable, heat-resisting (till 870°C) coatings were created. The interconnection between composition of PEO-coatings, their zone structure, semiconductive properties and electrochemical, corrosive

behaviour of coatings in the sea-water were investigated. Contemporary model's ideas about mechanism of microplasma oxidation were suggested.

2 MODEL OF PEO—METHOD

Plasma electrolytic oxidation (PEO) method is based on the anodic polarization of processing metals (valves metals: Ti, Al, etc.) and its alloys under microplasma discharges on the anodic surface by Gordienko & Gnedenkov (1997). These discharges arise after exceeding of critical polarization potential. In the result of this processing on the surface of metals and alloys the coatings possessed by important for practice physical-chemical properties are formed. For obtaining of anodic oxidation general picture, the polarization curve of given process with simultaneous registration of acoustic and light signals accompanying the PEO-process (Figure 1) was recorded. At the same time the visual observation for process on the anode took place. It was established that the current increasing and fall on the polarization curve on the initial stage of PEO-process is determined by origin, spreading and interrupting of microplasma discharges

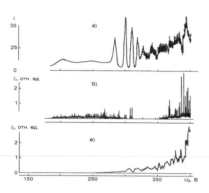

Figure 1. Dependence of density current alteration a), acoustic b), and light c) signals from the formation voltage during titanium oxidation in the phosphate electrolyte ($dU_f/dt = 80$ V/min).

Figure 2. View of industrial plant for testing of the marine equipment with PEO-coating in seawater.

area on the anode surface. The thickness of PEO-coatings (d) determined by electrical capacity method increases linearly with the increasing of formation tension time (U_f).

At the same the electrical field strength in the film decreases that is not able to explain the resumption of microplasma discharges area after it interruption.

Obtained experimental results are well explained if to take into account the semiconductive properties of MAO-structures. The coatings obtained in the phosphate electrolyte consist of TiO_2, which is attributed to the n-type conductivity semiconductors. By anodic polarization of electrode in the surface area of coating, the space charge region (SCR) impoverished by chargers is formed. Its thickness is none above of the thousandth centimeter parts. The electrical field strength in this area achieves to the critical values (10^7 V/cm). The powerful ionic current in the microplasma discharges channels destroys the lock layer, decreases the electrical resistance of system and intensifies the growth of film. As the film is growing the discharge intensity is decreased, the width of SCR is increased and conditions of passing next avalanche arise. According to suggested model, the electrical field strength, which is necessary for micro-discharges leaking, arises but not on the all film thickness and only in the SCR. The typical distribution of chemical elements along the film section is the confirmation of surface discharge model. The greatest element concentration presented in the film composition from electrolyte solution is observed in the coating surface field.

3 PLANT FOR TESTING OF THE MARINE EQUIPMENT

Industrial plant for testing of the marine equipment with PEO-coating was constructed at Shipyard "Zvezds" (Figure 2).

Scale-formation and corrosion research at seawater forced motion was carried out on the experimental complex intended for different types of heat-exchange apparatus tests carrying out on seawater in ship conditions. The experimental complex operates in the following way. The overboard seawater is fed through the pipeline by pump from kingston box to the heat-exchange apparatus. It consists of the heat exchanger "tube in tube" type, where heating vapour passes through internal tubes and they are washed by cooling seawater counter flow from the outside by Minaev & Lysenko (2001).

4 ANTICORROSIVE COATINGS

Anticorrosive coatings on the titanium and titanium alloys reduced the current of galvanic corrosion between steel and titanium details exploited in the sea-water in 200 times as compared with contact-couples without protective coatings. These coatings were created on the titanium samples in the solution containing hypophospite-ion by Gnedenkov (2000).

According to x-rays analysis results and electronic x-rays microscopy data, the anticorrosive coatings were x-rays amorphous and consisted of: 15 mass % Al; 5 mass % P; 3 mass % Ca; 14 mass % Ti; O—the rest. After heating till 1000°C in the inert atmosphere the amorphous phase turned into TiO_2 (rutile) and α-Al_2O_3, that allowed to propose the presence of small size crystals of these combinations in the coating composition. We investigated the electrochemical, semiconductive properties by capacity-voltage and polarization characteristics methods of some protective layers on the titanium: natural oxide film, thermal oxide, PEO-coating, created in the hypophospite containing electrolyte. We had ascertained, that for suitability of Mott-Schottky equation to investigation of surface layers, consisting of TiO_2, 1M HCl solution was best, since it provided high means of Helmholtz layer capacity on the phase boundary oxde/electrolyte.

In accordance with experimental C^{-2}, φ-curves, important parameters (flatband potential—φ_{FB} versus normal hydrogen electrode (NHE); concentration of charge carriers—N_D; Fermi level—F; energy of bottom of conduction zone—E_C; etc.) were calculated.

Statistic processed results are given in Tables 1, 2.

The experimental data allowed to conclude: coatings had the better protective properties the more R and smaller N_D; according to zonal theory, semiconductor must not be degenerated (E_C–$F \geq 3kT$).

Instability of aluminate-ion containing solutions during the time caused by the polymerization of aluminate complexes, including PEO, has essentially influence on the surface layers properties. In given work the properties of coatings on the titanium VT1-0, formed in hypophospite-aluminate electrolyte were investigated. The investigations were carried out through some time interval, beginning from the moment of solution preparation: 1-in 72 hours after solution preparation; 2-in 48 hours; 3-in 24 hours; 4-in 5 hours; 5-at once after solution preparation. Dependence of contact corrosion current from the work's time of galvanic couples titanium with PEO-coatings/steel are presented on the Figure 3.

The protective properties of oxidic layers are worse as a result of exposure of formed solution. With using of nuclear magnetic resonance (NMR)-method it was established that aluminium concentration with tetragonal coordination in the electrolyte solution is change in time and through twenty-four hours after solution preparation it is decreased in two times approximately. In the electrolyte the solid deposition is formed and according to x-ray analysis data consist of aluminium hydroxide. The process of electrolyte exposure caused the contact corrosion current increasing change essentially the phasal, elemental compositions of protective layers obtained in this electrolyte. Over twenty-four hours after solution preparation in the initially x-ray amorphous coating the TiO_2 Crystalline inclusions appear. In the coatings formed in the electrolyte bearing for two days and more the phase TiO_2 (rutile) is

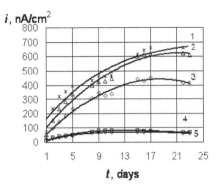

Figure 3. Dependence of contact corrosion current alteration from the work time of galvanic couple titanium/steel, titanium surface in which is processed in hypophospite-aluminate electrolyte, on the different time stages of solution exposure (the meaning are on the text).

fixed. After electrolyte bearing for three days aluminium concentration in the film is decreased in 8 times and calcium in 7.5 times. The outlook of cathodic polarization characteristics, taking in the potentials field, realized by galvanic contact with steel confirms the relaxation of PEO—layers protective proprieties in the result of decreasing of coatings polarization resistance that lead to the increasing of fixed current. On the Fig. 4 for investigated coatings the C^{-2} (φ)-dependencies with distinguished diapasons of polarization potentials: $-0.10 \ldots 0.35$ V; $-0.30 \ldots -0.10$ V; $-0.5 \ldots -0.30$ V are presented. Inside of these diapasons the C^{-2} (φ)-dependencies are described by straight-line equation (with the exclusion of curve 1, which in the diapasons $-0.10 \ldots 0.35$ V has two line sections).

The decreasing of angle tangents of inclination of linear plots for investigated coatings as compared with the same diapasons of polarization potentials testify about increasing of charges carriers concentration with increasing of time electrolyte exposure. The presence of two and more linear plots on the C^{-2} (φ)-dependencies may be cause by few reasons: the presence in semiconductor materials of two or more types of donor admixtures (for TiO_2—electrodes such as may be Ti^{3+} and O^{2-}, which are on the different energetic positions); by presence in the SCR of some layers with different composition and physical-chemical properties; impoverishment of semiconductor material by the main charge carriers upon the high polarization potentials; the presence of "parasite" capacities on the semiconductors surface; the presence of the charge carriers concentration gradient. According to above-mentioned results, the coatings formed on the different stages of electrolyte exposure with decreasing of inclusion elements concentration from the electrolyte (Ca, Al, Na) pass from x-ray amorphous state into crystalline, having the composition

Table 1. Some parameters of surface layers on titanium.

Layers type	$D, \mu m$	N_D, cm^{-3}	φ_{FB}, V (NHE)	F, eV
Natural oxide	0.013	$5.9 \cdot 10^{19}$	-0.23	-4.17
Thermal oxide	1.17	$3.4 \cdot 10^{15}$	-0.47	-3.93
PEO-coating	1.01	$9.6 \cdot 10^{15}$	-0.54	-3.86

Table 2. Some parameters of surface layers on titanium.

Layers type	E_C	R, Ohm	i, nA/cm^2 (after 96 hours)
Natural oxide	-4.19	50	12 100
Thermal oxide	-3.70	5000	1 500
PO-coating	-3.67	12800	64

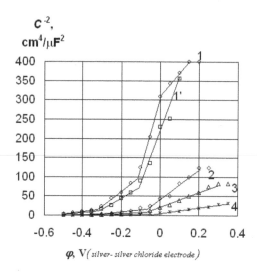

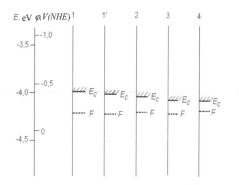

Figure 4. C^{-2} (φ)-dependencies for titanium samples with MAO-coatings obtained for different stages of hypophospite-aluminate electrolyte exposition (the meanings are in the text).

Figure 5. The energetic diagrams (for the case of flat band potentials) of PEO-layers on the titanium, obtained on the different provisional stages of hypophospite-aluminate electrolyte exposure (the meaning are in the text).

like TiO_2. The material of coatings, obtained in the electrolyte on the third day from it exposure is more homogeneous on the composition, but it is characterized by more high concentration of charge carriers as compared with the coating obtained at once after solution preparation. Under generalizing of obtained results the conclusion may be done: the investigated coatings are the heterostructure.

The surface part of none-porous under-layer, determining the electrochemical and corrosive behaviour of layer on the whole, contains the x-ray amorphous phase (glass-phase) which presence and influence accordingly decrease with the electrolyte exposure.

The inter part of non-porous under-layer, adjacent to the metal consist of small-dispersive TiO_2. It may to assume, that the inclination angle of $C^{-2}(\varphi)$-dependencies on the plot of potentials $-0.1 \ldots 0.35$ V is determined in the more great degree by state of surface non-porous under-layer (glass phase), just as on the two other plots the inclination angle is set by TiO_{2-}state. From the analysis of zone diagram (Fig. 5) constructed for potentials diapason of $-0.30 \ldots -0.10$ V it is follows that the energetic interval between the bottom of conduction zone and Fermi level (E_c–F) for coatings formed in the electrolyte on the measure of its exposure is decreased. So that the semiconductor is approached to the state of degeneration. This lead to the increasing of contact corrosion current and weakening of PEO-layer protective properties.

5 MECHANICAL PROPERTIES OF PEO-COATINGS

Plasma electrolytic oxidation unlike thermal oxidation did not change for worse the corrosive-mechanical properties of metals and alloys. According to results of our investigations PEO-coatings deposition carried out in phosphate- or hypophosphite-containing electrolytes did not changed plasticity and durability characterics of raw alloy (Table 3), while the thermal oxidation reduced diametrical narrowing till 7% and shock viscosity till $10 \div 15\%$. Small-cycles fatigue, determined in the sea-water and in air for titanium samples having fatigue crack and without it processed by PEO-method, did not changed also.

The wear-resisting coatings synthesis conditions on the aluminium alloys their phasal and elemental compositions were investigated. The analysis of obtained results, allows to propose that x-ray amorphous coatings with microhardness till 7000–10 000 MPa and thickness till 30 mkm consists of fine dispersity aluminium oxide modified by fluorine. The chemical formula may be presented as $Al_2O_{3-x}F_{2x}$, what is typical for glass-phase. However in the case of unhomogeneous coating's state it is impossible to except the presence of aluminium oxyfluorides and also aluminium fluorides. The main quantity on the volume of fluorine concentration, presented in these compounds composition, does not allow with confidence to identify the fluorides or oxyfluorides in the film. It is established that obtained coating possesses by heat-resistance till 870°C. The analysis of T, G-curves permits to conclude about absence till indicated temperature of heat's, weight's changes, connected with coatings material, which testify about invariability of composition and consequently the properties of surface layers by temperature exceeding the melting temperature of aluminium base (620°C) fixed on the

Table 3. Plasticity and durability characteristics of titanium alloys PT—1M (Al-1.2%); PT—3V (Al-3.5%, V-1.2%) non-oxidized (NO) and after PEO-processing.

| | Kind of alloy | | | |
| | 1M | | 3V | |
Characteristic	NO	PEO	NO	PEO
Limit of durability-σ_B, MPa	420	416	807	797
Limit of fluidity-$\sigma_{0.2}$, MPa	336	339	769	772
Relative lengthening-δ, %	38.2	41.3	16.2	14.4
Diametrical narrowing-ψ, %	71.5	75.2	45.6	47.2
Schock viscosity- a_H, MPa	1.02	1.14	1.06	1.04

location of endothermal effect on DTA-curves (differential thermal analysis). The coating does not melt in the studied temperature interval and plays the role of thermostable container to prevent the spreading and vaporization of metal (alloy). After annealing till 870°C and next sample cooling till room temperature the coating does not fall off and the sample with coating remains its initial shape. At the temperature equal to 870°C by annealing both in helium atmosphere and on air the heat (or totality of heats) which is accompanied by sample's mass changing fixed on the T, G-curve (termogravimetry) and DTG-curves (derivative thermogravimetry) is observed. After heat passing the mass decreasing is noted. With further temperature rising the mass of sample increased. Likely upon the temperature exceeding 870°C in the volume of the coating material structure alterations take place. This lead to formation of microcracks, which cause the evaporation and oxidation of fused aluminium alloy. It is established that obtained coatings in the presented electrolyte composition possess by significant adhesion to the metal surface elasticity without separating from the metal base by frequent bends till the angles exceeding 90°.

6 ANTISCALE PEO-COATINGS

Antiscale coatings may be formed in the electrolyte, containing phosphate-ions with help plasma electrolytic oxidation technologies (PEO) by Gnedenkov & Mashtalyar (2008). These layers decreased the scaling intensity on the surface of titanium heat-exchanger, working in the sea-water by values of thermal flow equal to $0.2 \div 0.5$ MW/m^2.

The samples under investigation were the plates with sizes $110 \times 80 \times 1$ mm on which the coatings in water solution of phosphate electrolyte ($Na_2PO_4 \cdot 12$ H_2O- 10 g/l) with using PEO method were formed.

Then the ultra dispersed polytetrafluorinethylene of average fraction composition (afc) and also its different fractions obtained under thermal destruction of the polymer: low temperature fraction (ltf) and high temperature fraction (htf) were applied on the surface. The original UPTFE and its fractions (ltf and htf) were investigated by the method of differential thermal analysis and thermal gravimetry. The investigations were performed on the derivathermo gravimeter DTG-60 H Shimadzu with the heating velocity of 2, 5 degree/min in the air atmosphere in corundum crucibles. As follows from the derivatogramms analysis the low temperature fraction of PTFE obtained under 70–90°C was heated until the temperature 150°C (full mass loss). On the DTA curve the endothermic effect with maximum under the temperature 83°C connected with sublimation of the polymer was observed. High temperature fraction of the polymer presents itself the residue obtained under heating of initial PTFE till 300°C. The presence on the DTA curve of pick under the temperature 227°C is connected with melting of PTFE sample. In the temperature interval (340–520°C) on the TG curve the stepped mass loss of the samples is observed that can be clearly seen on DTG curve. The noted discontinuity can be as a result of polymer boiling containing polymer chains of different lengths, which are in the composition of fraction investigated. The averaged fraction of PTFE was heated until the temperature 560°C when the full polymer mass loss was observed. On the DTA curve the wide peak with maximum under the temperature 274°C is observed and caused by the presence in the composition of the UPTFE of significant content of high temperature fractions with melting temperature fixed in the interval of 232–320°C. However the active sample mass loss fixed on the curve TG beginning from the temperature 50–60°C indicates about the presence in the given fraction of the certain content of oligomers possessing of low melting temperature or sublimation.

In whole the TG and DTG curves behavior for (afc) is more similar with TG and DTG curves for high temperature fraction of PTFE than for (ltf) which can characterize the fractions under investigation. The analysis of DTA- TG- DTG curves allows to forecast the differences of composition functional coatings in which structures the different thermal fractions of the PTFE are used. To establish the stability of composite layers containing in their structure the PTFE layer of different fraction composition the changing of coating weigh in a process of different time annealing was analyzed in the given investigation. The annealing conditions (T = 300°C, in muffle) designed the hard temperature effect realized on the heat exchanger's surface energetic machines. The weighting of samples till the annealing and on the different stages of the temperature effect carried out on the analytical scales AUW 120 D (Shimadzu, Japan) with the accuracy 0, 1 mg. Metal oxide structure before the PTFE processing was released from the presence of sorption water under the previous annealing during an hour until 100°C.

The evaluation of different time annealing on the electrochemical properties of hetero structures and also on the condition of the composite layer (PEO layer + polymer)/electrolyte interface was carried out by methods of polarization curves and electrochemical impedance spectroscopy with using of the electrochemical system 12558 WB (Solartron Analytical, England). The 3% water solution of sodium chloride NaCl (analog of sea water) was used as electrolyte. The measurements were performed under the room temperature.

The sample-exposing square was 32 cm² The potentiodynamic scanning were made with a scan rate of 1 mV/s. As a perturbation signal for the dimension of impedance the signal of the sinusoidal shape with amplitude 10 mV and frequency from 0,01 Hz till 1 MHz with logarithmic scanning of 10 points on period was used. The spectra recording was performed under the composition layer (PEO layer + polymer)/electrolyte under the constant potential mean. The experimental data are presented in Bode coordinates in which the changing of impedance module | Z | and phase angle Θ are shown in relation to frequency f. From the analysis of polarization curves it can conclude that the covering of the PTFE different fractions on the PEO coating surface influences positively on the protective properties of layers obtained reducing the currents of free corrosion as in active as well in passive areas and giving the ennobling of corrosion potential. For composite layers annealed for a hour the further reducing of free corrosion currents is observed, and the polarization curves oscillations decreases what can be explained by more homogeneous structure formed as a result of polymer particles melting and connecting them with each other. For the layers under investigation on the given annealing stage without the dependence from the PTFE fraction the

polarization curves are practically identical which can characterize the analogical or close to it the surface condition. Changes coming from polarization curves for composite layers which were annealed during 8, hours, namely: disenabling of the constant potential, the increasing of free corrosion currents may be caused by sublimation of low temperature fraction of PTFE. As a result on the surface of PEO coating the areas, which are not protected by polymer, appear and also can be caused by possible formation of defects in surface layer. The structure modeling of processes suggested on the base of the experimental data obtained by the method of impedance spectroscopy is found on the system way on which the object under investigation is examined as an equivalent electrical circuit, concluding elements characterizing the electrolyte/electrode interface. For the fitting of impedance data the model (Figure 6) was used. It can be applied for description of two electrochemical interfaces was used.

In this equivalent circuit instead of the electrical capacity the constant phase element CPE applied for description of non-ideal capacitors (heterogeneous surface layers which are non homogeneous on the composition and thickness, complicate morphology, the presence of the gradient of the charge carriers through the section of oxide layer) was used. The impedance CPE is described with formula

$$Z_{CPE} = \frac{1}{Q(j\omega)^n}, \qquad (1)$$

where Q—is pre exponential factor, which is as frequency independent factor, n—is the exponent determining the character of frequency dependence ($-1 \leqq n \leq 1$), $\omega = 2\pi f$ is a circular frequency and $j = \sqrt{-1}$ is an imaginary unit.

The constant phase shift element is widely used under modern electrochemical modeling of different complicated objects including the description of processes on the interface of anodic oxide layers. The resistor R_e is a resistance of the electrolyte ($R_e = 120 \div 140$ Ohm $\cdot$ cm²), CPE$_1$ is geometrical capacity

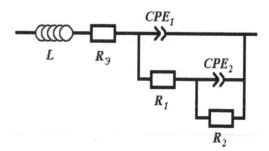

Figure 6. The equivalent electrical circuit in the connection with structure of oxide layer.

of all composite layers. The parallel with CPE_1 element R_1 is responsible for the electrical resistance of pores to ionic current. Elements of parallel circuit CPE_2-R_2 are destined for the description of the charge transfer process on the nonporous layer/electrolyte interface. The high values of exponent n for both of constant phase elements indicate about the capacitive character of suggested CPE.

The necessity of inductivity L concluding into equivalent electrical circuit is made by contribution in the high frequency part of spectra the cable for measurement (L = 1 ÷ 6 μH). With using of model suggested of electrode/electrolyte interface the resistive and capacitive characteristics of every layer before and after thermal processing were calculated.

The low values χ^2 (10^{-3} ÷ 10^{-5}) are in accordance with a model suggested to obtained experimental data. The analysis of impedance spectra permit to suggest and to conclude about the condition of surface of the samples under investigation. The presence and location of the time constants determined by the bends on the phase angle Θ vs. frequency dependence characterizes the morphological peculiarities (roughness, porosity) and heterogeneity of composite layers. Two time constants on this graphs for the sample with surface coatings indicate about two layers structure of electrode: there are the outer porous layer and inner non porous barrier layer.

The treatment of the surface with PTFE not only increases the value of impedance module $|Z|_{f=0,0,1}$ from $5 \cdot 10^5$ till $5 \cdot 10^6$ Ohm^{-1} $\cdot$ cm^{-2}, but also gives to the electrode/electrolyte interface more capacitive character (the tendency of phase angle to change to the higher values). It indicates that the processing with PTFE followed by thermal processing allows to fill up pores of the coating by polymer so that to make the surface more homogeneous. The transformation of location of time constant amplitude in the frequency range $5 \cdot 10^2$–$5 \cdot 10^5$ Hz is more clearly shown for the average and high temperature fractions.

PTFE under thermal processing indicates the changing of surface condition: porosity, homogeneity of composite layer. In accordance with tendency of changing $|Z|$ and the capacity of the composite layers decreases and therefore the thickness of protective layer increases that is possible as a result of PTFE spreading on the surface and forming of pseudo continuous complete polymer film on the surface of the PEO layer.

The treatment of PEO-coatings by polytetrafluorethylene powder was made for creating of hydrophobic layer on the surface, decreasing its roughness. The processing of non-oxidized surface by polymer powder is not effective because the powder has not good adhesion to metal. The PEO-coating, possessing some roughness, guarantees fine adhesion. The experiment for each type of surface layer was carried out during 110 hours.

In the Figure 7 is shown temporal changes power density (q, kWt/m^2) on titanium heat-exchanger with different surface layers: 1—natural oxide film; 2—PEO-coating after treatment by small size polytetrafluorethylene powder. Titanium heat-exchanger with natural oxide film is shown on Figure 8. Titanium heat-exchanger with PEO-coating after treatment by small size polytetrafluoroethylene powder is shown on Figure 9.

7 CONCLUSIONS

Realized investigation show that PEO-coating can be recommended as anticorrosion coatings; antiscale coatings; wear-resistant coatings; antifriction coatings; antifouling coatings; thermally stable coatings; decorative anticorrosion coatings (some colors and hues) improving the appearance and anticorrosion resistance of aluminum articles applied in the sea-climate; electro isolated coatings. Properties of composite coatings on the base of oxide layers

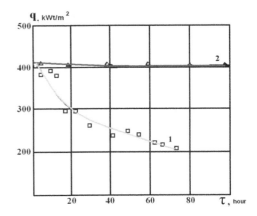

Figure 7. Alteration of specific thermal flow on titanium heat-exchanger with different surface layers: 1—natural oxide film; 2—PEO-coating after treatment by small size polytetrafluoroethylene powder.

Figure 8. Titanium heat-exchanger with natural oxide film.

Figure 9. Titanium heat-exchanger with PEO-coating after treatment by small size polytetrafluorethylene powder.

formed on the heating surface by plasma electrolytic oxidation method and processed with ultra dispersed polytetrafluorinethylene Forum® were investigated. The combination of the methods of electrochemical impedance spectroscopy, differential thermal analysis and thermal gravimetry allowed to establish the changing of condition of the surface as a result of heating determining the charge transfer mechanism on the hetero structure/electrolyte interface and caused by difference of thermo dynamic stability of polymer temperature fractions. The information obtained with the use of impedance spectroscopy method confirms the essential influence of PTFE on the process of charge transfer through the composite layer/electrolyte interface and therefore to explain the difference in the influence of various fractions obtained under the PTFE thermal destruction and used for producing of composite coatings on the scaling process. It was also established the significant influence both the method of the thermal processing and its duration on electro physical properties of coatings.

Now Shipyard "Zvezds" of Russia successfully used the PEO—technology in the fields of ship building and ship repair.

ACKNOWLEDGEMENTS

The autors would like to acknowledge that this work was supported by the Innovative educational program "Development of the Far East scientific—educational centre for formation Innovative terminal of Russia in Asian—Pacific region".

REFERENCES

Yerokhin, A.L., Nie, X., Leyland, A., Matthews, A. & Dowey. S.J. 1999. Surf. Coat. Technol; 122:73–93.
Yerokhin A.L., Nie, X., Leyland, A., & Matthews, A. 2000. Surf. Coat. Technol; 130:195–206.
Curran, J.A. & Clyne, T.W. 2005. Thermo-physical Properties of Plasma Electrolytic Oxide Coatings on Aluminium, Surface and Coatings Technology, Vol.199 (2–3), pp.168–176.
Curran, J.A. & Clyne, T.W. 2006. Porosity in plasma electrolytic oxide coatings, Acta Materialia, Vol. 54, pp. 1985–1993.
Gordienko, S.V. & Gnedenkov. 1997. Microarc oxidation of ti- tanium andits alloys. Vladivostok: Dalnauka.
Minaev, A.N. & Lysenko, L.V. 2001. Energotechnological processes using sea- or mineralized water. Moscow. Bauman Moscow State Technical University.
Gnedenkov, S.V. et al. 2000. Anticorrosion, Antiscale Coatings Obtained on the Surface of Titanium Alloys by Microarc Oxidation Method and Used in Seawater. 24–31// Corrosion No. 1 (56).
Gnedenkov, S.V. et al. 2007. Charge Transfer at the Antiscale Composite Layer-Electrolyte Interface. 667–673. Protection of Metals. Vol. 43, No. 7.
Gnedenkov, S.V. et al. 2008. Composite Polymer Containing Protective Layers on Titanium. 67–72. Protection of Metals. Vol. 44, No. 7.

Analysis and Design of Marine Structures – Guedes Soares & Das (eds)
© 2009 Taylor & Francis Group, London, ISBN 978-0-415-54934-9

Corrosion wastage statistics and maintenance planning of corroded hull structures of bulk carriers

Y. Garbatov & C. Guedes Soares
Centre for Marine Technology and Engineering (CENTEC), Technical University of Lisbon,
Instituto Superior Técnico, Lisboa, Portugal

ABSTRACT: An approach based on the statistical analysis of failure data leading to probabilistic models of time to failure and maintenance planning is presented here. The approach adopts the Weibull model for analyzing the failure data. Based on historical data of thickness measurements or corresponding corrosion wastage thickness of structural components in bulk carriers and the progress of corrosion, critical failure levels are defined. The analysis demonstrates how data can be used to address important issues as the inspection intervals, condition based maintenance action and structural component replacement. An effort is made to establish realistic decisions about when to perform maintenance on structure that will reach a failed state. Different scenarios are analyzed and inspection intervals are proposed.

1 INTRODUCTION

Corrosion is considered as one of the most important factors leading to age-related structural degradation of ships. Corrosion can take different forms of general attack, pitting corrosion, stress corrosion cracking, corrosion fatigue, fretting corrosion, filiform corrosion, weld corrosion, bimetallic corrosion and bacterial corrosion. General corrosion, which is a common form of corrosion, is spread over the whole surface of the metal and may lead to thickness reduction, and consequently facilitates fatigue cracks, brittle fracture and unstable failure.

The thickness measurements indicate the amount of wastage as a function of the age of structural elements and this data is used to develop relations that predict how corrosion wastage grows with time. Some databases are available worldwide as reported by TSCF, (1992), Yamamoto & Ikegaki, (1998) and Wang et al., (2003). These databases show that corrosion wastage has a rapid growth in the initial years and tends to level off later. Recently a new set of models have been proposed by several authors Guedes Soares & Garbatov, (1998, 1999), Paik et al., (1998) and Qin & Cui, (1992). The model of Guedes Soares & Garbatov, (1998, 1999) has been calibrated with data from tankers by Garbatov et al, (2007) and from bulk carriers by Garbatov & Guedes Soares (2008) showing a very good adjustment.

All of these models do not represent the succession of corrosion mechanisms as described by Melchers, (2003a,b), but they represent the overall trend that can be seen both in the Melchers model as well as in ship historical data. Therefore, they are the appropriate

models to be used by the Classification Societies and ship owners to predict the growth of corrosion in general and to plan inspections on that basis.

Experience has also shown that sister ships can experience different levels of corrosion, showing that the application of such models represent average situations but can have significant deviations when applied to one specific ship. Furthermore, experience has shown that even in the same ship corrosion rates very significantly from location to location. The reason is that the environmental conditions that are present in the different ship spaces and ships experience are different.

The first approach to solve this question has been adopted by separating the corrosion thickness measurements in areas of relatively constant environmental conditions, such as deck, tanks, hull etc. However the definitive way of dealing with this problem is to represent explicitly the effect that the various environmental factors have on corrosion.

Guedes Soares et al. (2005, 2008) studied the effect of the different marine immersion and atmospheric factors on the behaviour of marine corrosion wastage under immersion and atmospheric conditions respectively all over the ship's service life. The studies proposed a new corrosion wastage models based on the non-linear time dependent corrosion model accounting for various immersion and atmospheric environmental factors.

The decision about the preferred level of safety and consequently of the plate thickness to be included in the ship is more often made on the basis of reliability studies that incorporate a model of corrosion growth. Guedes Soares and Garbatov, (1996) presented a time variant formulation to model the degrading effect that corrosion has on the reliability of ships' hulls. They also considered one repair policy showing the effect of plate replacement when its thickness reached 75% of the as built thickness. Wirshing et al. (1997) presented a reliability assessment relative to the ultimate strength failure of a ship hull experiencing structural degradation due to corrosion.

The approaches discussed earlier were based on using structural reliability theory combined with models of corrosion growth with time. The approach in this paper deals with the statistical analysis of corrosion wastage data leading to probabilistic models of time to failure, which are used as basis for maintenance decisions.

Classical theory of system maintenance describe the failure of components by probabilistic models often of the Weibull family, which represent failure rates in operational phases and in the aging phases of the life of components as described in various textbooks (Moubray, 1997, Rausand, 1998 and Jardine & Tsang, 2005).

The present paper adopts that type of approach and demonstrates how they can be applied to structural maintenance of ships that are subjected to corrosion. The approach applied here is based on historical data of thickness measurements or corresponding corrosion wastage thickness in bulk carriers. Based on the progress of corrosion, critical corrosion levels are defined as "failure", which is modeled by a Weibull distribution. Existing formulations obtained for systems are applied to this case, leading to results that are in agreement with standard practice.

Corrosion data of plates of bulk carriers discussed by Garbatov & Guedes Soares, (2008) have been updated and analyzed here. The analysis demonstrates how this data can be used to address important issues such as inspection intervals, condition based maintenance action and structural component replacement. An effort is made to establish realistic decisions about when to perform maintenance on a structure that will reach a failed (corroded) state. Different scenarios are analyzed and optimum intervals and age are proposed.

The optimum intervals are based on statistical analysis of operational data using the Weibull model and some assumptions about the inspection and the time required for repair in the case of failure are considered here. The present analysis applies the general framework that was developed for failure of components in a system and adopts it to the corrosion deterioration problem by considering the different corrosion occurrence tolerances as failure criteria.

2 CORROSION DATA

Fourteen sets of corrosion data of structural components of balk carriers, partially included in the study in

Garbatov & Guedes Soares, (2008) and latter updated are analyzed here. These sets cover bottom (1), inner bottom (2), below top of bilge-hopper tank-face (3), lower slopping (4), lower wing tank-side shell (5), below top of bilge-hopper tank-web (6), between top of bilge, hopper tank, face (7), between top of bilge, hopper tank, web (8), side shell (9), upper than bottom of top side tank, face (10), upper deck (11), upper slopping (12), upper wing tank side shell (13), upper than bottom of top side tank, web (14) in a total of 8832 measurements. Using regression analysis, the mean value and standard deviation of corrosion wastage as a function of time are fitted to the following equation (Garbatov & Guedes Soares, 2008):

$$\text{Mean value}[d(t)] = d_\infty \left[1 - \exp\left(-\frac{t - \tau_c}{\tau_t} \right) \right]$$

$$\text{Stdev}[d(t)] = a \, \text{Log}(t - \tau_c - b) - c$$

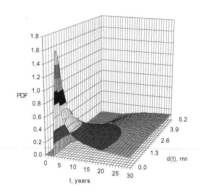

Figure 1. Probability density function of corrosion wastage (eqn. 1).

Table 1. Mean value and standard deviation of corrosion wastage regression equation descriptors.

Data set	d_∞ mm	τ_t years	τ_c	a	b	c	n
1	1.42	14.14	3.17	7.16	−11.60	7.41	684
2	3.50	18.29	0.00	1.47	−1.00	−0.23	556
3	4.68	3.51	9.11	2.08	−1.00	−0.46	421
4	2.75	19.75	0.00	16.09	−6.18	12.72	218
5	1.41	14.72	3.35	13.84	−9.16	13.22	152
6	3.55	3.17	8.73	11.24	−11.65	11.53	929
7	4.28	8.54	6.85	7.53	7.53	7.53	1380
8	3.35	5.66	7.26	17.93	−7.75	15.70	1901
9	2.25	25.46	0.00	1.91	−1.44	0.01	383
10	2.83	2.92	8.75	1.60	−1.00	−0.42	509
11	2.50	19.28	0.00	12.93	−9.83	12.62	362
12	1.19	15.67	0.69	12.23	12.23	12.23	432
13	1.58	20.56	0.00	9.88	−13.24	10.89	201
14	2.60	5.41	7.25	13.41	−16.42	15.72	704
						Total N	8832

216

where t is time, d_∞ is a parameter defining the long term corrosion depth, τ_c is the coating life, τ_t is the transition time and a, b and c are coefficients. The long-term probability density function as a function of time is defined as a truncated Normal probability density function.

The probability density function of corrosion wastage as a function of time for bottom 1 is shown in Figure 1, the mean value and the standard deviation of corrosion wastage regression equation descriptors for all sets of data are given in Table 1.

3 RELIABILITY ANALYSIS

A statistical parameter frequently used in reliability studies is the hazard rate, $h(t)$. An estimate of the hazard rate of a component at any point in time may be thought of as the ratio of a number of components that failed in an interval of time to the number of components in the original population that were operational at the start of the interval. Thus, the hazard rate of a component at time t is the probability that the component will fail in the next interval of time given that it is good at the start of the interval.

In general, the distributions of failure data of component life can be modeled by a Weibull probability density function of the following form:

$$f(t) = \begin{cases} \dfrac{\beta}{\eta}\left(\dfrac{t-\gamma}{\eta}\right)\exp\left(-\left(\dfrac{t-\gamma}{\eta}\right)^{\beta}\right) & \text{for } t > \gamma \\ 0 & \text{for } t \leq \gamma \end{cases}$$

(1)

where the three parameters of the Weibull distribution are β the shape parameter, γ the location parameter and η the scale parameter, β and η are greater than 0. The β value determines the shape of the distribution and when $\beta < 1$, the Weibull distribution has a hyperbolic shape with $f(0) = \infty$. When $\beta = 1$, it becomes an exponential function and when β exceeds 1, it is a unimodal function where skewness changes from left to right as the value of β increases. When $\beta = 3.44$, the Weibull distribution approximates the symmetrical Normal function.

The hazard rate, $h(t)$ of the Weibull distribution is:

$$h(t) = \frac{\beta}{\eta}\left(\frac{t-\gamma}{\eta}\right)^{\beta-1} \quad \text{when } t > \gamma, 0 \text{ otherwise.} \quad (2)$$

where $h(t)$ varies with the value of the independent variable t and in particular, when $\beta < 1$, $h(t)$ is a decreasing function of t. When $\beta = 1$, $h(t)$ does not vary with t, $h(t)$ becomes an increasing function of t when $\beta > 1$. When $t - \gamma = \eta$, $F(t) = 1 - \exp(-1)$, or approximately 63.2%, for all values of β. Thus, η

is also known as the characteristic life of the Weibull distribution.

By definition, the probability density function of the Weibull distribution is zero for $t < \gamma$. That is, there is no risk of failure before γ, which is therefore termed the location parameter or the failure-free period of the distribution. In practice, γ may be negative, in which case the component may have undergone a run-in process or it had been in use prior to $t = 0$.

A function complementary to the cumulative distribution function is the reliability function, also known as the survival function. It is determined from the probability that the structure will survive at least to some specified time, t. The reliability function is denoted by $R(t)$ and is defined as:

$$R(t) = \int_t^\infty f(t)dt = 1 - F(t) \quad (3)$$

Maintenance decision analysis requires the use of the failure time distribution of structural components, which may not be known. There may, however, be a set of observations of failure times available from historical records. One might wish to find the Weibull distribution that fits the observations, and to assess the goodness of the fit.

If the Weibull plot is a curve when the location parameter is considered as non-zero, then a three-parameter Weibull distribution has to be used to model the data set. The curvature of the plot suggests that the location parameter is $\gamma > 0$. Obviously, γ must be less than or equal to the shortest failure time, t_i. Finding the correct value of γ will produce a linear plot.

The failure data can be analyzed in order to estimate measures of reliability such as $R(t)$ from the Weibull plot. For more accuracy, a confidence interval can be determined on the estimation of $R(t)$.

In practice, not every structural component is observed up to failure. When there is only partial information about a components lifetime, i.e. when some of the components have not failed, the information is known as censored or to have suspended data.

The most common case of censoring is right censored data, or suspended data, which means that the life data sets have components that did not fail. For example, if five components are tested and only three have failed by the end of the test or of the observation time period, then the two unfailed components represent suspended data (or right censored data). The term "right censored" implies that the event of interest (i.e. the time of failure) is to the right of the reference period. If the components were to keep operating, the failure would occur at some time after that data point (or to the right on the time scale).

The second type of censoring is interval censored data and it reflects the times the components failed

Table 2. Weibull distribution function parameters and statistics descriptors for the 14 data sets.

		L	M	H	E			L	M	H	E
1	α	2.4	4.7	*	*	8	α	1.4	3.0	3.0	2.8
	β	12.0	14.1	*	*		β	4.7	5.8	6.3	6.6
	γ	5.0	5.0	*	*		γ	7.5	7.5	7.5	7.5
2	α	1.3	1.8	2.0	1.9	9	α	1.4	3.0	3.0	2.8
	β	9.0	13.8	15.6	21.2		β	4.7	5.8	6.3	6.6
	γ	5.0	5.0	5.0	5.0		γ	7.3	7.3	7.3	7.3
3	α	2.1	2.9	3.5	3.3	10	α	1.7	3.6	3.9	3.5
	β	4.4	4.6	5.0	5.2		β	5.8	6.2	6.7	7.1
	γ	8.3	8.3	8.3	8.3		γ	7.3	7.3	7.3	7.3
4	α	1.3	1.8	1.6	1.5	11	α	1.5	1.3	1.3	4.6
	β	9.6	11.6	15.5	18.0		β	10.3	12.0	17.6	14.6
	γ	5.0	5.0	5.0	5.0		γ	5.0	5.0	5.0	5.0
5	α	2.2	5.3	*	*	12	α	3.1	3.3	*	*
	β	13.2	15.5	*	*		β	11.8	15.3	*	*
	γ	5.0	5.0	*	*		γ	5.0	5.0	*	*
6	α	1.3	4.0	4.7	5.7	13	α	1.5	1.6	*	*
	β	3.8	5.0	5.1	5.3		β	11.0	19.4	*	*
	γ	7.5	7.5	7.5	7.5		γ	5.0	5.0	*	*
7	α	1.3	1.7	2.4	3.2	14	α	3.4	5.0	6.9	5.6
	β	5.7	6.9	8.5	7.5		β	5.1	5.3	5.6	5.9
	γ	7.3	7.3	7.3	7.3		γ	7.3	7.3	7.3	7.3

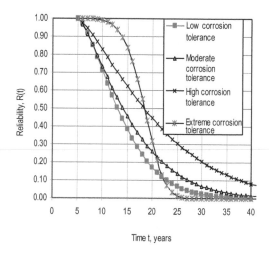

Figure 2. Reliability, lower wing tank.

the different levels of corrosion limits are given in Figure 2.

within an interval. This type of data comes from situations where the components are not constantly monitored. The only information that is recorded is that they failed in a certain interval of time.

The third type of censoring is similar to the interval censoring and is left censored data. In the left censored data, a failure time is only known to be before a certain time. This is identical to interval censored data in which the starting time for the interval is zero.

Right censored data will not cause complications in the Weibull analysis because all of them correspond to longer failure times than the time of observation. However, it is necessary to use a special procedure to take care of them and in this case suspended data are handled by assigning an average order number to each failure time.

The analysis presented here uses data sets of failure times of corroded structural components in bulk carriers. Four different levels of censoring related to the failure state of corroded plates are introduced: low (L) corrosion that is 0.5 mm corrosion degradation with respect to the plate thickness as built up, moderate (M) corrosion is 1.0 mm, high (H) corrosion is 1.5 mm and extreme (E) corrosion is 2.0 mm respectively. The corrosion levels are set up here as permissible corrosion levels, any time at which corrosion depths may reach them is classified as complete failure, and others are censored.

The completed failure times are described by the Weibull distribution and its statistical descriptors for the different corrosion levels can be seen in Table 2. The reliability estimates of a lower wing tank for

4 OPTIMAL REPLACEMENT INTERVAL— MINIMIZATION OF TOTAL COST

Metal structural components are subjected to corrosion and when failure occurs, they have to be replaced. Since failure is unexpected then it may be assumed that a failure replacement is more costly than an earlier replacement. In order to reduce the number of failures, replacements can be scheduled to occur at specified intervals. However, a balance is required between the amount spent on the replacements and their resulting benefits, that is, reduced failure replacements.

It is assumed that the problem is dealing with a long period over which the structure is to be in good condition and the intervals between the replacements are relatively short. When this is the case, it is necessary to consider only one cycle of operation and to develop a model for one cycle. If the interval between the replacements is long, it would be necessary to use a discounting approach, and the series of cycles would have to be included in the model to take into account the time value of money.

The replacement policy is one where replacements occur at fixed intervals of time; failure replacements occur whenever necessary. The problem is to determine the optimal interval between the replacements to minimize the total expected cost of replacing the corroded plates per unit time.

The total cost of a replacement before failure is defined as C_p, while C_f is the total cost of a failure replacement and $f(t)$ is the probability density function of the plate's failure times. The replacement policy is to perform replacements at constant intervals of time t_p, irrespective of the age of the plate, and failure replacements occur as many times as required in the interval (0, t_p).

To determine the optimal interval between replacements the total expected replacement cost per unit time is minimized. The total expected cost per unit time for replacement at the intervals of length t_p, denoted $C(t_p)$ equals to the total expected cost in the interval (0, t_p) divided by the length of a interval (Jardine & Tsang, 2005):

$$C(t_p) = \frac{C_p + C_f H(t_p)}{t_p} \qquad (4)$$

where $H(t_p)$ is the expected number of failures in the interval (0, t_p). To determine $H(t)$, the renewal theory approach is to be applied (Jardine & Tsang, 2005) leading to the recurrence relation:

$$H(T) = \sum_{i=0}^{T-1} [1 + H(T - i - 1)] \int_i^{i+1} f(t)\,dt, \quad T \geq 1$$

$$(5)$$

with $H(0) = 0$.

The expected cost of failure replacement considered here is modeled as $C_f = nC_p$, where C_p is assumed 10,000 units as an example here and n varies as 10, 25, 50 and 75 that results in low (L), moderate (M), high (H) and extreme (E) repair cost consequence respectively. The optimal replacement interval for corroded plates subjected to the replacement strategy is based on the normalized total repair cost with respect to the expected cost of failure replacement.

In this analysis, no account was taken for the time required to perform replacements since they were

considered very short, compared to the mean time between replacements of plates. When necessary, the replacement durations can be incorporated into the replacement model, as is required when the goal is the minimization of total downtime or, equivalently, the maximization of component availability. However, any cost that is incurred due to the replacement stoppages need to be included as part of the total cost before failure or in the total cost of a failure replacement.

Optimal replacement intervals for the sets of data accounting for different corrosion and repair cost consequence levels are given in Table 3. The columns, in

Table 3. Optimal replacement intervals for the 14 data sets.

Table 4. Optimal replacement intervals, (years) for the 14 data sets.

Table 5. Optimal inspection intervals, (years) for the 14 data sets.

219

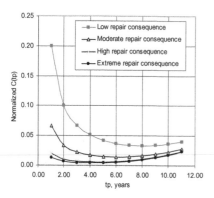

Figure 3. Replacement intervals, lower wing tank.

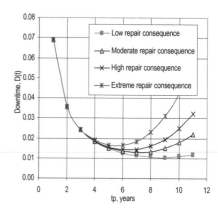

Figure 4. Replacement intervals, lower wing tank.

Table 3, with L, M, H and E show low (L), moderate (M), high (H) and extreme (E) corrosion degradation levels and rows with L, M, H and E show the resulting low (L), moderate (M), high (H) and extreme (E) repair cost consequences levels. These designations are also used for Table 4 and 5.

Figure 3 shows the normalized total cost as a function of replacement interval for lower wing tanks. It can be seen that the minimum inspection interval is achieved when there is a combination of lower corrosion and extreme total repair cost consequence, which leads to 1 year optimal replacement interval. The maximum inspection interval is achieved in a combination of lower total repair cost consequence and extreme corrosion, which results in 8 years. Different variations of Corrosion and total repair cost consequence result in different optimal replacements intervals defined as L ($6 < t_p$ years), M ($4 < t_p \leq 6$ years), H ($2 < t_p \leq 4$ years) and Extreme ($1 \leq t_p < 2$ years) frequency repairs (see Table 3 and Figure 3).

5 OPTIMAL REPLACEMENT INTERVAL— MINIMIZATION OF DOWNTIME

In some cases due to difficulties in costing or to the desire to get maximum throughput or utilization of the structures, the replacement policy required may be one that minimizes total downtime per unit time or, equivalently, maximizes availability. The problem is to determine the best times at which replacements should occur to minimize total downtime per unit time. The basic conflicts are that as the replacement frequency increases, there is an increase in downtime due to these replacements, but a consequence of this is a reduction of downtime due to failure replacements, and we wish to get the best balance between them has been defined.

The objective is to determine the optimal replacement interval t_p between replacements in order to minimize the total downtime per unit time. The total

downtime per unit time, for replacement at time t_p, is denoted as $D(t_p)$ and it is defined as the number of failures, $H(t_p)$ in the time interval ($0, t_p$) time the time required to make a failure replacement, T_f. plus the time required to make replacement before failure divided to the interval, $t_p + T_p$, results in (Jardine & Tsang, 2005):

$$D\left(t_p\right) = \frac{H\left(t_p\right) T_f + T_p}{t_p + T_p} \qquad (6)$$

The replacement interval to minimize total downtime for the sets of data of failure time studied here are modeled as $T_f = nT_p$.

Four different levels of downtime consequence are defined for n equals to 5, 10, 25 and 75 conditioning to $T_p = 4$ weeks as L, M, H and E downtime consequence respectively. The optimal replacement interval for different level of corrosion and downtime consequences result in different optimal replacement intervals defined as L ($6 < t_p$ years), M ($4 < t_p \leq 6$ years), H ($2 < t_p \leq 4$ years) and Extreme ($1 \leq t_p < 2$ years) replacement interval as can be seen in Table 4. The variation of the downtime as a function of replacement interval for lower wing tank is shown in Figure 4.

6 OPTIMAL INSPECTION INTERVAL TO MAXIMIZE THE AVAILABILITY

The basic purpose behind an inspection is to determine the state of the structure. One indicator, such as corrosion deterioration, which is used to describe that state, has to be specified, and the inspection is made to determine the values of this indicator. Then some maintenance action may be taken, depending on the state identified. The decision about when the inspection should take place ought to be influenced by the costs of the inspection and the benefits of the inspection, such as detection and correction of minor defects before major breakdown occurs.

The primary goal addressed here is that of making the structure more reliable through inspection because of establishing the optimal inspection interval for structures, and this interval is called the failure-finding interval.

The time required to conduct an inspection is T_i. It is assumed that after the inspection, if no major faults are found requiring repair or complete component replacement, the component is in the as-new state. This may be as a result of minor modifications being made during the inspection. T_r is the time required to make a repair or replacement. After the repair or replacement, it is assumed that the component is in the as-new state.

The objective is to determine the interval t_i, between inspections in order to maximize availability per unit time. The availability per unit time, denoted by $A(t_i)$, is a function of the inspection interval t_i and it is the expected availability per cycle/expected cycle length.

The uptime in a good cycle equals to t_i, since no failure is detected at the inspection. If a failure is detected, then the uptime of the failed cycle can be taken as the mean time to failure of the component, given that inspection takes place at t_i.

The expected uptime per cycle is calculated as (Rausand, 1998):

$$A(t_i) = \frac{t_i R(t_i) + \int_{-\infty}^{t_i} tf(t)\, dt}{(t_i + T_i) R(t_i) + (t_i + T_i + T_r)[1 - R(t_i)]}$$

(7)

The analysis here uses the data sets of failure times of corroded structural components and the optimal inspection interval to maximize the availability considering $T_i = 1$ week and $T_r = nT_i$, where n varies as 10, 25, 50 and 75 respectively, which results in L, M, H and E downtime consequence and it can be seen in Table 5 and Figure 5 shows the variation of availability as a function of inspection intervals for the lower wing tank. The inspection interval defined here are classified as L ($6 < t_p$ years), M ($4 < t_p \leq 6$ years), H ($2 < t_p \leq 4$ years) and Extreme ($1 \leq t_p < 2$ years) inspection interval.

The crucial assumption in the model here is that plates can be assumed as good as new after inspection if no repair or replacement takes place. In practice, this may be reasonable, and it will certainly be the case if the failure distribution of the component was exponential (since the conditional probability remains constant).

If the as-new assumption is not realistic and the failure distribution has an increasing failure rate, then rather than having inspection at constant intervals, it may be advisable to increase the inspection frequency, as the component gets older.

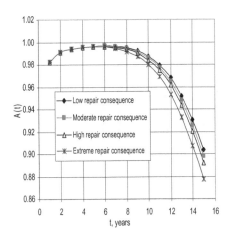

Figure 5. Inspection intervals, lower wing tank.

7 COMPARATIVE ANALYSIS

Fourteen sets of corrosion data are analyzed here and a comparison between different strategies for optimal inspection interval accounting for minimization of total cost, minimization of downtime and maximization of availability of different structures subjected to moderate level of corrosion and for the resulting moderate cost consequence is presented in Figure 6. As can be observed, ship structural components 3, 6–8, 10 and 14 require more frequent repair work with respect to the criteria of "total repair cost" and "downtime". When optimal replacement is defined to "maximize the availability" structural components 3, 7 and 11.

The structural components behave in a similar manner with respect to the criteria "minimization of the total repair cost" and "minimization of downtime". However, it has to be pointed out that after some years of service some components will deteriorate faster and because of that more intensive repair work will be required. Furthermore, different assumptions about the costs involved and the operational constraints for the time of inspections will result in different strategies of maintenance.

The analysis just presented is based on failure data collected during the entire life of corroded structural components. To make use of this analysis for a specific ship, it would be necessary to conduct a similar analysis for an appropriate data set. Additional constrains have to be applied with respect to the time of planned inspections for any class notation as defined by Classification Societies.

8 CONCLUSIONS

An analysis has been made, which predicts the optimum interval for maintenance, accounting for the

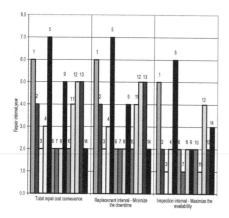

Figure 6. Optimum repair intervals for different criteria.

general corrosion deterioration. The analysis used the general framework that was developed for failure of components in a system and adopted it to the present problem by considering the different corrosion levels as failure criteria.

It is demonstrated that after fitting a Weibull distribution to corrosion data different criteria can be adopted and the corresponding optimal repair schedule can be determined.

Different criteria for corrosion detection and for the effect of repair can be easily incorporated. Various assumptions were made about different operational times and costs that are not essential to the method but are needed for the example calculation.

ACKNOWLEDGEMENT

This work has been performed within the project "MARSTRUCT—Network of Excellence on Marine Structures" (http://www.mar.ist.utl.pt/marstruct/) and has been partially funded by the European Union through the Growth programme under contract TNE3-CT-2003-506141.

REFERENCES

Garbatov, Y. & Guedes Soares, C., 2008, Corrosion Wastage Modeling of Deteriorated Ship Structures, *International Shipbuilding Progress*, Vol. 55, pp. 109–125.

Garbatov, Y., Guedes Soares, C. & Wang, G., 2007, Non-linear Time Dependent Corrosion Wastage of Deck Plates of Ballast and Cargo Tanks of Tankers, *Journal of Offshore Mechanics and Arctic Engineering*, Vol. 129, pp. 48–55.

Guedes Soares, C. & Garbatov, Y., 1996, Fatigue Reliability of the Ship Hull Girder Accounting for Inspection and Repair, *Reliability Engineering and System Safety*, Vol. 51, pp. 341–351.

Guedes Soares, C. & Garbatov, Y., 1998, Non-linear Time Dependent Model of Corrosion for the Reliability Assessment of Maintained Structural Components, *Safety and Reliability*, S. Lydersen, G. Hansen and H. Sandtorv, Rotterdam, Balkema, Vol. 2, pp. 929–936.

Guedes Soares, C. & Garbatov, Y., 1999, Reliability of Maintained, Corrosion Protected Plate Subjected to Non-linear Corrosion and Compressive Loads, *Marine Structures*, Vol. 12, pp. 425–445.

Guedes Soares, C., Garbatov, Y., Zayed, A. & Wang, G., 2005, Non-linear Corrosion Model for Immersed Steel Plates Accounting for Environmental Factors, *Transactions of the Society of Naval Architecture and Marine Engineering*, Vol. 13, pp. 48–55.

Guedes Soares, C., Garbatov, Y., Zayed, A. & Wang, G., 2008, Corrosion Wastage Model for Ship Crude Oil Tanks, *Corrosion Science*, Vol. 50, pp.3095–3106.

Jardine, A. & Tsang, A., 2005, *Maintenance, Replacement and Reliability, Theory and Applications*, Taylor & Francis.

Melchers, R., 2003a, Probabilistic Models for Corrosion in Structural Reliability Assessment Part 1: Empirical Models, *Journal of Offshore Mechanics and Arctic Engineering*, Vol. 125, pp. 264–271.

Melchers, R., 2003b, Probabilistic Models for Corrosion in Structural Reliability Assessment Part 2: Models Based on Mechanics, *Journal of Offshore Mechanics and Arctic Engineering*, Vol. 125, pp. 272–280.

Moubray, M., 1997, *Reliability Centered Maintenance*, Oxford, Butterworth Heinemann.

Paik, J. K., Kim, S. K., Lee, S. & Park, Y. E., 1998, A Probabilistic Corrosion Rate Estimation Model for Longitudinal Strength Members of Bulk Carriers, *Journal of Ship and Ocean Technology*, Vol. 2, pp. 58–70.

Qin, S. & Cui, W., 2002, Effect of Corrosion Models on the Time-Dependent Reliability of Steel Plated Elements, *Marine Structures*, pp. 15–34.

Rausand, M., 1998, Reliability Centred Maintenance, *Journal of Reliability Engineering & System Safety*, Vol. 60, pp. 121–132.

TSCF, 1992, *Condition Evaluation and Maintenance of Tanker Structures*, Tanker Structure Cooperative Forum, Whiterby & Co Ltd.

Wang, G., Spencer, J. & Sun, H., 2003, Assessment of Corrosion Risks to Aging Ships using an Experience Database, *Proceedings of the 22nd International Conference on Offshore Mechanics and Arctic Engineering*, ASME, pp. OMAE 2003-37299.

Wirshing, P., Ferensic, J. & Thayamballi, A., 1997, Reliability with Respect to Ultimate Strength of a Corroding Ship Hull, *Marine Structures*, Vol. 10, pp. 501–518.

Yamamoto, N. & Ikegaki, K., 1998, A Study on the Degradation of Coating and Corrosion on Ship's Hull Based on the Probabilistic Approach, *Journal of Offshore Mechanics and Arctic Engineering*, Vol. 120, pp. 121–128.

Analysis and Design of Marine Structures – Guedes Soares & Das (eds)
© 2009 Taylor & Francis Group, London, ISBN 978-0-415-54934-9

Numerical simulation of strength and deformability of steel plates with surface pits and replicated corrosion-surface

Md. Mobesher Ahmmad & Yoichi Sumi
Yokohama National University, Yokohama, Japan

ABSTRACT: Strength and deformability of steel plates for marine use are studied both experimentally and numerically. In order to accurately simulate this problem it is essential to have the precise information about the true stress-strain relationship of materials. In the present study, we first estimate the true stress-strain relationship of flat steel plate under uni-axial tension by introducing a vision sensor system to the deformation measurements. The experimental technique consists of the measurement of minimum cross-sectional area and the estimation of a correction factor for the tri-axial stress state after the necking phenomenon, where we use a flat specimen without any imperfections. The measured true stress-stain is then applied to assess the strength and deformability of steel plates with the use of non-linear, large deformation and three-dimensional finite element method. The relation between the final failure criterion and the finite element mesh size used for the analyses is clarified. Then the simulations are carried out for flat plates with surface pits and for replica specimens of general corrosion by a nonlinear implicit finite element code; LSDYNA. The numerical results are then validated by experimental counterparts. Two different steels having yield ratios, 0.657 and 0.841, respectively, are prepared to see the material effects for corrosion damage. It has been shown that the strength and deformability are not so dependent upon the yield ratios of the present two materials, while they depend on the surface configuration such as the minimum cross-sectional area of the specimens.

1 INTRODUCTION

Marine structures are subjected to age related deterioration such as corrosion wastage, fatigue cracking or mechanical damage during their service life. These damages can give rise to significant issues in terms of safety, health, environment and financial cost. It is thus of great importance to develop advanced technologies, which can allow for proper management and control of such age related deterioration (ISSC, 2006). In order to assess the structural performance of aged ships, it is of essential importance to predict the strength and absorbing energy during collapse and/or fracture of corroded plates.

Now-a-days numerical simulation is replacing the tedious experimental work. An exact simulation of tension test requires a complete true stress-strain relationship. Here we, at first, estimate the true stress-strain of steel plate with rectangular cross-section. A vision sensor system is employed to estimate the deformation field from the specimen surface by which an average least cross-sectional area and a correction factor due to tri-axial stress state can be estimated. The measured true stress-strain relationship is then applied to an

elasto-plastic material model of LS-DYNA to assess the strength and deformability of corroded steel plates.

A great number of research works have been carried out for the structural integrity of aged ships. Nakai et al. (2004) have studied the strength reduction due to periodical array of pits, while Sumi (2008) has investigated the self-similarity of surface corrosion, experimentally. Paik et al. (2003 & 2004) studied the ultimate strength characteristics of pitted plates under axial compressive load and in-plane shear load. In the present paper, we shall discuss the geometrical effect on strength and deformability for various pit size and for plates with general corrosion. The shape of the pit is conical and its depth to diameter ratio is 1:8. In the case of general corrosion, replica specimens are made to simulate corroded surfaces sampled from an aged actual ship. The geometry of corroded surfaces are generated by a CAD system, and mechanically processed by NC milling machine by a CAM system. Investigations are made for two different steels, with same ultimate strength, having yield ratios, 0.657 (steel A) and 0.841 (steel B), respectably, to identify the material effect for corrosion damage. Note that the former is commonly used in marine industries.

2 MESUREMENT OF TRUE STRESS-STRAIN RELATIONSHIP

The true stress-strain relationship including material response in both pre-and post plastic localization phase is necessary as input for numerical analyses. In some cases, structural analysts use power law stress-strain relationship. It has been demonstrated that power law stress-strain curve for certain examples of steel can overestimate the actual stress-strain curve at low plastic strain and underestimate at latter stage (Bannister et al. 2000). Strain measurement becomes complicated, especially, in case of flat tensile specimen due to inhomogeneous strain field and tri-axial stress state. Two practical difficulties can be mentioned here. The first problem is the measurement of instantaneous area of minimum cross section after necking. Figure 1 (a) depicts the necking phenomenon in the uniaxial tension test of a flat specimen. During the plastic instability, the cross-sectional area at maximum deformed zone shapes up as cushion like shape (Schider et al. 2004) as shown in Figure 1 (b). It makes difficult to measure the cross-sectional area at the neck. The second challenge is the measurement of a/R to estimate a correction factor e.g. Bridgman (1964) and Ostsemin (1992) correction factor, arising due to tri-axial stress condition after necking initiation, where a is half thickness and R is radius of curvature of the neck (see Fig. 3).

2.1 Theoretical background

For any stage of deformation, true stress-strain is defined by

$$\sigma_T = \frac{F}{A}, \quad \varepsilon_T = \ln\left(\frac{l}{l_0}\right), \tag{1}$$

where A, F, l_0 and l are the instantaneous area, applied force, initial length of a very small distance, say 1 mm, at probable neck zone and its current value, respectively. As long as uniform deformation occurs, it can be calculated in terms of engineering stress and strain by

$$\varepsilon_T = \ln(1 + \varepsilon_e); \quad \sigma_T = \sigma_e(1 + \varepsilon_e). \tag{2}$$

In the Scheider's procedure the effective strain, $\bar{\varepsilon}$, after bifurcation was calculated by

$$\bar{\varepsilon} = \sqrt{\frac{4}{3}(\varepsilon_I^2 + \varepsilon_I\varepsilon_{II} + \varepsilon_{II}^2)}, \tag{3}$$

where ε_I and ε_{II} are specimen's length and width directional true strain. Usually bifurcation phenomenon occurs soon after the maximum load (Okazawa et al.

2005). In our calculation, we shall use Equation (3) to measure the true strain. After initiation of necking the true stress can be calculated by

$$\sigma_T = \frac{F}{A} = \frac{F}{A_0}\exp(-\varepsilon_{II} - \varepsilon_{III}), \tag{4}$$

where ε_{III} is the thickness directional strain. Equation (5) can be obtained in case of uniform deformation

$$\sigma_T = \frac{F}{A_0}\exp(\varepsilon_I). \tag{5}$$

In practice, the axial strain over the cross-section, as shown in Figure 2 (a), is not uniform. An average true stress can be obtained from Equation (6) by measuring an average axial strain, $\bar{\varepsilon}_I$ (see Fig. 2(b)),

$$\sigma_T = \frac{F}{A_0}\exp(\bar{\varepsilon}_I). \tag{6}$$

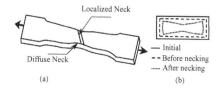

(a)   (b)

Figure 1. Illustration of (a) diffused and localized neck, (b) cross sectional area at various phases of deformation.

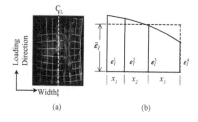

(a)   (b)

Figure 2. (a) Deformed grid on the surface of neck zone. (b) Estimation of average axial strain.

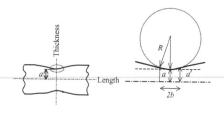

Figure 3. Illustration of neck geometry.

Stress correction due to triaxial stress state after the initiation of necking can respectively be given by the analytical solutions of Bridgman (1964)

$$C.F = \left[\left(1 + \frac{2R}{a} \right)^{1/2} \right.$$

$$\left. \times \ln \left\{ 1 + \frac{a}{R} + \left(\frac{2a}{R} \right)^{1/2} \left(1 + \frac{a}{2R} \right)^{1/2} \right\} - 1 \right],$$

(7)

and Ostsemin (1992)

$$C.F = \left(1 + \frac{a}{5R} \right).$$

(8)

In each case, the correction factor ($C.F$) depends on a parameter, a/R (see Fig. 3). The solid bold line in Figure 3 represents the upper boundary of centerline section at neck. The quantity a/R can be obtained by

$$\frac{a}{R} = \frac{2a(a' - a)}{(a' - a)^2 + b^2},$$

(9)

where, initially, a/b can be taken as 0.5~1.0 (Cabezas et al. 2004). The superscript '$\prime$' is used to indicate the measuring quantities estimated at a distance b from the center of neck (see Fig. 3).

The continuous value of a can be estimated in terms of on surface deformation measured by vision sensor:

$$a = a_0 e^{\varepsilon_{III}} = a_0 e^{-(\varepsilon_I + \varepsilon_{II})}$$

(10)

$$a' = a_0 e^{\varepsilon'_{III}} = a_0 e^{-(\varepsilon'_I + \varepsilon'_{II})}.$$

(11)

2.2 Vision sensor for measurement of true stress-strain relationship

The specimens are prepared according to Japanese Industrial Standards (JIS). The geometry of the flat specimen is shown in Figure 4 (a). In the present study

this model will be used for all cases. The surface is prepared as shown in Figure 4 (b): white dots are pointed on permanent black ink that is painted on the specimen. The wide range, 40 mm, of measuring length is designed so that necking occurs within this range without any imperfections.

Figure 5 shows the experimental setup. The monochromic vision sensor traces the white dots during the experiment. After reading the position of the dots on the steel specimen, its digital data is changed into analog data with a D/A converter. The analog data of deformation and load meet on a personal computer through voltage signal interface. A Programmable Logical Controller is used to synchronize the whole system. An extensometer is used to measure the strain of 100 mm length. Note that the white paint should have high deformability. Here we use correction fluid for white dots.

2.3 Results and discussions

In this study we observe that uniform deformation occurs until the first bifurcation (initiation of diffuse neck) at strain 0.25, while the strain at maximum load is 0.16 for steel A. The correction factor due to triaxial stress state becomes effective after the second bifurcation at strain 0.45 (see Fig. 6). The correction factor varies 1 to 1.03, which implies that the true stress is reduced by 0 to 3 percentages initiating from strain 0.45. Applying the procedure discussed in previous subsection, finally, the true stress-strain is obtained as shown in Figures 7a, b.

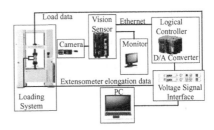

Figure 5. Vision system.

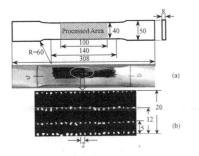

Figure 4. (a) The tensile specimen (all dimensions in mm). (b) White dots on specimen.

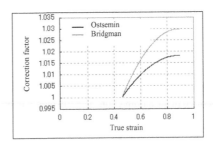

Figure 6. Correction factor due to tri-axial stress state (steel A).

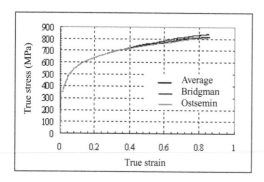

Figure 7a. True stress-strain relationship of steel A.

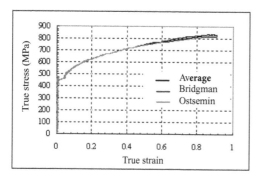

Figure 7b. True stress-strain relationship of steel B.

3 NUMERICAL ANALYSIS

Numerical analyses have been carried out with non-linear implicit finite element code of LS-DYNA, as the problem is a quasi-static type. The constitutive material model is an elasto-plastic material where an arbitrary stress verses strain curve can be defined. This material model is based on the J_2 flow theory with isotropic hardening (Hallquist, 1998).

3.1 Finite element model and material properties

The basic problem that has been analyzed is the quasi-static uniaxial extension of rectangular bar as shown in Figure 8 (a).

Due to symmetry condition only 1/8 of the model is analyzed. A constant velocity, $V(t) = 3$ mm/min, is prescribed in the X_1 direction. The FE-model is shown in Figure 8 (b), which is discretised by 8-node brick elements.

The material properties are listed in Table 1. The strain hardening is defined by the true stress-strain curves as shown in Figures 7a, b. Here fracture

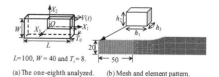

$L=100$, $W = 40$ and $T_s = 8$.

(a) The one-eighth analyzed.   (b) Mesh and element pattern.

Figure 8. Finite element model of flat specimen.

Table 1. Material properties.

Mat-erial	Yield strength N/mm²	Tensile strength N/mm²	Y/T ratio	E GPa	Poisson's ratio	Failure strain
A	344	523	0.657	206.5	0.3	0.92
B	440	523	0.841	204.5	0.3	0.90

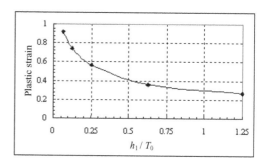

Figure 9. Mesh size effect on maximum plastic strain.

initiation is defined by strain to failure. The fracture strain, ε_f, is measured by

$$\varepsilon_f = \ln\left(\frac{A_0}{A_f}\right),\qquad(12)$$

where A_f is the projected fracture surface area measured after experiments.

3.2 Mesh size effect

Mesh size effects are important in the failure analyses of structures. It is universal that finer mesh size is needed for better results when large deformation occurs. However, a significant complication arises because of mesh size sensitivity whereby strain to failure generally increases with finer mesh.

In Figure 9, we compare the maximum plastic strain (failure strain) of four models (flat) at a nominal strain of 0.284. In these cases, we vary only loading directional element size, h_1, keeping the mesh size in the other two directions, h_2 and h_3, fixed, as their effect is less significant. From the figure, it can be stated

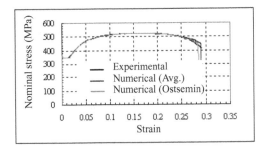

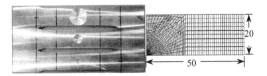

Figure 10. Verification of numerical nominal stress-strain by its experimental counterpart (steel A).

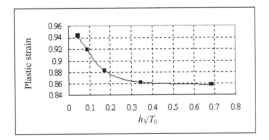

Figure 11. (a) Experimental specimens with single pit (left). (b) Mesh pattern of $^1/_4$ of processed zone ($50 \times 20 \times 8$ mm).

that the finer the mesh size the greater the maximum plastic strain. Figure 10 represents a comparison of experimental and numerical nominal stress-strain for flat plate of steel A. The element size is $0.5 \times 1 \times 2$ (mm). Similar results are obtained for steel B, where numerical result agrees with experimental counterpart.

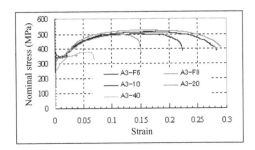

Figure 12. Mesh sensitivity due to surface pit.

4 SIMULAITON OF PITTED PLATES

4.1 *Effect of pit size on strength and deformability*

The true stress-strain relationship of steels A and B will be applied to specimen with a surface pit. To estimate the pit size effect on strength and deformability, we consider three different pit sizes of diameter 10 mm, 20 mm and 40 mm. Figure 11 shows the specimens (left) and mesh pattern (right) of 1/4 model.

The mesh sensitivity under the pit cusp has been analyzed by changing the mesh size along thickness direction while the mesh in other directions will remain unchanged. With the finer mesh under the pit cusp the plastic strain, measured at nominal strain 0.175, increases (Fig. 12).

The experimental and numerical results of nominal stress-strain are shown in Figures 13a, b, respectively. A good agreement can be observed between them. Similar results are found for steel B. Figures 14a, b show the quantity (experimental and numerical) of strength and deformability reduction for various pit sizes. From these figures it can easily be stated that with increasing pit size the deformability reduces considerably while the strength reduces moderately. As shown in figures the difference of the reduction of the strength and deformability due to steels A and B is insignificant. The significance of these results can be emphasized on the energy absorption by structural elements during deformation. The surface pits can be treated as highly stress concentrated zone.

Paik et al. (2003) and Nakai et al. (2004) have confirmed that the ultimate strength of a steel plate with pitting corrosion is governed by smallest cross-sectional area. Here we also consider the ultimate

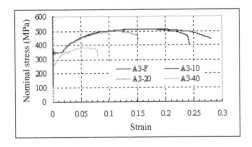

Figure 13a. Experimental stress-strain diagram for various pit sizes (A3–10: steel A, cross-head speed 3 mm/min, pit dia 8 mm; A3-F: steel A flat specimen).

Figure 13b. Numerical stress-strain diagram for various pit sizes.

strength reduction factor, R_u, as a function of damage. The damage value depends on smallest cross-sectional area, A_i, due to surface pits and can be defined as:

$$\text{Damage,} \quad D = \frac{A_0 - A_i}{A_0} \quad (13)$$

227

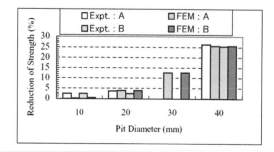

Figure 14a. Pit size effect on tensile strength.

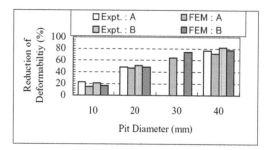

Figure 14b. Pit size effect on deformability.

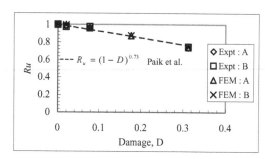

Figure 15. Strength reduction due to damage caused by pitting corrosion.

where A_0 is the initial undamaged area. R_u is defined as:

$$R_u = \frac{\sigma_{ui}}{\sigma_{u0}} \qquad (14)$$

where σ_{u0} and σ_{ui} are the ultimate tensile strength of undamaged plate and maximum tensile strength of pitted plate. Figure 15 shows the strength reduction factor decreases with increasing damage due to increase of pit size.

4.2 Simulation of plates with periodical pits

Sumi (2008) and Nakai et al. (2004) have experimentally investigated the strength and deformability of

Figure 16. Specimen with periodical pits and pit geometry.

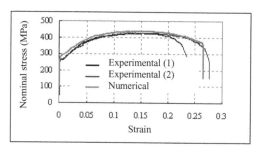

Figure 17. Mesh pattern of pitting model.

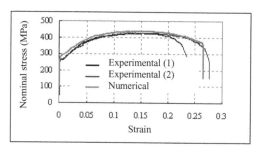

Figure 18. Numerical stress-strain curve agrees with experimental counterparts (steel A).

steel plates with periodical array of surface pits (e.g. Fig. 16). Periodical pits are made on both surfaces of plate and they are arranged asymmetrically with respect to middle plane of the specimen. To make a FE-model we, at first, make array of points, which describe the surfaces with pits. From the point data we can obtain NURBS surface. Now the surface can be discretized by iso-mesh. After obtaining both surfaces, solid elements (8-node hexahedron) can be obtained by sweeping action as shown in Figure 17.

The numerical and experimental nominal stress-strain curves of steel A are shown in Figure 18. Here the failure strain is defined as 0.7 as the element size is $1 \times 1 \times 4$ (mm). Similar results are obtained for material B, where a good agreement is observed among experimental and numerical nominal stress-strain curves.

5 GENERAL CORROSION AND ITS EFFECT

5.1 Replica specimen

In order to investigate the mechanical behavior of steel plate subjected to general corrosion, a steel plate

(250 mm × 100 mm) was sampled from the bottom plate of an aged heavy oil carrier, whose two surfaces have been contacting with heavy oil and seawater, respectively. Based on the result of self-similarity (Sumi, 2008), the replica specimen is reduced to 40% of the sample (100 mm × 40 mm), and the plate thickness before surface processing is 8 mm. Test specimen are made based on the following procedures:

a. The surface geometry of the sample plate is scanned by 0.5 mm interval by a laser displacement sensor, and the results are stored as the data for CAD system.
b. The specimen surfaces are processed by a numerically controlled milling machine. The experimental model is shown in Figure 19.

5.2 Finite element model

FE models are made by CAD software i.e. MSC. Patran. With the scanned data obtained by laser displacement sensor, two Non-Uniform Rational B-Splines (NURBS) surfaces are generated by CAD software, Rhinoceros. Rhinoceros generated surfaces are then imported to MSC. Patran. Noting that both surfaces should be topological congruent i.e. the points data should be arranged in same directional order during the surface generation. The surfaces are then discretized by iso-mesh. Now by sweeping action 3D solid element (8-node hexahedron) can be generated from these 2D elements. The 3D FE model is shown in Figure 20 (a). It consists of 17040 elements. The element size in the processed area is $0.5 × 1 × 4$ mm.

The accuracy of replica specimen and FE model is confirmed by comparing it with actual model.

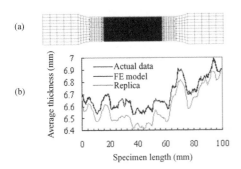

Figure 19. Replica specimen.

Figure 20. (a) FE-model. (b) Accuracy of replica specimen and FE-model.

Figure 20 (b) represents the average thickness distribution of actual plate, replica specimen and FE-model along the length, where a good agreement is observed. The average corrosion diminution can obtain it by

$$E[z_W(x_1)] = \frac{1}{W} \int_{-w/2}^{w/2} z(x_1, x_2)dx_2 \qquad (15)$$

where L and W are the length and width of the corroded plate, and $z(x_1, x_2)$ is the corrosion diminution of (x_1, x_2) surface. Equation (15) can be evaluated from discrete data.

5.3 Results and discussions

Figures 21a, b show experimental and numerical nominal stress-strain curves. In all cases the strength reduction is proportion to average thickness diminution, while the deformability is slightly less than that of flat plate.

Failure occurs by pure shear deformation, which is followed by a cross diagonal shear band. A less shear deformation (slip) is observed in FE-analyses. Hence, its deformability is less than that of experimental specimen. Note that, a plastic strain 0.92 is set as failure strain for FEM for steel A as the elements size are $0.5 × 0.5 × 4$ (mm). The fracture location is shown

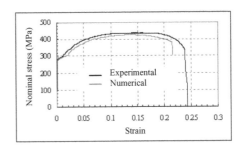

Figure 21a. Nominal stress-strain curves of general corroded plate (steel A).

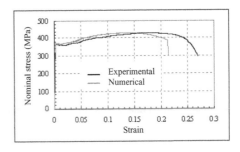

Figure 21b. Nominal stress-strain curves of general corroded plate (steel B).

Figure 22a.   Specimen of steels A and B after fracture.

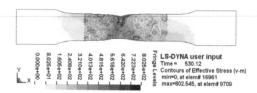

Figure 22b.   von Mises stress just before fracture (steel A).

in Figures 22a, b. The failure occurs in the zone of maximum thickness diminution.

In practical survey, the thickness diminution is measured by ultrasonic device. From the survey data we cannot obtain the actual corrosion surface. Moreover, it is very time consuming to analyze a replica model, due to vast number of degree of freedoms. It is worthy to find a suitable equivalent model in the near future.

6   CONCLUSION

Strength and deformability of corrosion degraded steel plate have been investigated both experimentally and numerically. At first, we have successfully measured the true stress-strain relationship by a vision sensor system. The pit size effect and general corrosion effect have been studied. Two steels i.e. yield ratio of 0.657, and 0.841, have been used in this study to investigate their integrity in aged structures. We can obtain the following conclusions.

1. The measurement of average axial strain and correction factor due to triaxial stress state improve the true stress-strain relationship after the bifurcation.
2. Mesh size effect after the bifurcation is significant. When maximum plastic strain is used as fracture criterion, it should be adjusted according to mesh size.
3. Deformability reduces considerably and strength reduces gradually with increasing pit size. This means, structures could lose significant amount of energy absorption capability under uniaxial tension.
4. A moderate strength and deformability reduction occur in case of general corrosion diminution.

5. Strength and deformability do not show significant dependence on yield ratio in the present study.

ACKNOWLEDGEMENT

The authors express their appreciation to Mr. Yamamura, Mr. Yamamuro, Mr. Shimoda, and Mr. Michiyama for their supports in the present work. This work has been supported by Grant-in-Aid for Scientific Research (No. A (2) 17206086) from the Ministry of Education, Culture, Sports, Science and Technology to Yokohama National University. The authors are grateful for the support.

REFERENCES

Bannister, A.C., Ocejo, J.R. & Gutierrez-Solana, F. 2000. Implications of the yield stress/tensile stress ratio to the SINTAP failure assessment diagrams for homogeneous materials. *Engineering Fracture Mechanics* 67 (6): 547–562.

Bridgman, P.W. 1964. Studies in Large Plastic Flow and Fracture. Cambridge, Massachusetts.

Cabezas, E.E. & Celentano, D.J. 2004. Experimental and numerical analysis of the tensile test using sheet specimens. *Finite Elements in Analysis and Design* 40 (5–6): 555–575.

Hallquist, J.O. 1998. *LS-DYNA Theoretical Manual.* Livermore Software Technology Corporation.

ISSC. 2006. Condition Assessment of Aged Ships. *Proceedings of the 16th International Ship and Offshore Structures Congress.* Committee V. 6: 255–307. Southampton, UK.

Nakai, T., Matsushita, H., Yamamoto, N. & Arai, H. 2004. Effect of pitting corrosion on local strength of hold frame of bulk carriers (1st report). *Marine Structures* 17: 403–432.

Okazawa, S., Fujikubo, M. & Hiroi, S. 2004. Static and dynamic necking analysis of steel plates in tension. *Proceedings of 3rd International Conference on Collision and Grounding of ships*: 276–284.

Ostsemin, A.A. 1992. Stress in the least cross section of round and plane specimen. *Strength of Materials* 24 (4): 298–301.

Paik, J.K., Lee, J.M. & Ko, M.J. 2003. Ultimate strength of plate elements with pit corrosion wastage. *J. Engineering for Maritime Environment* 217: 185–200.

Paik, J.K., Lee, J.M. & Ko, M.J. 2004. Ultimate shear strength of plate elements with pit corrosion wastage. *Thin-Wall Structures* 42: 1161–1176.

Schider, I., Brocks, W. & Cornec, A. 2004. Procedure for the determination of true stress-strain curves from tensile tests with rectangular cross-sections. *Journal of Engineering Materials and Technology* 126: 70–76.

Sumi, Y. 2008. Strength and Deformability of Corroded Steel Plates Estimated by Replicated Specimens. *Journal of Ship Production* 24–3:161–167.

Analysis and Design of Marine Structures – Guedes Soares & Das (eds)
© 2009 Taylor & Francis Group, London, ISBN 978-0-415-54934-9

Effect of pitting corrosion on the collapse strength of rectangular plates under axial compression

S. Saad-Eldeen & C. Guedes Soares

Centre for Marine Technology and Engineering (CENTEC), Technical University of Lisbon,
Instituto Superior Técnico, Lisboa, Portugal

ABSTRACT: The present study is focused on assessing the effects of localized pitting corrosion on the collapse strength of unstiffened plates. A series of non-linear (FE) analysis using ANSYS program are performed to explore the effect of pitting corrosion on the collapse strength of both rectangular and square steel plate subjected to axial compression, by varying the pitting distribution intensity as well as the plate geometry. A formula for predicting the collapse strength reduction due to pitting corrosion is derived by using regression analysis.

1 INTRODUCTION

Pitting is one of the most destructive and insidious forms of corrosion, as the attack is extremely localized resulting in holes in the metal and causing failures. It is often difficult to identify pits because of their small size and because the pits are often covered with corrosion products. The concept of ship structural design is to provide sufficient compressive strength to withstand the vertical bending moment of the ship hull.

Some studies on the influence of corrosion on plate elements have been carried out. Many of the studies deal with the effect of generalized corrosion as this is a type of corrosion more common. Initial studies considered simplified models of uniform corrosion increasing linearly with time, (Hart et al. 1986; Guedes Soares, 1988a; Shi, 1993) while more recent ones dealt with non-linear models (Guedes Soares, and Garbatov, 1999), and also with random fields (Teixeira, and Guedes Soares, 2008).

Pitting corrosion has also attracted the attention of researchers as it is a type of corrosion frequently found in ships. Daidola et al. (1997) proposed a mathematical model to estimate the residual thickness of pitted plates using the average and maximum values of pitting data, or the number of pits and the depth of the deepest pit, and presented a method to assess the effect of thickness reduction due to pitting on local yielding and plate buckling based on a probabilistic approach. They also developed a set of tools which can be used to assess the residual strength of pitted plates.

Paik et al. (2003, 2004) studied the ultimate strength characteristics of pitted plate elements under axial compressive loads and in-plane shear loads, and derived closed-form formulae for predicting the ultimate strength of pitted plates using the strength reduction (knock-down) factor approach. They dealt with the case where the shape of the corrosion pits is cylindrical and each pit is a through thickness for the tested rectangular plate which is really an idealization, however in the FEA of square plate the pitting depth and degree of pitting intensity were variable.

Nakai et al. (2006) studied the effect of pitting corrosion on the ultimate strength of steel plates subjected to in-plane compression and bending in bulk carriers hold frames. Corrosion pits were modeled with a circular cone shape and their effect on the local strength of hold frames was studied. A series of nonlinear FE analyses has been conducted with square plates having corrosion pits. They dealt with the case where the shape of the corrosion pits is cylindrical cone; the ratio of the diameter to the depth is in the range between 8:1 and 10:1 and the average pit is 25–30 mm. They studied a 14-years-old bulk carrier, and assumed that the corrosion pits exit on the both sides of the plate.

Duo et al. (2007) investigated the effects on the ultimate strength of localized corrosion, which concentrates at one or several possibly large areas, and calculated the strength reduction of unstiffened plates under uniaxial compression. Over 256 nonlinear FEA have been carried out to systematically investigate the effects of plate slenderness, locations, sizes and depths of pitting corrosion. They dealt with the case where the shape of corrosion is rectangular. New empirical formulae are derived to predict the ultimate strength and strength reduction of unstiffened plates with localized corrosion under uniaxial compression. The formula for single-side type pitting corrosion is slightly more accurate than that for both sides pitting corrosion.

Jiang and Guedes Soares (2008) studied the ultimate strength characteristics of pitted square plates elements subjected to in-plane compression. An eight-node shell element was used to model the plate with through pits, with a cylindrical shape. They concluded that when degree of pitting intensity increases much it starts dominating the ultimate strength, being more influential than plate slenderness. They also started looking at the effect of considering only partial depth pits and few calculations were made using a solid element.

The present study follows from that one and looks at the effect of pitting corrosion intensity on rectangular plate strength considering however that pitting damage is not through thickness. Both rectangular and square plate elements are modeled under compression loads as well as with pitting corrosion. A series of non-linear FE analysis are performed using the ANSYS program to explore the effect of pitting corrosion on the collapse strength of plate elements subjected to axial compression, by varying the pitting distribution intensity as well as the plate geometry.

2 PITTING CORROSION DAMAGE

The localized pitting corrosion can be concentrated in one or several possibly large areas in ballast tanks. Another type of localized pitting corrosion is of regularly pitted form and caused by microbiologically influenced corrosion (MIC), such as sulfate reducing bacteria (Duo et al. 2007).

2.1 Pitting degree intensity

To assess scale for breakdown due to pit corrosion, a parameter denoted by DOP (degree of pit intensity) is often used (Paik et al. 2003) as given in Equation 1, where DOP is defined in percentage as the ratio of the corroded surface area to the original plate surface area:

$$DOP = \frac{1}{ab} \sum A_{pi} * 100\% \qquad (1)$$

where n = number of pits; A_{pi} = surface area of the ith pit; a = plate length; and b = plate breadth.

Figure 1 shows different intensities of pitting damage.

2.2 Pitting corrosion shape

It is well known that shape of corrosion pits varies depending on the corrosive environment. Corrosion pits with a conical shape are typically observed in hold frames in way of cargo holds of bulk carriers which carry coal and iron ore. On the other hand for the bottom shell of a tanker, the shape of the corrosion pit is a part of a sphere (Nakai et al. 2007).

The relationship between depth and diameter of the corrosion pits is plotted in Figure 2. The ratio of the diameter to the depth is approximately constant (between 10 to 1 and 8 to 1 in the case of corrosion pits of the hold frames of the bulk carrier and between 4 to 1 and 6 to 1 in the case of the bottom shells of the tanker).

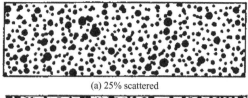

(a) 25% scattered

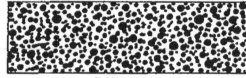

(b) 40% scattered

Figure 1. Pitting intensity diagram for a tanker (Daidola et al. 1997).

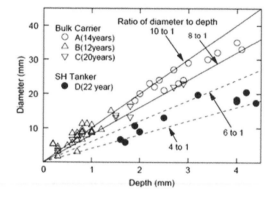

Figure 2. Relationship between diameter and depth of corrosion pit (Nakai et al. 2007).

It can be said that the corrosion pits on the bottom shells of the tanker have larger depth than that of hold frames of the bulk carrier when they have the same diameter. It is considered that corrosion pits grow in a different way depending on environment. For example, the difference of the amount of water existing in each environment would possibly affect the growth of corrosion pits (Nakai et al. 2004).

2.3 Smallest cross-sectional area based approach

The collapse of pitted plates is governed by the smallest cross-sectional area, and the ultimate strength reduction factor defined as the ratio of the ultimate strength of the plate with pit corrosion to that of the intact plate, also a formula derived based on ANSYS results and experimental data as given in Equation 2 (Paik et al. 2003):

$$R_{xu} = \frac{\sigma_{xu}}{\sigma_{xuo}} = \left(\frac{A_o - A_r}{A_o} \right)^{0.73} \qquad (2)$$

where R_{xu} = ultimate strength reduction factor for axial compressive loads; σ_{xu} = ultimate compressive strength for a plate with pit corrosion; σ_{xuo} = ultimate compressive strength for an intact plate; A_o = original cross-sectional area; and A_r = reduced cross-sectional area at the most heavily pitted location.

3 COMPRESSIVE STRENGTH OF PLATES WITH PITTING CORROSION

Plate elements are generally part of stiffened panels which are in fact the structural components whose strength one is finally interested to predict. However, global failure of the stiffened panels is usually avoided by design, so that the usual failure modes are inter-frame failure of the plate stiffener assembly or the plate elements between stiffeners. The different prediction methods assess the strength of the plate and stiffener assembly by properly accounting for the contribution of each one. Therefore, in addition to the interest in its own right, the prediction of plate strength is also a pre-requisite for assessing the strength of stiffened plates; this justifies the interest of having a complete under-standing and a prediction capability for its compressive strength (Guedes Soares, 1988b).

3.1 Model characteristics

The analyzed models are rectangular and square plates with pitting corrosion subjected to axial compression load as shown in Figure 3.

The distribution of pits is regular as shown in Figure 4. The geometry and material properties of the analyzed plates are shown in Tables 1 and 2.

3.2 Finite element simulation

The finite element code ANSYS has been used with both its geometrical and material nonlinear capabilities. To perform adequate simulation of the plate behavior, the solver has been used for large deflection analysis with the arc length method.

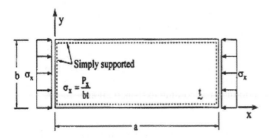

Figure 3. A simply supported rectangular plate under axial Compressive loads.

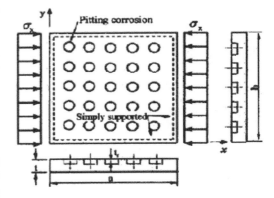

Figure 4. A simply supported rectangular plate with regular pitting corrosion under axial compressive load (Paik et al. 2003).

Table 1. Rectangular plate geometry and material properties.

Item	Geometry and material properties
Length (a)	3000 mm
Breadth (b)	1000 mm
b/t	40–70
Pit diameter (d)	30 mm
Pitting diameter to depth ratio	8:1
Distribution of pitting intensity	1.9%, 16.6%
Elastic modulus (E)	$210 * 10^3$ MPa
Poisson ratio (υ)	0.3
Material yield stress (σy)	240 MPa

Table 2. Square plate geometry and material properties.

Item	Geometry and material properties
Length (a)	500 mm
Breadth (b)	500 mm
b/t	25
Pit diameter (d)	30 mm
Pitting depth (both sides)	1.875
Distribution of pitting intensity	4.52%
Elastic modulus (E)	$210 * 10^3$ MPa
Poisson ratio (υ)	0.3
Material yield stress (σy)	240 MPa

The rectangular plate is modeled using solid element (Solid 95) from the ANSYS element library. Solid 95 is defined by 20 nodes having three degrees of freedom per node: translations in the nodal x, y, and z directions. The element may have any spatial orientation. The element has plasticity, creep, stress stiffening, large deflection, and large strain capabilities.

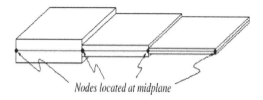

Figure 5. Position of nodes after reducing the thickness due to pitting (Duo et al. 2007).

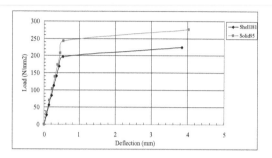

Figure 6. Average axial compressive load-deflection curve for plate with thickness 20 mm and DOP4.52%.

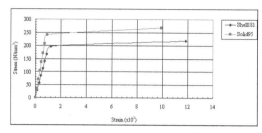

Figure 7. Average axial compressive stress-strain curve for plate with thickness 20 mm and DOP4.52%.

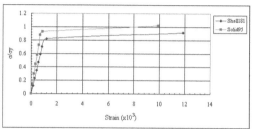

Figure 8. Effect of modeling methods on the average stress-strain relationship.

All the edges of the plate are assumed to be simply supported. The unloaded edges are allowed to deform in-plane but remain straight; this is achieved by coupling the deformation of nodes in that direction. This condition is to generate the actual situation of unstiffened plate between longitudinal and transverse stiffeners. The reaction edge is constrained to obtain an equal force caused due to loading edge.

Large displacement static analysis option is activated in geometric nonlinear analysis. Bilinear isotropic is used in material nonlinear analysis. The axial compressive load is applied as surface pressure on the loaded area.

In order to investigate the effect of changing the modelling method for pitted plates from shell elements as used by earlier authors (Paik et al. 2003, 2004, Nakai et al. 2006, Duo et al. 2007 & Jiang and Guedes Soares 2008) to a solid element, a square plate element with the characteristics given in Table 2 is modelled using two elements shell 181 and solid 95.

If a shell element is used to model pitting damage, the node in the pitted area will be located on the midplane as illustrated in Figure 5, and it is difficult to move this node to the right location (the bottom surface) for each pit, so the modelling is not accurate. For modelling the pitting damage as accurately as possible the pitting damage is presented on both sides of the plate. However, the use of solid elements allows the representation of pitting corrosion only in one side of the plate, representing its non symmetrical character with respect to the neutral axis.

Figures 6 and 7 show the load-deflection, stress-strain curves for the square plate with slenderness β of 0.85 and constant pitting diameter of 30 mm.

With the application of both solid and shell element for modelling the pitted square plate, there is a noticed deviation between them for the average stress-strain relationship as shown in Figure 8 with 11.23%.

As the use of shell elements does not allow the proper representation of pitting that is no symmetric with respect to the mid plate, solid modelling is adopted in this paper for simulating the pitted plate element.

4 ANALYSIS RESULTS

A series of nonlinear FE analysis has been conducted with rectangular plates having different corrosion pit intensity distribution and geometry.

Figures 9–12, show the load-deflection and stress-strain curves for the rectangular plate with different thicknesses, slenderness ratios, variable DOP, and constant pitting diameter 30 mm with the diameter to depth ratio of 8:1 according to (Nakai 2006).

The effect of the slenderness for the analyzed rectangular plate element using solid element modeling with different geometries on the plate element strength characteristics (load, deflection, stress, strain) are tabulated in Tables 3, 4, and shown in Figures 13.

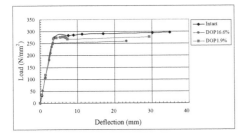

Figure 9. Average axial compressive load-deflection curve for plate with thickness 20 mm, β of 1.69 and pitting depth 3.75 mm.

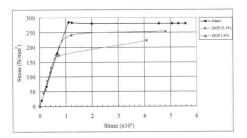

Figure 10. Average axial compressive stress-strain curve for plate with thickness 14.28 mm, β of 2.37, and pitting depth 3.75 mm.

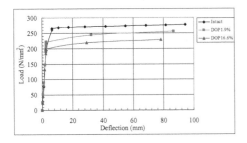

Figure 11. Average axial compressive load-deflection curve for plate with thickness 11.11 mm, β of 3.04, and pitting depth 3.75 mm.

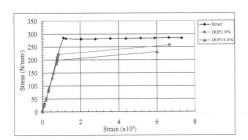

Figure 12. Average axial compressive stress-strain curve for plate with thickness 11.11 mm, β of 3.04, and pitting depth 3.75 mm.

Table 3. Percentage of changing the load and deflection with respect to slenderness.

| β | DOP1.9% | | DOP16.6% | |
	Load%	Deflection%	Load%	Deflection%
1.69	−7.26	−48.67	−13.08	−59.42
2.37	−6.10	−11.45	−12.89	−25.02
3.04	−7.94	−7.94	−17.15	−17.15

Table 4. Percentage of changing the stress and strain with respect to slenderness.

| β | DOP1.9% | | DOP16.6% | |
	Stress%	Strain%	Stress%	Strain%
1.69	−2.56	−20.55	−5.45	−27.82
2.37	−9.54	−13.60	−20.97	−26.50
3.04	−9.52	−8.55	−18.57	−17.69

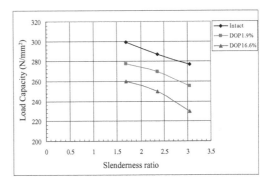

Figure 13. Relation between slenderness and load capacity for the rectangular plate.

From these analyses it is clear that as the degree of pitting increased, the plate compressive strength decreased and the percentage of this decrease increased with increasing the slenderness, for example if the slenderness increased form 1.69 to 3.04 with DOP of 1.9% then the plate strength decrease with 7.27% and 7.94% respectively. If the DOP increased to 16.6% the percentage of strength decease increase to 13.08% and 17.15% respectively.

From the finite element results in Figure 13, it looks as if the degree of pitting induces a reduction of strength that is approximately the same in all three slenderness plates. The limited number of results do bot allow definite conclusions to be drawn but at any rate a tendency can be extracted from the results by using a regression analysis to produce a formula that predicts the collapse strength reduction due to pitting corrosion for a rectangular plate is obtained as:

$$\Phi = \Phi_0 \left[\left(-0.7292 * \frac{\text{DOP}}{100} + 0.9759 \right) \right] \qquad (3)$$

with coefficient of determination $R^2 = 0.8144$, where Φ = pitted plate collapse strength; and Φ_0 = intact plate collapse strength.

5 CONCLUSIONS

The present study has addressed the effect of pitting corrosion on the collapse strength of unstiffened rectangular and square plate subjected to axial compressive loading, where the pits were not through thickness. A series of nonlinear FE analysis were carried out, using the ANSYS software and varying the degree of pitting intensity as well as the plate geometry.

It is evident that the collapse strength of a steel plate can be significantly decreased due to pit corrosion. For plate elements with slenderness of 1.69, as the degree of pitting intensity increased with 1.9% to 16.6% the plate load carrying capacity decreases by 7.3% and 13.1% respectively, also the ultimate compressive stress decreased by 2.6% and 5.4% of the intact plate strength.

A formula for predicting the collapse strength reduction of a rectangular plate due to pitting corrosion has been derived.

ACKNOWLEDGMENTS

The work has been performed in the scope of the project MARSTRUCT, Network of Excellence on Marine Structures (www.mar.ist.utl.pt/marstruct/), which has been financed by the EU through the GROWTH Programme under contract TNE3-CT-2003-506141.

REFERENCES

Daidola, J.C., Parente, J. & Orisamolu, I.R., et al. 1997. Residual strength assessment of pitted plate panels. *Report SSC-394*, Ship Structure Committee.

Duo, O.k., Yongchang P.u. & Atilla Incecik. 2007. Computation of ultimate strength of locally corroded unstiffened plates under compression. *Marine structures* 20: 100–114.

Guedes Soares, C. 1988a. Uncertainty modelling in plate buckling. *Struct Safety* 5:17–34.

Guedes Soares, C. 1988b. Design equation for the compressive strength of unstiffened plate elements with initial imperfections. *J. Construct. Steel Research* 9: 287–310.

Guedes Soares, C. & Garbatov, Y. 1999. Reliability of Maintained Corrosion Protected Plate Subjected to Non-Linear Corrosion and Compressive Loads. *Marine Structures* 12(6):425–446.

Hart, D.K., Rutherford, S.E. & Wickham, A.H.S. 1986. Structural reliability analysis of stiffened panels. *Trans Roy Inst Nav Architects (RINA)* 128:293–310.

Jiang, X. & Guedes Soares, C. 2008. Nonlinear FEM analysis of pitted mild steel square plates subjected to in-plane compression. *Proceedings TEAM 2008 Conference*, A Ergin (Ed), Istanbul Technical University, 6–9 October 2008.

Nakai, T., Matsushita, H., Yamamoto, N. & Arai, H. 2004. Effect of pitting corrosion on local strength of hold frames of bulk carriers (1st Report). *Marine Structures* Vol (17): 403–432.

Nakai, T., Matsushita, H. & Yamamoto, N. 2006. Effect of pitting corrosion on the ultimate strength of steel plates subjected to in-plane compression and bending. *J Mar Sci Technol*: 52–64.

Nakai, T., Matsushita, H. & Yamamoto, N. 2007. Visual assessment of corroded condition of plates with pitting corrosion taking into account residual strength—in the case of webs of hold frames of bulk carriers, *Proceedings OMAE Conference*, San Diego, California, USA Paper OMAE2007-29159.

Paik, J.K., Lee, J.M. & Ko, M.J. 2003. Ultimate compressive strength of plate element with pit corrosion wastage. *J. Engineering for the Maritime Environment* Vol. (217): 185–200.

Paik, J.K., Lee, J.M. & Ko, M.J. 2004. Ultimate shear strength of plate element with pit corrosion wastage. *Thin-Walled Structures* 42: 1161–1176.

Shi W.B. 1993. In-service assessment of ship structures: effects of general corrosion on ultimate strength. *Trans Roy Inst Nav Architects* (RINA) 135:77–91.

Teixeira, A.P. & Guedes Soares, C. 2008. Ultimate Strength of Plates with Random Fields of Corrosion. *Structure and Infrastructure Engineering*. 4, n. 5, 363–370.

Fatigue and fracture

Analysis and Design of Marine Structures – Guedes Soares & Das (eds)
© 2009 Taylor & Francis Group, London, ISBN 978-0-415-54934-9

Fracture mechanics procedures for assessing fatigue life of window and door corners in ship structures

M. Bäckström & S. Kivimaa
VTT Vehicle Engineering, Espoo, Finland

ABSTRACT: A comparative study of fracture mechanics procedures for assessing the fatigue life of ship structures is presented. The effect of some factors on life prediction equation is considered. Also emphasis is made to point out the usefulness of fracture mechanics for estimating the fatigue life of cracked and un-cracked ship structures. The procedures are verified against fatigue test data and demonstrated with an example calculation using data from onboard measurements and fatigue tests. The main interest is in the ship structures including large openings e.g. window and door corners. The example calculation concentrates in defining the remaining fatigue life and the critical crack size of the cracked door corners in a passenger ship. The investigated details contain longitudinal welds and gas cut edges.

1 INTRODUCTION

Side shells of passenger ship's superstructure have important part in shear force carrying capacity of hull girder. The importance of side shells is increasing in modern designs because the amount of longitudinal bulkheads is decreasing in superstructure. At passenger deck levels side shells are full of openings including windows and doors. Side shell of big cruise liner can include thousands openings. The trend is that the size of openings is increasing. Clearly the critical design detail exists at the window and door corner area due to stress concentration (Fig. 1). Detail design of the corner area has become increasingly important. It is necessary to check both the maximum stresses as well as the fatigue strength of this detail. (Bäckström & Kivimaa 2008)

The fatigue process can generally be broken into two distinct phases: initiation and propagation phase of the crack. The initiation phase is usually short in the welded structures because it can always be assumed that sharp-edged discontinuities exist in a welded structure. The fatigue life estimations for welded structures with discontinuities can be based on the crack propagation calculation by using fracture mechanics.

There are few procedures or standardized practices for fracture mechanics calculations for ships (DNV 1994, GL 2000, LR 1996, SSC-409 2000). Principles to fracture mechanics calculations of marine structures can be adapted from standards, recommendations and guidance for offshore and welded structures (BS 7608 1993, Hobbacher 2007, PD 6493 1991, RP-C203 2000). However, these procedures have been seldom applied to ship structures due to complexity of ship details, uncertainty of operational loads, uncertainty of welding residual stresses and the redundancy of ship structures. In addition, little attention has focused on development of fracture mechanics calculation procedures for longitudinal welds and gas-cut edges.

This study concentrates on development and utilization of fracture mechanics procedures in fatigue life assessment of ship structures. The main interest is in the ship structures including large openings e.g. window and door corners. The procedures are verified against window corner fatigue test data. The investigated details include longitudinal weld and gas-cut edge cracks. The example calculation concentrates in defining the remaining fatigue life and the critical crack size of the cracked door corner in a passenger ship. Emphasis is made to point out the usefulness of fracture mechanics for estimating the fatigue life of cracked details in ship hull.

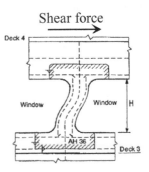

Figure 1. A sketch of the loading and deformation pattern at window and door area of the side shell.

2 REVIEW ON MAIN FACTORS IN FRACTURE MECHANICS OF WELDED STRUCTURES

2.1 Nominal, hot spot and local stress

To understand fatigue of welded structures it is essential to understand the difference between nominal, hot spot and local stress, see Figure 2. The nominal stress is the linear part of the stress distribution excluding the joint geometry. The macro-geometric effects e.g. large openings should be included to the nominal stress e.g. with stress concentration factors obtained from handbooks or with finite element calculations.

Hot spot is a term, which is used to refer to the critical point in a structure, where fatigue cracking can be expected to occur due to a discontinuity and/or a notch. Usually, the hot spot is located at the weld toe. The hot spot stress is the value of the structural stress at the hot spot (Niemi 1995). The hot spot stresses account only the overall geometry of the joint and exclude local stress concentration effects due to the weld geometry and discontinuities at the weld toe. The local stress is the total stress at the weld toe.

2.2 Stress-intensity factor solutions

2.2.1 Standard solution for flat plate

The deformation of the cracks is usually divided in three basic modes. These modes of deformation (not modes of cracking), are usually referred to simply by roman numerals I, II and III. However, if the loading of these modes is in phase, cracks will rapidly choose a direction of growth in which they are subjected to mode I. Thus, the majority of apparent combined mode cases are reduced to mode I by nature itself. There are few cases left which cannot be treated as pure mode I. In this report only the stress-intensity factors for the mode I are used in fatigue life prediction.

Stress-intensity factor (SIF) solution for an edge crack (Fig. 3a) has been given by Tada et al. (1973). The solution is given as a function of loading (tension and bending), crack depth and plate thickness. Crack shape is straight-fronted which means for a surface crack shown in Figure 3b that: crack depth/crack length = $a/2c \approx 0$.

Stress-intensity factor solution for a surface crack in a finite plate subjected to tension and bending loads (Fig. 3b) has been fitted to the finite-element results by Newman & Raju (1979). The solution is given as a function of loading (tension and bending), elliptic angle, crack depth, crack length, plate thickness and plate width.

2.2.2 M_k-factors for welded structures

Typically, the fracture mechanics calculations of welded structures are conducted with standard stress intensity factor solutions (Fig. 3) and hot spot stresses (Fig. 2). This considers the effects of overall geometry of the

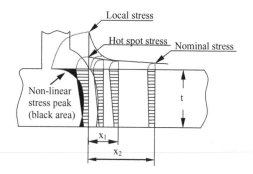

Figure 2.   Stress distributions across the plate thickness and along the surface in the vicinity of a weld toe.

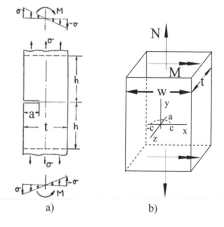

Figure 3.   a) Edge and b) surface crack under tension and bending loading.

joint and the crack. The stress-intensity factor K_I equation for welded structures in combined tension and bending loads is

$$K_1 = (f_{G.1} \cdot M_{K.1} \cdot F_1 \cdot \sigma_1 + F_{G.b} \cdot M_{k.b} \cdot F_b \cdot \sigma_b)$$
$$\times \sqrt{\pi \cdot a} \qquad (1)$$

where F_G = additional stress intensity correction factor for non-linear stress; M_k = stress intensity correction factor for weld geometry; F = stress intensity correction factor for flat plate; σ = normal stress; subscript t = tension; subscript b = bending; and a = surface crack depth or edge crack length.

The non-linear stress peak effects due to weld geometry are included to the standard stress intensity factor solution with a M_k-factor which is

$$M_K = \frac{K_1 \text{ for crack in welded detail}}{K_1 \text{ for same crack in flat plate}} \qquad (2)$$

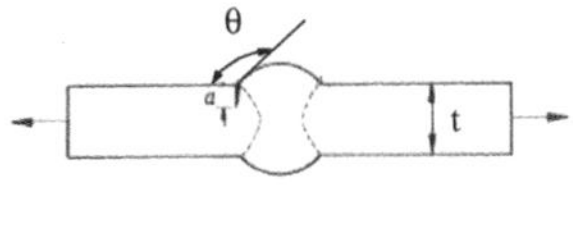

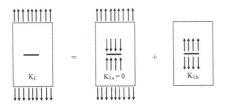

$$M_k = 2.5 \cdot \left(\frac{a}{t}\right)^{-q} \text{ where}$$

$$q = \frac{\log(11.584 - 0.0588 \cdot \theta)}{\log(200)} \text{ and}$$

θ is obtuse toe angle measured in degrees ($135 \leq \theta \leq 180°$).

Figure 4. M_k-factors for transverse butt welds with overfill (Gurney 1979).

Figure 5. Stress intensity factor calculation with superposition principle.

M_k-factor solution for a transverse butt welds with overfill has been given by Gurney (1979) which is shown in Figure 4 as an example. The solution is given for tension loading as a function of crack depth, plate thickness and obtuse toe angle. Crack shape is straight-fronted ($a/2c \approx 0$). More M_k-solutions can be found from references (Almar-Naess 1985, BS 7608 1993, Gurney 1979, Hobbacher 2007, PD 6493 1991). It should be noted that, in some references F_G is used instead of M_k.

2.2.3 Superposition principle and integration of Green's function

Superposition principle has been described as: "In the vicinity of the crack tip the total stress field due to two or more different mode I loading systems can be obtained by an algebraic summation of the respective stress-intensity factors" (Broek 1988). One very important result that can be derived with the superposition principle is presented in Figure 5. The figure shows that stress intensity factor calculation (K_I) can always be returned to a crack case which has loads/stresses on crack surface ($K_{I,b}$). The loads/stresses are determined from un-cracked structure in the location of crack surface. The stress intensity for any loading case is equal to stress intensity obtained by applying to the crack surfaces loads/stresses of un-cracked structure in the crack location when there was no crack.

One simple solution for obtaining correction function of non-uniform loading (F_G) is integration of

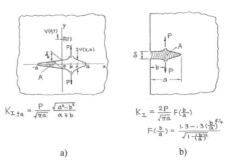

a)   b)

Figure 6. K_I solution for a pair concentrated splitting forces on crack surfaces a) infinite planes with a finite crack and one splitting force P and b) semi-infinite planes with an edge crack and one splitting force P (Tada et al. 1973).

Green's function. Integration of Green's function includes following steps:

– Determine actual stress distribution $\sigma(x)$ for crack surface loads/stresses of un-cracked structure in the crack location when there is no crack.
– Select standard stress intensity factor solution for the calculations that correspond best the actual stress distribution. Typically the reference stress distribution $\sigma_{\text{ref}}(x)$ is uniform bending or/and tension stress on the crack surface obtained from standard stress intensity factor solution with superposition principle.
– Select K_I solution for Green's function $K_{I,G}(a, b, P)$. Most conveniently are pair concentrated splitting forces on crack surfaces (Fig. 6).
– Calculate $F_G(a)$ with different crack lengths a:

$$F_G(a) = \frac{\int_0^a K_{I,G}(a, x, \sigma(x)dx)}{\int_0^a K_{I,G}(a, x, \sigma_{\text{ref}}(x)dx)} \quad (3)$$

where $K_{I,G}(a, x, \sigma(x)dx) = K_{I,G}$ for actual stress distribution; and $K_{I,G}(a, x, \sigma_{\text{ref}}(x)dx) = K_{I,G}$ for reference stress distribution.

2.2.4 Paris relation for crack growth

A simple power law is probably the most widely accepted method for describing the crack growth in constant or variable amplitude loading of welded structures. A power law relationship between the rate of the crack growth per cycle (da/dN) and the range of stress-intensity factor (ΔK_I) has been developed by Paris (Paris & Erdogan 1997):

$$\frac{da}{dN} = C(\Delta K_1)^m \quad \Delta K > \Delta K_{\text{th}}$$
$$\frac{da}{dN} = 0 \quad \Delta K \leq \Delta K_{\text{th}} \quad (4)$$

where C = crack growth coefficient; m = exponent in equation for crack growth rate; and ΔKth = threshold value of stress intensity factor range.

Most welded structures are in the as-welded condition and contain welding-induced tensile residual stresses of the order of the yield strength of the material. This means that the mean stresses have a minimal effect on the fatigue strength of welded joints and the fatigue failure can occur in welded joints under nominally compressive stresses. Stress range seems to be the main controlling parameter of the fatigue strength of the welded structures. (Almar-Naess 1985, Gurney 1979, Maddox 1991, Niemi 1995)

Stress-intensity factor range for as-welded structures is

$$\Delta K_I = K_{I.\max} - K_{I.\min} \qquad (5)$$

$K_{I \cdot \max}$ and $K_{I \cdot \min}$ is a sum of effects of the residual stress and external loading and it is assumed that $K_{I \cdot \min} > 0$. Because of this assumption the SIF range is $\Delta K_I = K_{I \cdot \max} - K_{I \cdot \min}$.

Fatigue life can be predicted using the Paris relation and the known stress intensity factor solution. Typically, the integration of the Paris relation is conducted numerically with a computer:

$$N = \int_{a_i}^{a_f} \frac{1}{C \cdot \Delta K_I^m} \cdot da \qquad (6)$$

where a_i = initial crack size; and a_f = final crack size.

3 FATIGUE ASSESSMENT OF WINDOW OPENING TEST SPECIMENS

3.1 Fatigue test data

Fatigue test data for fracture mechanics calculation was obtained from fatigue tests reported by Bäckström et al. (2000). Fatigue tests were conducted inFatHTS-project which was carried out by 14 partners from eight European countries and was partially funded by the European Union under the IMT Programme Brite Euram IV (project BE95-1937). Fatigue test specimens were T-shaped test structures simulating window openings in the ship (height = 2 m and width = 5.5 m). The vertical beam models the stiff deck and the horizontal beams model the connection strips between windows. The beams have a U cross section with additional flat bars at edges or stiffeners. The critical (hot spot) details were the corner areas and thus the structure had total four 'hot spots', which were loaded with combined bending and shear. Test specimens were fabricated using a thermo mechanically processed DH

36 steel. Fatigue tests included three constant amplitude and one spectrum fatigue test.

3.2 Calculation procedures

Five longitudinal (Fig. 7a) and three gas-cut edge (Fig. 7b) cracks were selected for fracture mechanics calculations. A total of 126 different crack cases were analyzed. Variables of calculation cases are summarized in Table 1.

Calculations were conducted for all the cracks with standard SIF solution of edge and surface crack

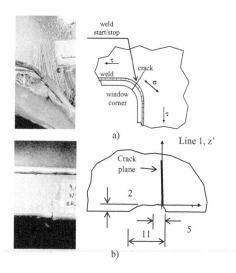

Figure 7. Analyzed crack cases: a) longitudinal weld crack (LW) and b) gas-cut edge crack (GCE).

Table 1. Input data of the calculation cases.

	LW1–LW3, LW5 GCE2–GCE3 LW4s, GCE1s*	LW1–LW3, LW5 LW4s*
SIF	Surface crack : (Fig. 3b) Edge crack : (Fig. 3a)	Surface crack with butt weld M_k : (Fig. 3b, 4)
a_i	0.10, 0.15 and 0.25 mm Surface crack : 1.5 mm	0.10, 0.15 and 0.25 mm 1.5 mm
c_i	Edge crack : ∞ Surface crack : T/2	T/2
a_f, c_f	Edge crack : 6 mm C Upper bound and mean	Upper bound and mean
$M_{k.a}$		
θ	–	LW1–LW3 : 160° LW4–LW5 : 175°
$M_{k.c}$	–	1.0
F_G	GCE2 : (Fig. 9)	–

*Spectrum fatigue test results.

242

(second column in Table 1). The longitudinal cracks were analysed using surface crack solution with $M_{k.a}$-function for transverse butt weld with overfill (third column in Table 1). Obtuse toe angle θ is a rough estimation from photos. The angle is perpendicular to crack growth direction. $M_{k.c}$ was 1.0 for longitudinal welds. Because it was assumed, that weld geometry has only small effects for crack growth in crack length direction of longitudinal welds (see surface point–c and c in Fig. 3b).

Initial crack size (a_i) and aspect ratio (a_i/c_i) for fatigue assessment of welded joints without detected cracks are given in following references BS 7608 (1993), Hobbacher (2007) and SSC-409 (2000). Methods and procedures of idealization and characterization of detected cracks or crack-like defects for fracture mechanics calculations are described in following references Hobbacher (2007), PD 6493 (1991) and SSC-409 (2000). The final crack size (a_f) can be obtained with so called Failure Assessment Diagram (FAD) when the integrity of a structure with discontinuities is limited by two limit states, i.e. plastic collapse and fracture (PD 6493 1991, SSC-409 2000).

Calculations were conducted with three different initial crack depth (a_i). Initial aspect ratio (a_i/c_i) varied from 0 to 0.17. The predicted fatigue life was calculated from (a_i, c_i) to (a_f and c_f). The predicted fatigue life is the shorter of the two predictions.

Material parameters C, m and ΔKth for welded structures can be found from following references Almar-Naess (1985), BS 7608 (1993), Gurney (1979), Hobbacher (2007), PD 6493 (1991) and SSC-409 (2000). However, it seems that most of the values are based on Gurney's (1979) mean values for welded joints. If the mean values are used in the design, it is recommended that the mean value of C should be multiplied by 2.0 to allow for scatter (BS 7608 1993). The upper bound C value means that, if different C values are obtained from different sources no value of C exceeds $C_{Upper\ bound}$. Upper bound (and characteristic) values are so-called design values. It should be noted that, the crack growth rate is approximately the same for all weld able structural steels, regardless of the yield limit of the material. Thus, the same fatigue strength curves can be applied to different weld able structural steels.

Crack growth rate was calculated with Paris relation, see Eq. 6. The crack growth increments were calculated separately for every load cycle.

In this study, exponent in Paris relation for crack growth rate was m = 3. Upper bound and mean values of crack growth coefficients were $C_{Upper\ bound} = 5.21 \cdot 10^{-13}$ (Hobbacher 2007) and $C_{Mean} = 1.83 \cdot 10^{-13}$ (Gurney 1979). Units for da/dN and ΔK are mm/cycle and MPa$\sqrt{}$/mm, respectively. ΔKth was zero. Material parameters are for high R-values. This means that, effects of residual stresses are considered with crack growth material parameters. The crack growth rates for surface cracks were calculated independently for crack depth and length according to Newman and Raju (1981).

The additional stress concentration which is not included in F and M_k-correction factors has been considered by integration of the Green's $K_{I.G}$ solution. F_G-correction is obtained according to procedures described in section 2.2.3.

3.3 Longitudinal weld cracks

Variables of longitudinal weld crack calculation cases are summarised in the second and third column of Table 1. Figure 8 shows the correlation between predicted fatigue life using fracture mechanics and cycles to failure obtained from fatigue test. The crack name is given on abscissa. The ratio between predicted fatigue life (N_{PRED}) and the cycles to failure obtained from fatigue test (N_{TEST}) are plotted on ordinate. N_{TEST} is cycles to failure when the fatigue test was stopped. The solid bars are calculated with C_{Mean}-values and texture bars with $C_{Upper\ bound}$-values.

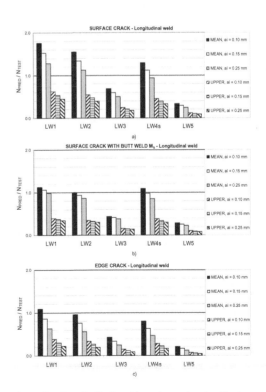

Figure 8. Longitudinal weld cracks analyzed with stress intensity factor solution for a) surface crack, b) surface crack with butt weld M_k factor and c) edge crack.

Calculations were conducted for longitudinal cracks with standard SIF solution of surface (Fig. 8a) and edge crack (Fig. 8c). The longitudinal cracks were also analysed using surface crack solution with $M_{k.a}$-function for transverse butt weld with overfill (Fig. 8b).

3.4 *Gas-cut edge cracks*

Variables of gas-cut edge crack calculation cases are summarised in the second column of Table 1. Figure 10 shows the correlation between predicted fatigue life using fracture mechanics and cycles to failure obtained from fatigue test. The crack name is given on abscissa. The ratio between predicted fatigue life (N_{PRED}) and the cycles to failure obtained from fatigue test (N_{TEST}) are plotted on ordinate. N_{TEST} is cycles to failure when the fatigue test was stopped. The solid bars are calculated with C_{Mean}-values and texture bars with $C_{Upper\ bound}$-values.

Calculations were conducted for gas-cut edgecracks with standard SIF solution of surface and edge crack. Surface crack was used instead of corner crack, because it was found out that the SIF of the surface crack is almost the same with the SIF of the corner crack.

It should be noted that one of the gas-cut edge cracks (GCE2) had an additional stress concentration.

The geometry and dimensions for the gas-cut edge crack with an additional stress concentration are given in Figure 7b. The additional stress concentration has been also considered by integration of the Green's K_I solution for semi-infinite planes with an edge crack and one splitting force P according to procedures described in section 2.2.3. The assumed actual stress distribution σ_y' in the un-cracked structure is shown in Figure 9. The reference stress distribution is σ_b. The F_G-function is calculated with Eq. 3.

4 CASE STUDY OF CRACKED DOOR OPENINGS IN A PASSENGER SHIP

Fracture mechanics calculations were applied in order to estimate growth of cracks at a passenger ship's door corner for two cases (Bäckström & Kivimaa 2008). First fatigue assessment was made for the case where no visible cracks have been detected in structures. In general the results obtained in the calculations were in line with the observations which have been made on board the target ship after 10 years of operation. The second assessment was conducted for the case where cracks have been detected during inspections. Figure 11a are calculated with $C_{Upper\ bound}$—material

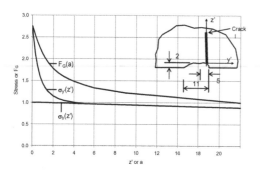

Figure 9. Non-uniform correction function of gas-cut edge crack with an additional stress concentration in the gas-cut edge (F_G) for non-uniform stress distribution (σ_y') with reference function of bending (σ_b).

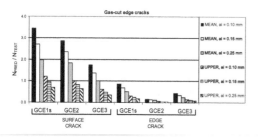

Figure 10. Gas-cut edge cracks analyzed with stress intensity factor solution for surface and edge crack.

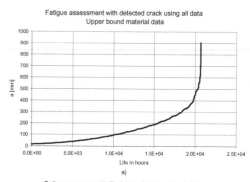

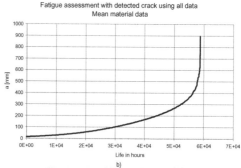

Figure 11. Fatigue assessment of welded joint in door corner with detected cracks using fracture mechanics and a) design material data and b) mean material data.

data and Figure 11b with C_{Mean}-values. The main purpose of this study was to get an overview of critical crack size and growth, and their influence on the remaining fatigue life of the door corner and the associated bulkhead.

Figure 11a gives the relation between number of hours and predicted crack length (a). The fatigue assessment with the detected cracks using design material data specifies the time period that ship can operate before maintenance is required for the cracked door and bulkhead. It was estimated that the critical crack size of 1000 mm was relevant to adjust the remaining fatigue life of the bulkhead before the crack grows to the deck. Thus the cracked door and bulkhead should be repaired in 2.1E4/24/365 = 2.4 years after the first visible cracks have been detected.

5 DISCUSSION

5.1 *Stress intensity correction factors*

According to Eq. 1, calculation of stress intensity factor for welded structures includes three different correction factors which are:

- standard correction factor F which considers effects of loading (tension and bending), elliptic angle, crack depth, crack length, plate thickness and plate width; it should be noted that overall geometry of the joint and the crack can be considered with standard correction factor and hot spot stresses
- weld M_k-factor for welded structures considers local stress concentration effects due to the weld geometry
- correction factor for non-uniform loading F_G considers additional local stress concentrations, which have not been included in standard correction factor F and M_k-factor solutions

The usage of correction factor for non-uniform loading F_G is demonstrated in Figure 9. As expected, F_G is equal to σ'_y when $z' = a \approx 0$. This indicates that maximum value of the non-uniform stress distribution can be used instead of F_G-correction function for short cracks. If the crack is longer, non-uniform stress distribution gives too low values compared to F_G-correction. It can also be noted that non-uniform stress distribution coincide with the reference stress distribution at $z' = 5$ mm. However, the F_G-correction is still 1.4 and approaches unity at $z' = 20$ mm. It should be noted that the usage of maximum σ'_y value instead of F_G-correction function can lead to unsafe fatigue predictions.

5.2 *Longitudinal weld cracks*

Most challenging task for longitudinal weld cracks (Fig. 7a) is to find appropriate M_k-factors for crack depth (deepest point) and length direction (surface point). It should be noted that these longitudinal cracks are growing perpendicular to maximum principal stress range and perpendicular to weld line. In addition, the main loading direction of these welds is in the direction of the welds, which indicates of a low local stress concentration due to the weld geometry.

Figure 8 summarises results for longitudinal weld cracks that has been obtained with stress intensity factors of surface crack, surface crack with butt weld M_k and $M_{k.c} = 1$, and edge crack. M_k-factor considers local stress concentration effects due to the weld geometry for crack depth direction and $M_{k.c}$ for crack length direction. $M_{k.c}$-factor in crack length direction is one, which means that the effects of the weld are ignored in the surface direction. It can be seen that surface crack gives un-conservative results when the results for surface crack with butt weld M_k are good. Surprisingly, edge crack results seem to be truthful even if the surface crack is a typical crack type in the longitudinal welds.

It seems that for the longitudinal welds the initial crack size could be taken as $a_i = 0.15$ mm (depth) and $c_i = 1.5$ mm (half-length). If there is no information of the obtuse toe angle, the obtuse toe angle could be $\theta = 160°$ for weld start/stop positions and $\theta = 175°$ for small ripples on the surface of the weld.

5.3 *Gas-cut edge cracks*

It seems that little attention has focused on fracture mechanics calculations of gas-cut edge cracks. Actually, there are no general rules for selection of material parameters, initial crack size and stress intensity factor solution (surface or edge crack) for gas-cut edge cracks in open literature (Almar-Naess 1985, Gurney 1979, Maddox 1991), standards (BS 7608 1993), or recommendations (Hobbacher 2007, PD 6493 1991, SSC-409 2000).

Figure 10 presents results for gas-cut edge cracks that has been analysed with stress intensity factors of edge and surface crack with three different initial crack sizes. Correction factor of non-uniform loading for GCE2-crack due to the additional stress concentration have also been included in the calculations. The present results suggest that the gas-cut edges could be assessed using material parameters for weld able structural steels, an initial crack size of $a_i = 0.15$ mm and the standard stress intensity factor solution of the edge crack.

6 CONCLUSIONS

This study presents fracture mechanics calculations results for longitudinal weld and gas-cut edge cracks. The material parameters are taken from a literature survey. The fatigue assessment was conducted with three different initial crack sizes 0.10, 0.15 and 0.25 mm covering the range of recommended design values. Reasonable results were obtained with all applied initial crack sizes. The calculation procedures are verified with window corner fatigue test data and demonstrated with onboard measurement data for a cracked door corner. The calculations concentrated in ship structures including large openings like window and door corners.

Good results were obtained, when longitudinal weld cracks were analysed with stress intensity factor of surface crack with butt weld M_k and $M_{k.c} = 1$. M_k-factor considers local stress concentration effects due to the weld geometry for crack depth direction (deepest point) and $M_{k.c}$ for crack length direction (surface point). It seems that initial crack size could be taken as $a_i = 0.15$ mm (depth) and $c_i = 1.5$ mm (length). If there is no information of the obtuse toe angle, the obtuse toe angle could be $\theta = 160°$ for weld start/stop positions and $\theta = 175°$ for small ripples on the surface of the weld.

Gas-cut edge cracks have been analysed with stress intensity factors of edge and surface crack with three different initial crack sizes. According to the results in this study, it seems that the gas-cut edges could be assessed with an initial crack size of $a_i = 0.15$ mm and the standard stress intensity factor solution of the edge crack.

Overall the application of fracture mechanics in estimating the fatigue life of welded ship structures seems promising. Especially, it is important to find new methodologies in order to analyse remaining fatigue life of structural components such as bulkheads after detection of first cracks. Fracture mechanics is a useful tool which can improve the planning of maintenance and dry-docking schedules.

ACKNOWLEDGEMENTS

The work has been performed in the scope of the project MARSTRUCT, Network of Excellence on Marine Structures, (www.mar.ist.utl.pt/marstruct/), which has been financed by the EU through the GROWTH Programme under contract TNE3-CT-2003-506141.

REFERENCES

Almar-Naess, A. 1985. Fatigue handbook, Offshore steel structures. Norge: TAPIR.
BS 7608 1993. Code of Practice for Fatigue Design and Assessment of Steel Structures. London: British Standards Institution (BSI).
Broek, D. 1988. The practical use of the fracture mechanics. Netherlands: Kluwer Academic Publishers.
Bäckström, M., Mikkola, T.P.J., Marquis, G. & Ortmans, O. 2000. Testing and analysis of window details for passenger ships. In Bache, M.R., Blackmore, P.A., Draper, J., Edwards, J.H., Roberts, P., Yates, J.R. (ed.), Fatigue 2000, Fatigue & durability assessment of materials, components and structures. U.K.: EMAS.
Bäckström, M. & Kivimaa, S. 2008. Estimation of crack propagation in a passenger ship's door corner. Glasgow: ASRANet Ltd.
DNV 1994. Fatigue assessment of ship structures. Norge: DNV Classification AS.
GL 2000. Guidlines for Fatigue Strength Analyses of Ship Structures, Chapter V-1-2, Section 1. Germanischer Lloyd.
Gurney, T.R. 1979. Fatigue of welded structures. Gambridge: Gambridge University Press.
Hobbacher, A. 2007. Recommendations for fatigue design of welded joints and components. International Institute of Welding (IIW).
LR 1996. Fatigue Design Assessment procedure—Structural Detail Design Guide—January 1996, FDASDDG: Ch1,3.1. Lloyd's Register.
Maddox, S.J. 1991. Fatigue strength of welded structures. Cambridge: Abington Publishing.
Newman, J.C. & Raju, I.S. 1979. Analyses of Surface Cracks in Finite Plates Under Tension or Bending Loads. National Aeronautics and Space Administration (NASA).
Newman, J.C., JR. & Raju, I.S. 1981. An empirical stress-intensity factor equation for surface crack. Engineering Fracture Mechanics 15(1–2): 185–192.
Niemi, E. 1995. Stress determination for fatigue analysis of welded components. IIW.
Paris, P. & Erdogan, F.A. 1997. Critical Analysis of Crack Propagation Laws. In: R.J. Sanford (ed.), Selected Papers on Foundations of Linear Elastic Fracture Mechanics, SEM Classic Papers, Volume CP 1. Bethel, Connecticut: Society for Experimental Mechanics (SEM).
PD 6493 1991. Guidance on methods for assessing the acceptability of flaws in fusion welded structures. London: BSI.
RP-C203 2000. Fatigue Strength Analysis of Offshore Steel Structures. Norway: DNV.
SSC-409 2000. Guide to damage tolerance analysis of marine structures. Ship Structure Committee (SSC).
Tada, H., Paris, P.C. & Irwin, G.R. 1973. The Stress Analysis of Cracks Handbook. St. Louis, Missouri: Del Research Corporation Hellertown.

Analysis and Design of Marine Structures – Guedes Soares & Das (eds)
© 2009 Taylor & Francis Group, London, ISBN 978-0-415-54934-9

Experimental and numerical fatigue analysis of partial-load and full-load carrying fillet welds at doubler plates and lap joints

O. Feltz & W. Fricke

Hamburg University of Technology (TUHH), Hamburg, Germany

ABSTRACT: Approaches for the fatigue strength assessment consider the loads transferred by fillet welds in different ways resulting in varying life predictions. This is particularly true for the structural stress approaches where modified stress distributions in thickness direction of the parent plate have recently been proposed. In order to clarify the real fatigue behaviour, some fatigue tests have been performed with doubler plates having partial-load carrying fillet welds and lap joints having full-load carrying fillet welds. Two throat thicknesses have been realized (3 and 7 mm) joining 12 mm thick plates. In the fatigue analyses, different approaches have been applied, including different structural stress approaches and the effective notch stress approach. From a comparison of the approaches with the fatigue lives obtained, conclusions are drawn with respect to a realistic consideration of the amount of load carried by fillet welds.

1 INTRODUCTION

Fatigue is one of the major design criteria for ship structures, due to high cyclic stresses mainly caused by the seaway and to relatively high notch effects in the welded structure. Different approaches exist for the fatigue assessment of welded structures, which have recently been summarized by Radaj et al. (2006).

Apart from the nominal stress approach, the so-called structural hot-spot stress approach is widely used in ship structural design. It has the advantage that the stress increase due to the structural configuration, which varies to a large extent, can be rationally considered (Niemi et al., 2006). It has been shown in round-robin studies that guidelines on stress analysis can limit the scatter of results (Fricke, 2002; Fricke et al., 2002 and 2007).

One disadvantage of the approach is that the structural hot-spot stress cannot catch the intensified local stress at the toe of load carrying fillet welds in comparison to non-load carrying fillet welds. Therefore, two different fatigue classes have been introduced in the Fatigue Design Recommendations of the International Institute of Welding (Hobbacher, 2007), i.e. FAT 90 for load carrying and FAT 100 for non-load carrying fillet welds.

Alternative approaches have been developed to consider this effect in the structural stress, which are summarized by Radaj et al. (2009). Xiao and Yamada (2004) utilize the structural stress at a location 1 mm below the weld toe in connection with the fatigue class FAT 100. Poutiainen (2006) proposes a modified through-thickness stress distribution across the plate in front of the weld, which is determined by the

stress transferred by the weld, together with a modified fatigue class.

Another approach which can consider the effect of load carrying welds is the effective notch stress approach (Hobbacher, 2007), using the elastic stress in the weld toe rounded by $r_{ref} = 1$ mm as fatigue parameter. The corresponding fatigue class is FAT 225.

Poutiainen and Marquis (2006) have shown that the four approaches mentioned yield quite different fatigue lives for load-carrying welds of different size in a doubler plate on a T-bar. As no fatigue tests have been performed for this example, some additional fatigue tests have been performed within the Network of Excellence on Marine Structures (MARSTRUCT) which can serve as a basis for the assessment of the different approaches.

In this paper, the fatigue tests and their results are described in detail. In addition, some stress analyses have been performed which allow to draw first conclusions regarding the above mentioned questions. Further analyses are in progress in MARSTRUCT.

2 FATIGUE TESTS

2.1 Test specimens

The objective of the tests was to investigate the fatigue behaviour of fillet-welded small scale specimens where the load carrying grade and the size of the weld are varied.

As the fillet welds of lap joints are full-load carrying, while those at doubler plates are partial-load carrying, it was decided to choose these two joint

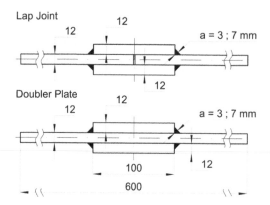

Lap Joint

12
12
a = 3 ; 7 mm

12

Doubler Plate
12
12
a = 3 ; 7 mm

100
12

600

Figure 1. Geometry of the specimens for the fatigue tests.

Figure 2. Weld toe failure in a specimen with doubler plates.

Figure 3. Weld root failure (left) in a specimen with lap joints.

types. Fig. 1 shows the geometry of the specimens with lap or doubler plates on both sides in order to avoid local bending. A plate thickness of 12 mm and two different weld throat thicknesses of $a = 3$ and 7 mm were chosen resulting in four different types of specimens, denoted L.3, L.7, D.3 and D.7.

The test specimens were fabricated from structural steel S355 having a nominal yield limit $R_{eH} = 355$ N/mm². 100, 300 and 600 mm wide plate strips were welded together by the MAG process. While the fillet welds with $a = 3$ mm were performed with one run, 3 runs were necessary for the throat thickness of 7 mm. The actual thickness varied between 2.8 and 3.7 mm and between 6.6 and 7.3 mm.

After welding, the plate strips were saw-cut to 50 mm wide specimens, having a length of 600 mm. In total, 10 specimens were fabricated for each geometry, i.e. for each of the four test series. Finally, the angular distortion of the specimens was measured at both ends of the lap and doubler plates. The angular distortion was relatively small with values between 0.1 and 0.7 degrees.

2.2 Performance of the fatigue tests

The fatigue tests were performed with constant load amplitudes in a resonance pulsator at a frequency of approx. 30 Hz and room temperature. A stress ratio $R \approx 0$ was chosen. The length between the grips was 480 mm.

Depending on the type of specimen, fatigue crack initiation was observed at one of the weld toes, running through the main plate, or at one of the weld roots, running through the weld throat. Fracture of the specimen was taken as failure criterion as usual for small-scale specimens. Figs. 2 and 3 show typical failures.

2.3 Fatigue test results

The fatigue tests were performed at different load levels, corresponding to nominal stress ranges in the

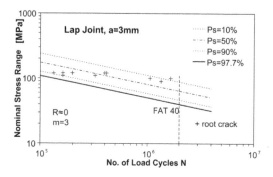

Figure 4. S-N data for the lap joints with $a = 3$ mm (L.3).

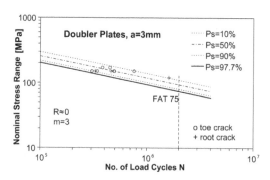

Figure 5. S-N data for the doubler plates with $a = 3$ mm (D.3).

plate in front of the stiffener between $\Delta\sigma_n = 90$ and 210 N/mm².

The fatigue lives of each test series was statistically evaluated in order to obtain characteristic data of the S-N curve such as the characteristic fatigue strength at two million cycles for a survival probability $P_s = 97.7\%$ (FAT-class), corresponding to two standard deviations below the mean. A forced slope

248

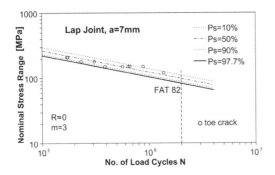

Figure 6.   S-N data for the lap joints with a = 7 mm (L.7).

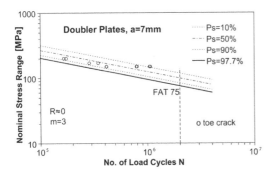

Figure 7.   S-N data for the doubler plates with a = 7 mm (D.7).

exponent $m = 3$ was assumed for the S-N curves being typical for welded joints.

The fatigue lives of the lap joints with throat thickness $a = 3$ mm are shown in Fig. 4. All specimens showed root cracks due to the small weld size and the transfer of the whole plate force. The evaluated fatigue class is FAT 40. It should be noted that the S-N curve seems to have a shallower slope causing a wide scatter band when $m = 3$ is assumed.

The fatigue lives of the doubler plates with throat thickness $a = 3$ mm are plotted in Fig. 5. Seven specimens showed weld toe cracks, while three caused weld root cracks. The evaluated fatigue class is FAT 75 when all specimens are taken into account.

Fig. 6 shows the fatigue lives of the lap joints with throat thickness $a = 7$ mm. All specimens showed weld toe cracks with a characteristic fatigue strength of FAT 82.

Also the doubler plates with throat thickness $a = 7$ mm failed at the weld toes. The evaluation of the SN data, displayed in Fig. 7, yields a fatigue class FAT 75, which is surprisingly lower than that of the corresponding lap joints, although the force transferred by the welds is reduced.

It is interesting to note that there seem to be no significant effects of the throat thickness nor of the

load carrying grade on the fatigue behaviour regarding weld toe failure. However, the occurrence of weld root failure depends highly on these factors.

3   FATIGUE ASSESSMENT WITH THE NOMINAL STRESS APPROACH

The details investigated can be found in the catalogues of codes and recommendations based on the nominal stress approach. In the case 611 of the IIW-Recommendations (Hobbacher, 2007), the lap joint is classified as FAT 63 for weld toe failure (parent metal) and FAT 45 for weld throat failure (based on the nominal stress in the weld throat). The above mentioned test results are generally above these values. Concerning the results in Fig. 4 it should be kept in mind that the nominal stress in the weld is twice as high as in the plate according to the areas for force transfer (Fig. 7 shows the plate stress).

The FAT value of the doubler plate in the recommendations depends on the structural configuration. On a continuous plate it is classified as FAT 71 (case 513), while on an I-beam or a rectangular hollow section it corresponds to FAT 50 (cases 711 and 713). The first case corresponds better to the test specimen as the latter includes additional structural stress concentrations. The test results are again on the safe side.

4   FATIGUE ASSESSMENT WITH STRUCTURAL STRESS APPROACHES

4.1   Structural hot-spot stress approach

The structural hot-spot stress approach, which is applicable only to weld toe failures, is based on the stress component at the weld toe which disregards the local stress increase due to the local weld toe notch. The structural hot-spot stress can be computed by surface stress extrapolation using extrapolation points depending on plate thickness or by linearization of the stress through the plate thickness (Niemi et al., 2006).

In the present cases, the structural hot-spot stress corresponds to the nominal stress as long as the stress increase due to misalignment is negligible (5% or less), which may be assumed here.

The characteristic fatigue strength is FAT 90 for load carrying and FAT 100 for non-load carrying fillet welds. Partial-load carrying welds, as in the doubler plate specimens, should be considered as load-carrying (Hobbacher, 2007), which means that FAT 90 generally applies here.

This assessment is obviously slightly non-conservative for the tested specimens showing weld toe failure (Figs. 4–6).

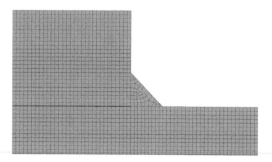

Figure 8. Finite element model (1/4 model) for the structural stress analysis acc. to the approach of Xiao/Yamada (2004).

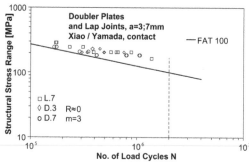

Figure 9. S-N data for weld toe failure based on the principal structural stress 1 mm below the weld toe.

Table 1. Computed structural stress concentration factors acc. to the approach of Xiao and Yamada (2004).

Specimen type	Contact		No contact	
	SCF (σ_x)	SCF (σ_1)	SCF (σ_x)	SCF (σ_1)
Lap joint $a = 3$ mm	1.40	1.88	1.46	2.27
Doubler pl. $a = 3$ mm	1.20	1.36	1.15	1.31
Lap joint $a = 7$ mm	1.20	1.37	1.30	1.64
Doubler pl. $a = 7$ mm	1.13	1.23	1.17	1.31

4.2 Structural stress approach of Xiao/Xamada

The structural stress approach of Xiao and Yamada (2004) takes the stress at a location 1 mm below the weld toe on the anticipated crack path as the relevant stress parameter for the fatigue assessment using an S-N curve according to FAT 100. A finite element analysis is necessary with a mesh having element sizes of 1 mm or less.

Simple finite element meshes were created using plane elements with quadratic displacement functions. The element size has been set to 0.5 mm (Fig. 8) to avoid stress underestimation which has been observed in previous analyses with 1 mm element size (Fricke and Kahl, 2005). It should be noted that contact elements (without friction) were introduced in the gap between parent and lap resp. doubler plate to avoid any overlap under loading. The gap width has been assumed to be zero. The analyses were performed for a nominal stress of $\sigma_n = 150$ N/mm^2 which is the average loading during the fatigue tests.

As it is unclear if the stress σ_x in longitudinal direction or the max. principal stress σ_1 is the relevant parameter, both stresses have been computed. Table 1 gives the resulting structural stress concentration factors referring to the nominal stress in the parent plate for contact and no contact. Both the throat thickness and the load carrying grade affect the SCFs.

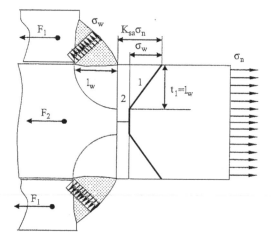

Figure 10. Modified through-thickness structural stress for axial loading according to Poutiainen (2006).

Using the principal structural stresses, which seem to represent better the concentrated force flow at the weld toe, all fatigue test data plotted in Fig. 9 are above the FAT 100 S-N curve.

4.3 Structural stress approach of Poutiainen

Poutiainen (2006) proposed a modified through-thickness structural stress to take into account the local stress increase due to the force transferred by the weld. Fig. 10 shows the tri-linear stress distribution for a plate with two-side fillet welds. The weld stress σ_w, which can be calculated from the force F_1 in the attached plate, determines the stress increase at weld toe.

As the two triangular stress distributions approach each other in case of relatively large weld leg lengths

Table 2. Computed structural stress concentration factors K_{sa} acc. to the approach by Poutiainen (2006).

Specimen type	σ_w	l_w	K_{sa}
Lap joint $a = 3$ mm	$1.41 \cdot \sigma_n$	4.24 mm	1.91
Doubler pl. $a = 3$ mm	$0.76 \cdot \sigma_n$	4.24 mm	1.49
Lap joint $a = 7$ mm	$0.61 \cdot \sigma_n$	9.90 mm	1.18
Doubler pl. $a = 7$ mm	$0.36 \cdot \sigma_n$	9.90 mm	1.11

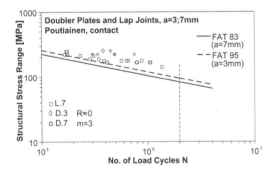

Figure 11. S-N data based on the modified structural stress (Poutiainen, 2006).

l_w, two cases have to be distinguished for the calculation of the structural stress concentration factor K_{sa}:

$$K_{sa} = 1 + \frac{\sigma_w}{\sigma_n}\left(1 - \frac{\ell_w}{t}\right) \quad \text{for } l_w \leq t/2 \quad (1)$$

$$K_{sa} = 1 + \frac{\sigma_w}{\sigma_n}\left(\frac{t}{4\ell_w}\right) \quad \text{for } l_w \geq t/2 \quad (2)$$

For the lap joints, $F_1 = \sigma_n \cdot t/2$ applies so that K_{sa} can easily be calculated. In order to calculate K_{sa} for the doubler plates, the above mentioned finite element models were used showing that each of the doubler plates carry 26.75% of the plate force for $a = 3$ mm and 30% for $a = 7$ mm. The resulting structural stress concentration factors are summarized in Table 2.

The differences between the specimens are larger compared with those acc. to the approach of Xiao and Yamada. The fatigue class proposed by Poutiainen (2006) depends on the plate and weld thickness, yielding FAT 95 for $a = 3$ mm and FAT 83 for $a = 7$ mm. Fig. 11 shows the fatigue test data based on the modified structural stress. The results are well on the safe side of the associated S-N curve, particularly those for the doubler plate with $a = 3$ mm (D.3).

5 FATIGUE ASSESSMENT WITH THE EFFECTIVE NOTCH STRESS APPROACH

The effective notch stress approach, using the elastic stress in the notches rounded by a reference radius $r_{ref} = 1$ mm, is able to assess both, weld toes and

weld roots (Radaj et al., 2006; Hobbacher, 2007). Weld roots at the end of non-fused plate surfaces as ocurring in the test specimens are usually rounded with a so-called keyhole notch.

Relatively fine-meshed finite element models have been created for the numerical analyses (Fig. 12) considering the recommendations by Fricke (2008). The keyhole notch has been placed such that the minimum distance between the rounded notch surface and the weld surface is exactly the throat thickness (3 mm resp. 7 mm).

The analysis has been performed with and without contact elements between the parent and the lap/doubler plate. Again, the analyses have been performed for a nominal stress of $\sigma_n = 150$ N/mm². Figs. 13 to 16 show the von Mises equivalent stress distribution in the finite element models with contact elements.

The resulting stress concentration factors, i.e. the max. principal stress in the notch related to the nominal stress, are given in Table 3. The differences between the analyses with and without contact are quite high. When contact is assumed, the SCFs computed for the 3 mm welds correspond better with the failure behaviour of the specimens. However, the notch stress

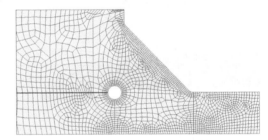

Figure 12. Typical mesh (1/4 model) for the effective notch stress analysis.

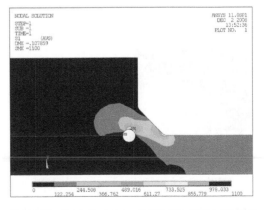

Figure 13. Equivalent stress distribution for the lap joint with a = 3 mm (L.3).

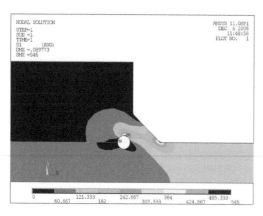

Figure 14. Equivalent stress distribution for the doubler plate with a = 3 mm (D.3).

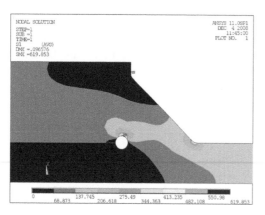

Figure 15. Equivalent stress distribution for the lap joint with a = 7 mm (L.7).

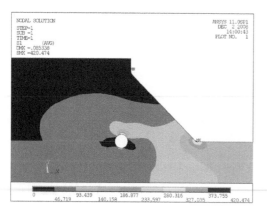

Figure 16. Equivalent stress distribution for the doubler plate with a = 7 mm (D.7).

Table 3. Computed notch stress concentration factors for reference radius r_{ref} = 1 mm.

	Contact		No contact	
Specimen type	SCF (toe)	SCF (root)	SCF (toe)	SCF (root)
Lap joint a = 3 mm	5.76	7.34	7.33	6.21
Doubler pl. a = 3 mm	3.54	3.60	3.44	2.23
Lap joint a = 7 mm	3.35	4.12	3.97	2.73
Doubler pl. a = 7 mm	2.82	2.24	3.03	1.38

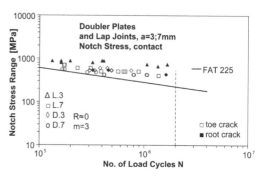

Figure 17. S-N data based on the effective notch stress.

at the weld root seems to be unrealistically high for the lap joint with $a = 7$ mm.

It can be stated that the effective notch stress is quite sensitive to the weld throat thickness, but also to the load carrying grade, particularly at the weld root.

Fig. 17 shows the fatigue test data, using the effective notch stresses for the actual crack initiation site. It can be seen that the data are generally above the S-N curve according to FAT 225 recommended by Hobbacher (2007), although the results for the root cracks are quite conservative.

6 SUMMARY AND CONCLUSIONS

Different approaches for the fatigue strength assessment of fillet welded joints consider effects of the weld throat thickness and the load carrying grade in different ways. In order to clarify this matter, fatigue tests were performed with 12 mm thick lap joints, having full-load carrying fillet welds, and doubler plates, where the welds carry only part of the load in the plate. Two different weld throat thicknesses were realized (3 and 7 mm).

Four approaches have been applied to the investigated specimen types. From the tests and the computations, the following conclusions are drawn:

- Joints with large weld throat thickness show crack initiation at the weld toe, while a small weld throat thickness promote crack initiation from the weld root, particularly with full-load carrying fillet welds.
- Three test series with weld toe failure showed similar characteristic fatigue strengths (FAT). Here, the load carrying grade seems to have no effect on the fatigue strength so that partial-load carrying fillet welds should be classified as full-load carrying welds.
- The results agree with the common joint classification according to the nominal stress approach which can also be applied to the cases with weld root failure.
- The structural hot-spot stress approach, applicable to weld toes, yields the same structural stress and strength for partial-and full-load carrying fillet welds, however being slightly on the non-conservative side.
- The structural stress approach of Xiao/Yamada, which is applicable to weld toes as well, shows slightly different structural stresses for the specimen types, which are on the conservative side. The max. principal stress yields better results than the directional stress.
- The modified structural stress approach by Poutiainen, again applicable to weld toes, shows larger differences between the different specimens, giving again conservative results.
- The effective notch stress approach, which applies also to the weld root, shows again significant differences between the different specimens. The calculated notch stresses correspond to the observed crack initiation site except for the lap joint with 7 mm throat thickness, where the stress at the keyhole notch representing the weld root seems to be over-estimated.

ACKNOWLEDGEMENT

The investigations were performed within the Network of Excellence on Marine Structures MARSTRUCT (http://www.mar.ist.utl.pt/marstruct/), being funded by the European Union through the Growth programme under contract TNE3-CT-2003–506141.

REFERENCES

Fricke, W. 2002, Recommended Hot-Spot Analysis Procedure for Structural Details of Ships and FPSOs Based on Robin FE Analyses. Int. J. of Offshore and Polar Eng., 12:40–47.

Fricke, W.: Guideline for the Fatigue Assessment by Notch Stress Analysis for Welded Structures. IIW-Doc. XIII-2240r1-08/XV-1289r1-08, Int. Institute of Welding 2008.

Fricke, W.; Bollero, A.; Chirica, I.; Garbatov, Y.; Jancart, F.; Kahl, A.; Remes, H.; Rizzo, C.M.; von Selle, H.; Urban, A. and Wei, L. 2007, Round Robin study on structural hot-spot and effective notch stress analysis. In: Advancements in Marine Structures (Ed. C. Guedes Soares & P.K. Das), 169–176, Taylor & Francis, London.

Fricke, W.; Cui, W.; Kierkegaard, H.; Kihl, D.; Koval, M. Mikkola, T.; Parmentier, G.; Toyosada, M. and Yoon, J.-H. 2002, Comparative Fatigue Strength Assessment of a Structural Detail in a Containership using Various Approaches of Classification Societies. Marine Structures 15:1–13.

Fricke, W. and Kahl, A. 2005, Comparison of different structural stress approaches for fatigue assessment of welded ship structures. Marine Structures 18:473–488.

Hobbacher, A. 2007, Recommendations for Fatigue Design of Welded Joints and Components, Final Draft, IIW-Doc. XIII-2151r1-07/XV-1254r1-07, International Institute of Welding.

Niemi, E.; Fricke, W. and Maddox, S.J. 2006, Fatigue Analysis of Welded Components—Designer's Guide to Structural Hot-Spot Approach, Cambridge: Woodhead Publ.

Poutiainen, I. 2006, A modified structural stress method for fatigue assessment of welded structures. Doctoral Thesis 251, Lappeenranta University of Technology.

Poutiainen, I. and Marquis, G. 2006, Comparison of local approaches in fatigue analysis of welded structures. IIW-Doc. XIII-2105-06, Int. Institute of Welding.

Radaj, D.; Sonsino, C.M. and Fricke, W. 2006, Fatigue Assessment of Welded Joints by Local Approaches. Cambridge: Woodhead Publishing (2nd Edition).

Radaj, D.; Sonsino, C.M. and Fricke, W. 2009, Recent developments in local concepts of fatigue assessment of welded joints. Int. J. Fatigue 31:2–11.

Xiao, Z.G. and Yamada, K. 2004, A method of determining geometric stress for fatigue strength evaluation of steel welded joints. Int. J. Fatigue 26:1277–93.

Analysis and Design of Marine Structures – Guedes Soares & Das (eds)
© 2009 Taylor & Francis Group, London, ISBN 978-0-415-54934-9

Global strength analysis of ships with special focus on fatigue of hatch corners

H. von Selle, O. Doerk & M. Scharrer
Germanischer Lloyd AG, Hamburg, Germany

ABSTRACT: Global analysis of ships using the finite element method is a widespread technique. The importance of this type of analysis has grown in recent years due to the fact that ship sizes are increasing rapidly, especially container vessels. This paper gives an overview of global FE-analysis and evaluation. The main focus is the fatigue assessment of hatch corners. Due to the stress concentration hatch corners show a high degree of utilization, thus sound fatigue behaviour is essential. Germanischer Lloyd (GL) developed a technique allowing fatigue assessment of hatch corner radii in a very effective way by means of local detail model computation. The result assessment according to GL rules is being presented here. Finally the paper describes the extension of the tool on hatch corners without any radii which are found frequently in lower decks. Here the hotspot or structural stress approach is used. The assessment is based on latest fatigue recommendations published by the International Institute of Welding IIW.

1 INTRODUCTION

Nowadays it is common practice to perform global strength analyses of ships as a powerful tool to design a well-balanced and utilized vessel. This analysis technique is recommended especially in cases where new ship designs differ significantly from those ships which are proven and already operating for a long time. Increased ship sizes, unconventional designs, use of new materials or materials with new properties, e.g. steel with increased yield strength are reasons for global finite element calculations. It enables the engineer to detect areas of poor design and it also allows design optimisation.

From a class point of view it has to be emphasized that such an analysis does not replace the plan approval procedure. This is carried out in addition and will lead to a better design.

This paper is mainly based on analysis performed by Germanischer Lloyd (GL) as engineering services.

2 OVERVIEW OF GLOBAL FE-ANALYSIS

Despite many efforts in the development of global analyses, this technique is still very time consuming. The following steps have to be taken:

- Generation of a global finite element model
- Generation of loading condition and design wave load cases
- FE-calculation
- Calculation evaluation

- Drawing conclusions (proposal of reinforcements if overstressed areas have been detected)

A more detailed description can be found in Payer & Fricke (1994).

2.1 Finite element model

Many software packages are available for generating finite element models. As well as others Patran, Ansys, Hypermesh and Poseidon are well known in the shipbuilding industries. GL is using GL Ship-Model, a Patran based tool, Poseidon as well as an in-house developed program ISG, which is an internal GL program. Not only the capabilities of a tool are of importance but also the support by the developer, the acceptance by the users as well as the in-house experience, especially in the case of trouble shooting.

Global FE-models consist of a big amount of input data and therefore it is important to keep an overview of the model. A well grouped model structure and proper model documentation is essential for checking the FE model, e.g. element scantlings as well as material properties. Figure 1 shows a sample FE-model of a large container vessel.

2.2 Load generation

The load generation consists of the definition of relevant loading conditions, e.g. ballast and fully loaded condition for tankers and bulk carriers or representative loading conditions for container vessels and other ship types.

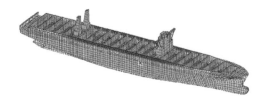

Figure 1.   Sample Global FE-model.

Figure 2.   Container vessel in wave sagging condition.

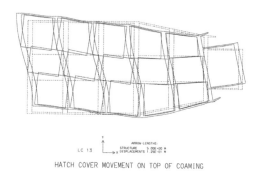

Figure 3.   Hatch cover movements.

For these loading conditions design wave load cases have to be created which need to meet the requirements given in rules or values coming from direct load calculation. For this purpose the software package GL ShipLoad was developed (Roerup et al., 2008). A graphical user interface facilitates the convenient application of ship and cargo masses to the finite element model and helps the selection of the relevant design wave situation. User defined selection criteria, such as the maximum values of sectional forces and moments, specify which waves have to be chosen for the global analysis. Scanning some thousands of wave parameter combinations leads to about 30 to 60 load cases which are taken as design load cases for the finite element calculation. The load cases are well balanced, which means ship acceleration forces are in equilibrium with forces from water pressure. For information on underlying theory see Hachmann (1991). Figure 2 shows a vessel in a typical wave sagging condition.

2.3   Calculation

During the last decades the demands on strength analyses have been increased rapidly but at the same time the computation capacity has been increased too. Therefore computation restrictions no longer exist in most cases. Most software packages include pre-processing tools, solvers as well as post processing tools.

2.4   Evaluation

Calculation evaluation according to the rules and standards should cover the following items:

- Deformations
- Stresses
- Buckling
- Ultimate strength
- Fatigue

Fatigue as the major topic of this paper is described in the next chapter separately.

2.4.1   Deformation

The checking of deformation is a precondition to gain confidence in calculated results. In general deformations do not influence strength in a negative way, but the functionality of structures must be ensured, e.g. ramps, door openings, clearance of car decks and hatch cover movement. The latter are demonstrated in Figure 3.

2.4.2   Stresses

To get a first impression of stress behaviour, equivalent stress plots are helpful. Principal stresses as well as shear stresses should be checked for further assessment. To keep the output as small as possible maximum stresses from all load cases can be stored for each element. This is a special GL-feature. Each stress type has to be checked separately against allowed values.

Figure 4 is a sample demonstration of equivalent stresses for a container vessel.

2.4.3   Buckling

Normally, a plate field buckling evaluation is one of the most time consuming tasks. This is due to the fact that the plate field and stiffener scantling information are not stored in FE-models, even though they are needed for the evaluation. However, the Poseidon software does store these information needed for dimensioning.

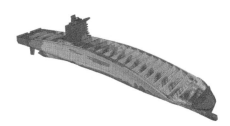

Figure 4. Equivalent stresses for a container vessel.

2.4.4 *Ultimate strength*

The Ultimate strength analyses by means of the finite element method are occasionally carried out for investigation of collision and grounding. The latest GL Rules cover also the ultimate checks but they are not FE-based.

3 FATIGUE

The fatigue of ship structures has become more and more important during the last decades. Reasons are trends to light weight structures, use of new materials as well as structural optimisation to a large extent.

Potential fatigue crack initiation points are welded components or free plate edges. In both cases the local design is the crucial factor. For the assessment of these, local design methods can be used, as described below. For typical cracks see Figure 5. The cracked welded detail (left) shows impressively the smooth surface of the fatigue crack propagation area and the rough surface of the final crack. The right Figure shows a crack starting from a plate edge. For fatigue assessment fine mesh calculations are usually needed.

3.1 *Local strength analysis*

The use of local strength analyses is a common practice to overcome the failure of global analyses with coarse meshes. The following methods exist:

- Stress concentration factors SCFs
- Sub-model technique
- Sub-structure technique
- Locally refined global FE-models

The use of stress concentration factors SCFs is a very fast and effective method but it works only if SCFs are available and if a well defined nominal stress state can be found.

Sub-models represent local details modelled with a fine mesh. The models are to be solved separately by applying the boundary displacements taken from the global analysis.

The sub-structure technique requires also a local fine mesh model but the internal degrees of freedom

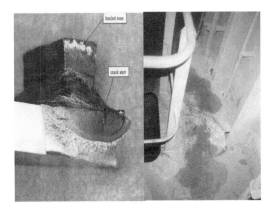

Figure 5. Fatigue cracks.

are condensed to boundary nodes and the resulting stiffness matrix is incorporated into the global finite element model.

For the last method the global FE-model has to be refined locally. If there is a lot of details, this method results in a large FE-system and corresponding computation time.

GL prefers the sub-model technique as it is a very simple post-processing method. Besides it does not require a re-calculation of the global model as would be necessary for the last two procedures. From a practical point of view it allows generation of the global model as well as the definition of details in parallel.

3.2 *Fatigue of hatch corners with corner radii (free plate edge)*

The sub-model procedure described here is an in-house development. Despite the fact that commercial programs such as ANSYS offer also the opportunity for local mesh refinement, the in-house development has been preferred. This is due to the fact that beside simple mesh refinement components have to be incorporated into the local model which are not part of the global model such as hatch corner radii and insert plates. Another reason is that the procedure fits well into GL's pre- and post-processing environment.

Steps are:

- Definition of details after global calculation run
- Set up of locally refined detail model by GL software
- Fatigue evaluation according to GL Rules

The definition of the details which are to be investigated can be done either interactively using the global model or by using an editor. In both cases the following data has to be defined:

- Geometry
  1. Location
  2. Size of radius (or ellipse)

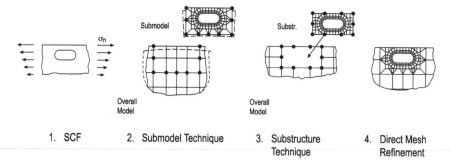

| 1. SCF | 2. Submodel Technique | 3. Substructure Technique | 4. Direct Mesh Refinement |

Figure 6.   Methods for local strength analyses.

3. Plate thickness
4. Size of insert plate (if arranged)
• Fatigue parameters
  1. Material yield strength
  2. Fat class or detail category (quality of cut surface)
  3. Slope of SN-curve
  4. Shape of load spectrum
  5. Number of cycles (design life time)
  6. Load cases to be considerd.

One additional parameter defines the size of the detail model, i.e. how many elements of the global model adjacent to the corner are included in the detail model.

Bottom of Figure 7 shows a typical local model. As can be seen, the mesh is not ideal. The transition gradient from small elements to large sized elements is very hard, but comparison calculations showed that the stresses along the corner radius are not affected by this. Top of Figure 7 shows all local models for the fore part of a container vessel. For each hatch corner a separate model is established. The models penetrate each other but they are not connected physically. A special feature can be used to shift the model so that it can be checked better.

Calculation results are tangential stresses along the corner radius which is subdivided into 10 elements (evaluation positions). For each position the maximum stress range from all load cases is listed. In case of symmetric ship structures it is sufficient to consider waves coming from one ship side only. In these cases stress results from port side and starboard have to be combined. In the end the acting stress ranges are compared with permissible stress ranges taking into consideration the mean stress effect.

The permissible stress ranges can easily be determined by using the GL Rules (Germanischer Lloyd 2008).

From the load side a straight line spectrum can be assumed (spectrum A acc. Figure 8) in combination with 50 million load cycles for 20 years design life, as is usual for seaway induced loads.

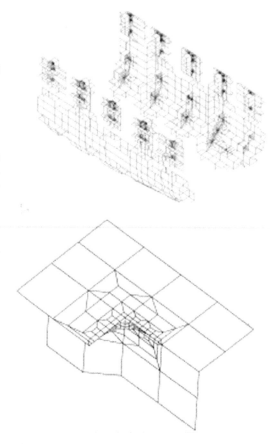

Figure 7.   Detail models of hatch corners radius.

The load carrying capacity is reflected by an appropriate SN-curve (Figure 9). The different curves are valid according to the surface quality.

Using these data a linear damage calculation according to Miner's Rule can be performed resulting in a damage sum D. Instead of the damage sum D the usage factor U is used frequently, as it is related to

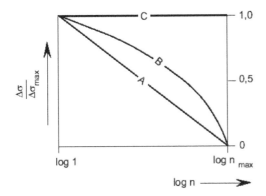

Figure 8.   Standard stress spectra.

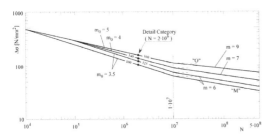

Figure 9.   SN-curve for cut edges of different quality.

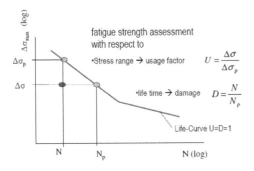

Figure 10.   Definition of damage D and usage factor U.

stresses and therefore more helpful for engineers. The difference is illustrated in Figure 10.

In case of a detection of overstressed hatch corners an Excel-sheet is provided to find appropriate countermeasures. It is based on regression curves which reflect the influence of:

- Increase of corner radius
- Arrangement of insert plate with increased thickness
- Use of higher tensile steel
- Improvement of surface quality by grinding

The best solution should be fined for together with the shipyard.

It should be emphasised that this procedure is not restricted to hatch corners but it can also be used for opening corners in general, e.g. doors and windows.

3.3   Fatigue of hatch corners without corner radii

The work flow for hatch corners without radii is similar to that with radii but the concept of assessment has to be changed as the potential crack initiation point is now a weld.

Differences are:

- Assessment of welds instead of plate edges
- No fatigue bonus for higher tensile steel
- Special requirement on meshing
- Different SN-curves

The method implemented at GL is the so-called hot-spot or structural stress concept (Hobbacher, 2007, Niemi & Fricke & Maddox 2006). The notch stress approach (Fricke, 2008) would be suitable too, but it is much more complicated from a modelling point of view and the results are not always reliable.

The hot-spot concept requires a pre-defined mesh arrangement of the FE-model close to the weld, e.g. element sizes equal to plate thickness for welds on a plate surface. The method works with shell elements as well as with solid elements (Figure 11). The hot-spot stress is taken as the linear extrapolated stress from the two elements adjacent to the weld. A sample local FE-model is shown in Figure 12.

Again a stress spectrum using the maximum hot-spot stress range has to be set up (e.g. straight line spectrum A acc. to Germanischer Lloyd 2007, see Figure 8). As the hot-spot stress includes all stress concentrations towards the weld, unique SN-curves with appropriate Fat classes are used (Figure 13). The reference stress range values or Fat-classes are 90 and 100 for load carrying and non-load carrying welds respectively.

As described for hatch corners with radii Damages or Usage factors have to be calculated. Countermeasures in case of a detection of overstressed corners are a local re-design or the arrangement of a radius bracket.

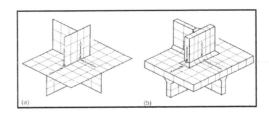

Figure 11.   FE-mesh according to hot-spot concept.

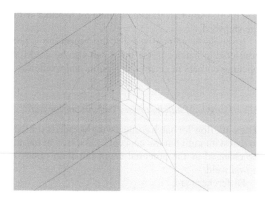

Figure 12. Sample FE-model for hot-spot concept.

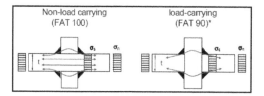

Figure 13. Unique fat classes for hot-spot concept.

4 CONCLUSIONS

With the fatigue assessment for hatch corners as described above, a powerful tool has been developed. It enables engineers to assess all hatch corners of a large container vessel in one day. The method is not restricted to hatch corners but is also suitable for other opening corners such as windows and door openings.

ACKNOWLEDGMENTS

This paper has been partly compiled within the Network of Excellence on Marine Structures (MARSTRUCT), which has been founded by the European Union through the Growth programme under contract TNE3-CT-2003-506141.

REFERENCES

Cabos, C., Eisen, H. & Krömer, M., 2006, GL ShipLoad: An integrated load generation tool for FE analysis. *Proc. 5th Int. Conf. on Computer Application and Information Technology in the Maritime Industries (COMPIT)*, Grimmelius, H.T. (ed.), Delft University of Technology, Oegstgeest, The Netherlands, pp. 199–210.

Fricke, W., 2008, Guideline for the Fatigue Assessment by Notch Stress Analysis for Welded Structures, *International Institute of Welding*, IIW-Doc. XII-WG3–03r7.

Germanischer Lloyd, 1997, *GL-Technology*, Fatigue Strength of Ship Structures, part I: Basic Principles.

Germanischer Lloyd, 1998, *GL-Technology*, Fatigue Strength of Ship Structures, part II: Examples.

Germanischer Lloyd, 2008, *Rules for Classification and Construction*, I—Ship Technology, Part 1—Sea-going Ships, Chapter 1—Hull Structures.

Germanischer Lloyd, 2007, *Rules for Classification and Construction*, V—Analysis Techniques, Part 1—Hull Structural Design Analyses, Chapter 1—Guidelines for Global Strength Analyses of Container Ships.

Germanischer Lloyd, 2009, *Tutorial* Poseidon Software.

Hachmann, D., 1991, Calculation of Pressures on a Ship's Hull in Waves. *J. Ship Tech. Res.*, Vol. 38, pp. 11–132.

Payer, H.G. & Fricke, W., 1994, Rational Dimensioning and Analysis of Complex Ship Structures, *SNAME Trans.* Vol. 102.

Roerup, J., Schellin, T.E. & Rathje, H., 2008.

Load Generation for Structural Strength Analysis of Large Containerships, *ASME 27th International Conference on Offshore Mechanics and Arctic Engineering*.

Analysis and Design of Marine Structures – Guedes Soares & Das (eds)
© *2009 Taylor & Francis Group, London, ISBN 978-0-415-54934-9*

Structural integrity monitoring index for ship and offshore structures

B. de Leeuw
Plant Asset Management, Houston, TX, USA

F.P. Brennan
School of Engineering, Cranfield University, Cranfield, UK

ABSTRACT: Recent years have seen enormous activity in the development of Structural Integrity Monitoring/Structural Health Monitoring equipment and systems for Ship and Offshore Structures. Systems based on technology that could in the past have only been used under laboratory conditions are now frequently deployed offshore very often claiming accuracy and reliability commensurate with laboratory measurements. Monitoring is certainly an exciting prospect and has many advantages over traditional NDT; there are however, some very fundamental issues that must be resolved to benefit fully from these new technologies. Not least of these is the development of objective measures to quantitatively assess the performance characteristics of monitoring technologies. This paper presents the background and development of such a measure of performance based on a fatigue failure model for Ships and Offshore Structures. This new measure, the Structural Integrity Monitoring Index or SIMDex can be similarly applied using any failure model and criterion and means that Structural Integrity Monitoring technologies can be objectively judged solely on their suitability for specific applications.

1 INTRODUCTION

Technological advances over the last decade have led to the development of an increased number of state-of-the-art Structural Integrity Monitoring (SIM) or Structural Health Monitoring (SHM) systems. Although the potential benefits of applying long term continuous monitoring are well publicized, thus far little attention has been paid to determining how well such systems perform their intended task. Different systems vary in terms of application and the parameters they record to determine the integrity of a particular structure. The level of performance of each monitoring technique will also vary depending on the type of structure being monitored. Whereas for NDT inspection systems, parameters such as Probability of Detection (POD) and Probability of Sizing (POS) are available, no such approach exists for stress or strain monitoring applications.

The design of state-of-the-art SIM systems must be led by an understanding of the requirements of maintenance and structural integrity calculations. This also provides a basis for the characterization of the monitoring functions and enables the objective evaluation of a particular system. Measurement of the performance characteristics of a SIM system is important as it enables the determination of its suitability for monitoring a particular application as well as allowing direct and objective comparisons to be made between different systems. When considering system applicability

to the integrity management of a particular structure, consideration must also be given to the combined performance of different systems. This includes NDT inspection systems as well as continuous monitoring systems.

2 DEFINING SYSTEM PERFORMANCE

Performance evaluations depend in part on the understanding of the requirements of maintenance and structural integrity calculations. These will not only dictate the data measurement requirements from a monitoring system but also provide an indication of the actual effect of measurement inaccuracies on structural safety estimates. Traditional NDT techniques can be evaluated using well publicized statistical approaches such as POD and POS (Packman et al, 1976) which rely on single point measurements taken at a particular point in time. With SIM, the continuous collection of data over a longer period of time is considered usually under varying environmental and operational conditions. The assessment is therefore complicated by the large number of variables involved. However, as for POD trials, the only way to effectively assess SIM systems is through controlled trials, where loading and environmental conditions can be specified and simulated accordingly. The three main areas of interest when considering SIM system performance include the Probability of

Detecting the onset of damage, the likelihood of accurate identification of the location of the damage and the Performance for Purpose of the system. The latter is a measure of the performance capability of a particular system in measuring and monitoring the relevant structural parameters. This is the most important measure because if this is not satisfactory it is unlikely that either of the first two measures can be met with any degree of confidence.

As with crack inspection techniques stress/strain-monitoring devices need to be evaluated in terms of accuracy of measurement with respect to material and environment. However, other additional variables of importance include static and dynamic loading parameters, repeatability and drift. Such systems are also likely to respond differently to different loading regimes. For example, some systems will respond well to low R-ratio cyclic stresses but may not perform as well under high mean stress or vice-versa. Similarly, the frequency of applied cyclic stresses may also affect the performance of some monitoring systems. These variables all need to be taken into account when planning the trial stages for the SIM system assessment.

For effective comparison between systems it is necessary to design a procedure that will enable a direct and easy assessment of each system under certain predetermined criteria or parameters. These criteria will vary depending on the type of structure and the conditions in which they operate. Ideally an assessment would result in a single indexed value indicating the performance of different stress monitoring technologies as a quantitative basis on which to compare different systems. Other factors that need to be taken into account when considering system performance are Repeatability, Sensitivity and Drift for the particular system.

Research efforts by the authors have led to a new objective approach to tackle this problem in the form of a new Structural Integrity Monitoring Index (SIMdex) (de Leeuw, 2004). In this paper the SIMdex is illustrated by its application to stress/strain monitoring systems.

3 STRUCTURAL INTEGRITY MONITORING INDEX (SIMDEX)

The main purpose of the SIMdex is to provide a measure of the performance of a particular monitoring system and is based on its ability to measure the actual damaging stresses within a component. This approach is applicable to a range of material and fatigue loading scenarios allowing comparative system assessments to be made under different operating conditions.

The assessment of system accuracy provided here directly relates the measured stresses to the level of damage present in the structure, highlighting the fact that as the structure accumulates damage, the accuracy of a monitoring system becomes progressively more important. This also means that the determination of the SIMdex is linked closely to an appropriate failure model. The analysis below presents an approach to the SIMdex calculation based on fatigue life defined by the Stress-Life damage model.

The SIMdex values are represented by a range of values between One and Three (i.e., $1 \leq \text{SIMdex} \leq 3$). These are the limit values and were chosen arbitrarily, Three representing the worst case, where the significance of the monitored stress range error is large. This indicates a less accurate system which is likely to provide an inaccurate assessment of the structure's condition. One is the target value for any monitoring system, indicating that the greatest proportion of measurements correspond closely to the actual Stress Range experienced by the component. A critical part of the analysis is the determination of these limit values as these ultimately determine the index weightings which relate to the spread of the data, and from which the SIMdex value is derived.

4 SIMDEX—STRESS-LIFE APPROACH

The approach illustrated considers a methodology for the determination of the SIMdex in terms of fatigue damage using a Stress-Life (S-N) failure model. The assessment process takes into account the stage in the lifetime of the structure at which it is being monitored. Consider the S-N Curve defined in Equation 1 below for welded steel plate joints [3]:

$$\text{Log}(N) = 11.705 - 3\text{Log}(S) \qquad (1)$$

The lower SIMdex limit (One) is linked to the actual applied stress range and is the target value for the monitoring system. When considering the Upper Threshold for the SIMdex two scenarios need to be considered in terms of deviation of measured stress range relative to actual applied stress range. A system could either over- or under-estimate the stress range and the significance of each of these needs to be taken into account. This is achieved by correlating the Upper limit value (Three) of the SIMdex to two different limiting stress ranges, one below the actual applied range and one above, each corresponding to either the over- or under-estimation scenarios.

As a starting point the SIMdex limit value is set for the case of over estimation of the monitored stress range. This is dependant on the point in time in the structure's life that the assessment is taking place and corresponds to the stress range that would cause its failure due to fatigue at that particular point in time. To illustrate this, consider the stress-life curve in Figure 1 below and assume that the system assessment is carried out after 130,000 stress cycles with an actual cyclic

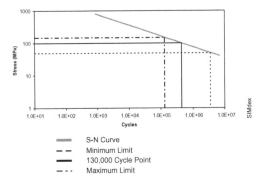

Figure 1. Stress-Life Curve and SIMdex Limit setting after 130,000 cycles on.

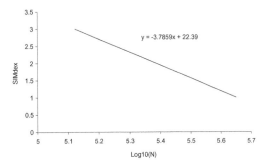

Figure 2. Determination of Index Weightings for the case of over-estimation of the applied Stress Range.

stress range of 100 MPa. The maximum threshold for the index (Three) would then be set at approximately 150 MPa which is the stress range at which fatigue failure would occur at approximately 130,000 cycles as predicted by the Stress-Life Equation (1). This point represents the scenario of a system for which the greatest proportion of stress data recorded deviates from the actual applied mean stress range by an amount that would conservatively predict instant failure. This means the system is so conservative in its overestimation of stress that it would be practically useless at this stage in the life of the structure.

Considering the underestimation of the actual stress range a lower threshold (minimum limit in Figure 1) of the stress range needs to be associated with the maximum SIMdex value (Three). In the first instance a logical choice may seem to set this to the material fatigue endurance limit. However, in practice this was found to result in the underestimation of stress range becoming a less significant factor in the system assessment at higher operating stress ranges, therefore leading to overconfidence and should therefore not be applied (de Leeuw, 2004).

Taking this into account, another approach was considered for the application of the SIMdex limit to underestimation of the applied stress range. The simplest approach is to maintain a constant ratio between maximum overestimation and underestimation of the stress range on the S-N Curve plotted on a log-log scale as illustrated in Figure 1 i.e. Maximum Limit is 150 MPa and Minimum Limit is 50 MPa. This method means the lower limit also varies with the remaining life of the structure, resulting in an increased penalty for inaccuracy as the fatigue life is approached. Also, due to the general shape of the S-N curve this relationship leads to a more sensitive SIMdex and a relatively larger penalty for an underestimation of stress range compared to over-estimation. This is discussed in greater detail following the explanation of the remaining methodology for calculating the SIMdex.

Having determined the limiting values of the index it is necessary to represent its variation at all stages between these two extremes. This is achieved by relating Log variation of cycles to failure to the limit values of the SIMdex, thus allowing a linear interpolation on a Log-liner scale to be carried out, providing a complete set of values of the SIMdex relating to stress range. Figure 2 illustrates this procedure for cases of over-estimation for an applied stress range of 100 MPa.

The Index Weightings (or penalty values) used to calculate the value of the SIMdex are based on the relationship between the inverse logarithm of the variation of the SIMdex Factor with Log(N) (shown in Figure 2) and the Stress Range distribution output from the SIM system being considered. Figures 3 and 4 below illustrate the correlation of the Index Weightings with monitored stress range for a particular S-N curve. The Index Weightings (or penalty values) used to calculate the value of the SIMdex are based on the relationship between the inverse logarithm of the variation of the SIMdex Factor with Log(N).

It can be seen that this not only yields a higher penalty the greater the variability in monitored data but also that the SIMdex becomes more sensitive to dispersion at higher stress levels and at later stages in a structure's life. This is less obvious for the case of underestimation of the actual stress range, which is penalised higher than the overestimation of stress at all times. This is a significant attribute of the SIMdex as underestimation is more serious as it could lead to erroneous over-confidence in the safety of the structure or component.

In order to estimate the ultimate SIMdex value representative of the performance of the monitoring system, the Index Weightings are multiplied by the corresponding proportions of the data across each stress range recorded by the system. An example of a single point 'j' on the probability distribution and its multiplication to the Index Weighting is illustrated

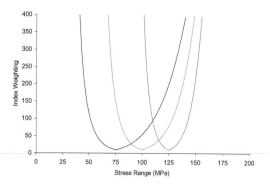

Figure 3. Effect of Stress Range on the SIM Index Weighting following 130,000 Cycles.

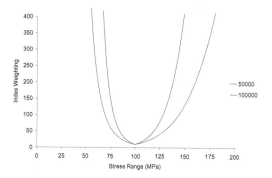

Figure 4. Change in the SIM Index Weighting with increased time in service of the component.

in Figure 5. Similar multiplications are carried out on each point on the probability distribution function.

The final SIMdex value is then obtained by taking the logarithm of the summation of all the individual results of the multiplication procedures. This can be defined by the following expression:

$$\text{SIMdex} = \text{Log10} \left[\sum_{1}^{j} \left(\text{IW}_j \times \text{D}_j \right) \right] \quad (2)$$

Where: IW_j = Array of Index Weighting Data, D = Array defining proportion of data points monitored within each Stress Range.

The resulting number will usually lie between 1 and 3, One indicating a perfect system and Three or above means the system is either over conservative in predicting failure or alternatively likely to be greatly underestimating the actual stress range applied. In the case of extremely underperforming systems the resulting SIMdex value may well exceed the value Three by a considerable margin.

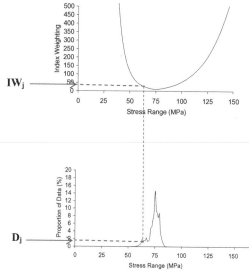

Figure 5. Example of correlation between Index Weightings and the monitored Stress distribution.

5 PERFORMANCE FOR PURPOSE RATIO (PFP)

The Performance for Purpose Ratio (PfP) provides an approach to normalise the SIMdex rather than presenting it in the form of the arbitrary index values. This should simplify comparisons between different monitoring systems, particularly in cases where different SIMdex weightings are applied.

The proposed performance ratio for stress monitoring systems is based on the SIMdex and defines the absolute deviation of the calculated SIMdex rating from the target value of One, which represents the best possible system performance. Given the term "Performance for Purpose (PfP)" this performance ratio can be calculated using the expression below.

$$\text{PfP} = 100 \times \frac{(2 - |1 - \text{SIMdex}|)}{2} \quad (3)$$

Presenting the information in such a manner provides a more understandable quantitative representation of system performance as well as enabling easy comparisons to be made between different systems.

Note that a PfP of zero represents the worst performance case, where the system erroneously predicts failure when none is imminent. A minimum performance threshold could be set for practical purposes below which a system would not be considered for the particular application.

6 SIMDEX SAMPLE CALCULATION— STRESS-LIFE APPROACH

The variation of the SIMdex with fatigue life of a tubular joint was determined using an S-N approach. The analysis assumes a constant amplitude Stress Range of 125 MPa. For comparative purposes three experimental stress range distributions from a stress monitoring system are applied to illustrate three different levels of performance. In order of decreasing performance, will be referred to as Eg1, Eg2, and Eg3 for the purpose of the study below. Sample calculations for the SIMdex relating to stress distribution "Eg3" after 1 million cycles are presented below.

$$\text{SIMdex} = \text{Log10} \left[\sum_{1}^{j} \left(\text{IW}_j \times D_j \right) \right] \quad (4)$$

$$= \text{Log10}(31.6227766)$$

$$\text{SIMdex} = 1.50$$

$$\text{PfP} = 75\%$$

Figure 6 summarizes the system performance results in terms of Performance for Purpose for three data distributions using the Stress-Life Approach.

Figure 6 clearly shows a drop in system performance as the quality of the stress distribution drops, "Eg1" being the best and "Eg3" the worst performer. The quality of system "Eg1" is such that it performs well, and thus provides accurate measurements of stress, over the whole fatigue life of the structure being monitored.

The Stress Life approach is conservative in its performance assessment and penalizes underperforming systems from the start.

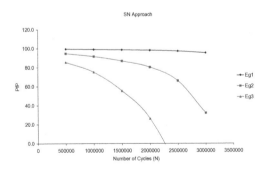

Figure 6. Results for Stress-Life approach to PfP.

7 CONCLUSIONS

Decisions regarding the safety of a particular structure rely heavily on information provided by NDT and structural monitoring. As such the need to quantify the performance capabilities of such systems is a priority. This is particularly true for SIM systems which, due to their relatively recent development, need to generate a level of confidence in their performance and capabilities before they become more widely used.

A new approach for the assessment of structural monitoring techniques in the form of the Structural Integrity Monitoring Index has been proposed and presented. As the SIMdex uses real monitored data it is takes into account accuracy or otherwise of measurements that might be influenced by applied stress ratio and frequency.

Overall the SIMdex provides a quick and robust approach to assessing the performance of stress monitoring systems under different conditions. Results presented by the single index value or in the form of Performance for Purpose means different systems can easily be directly compared to one another.

REFERENCES

Balladon, P. and Coudert, E., (1995), "TPG500 Structural Assessment", Rapport Technique 95072 C.
BS7910, (2006), "Guide on methods for assessing the acceptability of flaws in metallic structures", British Standards.
de Leeuw, B., (2004), "Analysis and Assessment of Structural Integrity Monitoring", PhD Thesis, University of London.
Etube, L., (2000), "Variable Amplitude Corrosion Fatigue and Fracture Mechanics of Weldable High Strength Jack-up Steels", Professional Engineering Publishing, ISBN: 1860583121.
Myers, P., (1998), "Corrosion Fatigue and Fracture Mechanics of High Strength Jack Up Steels", PhD Thesis, University of London.
Packman, P.F., Klima, S.J., Davies, R.L., Malpani, J., Moyzis, J., Walker, W., Yee, B.G.W. and Johnson, D.P., (1976), "Reliability of Flaw Detection by Nondestructive Inspection", ASME Metals Handbook, 8th Edition, Vol. 1, pp. 414–424.
Talei-Faz, B., (2003), "Fatigue and Fracture of Tubulars Containing Large Cracks", PhD Thesis, University of London.

Analysis and Design of Marine Structures – Guedes Soares & Das (eds)
© 2009 Taylor & Francis Group, London, ISBN 978-0-415-54934-9

Effect of uncertain weld shape on the structural hot-spot stress distribution

B. Gaspar, Y. Garbatov & C. Guedes Soares
Centre for Marine Technology and Engineering (CENTEC), Technical University of Lisbon,
Instituto Superior Técnico, Lisbon, Portugal

ABSTRACT: This paper presents a probabilistic study of the effect of uncertain weld shape on the structural hot-spot stress distribution along the weld toe using a Monte-Carlo simulation and the finite element analysis. A structural detail consisting of a plate strip with a transversal butt welded joint and a tapered thickness step is used as case study. The weld shape is modelled based on weld profile parameters defined as random variables allowing the simulation of typical imperfections as a function of the quality level of welding. The analysis uses a linear finite element model build with second order solid finite elements. A finite element analysis program is used to calculate the stress distribution on the vicinity of the weld toe and a standard extrapolation procedure is used to obtain the structural hot-spot stresses. Different quality levels are used to generate the weld shapes and its effect on the hot-spot stresses is calculated afterwards.

1 INTRODUCTION

Fatigue strength is an important requirement in the design of marine structures and welded structures and components in general. This requirement has been reflected is the design rules published by classification societies, where the fatigue strength assessment is mandatory for the hull structural design classification.

The most recent methods that are being used for fatigue assessment are based on the structural hotspot stress approach, using the finite element analysis to calculate the stress distribution on the vicinity of the hot-spots. Some research institutions have been publishing and proposing design recommendations for fatigue assessment of welded structures and components based on these methods. The International Institute of Welding (IIW) is one of those institutions, which has been publishing widely recognized guidelines. The designer's guide proposed by Niemi et al. (2004) is an example and was adopted for the present study. As recognized in this document, the fatigue assessment based on the structural hot-spot stress approach combined with the finite element analysis is nowadays the most accurate procedure for complex structures as the marine structures. However, these methods are still in development, and there is scope for new approaches that take more advantage of the potential of the finite element analysis (Niemi et al. 2004).

The fatigue assessment methods that are currently in use for design applications, including the most recent ones previously mentioned, still have uncertainties in important variables that govern the fatigue phenomena, and that are not usually quantified by means of numerical tools that may be used to rationally consider these uncertainties in the design process. These

uncertainties are related to various design variables, as the geometrical imperfections that are intrinsically present in the construction and repair processes as well as the numerical methods and calculation tools that are used to predict the loading and the structural response (Garbatov et al. 2006).

There are already publications where some of these uncertainties were quantified. Fricke et al. (2007) quantified the uncertainties related to different approaches that are now in practice for modelling and stress evaluation by means of round robin studies. It was concluded that this uncertainties may significantly affect the fatigue assessment results. Chakarov et al. (2008) studied the effect of the geometrical imperfections related to misalignments in the welded joints, intrinsically present in the construction and repair processes, and it was concluded that in some cases these geometrical imperfections may increase considerably the structural hot-spot stress.

The present study deals with the effect of the weld shape uncertainties in the structural hot-spot stress distribution. These uncertainties are related to the geometrical imperfections induced by the welding processes commonly used in the shipbuilding and repair industries. These imperfections are of random nature and probabilistic methods have to be used to address this problem. A Monte-Carlo simulation method combined with the procedure proposed in Niemi et al. (2004) are used to quantify the effect of the weld shape uncertainty in the structural hot-spot stress distribution. A simulation procedure is proposed in this paper, using a structural detail typical of containership's deck structures as case study. The weld shape imperfections are modelled based on weld profile parameters included in the Monte-Carlo simulation as random variables. Five classes of weld shape imperfections (i.e. quality levels of the welding process) are simulated using the proposed procedure and the effect of the weld shape uncertainty in the maximum stress concentration factor is assessed using fitted probability distributions and regression analyses.

2 STRUCTURAL DETAIL DESCRIPTION

The structural detail used as case study is a two plate deck strip, typical of containerships' deck structure, with two longitudinal stiffeners, one vertical plate at each side and a tapered thickness step at half-length, as shown in Figure 1. The material used is a common shipbuilding steel with typical values for the mechanical properties considered in linear finite element analysis, i.e. modulus of elasticity and Poisson's ratio. The structural detail region of interest for the present study is the deck plating region in the vicinity of the transversal butt welded joint between the two deck plates, as shown in Figure 2. The structural detail and butt welded joint geometries are described next.

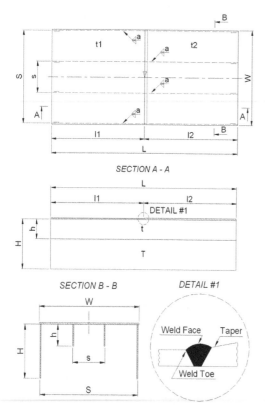

Figure 1. Structural detail geometry views: top, longitudinal section, transversal section and butt welded joint detail.

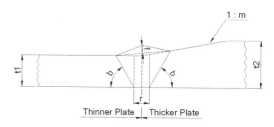

Figure 2. Butt welded joint and butt weld profile geometries.

Table 1. Structural detail geometric parameters.

Length, L	4000	Stiffeners, h^*	400
Width, W	2000	Stiffeners, t^*	20.0
1st spacing, S	1960	Thinner plate, l_1	2000
2nd spacing, s	660	Thinner plate, t_1	20.0
Side plates, H	1000	Thicker plate, l_2	2000
Side plates, T	20.0	Thicker plate, t_2	30.0

All dimensions in mm.
* Flat bar profile.

2.1 Structural detail geometry

The detail geometry was defined parametrically in order to simplify the geometrical modelling in the finite element analysis program and to allow future parametric studies. Figure 1 shows the detail geometry and the parameters adopted. The values fixed for this study are given in Table 1 and were selected from the work of Fricke et al. (2007), where the same structural detail was studied. The length is fixed equal to the hull structure frame spacing and the two end sections are considered coincident with two adjacent frame transverse sections. The deck plating width is imposed by the double hull width in the structural detail region, taking into account appropriate margins at each side to accommodate the side plates longitudinal welding. The two longitudinal stiffeners are equally spaced relatively to the detail longitudinal plane of symmetry and its cross section is the flat bar profile. The vertical extension of the side plates, which represent the side shell plating and the longitudinal bulkhead plating, was selected in order to model properly the interaction between the deck plating and this two attached vertical plates. The longitudinal welding between the deck plating and the stiffeners and side plates is tee type with 4 mm of weld throat depth.

2.2 Butt welded joint geometry

The butt welded joint and the butt weld profile geometries were also defined parametrically. In this case the parametrical description shown to be of extremely importance, as it allows an accurate description of the weld profile geometry as well as the geometrical

Table 2. Butt welded joint and butt weld profile geometric parameters.

| Face height, f | 5.0 | Bevel angle, b | 60.0 |
| Gap size, r | 5.0 | Taper ratio, m | 5.5 |

All dimensions in mm or degree, where applicable.

imperfections related to the welding process quality levels. Figure 2 shows the butt welded joint and the butt weld profile geometries with the parameters used in the geometrical description. The values adopted for the present study are presented in Table 2. For these specific parameters no information is given in Fricke et al. (2007). However, the parameters were properly defined taking into account shipbuilding and repair quality standards as well as common classification societies requirements for this type of welded joint, (IACS 2006, DNV 2001).

Usually this type of joint is built as shown schematically in Figure 2, where a perfect alignment of the lower surfaces of the two plates were assumed. However, there are very often misalignments due to irregularities in the deck moulded surface resulting from the construction process of the ship hull structure, which can increase the structural hot-spot stresses in some cases, (Chakarov et al. 2008). As already mentioned, this type of geometrical imperfections was not considered, since the aim is the study of the effect of the weld shape uncertainty on the structural hot-spot stress distribution.

The assessment was carried out considering simplified weld face geometry defined with two flat surfaces, as shown in Figure 2. This simplification permits an easier description and modelling of the joint geometry without introducing significant inaccuracies in the model. In this simplified geometry, the weld face height was defined as being the distance from the weld face edge to the reference line defined by the weld toes. Since this simplification of the real geometry may generate weld face geometries with slightly lower toe angles, the face height parameter may be used as an adjusting parameter.

This type of welded joint induce secondary bending stresses in the plating when it is axially loaded, due to the eccentricity between the thinner and thicker plates neutral surfaces, resulting in an increase of the structural hot-spot stresses along the thinner plate weld toe line. As shown in Chakarov et al. (2008) this secondary bending stresses vanishes near the longitudinal stiffeners and near the side plates, due to the increase in bending stiffness of the plates in these regions. The maximum values of the secondary stresses occur between this vertical structural elements, and the maximum structural hot-spot stress occur at the point defined by the intersection of the thinner plate weld toe line with the longitudinal plane of symmetry.

3 DETERMINATION OF STRUCTURAL HOTSPOT STRESSES USING EXTRAPOLATION

In the design phase of welded structures and components, the finite element analysis has been considered and accepted as the ideal tool for determining structural hot-spot stresses for fatigue assessment (Niemi et al. 2004). In the field of marine structures design, most of the recent design rules published by classification societies are using the finite element analysis and the structural hot-spot stress approach for fatigue assessment of critical hull structural details. A recent example is the common structural rules (CSR) for hull structural design of oil tankers and bulk carriers, where this approach was widely adopted. Various authors have been proposing different procedures for finite element analysis modelling and stress evaluation using the hot-spot stress approach, which has been recognised as its main disadvantage, since different procedures as well as different finite element types and programs can give different results (Fricke et al. 2007). In the present study, the procedure recommended by the IIW was adopted, taking as main reference the designer's guide proposed by Niemi et al. (2004), as already mentioned.

As known from the literature, the structural hotspot stress approach is only applicable to welded joints for which the potential fatigue crack will initiate at the weld toe, due to fluctuating stresses acting predominantly in the transverse direction when the structure or component is in service. The procedures used for finite element analysis modelling and stress evaluation usually vary with the type of hot-spot to be assessed, since the weld toe may be located on a plate surface or on a plate edge, and consequently different procedures need to be considered. According to the designer's guide adopted, these two possible locations for the hot-spots are classified as Type "a" and Type "b", respectively, and different procedures are used for each case.

In the welded joint used as case study, shown in Figures 1–2, the locations prone to fatigue crack initiation are all the points on the thinner plate surface defined by the left butt weld toe line. Consequently, the procedure to be used in the finite element analysis modelling and stress evaluation should be that proposed for Type "a" hot-spots, where two different types of finite elements and mesh densities can be applied according to the designer's guide. The choice of the ideal combination of element type and mesh density depends on many factors, as the type of welded joint to be analysed, but also on practical question related to the modelling complexity, computational time and sufficient accuracy of the results. For large structures like the marine structures, a practical, relatively coarse mesh is usually preferable instead of a relatively fine one due to the large number of structural details that have to be usually analysed in the design phase of this type of structures.

The strategy adopted for the finite element analysis modelling of the structural detail selected was, therefore, the implementation of a relatively coarse mesh with one layer of 20-node solid elements over the plate thickness. Usually, this type of elements yield reasonably accurate results in the assessment of Type "a" hot-spots, with more reliable results in complex details compared to those obtained with shell elements, as mentioned in the designer's guide. The relatively coarse mesh approach has been widely used by the classification societies, since it is suitable for design applications. This approach, combined with solid elements, is sufficient when the following conditions are satisfied: a) there are no other severe discontinuities in the vicinity; b) the stress gradient close to the hot-spot is not extremely high; and, d) stresses are resolved at the element's surface centre.

With this approach, the finite element mesh on the vicinity of the weld toe is characterised by element's edge sizes equal to the plate thickness along the direction perpendicular to the weld toe line, as shown in Figure 3. This requirement is to be applied to, at least, two adjacent elements on the vicinity of the weld toe, in order to properly apply the extrapolation procedure. However, more elements should be considered for a suitable mesh generation in the weld toe region and an accurate prediction of the stress distribution. The element's width should also be equal to the plate thickness, in order to generate elements with cube geometry. Although, in some cases, this requirement cannot be complied with exactly, as for example, in structural details with brackets, where sometimes the element's width has to be equal to the bracket's thickness plus two weld leg lengths.

The structural hot-spot stresses are obtained by linear extrapolation according to the procedure presented in the designer's guide. The extrapolation is performed based on the following equation,

$$\sigma_{HS} = \alpha \sigma_{0.5t} + \sigma_{1.5t} \qquad (1)$$

where $\sigma_{0.5t}$ and $\sigma_{1.5t}$ are the normal stresses perpendicular to weld toe line and calculated at the surface centre of the first and second elements adjacent to the weld toe respectively, as shown in Figure 3. With the mesh properly defined, the elements surface centre, or mid-point, should be at distances 0.5 t and 1.5 t from the weld toe modelled, for the first and second elements respectively, as represented in Figure 3. Consequently, in Equation 1 the constant α assumes the value 1.50 and the constant β the value -0.50, which may be derived based on geometrical considerations. The normal stresses to be used in the structural hot-spot stress extrapolation are defined in detail in the designer's guide. In the welded joint considered in the present study, the weld toe line is perpendicular to the largest principal stress direction along the thinner plate surface, so the normal stresses to be used in Equation 1 are interpreted directly as being that principal stresses evaluated at the extrapolation points.

4 FINITE ELEMENT ANALYSIS MODELLING DESCRIPTION

The finite element analysis modelling was carried out using the widely recognized capabilities of the commercial program ANSYS® (2007). Advanced modelling and analysis techniques were applied due to the scope of the simulation performed in the present study. The programming language ANSYS® Parametric Design Language (APDL) was used to write a set of macro files that can be used to automate all the modelling steps as a function of a set of parameters previously defined. These parameters are related to the geometry, material properties, meshing, loading and boundary conditions.

The modelling is performed accordingly to the following main steps, after all the modelling parameters are known: 1) generate the idealised geometry; 2) generate the finite element mesh; 3) generate the weld shape random imperfections; and, 4) apply the loading and boundary conditions. Since the analysis is based on the Monte-Carlo simulation method, all these modelling steps are applied sequentially in each simulation performed. For this purpose, an external pre-processing and solution program was written to drive the Monte-Carlo simulation, using the batch-mode capabilities of the finite element analysis program and the macro files developed to perform all the modelling steps.

The procedure used to simulate the weld shape random imperfections is presented in the next chapter for convenience. The structural hot-spot stresses are calculated along the weld toe line, for each simulation performed, accordingly to the extrapolation procedure presented in the previous chapter. Due to the amount

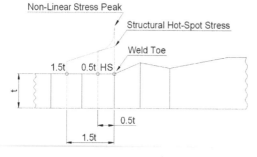

Figure 3. Determination of structural hot-spot stresses using extrapolation and the relatively coarse mesh approach combined with 20-node solid elements.

of output files that may exist in this type of analysis, an external program was written for the post-processing of the simulation results. This program reads all the output files generated in the solution phase and write a summary of the simulation results in a final output file. The first, second and fourth modelling steps are briefly described next.

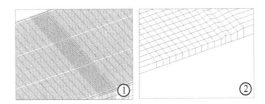

Figure 5. Finite element model mesh: 1) course and fine mesh regions; 2) mesh size on the vicinity of the weld toe.

4.1 Idealised geometry

The geometry is generated using the parameters presented in the structural detail description and the solid model method of the finite element analysis program ANSYS® (2007). The parameters are used to define positions of reference keypoints, which are used afterwards to define volume entities. These volume entities represent the plates and welding domain, as shown in Figure 4. The symmetry properties of the structural detail geometry are taking into account in the model generation. However, the whole geometry has to be represented since the butt weld shape random imperfections destroy the geometry transversal symmetry.

As already mentioned, the butt weld profile was idealised as shown in Figure 2, using two flat surfaces to represent the weld face. The volume entities representing the thinner plate domain, defined as Plate 1 in the previous figure, are generated taking into account the meshing step in order to ensure that the requirements imposed by the structural hot-spot stress extrapolation method are complied with.

4.2 Finite element mesh

The finite element mesh is automatically generated using the meshing capabilities of the finite element analysis program. A mapped mesh approach (i.e. mesh with a regular pattern of elements with similar geometry) with coarse and fine mesh regions was adopted. The solid element Solid 95 was selected for the modelling, since it complies with the recommendations given in Niemi et al. (2004) and it was also used by Fricke et al. (2007) and Chakarov et al. (2008) to study identical structural details. This element is typically used for 3-D structural solid modelling. The element is defined by 20 nodes, with three degrees of freedom

per node, i.e. translation in the x, y and z directions. It has second order displacement shape functions and it can tolerate irregular geometries without much loss of accuracy. More features of this element can be found in the elements reference of the finite element analysis program user's manual, (ANSYS® 2007).

The fine and course mesh regions can be seen in Figure 5. The fine mesh region was defined in the thinner plate domain, on the vicinity of the butt weld toe line, where the hot-spots are located. In the transverse direction, the extent of the fine mesh region is equal to the plate width. In the longitudinal direction, the extent was fixed proportional to the plate thickness, i.e. extent fixed equal to $\eta \cdot t_1$, where t_1 is the thinner plate thickness and η a constant to be specified. In the present study, this constant was assumed equal to 10 in order to allow an accurate calculation of the stress distribution on the vicinity of the weld toe line. The element size in the course and fine mesh regions was imposed directly by lines division using uniformly distributed division lengths along each line. This procedure allows an accurate control of the mesh size along the model domain. The mesh generated for the present study has an element edge length equal to t_1 (i.e. edge length in the three directions of the element's coordinate system) in the refined mesh region. In the coarse mesh region, the element edge length is t_1 along the thickness and transverse directions and approximately $5t_1$ along the longitudinal direction in order to reduce the number of degrees freedom of the model. In the thicker plate domain, the mesh was generated with the same parameters used for the coarse mesh region of the thinner plate. In the stiffeners and side plates domain a coarse mesh is sufficient and a reference edge length of $5t_1$ was adopted, as can be seen in Figure 6.

4.3 Loading and boundary conditions

The loading and boundary conditions were defined accordingly to Fricke et al. (2007). The boundary conditions are given in Table 3 and Figure 6. All nodal displacements at the model end sections and at the lower edges of side plates are constrained in the vertical direction with zero displacement conditions. The longitudinal nodal displacements at the right end sections

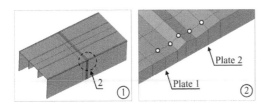

Figure 4. Finite element model geometry: 1) geometry of the structural detail; 2) idealized geometry of the butt weld profile.

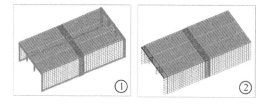

Figure 6. Finite element model loading and boundary conditions: 1) boundary conditions at the end sections and at the side plates lower edges; 2) longitudinal nodal forces at the left end section.

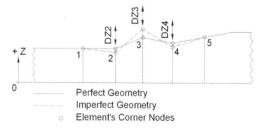

Figure 7. Longitudinal cross section of the butt-welded joint. Convention adopted for the reference nodes numbering and vertical position for perfect and imperfect geometries.

Table 3. Finite element model boundary conditions.

	UX*	UY*	UZ*
Left end section		s	A
Right end section	A	C	A
Side plates lower edge			A

* (C)—Only nodes at half-model width are constrained.
* (A)—All nodes are constrained.

are also constrained using the same condition. In order to avoid rigid body motions in the transverse direction, the displacement of the right end section nodes at half-model width are constrained in that direction imposing zero displacement.

The loading is a uniformly distributed longitudinal stress applied on the left end section of the model, as shown in Figure 6. This stress distribution is modelled first on the solid model surface as a pressure distribution with negative value. Afterwards, it is converted on equivalent nodal forces and applied at the finite element nodes. In order to simplify the calculation of the stress concentration factors the stress distribution was defined with unit magnitude. With this procedure the nominal stress distribution along the model has unit magnitude and, consequently, the structural hot-spot stresses are interpreted directly as stress concentration factors.

5 WELD SHAPE UNCERTAINTY ANALYSIS

5.1 *Monte-Carlo simulation of the weld shapes*

The weld shape imperfections are modelled as random deviations on the vertical positions of a mesh of equally spaced reference points positioned along the weld toe and along the weld face edge, as shown in Figure 7. These reference points are defined in the simulation as being the corner nodes of the finite element mesh on the weld toe and weld face edge lines. For convenience, these nodes are now one designated as reference nodes.

The convention adopted for the reference nodes numbering and vertical position, for perfect and imperfect geometries or weld shapes, are shown in Figure 7. This convention was used to derive the formulation that describes the vertical position of the reference nodes in the Monte-Carlo simulation of the weld shapes. As shown in Figure 7 the weld shapes simulation is performed considering that the reference nodes 1 and 5 are fixed, and that the positions of the reference nodes 3 to 4 describe the weld profiles along the butt weld for perfect and imperfect weld shapes. The vertical position of these reference nodes are given by,

$$Z_i^I = Z_i^P + Z_i^D \qquad (2)$$

with $i = 2$ to 4 the reference node index. According to Equation 2, the vertical coordinates for imperfect weld shapes, Z^I, are given by the vertical coordinates for perfect weld shapes, Z^P, affected by random vertical deviations, Z^D. For perfect weld shapes the vertical coordinates of the reference nodes can be easily obtained from the structural detail geometric parameters presented in Tables 1–2. It should be noted that the perfect weld shape is an idealization used in this study as a reference geometry for weld shapes simulation purposes, since it is not possible in practice to reproduce such geometry due to the imperfections induced by the welding process.

The vertical random deviations are given by a multivariate normal distribution. This probabilistic distribution is widely known and can be found in the literature about statistics or reliability, as for example, in Melchers (1999). This distribution is given by,

$$f_Z(Z, C_Z) = (2\pi)^{-N/2} |C_Z|^{-1/2}$$

$$\exp\left\{ -\frac{1}{2}(Z - \mu_Z)^T C_Z^{-1}(Z - \mu_Z) \right\} \qquad (3)$$

where N is the dimension of the distribution, Z the vector of random variables (i.e. considering $Z = Z^D$ for convenience), μ_Z the vector of mean values and

272

C_Z the covariance matrix of the random vector Z. The simulation was performed considering the dimension N equal to three, since the weld profiles are defined by the vertical coordinates of the three reference nodes in Figure 7, as already mentioned.

The random deviations Z^D are positive or negative increments in the vertical position, considering that they are measured from the perfect geometry and have zero mean value μ_Z. The covariance matrix C_Z can be written as a function of the correlations and standard deviations of the random variables, as known from statistics. The standard deviation of each random variable is defined by the coefficient of variation, considering as mean value the vertical coordinate in the perfect geometry,

$$\delta_i = \frac{\sigma_i}{\mu_i} = \frac{\sigma_i}{Z_i^P} \qquad (4)$$

with $i = 2$ to 4 the reference node index, or random variable index, δ_i the coefficients of variation and Z^P the vertical coordinates in the perfect geometry. The coefficients of variation are defined for each Monte-Carlo simulation using the same value for all random variables, i.e. $\delta_i = \delta$ with $i = 2$ to 4. This parameter is used as a global measure of the intensity of the weld shape imperfections. The vertical coordinates Z^P are known from the structural detail geometry, as already mentioned.

The correlations between the random variables are defined assuming that the random deviations in the vertical position of the reference node 2 and reference node 4 are independent, which means that $\rho_{24} = \rho_{42} = 0$. The correlations between the vertical deviations in the weld face node and weld toe nodes, described by the coefficients $\rho_{23} = \rho_{32}$ and $\rho_{34} = \rho_{43}$, may be high due to the welding process characteristics, i.e. correlations of 0.70 and even higher may have to be used in the simulation.

The random deviations are defined by the previous probability distribution with the proposed procedure to define the covariance matrix. However, generating random vectors according to a multivariate distribution is very often a difficult task. There are some methods in the literature that may be used to perform this kind of simulations, most of them frequently applied in reliability calculations, (Melchers 1999). In the simulation performed in this study, the random vectors were generated using a subroutine available in numerical libraries for statistical applications, (IMSL® 1997). This subroutine is called RNMVN and can be used to generate pseudorandom vectors from a multivariate normal distribution. The numerical procedure adopted in this subroutine can be found in the numerical library.

The Monte-Carlo simulation of the weld shapes is performed according to the previous procedure, in which a random vector is generated for each set of reference nodes. It is assumed that the random deviations along the butt weld direction are independent, i.e. the random deviations of a set of reference nodes with $y = y_1$ are independent of the random deviations of a set of reference nodes with $y = y_2$, where y is the transversal coordinate aligned with the butt weld and y_1 is different than y_2. This assumption is reasonable since the transversal distance between adjacent nodes is equal to the plate thickness.

The modelling of the imperfect weld shapes in the finite element analysis model is made according to the previous procedure, using the simulated random deviations to adjust the vertical position of the reference nodes according to Equation 2, as exemplified in Figure 7. The position of the finite elements midside nodes is also adjusted. The modified positions are given by the midpoint of the line segments connecting the adjusted corner nodes.

5.2 Weld shapes as a function of the quality levels

The simulation of the weld shapes can be performed with any reasonable value defined for the parameter δ. This parameter is used to define the intensity of the weld shape random imperfections; however, it does not permit by itself an easy interpretation of the weld shapes, since there is not any correlation between this parameter and geometrical parameters that can be used to describe the weld profile. Thus, it is important to quantify the imperfect weld shapes generated with a range of δ values by means of geometrical parameters usually applied in practice to characterize the weld profile. These parameters can be found in guidelines related to shipbuilding and repair quality standards, as for example, IACS (2006). According to this quality standard, there are three main parameters for this type of welded joint: 1) the weld face height, H; 2) the weld toe angle, θ; and, 3) the weld toe position, related to the weld undercut, D.

The three geometrical parameters were used to quantify statistically five random samples of imperfect weld shapes, each of it corresponding to a different value of the parameter δ. In the present study the values considered were all integers in the interval $[1, 5]$, which define five classes of weld shapes, designated from now on as quality levels for convenience, since they are related to different quality levels of the welding process.

For each quality level, 100 simulations of the weld shape were performed, obtaining for each simulation one random sample with 197 random vectors with coordinates of the weld profile reference nodes, since the number of nodes in the transversal direction of the finite element mesh is 197. For each quality level, the 100 random samples were combined and the three geometrical parameters $\{H; \theta; D\}$ calculated afterwards. The sample statistics of these three parameters for the five quality levels considered are presented next by means of histograms, fitted probability distributions and tolerance intervals.

Figure 8 shows some examples of weld shapes generated with the five quality levels considered, where can be seen the effect of the parameter δ on the weld shape imperfections. The perfect weld shape is presented for comparison purposes, since it is not representative of real weld shapes. As expected, the intensity of the weld shape imperfections increases with the parameter δ.

Figure 10 presents fitted probability distributions to the geometrical parameters sample data for the five quality levels considered. It was concluded that the random deviations in the three parameters are normally distributed with zero mean, as can be seen in Figure 9, where the histograms and the fitted distributions are shown for the quality level three as example. This conclusion was expected since the random deviations in the vertical position of the weld profile reference nodes are generated according to a normal distribution.

The parameters of the fitted distributions are given in Tables 4–6 for each quality level. The tolerance intervals presented for the three geometrical parameters were calculated considering 95% of the weld profile population and assuming the same percentage for the confidence level. These parameters are easily interpreted; however, the definition adopted for the weld toe angle should be clarified. This parameter was defined as the angle between the tangent to the plate

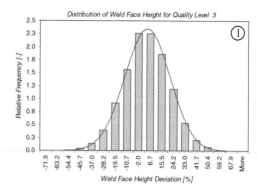

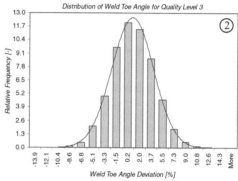

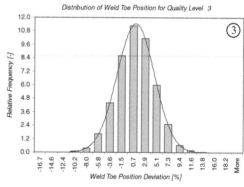

Figure 9. Histograms and fitted distributions of random deviations in the geometrical parameters with quality level three: 1) weld face height, H; 2) weld toe angle, θ; and, 3) weld toe position, related to the weld undercut, D.

surface and the tangent to the weld face, both evaluated at the weld toe. The deviations referred in the sample statistics are measured in relation to the perfect weld shape.

5.3 Effect of the weld shape uncertainty on the structural hot-spot stress

The study of the effect of the weld shape uncertainty on the structural hot-spot stresses was carried out using

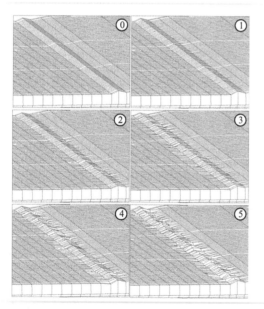

Figure 8. Weld shape examples for each quality level considered: 0) perfect case; 1) quality level 1; 2) quality level 2; 3) quality level 3; 4) quality level 4; and, 5) quality level 5.

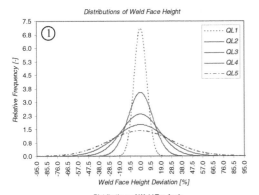

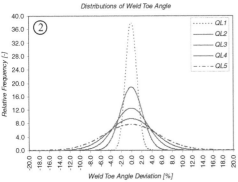

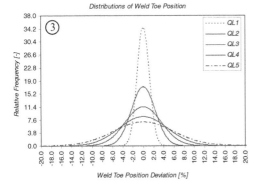

Figure 10. Distributions of random deviations as a function of the quality levels: 1) distributions weld face height, H; 2) distributions weld toe angle, θ; and, 3) distributions weld toe position, related to the weld undercut, D.

Table 4. Distribution parameters and tolerance intervals for weld face height deviation as a function of the quality levels.

Quality Level		1	2	3	4	5
μ	mm	0.000	0.000	0.000	0.000	0.000
σ	mm	0.281	0.561	0.847	1.130	1.411
H_-	mm	−0.6	−1.1	−1.7	−2.2	−2.8
	%	−11.1	−22.1	−33.3	−44.5	−55.5
H_+	mm	0.6	1.1	1.7	2.2	2.8
	%	11.1	22.1	33.3	44.5	55.5
ΔH	mm	1.1	2.2	3.3	4.4	5.5
	%	22.1	44.2	66.7	89.0	111.1

Tolerance interval for 95% of the weld profile population with 95% confidence.

Table 5. Distribution parameters and tolerance intervals for weld toe angle deviation as a function of the quality levels.

Quality Level		1	2	3	4	5
μ	Deg.	0.000	0.000	0.000	0.000	0.000
σ	Deg.	1.641	3.293	4.952	6.601	8.163
θ_-	Deg.	−3.2	−6.5	−9.7	−12.9	−16.0
	%	−2.1	−4.1	−6.2	−8.3	−10.3
θ_+	Deg.	3.2	6.5	9.7	12.9	16.0
	%	2.1	4.1	6.2	8.3	10.3
$\Delta\theta$	Deg.	6.4	12.9	19.4	25.9	32.0
	%	4.1	8.3	12.5	16.6	20.6

Tolerance interval for 95% of the weld profile population with 95% confidence.

Table 6. Distribution parameters and tolerance intervals for weld toe position deviation as a function of the quality levels.

Quality Level		1	2	3	4	5
μ	mm	0.000	0.000	0.000	0.000	0.000
σ	mm	0.232	0.463	0.697	0.934	1.157
D_-	mm	−0.5	−0.9	−1.4	−1.8	−2.3
	%	−2.3	−4.5	−6.8	−9.2	−11.3
D_+	mm	0.5	0.9	1.4	1.8	2.3
	%	2.3	4.5	6.8	9.2	11.3
ΔD	mm	0.9	1.8	2.7	3.7	4.5
	%	4.5	9.1	13.7	18.3	22.7

Tolerance interval for 95% of the weld profile population with 95% confidence.

the pre-processing and solution program. For each quality level considered 1000 simulations were performed, obtaining for each simulation the structural hot-spot stress distribution along the weld toe line. The post-processing program was used afterwards to obtain samples of maximum structural hotspot stress for each quality level.

For convenience, these stresses are presented as maximum stress concentration factors normalized by the value obtained for the perfect case, i.e. normalized by the maximum stress concentration factor that is obtained if the weld shape is considered perfect.

For this reference case the value obtained was $K_P = 1.613$, which is similar to the result presented in Fricke

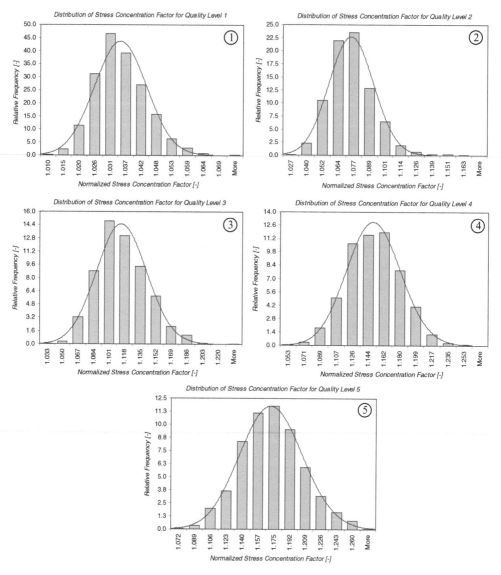

Figure 11. Histograms of the normalized maximum stress concentration factor and fitted distributions for each quality level considered: 1) quality level 1; 2) quality level 2; 3) quality level 3; 4) quality level 4; and, 5) quality level 5.

et al. (2007), where the same structural detail was studied. The sample statistics of this parameter for the five quality levels considered are presented next by means of histograms, fitted probability distributions and tolerance intervals.

Figure 11 presents histograms and fitted probability distributions to the simulation results for the five quality levels. The histograms show that the results obtained for the maximum stress concentration factor may be described by a log-normal probability distribution or, for some quality levels, by a normal probability distribution. The best distribution to describe the simulation results was selected using probability paper plots and Kolmogorov-Smirnov goodness-of-fit tests. These methods are widely used and can be found in classical texts, as for example, Ang & Tang (1975).

It was concluded that the log-normal distribution is more appropriate than the normal distribution, since it can describe the simulation results for the five quality levels considered with better results in the Kolmogorov-Smirnov goodness-of-fit test, where a 5% significance level was adopted. The parameters of

Table 7. Regression coefficients for the normalized maximum stress concentration factor mean values and tolerance limits.

	δ^2	δ	1	R^2
K_+/K_P	−2.34E-03	5.89E-02	1.00	9.97E-01
K_0/K_P	0.00	3.40E-02	1.00	9.96E-01
K_-/K_P	9.22E-04	1.56E-02	1.00	9.95E-01

Normalization factor, $K_P = 1.613$.

Table 8. Distribution parameters and tolerance intervals for the normalized maximum stress concentration factor as a function of the quality levels.

Quality Level		1	2	3	4	5
λ	–		0.032	0.066	0.101	0.132	0.151
ξ	–		0.009	0.016	0.025	0.027	0.029
K	–		1.015	1.035	1.053	1.082	1.099
	%		−1.7	−3.2	−4.8	−5.2	−5.6
K_+	–		1.050	1.104	1.161	1.203	1.232
	%		1.7	3.3	5.0	5.4	5.8
ΔK	–		0.036	0.069	0.108	0.121	0.133
	%		3.5	6.4	9.8	10.6	11.4

Tolerance interval for 95% of the weld profile population with 95% confidence.

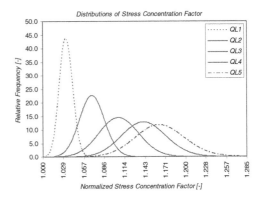

Figure 12. Distributions of the normalized maximum stress concentration factor as a function of the quality levels.

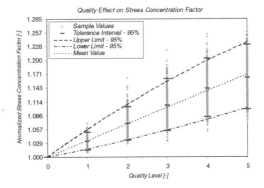

Figure 13. Effect of the weld shape quality levels on the normalized maximum stress concentration factor.

the fitted log-normal distributions are given in Table 8. Figure 12 shows the fitted probability density functions for the five quality levels. The effect of the weld shape imperfections on the maximum stress concentration factor is given in Figure 13. This figure shows the sample values of normalized maximum stress concentration factor obtained with the simulations performed. The tolerance intervals were calculated considering 95% of the weld profile population and assuming the same percentage for the confidence level. The tolerance limits are given in Table 8 as a function of the quality level parameter, in absolute values as well as in relation to the mean values.

Simple linear and quadratic regression analyses were carried out in order to find appropriate polynomial functions that may be used to define the mean value and the tolerance limits as a function of the quality level parameter. Based on these analyses, it was concluded that these variables increase with these quality level parameter. The increase is approximately linear for the mean value and approximately quadratic for the upper and lower tolerance limits, as can be seen in Figure 13. The distance between the upper and lower tolerance limits tends to increase due to the increase in the simulation results variance, which reflects the effect of the weld shape random imperfections variance on the maximum structural hot-spot stress.

The regression coefficients obtained for each case are given in Table 7 as well as the coefficients of determination. In this table the quality level parameter is denoted by δ and the regression coefficients associated with the quadratic, linear and constant terms of the polynomials are identified by the order of this parameter in the first line of the table.

6 CONCLUSIONS

It has been recognized nowadays that the fatigue assessment methods commonly used in the design of welded structures and components still have uncertainties in important variables that are not usually quantified by means of numerical tools that may be efficiently used for design purposes.

In this study an attempt has been made to quantify the effect of the weld shape uncertainties in the structural hot-spot stress distribution. A simulation procedure has been proposed to address this problem, using the structural hot-spot stress approach combined with the finite element analysis and the Monte-Carlo simulation method.

The weld shape uncertainty is related to the geometrical imperfections induced by the welding processes commonly used in the shipbuilding and repair industries. These imperfections were simulated by means of weld profile parameters included in the Monte-Carlo simulation as random variables.

It can be concluded that it is possible to quantify the effect of the weld shape uncertainty in the structural hot-spot stress distribution using a simulation procedure that is computationally efficient. The results of the simulation performed with the case study have been shown that the maximum stress concentration factor increases considerably if the weld shape uncertainty is taken into account. As a numerical example, if a weld shape with moderate random imperfections is assumed (i.e. quality level three in the present study), and the 95% upper limit polynomial is considered to describe the simulation results, the increase in the maximum stress concentration factor may be 15.6% comparatively with the value obtained if a perfect weld shape is considered.

It can also be concluded that the proposed procedure may be applied to study the effect of the weld shape uncertainty in different welded joint configurations. This flexibility may be used to derive parametric formulae for the stress concentration factor of different classes of welded joints, taking into account the weld shape uncertainty and adequate safety margins for fatigue assessment.

ACKNOWLEDGMENTS

This work has been performed within the project "MARSTRUCT"—Network of Excellence on Marine Structures" (http://www.mar.ist.utl.pt/marstruct/) and has been partially funded by the European Union through the Growth programme under contract TNE3-CT-2003-506141.

The first author has been funded by the Portuguese Foundation for Science and Technology (Fundação para a Ciência e Tecnologia—FCT) under contract SFRH/BD/39106/2007.

REFERENCES

Ang, A.H.-S. & Tang, W.H. 1975. *Probability Concepts in Engineering Planning and Design. Volume 1. Basic Principles*. New York: John Wiley & Sons Ltd.

ANSYS®. 2007. *Release 11.0 Documentation for ANSYS*. United States of America: ANSYS Inc.

Chakarov, K., Garbatov, Y. & Guedes Soares, C. 2008. Fatigue analysis of ship deck structures accounting for imperfections. *International Journal of Fatigue* 30: 1881–1897.

DNV. 2001. *Hull Structural Design. Ships with 100 Meters and Above. Rules for Classification of Ships*. Hovik: Det Norske Veritas.

Fricke, W., Bollero, A., Chirica, I., Garbatov, Y., Jancart, F., Kahl, A., Remes, H., Rizzo, C.M., von Selle, H., Urban, A. & Wei, L. 2007. Round Robin study on structural hot-spot and effective notch stress analysis. In C. Guedes Soares & P.K. Das (ed.). *Advancements in Marine Structures*. London: Taylor & Francis.

Garbatov, Y., Guedes Soares, C., OK, D., Pu, Y., Rizzo, C.M., Rizzoto, E., Rouhan, A. & Parmentier, G. 2006. Modelling strength degradation phenomena and inspections used for reliability assessment based on maintenance planning. *Proceedings of the 23rd International Conference on Offshore Mechanics and Arctic Engineering*. Paper OMAE2006-92090: ASME.

IACS. 2006. *No.47 Shipbuilding and Repair Quality Standard. Revised Document 3*. London: International Association of Classification Societies Ltd.

IMSL®. 1997. *Fortran Subroutines for Statistical Applications. Statistical Library. Volume 2*. United States of America: Visual Numerics Inc.

Melchers, R.E. 1999. *Structural Reliability Analysis and Prediction*. Sussex: John Wiley & Sons Ltd.

Niemi, E., Fricke, W. & Maddox, S.J. 2004. *Structural Hot-Spot Stress Approach to Fatigue Analysis of Welded Components: Designer's Guide. IIW-Document XIII-1819-00/XV-1090-01*. Finland: International Institute of Welding.

Analysis and Design of Marine Structures – Guedes Soares & Das (eds)
© 2009 Taylor & Francis Group, London, ISBN 978-0-415-54934-9

A study on a method for maintenance of ship structures considering remaining life benefit

Y. Kawamura & Y. Sumi
Faculty of Engineering, Yokohama National University, Yokohama, Japan

M. Nishimoto
Graduate School of Engineering, Yokohama National University, Yokohama, Japan

ABSTRACT: The failure of ship structures usually causes serious disaster such as human loss or oil spill from tankers that may cause catastrophic environmental destruction by the pollution of the sea. To avoid such casualties and disasters, it is necessary to make a proper maintenance plan for the ship by rationally evaluating the condition of the ship structures. However at present, the strategy of maintenance of a ship structures is different by different ship-owners. Some ship-owners may well maintain the ship structures and operate it over a long period. Whereas other ship owners may leave a ship in bad structural condition without any repair action in order to reduce the maintenance cost from the economical viewpoint. From the rational point of view, the strategy of the maintenance of a ship must be decided by considering both the allowable level of the safety and the cost of the maintenance. To overcome this problem, we propose a new method for rational decision-making of a maintenance plan for a target ship. In the proposed method, the Remaining Life Benefit (RLB) is computed considering the survey results and the risk of failure of the ship. Then, a proper maintenance plan can be selected by maximizing the RLB. In this paper, formulation of the RLB and evaluation of the cost and the risk is firstly presented. Secondly, calculation of the risk of failure based on the structural reliability analysis for hull girder strength is described in which deterioration of the structure by corrosion and fatigue is accounted for. Finally, an example of the calculated RLB of a bulk carrier for different maintenance plans is displayed to show the validity of the proposed method.

1 INTRODUCTION

The failure of ship structures usually causes serious disaster such as human loss or oil spill from tankers that may cause catastrophic environmental destruction by the pollution of the sea. One of the major causes of such accidents is poor maintenance strategy of ship structures during its service period. Usually, the maintenance schedule of a ship is decided by the ship owner. Though some ship owners may use their ships for a long period by carrying out good maintenance of structures, some ship owners may intentionally reduce the cost of the maintenance considering the short-term benefit. As a result, a number of substandard ships may increase which may cause serious accidents.

If the cost of maintenance of the structures is reduced, the ship hull may become more damageable than well-maintained ships. And if the serious failure occurs in the ship structure, expense for repair of the ship is necessary. Moreover, if a serious disaster such as human loss or catastrophic environmental destruction occurs caused by the accident, the ship owner may have to compensate for that accident. In that case, it is possible to say that the total operational cost with the reduction of the maintenance cost and with the serious accident is much greater than the operational cost with good maintenance without any accidents. If the ship owner wants to reduce the true operational cost, the life cycle cost, including not only the operational cost and the maintenance cost but also the risk of the accident, should be evaluated.

From the above reasons, we propose a new method for maintenance of ship structures considering remaining life benefit (RLB) and risk of the accident. In the proposed concept, the estimation method of RLB is formulated considering the operational cost, the insurance cost, the maintenance cost, the revenue and the risk of failure. Then availability of the formulation is shown with a calculation example of the RLB for a bulk carrier operated on the shipping route of North Atlantic Ocean.

2 METHOD FOR MAINTENANCE OF SHIP STRUCTURES BASED ON THE REMAINING LIFE BENEFIT

2.1 Design of structures considering life cycle cost

The concept of design considering the life cycle cost (LCC) has attracted attention in recent years. For example in ship structural design, the relationship between the probability of failure (p_f) and each component of LCC can be generally represented as shown in Figure 1. In this figure, the initial cost (construction cost, C_I) is high if the probability of failure (p_f) is low, because a lot of steel is necessary to construct a ship with high strength. And the initial cost becomes lower as the probability of failure becomes higher. On the other hand, the maintenance cost (C_M) and the risk of failure (C_{RISK}) may be low if the probability of failure is low, and may become higher if the probability of failure becomes higher. Here, the risk of failure (C_{RISK}) can be represented by the product of the probability of failure (p_f) and the expected value of the amount of damage (C_F).

$$C_{RISK} = p_f \times C_F \qquad (1)$$

As shown in Figure 1, the life cycle cost (LCC) can be regarded as the sum of these costs. And the optimal structural design (optimal p_f) can be decided by minimizing the LCC in the structural design considering the lifecycle cost (Melchers 1999, ISSC 2006).

2.2 Concept of remaining life benefit

Based on the above concept of structural design considering LCC, we propose a new method for maintenance of ship structures considering remaining life benefit (RLB). In this method, a maintenance plan of a certain ship can be decided by predicting the life cycle revenue and the cost, considering the result of inspection of the ship structure. An equation for estimation of RLB is defined as follows.

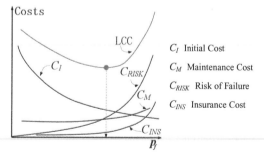

Figure 1. Concept of structural design considering life cycle cost.

$$C_{RL}(P_L) = \sum_{i=t_{cur}}^{t_{retire}} \left[\left\{ \prod_{j=t_{cur}}^{i-1} (1 - p_{fj}) \right\} C_{ALLi}(P_L) \right] \qquad (2)$$

$$C_{ALLi}(P_L) = C_{OPEi}(P_L) + C_{INSi} + C_{Mi}(P_L) + C_{RISKi}(P_L) \qquad (3)$$

$$R_{RL}(P_L) = \sum_{i=t_{cur}}^{t_{retire}} \left[\left\{ \prod_{j=t_{cur}}^{i-1} (1 - p_{fj}) \right\} R_{OPEi}(P_L) \right]$$
$$+ \prod_{j=t_{cur}}^{t_{retire}} (1 - p_{fj}) R_{RE}(P_L) \qquad (4)$$

$$RLB = R_{RL}(P_L) - C_{RL}(P_L) \qquad (5)$$

In these equations, t_{cur} is the current ship-year when the estimation of RLB is carried out. It is assumed that the target ship will be used from t_{cur} to t_{retire} and the ship will be scrapped after t_{retire}. In the equation (2), the remaining life cost (C_{RL}) is defined as sum of the necessary annual expense of i-th ship-year (C_{ALLi}) from the current ship-year (t_{cur}) to the end year (t_{retire}) just before the scrapping. As annual expense (C_{ALLi}), operational cost (C_{OPEi}), insurance cost (C_{INSi}), maintenance cost (C_{Mi}) and the risk of failure (C_{RISKi}) are considered as indicated in the equation (3). The reason why these costs are evaluated year by year is that the risk of failure (C_{RISKi}) changes every year because of structural deterioration such as corrosion and fatigue cracking, and that such structural deterioration usually affects to the maintenance plan and the various costs. In the above equations, p_{fj} shows the probability of failure with total loss at j-th ship-year and can be computed by using structural reliability analysis for the ship structure. In the equations (2) and (3), $\prod(1 - p_{fj})$ means the probability that the ship does not meet total failure before the i-th ship-year. If the ship gets accident of total loss, the costs and the risk will not occur any more after the accident.

Generally, the maintenance plan of the ship structure affects not only the remaining life cost but also the remaining life revenue. For this reason, we propose an equation of the remaining life benefit (RLB) shown in the equation (5), in which the RLB is defined as the difference of the remaining life revenue (R_{RL}) and the remaining life cost (C_{RL}). For the remaining life revenue (R_{RL}), operational revenue for each ship-year (R_{OPEi}) and selling benefit (R_{RE}) of the scrapped ship are considered (Equation (4)).

As shown in the equation (5), the remaining life benefit (RLB) can be regarded as a function of the maintenance plan (P_L) which should be decided by the ship owner. In other words, by using the proposed equations, it becomes possible that the optimal maintenance plan with low cost considering the risk of failure can be selected by maximizing the RLB estimated by the equations from (2) to (5).

3 ESTIMATION OF ANNUAL COSTS AND REVENUE

In this paper, RLB for a Bulk Carrier on North Atlantic Ocean route is computed in order to select a proper maintenance plan for the Bulk Carrier. The costs and revenue indicated in equations (2)–(5) are estimated by the following procedures.

3.1 Estimation of operational cost (C_{OPEi}), operational revenue (R_{OPEi})

One of the criteria that is generally used to estimate reliability or maintainability of a system is the availability (AVL). In this study, the AVL for operation of a ship is defined as the rate of a number of operating days in a certain ship-year by the following equation,

$$AVL = 1 - \frac{d_M}{365}, \qquad (6)$$

where d_M is a number of days for repair of the ship in the ship-year. Then, the number of navigation days in the ship-year, N_R, can be defined as follows.

$$N_R = \frac{f \times V \cdot 24}{drp} \cdot 365 \times AVL \qquad (7)$$

In this equation, f is the rate of sailing days (assumed as 0.7), V is the velocity of the ship, and drp is the round-trip distance of the shipping route.

By using the above definitions and by using the method in Kuniyasu et al. (1968), we formulate the equations in Table 1 to estimate the annual operational cost of a bulk carrier. That is, the operational cost (C_{OPEi}) is defined as sum of the costs for port dues (C_1), the cost for fuel oil (C_2), the cost for lubricants (C_3), the crew cost (C_4), the cost of stores (C_5) and the other miscellaneous cost (C_6). The factors (unit values) shown in these equations are estimated based on the statistical data (Opcost2004) and etc.

Also, we estimate the operational revenue (R_{OPEi}) by using the following equation,

$$R_{OPE_i} = DWT \times R_C \times N_R \qquad (8)$$

where R_C is the charter rate (the income by the cargo per weight) estimated by the data of charter fee for the bulk carriers in the route of North Atlantic Ocean (NYK LINE 2006). It is noted that as the value of the cargo becomes higher, the difference of the maintenance plan affects more to the remaining life benefit (RLB), because the total revenue indicated in Table 1 is affected by the AVL.

3.2 Estimation of maintenance cost (C_{Mi})

It is necessary for bulk carriers to have close-up survey in the periodical inspections by a classification society based on the regulations by IMO. For these periodical inspections, the cost of inspection is necessary. And if damages are detected by the inspection, the repair cost is also necessary. Generally, method for repair is decided based on the kind of the damage and the degree of the damage. For example, the plate damaged by fatigue cracks might be replaced to a new plate. If a thickness of a plate is decreased by corrosion, it might be replaced to a new plate or a new structural member such as a stiffener might be added to strengthen the structure. Also, if the decrease of the thickness of the plate by corrosion is relatively small and the coating condition is poor, re-coating is often carried out as the repair work. In this paper, the fatigue crack and the uniform corrosion are considered as deterioration of the ship structure, and the replacement of the plate and the re-coating are considered as the repair work. Then the maintenance cost (C_{Mi}) and the number of days for repair (in i-th ship-year) can be estimated by the equations,

$$C_{Mi} = C_0 + C_r w + C_c A, \qquad (9)$$

$$d_M = d_0 + \frac{w}{w_{dm}} + \frac{A}{A_{dm}}, \qquad (10)$$

Table 1. Operating cost and estimation ((Japanese) yen*).

Total operational cost	$C_{OPE} = \sum_{i=1}^{6} C_i$
Port dues	$C_1 = 2 \cdot N_R C_{p1} + day_P C_{p2}$ C_{P1}: Port due per anchorage C_{p2}: Port due per day day_P: Number of days in harbor per year
Fuel cost	$C_2 = C_c (who \times day_n \times N_R + whp \times day_P)$ C_c: Unit price who: Consumption per day day_n: Number of days per round trip whp: Consumption per day (anchorage)
Lubricants cost	$C_3 = C_{oil} \cdot sfo \cdot day_n \cdot N_R$ C_{oil}: unit price sfo: consumption per day
Crew cost	$C_4 = p \times C_w \times AVL$ p: Number of crews C_w: crew salary
Stores other	$C_5 = 28.993 \cdot DWT + 9 \times 10^6$
Other cost	C_6 other cost per year

* 1 US dollar is about 120 Japanese yen when the study is carried out.

where w is the weight of plates for replacement and A is the area for re-coating. In the equation (9), C_0 is the cost for inspection and the cost for repairing engine and rigs, C_r is the cost for replacement of plates per weight, and C_c is the cost of re-coating per unit area. Also in the equation (10), d_0 is the number of days for repairing engine and rigs, w_{dm} is the weight of the plate for replacement per day, and A_{dm} is the area of re-coating per day. Usually, these factors may be different if the location of the repair dock is different. In this study, we set up these factors based on the interview to a repairing yard in Japan.

In the above equations for estimation of the maintenance cost, the cost for repair not for the structure, C_0, is assumed to be constant in the life time of the ship. However, if the high-valued ship such as a LNG carrier is used for a long period, the cost (C_0) should increase as the ship ages because replacement of fitting equipments is usually necessary for aged ships (Yuasa et al. 2004).

3.3 Estimation of insurance cost (C_{INSi})

Generally, in order to decide the insurance cost of a ship, constant insurance rate is considered for total loss of the ship, while insurance rate for partial loss is decided proportional to the tonnage of the ship. Here in this study, it is assumed that the insurance cost is represented by linear polynomial with respect to the deadweight capacity (DWT) of the ship. And the coefficient is estimated based on the statistical data in the reference (Opcost 2004).

$$C_{INSi} = 84.991 \times DWT + 10^7 \text{ [Japaneseyen]} \quad (11)$$

3.4 Estimation of profit on sale of the ship (B_{RE})

As a gain of selling the ship, it is possible to consider selling the ship as a second hand ship, or selling the ship as scrap. In this study, it is assumed that the ship is sold as scrap after the retirement at 30 ship-years. Then the profit on sale of the ship can be estimated by the following equation.

$$B_{RE} = C_{os} \times W_{st} \text{ [Japaneseyen]} \quad (12)$$

where (C_{OS}) is unit price of the scrap steel (per ton) and W_{st} is the total weight of the steel of the discarded ship.

4 ESTIMATION OF THE RISK OF FAILURE (C_{RISK}) BY USING STRUCTURAL RELIABILITY ANALYSIS

In this study, the risk of failure of the ship (C_{RISKi}) at i-th ship-year is estimated as the product of probability of failure (p_f) and the expected amount of loss (C_F) as shown in the equation (1).

4.1 Probability of failure of the ship (p_f)

One of the methods to estimate the probability of failure of a structure is structural reliability analysis. In the field of naval architecture and ocean engineering, many studies about structural reliability for longitudinal strength of ships are carried out recently. Mansour et al. computed the reliability index by proposing limit state functions for the longitudinal strength with estimation of the ultimate strength and the wave load (Mansour et al. 1997a,b, Mansour & Hoven 1994). Guedes Soares & Garbatov (1997, 1998), Akpan et al. (2002) and Sun et al. (2003) carried out structural reliability analysis for the longitudinal strength considering deterioration by fatigue and corrosion. In this study, based on these concepts, we calculate the failure probability for the longitudinal strength of the ship, p_f^{hul}, and estimate the total failure probability (p_f) of the ship.

4.1.1 Definition of limit state function

When structural reliability analysis is carried out, it is necessary to define a limit state function ($g(t)$) to compute the failure probability. In this study, we define the following limit state function,

$$g(t) = M_U - M_L, \quad (13)$$

where M_U is the longitudinal strength of the ship, and M_L is the load (longitudinal bending moment). This limit state function is represented by multiple probability variables. If this function takes negative value ($g(t) < 0$), the ship is regarded as in failure condition. By using this limit state function, the failure probability per year related to the longitudinal strength is computed for each ship-year.

4.1.2 Modelling of load (M_L)

In this study, the still-water bending moment (M_{SW}) and the wave bending moment (M_W) are considered as the load for the longitudinal strength of the ship.

$$M_L = x_{sw}M_{SW} + x_w M_W \quad (14)$$

In this equation, x_{sw} and x_w are the probability variables which represent uncertainty of modeling (Mansour & Hoven 1994, Akpan et al. 2002, Sun & Bai 2003, Guedes Soares & Moan 1991).

1) *Stillwater bending moment (M_{SW})*
Generally, the still-water bending moment depends on the deadweight and the cargo weight of the ship so that it varies for each voyage. For this reason, the still-water bending moment (M_{SW}) should be treated as a probability variable. In the rule of classification societies,

the still-water bending moment for design criteria is estimated by the following equation.

$$M_{sw,max} = f C_w L^2 B (C_b + 0.7) \qquad (15)$$

In this equation, L is the length of the ship, B is the breadth, C_b is the block coefficient, C_w is the coefficient depending on the length of the ship, and the coefficient f is 0.072 for sagging. This still-water bending moment can be regarded as the maximum bending moment for the ship to undergo during its whole life. Because the limit state function should be evaluated for every ship year, the still-water bending moment should be evaluated as the maximum value in one year. In this study, based on the statistical investigation about still-water bending moment (Guedes Soares & Moan 1988) and the concept of the combination factor (Ψ_S) (ISSC 1991), we evaluate the still-water bending moment (M_{SW}) by the following equation.

$$M_{SW} = \Psi_S M_{sw,max} \qquad (16)$$

2) *Wave bending moment (M_W)*
In order to predict the wave bending moment applied to ship structures, short-term prediction and long-term prediction are generally carried out (Owen 1988). In the short-term prediction, ship motion and structural response in regular waves (frequency response function) are firstly obtained by using the numerical method such as the strip method. Then, the response spectrum for short-term sea condition can be estimated by using the frequency response function and the wave spectrum. Then, the maximum structural response during the service life of the ship can be estimated by using the long-term statistical data of wave for the sea areas. By using these methods, it is possible to predict the load applied to the structure for the specific route of the ship as a probabilistic variable. However in this study, we assume for simplicity that the target ship operates on North Atlantic Ocean route, and that the following design wave load (M_w^d) can be used to estimate the probability distribution of the wave bending moment (M_W).

$$M_w^d = \begin{cases} -0.11 C_w L^2 B (C_b + 0.7) \text{ (kNm) } \textit{for sagging} \\ \\ 0.19 C_w L^2 B C_b \text{ (kNm) } \textit{ for hogging} \end{cases} \qquad (17)$$

It is known that the wave bending moment obtained by the long-term prediction generally conforms to the

following Weibull distribution function,

$$P_{MW}(x) = 1 - \exp\left[-\frac{x}{\beta}\right]^{h_w}, \qquad (18)$$

where h_w and β are the shape parameter and the scale parameter respectively. If we assume that the design load defined by IACS (the equation (17)) corresponds to the load of 10^{-8} level, the scale parameter β can be obtained as follows.

$$\beta = \frac{M_w^d}{\ln 10^8} \qquad (19)$$

Also in the equation (18), we assume that the shape parameter h_w is equal to 1.0. Next, in order to estimate the remaining life benefit (RLB) for each ship-year, probability density function of maximal wave bending moment in a certain ship-year is computed as follows.

$$f_{Mw}(x) = p_{MW}(x) \nu_0 [P_{MW}(x)]^{\nu_0 - 1} \qquad (20)$$

In this equation, $p_{MW}(x)$ is the probability density function for the long-term wave bending moment defined in the equation (18), and ν_0 is the number of waves that the ship meets in one year.

4.1.3 *Estimation of strength (M_U)*
1) *Estimation of ultimate strength*
For estimation of the ultimate longitudinal strength of ships, numerical methods such as FEM and ISUM, or simplified formulas are studied in recent years (Yao 2003). In this study, it is necessary to estimate the ultimate strength considering the deterioration of structures by aging. For this reason, we adopt the simple formula in which the strength is represented by the product of the section modulus ($Z(t)$) and the yield strength of the material (σ_Y) as follows.

$$M_U = x_u \phi \sigma_Y Z(t) \qquad (21)$$

In this equation, x_u is the probability variable that represents model uncertainty of the ultimate strength. And ϕ is estimated as the ratio between the initial yield moment and the ultimate strength for the intact ship-hull obtained by the modified Caldwell's formula (Caldwell 1965, Paik & Mansour 1995).

2) *Effect of structural deterioration*
In this study, uniform corrosion and fatigue cracking is considered as the effect of structural deterioration by aging of the ship. As shown in the deterioration model in Figure 2, it is assumed that the thickness of the plate is decreased as the ship ages, and the width or the height of the plate reduces as the fatigue crack is extended, so that the section modulus ($Z(t)$) decreases gradually by aging.

283

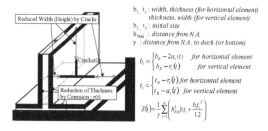

b_i, t_i : width, thickness (for horizontal element)
thickness, width (for vertical element)
b_i, t_i : initial size
h_{NAi} : distance from N.A.
y : distance from N.A. to deck (or bottom)

$$b_i = \begin{cases} b_{Ii} - 2a_i(t) & \text{for horizontal element} \\ b_{Ii} - r_i(t) & \text{for vertical element} \end{cases}$$

$$t_i = \begin{cases} t_{Ii} - r_i(t) & \text{for horizontal element} \\ t_{Ii} - a_i(t) & \text{for vertical element} \end{cases}$$

$$Z(t) = \frac{1}{y} \sum_{i=1}^{n} \left(h_{NAi}^2 b_i t_i + \frac{b_i t_i^3}{12} \right)$$

Figure 2. Deterioration of plate structures considering the corrosion and the fatigue crack.

For the estimation of corrosion wastage by aging, many corrosion wastage models are proposed by many researchers. Yamamoto & Ikegami (1998) proposed the corrosion model which accounts for the life of coating. Also, Paik et al. (2004) considered the life of coating and carried out the statistical investigation of structures of bulk carriers to propose the following formula for the estimation of corrosion wastage (thickness reduction).

$$r(t) = \begin{cases} 0 & t \leq t_1 \\ C_1 (t - t_C)^{C_2} & t > t_1 \end{cases} \tag{22}$$

In this equation, t_c is the life of coating defined as a probability variable (normal distribution), and t is the ship-year, C_1 is the probability variable assumed as Weibull distribution that represents the corrosion rate, and C_2 is assumed to be constant variable. In the reference (Paik et al. 2004), the values of parameters in the probability distribution of C_1 are decided from the statistical investigation for some locations in the bulk carrier. In this paper, this probabilistic model is used assuming that $C_2 = 1$ in order to estimate the corrosion wastage for structural members.

For a fatigue crack propagation model, fracture mechanics based on the Paris's law is used to estimate the crack length, $a(t)$, by using the equation (20).

$$a(N) = \left[a_0^{1-m/2} + \left(1 - \frac{m}{2}\right) C \Delta \sigma^m Y^m \pi^{m/2} v_0 t \right] \\ \times 1/1 - \frac{m}{2} \tag{23}$$

In this equation, N is the number of load cycles, and C and m are the material constants. Also, $\Delta \sigma$ is the stress range, Y is the geometric constant and a_0 is the initial crack length. As shown in Figure 2, crack propagation estimated by the equation (23) affects the section modulus by using this model (Guedes Soares and Garbatov 1998, Akpan et al. 2002, Sun & Bai 2003). We think that this model is not strictly proper from a physical viewpoint. However, we adopt this

model because it qualitatively represents the effect of deterioration of the strength of the ship structure.

By representing the load and the strength by probability variables, the limit state function, $g(t)$, can be evaluated. In this study, First Order Reliability Method (FORM) with the iterative numerical algorithm is used to calculate the reliability index, β Ditlevesen & Madsen 1996). Then, the failure probability, p_f^{hull}, for the limit state function can be approximately derived by using the standard normal distribution function, Φ, as follows.

$$P_f^{hull} = \Phi(-\beta) \tag{24}$$

4.2 Risk of failure of the ship (C_{RISKi})

In the previous section, we show the method to calculate the failure probability for the failure mode related to the longitudinal strength. However, in the case of a ship in service, the cause of the accident is not only the shortage of the longitudinal strength, but also the failure of the local structure and subsequent flooding, or the human error and etc. In this study, we assume that the total failure probability (p_f) is proportional to the failure probability for the longitudinal strength (p_f^{hull}) as follows.

$$p_f = k \times p_f^{hull}, \tag{25}$$

where we define the factor, k, as the system failure factor. In the Formal Safety Assessment (FSA) for bulk carriers (IMO 2001), scenarios of the accident are analyzed by using ETA and FTA with the statistical database of accidents in 20 years. By referring the results of the analysis, we assume $k = 30.0$ as the rate between the total failure probability and the failure probability for the longitudinal strength. After the failure probability (p_f) is computed, the risk of the accident can be estimated by using the equation (1). For the expected amount of damages (C_F), it is necessary to account for not only the loss of the cargo and the ship, but also human loss or the compensation of environmental destruction. In this study, the estimation of amount of damages for an accident of a bulk carrier (IMO 2001) is used. Also we account for the loss of human lives where we assume that the amount of the compensation is 300 million Japanese-yen per person.

5 COMPUTATION OF RLB (REMAINING LIFE BENEFIT) FOR A BULK CARRIER

In order to study the availability of RLB, we computed the RLB for a bulk carrier that sails on North

Atlantic Ocean route for four different maintenance plans. Here, it is assumed that maintenance plan is decided just after the construction of the ship ($t_{cur} = 0$), and that the ship will be scraped after 30 years operation ($t_{retire} = 30$).

5.1 Target ship

The specification of the target bulk carrier and the values of factors used for calculation of RLB are shown in Figure 3. To decide an optimal maintenance plan for the bulk carrier, we consider four candidate plans shown in Table 2 In these plans, fatigue cracks are repaired (plates with cracks are replaced) if the crack length is more than 10 mm. The interval of the period of inspection is assumed to be 2 or 4 years. The re-coating for corrosion prevention is not carried out in

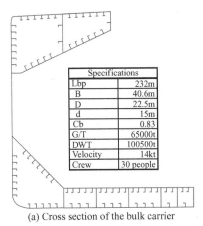

(a) Cross section of the bulk carrier

Items			Factor	Value (Formula)	Unit
			R_c	1,920	[yen/(ton · voyage)]
	Port		C_{p1}	13.2×G/T	[yen/entry into port]
			C_{p2}	13.4×G/T+14000yen×2	4[yen/day]
	Fuel		who	39.254	[ton/day]
			whp	0.297	[ton/day]
Operational Cost			C_{co}	46,800	[yen/ton]
	Lubricants		Coil	235,500	[yen/ton]
			sfo	1.2×10a^{-6}	[ton/PSh]
	Crew		C_w	180,000	[yen]
	Other		C_{fi}	35,913,681	[yen]
	Cost		C_0	600,000,000	[yen]
			C_r	1000,000	[yen/ton]
Maintenance Cost			C_c	30,000	[yen/m²]
	Days		d_o	10	[day]
			W_{dm}	5	[ton/day]
			A_{dm}	3000	[m²/day]
Selling Benefit			C_{os}	24,000	[yen]
			Wst	19,701	[ton]
Risk			Cf	7,467,900,000	[yen]

(b) Values of the factors used in the calculation*

Figure 3. Cross section of the bulk carrier and values of factors for computation.
Source: * [yen] = [Japaneseyen]: 1 US dollar is about 120 japanese yen when this study is carried out.

Table 2. Maintenance plans for evaluation of the RLB.

Plan	Period of inspection	Crack size for repair	Re-coating action
Plan1	every 4 years	more than 10 mm	no
Plan2	every 4 years	more than 10 mm	every 8 years
Plan3	every 4 years	more than 10 mm	every 12 years
Plan4	every 2 years	more than 10 mm	every 8 years

the Plan1, but carried out every 8 years in the Plan2 and Plan4, or every 12 years in Plan3.

5.2 Structural reliability analysis

For each maintenance plan, structural reliability analysis is carried out considering the structural deterioration by aging. Probability distribution, mean and coefficient of variation of the probability variables are shown in Table 3 and Table 4, some of which are decided based on the data in the references (Akpan et al. 2002, Sun & Bai 2003, ISSC 1991). For the probability variables indicated in the equation (22), different corrosion rates are used for different structural members based on the Paik's investigation (Paik et al. 2003), and the mean value of the life of coating (t_c) is assumed to be 7.5 years for most of the structural members. Next, about estimation of fatigue cracks by the equation (23), it is assumed that cracks initiate at the intersection between the longitudinal stiffeners and the bottom, deck or side plates at mid-ship. And also it is assumed that the amount of repair (amount of replaced plates) at these locations is representative of the amount of repair for the whole ship. Though this assumption in the repair model seems to be very crude, it is possible to say that the model can qualitatively represents the effect of structural deterioration by fatigue cracks.

Figure 4 shows the change of the failure probability, p_f^{hull}, by aging for four maintenance plans (Plan1–4). As shown in this figure, the probability of failure is reduced by the repair actions at the periodical inspections. It is noted that the probabilities of failure are very different at 30 ship-year for different maintenance plans. It can be said that the probability of failure becomes smaller as the maintenance and repair plan becomes careful (Plan4 < Plan3 < Plan2 < Plan1).

Next, the risk of failure is evaluated based on the obtained failure probability (p_f), and the remaining life benefit (RLB) is calculated by using the equations (2)–(5) as shown in Figure 5. It is noted that RLB for plans with re-coating (Plan2, Plan3, Plan4) is much more greater than that for the Plan1 without re-coating. It can be concluded that the maintenance plan with re-coating is advantageous from an economic viewpoint. Also as shown in Figure 5, the maintenance plan with the highest RLB is not the most elaborate plan (Plan4), but is Plan2. This is because the mean value of the

Table 3. Probabilistic variables for the strength and the load.

Variables	Definition	Mean	COV	Distribution
σy	yield stress [MPa]	355	0.066	Lognormal
φ	strength factor	1.154	–	Fixed
xu	uncertainty parameter	1	0.15	Normal
xsw	uncertainty parameter	1	0.1	Normal
xw	uncertainty parameter	1	0.24	Normal
Ψ Msw	still water bending moment [MNm]	1310	0.4	Normal
Mw	wave-induced bending moment [MNm]	3255	0.068	Weibul

Table 4. Probabilistic variables for fatigue crack model.

Variables	Definition	Mean	COV	Distribution
Y	geometry function	1.05	0.02	Normal
C	material parameter	$6.94 * 10^{-9}$	–	Fixed
m	material parameter	2.75	–	Fixed
a_0	initial crack size[mm]	1.00	0.18	Normal
v_0	upcrossing rate [/ year]	$5 * 10^6$	–	Fixed

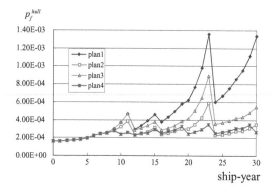

Figure 4. Time variance of the probability of failure for each maintenance plan.

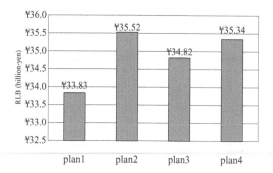

Figure 5. Estimated RLB (Japanese yen) for each maintenance plan.

coating life (t_c), which appears in the equation (22), is assumed as 7.5 years. Though the temporary repair cost with re-coating is generally very high, the total cost for the remaining life of the ship becomes lower if re-coating is carried out just after the coating reaches the end of its life.

As discussed above, by using the proposed concept of the remaining life benefit (RLB), it becomes possible to evaluate reasonably whether or not the maintenance plan is effective. Moreover, by using this concept, it might be possible to show that the ship with poor maintenance is disadvantageous from a viewpoint of safety as well as from an economic viewpoint.

6 ESTIMATION OF REMAINING LIFE BENEFIT BASED ON SURVEY RESULTS

In the previous section, for the prediction of reduction of the plate thickness by uniform corrosion, the probability distribution of the corrosion rate for each structural member is assumed as the distribution obtained from the statistical data. Therefore, the computed remaining life benefit (RLB) should be considered as the mean value using the average (original) probability distribution of the corrosion rate. However in the practical application of this concept to a certain ship, the rate of corrosion is different from the average rate, and is depending on its environmental condition. In this study, we make an attempt to evaluate the remaining life benefit (RLB) for a certain ship considering its condition by modifying the corrosion model based on the Bayesian inference (Itagaki & Yamamoto 1985) using the inspection results.

6.1 Modification of corrosion model based on bayesian inference

In the corrosion model in the equation (22), there are two probability variables, coating life, t_c, and corrosion rate, C_1. In this section, in order to improve the corrosion model effectively at each periodical survey, the Bayesian inference method is applied to the corrosion rate. In the Paik's corrosion model, corrosion rate, C_1, is represented by Weibull distribution (probability distribution function) (Paik et al. 2004),

$$F_{C_1}(C_1) = 1 - \exp\left\{-\left(\frac{C_1}{\beta}\right)^\alpha\right\} \quad (26)$$

where α is a shape parameter and β is a scale parameter. Here, it is assumed that periodical survey of the ship is carried out every n_f years, and that the survey data of corrosion depth at certain five locations $j(j = 1, 2, \ldots, 5)$ of a structural member, $\mathbf{R}_k(= r_1^k, r_2^k, \ldots, r_5^k (mm))$, are obtained at k-th periodical survey (ship-year t_k) as survey results. In this case, by using the information about results of 'k-1'-th survey, $\mathbf{R}_{k-1}$, the corrosion rate $\mathbf{C}_1^k (= C_{11}^k, C_{12}^k, \ldots, C_{15}^k (mm/year))$ observed at the locations $j(j = 1, 2, \ldots, 5)$ of the inspection can be defined as follows.

$$C_{1j}^k = \frac{(r_j^k - r_j^{k-1})}{n_{int}} \quad (27)$$

Then, the occurrence probability, Pr_j, in which the corrosion rate C_{1j}^k at the location j is observed, can be represented by the following equation,

$$Pr_j[Data(k : j) | \alpha, \beta] = f_{C_1}(C_{1j}^k) \times dr, \quad (28)$$

where $Data(k : j)$ means the event that C_{1j}^k is observed, f_{C1} is the probability density function of the equation (23), and dr is the interval of partition of corrosion depth. In this study, dr is taken as 0.5 mm, which is comparable level with the accuracy of thickness measurement.

If we assume that the corrosion rates for each location are independent variables, the probability, Pr, of which the k-th survey result occurs can be obtained by the following equation.

$$Pr[Data(k) | \alpha, \beta] = \prod_{j=1}^{5} Pr_j[Data(k : j) | \alpha, \beta] \quad (29)$$

Then, the parameters (α, β) of the Weibull distribution of the corrosion rate C_1 can be deduced by the Bayesian inference. In the Bayesian inference, prior distributions of the probability variables are firstly assumed from a subjective viewpoint before the inference. In this paper, as initial prior distributions of α and β, the probability density function, $f^{(0)}(\alpha, \beta)$, is assumed so that the mean value of the statistical data obtained by Paik's investigation is coincident with the mean value of the prior distribution (Fig. 6a). When the inspection result for k-th survey at certain locations of the structural member is obtained, improved (posteriori) distribution of the parameters (α, β) can be derived based on Bayes' theorem (Melchers 1999) as follows.

$$f^{(k)}(\alpha, \beta) = \frac{\prod_{r=1}^{k} Pr[Data(r) | \alpha, \beta] f^{(0)}(\alpha, \beta)}{\int\int\left[\prod_{r=1}^{k} Pr[Data(r) | \alpha, \beta] f^{(0)}(\alpha, \beta)\right] d\alpha d\beta} \quad (30)$$

By this equation, it is possible to improve the unknown parameters (α, β) at least once every periodical survey. Then, more accurate estimation of the structural deterioration and the remaining life benefit (RLB) is possible after the survey by using the improved parameters.

6.2 Computation of RLB

To examine the validity of this method, we carried out a trial calculation of the remaining life benefit (RLB). For this calculation, we prepared the hypothetical data of survey results for every two years, where the corrosion progresses at a faster rate than the average rate of the original distribution. Figure 6 shows the history of improvement of the distribution of the parameters (α, β) at the bottom plate. By using the Bayesian inference using the survey results, it is possible to estimate more accurate parameters than the initial values.

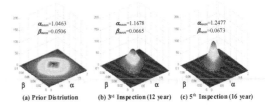

(a) Prior Distribution   (b) 3rd Inspection (12 year)   (c) 5th Inspection (16 year)

Figure 6. Estimated (improved) distribution of the parameters at outer bottom plate ($\alpha_{mean(true)}$ = 1.2719, $\beta_{mean(true)}$ = 0.0806).

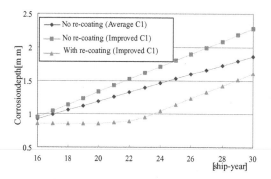

Figure 7. Predicted corrosion depth.

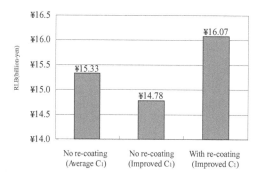

Figure 8. Estimated RLB (Japanese yen) for each case.

Next, remaining life benefit is evaluated by using the improved parameters. Figure 7 shows the estimated change of corrosion depth predicted at the 5th survey on 16 ship-years. The case "♦" is predicted by using the original distribution of corrosion rate, while the case "■" is predicted based on the distribution of (α, β) improved by Bayesian inference. It can be seen that the corrosion rate predicted by using the Bayesian inference is faster than that based on the original probability distribution. Also in Figure 8, estimated remaining life benefits (RLB), predicted by using the original distribution and by the improved distribution, are shown. It is shown that the RLB predicted by the original corrosion rate is much greater than the RLB based on the Bayesian inference. Also, as shown in Figure 7(▲), it is possible to retard the progress of corrosion by re-coating the structures at the survey of 16 ship-years. Then the RLB with re-coating (Fig. 8-right) becomes greater than that without re-coating. As shown the above, by using the result of survey, it becomes possible to evaluate the RLB for actual status of the ship, so that reasonable operation of the ship with a proper maintenance plan might become possible.

7 CONCLUSIONS

In this study, we arrive at the following conclusions.

1. In this study, we proposed a concept for decision making of a maintenance plan of a ship by evaluating the remaining life benefit (RLB). In particular, the method to estimate the revenue and the cost during its remaining service life is shown, and the formulation to estimate the remaining life benefit is proposed.
2. In the proposed method, the risk of failure is considered to estimate the remaining life benefit. In the computation of the risk, total failure probability is estimated by multiplying the system failure factor to the failure probability related to the longitudinal strength, which is derived by using the structural reliability analysis. Moreover, the deterioration by corrosion and fatigue is considered for evaluation of the longitudinal strength.
3. According to the estimated results of the remaining life benefit (RLB) for a bulk carrier, it is not necessarily the case that the RLB might become higher as the maintenance and repair is carried out more carefully. Also, if repair is not carried out to save the maintenance cost, the RLB becomes lower, because the risk of failure becomes higher without repair of the ship. It can be said that the optimum maintenance plan might exists for a certain ship, and that by using the concept of the proposed RLB, it is possible to find out the most suitable maintenance plan in which safe operation at low cost might be possible.
4. In this study, the concept to improve the parameters of the corrosion rate based on the Bayesian inference by using the survey result is shown, by which the proper estimation of the corrosion rate for actual ship condition is possible. By using this method, it becomes possible to estimate the remaining life benefit more accurately, and to plan a proper maintenance strategy considering the actual status of deterioration for a certain ship.

ACKNOWLEGEMENT

This work is supported by the Ministry of Education, Science, Sports and Culture, Grant-in-Aid for Scientific Research (A) 17206086, and Grant-in-Aid for Scientific Research (B) 20360391. Also this work is supported by Fundamental Research Developing Association for Shipbuilding and Offshore (REDAS). The authors are grateful for these supports.

REFERENCES

Akpan U.O. et al. 2002. Risk assessment of aging ship hull structures in the presence of corrosion and fatigue, *Marine Structures*, 15: pp. 211–231.

Caldwell, J.B. 1965. Ultimate longitudinal strength, *Trans RINA*, 107: pp. 411–430.

Ditlevesen, O. & Madsen, H.O. 1996. *Structural Reliability Method*: John Wiley& Sons.

Guedes Soares, C. & Moan, T. 1988. Statistical analysis of stillwater load effects in ship structures, *Trans. SNAME*, Vol. 96: pp. 129–156.

Guedes Soares, C. & Moan, T. 1991, Model uncertainty in the long-term distribution of wave-induced bending moments for fatigue design of ship structures, *Marine Structures*, 4: pp. 295–315.

Guedes Soares, C. & Garbatov, Y. 1997. Reliability assessment of maintained ship hulls with correlated corroded elements, *Marine Structures*, 10: pp. 629–653.

Guedes Soares, C. & Garbatov, Y. 1998. Reliability of maintained ship hull girders subjected to corrosion and fatigue, *Structural Safety*, 20: pp. 201–219.

IMO. 2001. Bulk Carrier Safety, Formal Safety Assessment of Life Saving Appliances for Bulk Carriers FSA/LSA/BC, *MSC 74/5/5*.

ISSC. 1991. Committee V.1 Applied Design, *Proc. 11th International Ship And Offshore Structures Congress (ISSC)*, Vol. 2: p. 14.

ISSC. 2006. Special Task Committee VI.1, Reliability-based structural design and code development, prepared by Moan T., et al., *Proceedings of 16th international ship and offshore structures congress*, 20–28 August 2006, Southampton, UK.

Itagaki, H. & Yamamoto, N. 1985. Bayesian reliability analysis on ship structural members: An application to the fatigue failures at the lower end of hold frames of ship, *Journal of the Society of Naval Architects of Japan*, No.158: pp. 565–570 (in Japanese).

Kuniyasu T. et al. 1968, A Computer Application to Principal Dimensions Study of Ships, *J. Soc. Naval Architectures of Japan*, Vol. 123: pp. 332–346 (in Japanese).

Mansour, A.E. & Hoven, L. 1994. Probability-based ship structural safety analysis, *J. Ship Research*, 38, 4: pp. 329–339.

Mansour A.E. et al. 1997a, Assessment of Reliability of Ship Structures, *SSC-Report-398*: Ship Structure Committee.

Mansour, A.E. et al. 1997b. Structural safety of ships, *Trans. SNAME*, Vol. 105: pp. 61–98.

Melchers, R.E. 1999, Structural Reliability Analysis and Prediction: John Wiley& Sons.

NYK LINE. 2006. *Monthly Report*, March 2006: pp. 81–82. (in Japanese)

OpCost2004. MOORE STEPHENS London: http://www.moorestephens.co.uk/

Owen F. Hughes. 1988. *Ship Structural Design*, Chapter 4: Wave loads-statistical, dynamic, and nonlinear aspects: The Society of Naval Architects and Marine Engineers.

Paik, J.K. & Mansour, A.E. 1995. A simple formulation for predicting the ultimate strength of ships, *Journal of Marine Science and Technology*, 1: pp. 52–62.

Paik, J.K. et al. 2003. Time-dependent risk assessment of aging ships accounting for general/pit corrosion, fatigue cracking and local denting damage, Presented at SNAME San Francisco, CA., October, *ABS Technical Papers—Marine Issues*, 2003.

Paik, J.K. et al. 2004. A time- dependent corrosion wastage model for seawater ballast tank structures of ships, *Corrosion Science*, 46, pp. 471–486.

Sun, H.H. & Bai, Y. 2003. Time-variant reliability assessment of FPSO hull girders, *Marine Structure*, 16: pp. 219–253.

Yamamoto, N. & Ikegam, K. 1998. A Study on the degradation of coating and corrosion of ship's hull based on the probabilistic approach, *J. OMAE, Trans. ASME*, 120, 3: pp. 121–128.

Yao. T. 2003. Hull girder strength, *Marine Structures*, 16: pp. 1–13.

Yuasa, K. et al. 2004, Refurbishment works for life time extension of LNG carriers to enhance safety and reliability, *The 14th Liquefied Natural Gas Conference and Exhibition*, March 21–24, Doha, Qatar.

Impact strength

Impact behaviour of GRP, aluminium and steel plates

L.S. Sutherland & C. Guedes Soares
Centre for Marine Technology and Engineering (CENTEC), Technical University of Lisbon,
Instituto Superior Técnico, Lisboa, Portugal

ABSTRACT: Comparative experimental data has been obtained for GRP laminates, and Aluminium 5083 and steel circular fully-clamped plates subjected to lateral impact. The metal plates were much more resistant to perforation than the laminated plates, but the behaviour up to perforation was more complex. A decrease in maximum force and an increase in maximum and final deflections, and absorbed energy was seen when fibre damage of the composite materials became significant. Both composite material and metal plates suffered damage at very low incident energies. However, the composite delamination damage would in practice be much less visible than that for metal plates. Care must be taken to ensure that relevant comparisons are made; flexural stiffness equivalence has been used here, but if strength, thickness, or weight equivalence would have been considered the results would have differed in each case.

1 INTRODUCTION

Three of the main building materials used in the construction of modern small craft are composites, steel and aluminium. Of these, composites are by far the most ubiquitous, almost exclusively as glass reinforced plastic (GRP) in the form of Chopped-Strand Mat (CSM) or Woven Roving (WR) reinforced polyester resin. Metal hull construction uses mild steel or 5000 series aluminium alloys. Wood is also still used, but since there are such a large number of combinations of species and forms and because this is a relatively specialised industry, this material is not considered in the current study.

The main advantages and disadvantages of each material are listed in Table 1 (Gerr 2000, Brewer 1994, Trower 1999, Du Plessis 1996). The numerous advantages of GRP explain to a large extent why it has become so popular. However, a major disadvantage of composites in comparison with metals is the lack of plastic deformation to failure, which means that metal boats are seen as more durable or damage resistant to groundings or collisions.

The susceptibility of composite materials in general to impact damage of has attracted great attention in the literature (Abrate 1998). The authors' previous work (Sutherland & Guedes Soares 1999a, b, 2003, 2004, 2005a, b, c, 2006, 2007) has shown the impact behaviour of GRP to be complex. Internal delamination occurred at very low incident energies, only followed at much higher energies by visible fibre failure leading to penetration and finally perforation. Membrane and bending effects dominated for thinner laminates until back-face fibre damage initiated final failure. Thicker laminates were dominated by shear effects due to delamination, with indentation damage

leading to final failure. The response was seen to be dependent on many impact and material parameters.

Table 1. Construction materials comparisons.

Advantages	Disadvantages
COMPOSITES:	
Lightweight	Little plastic deformation to
Easy to mould complex	fail (for impact energy
shapes	absorption).
Cheap more so for series	Low fire resistance, toxic
production	fumes.
Ability to tailor properties	Production working
No painting or fairing	environment
Low wastage	
No rot or corrosion	
Low maintenance	
Easy to repair	
Non-magnetic	
ALUMINIUM:	
Lightweight	Expensive
Easy to work	Welding distortion
No corrosion	Costly to paint
Non-magnetic	Less readily available
High plastic deformation to	Requires heat insulation
fail (for impact energy	Electrolysis
absorption)	Fatigue
	Less abrasion-resistant
	Can melt at temperatures
	seen in fires
STEEL:	
Cheap Simple to fabricate	Heavy
Easy to repair Fire resistant	Hard to Shape
High plastic deformation to	Welding distortion Requires
fail (for impact energy	heat insulation
absorption)	Corrosion
	Electrolysis

The aim of this study is to further explore and elaborate on the relative impact performance of GRP, steel and aluminium through a experimental comparison of transversely impacted plates.

However, since the impact response of a given material is not, in the pure sense of the term, a material property but will depend on the structural use to which it has been put, in order to make any meaningful comparisons from a design point of view, a pertinent equivalence between plates of different materials must be selected. For example, should the plates be of the same thickness, or the same stiffness, or of the same strength? Which criterion is selected will strongly influence the relative impact performance of the various materials considered, both quantitatively and, in all probability, qualitatively.

Here, the approach taken is to consider the design case of plates of equivalent bending stiffness. The aim is to provide a simple, but relevant and useful comparison of the impact behaviours of the materials considered, not to attempt to exactly predict the very complex response in each case.

2 EXPERIMENTAL DETAILS

As representative of the materials used in the marine industry, WR & CSM hand laid-up laminates, structural carbon or 'mild' steel and Aluminium 5083-H111 were considered. The tests on the two laminates were those as reported in previous work (Sutherland & Guedes Soares 2006).

An orthophthalic polyester resin was used throughout to laminate 1 m square panels by hand on horizontal flat moulds. A fibre mass-fraction of 0.5 and 0.3 (equivalent to fibre volume fractions of approximately 0.35 and 0.2) for WR and CSM plies respectively was stipulated as representative of the values commonly achieved under production conditions in the marine industry. 1, 2 and 3% by mass of accelerator, catalyst and paraffin respectively were added at an ambient temperature of between 18 and 21°C to cure the resin. In order to ensure sufficient cure, laminates were stored at room temperature for two months before testing.

A plate laterally impacted by a hemi-spherically ended projectile was considered as a commonly studied and relatively severe impact event. In order to make meaningful comparisons the thicknesses of each material were selected to give plate bending stiffness equivalence, using the appropriate WR laminate thickness as the reference value. The closest plate thicknesses available for the metals were used, but since bending stiffness varies as thickness cubed, exact equivalence was not possible (Table 2). However, allowance for these differences was made as discussed in the next section.

Table 2. Specimen details.

	Thin	Thick
WR:		
Panel code (No. plies)	W5	W15
Thickness (mm)	3.17	9.29
Diameter/thickness	32	11
CSM:		
Panel code (No. plies)	C4	C12
Thickness (mm)	3.75 (3.78)*	10.92 (11.06)*
Stiffness : WR stiffness	−2%	−4%
Diameter/thickness	27	9
ALUMINIUM:		
Panel code	AL1	AL2
Thickness (mm)	2.00 (1.82)*	5.92 (5.32)*
Stiffness : WR stiffness	+33%	+38%
Diameter/thickness	50	17
STEEL:		
Panel code	FE1	FE2
Thickness (mm)	1.37 (1.28)*	4.00 (3.74)*
Stiffness : WR stiffness	+23%	+23%
Diameter/thickness	73	25

* Thicknesses in parenthesis are those which would give equivalent bending stiffness to the appropriate WR laminate.

Two series of tests were run; 'thin' specimens to induce a bending/membrane controlled event such as would occur at the centre of a panel, and 'thick' specimens to induce a shear/indentation controlled event such as would occur near to a stiffener. 200 mm square specimens were cut from the panels and thickness measurements then taken at four points on each specimen prior to testing. Specimen thicknesses are given in Table 2. The specimens were fully clamped between two thick annular circular steel plates of 100 mm internal diameter. Four bolts that passed through holes in the clamp plates (and specimen) applied the clamping force.

Impact testing was performed using a fully instrumented Rosand IFW5 falling weight machine. A small, light hemispherical ended cylindrical projectile (of diameter 10 mm for these tests) is dropped from a known, variable height between guide rails onto a clamped horizontally supported plate target. A much larger, variable mass is attached to the projectile and a load cell between the two gives the variation of impact force with time. An optical gate gives the incident velocity, and hence the projectile displacement and velocity and the energy it imparts are calculated from the force-time data by successive numerical integrations. A pneumatic catching device prevents further rebound impacts. Since the projectile is assumed to remain in contact with the specimen throughout the impact event, the projectile displacement is used to give the displacement and velocity of the top face of the specimen, under the projectile. Assuming that frictional and heating effects are negligible the energy imparted by the indenter is that absorbed by the specimen. Thus, this energy value at the end of the test is that irreversibly absorbed by the specimen.

A series of tests were performed for a range of increasing incident energies for each material. Two sets of nominal incident energies were used, one for 'thick' and one for 'thin' specimens. As far as possible exactly the same incident energies were used for all materials but due to guiding rail friction effects this was not always possible, especially at low drop heights. The impact mass was 3.103 kg and 4.853 kg for the thin and thick specimens respectively.

3 THEORY

As noted in the previous section, it was not possible to arrange plates of the various different materials with exactly the same bending stiffness. Hence, a simplified model is developed here to allow for these differences through normalisation of the various experimental impact responses recorded. The 'ideal' case of a spring-mass model (Abrate 2001) with no damage or membrane or indentation effects will be considered. This simplified model can also be used as a 'baseline' reference to make useful comparisons between the different material impact behaviours seen.

However, the model is not expected to be a good predictor of behaviour since damage, which is not considered, occurs even at low incident energies, and membrane and indentation effects are large for thinner and thicker specimens respectively. Also, for simple comparative purposes the behaviour of the woven roving plates is simplified to that of an isotropic material.

The two degree-of-freedom spring-mass model of the impact event, for no damage, is shown schematically in Figure 1.

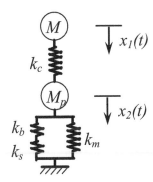

Figure 1. Two degree-of-freedom spring-mass model of impact event.

The bending, membrane, contact and shear stiffness' of the plate are represented by the springs k_b, k_s, k_m and k_c respectively. M is the mass of the projectile and M_p that of the plate. The displacements of the projectile and plate are given by the functions of time x_1 and x_2 respectively.

The free body diagrams of M and M_p give the equations of motion,

$$M\ddot{x}_1 + F = 0$$
$$M_p\ddot{x}_2 + k_{bs}x_2 + k_m x_2^3 - F = 0 \qquad (1)$$

where F is the contact force which is a non-linear function of the indentation,

$$F = f(\alpha) = f(x_1 - x_2) \quad x_1 > x_2 \qquad (2)$$

When $x_2 > x_1$ the projectile leaves the plate surface and $F = 0$.

The initial conditions are given by incident impact velocity and that the initial displacements are both zero,

$$\dot{x}(0) = v_i, \quad x_1(0) = x_2(0) = 0 \qquad (3)$$

The problem must be studied numerically, but assuming indentation and membrane effects are negligible, that shear stiffness is much greater than bending stiffness, and that in comparison to that of the projectile the mass of the plate is very small, leads to a simple one degree of freedom system where $x_1 = x_2 = x$ with the equation of motion,

$$M\ddot{x} + k_b x = 0 \qquad (4)$$

Solving the general solution to (4) using the initial conditions (3) gives,

$$x = v_i\sqrt{\frac{M}{k_b}} \cdot \sin\left(\sqrt{\frac{k_b}{M}} \cdot t\right) \qquad (5)$$

Hence,

$$x_{\max} = v_i\sqrt{\frac{M}{k_b}} \qquad (6)$$

The impact force is now given by that in the spring k_b,

$$F = k_b x = v_i\sqrt{k_b M} \cdot \sin\left(\sqrt{\frac{k_b}{M}} \cdot t\right) \quad \omega t < \pi \qquad (7)$$

Hence,

$$F_{\max} = v_i\sqrt{k_b M} \qquad (8)$$

The impact duration is given when the projectile leaves the plate at $t = T_c = \pi\omega$,

$$T_c = \pi\sqrt{\frac{M}{k_b}} \qquad (9)$$

From Young (1989), for a uniform load over a very small central circular area of radius r_o on a circular plate of radius a and flexural rigidity D with fixed edges,

$$F = \frac{16\pi D}{[a^2 - r^2(1 + 2\ln\frac{a}{r})]} \cdot x \quad r > r_o \qquad (10)$$

where,

$$D = \frac{h^3 E}{12(1 - \upsilon^2)} \qquad (11)$$

where h is thickness, E is Young's modulus and υ is poisson's ratio.

Since the contact radius here is much smaller than the plate radius then it is valid to approximate to a point load at the centre. Hence the load at the plate centre is given by,

$$F = \frac{16\pi D}{a^2} \cdot x \qquad (12)$$

Comparing Equations (7) and (12) gives,

$$k_b = \frac{16\pi D}{a^2} \qquad (13)$$

Which is the same as that given in Shivakumar et al (1985).

This simple model will now be used to develop relationships that normalise for the differences in flexural rigidities of the plates of each material, using the rigidity of the WR plates as a reference value. Since a is constant, equation (13) shows that plate stiffness k_b is proportional to flexural rigidity D.

From equation (6), and since for these tests v_i and M are constant, in order to reduce the impact displacement for direct comparison with that of the WR plate,

$$x_{\text{max (normalised)}} = x_{\text{max}}\sqrt{\frac{D}{D_{WR}}} \qquad (14)$$

Where D_{WR} is the flexural rigidity of the equivalent woven roving platefrom equation 11. Equation (14) may also be used to normalise for x_{end}.

Similarly, from equation (8) for maximum impact force, and since for these tests v_i and M are constant,

in order to normalise the maximum force for direct comparison with that of the WR plate,

$$F_{\text{max (normalised)}} = F_{\text{max}}\sqrt{\frac{D_{WR}}{D}} \qquad (15)$$

From equation (9), and since for these tests M is constant, in order to reduce the impact duration for direct comparison with that of the Woven Roving (WR) plate,

$$T_{c(\text{normalised})} = T_c\sqrt{\frac{D}{D_{WR}}} \qquad (16)$$

4 RESULTS

The damage incurred by the GRP specimens is summarised in Table 3 and may be categorised into three main stages:

i. At very low impact energies the damage was limited to slight permanent indentation and/or matrix cracking.

ii. At *very low incident energies* an internal delamination suddenly occurred. This delamination then grew and multiple delaminations at different ply interfaces were seen.

Table 3. GRP damage.

WR			CSM		
IE (J)	Delamination (mm²)	Fibre Damage	IE (J)	Delamination (mm²)	Fibre Damage
Thin specimens					
1.3	41	–	1	–	–
3.5	219	–	3.4	110	–
5.4	257	–	5.4	177	–
7.5	328	–	7.4	257	i*
10	426	–	10	315	i/b*
15	448	i*	16	358	f/b sp
20	590	i/b*	20	455	f/b sp
31	790	f/b sp	30	400	p
Thick Specimens					
2.7	29	–	2.2	–	–
5.5	152	–	5.8	–	–
11	702	i*	11	870	i*
16	1008	i	15	1106	i
31	1417	f*	30	1466	f*
55	2607	f	55	1876	f
74	2769	f	75	2757	f sp
90	3743	f	90	2782	f sp

f = front-face, b = back-face, i = indent, * = very slight, sp = start of penetration, p = penetration.

iii. At high incident energies fibre failure occurred (on the back face for thin specimens and on the impacted face for thick specimens), leading to penetration.

Typical contact indentation and delamination damage to a CSM laminate are shown Figure 2(a) and (b) respectively. Further details of the impact damage on marine composites, including the laminates considered here may be found in Sutherland & Guedes Soares (2006).

The metal plates also suffered impact damage at all but the very lowest incident energies. Only 2 specimens, the thin and thick steel plates at 1 and 2J respectively, showed no visible damage. The thick aluminium plate subjected to 2J and the thick steel plate subjected to 5J showed only slight indentation damage under the projectile. All the remaining metal plates suffered both indentation damage (Figure 3a) and plastic deformation of the plate itself (Figure 3b), both of which became more severe with increasing incident energy.

In previous work the load cell force data filtered with a low pass filter (second-order discrete Butterworth). However, this often led to a significant phase lag. Hence, the moving average method was used here. Averaging over 5% of the data span resulted in very good filtering with negligible phase lag as can be seen in Figure 4(a). However, as can be seen in Figure 4(b), this method did not filter the initial oscillations satisfactorily in some cases when an initial peak was severe (mostly for the thick metal specimens). Despite this, smoothing was used throughout since the advantages of avoiding phase lag were great.

The force-displacement plots for the thin and thick specimens are presented in Figure 5 and Figure 6 respectively. All tests for that material are shown on each plot, and the incident energies are labelled accordingly.

The impact responses of maximum force, maximum displacement, displacement at end of the impact event and energy irreversibly absorbed are presented in Figure 7 and Figure 8. Impact durations are given in Figure 9. Where appropriate the values have been normalised, as described in the previous section, to give comparability with the WR results.

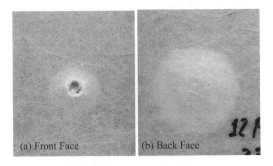

Figure 2.   Damage to thick CSM plate (90J).

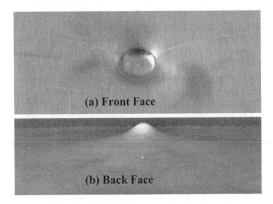

Figure 3.   Damage to aluminium thin plate (55J).

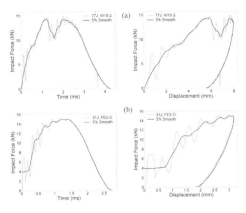

Figure 4.   Typical smoothed plots.

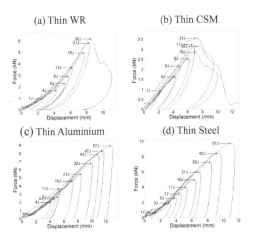

Figure 5.   Thin specimens force-displacement plots.

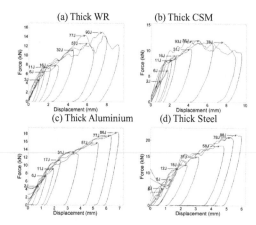

Figure 6. Thick specimens force-displacement plots.

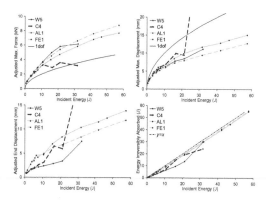

Figure 7. Thin specimens results.

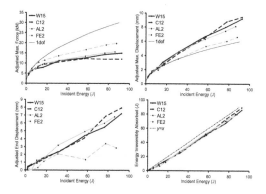

Figure 8. Thick specimens results.

5 DISCUSSION OF RESULTS

The impact behaviour was seen to be complex, and differed between the thin and the thick plates, and hence these two cases are discussed separately in this section.

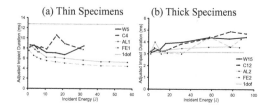

Figure 9. Impact durations.

5.1 *Thin Plates*

Although the thin composite plates were perforated at higher incident energies, the metal plates were not perforated even at the maximum incident energy attainable by the impact machine with the impact mass used in these tests. However, from the damage seen it was possible to predict that the order of increasing resistance to perforation was CSM, WR, aluminium, steel. Additional tests were also made on 1050 alloy aluminium, which was found to perforate at a much lower incident energy than the 5083 alloy (1050 at 35 J, 5083 > 60 J). Hence in terms of resistance to perforation, for this impact event, the metal plates outperformed the composite materials considered here.

Only the thin WR plates show a strong non-linear membrane stiffening effect right up to fibre damage (Figure 5(a)). The metal plates show a slight effect only at low incident energies (Figure 5(c) and (d)) and the CSM results exhibit linear stiffness (Figure 5(b)). However, Figure 7 and Figure 9 show that compared to the one degree of freedom model (which ignores membrane effects), maximum force is greater and maximum deflection and impact duration are both lower in all cases where there is no perforation. Figure 9 also shows that the impact duration of the metal plates decreases with increasing incident energy. Hence membrane forces are present (which would be expected from the large deflections seen), but their effects have been modified, almost certainly by plastic deformation of the metals, and because the CSM has no continuous fibres to resist such forces and indentation damage is more severe.

The maximum force is very similar for all materials at lower incident energies until fibre failure and perforation causes the CSM values to drop significantly. Plastic deformation leads to lower maximum forces for the metal plates, with the aluminium values lower than those for steel. The WR plates give the highest impact forces of all until fibre damage is suffered.

Very similar maximum displacements are seen for all plates until perforation of the CSM, the order of increasing displacement for the other specimens being aluminium > WR > steel. The final displacement gives some measure of the permanent deformation suffered, a higher value indicating more permanent deformation, and, until fibre damage, this is much lower for the composite materials studied here. When plastic deformation becomes significant, the end displacement of the aluminium becomes greater than that of steel.

It is clear from Figure 5 that the areas under the curves are much lower in the case of the composite materials, and this is reflected in the absorbed energy values of Figure 7. Only 60 to 70 percent of the incident energy is irreversibly absorbed by the composite plates as compared to 90 percent for the metal specimens, until fibre damage raises the composite value to 90 percent.

5.2 Thick plates

Figure 6(a) and (b) show the typical bi-linear force-displacement behaviour with a sudden drop in stiffness due to delamination. A reduction in stiffness at higher incident energies for metals is also seen in Figure 6(c) and (d), although this is not as sudden in these cases. This behaviour is thought to be due to local indentation followed by plastic deformation of the plate at higher incident energies.

The maximum forces of Figure 8 follow the one degree of freedom model at low incident energies, but then drop away as indentation fibre damage or plastic deformation occurs. Aluminium and composite plates produce very similar maximum forces, until indentation damage of the CSM becomes significant. The forces on the steel plates are greater than those on the other materials.

The one degree of freedom model predicts well the maximum displacements for steel, and also for the other materials before fibre damage and plastic deformation occur (Figure 8). At higher incident energies the deflections of the aluminium and composite plates are very similar and greater than those of steel plates. The final displacements are very similar for all materials except for the CSM plates which give lower values (Figure 8). This is contrary to what would be expected since Table 3 indicates that these laminates incurred significant indentation damage. This is thought to be because when CSM suffers damage the composite can delaminate causing an increase in laminate thickness.

Figure 8 also shows that the energy absorbed by steel plates at low energies is approximately 50% of the incident energy, but that for the other materials this value is approximately 75% indicating that steel is better at resisting indentation damage. Above an incident energy of approximately 15 J, the energy absorbed by all materials is very similar at around 5 J less than the incident energy, showing that almost all energy is absorbed by damage.

The impact durations (Figure 9) are generally longer than those predicted by the one degree of freedom, and increase with increasing incident energy due to damage and permanent deformation. However, at very low incident energies the one degree of freedom model predicts well the impact duration, and this model also applies up to approximately 30J for the steel plates.

6 CONCLUSIONS

Comparative data for approximately equal flexural stiffness has been obtained for woven roving and chopped-strand mat glass-polyester laminates, for Aluminium 5083 and for steel circular fully-clamped circular plates subjected to central lateral impact with a hemispherical ended dropped weight. The metal plates gave a much higher resistance to perforation than the laminated plates, but the behaviour up to perforation was more complex.

The behaviour with respect to maximum force, maximum deflection, end deflection and absorbed energy are summarised in Table 4. When fibre damage of the composite materials became significant this was accompanied by a corresponding decrease in maximum force and an increase in maximum and final deflections, and absorbed energy.

A simple one degree of freedom model was useful in making pertinent comparisons between plates of slightly differing flexural rigidity, but only predicted the impact response for very low incident energies on thick plates.

Both composite material and metal plates suffered damage at very low incident energies. However, the composite delamination damage would not be visible if gel-coated or painted as is normal in practice; the first visible damage would be fibre damage which only occurs at much higher incident energies. Damage to metals is visible even at low incident energies.

Table 4. Summary of impact responses.

	Thin plates		Thick plates	
	Low IE	Higher IE	Low IE	Higher IE
Maximum force	Little difference	WR > Fe > Al > CSM	Little difference except	Fe greater
Maximum deflection	Very similar	Fe < WR < Al < CSM	Little difference	Fe lower
Final deflection	Metals greater	Similar	Similar except	CSM lower
Absorbed energy	Metals greater	Similar	Fe lower	Little difference

Care must be taken to ensure that relevant comparisons are made; flexural stiffness equivalence has been used here, but if strength, thickness, or weight equivalence would have been considered the results would have differed in each case. It is not only material choice but also the plate design that will influence the impact behaviour. For example exchanging steel for GRP based on a stiffness design criterion will give a much thicker plate, and the change in plate thickness may well influence the impact behaviour more than does the change in material.

ACKNOWLEDGEMENTS

This work has been performed within the project "MARSTRUCT —Network of Excellence on Marine Structures" (http://www.mar.ist.utl.pt/marstruct/) and has been partially funded by the European Union through the Growth programme under contract TNE3-CT-2003-506141. The first author was financed by the Portuguese Foundation of Science and Technology under the contract number SFRH/BPD/20547/2004.

REFERENCES

Abrate, S. 1998. *Impact on Composite Structures*. Cambridge: Cambridge University Press.
Abrate, S. 2001. Modelling of impacts on composite structures. *Composite Structures* 51: 129–38.
Brewer, T. 1994. *Understanding Boat Design (4th Edition)*. Camden Maine: International Marine.
Du Plessis, H. 1996. *Fibreglass Boats (3rd Edition)*. London: Adlard Coles Nautical.
Gerr, D. 2000. *The Elements of Boat Design*. Camden Maine: International Marine.

Shivakumar KN., Elber W. and Illg W. 1985. Prediction of impact force and duration due to low-velocity impact on circular composite laminates. *Journal of Applied Mechanics* 52: 674–680.
Sutherland, L.S. & Guedes Soares C. 1999a. Impact tests on woven roving E-glass/polyester laminates. *Composites Science and Technology* 59: 1553–1567.
Sutherland, L.S. & Guedes Soares C. 1999b. Effects of laminate thickness and reinforcement type on the impact behaviour of E-glass/polyester laminates. *Composites Science and Technology* 59: 2243–2260.
Sutherland, L.S. & Guedes Soares C. 2003. The effects of test parameters on the impact response of glass reinforced plastic using an experimental design approach. *Composites Science & Technology* 63: 1–18.
Sutherland L.S. & Guedes Soares C. 2004. Effect of laminate thickness and of matrix resin on the impact of low fibre-volume, woven roving E-glass composites. *Composites Science and Technology* 64: 1691–1700.
Sutherland L.S. & Guedes Soares C. 2005a. Impact characterisation of low fibre-volume glass reinforced polyester circular laminated plates. *International Journal of Impact Engineering* 31: 1–23.
Sutherland, L.S. & Guedes Soares C. 2005b. Impact on low fibre-volume, glass/polyester rectangular plates. *Composites Structures* 68: 13–22.
Sutherland, L.S. & Guedes Soares C. 2005c. Contact indentation of marine composites. *Composites Structures* 70(3): 287–294.
Sutherland, L.S. & Guedes Soares C. 2006. Impact behaviour of typical marine composite laminates. *Composites Part B: Engineering* 37(2–3): 89–100.
Sutherland, L.S. & Guedes Soares C. 2007. Scaling of impact on low fibre-volume glass-polyester laminates. *Composites Part A: Applied Science and Manufacturing* 38: 307–317.
Trower, G. 1999. *Yacht and Small Craft Construction: Design Decisions*. Marlborough Wiltshire: Crowood Press.
Young, W.C. 1989. *Roark's Formulas for Stress and Strain*. Singapore: McGraw-Hill.

Analysis and Design of Marine Structures – Guedes Soares & Das (eds)
© 2009 Taylor & Francis Group, London, ISBN 978-0-415-54934-9

Impact damage of MARK III type LNG carrier cargo containment system due to dropped objects: An experimental study

J.K. Paik, B.J. Kim & T.H. Kim
Pusan National University, Busan, Korea

M.K. Ha, Y.S. Suh & S.E. Chun
Samsung Heavy Industries Co. Ltd., Geoje, Korea

ABSTRACT: The objective of the present study is to investigate the structural damage characteristics on MARK III type LNG carrier cargo containment system due to dropped object impacts. A series of experimental studies on a full scale structure model including primary barrier sheet (stainless steel), RPUF (reinforced poly-urethane foam) insulation panels and secondary barrier are undertaken under impacts arising from dropped objects such as pipe supports, bolts and nuts. A gun type impact machine is employed for the tests. The insights and conclusions developed from the present study will be useful for impact damage assessment and management on LNG cargo containment system due to dropped objects.

1 INTRODUCTION

For structural design and strength assessment, the application of the limit states based approaches tends to be mandatory today (Paik & Thayamballi 2003, 2007).

While in service, ships and offshore structures are subjected to impacts arising from dropped objects, which cause mechanical damage such as local denting. In some cases, the structural impacts cause more serious damage such as perforation which can lead to leakage of oil or gas.

Useful observations on this issue may be found in the literature (Jones 1989, Paik & Melchers 2008) in terms of damage assessment and management. Particular interests in this areas are structural damage caused by dropped objects (Backman & Goldsmith 1978, Caridis & Samuelides 1994, Corbett et al 1996, Jones & Birch 2008, Jones et al 2008).

The present study deals with impact damage on a MARK III type LNG carrier cargo containment system due to dropped objects. A series of experimental studies using a gun type impact machine are undertaken. Test data and insights obtained from the present study are documented.

2 IMPACT SCENARIO ON THE TARGET STRUCTURE

The object structure is the cargo containment system of a 150 km³ class MARK III type LNG carrier. Figure 1 shows a typical structure inside the MARK III type LNG carrier cargo tanks.

It is seen from Figure 1 that the cargo containment system is composed of primary barrier (stainless steel sheet), RPUF (reinforced poly-urethane foam) panels, secondary barrier, and plywood panels. The structure is then connected with inner hull via a set of mastic stripes (ropes). Figure 2 shows the target structure taken from the real cargo containment system tank. Figure 3 shows a zoomed-up picture of the membrane sheet (primary barrier) with small and large corrugations.

As an impact scenario, it is considered that an object freely falls onto the structure from the height of 27 m. The dropped objects are an assembled bolt and nut

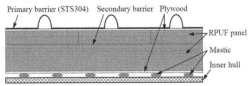

Figure 1. A typical structure inside the MARK III type LNG carrier cargo tank.

Figure 2. A real structure of MARK III type LNG cargo containment system on the full scale tested in the present study.

Figure 3. Zoomed-up picture of the membrane sheet (primary barrier) with small and large corrugations.

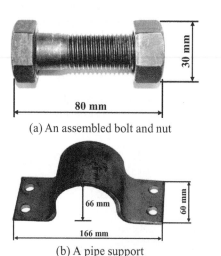

(a) An assembled bolt and nut

(b) A pipe support

Figure 4. The dropped objects and their geometrical dimensions considered in the present study.

with the mass of 0.280 kg and a pipe support with the mass of 0.555 kg, as shown in Figure 4.

The speed of the free-falling object is approximately predicted by

$$v = \sqrt{2\,gh} = 23.02 \text{ m/s} \tag{1}$$

where v = speed of falling object, h = height = 27 m, g = acceleration of gravity.

3 EXPERIMENTAL STUDY

3.1 Test facility

A gun type impact machine as shown in Figure 5 is employed for the tests. The chamber is pressurized by air compressor via air inlet. The level of pressure inside the chamber is measured by pressure gauge. A valve is accommodated to open or close the pressure chamber and subsequently control the shooting of the object located at the outlet. It is evident that the speed of the object movement depends on the level of pressure in the chamber.

The speed of the object movement after being shot is obtained by the measuring time which is taken while an object (projectile) passes through a designated distance between two positions, e.g., 150 mm, near the outlet, as shown in Figure 6.

The time of passing through the two reference positions can be measured by the observation of voltage changes in an electric wire inherently attached to each position due to wire cutting when the object passes through, namely

$$v = \frac{L}{\Delta t} \tag{2}$$

Figure 5. A gun type impact test machine.

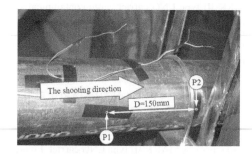

Figure 6. The speed measurement of the impact object movement after being shot.

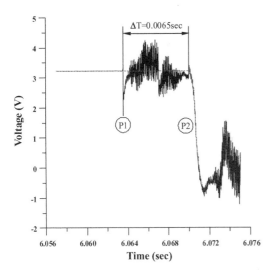

Figure 7. An example of the time measurement taken for passing through the two reference positions.

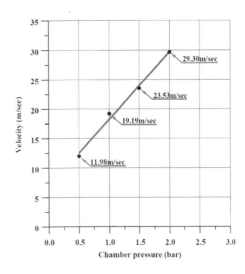

Figure 8. The relationship between the chamber pressure versus the object speed.

where L = distance between two positions which is taken as 150 mm in the present study, Δt = measured time taken for passing through the two positions.

Figure 7 shows an example of the measured time during passing through the two reference positions. Figure 8 shows the relationship between the chamber pressure versus the velocity of the moving object at the test machine. It is evident that the chamber pressure must be about 1.5 bar to achieve the target speed (i.e., 23 m/s) equivalent to the impact scenario indicated in Equation (1).

3.2 Impact tests

A total of 8 locations are considered as impact positions as shown in Figure 9. The impact locations for the assembled bolt and nut include intersection of large and small corrugations (B1), top of small corrugation (B2), membrane sheet floor (B3), membrane welded sheet floor (B4), top of small welded corrugation (B5) and top of large corrugation (B6). Additionally top of large corrugation (S1) and membrane sheet floor (S2) are chosen for the impact of pipe support.

The assembled bolt and nut, and the pipe support shown in Figure 4 are shot toward the target structure (i.e., cargo containment system) in the impact position noted above with the constant speed of 23 m/s. The objects were positioned in the longitudinal direction in the tests so that the heating point (impact location) with the test structure was the longitudinal tip of each object.

Figure 10 shows the impact damages on the object structure after testing. Figure 11 shows detailed pictures of the impact damages on various locations.

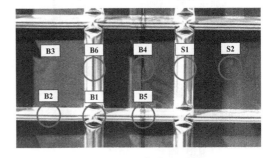

Figure 9. Impact locations considered in the present study.

3.3 Discussions

As shown in Figure 11, the impact damage on the top of the small or large corrugation is larger than those on the other locations. Even though the maximum penetration is 11 mm for B2 and B6 and 18 mm for S1, the damage is restricted within the primary barrier (membrane sheet), and serious damage such as perforation which can lead to leakage of gas was not observed. In fact, the impact perforation was the primary concern as an acceptance criterion in the present study.

However, for B3, B4 and S2 (i.e., with the impact location on the membrane sheet floor), it is observed that the plywood panel must be damaged to some extent, considering that the dent depth is 2–3 mm which is larger than the thickness of the membrane sheet (primary barrier).

In the present paper, one particular scenario of impacts in terms of shapes and height of dropped objects was examined, as per the concern of the ship

303

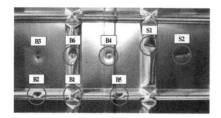

Figure 10. Impact damages after testing.

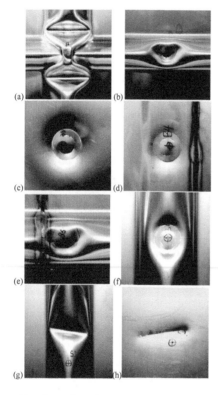

Figure 11. Zoomed-up pictures of impact damages on various locations: (a) B1, (b) B2, (c) B3, (d) B4, (e) B5, (f) B6, (g) S1 and (h) S2.

owner at the stage of front-end engineering and design. However, it is readily surmised that the impact damages may depend on different scenarios, and therefore further study is necessary.

4 CONCLUDING REMARKS

The aim of the present paper has been to perform an experimental investigation of the impact damage characteristics on a MARK III type LNG carrier cargo containment system due to dropped objects such as a pipe support and an assembled bolt and nut.

Serious damage such as perforation in the primary barrier which can lead to leakage of gas was not observed in the free falls from the height of 27 m onto the MARK III type LNG carrier cargo containment system.

Further study is now ongoing with different impact scenarios such as types of objects and heights of free falls. Also, a technology for nonlinear finite element method simulations on impact damage is also being undertaken.

ACKNOWLEDGEMENTS

The present study was undertaken at the Lloyd's Register Educational Trust (LRET) Research Centre of Excellence at Pusan National University, Korea. The authors are pleased to acknowledge the support of Samsung Heavy Industries Co., Ltd.

REFERENCES

Backman, M.E. and Goldsmith, W. 1978. The machanics of penetration of projectiles into targets. International Journal of Engineering Sciences, Vol.16, pp. 1–99.

Caridis, P.A. and Samuelides, E. 1994. On the dynamic response of ship plating under lateral impact. International Journal of Impact Engineering, Vol.15, Issue 3, pp.149–164.

Corbett, G.G., Reid, S.R. and Johnson, W. 1996. Impact loading of plates and shells by free-flying projectiles: A review. International Journal of Impact Engineering, Vol.18, No.2, pp.141–230.

Jones, N. 1989. Structural impact. Cambridge University Press, Cambridge, UK.

Jones, N. and Birch, R.S. 2008. On the scaling of low-velocity perforation of mild steel plates. Journal of Pressure Vessel Technology, Vol.130, Issue 3, 031207.

Jones, N., Birch, R.S. and Duan, R. 2008. Low-velocity perforation of mild steel rectangular plates with projectiles having different shaped impact forces. Journal of Pressure Vessel Technology, Vol.130, Issue 3, 031206.

Paik, J.K. and Melchers, R.E. 2008. Condition assessment of aged structures, CRC Press, New York, USA.

Paik, J.K. and Thayamballi, A.K. 2003. Ultimate limit state design of steel-plated structures. John Wiley & Sons, Chichester, UK.

Paik, J.K. and Thayamballi, A.K. 2007. Ship-shaped offshore installations: Design, building, and operation. Cambridge University Press, Cambridge, UK.

Paik, J.K. and Won, S.H. 2007. On deformation and perforation of ship structures under ballistic impacts. Ships and Offshore Structures, Vol.2, No.3, pp. 217–226.

Analysis and Design of Marine Structures – Guedes Soares & Das (eds)
© 2009 Taylor & Francis Group, London, ISBN 978-0-415-54934-9

Simulation of the response of double bottoms under grounding actions using finite elements

I. Zilakos, M. Toulios & M. Samuelides

National Technical University of Athens (NTUA), School of Naval Architecture and Marine Engineering, Athens, Greece

Tan-Hoi Nguyen & J. Amdahl

Department of Marine Technology, Norwegian University of Science and Technology (NTNU), Trondheim, Norway

ABSTRACT: Although simulations of impact on structures have been extensively performed using finite element codes, their use still presents challenges to researchers and designers. The present paper reports on work that has been carried out within the MARSTRUCT network of excellence, aiming on one hand to define simulation techniques that may be used to perform realistic simulations and, on the other hand, to determine the reactions of the double bottom structures of tankers to grounding actions. An extensive series of simulations have been performed independently by the MARTSRUCT partners NTUA and NTNU in order to investigate the effect of the user selected parameters on the results and, ultimately, to examine if widely accepted modeling techniques may be used to produce reliable results.

1 INTRODUCTION

The analysis of the structural response due to grounding actions involves the overall behavior of the vessel relative to the obstruction as well as the global and local structural behavior. The latter is primarily concerned with the prediction of the structural resistance to crushing and tearing. There are two main approaches used in these predictions which have received considerable attention in the literature and are still the subject of ongoing investigation. The first approach is a simplified methodology that uses admissible plastic deformation mechanisms in order to analytically obtain estimates of the structural resistance, e.g. Simonsen & Wierzbicki (1998), Hong & Amdahl (2008). The second approach is based on the application of the finite element (FE) method, e.g. Samuelides et al. (2007), Naar et al. (2002). Despite the availability of a number of FE codes which have the algorithms to carry out these complex simulations, such as LS-DYNA, ABAQUS, DYTRAN, there are some key issues that require further study. One such issue concerns the FE model used in the simulation since this should include key structural features of the hull structure without leading to unrealistic computational times. A more fundamental issue is the need to develop computationally efficient criteria to reliably predict the onset of fracture and the subsequent tearing of large scale shell structures. The present paper addresses a more practical topic that resides in the very nature of the problem in hand. Grounding

and collision simulations involve high nonlinearity, geometric and material, contact/friction and shell element formulations that are capable to adequately represent the complex deformation and failure modes at a reasonable computational cost. Accordingly, it is highly desirable to compare the predictions of FE codes when simulating hull structures, despite the lack of experimental evidence to validate the results of the simulations. Moreover, in order to reduce the uncertainty of the comparison, at least in the selection of user defined parameters, it is advisable to analyze the same model geometry, imposing identical boundary conditions and using the same material deformation model. In this way variability in the prediction of key results can be attributed to algorithmic features of the FE codes used.

In the present work the FE codes ABAQUS and LS-DYNA were used to simulate a section of the double bottom structure of a tanker subjected to a grounding action. The vessel considered is a 265 m long tanker with a displacement of 150000 tons. A three bay section of the double bottom has been modeled, including the transverse floors in order to investigate their effect on the overall structural response. The model spans the entire breadth at the middle section of the vessel. Three separates models have been constructed in order to examine the effect of the longitudinal stiffening to the grounding action. The first model excludes all longitudinal stiffeners, with only the main girders included in the structure, while the final model is fully strengthened with flatbar stiffeners in the plating and girders. The load has been imposed to the double bottom through the motion of a conical shaped rigid indenter in the longitudinal direction of the ship. Two scenarios of grounding have been studied: impact on a longitudinal girder and also on the plating between two girders. Following the guidelines outlined in the previous paragraph, the same model geometry of the double bottom has been used in the simulations with ABAQUS and LS-DYNA, carried out at NTUA and NTNU respectively. The paper initially discusses the modeling techniques used and the selections made in the construction of the FE models. It then concentrates on the computed reaction forces obtained from the simulations of the longitudinally unstiffened and stiffened structure. Finally, a few additional runs are discussed which were intended to assist in the understanding of the differences found in the force prediction from the two FE codes.

2 FE MODELLING OF THE DOUBLE BOTTOM

2.1 Geometry details

The simulations were performed with the double bottom geometry shown in Figure 1. All structural elements shown in this Figure, i.e. outer/inner bottom and longitudinal girders are made of mild steel plating having a thickness of 15 to 19 mm. The span between two transverse floors is a = 4 mm and the floor plating has a thickness of 18 mm. All the longitudinal stiffeners on the plating are flatbars with a height of 400 or 430 mm and their thickness at mid-ship is 16 mm decreasing to 14 mm in the side shell region. The longitudinal girders are also strengthened with flatbars having a 200 mm height and an 18 mm thickness. Mild steel was modeled as an elastic perfectly plastic material with a yield stress of 245 MPa and a Young's modulus of 2.06 E + 5 MPa.

The rigid obstacle that comes in contact with the double bottom is presented in Figure 2. An identical impactor was used in Samuelides et al. (2007). It has a conical shape with a right angle apex which is replaced by a spherical tip having a radius of 241 mm. The impactor penetrates the outer shell of the double bottom up to a total distance of 500 mm. Two locations of impact have been considered, directly on a longitudinal girder, marked A in Fig. 1, and on the bottom plating between two girders, marked B in Fig. 1.

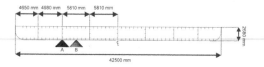

Figure 1.   Geometry of the double bottom.

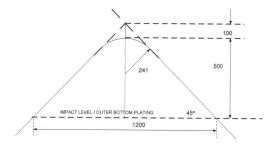

Figure 2.   Model or rock—conical obstacle.

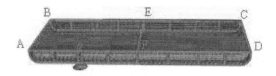

Figure 3.   FE model in the case of impact on girder highlighting regions edges and faces were the boundary conditions were applied.

The FE model of the three bay structure, including all the longitudinal stiffening, is shown in Fig. 3 for the case of impact on the girder. The total span of the model is a/2 + 3a + a/2, giving a total length of 14 m. The applied boundary conditions are as follows: (i) the front and rear faces of the model, bounded by the edges AB and CD respectively in Fig. 3, are fixed in the longitudinal direction; (ii) along the middle line EF and the longitudinal edges AB and CD the vertical displacement is set to zero and, (iii) the transverse displacement at the point E along the middle line is set to zero.

The impactor is slowly raised vertically in the forward a/2 span of the plating and is then moved horizontally in the longitudinal direction with a constant speed of 10 m/s until it reaches the aft a/2 s pan. Previous work on a simplified one bay model without transverse floors (Samuelides et al. 2007), has found small to negligible dependence of the reaction forces on penetration speed in the range of 20 to 1 m/s. This conclusion has also been confirmed here, albeit from a limited number of runs at speeds of 10 m/s and 5 m/s, see Section 4.

2.2 FE mesh details

A 200 mm square shell element was used to construct the FE mesh remotely from the impact region. In the impact zone the mesh was refined down to a 50 mm square element. Four node shell elements were used in both regions. Although the same mesh topology was used in both the ABAQUS and LS-DYNA models, the transition from the 200 mm to the 50 mm regions was constructed differently in the two codes. In LS-DYNA a traditional "fan-out" technique was used to coarsen the mesh leading to a somewhat wider fine region in the case of impact on the girder, Fig. 4a, while in ABAQUS constraint equations were used to eliminate nodes that fall on the mid-side of adjacent elements due to coarsening, Fig. 4b. This is a built-in feature of ABAQUS that leads to right angle elements in the transition region and can be applied with reasonable ease.

Figure 5a and 5b show the corresponding regions of the double bottom where the rigid cone impacts the plating amidst two longitudinal girders. Details of the mesh and the differences in the transition zone from the fine to the coarse region are clearly visible.

The selection of a 50 mm element size in the impact zone is based on previous work reported in Samuelides et al. (2007). The structure used was a simplified model of a double bottom of 5.2 m length and no transverse floors, also shown here in Figure 6. The mesh was refined in the central region from a coarse mesh of 100 mm square elements to a 50 mm element side

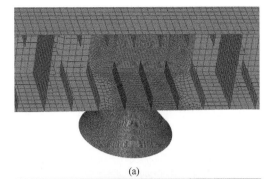

(a)

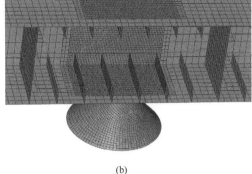

(b)

Figure 5. Mesh design in the vicinity of the impact zone for the impact on plating between two longitudinal girders: (a) NTNU model, (b) NTUA model.

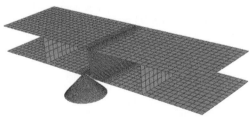

Figure 6. FE model of the single bay of the double bottom used in the mesh convergence study.

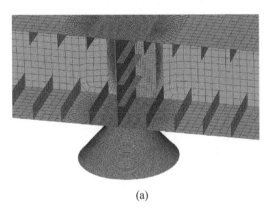

(a)

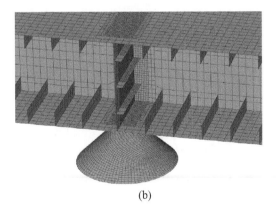

(b)

Figure 4. Mesh design in the vicinity of the impact zone for the impact on girder cased: (a) NTNU model, (b) NTUA model.

and, finally, down to a 25 mm element. The results in Samuelides et al. (2007) showed that, excluding a relatively short span of approximately 0.6 m, corresponding to the distance required for the rigid indenter to fully contact the bottom plating, the "steady-state" reaction forces obtained with the fine and medium meshes were in good agreement. Although the deformation mode of the girder was found to change with the element side, contours of the out-of-plane displacement exhibited similar distributions for the 25 and 50 mm element meshes. More detailed results are presented in Figure 7, which shows the in-plane and

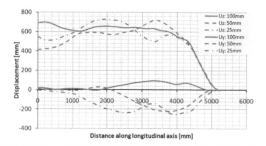

Figure 7. Variation of in-plane (Uz) and out-of-plane (Uy) displacement on the plate-girder intersection with element size.

out-of plane displacement variation along the central girder to the plate intersection. There is once more closer agreement in the 50 mm and 25 mm results, although some differences remain.

3 SOLVER AND ELEMENT ISSUES

The simulations were carried out with explicit solvers, LS-DYNA is primarily an explicit code and in ABAQUS the explicit procedures were specified. Both codes use a central difference scheme to integrate the equations of motion and diagonal (lumped) element mass matrices, see ABAQUS (2008), Hallquist (1998). They are therefore efficient solvers which do not require the solution of a system of equations, as implicit solvers do, but need a large number of small time increments dictated by the algorithm's stability limit.

Two types of shell elements were used in the FE models. The first type is based on a corotational velocity-strain formulation embedding a local coordinate system in each element. It was originally developed by Belytschko and his workers (e.g. Belytschko et al. 1984) and leads to a computationally efficient formulation, see the LS-DYNA Theory manual (Hallquist 1998), where the element is referred to as the Belytschko-Lin-Tsay (BLT) shell element. A similar element formulation is available in ABAQUS, referred to S4RS, however this element is limited to small membrane strains. The second type of elements, used only in the analysis with the ABAQUS code, does not have the small strain limitation of S4RS and defines the membrane strain increment through a logarithmic strain expression. These elements are referred to in ABAQUS as S4 or S4R, depending whether full integration or reduced (R) one point integration is used to define the element stiffness. In the simulations reported here one point quadrature elements were used, that is the BLT element in LS-DYNA and the S4R/S4RS elements in ABAQUS, due to their superior computational efficiency. Selective checks have

been carried out to ensure that the hourglass energy in the model is very small compared to the internal energy.

4 RESULTS AND DISCUSSION

The results of this work are presented here in the following three subsections. The first subsection concerns the simulations carried out for the impact on the longitudinal girder while the second discusses the results obtained for the impact on the plating between two longitudinal girders, Figure 1. The third subsection discusses additional analyses that were carried out in order to clarify certain differences in the ABAQUS and LS-DYNA results.

4.1 Impact on a longitudinal girder

Prior to examining the predictions from the two codes, it is pertinent to examine the effect of the longitudinal stiffening on the plating and girders. Accordingly, the longitudinal and vertical reaction forces are shown in Figures 8 and 9 respectively. It can be seen that there is a gradual increase of the longitudinal force as

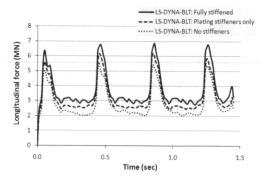

Figure 8. Longitudinal force versus time—effect of stiffening.

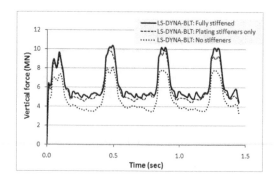

Figure 9. Vertical force versus time—effect of stiffening.

stiffening is added to the model, leading to an approximately fifty percent increase on the reaction force between the floors when comparing the unstiffened and fully stiffened structure.

The reaction forces obtained from the two FE codes are compared in Figures 10 and 11. For clarity, only the longitudinal and vertical forces for the unstiffened and fully stiffened models are presented. It is evident that the peaks in the forces, that is corresponding to the short period when the impactor crosses the transverse floors, are in good agreement. However, both the longitudinal and vertical forces computed with ABAQUS in the longer period between two floors are significantly lower than the LS-DYNA values, in the range of 15–30%. The curves obtained with ABAQUS also show a higher degree of fluctuation. This can be attributed to a continuous change of the contact points between the impactor and the structure, with the tip of the cone alternating in contact with elements on either side of the girder to plane intersection.

Figure 10 also shows the computed horizontal force from the ABAQUS simulation with the impactor moving at a constant speed of 5 m/s, in contrast to all the other results presented here where the impactor speed is 10 m/s. As discussed at the end of sub-section 2.1, there is negligible difference in the reaction forces. The corresponding data for the vertical force are also in very close agreement.

4.2 Impact on plating between two longitudinal girders

As in the case of impact on the girder, it can be seen from Figures 12 and 13 that the inclusion of the stiffeners leads to significantly higher reaction forces. However, when comparing the fully stiffened and plating only stiffened models, it is evident that the reaction forces are very similar. This is attributed to the 'localized' deformation induced by the impactor without spreading to the neighboring girders which remain almost intact. Clearly this result will change if a wider impactor was used in the simulations or if the penetration depth was larger. In the present work the width of the conical impactor that contacts the outer plating is approximately 1200 mm while the distance between the two adjoining girders is 5810 mm, see Figure 1.

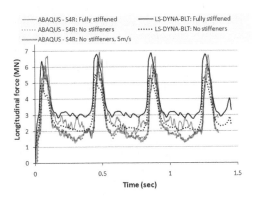

Figure 10. Comparison of the longitudinal force versus time data computed with the two FE codes.

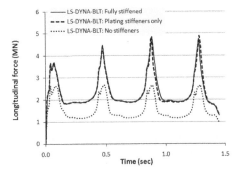

Figure 12. Longitudinal force versus time—effect of stiffening.

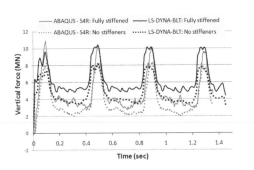

Figure 11. Comparison of the vertical force versus time data computed with the two FE codes.

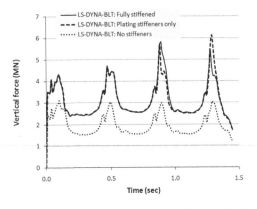

Figure 13. Vertical force versus time—effect of stiffening.

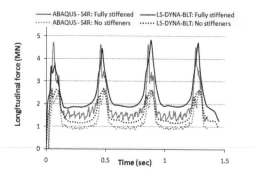

Figure 14. Comparison of the longitudinal force versus time data computed with the two FE codes.

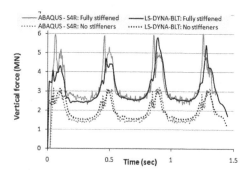

Figure 15. Comparison of the longitudinal force versus time data computed with the two FE codes.

The reaction forces obtained from the analyses with the two FE codes are presented in Figures 14 and 15. Again only the results from the unstiffened and fully stiffened models are shown. There is closer agreement here in the vertical forces, at least during the periods when the impactor is moving between the transverse floors, although there are some discrepancies in the peaks for the fully stiffened case.

The differences in the horizontal forces are observed only for the periods between the floors, with LS-DYNA giving again higher values. However these are in this case somewhat smaller, in the range of 20–25%.

4.3 Additional analyses

A few additional analyses have been performed in order to hopefully understand better the differences in the reaction forces computed with the two codes. Only the impact on the longitudinal girder has been re-analyzed since the discrepancies were higher in this case. In the first instance the Belytschko-Lin-Tsay type elements (S4RS) available in ABAQUS were used in a re-analysis of the NTUA model, instead of the

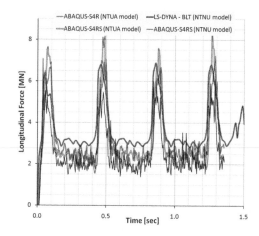

Figure 16. Comparison of the longitudinal force versus time curves obtained with ABAQUS using the S4RS element and the NTNU model with previous results. Only the impact on girder with the fully stiffened structure is shown.

S4R element type used in all the simulations reported above. A further run was carried out with ABAQUS using the NTNU model and the S4RS elements. The boundary conditions were identical in all models.

The longitudinal reaction forces computed from these analyses are compared in Figure 16 to the corresponding LS-DYNA results and the ABAQUS run with the S4R shell element. The use of the S4RS element has somewhat reduced the differences in the force values computed for the period between two floors, particularly in the second floor. However it has clearly resulted in higher peak force values when the impactor is crossing a floor, suggesting that the element leads to additional stiffening. The analysis of the NTNU model with ABAQUS and the S4RS element has again resulted in horizontal forces, particularly in the in-between floor periods, which are in closer agreement with other simulations carried out with ABAQUS. Accordingly the discrepancies in the reaction force predictions cannot be attributed to differences in the mesh topology in the vicinity of the impact zone. Further, the use of Belytschko-Lin-Tsay type elements in the simulations performed with both programs has also been unable to limit the differences observed. Although there may well be variations in the implementation of the BLT elements in the two FE codes, the discrepancies in the force predictions are probably attributed to other algorithmic features, for example the contact algorithms.

5 CONCLUSIONS

A key prerequisite for the wider application of the Finite Element Method in the design of ships against grounding and other collision accidents, is the reliability of the predictions. In this context, a series of grounding simulations of the double bottom structure of tankers have been performed independently by NTUA and NTNU using two commercially available FE codes, ABAQUS and LS-DYNA. The emphasis has been to model a full breadth section of the double bottom and to use identical model geometries in the FE simulations. Further, in order to reduce the uncertainty of the comparison, the same mesh topology and element sizes were used overall, although certain differences in the topology remained in the transition from the fine to coarse mesh regions due to differing features in the two codes. In order to achieve practical computational times, one point quadrature shell elements were used in all simulations. However the element formulations differed, using in LS-DYNA a popular shell element developed by Belytschko and co-workers while selecting in ABAQUS a finite (logarithmic) membrane strain alternative. The simulations considered two possible scenarios for grounding, that is impact on a longitudinal girder and on the plating, in-between girders. The reaction forces computed with the two FE codes for the periods between two transverse floors were found to differ in the range 15–30%. There was, however, closer agreement in the force peaks obtained as the impactor crosses each floor present in the model. Additional analyses were performed with ABAQUS in order to examine possible reasons for these discrepancies, in particular using the available Belytschko type shell element and repeating a run with the NTNU FE model. However the differences were not markedly reduced and this suggests that these may be due to other algorithmic features in the two codes. Accordingly, the present paper offers some useful insight into the variability of grounding predictions obtained with two widely used FE codes.

ACKNOWLEDGEMENTS

The work reported in this paper has been performed within the scope of the project MARSTRUCT—Network of Excellence on Marine Structures, (www.mar.ist.utl/marstruct), which is financed by the EU through the GROWTH Programme under contract TNE3-CT-2003-506141.

REFERENCES

Simonsen, B.C. & Wierzbicki, T. 1998. Plasticity, fracture and friction in steady state plate cutting. *International Journal of Impact Engineering* 21: 387–411.

Hong, L. & Amdahl, J. 2008. Plastic mechanism analysis of the resistance of ship longitudinal girders in grounding and collision. *Intl. J. Ships and Offshore Structures*: in press.

Samuelides, M.S., Voudouris, G., Toulios, M., Amdahl, J. & Dow, R. 2007. Simulation of the behavior of double bottoms subjected to grounding actions. In: *International Conference on Collision and Grounding of Ships*; *International Conference for Collision and Groundings ICCGS2007, September 2007*. Hamburg, Germany.

Naar, H., Kujala, P., Simonsen, B.C. & Lundolphy, H. 2002. Comparison of the crashworthiness of various bottom and side structures. *Marine Structures* 15: 443–460.

Belytschko, T., Lin, J.I. & Tsay, C.S. 1984. Explicit Algorithms for the Nonlinear Dynamics of Shells. *Computer Methods in Applied Mechanics and Engineering* 43: 251–276.

Hallquist, J.O. 1998. LS-DYNA Theoretical Manual. Livermore Software Technology Corporation. Livermore: California, USA.

ABAQUS. 2008. Theory Manual, Version 6.8. Dassault Systèmes Simulia Corp. Providence: RI, USA.

Fire and explosion

Analysis and Design of Marine Structures – Guedes Soares & Das (eds)
© 2009 Taylor & Francis Group, London, ISBN 978-0-415-54934-9

CFD simulations on gas explosion and fire actions

J.K. Paik, B.J. Kim, J.S. Jeong & S.H. Kim
Pusan National University, Busan, Korea

Y.S. Jang & G.S. Kim
Hyundai Heavy Industries, Ulsan, Korea

J.H. Woo, Y.S. Kim & M.J. Chun
Daewoo Shipbuilding & Marine Engineering, Geoje, Korea

Y.S. Shin
American Bureau of Shipping, Houston, USA

J. Czujko
Nowatec AS, Asker, Norway

ABSTRACT: On offshore installations, hydrocarbon (gas) explosion and fire are the two most important hazardous events which cause the highest potential risk. Within the framework of quantified risk assessment and management for offshore installations, therefore, more refined computations of consequences or hazardous action effects due to both explosion and fire are required. While the action characteristics must be first identified prior to the consequence analysis, the objective of the present paper is to develop a modeling technique of CFD simulations on hydrocarbon explosion and fire actions. The results of the present paper are part of Phase I in association with the joint industry project on explosion and fire engineering of FPSOs (EFEF JIP) which is composed of three Phases.

1 INTRODUCTION

For structural design and safety assessment, the characteristics of both actions and action effects should be identified, where the term action is an external load applied to the structure (direct action) or an imposed deformation or acceleration (indirect action), while the term action effect is an effect of actions on a global structure or structural components (ISO 2007). The present paper is focused on the characteristics of actions due to gas explosion and fire.

Hydrocarbon explosion and fire are different phenomena by nature, although they are often accompanied by in offshore installations.

Hydrocarbon is an organic compound containing only hydrogen and carbon (Nolan 1996). The hydrocarbons typically take the form of gases at ordinary temperatures, but they change the form to the liquid or even the solid state with increase in the molecular weight. In fact, the principal constituents of petroleum and natural gas are the hydrocarbons.

Hydrocarbons can be exploded by ignition with the mixture of an oxidizer (usually oxygen or air) when the temperature is raised to the point where the molecules of hydrocarbons react spontaneously with

an oxidizer and subsequently combustion takes place. The hydrocarbon explosion causes a blast or a rapid increase in pressure.

On the other hand, fire is a combustible vapor or gas combining with an oxidizer in a combustion process manifested by the evolution of light, heat, and flame.

Of primary concerns in terms of actions due to hazards within the framework of risk assessment and management is impact pressure on explosion and elevated temperature on fire.

Topsides of offshore platforms are likely exposed to hazards such as hydrocarbon explosion and/or fire. A number of major accidents involving hydrocarbon explosion and fire have reportedly happened, see Figure 1. Some details of such accidents may be found in Vinnem (2007).

Within the framework of quantified risk assessment and management, it is required to analyze consequences due to hazards in a more refined way.

Before starting the consequence analysis, the characteristics of actions arising from hazards must be first identified (Czujko 2001, 2005, 2007, Paik & Thayamballi 2003, 2007, Paik & Melchers 2008).

It is recognized that CFD (computational fluid dynamics) method is one of the most powerful tools for

Figure 1. Piper Alpha explosion and fire accident, 6 July 1988 (Cullen 1990).

identifying action characteristics relating to hydrocarbon explosion and fire, among various other methodologies.

The aim of the present study is to develop a modeling technique for CFD simulations in terms of hydrocarbon explosion and fire actions in association with the joint industry project on explosion and fire engineering of FPSOs (abbreviated as EFEF JIP) which was started from March 2008 (Paik & Czujko 2008). The main objective of the project has been to develop a fully quantified risk assessment and management against hydrocarbon explosion and fire.

The EFEF joint industry project is composed of the following three Phases, namely

- Phase I (March 2008–February 2009)

  – Literature survey of previous JIPs
  – Technologies for frequency analysis of gas explosion and fire on FPSOs
  – Technologies for action analysis of gas explosion and fire

- Phase II (March 2009–February 2010)

  – Hypothetical design of FPSO topsides
  – Wind tunnel tests on FPSO topside modules in association with fire heat dispersion with varying direction and speed of wind actions
  – Fire tests on plated and framed structural models in association with actions and action effects
  – Technologies for action effect (consequence) analysis of gas explosion and fire

- Phase III (March 2009–February 2011)

  – Quantified risk assessment and management of FPSO topsides against gas explosion and fire
  – Acceptance criteria
  – Guidelines and engineering handbook

The results of the present paper are part of Phase I with the focus on the development of a modeling technique of CFD simulations to identify action characteristics due to hydrocarbon explosion and fire accidents. ANSYS CFX (2008) code is employed for that purpose.

2 ANALYSIS OF GAS EXPLOSION ACTIONS

2.1 Experimental study

Gieras et al. (2006) carried out a laboratory test on methane gas explosion. In the present study, their experimental study is reanalyzed by ANSYS CFX code in order to develop a modeling technique of CFD simulations in terms of identifying the explosion pressure actions.

Figure 2 shows the layout of the test. A 40 dm^3 volume chamber was used for the test. While the concentration of methane gas was varied in the original test, the present study adopts the test results on the methane concentration of 9.5%.

The initial gas temperature inside the explosion chamber was also varied at 293, 373 and 473 K while the ambient temperature outside the chamber was supposed to be 293 K. Pressure gauge and temperature sensor were attached to the outer surface of the chamber.

Figure 3 shows a photo of the chamber with the diameter of 140 mm and the height of 441 mm, used for the test.

2.2 CFX simulations

The chemical reaction in association with methane gas explosion with the mixture of air can be represented by

$$CH_4 + 2(O_2 + 3.76N_2) \rightarrow CO_2 + 2H_2O$$
$$+ 2(3.76N_2) + Energy \qquad (1)$$

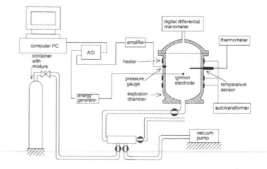

Figure 2. The layout of the laboratory test facility on methane gas explosion (Gieras et al 2006).

Figure 3. A photo of the chamber used for the explosion test (Gieras et al 2006).

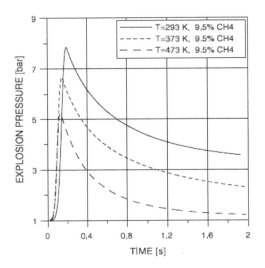

Figure 4. Explosion pressure versus time history obtained from the test with the methane concentration of 9.5% (Gieras et al 2006).

After the explosion, the energy is released causing a blast as shown in Figure 4. The peak pressure is reached in a very short rise time and the pressure decays with time.

ANSYS CFX code is now employed for the analysis of explosion actions. Figure 5 shows the CFX model for the Gieras explosion test, using tetrahedral elements. A half chamber is taken as the extent of analysis considering a symmetric action feature. The total number of nodes and elements are 32,364 and

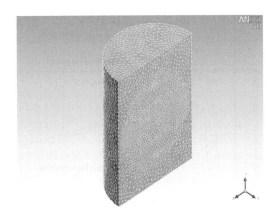

Figure 5. ANSYS CFX model for the Gieras explosion test using the tetrahedral elements.

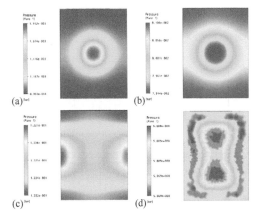

Figure 6. Variations of explosion pressure with time, (a) 0.002 s, (b) 0.012 s, (c) 0.02 s and (d) 0.132 s, obtained by ANSYS CFX simulations.

180,315, respectively, which were decided based on some convergence study.

In the CFX modeling, the turbulence is modeled by the k-epsilon model, and the radiation is modeled by the P1 model. The total energy model is used for modeling heat transfer inside the chamber. The CFX code facilitates four types of wall heat transfer options, namely adiabatic, temperature, heat flux, and heat coefficient. The present study uses the temperature option for the wall heat transfer behavior so as to reflect the ambient temperature at 293K.

Figure 6 shows the variations of explosion pressure with time at (a) 0.002 s, (b) 0.012 s, (c) 0.02 s and (d) 0.132 s, and Figure 7 shows the explosion pressure versus time history, as those obtained by CFX simulations. It is seen from Figure 7 that the CFX

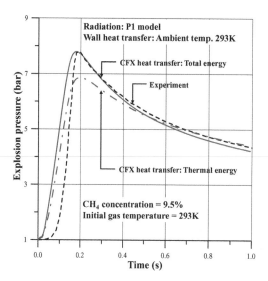

Figure 7a. Effect of heat transfer model on explosion pressure versus time history for the Gieras explosion test model, obtained by CFX simulations.

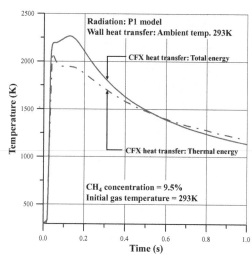

Figure 8a. Effect of heat transfer model on temperature versus time history for the Gieras explosion test model, obtained by CFX simulations.

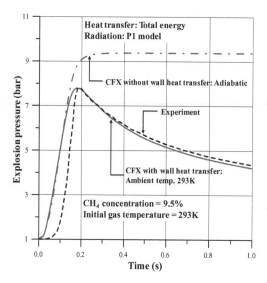

Figure 7b. Effect of wall heat transfer model on explosion pressure versus time history for the Gieras explosion test model, obtained by CFX simulations.

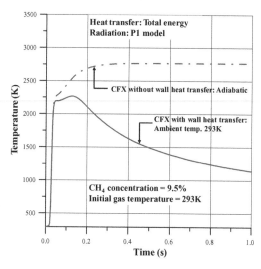

Figure 8b. Effect of wall transfer model on temperature versus time history for the Gieras explosion test model, obtained by CFX simulations.

simulations are in good agreement with experimental results in terms of pressure versus time history which can be characterized by four parameters, namely rise time, peak pressure, type of decay after reaching the peak pressure and duration time of impact pressure (Paik & Thayamballi 2007).

Figure 7(a) represents the effect of heat transfer modeling in terms of thermal or total energy models.

It is evident from Figure 7(a) that the total energy model taking account of the kinetic energy effect is more relevant for the analysis of explosion actions.

Figure 7(b) represents the effect of chamber wall heat transfer modeling. It is seen that the ambient temperature model equivalent to the test gives accurate results, while the adiabatic model does not transfer the heat energy to the outside of the chamber.

Although time-variant temperatures were not measured at the test, the CFX simulations can provide the time-variant temperature related information as shown in Figure 8. The peak temperature reaches over 2,200 K at the time of around 0.1 s after the ignition, which is meant to be prior to the time when the peak pressure occurs, i.e., at 0.2 s. Figure 8 also indicates that the adiabatic model of the wall heat transfer behavior can increase the peak temperature as well.

3 ANALYSIS OF FIRE ACTIONS

3.1 Experimental study

HSE (1995) carried out a laboratory test on jet fire accident in association with the combustion of propane gases.

A thermally insulated compartment was used for the fire test as shown in Figure 9. The compartment has an open door for the purpose of ventilation, with the breadth of 2.5 m and the height of 2.0 m. The heat and flame arising from fire can then flow to the outside via the ventilation door. The volume of the compartment is approximately 135 m^3.

The jet fire is activated through a nozzle located at the bottom centre of the compartment. The height of the nozzle is 380 mm and the diameter is 90 mm. It is supposed that the mass flow of jet fire at nozzle is 0.33 kg/s which may be equivalent to the flow velocity of 190 m/s. The ambient temperature outside the compartment is 281 K at the tests.

In the test, the temperatures during fire were measured by temperature sensors at various locations or elevations, i.e., at A-B, C-D and opening as indicated in Figure 9. Figure 10 shows a photo of jet firing inside the test compartment.

Figure 11 shows the measurements of temperatures at various locations sometimes after starting of the jet firing. It is seen that the maximum temperature reaches up to 1400 K. The temperature at the location of C-D tends to be higher than that at A-B. This is because the heat flow moves out via the opening.

Figure 10. A photo of the jet fire test (HSE 1995).

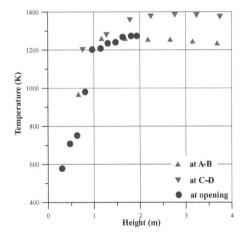

Figure 11. Variations of temperatures at locations or elevations as those obtained by the test (HSE 1995).

3.2 CFX simulations

While an imperfect combustion of propane gases may produce some CO and C_2H_2, the chemical reaction in association with the perfect combustion of propane gases with the mixture of air can be represented by

$$C_3H_3 + 5O_2 \quad \rightarrow \quad 3CO_2 + 4H_2O$$

ANSYS CFX code is now used for the analysis of the jet fire actions. Useful CFD simulations have also been performed on this test in the literature (e.g., Wen & Huang 2000).

Figure 12 shows the CFX model of the jet fire test, developed in the present study using the tetrahedral elements. Considering a symmetric action feature, a half the compartment is taken as the extent of analysis.

It is fairly considered that the heat flow actions outside the compartment may also affect the action characteristics inside the compartment. In this regard, some extent of the open air around the ventilation door was also included in the CFX model as shown in

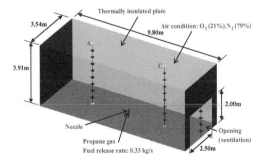

Figure 9. The jet fire test facility (HSE 1995).

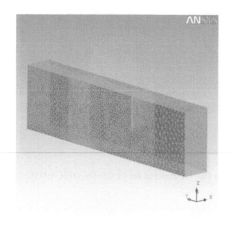

Figure 12a. ANSYS CFX model for the jet fire test, using the tetrahedral elements.

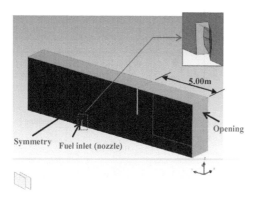

Figure 12b. Extent of the ANSYS CFX model analysis for the jet fire test.

Figure 12(b). The total number of nodes and elements are 30,918 and 165,353, respectively.

In the CFX modeling, the thermal energy model for heat transfer behavior, the eddy dissipation model for combustion behavior, the P1 model for radiation behavior, the k-epsilon model for turbulence behavior were applied.

In terms of wall heat transfer modeling, three types of options, i.e., with adiabatic, temperature, and transfer coefficient models were discussed in the present study. The adiabatic model represents that the heat flow does not transfer at all to the outside of the compartment via the wall, but the temperature model allows the transfer of heat flow directly to the ambient temperature. The transfer coefficient model can deal with a partial heat transfer behavior.

The test compartment is thermally insulated and it is thus considered that the wall heat transfer is not fully allowed but partially possible via the wall of the compartment. The total heat flux via the wall may be given by

$$q_w = h_c(T_o - T_w)$$

where q_w = total heat flux, h_c = heat transfer coefficient which is taken as $h_c = 30$ W/m²K for the present test, T_o = external or outside (ambient) temperature which is equivalent to 281 K in the present fire test, T_w = internal temperature.

Figure 13 shows the variations of temperatures with time at (a) 0.02 s, (b) 1.0 s, (c) 2.0 s, and (d) 9.0 s after starting of the jet firing, as those obtained by CFX simulations. It is evident from Figure 13 that the heat flow moves out via the opening with time. Figure 14 compares the temperatures at various locations or elevations between test measurements and CFX simulations.

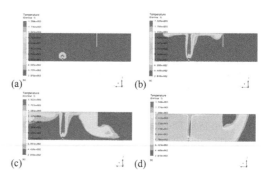

Figure 13. Variations of the temperatures with time at (a) 0.02 s, (b) 1.0 s, (c) 2.0 s and (d) 9.0 s.

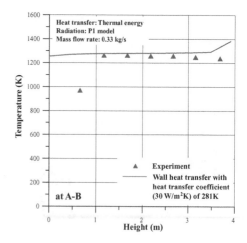

Figure 14a. Comparisons of temperatures between test measurements versus CFX simulations at A-B.

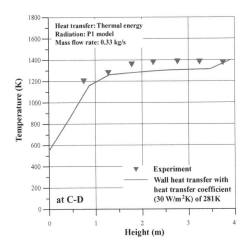

Figure 14b. Comparisons of temperatures between test measurements versus CFX simulations at C-D.

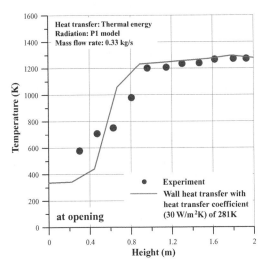

Figure 14c. Comparisons of temperatures between test measurements versus CFX simulations at opening.

Although the time-variant temperature characteristics were not measured at the test, CFX simulations provide such an information as shown in Figure 15.

Figure 16 shows the effects of wall heat transfer on temperature versus time history at the location of C-D with the elevation of 1.7 m, as obtained by CFX simulations. It is seen from Figure 16 that the adiabatic model overestimates the fire temperatures, while the ambient temperature model underestimates the fire temperatures. The proposed CFX model with wall heat transfer coefficient of 30 W/m²K gives very accurate solutions compared to the experimental results.

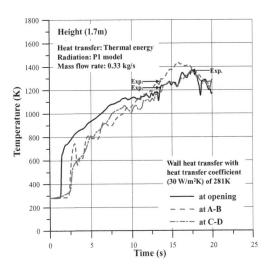

Figure 15. Temperature versus time history at various locations as obtained by CFX simulations at elevation of 1.7 m.

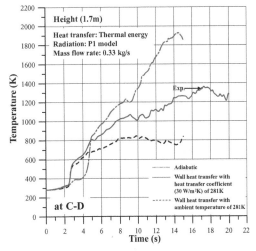

Figure 16. Effects of wall heat transfer models on temperature versus time history at C-D as obtained by CFX simulations at elevation of 1.7 m.

4 CONCLUDING REMARKS

The aim of the present paper has been to develop a CFD modeling technique for the analysis of hydrocarbon explosion and fire actions. Existing test data on methane gas explosion and propane gas jet fire was reanalyzed by ANSYS CFX code.

It is concluded that the CFD simulations proposed in the present study are in good agreements with experimental results.

The present study has been limited to laboratory test data obtained in closed chamber or compartment, although topsides of FPSOs are likely in open or partially closed environments, where the dispersion behavior of blast or heat may play an important role. In this regard, the related study is on-going in association with the EFEF JIP.

ACKNOWLEDGEMENTS

The present study was undertaken at the Ship and Offshore Structural Mechanics Laboratory (National Research Lab., Grant No. ROA2006000102390, funded by Korea Science and Engineering Foundation), Lloyd's Register Educational Trust (LRET) Research Centre of Excellence at Pusan National University, Korea.

The results of the present paper is part of Phase I in association with EFEF JIP (Joint Industry Project on Explosion and Fire Engineering of FPSOs). The authors are pleased to acknowledge the support of partners involved in the EFEF JIP, which include Pusan National University (Korea), Nowatec AS (Norway), Hyundai Heavy Industries (Korea), Daewoo Shipbuilding and Marine Engineering (Korea), American Bureau of Shipping (USA), and Health and Safety Executive (UK).

REFERENCES

ANSYS Inc. 2008. ANSYS CFX release 11.0, Canonsburg, PA, USA.

Cullen, L. 1990. The public inquiry into the Piper Alpha disaster. HMSO, London, UK.

Czujko, J. 2001. Design of offshore facilities to resist gas explosion hazard: Engineering handbook. CorrOcean, Oslo, Norway.

Czujko, J. 2005. Computational methods in the analysis of explosion resistant barriers. Proceedings of International Conference on Computational Methods in Marine Engineering, MARINE 2005.

Czujko, J. 2007. Consequences of explosions in various industries. Safety and Reliability of Industrial Products, Systems and Structures, SAFERELNET 2007.

Gieras, M., Klemens, R., Rarata, G. & Wolanski, P. 2006. Determination of explosion parameters of methane-air mixtures in the chamber of 40 dm^3 at normal and elevated temperature. *Journal of Loss Prevention in the Process Industries*, 19: 263–270.

HSE. 1995. Large scale compartment fires: Experimental details and data obtained in test comp-27. Report-OTO94024, Health and Safety Executive, London, UK.

ISO. 2007. International standards ISO 18072–1. Ships and marine technology—Ship structures—Part 1: General requirements for their limit state assessment, Geneva.

Nolan, D.P. 1996. Handbook of fire and explosion protection engineering principles for oil, gas, chemical, and related facilities. Noyes Publications, Westwood, NJ, USA.

Paik, J.K. & Czujko, Z. 2008. Joint industry project on explosion and fire engineering of FPSOs. Proceedings of the Workshop on Design of FPSO Structures, 20–21 November, Geoje, Korea.

Paik, J.K. & Melchers, R.E. 2008. Condition assessment of aged structures. CRC Press, New York, USA.

Paik, J.K. & Thayamballi, A.K. 2003. Ultimate limit state design of steel-plated structures. John Wiley & Sons, Chichester, UK.

Paik, J.K. & Thayamballi, A.K. 2007. Ship-shaped offshore installations: Design, building, and operation. Cambridge University Press, Cambridge, UK.

Vinnem, J.E. 2007. Offshore risk assessment. Springer-Verlag, London, UK.

Wen, J.X. & Huang, L.Y. 2000. CFD modeling of confined jet fires under ventilation-controlled conditions. *Fire Safety Journal*, 34: 1–24.

Analysis and Design of Marine Structures – Guedes Soares & Das (eds)
© 2009 Taylor & Francis Group, London, ISBN 978-0-415-54934-9

The effects of reliability-based vulnerability requirements on blast-loaded ship panels

S.J. Pahos & P.K. Das
Universities of Glasgow and Strathclyde, Department of Naval Architecture & Marine Engineering, Glasgow, UK

ABSTRACT: This paper describes the structural response of typical ship panels subjected to blast loads from asymmetric threats, commonly known as acts of terrorism. Three types of asymmetric threats will be investigated, namely: the blast load from a 4 kg Improvised Explosive Device (IED), the blast load from a Vehicle Borne IED (VBIED), and the coupled load of blast and impact from a Rocket Propelled Grenade (OG-7V). The investigation of blast loads on ship structures is a hot topic as maritime piracy and acts of terrorism are on the rise since the attacks of September 11, 2001. Previous work by Pahos & Das (2009) put forward a minimum increase of specific scantlings in order to raise the reliability index of ship panels subjected to blast loads. This proposal of Reliability-Based Vulnerability Requirements (RBVRs) is applied on the target panels in each of the above-mentioned threats; the benefits are discussed in terms of reduced fragmentation and the blast-induced structural damage. The non-linear analysis of blast and ballistic loading is carried out in LS-DYNA where the appropriate keywords of blast loads and contact are invoked.

1 INTRODUCTION

1.1 Contemporary events

Following the September 11, 2001 attacks, the Global War on Terror (GWOT) was launched by U.S. and allied forces. The post-9/11 years met a steady increase in the cost of manpower to patrol our coasts, ports, and further increase safety at sea. Material and financial recourses have also adapted to this demanding environment to respond adequately and effectively to the modern needs of this war. Various political, economical, and social factors have shaped the outcome of the GWOT in every place of the world. To date, all-above measures have shaped modern life and particularly transportation infrastructure, under the spectrum of preemptiveness and safety. Maritime terrorism has denigrated the market of leisure (cruise ships), and function (ROPAX vessels) with numerous deadly attacks around the world. Well-documented events such as that of SuperFerry 14, Seabourn Spirit, Limburg, Al Salam Boccaccio 98, are there to remind us the stark reality that there the GWOT has a long way ahead. Despite a series of regulations and amendments to cultivate a culture of safety and preemptiveness, it is acknowledged that all measures hitherto cannot prevent determined terrorists or other acts of extremism. The authors also recognize that little has been done in the maritime community to encourage the adoption of structural measures that would mitigate blast effects originating from acts of terrorism. No structural requirements exist to improve the blast response

of ship structures and mitigate risk at sea. Risk originating from fire, flooding, loss of structural integrity, stability, or collision has been addressed adequately through IMO and Classification rules. The present paper demonstrates the effect of RBVRs against a series of design threats, and presents the benefits from their application in reducing blast effects and enhancing safety at sea.

1.2 Background

Previous work on blast-loaded ship panels, from an IED of just 4 kg TNT equivalent, revealed a negative reliability index value of -0.05 (Pahos & Das, 2008). As this is deemed to be unacceptable given today's safety standards for ship structures, it was to an undetermined degree anticipated, considering that traditional ship panels commonly found throughout merchant vessels are not designed to withstand blast loads. Moving along the lines of preemptiveness and risk-based design, the next step forward was to identify what decision-makers need to do in order to raise the reliability index of ship panels, and subsequently mitigate blast effects. Taken into consideration the unavoidable manufacturing-induced distortions that degrade the strength of ship panels, a sensitivity analysis of the blast-loaded system revealed the most contributive strength variables considering the arisen stresses (Pahos & Das, 2008). Solicitous attention to detail was given so that existing scantlings would be increased, and consequently raise the reliability index. Further research concluded that for a typical stiffened

Table 1. The required scantlings increase for positive ($\beta > 0$), and satisfactory ($\beta = 1.5$), reliability index of blast-loaded ship panels from 4 kg IED (Pahos & Das, 2009).

Reliability index variable	$\beta > 0$ Req'd increase	$\beta = 1.5$ Req'd increase
Plate thickness	10%	20%
Flange thickness	N/C	25%
Yield stress	N/C	13%

Table 2. The considered design threats of this work.

ΘI:	4 kg PETN IED
ΘII:	227 kg VBIED[†]
ΘIII:	OG-7V

[†] 500lbs standard deliverance of sedan vehicles (ATF).

panel, loaded by the impulse of a 4 kg PETN (Pentaerythritol TetraNitrate) IED, the required increase for positive, and what is deemed to be a satisfactory reliability index of 1.5, is given in Table 1 known as RBVRs (Pahos & Das, 2009).

The application of RBVRs on ship panels should have a selective character in the sense that RBVRs could be applied in sensitive areas like compartments below the water line, or panels at the waterline level. These areas could be identified using total ship survivability assessment methods, or results from previous studies that reveal high vulnerability in the design. It should be mentioned here that sensitive areas are not necessarily below the waterline as one may assume. Previous work that looked at the survivability of a ROPAX vessel following the detonation of a VBIED on its car deck revealed that in certain sea states the vessel capsized (Pahos, 2008).

2 DEFINITIONS

2.1 The concept of design threats

Doerry (2007) credits the term design threat as:

"Any threat for which a design threat outcome has been defined"

Three, out of the seven, design threats Θ previously identified by Pahos (2008), are considered in this work as detailed in Table 2.

2.2 The Concept of design threat outcomes

Again, Doerry (2007) defines a design threat outcome as:

"The acceptable performance of the ship in terms of the aggregate of susceptibility, vulnerability, and recoverability when exposed to a design threat".

From then on, what decision-makers define as "design threat" and "design threat outcome" is only limited by human imagination. A lot of progress is

being made nowadays in reducing the susceptibility of ROPAX vessels; sensor technologies to detect explosives on the ramps, or piers, have appeared in many U.S. ports. Since recoverability is primarily an operational aspect relying on adequate training of the crew and perhaps on efficient instructions of passengers, from a structural point of view the next element to be considered is that of vulnerability. Vulnerability is defined as (Reese et al., 1998):

"The inability to withstand a man-made hostile environment after being hit"

From the above three terms, vulnerability reduction is closely tied to reducing human casualties and further mitigating blast effects.

3 DESIGN THREAT I (4KG PETN IED)

3.1 Mechanical properties of the FE model

The aspect of high tensile steel (HTS) to provide for lighter structures without loss of strength, the absence of transverse bulkheads on the car deck, and the passage of piping through structural members justify the use of HTS in the hull of ROPAX vessels. The same design principle is adopted herein; the FE model is assumed to have the mechanical properties of Table 3.

The effect of strain rate on yield strength is taken into account with the Cowper-Symonds strain rate model with $C = 3200$ s^{-1} and $q = 5$, being widely used and accepted values for HTS (Paik et al., 1999). The *MAT_PLASTIC_KINEMATIC card is used to model the geometry.

3.2 Geometry of the FE model

The blast-loaded system considered in this paragraph is assumed to be free of any imperfections with nominal scantlings as given in Table 4. SHELL163 elements were used throughout the model with five integration points through the element.

3.3 Boundary conditions

For high laterally distributed pressures the edge rotational constraints of the panel can become large enough to correspond to a clamped condition. In most practical situations, an idealised condition where the panel

Table 3. Material properties of the stiffened panel.

Property	Value
Density (kgm^{-3})	7800
Modulus of Elasticity (Pa)	206E + 09
Poisson's Ratio (Pa)	0.3
Yield Stress (Pa)	355E + 06
Tensile Stress (Pa)	510E + 06
Failure Strain	0.22

Table 4. The scantlings of the stiffened panel.

Structural member	Dimensions in (mm)
Panel span	8800
Panel depth	5300
Plate thickness	12.5
Web frame	W800 × 10 + F250 × 20
Web frame Spacing	2400
Frame profile	HP200 × 12
Frame spacing	800

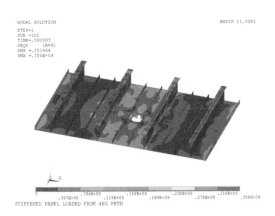

Figure 1. FE prediction of equivalent stresses (Pa) of the stiffened panel following the detonation of 4 kg PETN at point blank distance.

edges are considered to remain almost straight with zero rotational constraints along four edges, has been found to yield adequate results (Paik & Thayamballi, 2003). The same idea is adopted herein.

3.4 Structural response

The FE prediction of the arisen stresses when the IED is at point blank distance (0.2 m) is given in Figure 1. The resulting gash of 0.2 m^2 is the consequence of out-of-plane loads associated with global membrane and shear response known as direct (dynamic) shear. It can be seen that most of the blast effect is concentrated within the central zone between the stiffeners where the points of inflection fell along the circumference of a circle, known as inflection circle, with a diameter slightly greater than the stiffener spacing. The simulation in LS-DYNA ran for 0.5s allowing enough time for any oscillations to dampen down. Detonation was modelled with the *LOAD_BLAST card.

4 DESIGN THREAT II (VBIED)

4.1 Mechanical properties of the FE model

The second design threat is that of a Vehicle Borne IED (VBIED) onboard a ROPAX vessel. Such vessels usually shuttle between ports on a very tight schedule as passengers drive their vehicles on board essentially leaving no time to vet what is being brought onboard. A typical loading profile of a ROPAX vessel is shown in Figure 2. The booby-trapped vehicle is designated by bold lines beside the side shell on deck 3.

The same mechanical properties as the ones in Table 3 are considered in the FE model.

4.2 Geometry of the FE model

The parallel middle body of the FE model is approximately 1/10th of the LoA, i.e. 19.4 m and it extends up to deck 5 at 15.0 m. The scantlings at the immediate vicinity of the VBIED are given in Table 5 Frame spacing is 0.8 m.

4.3 Boundary conditions

Since the geometry in this design threat represents the parallel middle body of the vessel, symmetric boundary conditions were applied, i.e. longitudinally and transversely the out-of-plane translations and in-plane rotations are set to zero.

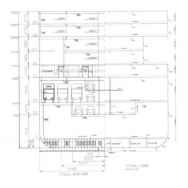

Figure 2. A typical loading profile of a ROPAX vessel with trucks and sedan vehicles parked alongside.

Table 5. The scantlings around the VBIED.

Structural member	Dimensions (mm)
Deck 3	
Deck thickness	10.5
Web frame	W550 × 10 + F250 × 15
Frame profile	HP180 × 8
Side shell	
Side shell thickness	12.5
Web frame	W450 × 10 + F250 × 20
Frame profile	HP140 × 7
Deck 5	
Deck thickness	6.0
Web frame	W650 × 10 + F250 × 10.5
Frame profile	HP100 × 6

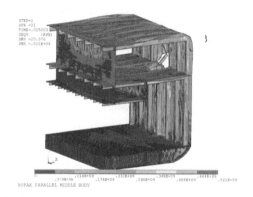

Figure 3. FE prediction of equivalent stresses (Pa) following the detonation of 227 kg TNT equivalent on deck 3.

4.4 Structural response

The predicted structural response from the detonation of 227 kg TNT equivalent at 0.4 m above deck 3 (average clearance of sedan vehicles), and at 1.9 m from the side shell (assuming the explosive charge to be in the centre of the vehicle) is depicted in Figure 3. Deck 3 suffered a gash of 3.41 m² while the side shell blown-off area is equal to 1.62 m². Secondary and tertiary stresses arose in the rest of the FE model as a result of stiffener bending, and plate bending between the stiffeners respectively. Simulation time in LS-DYNA was 2.5E-02 s allowed to adequately capture the blast effect. Detonation was modelled with the *LOAD_BLAST card.

5 DESIGN THREAT III (OG-7V)

5.1 Mechanical properties of the FE model

The third design threat is that against blast and ballistic loading. The attack against Seabourn Spirit on

November 5, 2005 off the Somali coast involved the firing of an RPG-7 that miraculously failed to explode when the warhead hit the side of the cruise liner. The RPG-7, although the name is a misnomer, is a point-detonated, anti-tank shoulder-launched, recoilless gun initially adopted in 1961. It can fire up to 4–6 rounds per minute, and its projectile can travel as far as 900 m where it self-destructs (approximately 4.5 s after firing). The grenade is carried in two parts, namely the warhead with the sustained motor, and the booster charge. The projectile being used here resembles the geometry of OG-7V; a high explosive anti-personnel grenade intended for use against soft-skinned vehicles and structures. It carries a 2 kg hexogen (relative effectiveness of 1.6) warhead with a point-detonating fuse. Once launched, the OG-7V travels at 117 ms⁻¹ and at approximately 11 metres from the launcher a sustainer rocket ignites and propels the warhead to a cruising speed of 180 ms⁻¹ (Jane's Infantry Weapons, 2007). Once airborne, accuracy is enhanced by the deployment of a set of fins that open after initial launch and provide improved stability in flight by imparting a slow rotational spin. The OG-7V is assumed to impact the panel at zero yaw in this analysis. Its light weight, low acquisition cost, simplicity, ruggedness, and reusability are some of the key-aspects that make RPGs a weapon of choice in more than forty countries (Grau, 1998). As of 2002, is it estimated that at least 9 million RPGs had been produced around the world (O'Sullivan, 2005).

Cruise ships make extensive use of AH36 marine steel; its mechanical properties are given in Table 6.

Due to lack of data in the open literature regarding the material properties of the projectile in question, the OG-7V is assumed to be made of aluminium alloy Al2024, widely used in missile parts, aircraft fittings and munitions due to its high strength and fatigue resistance. The mechanical properties of Al2024 are given in Table 7 (Boyer & Gall, 1985).

The foremost tip of the OG-7V is modelled as a blunt cylindrical geometry where the piezoelectric fuse is housed. The round diameter is 0.04 m and has a total length of 1.5 m including the stabilising pipe. SHELL163 elements were used to model the entire projectile. The detonation of the explosive yield, approximately 2 kg TNT equivalent, is modelled with

Table 6. Mechanical properties of AH36 marine steel.

Property	Value
Density (kgm⁻³)	7800
Modulus of elasticity (Pa)	210E + 09
Poisson's ratio (Pa)	0.3
Yield stress (Pa)	380E + 06
Tensile stress (Pa)	560E + 06
Failure strain	0.21

Table 7. Mechanical properties of Al2024 (Boyer & Gall, 1985).

Property	Value
Density (kgm^{-3})	2780
Modulus of elasticity (Pa)	73E + 09
Poisson's ratio (Pa)	0.33
Yield stress (Pa)	97E + 06
Tensile stress (Pa)	210E + 06
Failure strain	0.12

Table 8. The scantlings of a cruise ship's cross-stiffened panel.

Structural member	Dimensions (mm)
Panel span	6000
Panel depth	2720
Plate thickness	11.0
Transverse web	W300 × 11 + F130 × 12
Web spacing	2800
Longitudinal stiffener profile	HP180 × 8
Longitudinal stiffener spacing	680

the *LOAD_BLAST card upon contact with the target panel. The contact coefficients needed in the *CONTACT cards are taken from ship collision studies for mild steel/mild steel (Wu et al., 2004).

5.2 Geometry of the FE model

The FE model considered under the ballistic loading from the OG-7V is a typical cross-stiffened panel structure with the structural characteristics of Table 8.

5.3 Boundary conditions

The same principle as the one applied in §3.3 is adopted herein.

5.4 Structural response

The multi-faceted problem of the response of structures subjected to dynamic loading, and more precisely impact loading, entails elements of blast, friction, contact, and failure under shear and strain. LS-DYNA's capabilities and robustness helped in carrying out the explicit analysis. The LS-DYNA analysis ran for 3.0E-03 s that captured adequately the structural response of the target panel. The projectile was given an initial velocity of 180 ms^{-1} equal to the cruising speed of OG-7V. The blast pressure upon impact was found to be 1.25E + 09Pa at 5.63E-05 s. The anti-personnel properties of the OG-7V can be seen in the rather symmetric damage on the panel as a result of the

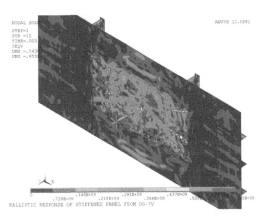

Figure 4. FE prediction of equivalent stresses (Pa) at 3.0 ms following the impact and blast from an OG-7V.

fragmentation. The panel is characterised by dishing that surrounded the impact point. For thin to moderately thick ductile metal plates, struck by blunt missiles, dishing precedes all other responses, and the subsequent failure mechanism is either plugging or discing (Stronge, 1985). The former was found to be caused by shear rupture, while the latter by tensile rupture in the model. Petalling is also identified as a result of high circumferential strains around the impact point that cause radial cracking and subsequently force the rotation of the affected material noted at the back of the panel (Wierzbicki, 1999). Overall, dishing dominates the structural response of the side shell panel as shown in Figure 4 where the side shell bends out of plane. From a structural point of view, it appears that the OG-7V has a mild effect on the target panel since it is an anti-personnel round. Nevertheless, its primary fragmentation pattern can be lethal in such cases. The longitudinal stiffener HP180x8 sustained most of the damage immediately behind the side shell, and decelerated the projectile later in the analysis. The resulting blown-off area was found to be equal to 0.04 m^2 while secondary fragmentation from the side shell plate, equal to 3.44 kg, found to travel at velocities as high as 595 ms^{-1}.

6 THE EFFECT OF RBVRS ON THE STRUCTURAL RESPONSE

6.1 Basics of reinforcing blast-loaded structures

The above three design threats were reconsidered, this time with the selected scantlings of Table 1 subjected to the corresponding increase. Prior to any FE analysis, one can predict that the blast effect will be mitigated as the scantlings are reinforced, this is happening as the structural response of impulsively-loaded structures is influenced by:

a. The ratio between blast duration and the natural period of the structure.
b. The rise time and natural period of the structure.

Thus, by increasing the denominator in the above ratios one decreases the dynamic amplification factor (Czujko, 2001).

6.2 FEA results

The structural response in the "as-built" condition following the blast load from the design threats of Table 2 is given in Table 9.

At this point, a metric of benefit needs to be defined in order to quantify the effect of RBVRs. It is perceived that for impulse-related problems such metric is the work done by the applied pressure on the structure. Intuitively, as the scantlings increase, the external work, essentially the amount of energy being transferred by a force, should take lower values. As a node follows a certain path of deformation (AB) in a three-dimensional FE model under the action of a force F, essentially a vector, and its own inertia, the work done along a path AB is defined as

$$W_{(AB)} = \int_{(AB)} F \cdot dr = \int_{(AB)} (F_1 dx + F_2 dy + F_3 dz) \quad (1)$$

where r is a position vector from a reference point in space.

In impact-loaded structures it is customary to assess any reinforcing effects by measuring the kinetic energy of the structure. The initial kinetic energy of a projectile is partly converted into plastic deformation to both the projectile and the target. The plastic strain energy

of the target is equal to (Calder & Goldsmith, 1971):

$$E_P = \int_V \left(\int [\sigma_r d\varepsilon_r + \sigma_\theta d\varepsilon_\theta] \right) dV \quad (2)$$

where σ_r is the radial stress (Pa); ε_r is the radial strain; σ_θ is the circumferential stress (Pa); and ε_θ is the circumferential strain.

Eq. (2) can be further manipulated to give the approximate plastic work done. If the loss of the kinetic energy of the projectile ΔE_{KIN} is set equal to the plastic energy, i.e.

$$\Delta E_{KIN} = E_P \quad (3)$$

the principle of energy conservation upon impact dictates that ΔE_{KIN} at any time should equal the total strain energy dissipated by the target panel, and the projectile, in the form of structural damage. One other way to confirm the principle of energy conservation is to check the kinetic energy equilibrium, that is:

$$E_{KIN.IN} = E_{KIN.TARGET} + E_{KIN.PROJ} \quad (4)$$

Strictly speaking, the initial kinetic energy E_{KININ} of the projectile is further dissipated due to friction, elastic strain energy, fragmentation etc.

In the context of ballistic loading, the metric of benefit can be mirrored in the projectile's kinetic energy difference. As the structure is reinforced, the ΔE_{KIN} value of the projectile should take higher values as the deceleration rate is higher. The metric used in ΘIII is the internal energy of the target which includes elastic strain energy and work done in plastic deformation.

The effect of RBVRs, for each one of the considered design threats prior to any reinforcement, and after the compliance of the influential structural members with the required increase of Table 1, are given in Tables 10

Table 9. The structural response in the "as-built" condition.

Design threat	δ (m)	Work Done (J)	Gash (m²)	Fragmentation mass (kg)
ΘI (4 kg PETN) Side shell				
ΘII (227 kg VBIED)	0.25	922E + 03	0.2	19.5
Deck 3	0.25	91.2E + 06[†]	3.41	279.3
Side shell	1.20		1.62	157.9
Deck 5	0.56		–	–
Design threat	δ (m)	E_{INT} (J)	Gash (m²)	Fragmentation Mass (kg)
ΘIII (OG-7V) Side shell	0.2	2.64E + 05	0.04	3.44

[†]Of the entire model.

Table 10. The effect of RBVRs for the requirements of positive reliability index ($\beta > 0$).

Design threat	δ (m)	Work Done (J)	Gash (m²)	Fragmentation Mass (kg)
ΘI (4 kg PETN) Side shell				
ΘII (227 kg VBIED)	0.24	811E + 03	0.05	5.5
Deck 3	0.24	88.4E + 06[†]	3.23	302.3
Side shell	0.85		1.19	130.0
Deck 5	0.53		–	–
Design threat	δ (m)	E_{INT} (J)	Gash (m²)	Fragmentation Mass (kg)
ΘIII (OG-7V) Side shell	0.14	2.74E + 05	0.028	2.62

[†]Of the entire model.

Table 11. The effect of RBVRs for the requirements of satisfactory reliability index ($\beta = 1.5$).

Design threat	δ (m)	Work Done (J)	Gash (m^2)	Fragmentation Mass (kg)
ΘI (4 kg PETN) Side shell				
ΘII (227 kg VBIED)	0.18 7	49E + 03	0.05	5.85
Deck 3	0.21	85.5E + 06[†]	3.09	301.3
Side shell	0.68		0.84	98.3
Deck	50.42		–	–

Design threat	δ (m)	E_{INT} (J)	Gash (m^2)	Fragmentation Mass (kg)
ΘIII (OG-7V) Side shell	0.13	3.62E + 05	0.024	2.51

[†]Of the entire model.

and 11. Table 10 paints the picture of the effects of RBVRs for positive reliability index, while Table 11 gives an idea about the structural response when reinforcement for what is deemed to be a satisfactory reliability index of 1.5 is applied.

7 DISCUSSION

The effect of RBVRs in reducing secondary fragmentation in ΘI is significant between the "as-built" design and the designs subjected to RBVRs. In ΘII the increase in fragmentation mass between the "as-built" design and that subjected to RBVRs for $\beta > 0$ is due to the sturdier scantlings being used. In addition to that, one should not neglect that primarily the development of RBVRs aimed at reducing the arisen stress levels in a model based on Quality of Service (QOS) and not secondary fragmentation.

$$QOS = f(\beta, \Theta) \qquad (5)$$

where β is the reliability index; and Θ is the design threat (s)

In ΘIII, the internal energy of the side shell plating between the "as-built" condition and the reinforced designs increases as a sign of higher levels of elastic strain energy being stored in the structure. The projectile's kinetic energy loss is also depicted in Figure 5 where one can see that as the target panel becomes sturdier the kinetic energy loss increases.

The effect of RBVRs was also found beneficial in reducing the blown-off area and secondary fragmentation. It is believed that even if a more destructive RPG round had been used (e.g. PG-7VR), the structural integrity of a highly redundant structure as a ship would not be compromised. The benefit from applying RBVRs is in reducing secondary fragmentation which can be particularly threatening. Primary and secondary fragments from RPG rounds have been reported to fly at distances as far as 150 m (TRADOC, 1976).

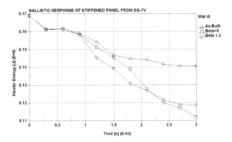

Figure 5. A comparison of the projectile's kinetic energy loss following impact.

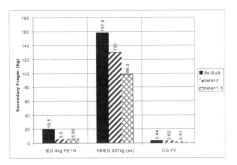

Figure 6. A comparison of the accounted design threats against secondary fragmentation (kg).

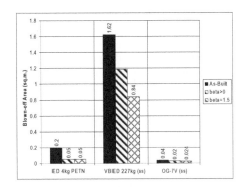

Figure 7. A comparison of the accounted design threats against the resulting blown-off area (m^2).

For the sake of clarity and comparison the above findings are plotted in Figures 6 and 7.

Since any reinforcement of the primary structure is to increase the hull weight with all the subsequent consequences in machinery etc, the application of RBVRs should be selective in terms of the vulnerable areas of the vessel that have been identified during the design stage.

For design threats above the water line, and for a typical ROPAX vessel, such vulnerable locale is the connection of the side shell and the car deck right above a bulkhead (Pahos, 2008). In a 194 m ROPAX vessel there are fourteen transverse bulkheads from which eleven span along the car deck. Since no statistical results regarding the probabilistic density distribution, and the cumulative probabilities related to the locale of IEDs or VBIEDs onboard vessels do not exist, the application or RBVRs can only be based on a qualitative approach. A compromise between the blown-off area and secondary fragmentation reduction is something to be decided upon the design stage. Perhaps at areas where passengers and crew are unlikely to be found during operational time, e.g. car deck, fragmentation should not be the primary concern. Likewise, in areas where passengers and crew are likely to be found, decision-makers should opt for reduced fragmentation. This could give food for thought in the years to come as to how many panels, and where, need to be reinforced.

The development of RBVRs as they appear in Table 1 was based on the detonation of a 4 kg TNT at point blank distance. Another thought to prevent the gash (reduce vulnerability, or achieve the design threat outcome of no-holing) would be to prevent an IED from getting within 0.2 m of the plate. This "specific keepout requirement" could be accomplished by non-structural bulkheads, screens, etc. at some loss of arrangeable area.

8 CONCLUSIONS

The present paper applied previous findings of the required reinforcement in order to reduce primary damage from a series of design threats. The key findings from the application of RBVRs on typical ship panels with slenderness ratio 3.0 are as follows:

- Secondary fragmentation from a 4 kg PETN IED could be reduced by 72%, the blown-off area by 75%, and the permanent set up by 28%.
- Secondary fragmentation from a 227 kg TNT equivalent VBIED on the car deck of a ROPAX vessel was found to increase up to 8%, while the blown-off area could be reduced by 9%. Secondary fragmentation on the side shell could be reduced by 37%, while the blown-off area could be reduced by 48%.
- For a ballistic attack from an OG-7V secondary fragmentation could be reduced by 27%; the blown-off area by 40%, while the permanent set by 35%.

ACKNOWLEDGMENT

The authors would like to acknowledge the support provided by the MARSTRUCT project, Task 2.6— Structural Strength under Fire and Explosion (http:// www. mar. ist. utl. pt/marstruct) Network of Excellence on Marine Structures, funded by the E.U.

REFERENCES

Boyer, H.E., and Gall, T.L. 1985 "Metals Handbook" *American Soc. Metals, Materials Park, OH*

Calder, C.A. and Goldsmith, W. 1971 "Plastic deformation and perforation of thin plates resulting from projectile impact" *Int. J. Solids & Structures, Vol. 7, No. 7, pp. 863–881*

Czujko, J. 2001 "Design of Offshore Facilities to Resist Gas Explosion Hazard" CorrOcean ASA, Oslo, Norway

Doerry, N. 2007 "Designing electrical power systems for survivability and quality of service" *Nav. Eng. J., Vol. 119, No. 2, pp. 25–34*

Grau, L.W. 1998 "The RPG-7 on the battlefield of today and tomorrow" *Infantry, May-August, pp. 6–8*

Jane's Infantry Weapons 2007 *Jane's Group of Information, New York, NY*

O'Sullivan, T. 2005 "External Terrorist Threats to Civilian Airliners: A summary of risk analysis of MANPADS, other ballistic weapon risks, future threats, and possible countermeasure policies" *CREATE Report #05-009, Los Angeles, CA*

Pahos, S.J. 2008 "An explicit dynamic analysis of a ROPAX vessel under impulsive load from a VBIED" *Proc. ASNE Symposium: Shipbuilding in Support of the GWOT, 14–17 April, Biloxi, MS*

Pahos, S.J. and Das, P.K. 2008 "A probabilistic risk assessment approach to decision-making for explosively-loaded ship panels" *Proc. 4th Int. ASRANET Colloquium, 25–27 June, Athens, Hellas*

Pahos, S.J. and Das, P.K. 2009 "Reliability-Based Vulnerability Requirements in the design of ship panels for impulsive loads" (in press) *Proc. 10th Int. Conf. on Structural Safety & Reliability, 13–17 September, Osaka, Japan*

Paik, J.K., Chung, J.Y., and Paik, Y.M. 1999 "On dynamic/impact tensile strength characteristics of thin high tensile steel materials for automobiles" *Trans. Korean Soc. Automotive Engineers, Vol. 7, No. 2, pp. 268–278 (in Korean)*

Paik, J.K. and Thayamballi, A.K. 2003 "Ultimate Limit State Design of Steel Plated Structures" *John Wiley & Sons, Chichester, UK*

Reese, R.M., Calvano, C.N., and Hopkins, T.M. 1998 "Operationally-oriented vulnerability requirements in the ship design process" *Naval Engineers Journal, Vol. 110, No. 1, pp. 19–34*

Stronge, W.J. 1985 "Impact and perforation of cylindrical shells by blunt missiles" *In S.R. Reid's Metal Forming and Impact Mechanics, pp. 289–302, Pergamon Press, New York, NY*

TRADOC BULLETIN 1976 "Soviet RPG-7 Antitank Grenade Launcher" *United States Army Training and Doctrine Command, Fort Monroe, VA*

Wierzbicki, T. 1999 "Petalling of plates under explosive and impact loading" *Int. J. Impact Engring, Vol. 22, No. 9–10, pp. 935–954*

Wu, F., Spong, R., and Wang, G. 2004 "Using numerical simulation to analyse ship collision" *Proc. 4th Int. Conf. on Collision and Grounding of Ships, Hamburg, Germany*

Structural monitoring

Analysis and Design of Marine Structures – Guedes Soares & Das (eds)
© 2009 Taylor & Francis Group, London, ISBN 978-0-415-54934-9

Structural monitoring of mast and rigging of sail ships

G. Carrera & C.M. Rizzo
DINAV, University of Genova, Italy

M. Paci
Perini Navi S.p.A., Viareggio, Italy

ABSTRACT: This paper describes still ongoing research activities carried out in cooperation by the Dept. of Naval Architecture and Marine Technologies (DINAV) of the University of Genova and the Perini Navi shipyard, one of the few ones in the world launching very large sailing yachts. Structural monitoring systems of mast and rigging performances were designed, built and installed onboard large sail ships during several sea trials. The measurements were used for the calibration of analytical and numerical models currently used in the design process resulting in a significant improvement of both the mast and the rigging performances.

1 INTRODUCTION

Mast and rigs are essential parts in sail yachts for various reasons; not least the fact that any rig failure may have catastrophic consequences for the ship and her crew. From a structural viewpoint it is a complex and coupled mechanical system, with slender shaped components subject to high compressive loads; i.e. to buckling problems.

Acting loads are uncertain, difficult to estimate and even to measure, largely variant with sailing conditions and sea state. Simplified methods to accurately evaluate the structural behavior under loading are mostly based on empirical loads distribution.

Mast and rig design significantly influences ship performances because moving up the center of gravity decreases ship's stability but mast deflection influences flying shapes of the sails and consequently sails efficiency. The mast shape is studied to reduce drag and disturbance of airflow around the sails. On the other hand the rig should have a certain ability to deform in a controllable manner, to trim the sails without collapsing. Therefore, a balanced solution should be looked for.

The mast and rig of the majority of Perini Navi's ships is traditionally Marconi (or Bermudian) shaped, either in sloop or ketch configurations, but their large sizes raises a number of specific design issues (Fig. 1). Moreover, these ships are leisure ones, thus requiring not only safe sailing performances but also high level comfort and fashioned aesthetic design.

With the aim of a structural optimization of mast and rigging designs, a research project started in the mast and rigging division of the shipyard. The structural scantling procedure, originally based on prescriptive analytical procedures, was improved and

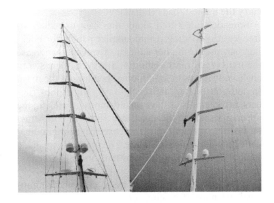

Figure 1. Masts of a 45 m ships and of a 56 m ship.

finite element (FE) models are now built up for a complete 3D simulation including global and local buckling analyses of the whole structure as well as of subcomponents, under both self-weight and dynamic loads. Indeed, design loads applied in current scantling procedures are still affected by rather large uncertainties and semi-empirical to a certain extent, even if they are now derived by more complex procedures and checked with rules requirements.

Within a strict cooperation framework aimed at numerical model calibrations, DINAV was mainly involved in experimental campaigns as well as in data analysis, focusing on the design of structural monitoring systems and on the definition of representative distributions of acting loads while structural design procedures, including FE analyses, were mainly developed by the Perini Navi mast division.

The parallel development of experimental and numerical activities was crucial for the success of the project, i.e. the optimization of the sail system and the improvement of ship sailing performances and as well as sea-keeping ones when motoring.

2 MAST AND RIGGING OF LARGE SHIPS

The mast and rigging of sail ships is generally built up by a set of beams carrying tension, compression, bending and torsion loads (mast and spreaders) and by cables providing lateral and longitudinal support to the mast (shrouds and stays). These may become slack under compression. The column mast parts extending between spreaders (but lowest one starts from deck or keel) are generally named panels.

Spreaders pull vertical shrouds (or V's) aside the mast increasing the angle between the cable and the mast itself. Diagonal shrouds (or D's) distribute internal loads along the mast. The lower shrouds usually carry the highest loads (Fig. 2).

Spreaders are fitted along the mast to reduce its free length, dividing it into smaller panels for buckling purposes. A higher number of spreaders increase the transversal mast stability. However, their sweep angle contributes to longitudinal bending of the mast. In addition to fore and back stays, sometimes runners, i.e. lateral backstays linked approximately in the upper midspan of the mast, are fitted to counteract the bending spreader effect and excessive longitudinal deflection due to inertial loads. The forestays are made straight under foresails loads by tensioning backstay(s) or runners (Fig. 3).

It is common practice that transversal and longitudinal behavior of rigging are analyzed separately in simplified design procedures, even if this is not the case because of spreaders' effects and others.

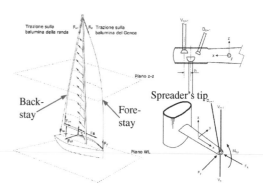

Figure 3. Sail forces distribution along mast and on spreaders (Bruni, 2007 and GL, 2002).

Boundary conditions of the structure should be carefully defined. The mast may be simply supported or clamped at deck level, depending on where it is stepped, i.e. directly on deck or on keel passing through the deck. Alternatively, keel support and deck horizontal motions can be fixed if the mast panel under deck is considered. Bringing the mast step down to keel increase its stability with respect to deck stepped masts.

Shrouds and stays are connected to the hull. In principle, also hull deflections should be considered as shrouds or stays ends may become closer, loosing pretensioning. This effect is relatively larger for rods than for usual wires because of their different stiffness but it is generally neglected, at least in the first phases of design.

A few analytical design approaches are described in open literature; among others Larsson & Eliafsson (2007), Marchaj (2000, 2003), Claughton et al. (1998) summarize the state of the art.

Indeed, various empirically derived, sometimes incomprehensible, safety factors account for uncertainties in strength assessment and loads definitions.

The Nordic Boat Standard (NBS, Larsson & Eliafsson, 2007) is one of the few yacht scantling standards taking rigging into account. It is applicable for normal masthead and fractional rigs but limited to 2 spreader pairs. Classification societies' rules generally provide simplified analytical scantling methods derived from NBS and applicable to relatively small boats. However, the Germanisher Lloyd (GL) rules (2002) provide modern scantling procedures for larger sail systems.

Bruni (2007) summarizes the evolution of rules and of literature on this topic. Figures 2 and 3 show main actions and forces on shrouds as considered in analytical procedures, based on beam/cable theory. Alternatively to current analytical approach, non linear FE analyses according to GL rules may be used, (Bruni, 2007; Janssen, 2004).

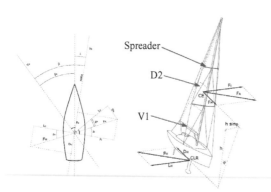

Figure 2. Main actions on a sail ship (Bruni, 2007).

3 LIMIT STATES AND ACTING LOADS

The sail ship behavior is a typical fluid structure interaction (FSI) problem, being the acting loads dependent on the structural deformations. Moreover, it is complicated by the fact that two fluids are in principle involved: water and air.

Limit states are different than in usual shipbuilding scantling as buckling is the governing one for such slender structures. However, yielding and ultimate strength should be considered as well, taking into account that weight optimization is of paramount importance because of its effects on ship's stability and sea-keeping performances.

The cornerstone of the mast design is the ability of the mast designer to define loads and corresponding distributions representative for critical/extreme structural conditions. Close hauling and running/broaching under spinnaker are generally considered, though others must be accounted for. Moreover, loads of sheets and halyards, basically depending on tension of sails' leech (i.e. the aft edge of sails), need to be estimated.

Traditionally, the equilibrium of the heeled ship is assumed as the starting basis for a quasi static approach (Fig. 2), hence separating the hydrodynamic actions from the aerodynamic ones. The maximum righting moment, generally estimated at heel angles of about 25°–30°, should be considered.

A rational engineering approach is applied to distribute the transverse rig loads due to sail forces on mast and rigging in different sailing conditions and equilibrating the corresponding righting moment: the usual simplifying assumption is that the ratio between main sail force and genoa force is the same as between their sail areas. Moreover, for triangular sails 3/7 of the sail force acts at the head and 2/7 each at tack and clew; for spinnakers 4/10 at the head and 3/10 each at the clews (see e.g. GL, 2002).

Leeward shrouds may become slack if not pre-tensioned, leading to mast buckling failure. However, excessive pre-tension causes unnecessary mast compression, possibly resulting in mast instability.

The compression at mast step is estimated by rule of thumb as twice the ship's displacement (Larsson & Eliafsson, 2007; GL, 2002), approximately distributed as 100% due to shrouds, 85% due to stays and 15% due to halyards. Therefore, mast tuning is of paramount importance for the structural behavior; actually, on large ships this is obtained by moving up the mast step by jack cylinders. As a matter of facts, shrouds turnbuckles cannot be moved when mast is tuned, i.e. if it is compressed by pre-tensioned shrouds, but only when shrouds and mast are unloaded. Sometimes, mast tuning is used for adjusting the mainsail profile in association with stays.

It is worth noting that inertial loads due to ship motions are not explicitly considered in analytical procedures, which are indeed fairly significant in large sail ships. Even when motoring without sails, ship motions induce pretty large accelerations, especially for components far from center of gravity of the ship. Moreover, geometrical non linearities, i.e. large deflections, implying change of loads directions, are exaggerated by large dimensions.

So far, factors of safety of 3 and more against breaking loads of components are primarily justified by uncertainties in the determination of actions and of their distribution on the rigging.

Without a doubt, FE analyses are powerful tools for structural strength modelling and account for the non linear behaviour of rigs but their reliability heavily depends on the accuracy of the loading and boundary conditions input. Rig loading caused by sails shall be a designer input to FE software.

4 THE PERINI'S DESIGN PROCESS

All rig design calculation procedures found in literature are more or less based on the well known Skene's method (Kinney & Skene, 1962), i.e. on ship righting moment, with variations to take into account the modeling assumptions and effects as distributed forces from sails, halyard forces, and longitudinal forces from stays.

The Perini mast design process starts from input information such as sail plan, ship's stability characteristics and load cases corresponding to different sail settings. In rig design, FE packages can be used to test an existing design, but they are difficult to apply to create a new design or tune an existing one. Therefore, in-house software has been developed by Perini Navi in order to assess the size of all shrouds and stays, strength sections of spreaders and mast panels using an analytical approach, adapted from currently available methods and rules.

External aerodynamic loads acting on rig are estimated on the basis of above mentioned general input information for several different sailing conditions, named load cases, using appropriate pressure distributions. Self-weight loads are applied to the structural model after a first estimate of mast and rig size and iteratively checked. Indeed, this procedure is an iterative loop which is automatically run in order to achieve a satisfactory convergence; in fact weight loads become quite influencing parameters at high heeling angles because of large dimensions.

At this stage the rig is analyzed as a statically determined structure by virtually inserting a hinge between shrouds and spreaders' tips. After a first estimation of rod sizes and strength sections of mast panels, the in-house code run some statically indeterminate simulations by adding constraints on the hinges connecting rigging components in order to evaluate lateral and longitudinal bending moment and corresponding displacements of mast shells. Moreover, the mast compression loads to be achieved during the pre-tension dock tuning steps are verified. Criteria of sizing rig structures are based on safety factors on stress level and buckling collapse, both global and local for thin walled section of the mast tube, according to GL (2002) recommendations.

Eventually, a complete 3D FE analysis is carried out on models including beam and cable (link) elements as shown in Figure 4: particularly it is now possible simulating the non-linear effects of tension-only cable elements used for shrouds and stays and the large deformation influence as the weight and inertial loads are included in the FE models as well. Sag fraction of stays and stability effect of runners or baby-stay is also predicted (Fig. 5).

Evaluation of buckling safety factors and of pre-tension steps loads are much more precise at this stage and the rig size or mast panels' inertia can be optimized

saving weight. It is worth noting that no analytical formulae exist for global longitudinal buckling of mast (Fig. 6), important for large sizes masts, as well as for local buckling of mast panels and spreaders (Fig. 7).

However, the same model of external loads used in the first phase is applied to the FE model, except for self-weight loads which are automatically introduced considering the gravity acceleration.

Dynamic loads on structures in motoring condition can be calculated in order to avoid stays slack in harsh sea conditions (Fig. 8), again using a quasi static loading approach.

Rig design is submitted to GL for approval, which carries out again a FE analysis according to load cases defined in their applicable rules.

Later on, small components of the mast such as mast head, spreader fittings, stays fitting, doublers, mast base are checked, generally running additional FE simulation whose boundary condition are properly derived from previous global analyses (Fig. 9).

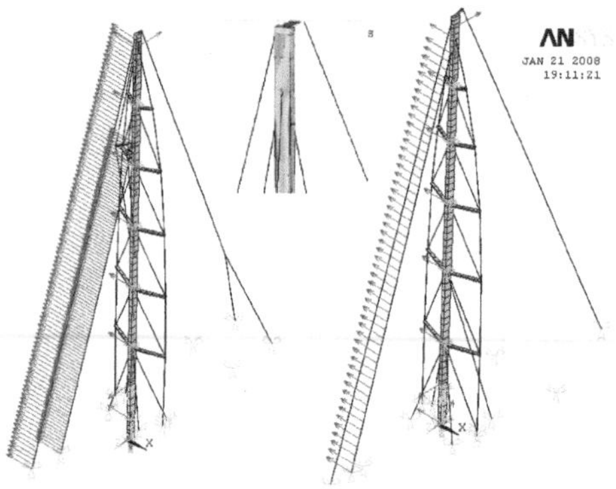

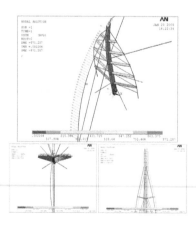

Figure 4. Examples of FE models with loads and boundary conditions, mainsail loads are here modeled as forces acting at mast/spreaders connections.

Figure 6. Global longitudinal buckling check of the mast.

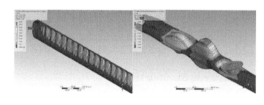

Figure 7. Local buckling checks of a mast panel and of a spreader, Paci (2008).

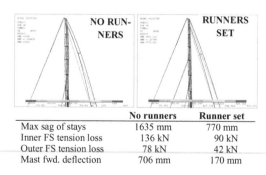

	No runners	Runner set
Max sag of stays	1635 mm	770 mm
Inner FS tension loss	136 kN	90 kN
Outer FS tension loss	78 kN	42 kN
Mast fwd. deflection	706 mm	170 mm

Figure 8. Slamming load simulation with and without runners.

Figure 5. Assessment of forestay sag, Paci (2008).

Figure 9. Mast head FE simulation, stress analysis.

5 THE MEASUREMENT SYSTEMS

The following features were stated by Perini Navi at the beginning of the project, having in mind to carry out measurements on ships during mast tuning in dock and during delivery sea trials:

1. Robustness and reliability of the system,
2. Easy to install on different ships,
3. User-friendly,
4. Digital acquisition of signals and visualization in real time during trials,
5. Appropriate definition of the frequency of acquisitions aimed at catching peaks in the saved files,
6. Resistance to sea environment,
7. Possibility to expand/improve the system,
8. Compatible with onboard electronics systems,
9. Last but not least, cheap.

So far, after discussions about the parameters to be measured, the project started selecting the appropriate sensors for measuring shrouds elongation. Ship's motions were measured by traditional accelerometers, gyroscope, rate-gyro and clinometers.

5.1 The shrouds sensors

The measurement of strains of a rod can be carried out using, among others, two transducers: electric strain gages and LVDTs. The strain gage is widely used as it allows a quasi punctual measurement being very small in size and as it's relatively cheap. However, its installation requires skilled personnel and preparation of surfaces. In case of stainless steel materials, as the rods of the shrouds, some concerns exist about duration of the glue, as demonstrated by the tests carried out in the DINAV laboratory.

A comparative analysis of other available sensors applicable to the problem was carried out and finally LVDT (Linear Variable Displacement Transducer Linear Voltage Differential Transformer) were considered satisfying all requirements, if properly engineered and carefully installed onboard.

They are rather easy to install and their high level outputs are less sensitive to noise. It is worth noting that LVDTs are contactless and therefore no friction occurs between moving parts, then minimum wear and long-life result. On the other hand, the sensor needs to be installed appropriately, with minimum geometrical tolerances. The clamping devices should be rigid and should not distort or slip neither wear the shrouds. The sensor has a high resolution, practically limited by the acquisition system only. The sensors bandwidth (200 Hz) is relatively large so their delay is practically negligible since it depends on the inertia of moving masses involved, on the frequency used for the LVDT coils supply and on the delay of the amplification system. Even if the sensor limits are exceeded, the sensor itself is not damaged, at least within a reasonable range.

Transducers with an electronic circuit inside and a single unregulated power supply were chosen because of their easier installation and compatibility with the majority of acquisition systems.

For strain measurements it is important to know the baseline length with minimum geometrical tolerances, so the sensor needs to be installed appropriately. The clamping devices should be rigid and should not distort or slip neither wear the shrouds.

Many laboratory tests were carried out and several installation systems where experienced, first in the lab and then during sea trials. Figure 10 shows the final solution installed on shrouds: clamping device geometry was studied to be intrusive as less as possible but easy to install and fully effective. A good compromise between sensitivity and overall dimensions was found with a baseline of about 400 mm and a sensor range of $2 \div 2.54$ mm corresponding to a deformation range of $5000 \div 6350 \ \mu\varepsilon$.

Being made of the same metal of shrouds, temperature effect should be limited. An intermediate support with a linear bearing was fitted to increase the stiffness

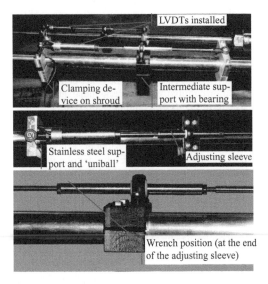

Figure 10. Particulars of the sensors installation and clamping devices on shrouds (Rizzo & Carrera, 2007).

Figure 11. Temperature and vibration tests on a piece of shroud in tension in the DINAV lab. (load up to 40 tons by a jack).

of the device and to reduce vibrations on moving parts. However, many laboratory tests were carried out on a piece of shroud assessing these aspects (Fig. 11). At the same time, several clamping devices were tested to reach the final solution.

The smoothed sleeve bar, shown in Figure 10, is utilized as bearing element and also for adjusting the mechanical 'zero' during installation and after dock tuning. The sensors were equipped with 'uniball' supports at ends for easier installation.

5.2 The acquisition system

The first step of the project was the choice between a single acquisition system, located near the PC, or a distributed acquisition system, with remote units placed near grouped sensors. A distributed acquisition system, more suitable for many sensors spread in a large area and therefore implying rather long wires, was preferred since the beginning of the project, also accounting for its modularity.

The second challenge to face was the selection of the communication medium: a traditional cable or an easy to install wireless communication system? Actually, the hardware and software configuration of the developed acquisition system can use both of them. Accounting for the system reliability, a traditional wired system was preferred because of radio interferences caused by several microwaves sources that are installed near the sensors and the boxes containing the acquisition terminal, e.g. radar and Inmarsat antennae.

The third step is the communication hardware protocol: a RS422 multipoint bus was preferred to a faster and more common Ethernet interface mainly because it is applicable for long lines, being the Ethernet, limited to about 90 m. Moreover, a bus topology is more flexible, for this specific application, in comparison of a 'star' configuration.

Figure 12 shows a simplified scheme of the DINAV-MDS (Measurement of Deformation of Shrouds). It basically use the National Instruments FieldPoint system, distributed in three boxes mounted on the spreaders. The boxes are chained together by a special cable that supplies also the power. A line terminator must be used on the last box in order to close the line. The hardware system can be easily implemented adding sensors and boxes. So also the software can be configured

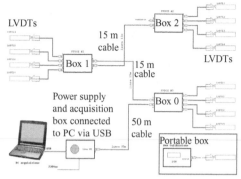

Figure 12. DINAV-MDS system scheme.

Figure 13. Boxes of the MDS system ready for installation.

accordingly. For the shown configuration, the MDS system can acquire with a maximum sample frequency of 20 Hz and with a maximum of eight channels per terminal (box), but these data can be easily modified adding new boxes and sensors.

Figure 13 shows the boxes, opened for tests. The boxes and connectors have an IP67–68 protection.

The acquisition software was developed in Lab-View environment: a user friendly graphical interface is included. During acquisition it is possible to see the data graphical trends and also numeric values (Fig. 14).

5.3 The development of the DINAV-MDS system

The first sea trials were carried out mainly aiming to have a look into practical applications of some sensors at the very beginning of the project, in parallel with laboratory tests.

During first onboard installation of the LVDT sensors only two sensors were used measuring dock tuning steps and delivery trials of a ship of about 45 m in length (Fig. 15, left). The acquisition system was a provisional one for prototyping purposes even if some very interesting data were collected. It was decided to install two different LVDT sensors, having a maximum displacement of 5 mm and 1 mm respectively. Two couples of stainless steel supports were provided having the same diameters of the two lowest shrouds, i.e. the V1 and the D1 shrouds.

In order to avoid problems with the deflection of the unloaded diagonal shroud (Fig. 15, right), the 1 mm sensor was installed on the D1, thus limiting as far as possible the reference base length, while the 5 mm

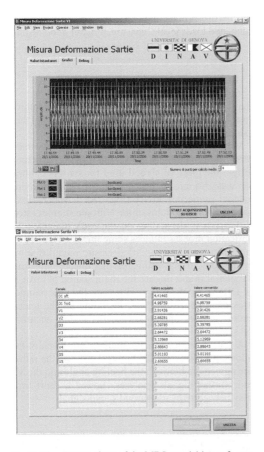

Figure 14.   Screen-shots of the MDS acquisition software.

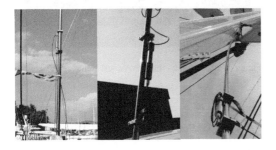

Figure 15.   First installation of LVDTs onboard a 45 m ship.

sensors was installed on the V1, thus requiring a longer measurement base.

The measuring base lengths for the sensors ranged respectively between abt. 1250 mm and 330 mm. This choice allowed selecting both the LVDT maximum displacement and the reference length for the measurements, best compromising precision and sensitivity, as previously said.

Moreover, some experience has been acquired on this occasion about assembling parts and other practical problems for the next trials and for the development of the complete system. Particular care was paid to sensors alignment in order to avoid erroneous measures of the shroud lengthening, e.g. because of bending or because of torsion of the rod. The laboratory tests were particularly useful as they allowed appropriately designing adjustable supports, clamping the sensors to the shroud in a non-intrusive way and defining an installation and calibration procedure, described in very detail in a working instruction provided with the MDS system. Actually, clamping devices should have very high mechanical tolerances, should not be too stiff and outsized in order to be not intrusive, i.e. to not modify locally the shroud's stiffness, but not too deformable as measurements resolution should be of the order of 10 μm.

The reference condition (or the 'zeros') for these measurements is difficult to be defined. As a matter of facts, it is practically difficult to install the sensors when the shroud is completely unloaded and undeformed. Therefore, it was decided to assume a specific condition related to the distance between the mast step and its basement on the ship keel as the reference condition for the measurements on this occasion.

The first application of the complete DINAV-MDS monitoring system occurred on an even larger sail ship, of 56 m in length. Further to the measurements of the mast during dock tuning, ship motions measurements were also carried out during sea trials aiming to check the interactions between the heel, the pitch and the other ship motions in general and the corresponding loads on shrouds. This measurement campaign was also direct toward a further development of the system, candidate to become a continuous monitoring system integrated with the onboard electronic equipment (i.e. GPS, wind instrument, log, etc.). However, structural rigging problems were considered as well since almost all Vs and Ds of a side were instrumented, installing total of 16 LVDTs sensors on shrouds and jumper of both main and mizzen masts.

Since the time between dock tuning and sea trials was rather long, it was decided to install the LVDTs with the mast ashore, before rigging, for dock tuning measurements and to dismantle them until sea trials, just leaving clamping devices on the shrouds (Fig. 16). It should be noted that installing sensors before placing the mast onboard allows measuring the 'zero', i.e. measuring the only occasion with nominally unloaded shrouds; as a matter of facts mast and rigging are tensioned since the first dock tuning steps and will be not unloaded during ship's lifetime unless the mast is disembarked.

After tuning operations, LVDT sensors were mounted again before sea trials, checking the stainless steel support position marked before dismantling for reference purposes. It is worth noting that some sensor

Figure 16.   LVDTs installed before rigging and during trials.

Figure 17.   LVDTs installed and protected by PVC pipes.

readings after re-installation were out of scale because some clamping devices were found moved. Therefore, it needed to calibrate them again with the calibration box but their zero reference condition (shroud unloaded) went lost.

Although the procedures and the design of the system have been carefully followed, some remarks raised in order to make easier the use of the system and to avoid or minimize uncertainties of the readings. High accuracy is required in the installation of sensors, adopting methods and techniques of high precision mechanics; however the large size of the structure and its vertical position make it rather difficult to realize in practice.

Moreover, the zero load reference condition is also difficult to measure, even if sensors are installed before tuning. A suggestion for the next trials was to install the sensors and to leave them until completion of measurements, appropriately protected against weather and mechanical damages up to completion of construction and later during sea trials.

As shown in Figure 17, tuning and sea trials measurements were carried out on the following ship, a 50 m ketch, protecting the sensor by a PVC pipe installed around the shrouds. The clamping devices of the sensors were adapted to fix the pipes.

The system DINAV-MDS was eventually settled and delivered to Perini Navi together with instruction procedures and upgraded versions of the acquisition software: it started to provide data for calibration of numerical models, without the assistance of DINAV technicians and researchers.

So far different types of large sailing yachts, including a 56 m sloop equipped with a 74 m mast claimed to be the world's tallest one, have been equipped with sensors on shrouds and stays and comparisons with calculated loads levels have given rather good results.

6   COMPARISONS

Really, a very large amount of data has been acquired since the beginning of the project in 2006 on more than 5 ships.

Moreover, laboratory tests also provided very interesting information for mast and rigging structural design: the piece of shroud used for sensors calibration was also tension tested in full scale, thus giving material properties advice (Fig. 18). It was realized that the Young modulus of the Nitronic stainless steel of the rods is depending on the rod's diameter. Its fabrication process induces some metal surface hardening resulting in a modified elastic modulus value for the rods. It was estimated that uncertainties due to material properties are the main source of error in the DINAV-MDS system. However, the uncertainties assessment confirmed that, including all the uncertainties sources, the measurements system accuracy and sensitivity is fully satisfactory.

A few examples of data analyses are given in the following Figures 19–22. Figure 19 shows the rather nice agreement between the heeling angle and the V1 load; further signals analyses showed a short time delay between roll and shroud's load signal.

Figure 18.   Shroud on the DINAV test frame.

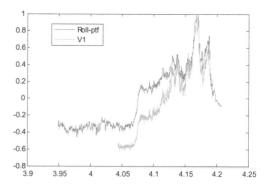

Figure 19.   Example of analysis of roll vs. load on V1 shroud, normalized values in time, Rizzo (2008).

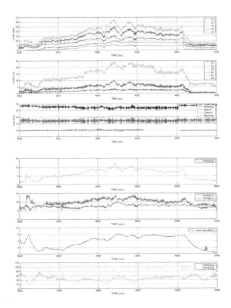

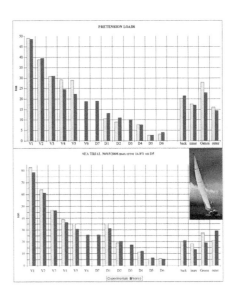

Figure 22. Comparison btw. measurements (yellow) and calculations (blue) of the 56 m sloop; dock tuning max loads (up) and sea trials max loads (down, max error abt. 17%).

Figure 20. Time history measurements of a 56 m ketch, including all Vs, all Ds, head stay, inner fore stay, outer fore stay, backstay, heel angle, apparent and true wind speed and angle, ship's speed.

Table 1. Weight reductions obtained after optimizations.

Ship type	Mast height [mm]	Weight reduction	
		[Kg]	%
45 mt sloop	49445	744	21
50 mt ketch (main)	48640	706	20
56 mt ketch (main)	59140	1543	23
(mizzen)	43470	471	17
56 mt sloop	72747	2202	24

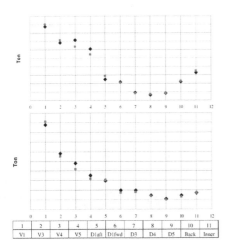

Figure 21. Comparison btw. measurements (blue) and calculations (magenta) of the 56 m ketch; dock tuning max loads (up) and sea trials max loads at heel 23° (down, max error abt. 13%).

Moreover, the V1 signal can be considered as a low-pass filter with respect to the input action, i.e. the roll. Figure 20 is a time history of several measurements, useful for correlation purposes. Figures 21–22 show comparisons between measured and FE calculated values: the max loads measured during mast tuning and

sea trials are in substantial agreement with the ones calculated by FE analysis. Sea trials loads were estimated by FE considering a heeling angle corresponding to the measured experimental heel angle for the 56 m ketch and the 56 m sloop.

Table 1 shows the weight reduction, for mast shell only, achieved by calibrating the FE models with experimental data.

7 CONCLUSIONS

Designing yacht rigs using empirical rules of thumb and large margins of safety can result in rigs that are substantially heavier than they need to be.

Even if the DINAV-MDS system can be further improved, especially as far as installation procedures, mechanical calibration and protection against weather and other damages of sensors is concerned, a set of very meaningful results was obtained.

Some sensors had some malfunctioning and they were needed to be installed and calibrated again after rigging: indeed, working difficulties with the mast onboard caused installation errors, inaccuracies and some damage to sensors. On the other hand, it can be easily understood that it is not simple for a technician to work with very high mechanical precisions at a height of several tens of meters, hung to a halyard in the bosun's chair, subject to ship's motions, rather large even when ship is berthed.

A protection with PVC pipe was designed and installed for shrouds adapting the clamping devices of LVDTs that will be made even fashioned in the future installations by a shaped stainless steel cover.

Diagonal shrouds are affected by deflection before tuning: this was recognized and it is now accounted for when installing, calibrating and recording the reference value of the sensor.

Unfortunately sea conditions during sea trials were relatively calm. Therefore, no extreme and representative critical loads could be measured, including significant inertial loads. For the time being, FE models have been calibrated considering for the heel angle as the main governing parameter, in calm sea and winds up to about 20 kt. More challenging loading conditions have been only extrapolated based on the measurements carried out.

8 FUTURE DEVELOPMENTS

Owing the difficulties experimented in defining representative critical design loads only relying on dock tuning and sea trials measurements, it was decided to develop a monitoring system to be permanently installed onboard.

The target is to acquire as much experimental data as possible in terms of static and dynamic loads and structural response during ships service in order to develop realistic load models by statistical analyses of experimental data. Despite the fact that such a strategy needs significant efforts and long time, it should be able to provide representative design scenarios and relevant distributions of acting loads.

Then, a new system was designed and built by the DINAV Ship Structures Laboratory, able to acquire ship motions and mast compression in addition to shrouds and stays loads; it is suited to be permanently installed onboard and save all data in a useful format for subsequent analysis.

This system, named DINAV-SSM (Sail Ship Monitoring), was additionally connected to the ship's navigation logging system (the Perini DLS2) to obtain the environmental and navigation condition and the sails status, thus providing a more complete picture of the loads and of the relevant structural behavior.

For the time being, a complete DINAV-SSM was developed by DINAV and tested in the laboratory. It applies the same communication system to acquire now five boxes on two different lines, as shown in Figure 23.

Inside the new two boxes of Line 2 tri-axial accelerometers and inclinometers are fitted, monitoring ship's sea keeping performances: Box 3 should be located in way of the center of gravity and Box 4 at main mast step and receives also fore stays' and strain gages bridges' signals. The mast compression is acquired by means of two strain gage bridges mounted around two holes in way of the mast step, on the opposite sides of the mast panel. Actually, the holes were used as strain amplifier, otherwise too small variations result in a not acceptable sensitivity and accuracy of the measurement system. Also, monitoring two holes on opposite sides of the mast allows assessing its transversal bending.

Again a parallel calibration of experimental and numerical activities was carried out, as shown in

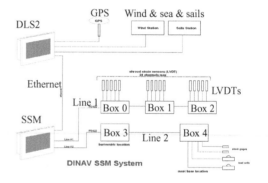

Figure 23. Scheme of the DINAV-SSM system.

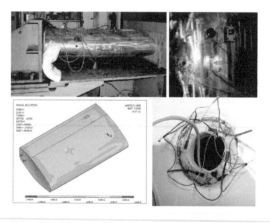

Figure 24. Mast specimen under compression testing in the lab (up), relevant FE model for stress concentration estimates and installation of gages on a mast during dock tuning and sea trials.

Figure 24. A mast specimen is under compression testing in the DINAV laboratory in order to estimate stress concentrations in way of holes and to calibrate both the measurement system and the FE models. A preliminary test on a real mast of a 56 m ship during dock tuning and sea trials was carried out, getting useful information for subsequent permanent installation on a sister-ship, still under analysis.

An industrial panel PC is connected to the Perini Data Logger (DLS2) via the ship Ethernet LAN. The acquisition software, developed on purpose by DINAV, monitors all data but start storing data on the PC hard disk only when defined scenarios occur (e.g. ship's speed higher than 7 kt and mainsail open). All data are synchronized with GPS time. The collected data are made available on the ship LAN.

The system is ready now for onboard installation, after laboratory testing. Its installation is foreseen in the beginning of 2009.

ACKNOWLEDGMENTS

The work reported in this paper was funded by Perini Navi S.p.A. and partially developed within the Network of Excellence on Marine Structures (MARSTRUCT), financed by the E.U. through the 6th FP under Contract No. TNE3-CT-2003-506141.

REFERENCES

Bruni L., 2007. Metodi di dimensionamento del sistema velico di grandi navi a vela, MSc thesis, Tutors: C.M. Rizzo & M. Paci, University of Genova, Italy (in italian).

Claughton A., Wellicome J. & Shenoi A., 1998. Sailing yacht design: theory & practice, Longman Pub Group, ISBN 0582368561 and ISBN 058236857X.

Germanisher Lloyd, 2002. Rules & Guidelines I Ship Technology—Part 4, Rigging Technology, Hamburg, Germany.

Janssen R., 2004. Best Mast: a new way to design a rig, The 18th International HISWA Symposium on Yacht Design and Yacht Construction, Amsterdam.

Kinney F.S. & Skene, N.L., 1962. Skene's element of yacht design, New York Dodd, Mead & Co ISBN: B000LAVWT2.

Larsson L. & Eliasson R.E., 2007. Principles of Yacht Design, International Marine McGraw-Hill, Camden, Maine, ISBN-10: 0071487697.

Marchaj C.A., 2000. Aero-hydrodynamics of sailing, Tiller ISBN 1888671181, 9781888671186.

Marchaj C.A., 2003. Sail Performance, McGraw-Hill, ISBN: 0071413103, 9780071413107.

Paci M., 2008. Analisi FEM su alberi di grandi dimensioni e confronto con dati sperimentali, Presentation at the technical meeting of the Royal Institution of Naval Architects, Genoa Branch on 02 July 2008 (in italian).

Rizzo C.M., 2007. Report MARSTRUCT MAR-W2-7-DINAV-38 (1), Monitoring sensors for shrouds deformation of a large sail ship, October 2006, MARSTRUCT NoE.

Rizzo C.M. & Carrera G., 2007. Report MARSTRUCT MAR-W2-7-DINAV-39 (1), Measurements of shrouds deformation of a large sail ship, September 2007, MARSTRUCT NoE.

Rizzo C.M., Il monitoraggio strutturale dell'alberatura di grandi navi a vela, Presentation at the technical meeting of the Royal Institution of Naval Architects, Genoa Branch on 02 July 2008 (in italian).

Analysis and Design of Marine Structures – Guedes Soares & Das (eds)
© 2009 Taylor & Francis Group, London, ISBN 978-0-415-54934-9

Assessment of ice-induced loads on ship hulls based on continuous response monitoring

B.J. Leira, L. Børsheim*, Ø. Espeland* & J. Amdahl

Department of Marine Technology, NTNU, Trondheim, Norway

ABSTRACT: The increased activity related to the oil and gas industry in polar waters implies that proper operation of ships in such areas also will be in focus. The loading on a particular ship hull depends strongly on the route selection and vessel speed. Lack of information about the actual ice condition and the corresponding loads acting on the hull is identified to be among the most critical factors when operating in Arctic waters. This implies that there is a challenging interaction between strength-related design rules and schemes for operation of ships in arctic regions. In particular, the possibility of monitoring ice-induced stresses in order to provide assistance in relation to ship manoeuvring becomes highly relevant.

The present paper is concerned with estimation of ice loads acting on the hull of the coast guard vessel KV Svalbard based on strains that were measured during the winters of 2007 and 2008. Application of a finite element model of the bow structure is also applied in order to correlate the loading with the measured strains. The influence of ice thickness and vessel speed on the measured strain levels is also investigated. Methods for extrapolation of experienced hull response into the future are also addressed.

1 INTRODUCTION

Expected growth in maritime and offshore activity in Polar areas has increased the focus on possible risks connected to operation in ice infested waters. Models for ice-induced loads on the shell structures of ships have a long history. For statistical analysis of ice loads on ship hulls, see e.g. Kheisin et al. (1973), Korri et al. (1979), Vuorio et al. (1979), Varsta (1983), Varsta (1984), Daley (1991) and Kujala (1994).

However, accurate models for computation of prevailing loads on the hull is still an identified challenge, and is the background for the Ice Load Monitoring (ILM) project managed by Det Norske Veritas (DNV) with partners. The overall aim of the ILM-project is to increase the knowledge about the actual ice conditions a vessel will meet, and their effect on the stresses induced within the hull structure. Relevant information obtained from the monitoring system should then be communicated to the bridge in order to provide decision support for the navigators.

A specific secondary goal for the ILM-project was to use instrumentation to assess a credible ice pressure distribution on the hull. With an assumed load distribution the hull utilisation can be estimated from Finite Element Analysis. This approach reduces the number of sensors involved compared to approaches with full instrumentation of all the hull where utilization

is considered directly from a complete coverage by strain sensors. A reduced number of sensors results in cheaper instrumentation, or alternatively monitoring on larger areas of the hull. A prototype of the monitoring system has been tested on Norwegian Coast Guard vessel KV Svalbard, resulting in large quantities of collected data on the experienced ice actions. The vessel is 103 m long with a displacement of 6500 tons. The beam is 19.1 m and the draft is 6.5 m. The vessel operates as a coast guard vessel in the Barents Sea and around the Svalbard islands. Measurements were performed during the winters of 2007 and 2008.

Some of the results obtained from these measurements are summarized in the present paper. Further details are given in Børsheim (2007), Espeland (2008), Leira et al. (2008). A brief description of the monitoring system is first given in the following.

2 ICE LOAD MONITORING SYSTEM

2.1 *Fiber optic strain sensors*

The applied fiber optic strain sensors are based on Fiber Bragg Gratings. Their relatively small size means that they are easily installed on girders and stiffeners in all parts of the hull. Mounting the sensors on girders and stiffeners ensures that the measurements are not contaminated by local vibrations (which would be the case for instrumented plating).

---

*Presently employed by DNV, Høvik, Norway.

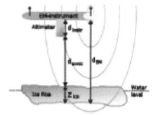

(a) Principle of EM ice thickness measuring device

(b) EM device mounted in the bow of KV Svalbard

Figure 1.   Measurement of ice thickness.

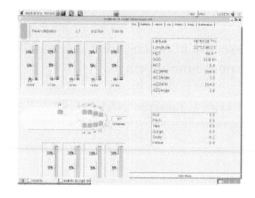

Figure 2.   Display on bridge.

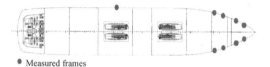

• Measured frames

Figure 3.   Location of instrumented frames.

2.2   Electromagnetic ice thickness measurements

An important difference between the ILM system and other monitoring systems is the ice thickness sensor sketched in Figure 1. In the prototype of the system, the sensor was mounted on a wooden beam ahead of the vessel. Ice thickness data were collected with a sampling frequency of 15 Hz. The sensor combines an electro-magnetic (EM) conductivity meter with a sonic distance-meter, see e.g. Haas (1997). The EM instrument measures the distance between the instrument and the water below the ice, d_{EM}. The sea ice has a negligible electrical conductivity, while the water has a high value. A magnetic field will therefore be induced in the water by the EM instrument, while the influence from the ice is negligible. The distance to the water surface is calculated from the strength of the induced field. The altimeter measures the distance to the top of the ice sheet, d_{sonic}, while d_{instr} expresses the vertical difference between the distance-meter and EM instrument. Now the ice thickness Z_{ice} can be obtained as $Z_{ice} = d_{EM} - d_{sonic} - d_{instr}$. It is noted that the sensor motions which are induced by those of the ship may introduce errors in the measurements in some cases.

Some properties of the ice thickness sensor are important to note. Ridge keels will not be measured correctly as they consist of a mixture of unconsolidated ice and water. Their depth will then be underestimated. It is also important to note that the EM-device measures the distance to the water surface over an area about 12 m^2, while the sonic device is capable of

detecting more local effects. For further information about the ice thickness sensor, reference is made to Mejlænder-Larsen et al. (2007).

2.3   Sattelite and meteorological information

During the 2007 voyage, WeatherViewTM from C-MAP Marine Forecast was used. Information about the ice conditions was supplied from the Norwegian Meteorological Institute in addition to satellite images from other suppliers. This information was used for route planning and validation.

2.4   Display at bridge

A screen is located at the bridge to display the estimated utilisation of the hull structure as well as all the other measured parameters. Both the instant values and the statistical values are available. Figure 2 shows an example of such a display.

2.5   Mounting of strain sensors

A total of 66 optic sensors have been mounted onboard KV Svalbard. The strain measuring arrangement is based on spot checks of critical frames mainly in the bow area. The sensors have been mounted on a total of 9 frames. Figures 3 and 4 show the location of the instrumented frames and how the sensors are mounted on a frame.

Figure 4. Mounting of sensors on a frame.

2.6 Other measurements

In addition to ice-thickness and strain measurements, e.g. the following data were also recorded: Temperature, ship motions, navigational data (includeng vessel heading) and ship settings (including propulsion).

3 FINITE ELEMENT ANALYSIS

3.1 Introduction

In the following, a finite element analysis of the bow area is outlined. The main task of the finite element study is to determine the area in which the mounted sensors can recognize that a load is present. This will be useful when the loads are to be estimated based on the measured strains. This task is also important in order to be able to limit the area where an acting load can influence the different strain sensors. If a load far away from an instrumented stiffener influences the measurements at the sensor locations, it is difficult to determine which load that gives rise to the specifice measurements observed in each case. For the analysis, the computer program ABAQUS.CAE was applied.

3.2 The model

The area that is modelled is shown in Figure 5. The rear part of the model is limited by the bulkhead at frame 9 in horizontal direction and by the 1st and 4th deck in the vertical direction. The model covers the sensors located at frame 1 to 4.

The complete finite element model is shown in Figure 5. To avoid any interference with the boundary

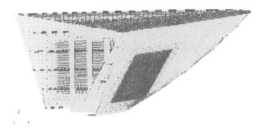

(a) The complete model

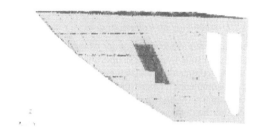

(b) Inside the model

Figure 5. The Finite Element model of the hull.

conditions, both the port and the starboard side of the ship hull are modeled.

The model is fixed in all translation directions along the rear edge of the model. This assumption is plausible because the model is ended at the bulkhead at frame 9. Since the model covers the entire bow area it is no need for a boundary at the bottom of the model. Initially FE trail was performed on only half of the model. The model was then also fixed along the centreline of the ship. The results within the instrumented area was the same for the two models, but a small difference was detected when the load was applied at the front of the model. This is the reason for making the complete model of both the port and starboard side.

Because this is a ship, the global motion could have an influence on the applied loads. However, since ice loads mostly are impulse loads, and will therefore not contribute to a global motion of the ship, this effect is not considered when applying the boundary conditions.

3.3 Location of sensors within the model

Figure 6 shows the numbering of the sensors. KV Svalbard has been equipped with sensors on both port and starboard side, but due to the symmetry in the model, the stresses have only been extracted on the starboard side.

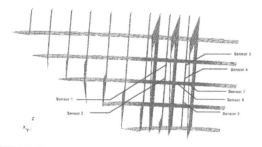

Figure 6.   Location of sensors.

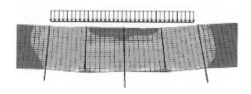

Figure 7.   Von Mises stress distribution on instrumented stringer for a specific load case.

3.4   *Load outside instrumented area*

One important aspect to consider when analysing the measured results from KV Svalbard is to what extent a load applied outside the instrumented area will affect the measurements. The instrumented area is defined as the distance between two stringers in the vertical direction, and the distance between two partial bulkheads in the horizontal direction, assuming an instrumented stiffener is located in between. If a load acting outside this area were to affect the measured results, it would be difficult to separate the different loads acting on the hull.

When considering the calculated stress in the stiffeners the difference between a load applied outside the instrumented area and a load applied within the instrumented area is large. For the load case which is shown in Figure 5 the stress is close to 0.5 MPa. When the load is moved one frame closer to the sensor location, the stress at sensors 1 and 2 increases to approximately 1 MPa.

When the load is applied directly on the stiffener, the von Mises stress in sensors 1 and 2 have increased to more than 20 MPa. Thus the effect that a load applied outside the instrumented area will have on the sensors on a stiffener is negligible. When considering the stringers only, the difference between a load applied outside the instrumented area and a load applied on the instrumented area is not that evident.

A total of 13 load cases and their respective results have been considered. In general the trend is the same as discussed above. When the load is outside the instrumented area, the stresses in the stiffeners are small. The stresses in the stringers are somewhat larger, but as explained above this effect can be neglected. Thus a load applied more than one partial bulkhead away from a instrumented stiffener or stringer will not affect the measurements in the sensors.

Figure 7 shows the von Mises stress distribution on the stringer subjected to one of the load cases. It it observed that the stress distribution when subjected to load 1 is not symmetric around the stiffener at frame 4. However, when considering e.g. load case 4 the stress is symmetric around the stiffener at frame 4, and the strain measured at sensor 4 is equal to the stain at sensor 5.

The magnitudes of the "ice load pulses" acting on the hull were obtained based on the measured shear strains, see Nyseth (2006), Børsheim (2007) and Espeland (2008) for further details.

4   MEASUREMENTS FROM KV SVALBARD

4.1   *General*

The following contains an analysis and a discussion of the measured data from KV Svalbard. The test voyage was carried out in March 2007. The raw data files contained the strain measured at each sensor. The data were sampled at a frequency of 600 Hz.

The statistics files were divided into 30 seconds, 2 minutes and 5 minutes statistics. In the present study, the 30 second statistics files are applied as this gives the best time resolution. Within each file one could find statistical data for 157 different parameters. The most important parameters used in this study are: (i) Shear stress at each cross section (ii) Vessel speed (iii) Heading and (iv) Ice thickness. Other parameters such as sea water temperature, thrust power and ship motion are also found in the statistic files.

For the different parameters the following statistical values were available: (i) Number of values used to calculate statistical parameters (ii) Mean value (iii) Standard deviation (iv) Skewness (v) Kurtosis (vi) Minimum (vii) Maximum (viii) Mean up-crossing count (ix) Max Peak-Peak value

The measurements provide a comprehensive data-based of statistical information. Some of the main observations are discussed below. Based on further study of the data, it was also decided to look at more detailed time series for a particular time segment. Figure 8 shows an example of the ice conditions during a particular period. To get a good graphical presentation of the results, the selected period has been divided into periods of one hour each.

4.2   *Ice thickness measurements*

In addition to consideration of the time sequence from 18:00 to 19:00 of March 25, a selection of other measurements sequences was made in order to obtain more comprehensive statistical information. These sequences are summarized in Table 1 below.

Ice thickness data presented by Kujala (1994), argues for a log-normal distribution for ice thickness in level ice conditions. Results obtained from the measurements from KV Svalbard are shown in Figure 9. The data are mean values over 30 second intervals between March 23th and March 28th, read from statistical files stored by the monitoring system. (Negative values below −0.5 m and above 10 m are considered unphysical, and therefore disregarded. In total they add up to less than one percent of the data. Small negative values may be due to inaccurate measurements or calibration errors, and values in the range from −50 to 0 centimeters are therefore included.) The largest group is found below 2 meter thickness, and is probably related to the pack ice without ridges. The mean thickness is around 40 centimeters, which agrees well with observations during the cruise.

The other group is located around 3 meters, and may represent ridges. A third group of data is found between 6 and 10 meters thickness. This is considered as high, but it is unclear wether they are physical or not. A ridge encountered during the test voyage was measured to be eight meters by the ice thickness sensor,

but the true ridge thickness was not measured and is hence unknown, Nyseth (2006).

The selection of recording periods for statistical analyses are not grouped in relation to ridging. Instead, they were chosen due to stable vessel speed, which is not expected in ridged areas. Figure 10 shows all ice thickness samples done between 19:00 and 21:00 on March 26th. The mean thickness is 0.9 m, with a standard deviation of 0.6 m. It should be noted that the negative values clearly deviates from the rest of the samples, which may invalidate the assumption of a calibration error. A log-normal distribution is fitted to the positive data in compliance with the findings of Kujala (1994) for level ice observations. The resulting fitted distribution is somewhat skewed to the left as compared to the observation histograms.

4.3 Consideration of response time histories

4.3.1 Ice load versus ice thickness
The simultaneous variation of the ice load and ice thicness is first considered. Regarding the mean load on a specific frame versus the ice thickness there does not seem to be any reasonable correlation. However, if we compare the peak loads for this particular time sequence, the correlation is more clear as shown in Figure 11. A sudden increase in the ice thickness results in a high load peak. This could indicate that the ice thickness influences the maximum load but not the mean value of the load.

However, an increase in the ice thickness did not always result in an increase in load. This could partly be explained by the fact that there is a spatial distance between (any of) the hull frames and the area where the ice thickness is measured (i.e. ahead of the bow). Clearly, the loads acting on the different frames will not occur at the same point in time. To be able to incorporate all the peaks in the foremost frames, Figure 12 shows the maximum peak load for frames 1 to 6.

Figure 12 shows a more distinct connection between the load and ice thickness than what we observed when only considering one frame at the time.

Figure 8. Ice conditions 03.25.2007 at 17:00.

Table 1. Selected measurement sequences for the purpose of statistical analysis.

Date	Interval	Mean ice th. [m]	St. dev. th. [m]	Mean speed [knots]	St. dv. speed [knots]
24.3.2007	05:10–05:40	0.7 m	0.5 m	3.9	0.8
25.3.2007	16:30–17:00	1.0 m	0.6 m	4.6	0.8
25.3.2007	21:00–21:30	0.8 m	0.7 m	3.6	0.7
26.3.2007	19:00–19:30	1.1 m	0.6 m	4.7	0.8
26.3.2007	19:35–20:05	0.5 m	0.3 m	6.0	0.3
26.3.2007	20:30–21:00	1.1 m	0.7 m	4.9	0.9

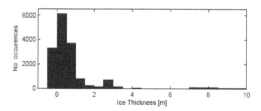

Figure 9. Histogram of ice thickness from 30 second statistics, March 24 to March 28.

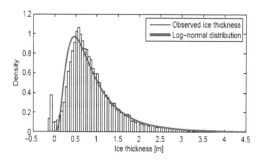

Figure 10. Ice observations at March 26 compared to fitted Lognormal distribution.

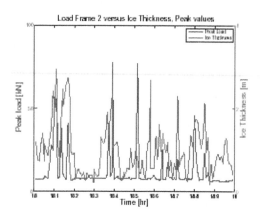

Figure 11. Variation of peak load on Frame 2 and ice thickness between 18:00 and 19:00.

4.3.2 Ice load versus vessel speed

Figure 13 shows the time variation of the peak load versus the vessel speed. When considering the mean value of the load, a distinct positive correlation is observed. An increase in the speed results in an increase of load (However, it should be noted that the mean load is quite small and does not induce any stresses of significance). This is according to what we would expect. On the other hand, as seen in the figure the peak load seems to increase when the speed is decreasing as shown in the figure. This could be due to the vessel experiencing a higher resistance from the ice sheet,

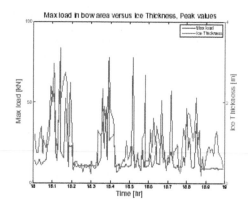

Figure 12. Time variation of max load on bow and ice thickness.

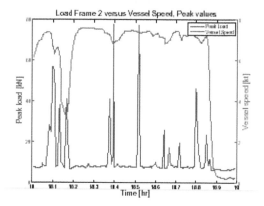

Figure 13. Time variation of load on Frame 2 and speed between 18:00 and 19:00.

hence leading both to reduced speed and increase of peak load.

This is also supported by Figure 14, where the corresponding time variation of the vessel speed versus ice thickness is shown. It is observed that for segments where peak values of the ice thicknss occur, there is typically an associated reduction of the mean value of the vessel speed. This will be even more pronounced if instantaneous velocities were applied instead.

4.4 Statistics of ice loads based on measured strains

One of the overall goals for the ice load monitoring system is to communicate information about the prevailing ice conditions to the bridge. Expected force maxima are relevant in this setting, as they give indications on future loads. Short term predictions are focused upon here, aiming at predictions for the next 30 minutes of operation. In general, long term distributions for expected loads during a season of operation,

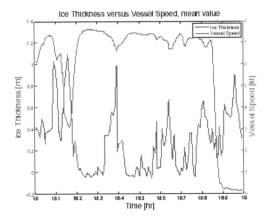

Figure 14. Example of ice thickness versus vessel speed record, 18:00 to 19:00 hrs.

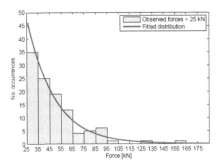

Figure 15. Peaks above 25 kN on Frame 2, March 26 at 19:00 to 9:30.

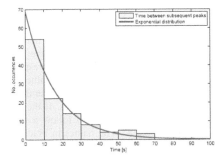

Figure 16. Observed time between subsequent peaks, March 26 at 19:00 to 19:30.

or the lifetime of the vessel may also be established, e.g. as described by Kujala (1994). In Lensu et al. (2003) two different methods are presented for the purpose of establishing short term ice load extreme distributions. These methods are referred to as the peak amplitude approach and the time window approach, respectively. Homogenous conditions are assumed by both approaches, which is not necessarily valid for pack ice conditions.

4.4.1 Peak amplitude approach

The peak amplitude approach is also referred to as the 'peak over threshold' method, as load peaks (i.e. of type seen in Figures 11 to 13) are defined as observations exceeding a predefined threshold value. The maximum value in a section is taken as the peak amplitude. A distribution model is then fitted to the observed maxima, see Figure 15 for an example of a fitted exponential distribution. A separate distribution is also fitted to the observed number of load peaks during an interval, see Figure 16 for an example. An exponential distribution fits the data quite well also for this cases. These models are combined to describe the distribution of interval maxima.

4.4.2 Time window approach

The time window approach differs from the previous method in the way peaks are identified. In this approach, the time history is divided into shorter subintervals, and the maximum for each interval is considered as a peak value. Instead of finding a distribution for the interval maxima, the method describes the relation between the mean maximum value from a set of sub intervals. The interval maxima is then predicted by linear extrapolation of the resulting relationship until a sub interval length equal to the whole interval. The approach is easy to implement, and shows good results in examples presented in Ref. [11].

The following advantages of this method was listed:

- No threshold value is needed. A threshold value will influence the duration between load peaks, and there are no physical reasons to assume that load peaks have a minimum amplitude.
- The time window approach can generate statistical models from shorter time histories.
- The length of the subintervals is treated as a variable, and no parameters are then chosen by the analyser. Results from different ships, ice conditions and locations are therefore always comparable.

A distribution is established from two variables, T and x_T. T is the length of the time windows the load time history is divided into. For a given value of T this corresponds to N_T time windows on a time history of length h:

$$N_T = \frac{h}{T}$$

The other variable x_T is calculated as the mean of all time window maxima. The maxima inside time window number i, x_i, is used to define x_T as

$$x_T = \frac{1}{N_T} \sum_{i=1}^{N_T} x_i$$

A number of values for x_T is then calculated from the load time history by varying the time window length T. Determination of suitable distributions for the variables, and prediction of future loads is based on linear extrapolation of the distribution on an appropriate probability paper, referred to as (T, x_T)-plots by the author. The variables are found to follow one of two principal linearised equations, depending on the time window length. For short window periods, x_T tends to follow equation the former equation below, while for longer time windows the relation in the second equation is more common, Ref. [11].

$$\ln(x_T) = H \ln(T) + c$$

$$x_T = H_0 \ln(T) + c_0$$

It is important to be aware that the extrapolated variable x_T describes the mean value of the maxima in several time windows, and not the respective window maxima x_i. According to Ref. [11], the corresponding distribution for x in intervals with mean value x_T will be a type-II extremal value distribution (Gumbel II) if x_T the first relationship above applies. If the second relationship applies for x_T, x will follow a Gumbel (Gumbel I) distribution. (The names in brackets are those applied by Lensu et al. (2003)).

This second approach typically leads to conservative estimates, and may call for an adjustment of the method to obtain better results. The adjustment will depend on the extent of extrapolation, and possibly ice conditions and speed. The correction proposed in Lensu et al. is to remove certain large loads based on statistical considerations, but no further comments about how this should be performed are given.

The findings from KV Svalbard are more ambiguous with respect to relation, as the different frames and time intervals seems to follow different trends. The time window plots indicate that bow section frames 2 and 4 follows the second relationship for time windows longer than eight seconds, while the curves for frames 6 and 8 (which are closer to the shoulder) follows the other relation. Figures 17 and 18 may serve as examples for frame 2 and frame 8, respectively. For other investigated intervals the conclusions are different. In the interval with lowest mean ice thickness and highest speed, between 19:35 and 20:05 on March 26th, all frames were found to fit the first relationship best.

4.4.3 Comparison of methods
Both approaches are able to predict loads in the right order of magnitude for the observed maxima, and may serve as a basis for prediction of future loads. Conservative choices of percentile value or extrapolation curves should be chosen to envelope the observed maxima.

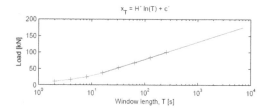

Figure 17. Time window approach applied to frame 2, March 26 at 19:00 to 19:00.

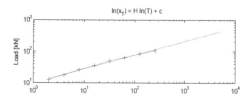

Figure 18. Time window approach applied to frame 8, March 26 at 19:00 to 19:00.

The time window approach is somewhat ambiguous with respect to relation between T and x_T, with fairly good fit for both distributions. The first relationship is generally very conservative, but is the only relation able to predict the highest observed value of 345 kN for frame 6. The second relationship was found to be comparable with the mean value of the maxima obtained by means of the peak amplitude approach. It should be noted that the 345 kN for frame 6 was well above the 99 percentile in the classical extreme value distribution from the peak amplitude approach. (Other high loads were experienced during the same interval, and it is hence not regarded as a single special event.)

From the few cases tested with the prediction routines, the peak amplitude approach seems most promising. However, further investigations for extended amounts of data are required for strong conclusions to be drawn.

4.4.4 Correlation of ice force versus ice thickness and vessel speed
Registered force versus ice thickness is shown in Figure 19(a). Reports presenting collected data from Nathaniel B. Palmer (SSC (1995), p. 93) and Polar Sea (SSC (1990), p. 88) shows comparable results, and explains the behaviour with the random nature of ice loads. The highest observed values of the force may then be expected to occur in areas with the highest number of registered events. A few loads are registered with a negative ice thickness, and this is expected to be due to calibration errors or measurement uncertainties.

Figure 19(b) leads to similar conclusions for the relation between ice force and vessel speed. The absence of observations for low speed is partly

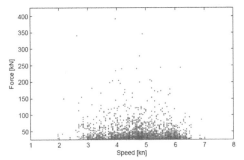

(a) Force (all frames) versus ice thickness.

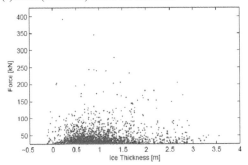

(b) Force (all frames) versus vessel speed.

Figure 19. Scatter plots relating (a) force to ice thickness and (b) force to vessel speed.

explained by the selection of intervals for analysis, where higher speeds were focused upon.

5 DISCUSSION AND CONCLUSIONS

The ice load monitoring system on KV Svalbard was presented. A correlation study between the ice load and the ice thickness and vessel speed has been performed. The study is based on 30 seconds statistical values. Some of the results were somewhat inconclusive, but in general it is possible to state that the maximum peak loads (i.e. the maximum peak for all frames at any time) depend on the ice thickness (for a given, constant value of the ship speed). However, Because KV Svalbard is a small vessel, the speed was reduced considerably when the ice thickness increased. (Furthermore, the mode of operation for the vessel also varied significantly due to other types of research missions during the voyage.) This will "disturb" the relationship between the ice load and the ice thickness for the highest values. Scatter-plots of peak ice loads versus ice thickness and vessel speed were also found to have a significant scatter.

The ice load monitoring system is able to inform about the instantaneous stress levels experienced by the hull. Furthermore, procedures are available (i.e. the

peak amplitude and the time window approaches) to extrapolate these stress levels into the future provided that parameters related to ice loading are constant (i.e. stationary conditions). For the case of nonstationary conditions, such extrapolation is more challenging and will require a comprehensive calibration.

A statistical evaluation of the time between the load peaks as well as the peak amplitudes has been performed. Both these parameters seem to be exponentially distributed, indicating that the peaks of the ice loading can be represented by a Poisson process with exponentially distributed amplitudes.

ACKNOWLEDGEMENT

Det Norske Veritas (DNV) is greatly acknowledged for giving access to the data from the measurements on KV SVALBARD and to publish the results which are reported in the paper. Parts of this work were performed within the project MARSTRUCT, Network of Excellence on Marine Structures (http://www.mar.ist.utl.pt/marstruct/), which has been financed by the EU through the GROWTH Programme under contract TNE3-CT-2003-506141.

REFERENCES

Børsheim, L. (2007) "Ship Hull Monitoring of Ice-Induced Stresses", Master Thesis, Dept. Marine Technology, NTNU, Trondheim.

Daley, C. (1991): "Ice Edge contact, a brittle failure process model", Dissertation, Acta Polytechnica Scandinavia, Me 100, Helsinki.

Espeland, Ø. (2008): "Ice Action and Response Monitoring of Ships". Master Thesis, Dept. Marine Technology, NTNU, Trondheim.

Haas, C., Gerland, S., Eicken, H. and Miller, H. (1997): "Comparison of sea-ice thickness measurements under summer and winter conditions in the Arctic using a small electromagnetic induction device", Geophysics, 62(3), pp. 749–757.

Kheisin and Popov (1973) (Eds.): "Ice Navigation Qualities of Ships", Cold Regions Research and Engineering Laboratory (CRREL), Draft translation 417. Hanover, New Hampshire.

Korri, P. and Varsta, P. (1979): "On the Ice Trial of a 14500 dwt tanker on the Gulf of Bothnia", NSTM-79, Society of Naval Architects in Finland (LARADI), Helsinki.

Kujala, P. (1994): "On the Statistics of Ice Load on Ship Hull in the Baltic", Acta Polytechnica Scandinavica, Mechanical Engineering Series No. 116, Helsinki, Finland.

Leira, B.J. and Børsheim, L. (2008): "Measured Ice-loading on KV Svalbard", Proc. OMAE 2008, Estoril, Portugal.

Lensu, M. and Hanninen, S. (2003): "Short Term Monitoring of Ice Loads Experienced by Ships", Proc. POAC '03, Trondheim, Norway.

Mejlænder-Larsen, M. and Nyseth, H. (2007): "Ice Load Monitoring" in Proceedings of the Royal Institution of Naval Architects—RINA, Design and construction of vessels operating in low temperature environments", London, UK.

Nyseth, H. (2006): "Strain measurements on board KV Sval-Bard with respect to ice loading", Technical report, Det Norske Veritas, 2006.

SSC (1990): "Ice Loads and Ship Response to Ice", Technical Report 329, Ship Structure Committee.

SSC (1995): "Ice Load Impact Study on NSF R/V NA THANIEL B. PALMER", Technical Report 376, Ship Structure Committee.

Varsta, P. (1983): "On the Mechanics of Ice Load on Ships in Level Ice in the Baltic Sea", Dissertation, Technical Research Centre of Finland, Publication 11, Espoo.

Varsta, P. (1984): "Determination of ice loads semiempirically", in Ship Strength and Winter Navigation", Technical Research Centre of Finland, VTT Symposium 52, Espoo.

Vuorio, J., Riska, K. and Varsta, P. (1979): "Long-term measurements of ice pressure and ice-induced stresses on the icebreaker Sisu in winter 1978", Winter Navigation Research Board, Report 28, Helsinki, Finland.

Materials and fabrication of structures

Welded Structures

Analysis and Design of Marine Structures – Guedes Soares & Das (eds)
© 2009 TWI Ltd, ISBN 978-0-415-54934-9

The importance of welding quality in ship construction

P.L. Moore
TWI Ltd, Cambridge, UK

ABSTRACT: Welding is one of the most critical operations within ship construction. When welds fail, often the whole structure fails. Fortunately, over sixty years of research and development in the field of welding has provided current ship builders with fabrication processes that are readily automated, can produce consistent welds reliably, and/or can weld thick sections in a single pass for controlled distortion. The expectation of weld quality has never been higher. This paper summarises how consistent weld quality is achieved in practise, through classification society rules, welding procedure qualification, welder certification and weld monitoring and control. However, no matter how rigorous the quality control, sometimes things go wrong. There are also a number of methods that can help to remedy the effects of poorer quality welding. For shipbuilding, weld repair is usually the main strategy, and the importance of repair welding will be discussed. However, it may not always be necessary to cut out the weld and start again. Methods such as fitness-for-service assessment and fatigue improvement techniques can be applied to justify whether imperfect welds might still be adequate.

1 INTRODUCTION

Welding is one of the most critical operations within ship construction. When welds fail, often the whole structure fails (Apps et al. 2002). Over the past sixty years of developments in welding technology, the current expectation of weld quality has never been higher. Ensuring good quality welding forms an important part of all ship classification societies' rules. Rules for shipbuilding are all written with the expectation of achieving safe shipping, including Lloyds Register (1999), Det Norske Veritas (DNV 2008), the American Bureau of Shipping (ABS 1997), and also the International Association of Classification Societies, IACS (2008). The approach is similar to that for many safety critical construction applications—using qualified welders and welding procedures.

2 THE PROBLEMS WITH WELDING

2.1 *Distortion*

Shipbuilders may consider the main problem with welding is the resultant distortion that occurs. Any process that uses a localised heat source, such as welding, is likely to result in some distortion. However, distortion can be minimised in welds that use low heat input and avoid excessive weld bead sizes. Using jigs and fixtures or pre-setting the components to offset the eventual distortion can also help. Most distortion is corrected after welding using localized flame spot heating to restore the required dimensions.

Distortion associated with welds may cause problems for the ship design strength and stiffness, and for the appearance of the finished vessel, and preventing and remedying distortion can be a major cost of shipbuilding. Despite it being a major subject in itself, distortion is associated with even the best welds and is not a weld quality issue as such, so will only be touched on briefly in this paper.

2.2 *Weld flaws*

Typical welding flaws can be classified into three categories to assess their significance; planar flaws (cracks, lack of fusion or penetration, undercut or overlap); volumetric flaws (porosity and slag inclusions); and weld shape flaws (misalignment, or incorrect profile). The least severe are gas and slag inclusions, and weld shape flaws. In terms of structural integrity, the most critical flaws are cracks, lack of fusion or penetration, and undercut. The sharp profile of these flaws makes them strong stress concentrators, and it is these flaws that most weld quality codes and standards are seeking to avoid.

3 GETTING WELD QUALITY RIGHT

3.1 *Approved welding procedures*

The weld procedure specification (or WPS) is the 'recipe' for carrying out a particular weld. It specifies the welding parameters that must be adhered to for compliance to that welding procedure. For arc welding these 'essential variables' include;

1. Welding process(es).
2. Voltage and current ranges, and polarity.

3. Welding travel speed and/or heat input.
4. Welding consumables (wire, rods and gases). Sometimes lists of approved welding consumables are issued by the Classification Society (such as ABS, Lloyds, DNV and RINA).
5. Parent materials type, grade and thickness range.
6. Joint design, welding sequence and welding position.
7. Preheat and inter-pass temperatures, and post-weld heat treatment (if required).

The ship classification societies all require approved welding procedures to be used. For example, DNV allows welding procedures to be approved in one of three different ways. Firstly, if the WPS is based on another approved welding procedure, secondly, from verification of documentation showing successful application of the WPS over a prolonged period of time. Thirdly, and most commonly, weld procedure approval is done through review of weld procedure qualification records (WPQRs), collected from the assessment of a test weld made using the WPS. IACS shipbuilding and repair quality standard (1999) also requires that a welding procedure should be supported by such a welding procedure qualification record.

3.2 Weld procedure qualification testing

Qualification of a welding procedure through testing generally consists of firstly preparing the written preliminary weld procedure specification (pWPS), then welding up a test piece following that pWPS. The test piece is then subjected to inspection and a range of non-destructive testing (NDT) and mechanical tests to demonstrate the strength, ductility, toughness, and the presence or absence of defects. If the results of the NDT and all the mechanical tests are satisfactory then the welding procedure specification (WPS) is qualified. Documentation such as the record of welding parameters used for the test weld, the NDT and mechanical testing certificates, and the signed weld procedure approval certificate, collectively form the weld procedure qualification record (WPQR). It is usually the responsibility of the ship builder to ensure that the welding procedures (and the welders) are qualified.

The NDT used on the test weld includes visual examination, a surface flaw detection method such as dye penetrant or magnetic particle (MPI) testing, and for butt welds, a method to detect buried flaws, usually radiography or ultrasonic testing. The standard mechanical tests for qualification of a test weld piece are:

1. Macro and hardness test to show the weld shape, and the maximum hardness of the weld and heat affected zone (HAZ).
2. Cross-weld tensile test to demonstrate that the weld is overmatching to the parent metal.

3. Guided bend tests; face bend and root bend, or two side bends (depending on material thickness), to demonstrate the weld ductility, and to open any flaws present for detection, shown in Figure 1.
4. All-weld metal round tensile to test the weld metal strength.
5. ABS also requires fillet weld fracture tests for WPS qualification.
6. Sometimes Charpy vee notch tests are carried out, notched in weld metal, HAZ fusion line, and HAZ+2mm, HAZ +5mm, to determine the impact toughness of different regions of the weld and HAZ. Charpy tests are carried out at a temperature dependant on the grade of steel being used, e.g. $-10°C$ for Grade D and $-40°C$ for Grade E (ABS 1997). The required impact energy is stated, usually between 27J and 47J at the given temperature.

Generally the Classification Societies require that the weld procedure qualification, including the welding of the test piece, the inspection, and subsequent mechanical testing is witnessed by an approved surveyor—this is a requirement of Lloyds Register for new welding procedures. It is uncommon for a Classification Society to specify an external standard to which the weld procedure needs to be qualified; ABS and DNV lay out their own requirements for test piece qualification. The most common general standards for qualification of welding procedures are ISO 15614-1 (2004a), ASME IX (2007), and ANSI/AWS D1.1 (2008) for arc welds, and ISO 15609-4 (2004b) for laser welds.

Figure 1. Guided bend test.

3.3 *Welder qualification*

The qualification of welders or weld operators often goes hand-in-hand with the weld procedure qualification. It is common for the welder who produced the test weld to also be approved for that particular welding procedure as well. All the Classification Societies require that welders and weld operators are properly skilled and qualified for the job; welders are normally required to be certified.

Qualification (and/or certification) of a welder is obtained following satisfactory assessment of a test weld, the production of which is usually witnessed by an examiner, surveyor or test body. The approval test examines a welder's skill and ability to produce a test weld of satisfactory quality. Limits are given on flaws associated with the shape of the weld bead such as, for example, excess weld metal, concavity, excess throat thickness and excess penetration. In a welder qualification, these features reveal the welder's competence and skill (Fig. 2). The welder's test piece is subjected to visual examination, possibly NDT (penetrant testing, MPI and radiography), and some mechanical testing which may include a macro-section, fillet break tests (where the fillet weld is fractured through the root and examined for defects), cross weld tensile tests, and butt weld fracture or bend tests (possibly using shallow notches cut into the weld metal of the side bend specimens).

DNV (2008) and IACS (1999) require that welders shall be qualified to a standard recognised by the society (i.e. EN 287, ISO 9606, ASME Section IX, ANSI/AWS D1.1). Lloyds Register requires that shipbuilders test the welders and weld operators to a suitable National standard, but is not specific, while ABS outlines its own programme of welder qualification testing. BS EN 287 (2004) is specifically for fusion welding of steels, while EN ISO 9606 (2004c) covers fusion welding of aluminium alloys. The challenges for manual arc welding of steel and aluminium are rather different, and hence it is important for welders to be qualified and trained specifically for the material that they will be welding. Within the remit of EN 287 there are also different categories of steels, depending on their composition and hence their likelihood of cracking during welding. Welding operators using fully mechanised or automated equipment are not usually subject to approval testing. However, they are usually required (e.g. by DNV and IACS) to have records of proficiency showing that operators are receiving regular training in setting, programming and operating the equipment.

3.4 *Range of approval*

Both weld procedure approval and welder approvals are not limited solely to fabrications using the exact welding parameters used for welding the test piece, but come with a given range of approval.

The objective of setting this range is to allow a weld procedure to also cover welds expected to give similar (or better) mechanical properties. For instance, although the material type cannot change (as it is an essential variable), the plate thickness for which the WPS is approved might range from half up to twice the test weld plate thickness. Another example is that DNV qualification allows welding of grades of steel with lower toughness requirements, but not those with higher toughness requirements.

The range of approval for the welder qualification is also intended to cover welds that are considered 'easier' to weld, so a welder that is qualified for butt welds may also weld fillet welds, but not vice versa; positional welding will qualify welding in the flat position but not vice versa. It is important to check the permitted range of approval before conducting the test welding as it might be possible to cover all the production welding with fewer weld procedure qualification tests given careful selection of the test weld parent material thickness, welding position, access to one or both sides etc.

3.5 *Welding co-ordination*

Often the requirement for qualification within welding ends with the welding procedure and the welders. However, it is also important that the staff who write the welding procedures and supervise the fabrication are competent and qualified for their roles. DNV

Figure 2.   Gas metal arc (GMAW) welder.

is one of the few Classification Societies to refer to this specifically, and gives a note that it bases quality requirements for welding on the EN 719 and 729 series of documents. ISO 3834 (1994) 'Quality requirements for welding' has now superseded EN 729 and ISO 14731 (2006) 'Welding co-ordination' has superseded EN 719. ISO 14731 is about people and responsibilities, and ISO 3834 is about companies implementing systems for qualified weld procedures. These two standards require companies to show that all their welding operations are under appropriate and technically competent control. The company must show that people with welding responsibilities possess relevant competence. The qualifications of the International Institute of Welding (i.e. International Welding Specialist/Technologist/Engineer diplomas amongst others) are mentioned as examples of such suitable qualifications.

4 MAINTAINING WELD QUALITY DURING FABRICATION

4.1 Production welding

Many Classification Societies provide recommendations on measures that can also be taken during welding to help improve the quality of welded structures, for example by minimising distortion. Welding from the centre outwards of a seam and selecting an appropriate sequence for welding different joints can minimise the distortion when welding stiffened panels. Fit-up using clamps, jigs, and tack welds (provided they are in accordance with the WPS) can also be used to ensure good alignment or to minimise residual stresses and distortions. Run-on and run-off plates can help to avoid welding defects and end craters in the actual structure.

Sometimes weld quality is affected by the adverse conditions of the environment during welding on site. Construction of vessel hulls is usually done in stages involving assembly and fabrication of sub-assemblies which are then brought together in the dry dock for final erection and welding. This final stage of fabrication is potentially more exposed to the weather, so where possible the welding environment should be kept free from moisture and draughts. The controlled welding consumables need to be kept dry and clean. A study of weld quality in a new-build FPSO was carried out to address concerns that welding quality was poorer when inspection was not required or if inspection equipment was unavailable (Still et al. 2004). The marine piping systems of the FPSO were not required to be inspected by the Class Rules, but the operator radiographed them for the study, and found 60% of the butt welds examined contained lack of root penetration defects. It seemed that when there was no expectation that the work would be inspected, and/or that the Class surveyor's workload is such that they may not have

adequate equipment or the time to carry out a thorough inspection, the weld quality reduced. This illustrates the importance of quality control by welding supervisors, since even an approved welding procedure can produce poor welding if not followed correctly during fabrication.

4.2 Joint or seam tracking

For mechanised and automated welding processes, including laser welding, joint tracking offers the potential for improved quality assurance by allowing on-line adjustment of the welding head position with respect to the workpiece to compensate for small variations in joint position and fit-up. Real-time joint tracking is performed by a sensor that first detects the position of the joint and then guides the automatic welding equipment. The sensor communicates with the welding head to send trajectory corrections, maintaining the tool centre point at the optimum position in the joint. The high degree of accuracy achieved while welding improves productivity by significantly decreasing the amount of operator monitoring and intervention required, increasing the welding travel speed, and reducing rework costs.

Vision-based sensors are currently used for the majority of joint tracking systems for both arc and laser welding applications, due to their high accuracy and signal update rate (Shi et al. 2007). Some other tracking systems used for arc welding processes such as tactile sensors or through-arc sensors are not suitable for laser welding applications due to their general lack of accuracy and low signal update rate. Faster data handling is required for laser welding due to the high welding speed, which increases the system response and processing speed needed. Vision-based sensors also use an additional low-power infra-red or visible laser which is focused to form a line on the workpiece surface, which detects the joint profile and gap.

Limitations of seam tracking come from trying to detect abrupt changes in direction whilst welding at high speeds, and detecting closely butting edges for laser welded joints, as these may not produce sufficient reflected signals to allow the seam to be detected. The effect is worse on highly reflective materials such as stainless steel or aluminium.

4.3 Laser welding and mechanised welding processes

The reproducibility of automated welding such as laser welding means that once welding parameters are set correctly, high quality welds can be reliably produced time and again. Laser beam welding (Fig. 3), as a substitute for arc welding, provides advantages in terms of high productivity (in terms of speed and/or thickness) and low heat input (and thus low distortion). Weld procedure qualification can also be used for laser processes, and is covered by ISO 15609-4 (ISO 2004b). Ship building is one of the main industries where laser welding is applied, and although welding lasers are not commonplace in shipyards, the process is mentioned specifically with respect to qualification of laser welding operators by DNV and IACS. The main issue is to ensure that the tight tolerances on joint fit-up required for laser welds are met, which can be difficult to achieve in shipyards. The solution is either to use a welding technique or procedure that is more tolerant to joint fit-up and/or include process control measures to allow accurate seam tracking to be integrated into automated welding systems to maintain weld quality. The combination of an electric arc with the laser beam in the same weld pool in a hybrid process (Fig. 4) can significantly improve the gap bridging capability and tolerance to misalignment of the laser process alone.

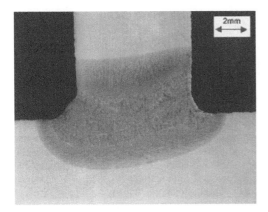

Figure 3.   Laser welded stiffener.

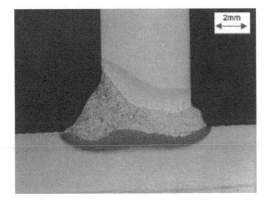

Figure 4.   Laser/arc hybrid welded stiffener.

For in-process monitoring of laser welding, most current techniques employ a variety of sensors to monitor electromagnetic signals from the molten pool during welding, with the objective of correlating the output from the sensor to features such as weld penetration, weld pores, and the weld shape.

5   WHAT CAN BE DONE WHEN THINGS GO WRONG?

5.1   Repair welding

Repairs to welds can be very expensive; up to 10 times the cost of getting it right the first time. Most classification societies, and the IACS shipbuilding repair quality standard (1999), do not allow any weld cracks to be left unrepaired irrespective of size. For shallow surface flaws it may be possible to grind out the flaw without any further weld repair. Methods of flaw removal include grinding or machining a smooth profiled groove, and then inspection to ensure the flaw has been entirely removed. Before repair welding it is important to know why the defects occurred, so that they can be avoided in the repair weld. Many problems are caused by either the welder, or from issues with the fit up. Therefore it might be necessary to re-train and re-certify the welder, and/or improve the fit-up of the joint. Repair welding must be done using the same quality controls as the original welding (i.e. approved WPS, qualified welder and certified consumables), but if following the welding parameters of the original WPS has resulted in defects, the WPS might need to be revised.

There is usually a limit to the number of times it is permitted to re-weld, for example DNV allows only two cycles of repair. The repair weld will usually be inspected using the same NDT methods as originally applied. Some additional difficulties for repair welding which are not usually experienced during production welding are the provisions required for proper access, lighting and ventilation, while being sheltered from the weather or seawater (for hull repairs).

5.2   Fitness-for-service assessment

Sometimes repairing a flaw might not be necessary with respect to ensuring structural integrity, e.g. a weld containing a small flaw could still be fit for service. This is where a fitness-for-service assessment can be invaluable. Most welding fabrication codes and classification society rules specify maximum tolerable flaw sizes based on good workmanship, i.e. what can reasonably be expected within normal working practices. These requirements tend to be somewhat arbitrary, and failure to achieve them does not necessarily mean that the structure is at risk of failure.

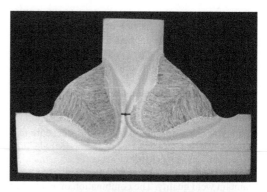

Figure 5.   Overview of the FFS assessment approach.

A fitness for service (FFS) assessment (also called engineering critical assessment or ECA) is an analysis, based on fracture mechanics principles, of whether or not a given flaw is safe from brittle fracture, fatigue or plastic collapse under specified loading conditions. An overview is shown in Figure 5. FFS assessment procedures are issued in a number of national standards. The results allow decisions to be made so that a flaw that is tolerable may be left safely in service, whereas an unacceptable flaw needs to be removed and/or repaired.

Although FFS assessment can be used to assess the significance of fabrication defects that have been found to be unacceptable to a given code, its greatest potential within shipping is for assessing flaws such as fatigue cracks that have grown during service. It can also be used to justify life extension of a vessel's service for instance. The FFS assessment concept is widely accepted by a range of engineering industries such as oil and gas, and power generation. However, it has yet to find wide acceptance by classification societies, despite its long track record in preventing unnecessary repairs, cost and delay while ensuring safety in service.

5.3   *Fatigue improvement*

Fatigue cracking during service is a major concern for ongoing structural integrity of welds in ships. Fatigue failures usually initiate at changes in cross section; machined grooves, bolt holes, sharp flaws or at welds. The sharper the notch, the higher the stress concentration, and the greater the limitation on fatigue life. The risk of fatigue can be reduced in structures under cyclic duty, by either reducing the loading on the structure, or by reducing the local stress concentrations by improved design. Misalignment and distortion of welded joints will increase the applied stress further, which reduces the expected fatigue life. A poorly shaped weld cap with a sharp transition between the weld and the parent metal will also have an adverse

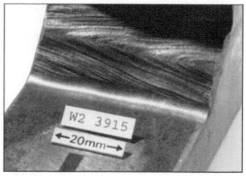

Figure 6.   Fatigue improvement by weld toe grinding.

effect on fatigue performance. In addition to these geometrical features affecting fatigue life there is also a small intrusion at the weld toe. In a welded component these intrusions act as initiators for fatigue and hence the bulk of the fatigue life is spent in propagating the fatigue crack without any need for a period of time to initiate the starter crack.

There are fatigue improvement techniques that can be applied to welds. The dual objectives of these are the removal of the intrusions at the weld toe and the achievement of a smooth transition between weld metal and parent plate, shown in Figure 6. Different welded joint configurations are given a classification according to their fatigue performance (BSI 1993). Full penetration butt welds made from both sides can be promoted from fatigue class D or E (depending on the welding procedure) to class B or C (depending on whether they are longitudinal or transverse welds) by machining the excess weld metal flush with the surface. Distortion of the joint can make this treatment difficult to apply. For fillet welds it is possible to dress the weld toes by grinding away material along the weld toe to remove the toe intrusion while maintaining a smooth weld profile. The toes may be dressed by the careful use of a disc grinder, making sure any machining marks are parallel to the axis of the main stress, to avoid forming initiation sites for further fatigue. For best results the toe should be machined with a fine rotary burr, even though this is slower. Toe grinding needs to be done with care to prevent too much metal being removed and hence thinning the component below its minimum design thickness. Ideally the dressing should remove no more than 0.5 mm depth of material for full burr grinding and 0.8 mm for disc grinding.

6 CONCLUSIONS

The shipbuilding industry has a history of using Classification Societies Rules for ensuring good quality ship fabrication. The qualification of welders and weld procedures is an effective way to ensure these quality requirements are met. However, supervision is needed to ensure that the procedures are followed during production welding. When flaws are found in welds, the usual requirement by Classification Societies is for removal and/or repair. However, it can be desirable for unnecessary repair welding to be avoided if possible, and the Classification Societies may wish to exploit the proven abilities of fitness-for-service assessment procedures for justifying this. In-service fatigue damage at welds can be prevented or remedied by a number of fatigue improvement techniques.

REFERENCES

ABS. 1997. Rules for Building and Classing Steel Vessels, 1997. Part 2, Section 3, Welding and Fabrication. American Bureau of Shipping, 1997.

Apps, B., Crossland, B., Fenn, R. & Evans, C. 2002. 'Killer consequences of defective welds—a plan for prevention', TWI Bulletin, January/February 2002.

ASME. 2007. Boiler and Pressure Vessel Code, Section IX: Welding and Brazing Qualifications.

AWS. 2008. D1.1/D1.1M:2008 Structural Welding Code—steel, American Welding Society.

BSI. 2004. BS EN 287-1:2004 Qualification test of welders—fusion welding—Part 1: steels, British Standards Institution.

BSI. 1993. BS 7608:1993 Code of practice for fatigue design and assessment of steel structures.

DNV. 2008. Rules for the Classification of Ships/High Speed, Light Craft and Naval Surface Craft. Newbuilding, January 2005, incorporating amendments 2008. Materials and Welding, Part 2 Chapter 3, Fabrication and Testing of Structures. Det Norske Veritas.

IACS. 2008. Common Structural Rules for Bulk Carriers, Consolidated edition of 1 July 2008.

IACS. 1999. Shipbuilding and Repair Quality Standard, Part A: Shipbuilding and Repair Quality Standard for New Construction, & Part B: Repair Quality Standard for Existing Ships. No. 47, Rev. 1, August 1999.

ISO. 1994. ISO 3834-1:1994 Quality requirements for welding—fusion welding of metallic materials.

ISO. 2004a. BS EN ISO 15614-1:2004 Specification and qualification of welding procedures for metallic materials—welding procedure test. Part 1: Arc and gas welding of steels and arc welding of nickel and nickel alloys (Supersedes EN 288 Part 3).

ISO. 2004b. BS EN ISO 15609-4:2004 Specification and qualification of welding procedures for metallic materials—Welding procedure specification. Part 4: Laser beam welding (Supersedes ISO 9956).

ISO. 2004c. BS EN ISO 9606-2:2004 Qualification test of welders. Fusion welding—Part 2: aluminium and aluminium alloys.

ISO. 2006. BS EN ISO 14731:2006 Welding coordination—Tasks and responsibilities (Supercedes BS EN 719:1994).

Lloyds. 1999. Rules and Regulations for the Classification of Ships, July 1999. Part 3, Ship Structures, Chapter 10, Welding and Structural Details. Lloyds Register.

Shi, G., Hilton, P. & Verhaeghe, G. 2007. 'In-process weld quality monitoring of laser and hybrid laser-arc fillet welds in 6–12 mm C-Mn steel' in Proceedings of the Fourth International WLT-Conference on Lasers in Manufacturing 2007 (LIM2007), 18–22 June 2007. Munich, Germany.

Still, J., Speck, J. & Pereira, M. 2004. 'Quality requirements for an FPSO hull and marine piping fabrication', in Proceedings of OMAE-FPSO 2004, OMAE Specialty Symposium on FPSO Integrity, 30 Aug.–2 Sept. 2004, Houston, USA.

Analysis and Design of Marine Structures – Guedes Soares & Das (eds)
© 2009 Taylor & Francis Group, London, ISBN 978-0-415-54934-9

A data mining analysis to evaluate the additional workloads caused by welding distortions

N. Losseau
University of Liège, ANAST, Fund for Training in Research in Industry and Agriculture of Belgium (F.R.I.A.), Belgium

J.D. Caprace & P. Rigo
University of Liège, ANAST, National Fund of Scientific Research of Belgium (F.N.R.S.), Belgium

F. Francisco Aracil
University of Liège, ANAST, Belgium

ABSTRACT: This paper presents a way to minimize costs in shipbuilding industry by using the results of a data mining analysis aiming to improve knowledge of the additional costs engendered by welding distortions. The presented analyses have exploited production data to establish formulations of the supplementary costs of steel working and straightening operations in function of welding distortions. Moreover, a complementary analysis based on the previous formulas has permitted to realise an economic comparison of post welding techniques aiming to reduce residual deformations.

1 INTRODUCTION

Since several years, the big shipyards use more and more plates of small thickness to build up the stiffened panels in order to decrease the structural weight of ships. The major problem relating to the utilization of thin plates is the appearance of welding distortions (see Figure 1) that have to be eliminated for fabrication, esthetical and service reasons. The supplementary operations to counter the problems of misalignment and lack of flatness involve non negligible costs and it seems thus important to characterize their economical impact on the hull production. To reach this goal, the idea is to establish assessment formulas of the supplementary workloads in function of scantlings and welding distortions. Those formulas could be useful to improve the research in the following domains: assessment of new welding equipments profitability, production simulation, cost assessment of hull fabrication, structure optimization, design for production, etc.

We distinguish several supplementary operations involved by the welding distortions:

- The steel working operations during the fixing of girders and bulkheads on plate's panels to constitute sections.
- The sections edges' straightening that consists in adjusting the plate's extremities of neighbour sections to constitute blocks.
- The desks straightening that permits to recover a certain flatness of plate's decks by heating techniques (see Figure 2).

The idea used to establish relations between those supplementary workloads and the residual welding distortions was to lead a statistical analysis based on the production data from a shipyard. This paper gathers the results of several analyses realised with the so called data mining technique.

Figure 1. Example of distorted panel.

Figure 2. Deck straightening by induction.

This method aims to explore the production databases, to find correlations between their attributes with the help of statistical tools and finally to establish an estimation formula of a particular attribute in order to understand its dependency with the others production parameters. For example, the data mining technique has been exploited in different studies relating to fabrication, exploitation or production: Bruce (2006), De Souza (2008).

The first analysis (Caprace (2007)) has exploited workload data of 13 passengers' ships in order to establish a formula linking the scantling (geometrical characteristics of stiffened panels) to the straightening cost [hour/m^2]. The second analysis (Losseau (2008)) was led further to a measure campaign gathering the welding distortions of one cruise ship and permitted to estimate the residual deformations in function of scantling. The last data mining studies realised recently had exploited production data and measure campaign data in order to generate relations between distortions and supplementary works of adjusting and straightening.

2 DECKS STRAIGHTENING IN FUNCTION OF SCANTLING

The first analysis aimed to generate a formula to estimate the decks straightening in function of scantling. This study was lead following the Data Mining methodology that is, by definition, the non-trivial process of extracting valid, previously unknown, comprehensible, and useful information from large databases. The successive steps of this data mining analysis are presented here after.

2.1 Database creation

A data base was first of all constituted; it gathered, for 13 passenger's ships, the characteristics of each section (global geometry, deck thickness, dimensions, interdistance of stiffeners, section family, deck number, steel grade, section weight, etc.) and the associated straightening workload [hour].

2.2 Data description stage

This step consisted in a presentation of the attributes (fields of the data base), with their distribution and other statistical parameters (minimum, maximum, mean and variance). One of difficulties which arose during the database analysis is that the most structural attributes show a discrete distribution with one or few dominant modes (see Figure 3 (3) and (4)); for instance, the distance between stiffeners has very often the same value. Those attributes are almost "constant" parameters and thus don't constitute a conclusive information source. In order to minimize this effect, we have replaced some attributes. In this scope, we have divided for example the plate weight

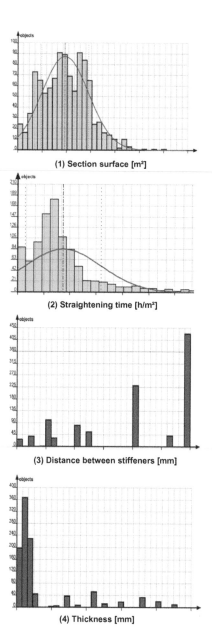

Figure 3. Distribution histograms of attributes.

by the section surface to obtain information similar to the thickness, but having the advantage to present a distribution much less discrete.

2.3 Data quality stage

This step listed the problematic recordings (strange distribution, missed values, data in conflict with their

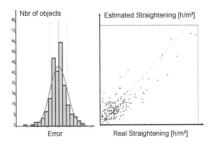

Figure 4. Errors diagrams relating to straightening work estimation (from PEPIT).®

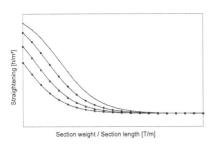

Figure 5. Decks straightening in function of section weight/section length, with the section family parameter.

physical meaning) in order to take care of them in the next stages.

A particular point has been noticed at this stage; the values related to the straightening work realised by sub-contractors are not reliable since those workloads correspond to estimated times and not times of strictly achieved work. Unfortunately this case concerns more than two third of the records and decreases thus the quantity of exploitable data.

2.4 Data exploration stage

This work stage consisted in using different approaches to visualize the correlations existing between the attributes and the straightening workload in order to finally select the parameters having the most relevant influence on the straightening assessment. In order to fulfil this stage, several techniques from PEPITo® software were used: a linear correlation analysis through dendrograms development, conditioned histograms, conventional dots clouds diagrams and decision trees analyses. PEPITo® software is developed by PEPITe society (Liège-Belgium).

2.5 Establishment of the formula

This stage consisted in building the relation between the straightening cost [hour/m²] and the sections characteristics. The used method was the Artificial Neural Networks (ANN) method which is a powerful technique permitting to establish non linear relations (i.e.

hyperbolic tangents) linking several inputs to a unique output. The formulas generated by the analyses thus are mathematical relations; they are not given analytically for confidential reasons by some schematic representations are given in the paper.

The input attributes selected to generate the formula were the following ones: thickness, longitudinal stiffeners spacing, transversal girders spacing, ratio stiffeners spacing/girders spacing, section family, section weight/m², section weight/section length. After having chosen the input parameters, it was necessary to restrict the number of records in order to ignore the sections carrying disruptive information. In this optic, we have ignored the sections whose straightening work was done by sub-contractors because time measurements of straightening were less reliable.

The formula was built exploiting 273 records and the correlation between the real values of straightening cost and the values estimated by the formula was 0.838 (see Figure 4).

2.6 Limitation of the formula

We have to notice that the generated formula (see Figure 5) has a limit. Firstly, since the recordings were restricted to the works realised by the shipyard workers, the quantity of data exploited was small and thus the robustness of the formula was not excellent. Moreover, when we have constructed the error diagrams, we have voluntarily tested the equation on the same data set than the one used to establish the relation. A consequence of this choice is that the precision given is not representative. Those precisions are optimistic in comparison to the precision obtained when the test set is different than the learning set.

3 DISTORTIONS DATA

A measure campaign gathering the deck distortions of a passenger ship was realised recently at the Saint Nazaire shipyard (see Figure 6). This brought new data and permitted to progress in the assessment of supplementary workloads caused by welding distortions. The distortions were measured after the blocks assembly stage and before the straightening operations. Each recorded value was related to the distortion occurring at the middle of rectangles (mesh division) delimited by consecutive longitudinal stiffeners and consecutive transversal girders. The database contained information about: scantling, coordinates of the mesh division in the deck, steel grade, neighbour structure element (bulkhead, longitudinal girder), sections characteristics (weight/m², fabrication workshop), etc. A particular attribute called "additional process" described the use of a distortion reduction technique (such as application of weld seams onto the plate between stiffeners).

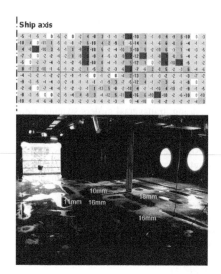

Figure 6. (a) Visualization of a distortions field, (b) Photo of a distorted deck.

Indeed the initial scope of the measure campaign was to evaluate the impact of such distortions reduction techniques.

This database represents an interesting source of information and permitted to lead researches in two domains: firstly the distortions evaluation in function of scantling and secondly the analysis of distortions impact in the production.

3.1 Distortions assessment

The distortions assessment has followed the successive data mining stages: investigate the correlation between attributes, extract meaningful parameters, fix exploitable data sub-sets and finally set up a formula through an ANN analysis.

Numerous attempts have been done to generate an effective estimation formula. It revealed that the exploitation of each distortion measure taken as unique record generates too much noise and involves thus a poor quality estimation (correlation around 0.5 for 23,000 elements). By gathering the data per plate (1400 elements) the correlation increased to 0.7 and by gathering the data per section (108 elements) the correlation reached 0.9. This last solution was thus utilised to evaluate the welding deformations in absolute value.

To generate the formula, the 108 sections concerned by the measure campaign were used and the following attributes were introduced as inputs: plate thickness, longitudinal stiffeners spacing, transversal girders spacing, ratio stiffeners spacing/girders spacing, section weight per m², workshop of the section, additional process. The correlation between the real distortions and the values estimated by the

formula was 0.947. This value is optimistic because the data set retained to test the formula was the same than the set selected to generate the formula.

Here again, the formula has a limitation. The estimation is quite good inside the range values encountered by the attributes in the database but is not excellent outside. For instance, an estimation relating to a thickness lower than 5 mm or higher than 8 mm becomes rapidly approximate.

4 ASSESSEMENT OF SUPPLEMENTARY WORKLOADS

4.1 Decks straightening in function of distortions

This analyse corresponds exactly to the scope fixed before: assess the impact of welding distortions on the production. This analysis was realised by exploiting the database from the distortions measure campaign. The idea was to generate a relation between the straightening workload and the level of welding deformations, with a more particular interest for the slope of the diagram curves than the precise estimation value of the straightening.

Several attempts have been realised on dataset before obtaining relevant results. In the first attempt, each record of the database represented a mesh division (i.e. rectangle delimited by consecutive stiffeners and consecutive girders) and gave the precise value of distortion, the straightening workload expressed in hour/m² and other characteristics. The results given by the neural network technique were not good. In the second attempt, we gather the data by sections and thus each record of the database gave for a section: the average value of distortions, the straightening workload expressed in hour/m² and the section characteristics. The results were once again not relevant.

Finally, we decided to continue to exploit the data by sections but we utilised the following attributes: total straightening workload on the section, sum of distortions of the section, average scantling (thickness, distance between stiffeners, and distance between girders). The results given by the neural network technique were quite good. Indeed the sensitivities of the generated formula (see Figure 7) show logic behaviour inside the validation domain:

– For a given thickness, the straightening increases when the distortions increase.
– For a given level of distortions, the straightening increases when the thickness increases. Indeed, it is necessary to heat longer a thick plate to reduce distortions.

One of the difficulties encountered here to generate a reliable function is that few data were exploitable since the straightening of decks was often realised

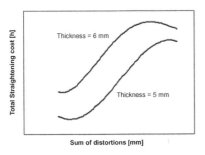

Figure 7. Decks straightening in function of the sum of distortions on a deck with thickness parameter.

Figure 8. Tools for steel working operations.

by subcontractors. The data of subcontractors corresponds to contract work times and not really executed work times.

4.2 Steel working assessment

The sections are generally constituted by a stiffened panel on which girders and bulkheads are welded. The steel working operations correspond to the operations of adjusting and fixing of those elements on panels in order to realise perfectly regular welds (see Figure 8). The idea of this analyse was to correlate the increase of steel working operations in function of distortions. Indeed more deformed are the plates, longer the workers will adjust steel elements to impose a precise contact between them.

The duration of steel works being highly dependant on section size, bulkheads number, etc. it was necessary to lead the analysis on similar sections and to introduce on top of the preponderant attributes (distortions, thickness, distance between stiffeners, etc.) the section weight by m^2 and also the bulkheads number divided by the mesh divisions number.

This analysis seems satisfactory because the correlation between estimated value and real values of steel working is quite high and the sensitivities respect quite well the observations in situ.

4.3 Sections edges straightening assessment

Similarly to the previous analysis, the sections edges straightening during the fabrication of block has been estimated. Only the distortions corresponding to the sections extremities have been introduced in the input "distortion" of the database. The results were not so perfect because few data about duration of those operations were exploitable.

5 ECONOMIC ANALYSIS OF TECHNIQUES REDUCING THE WELDING DISTORTIONS

At the Saint Nazaire shipyard, tests have been done to assess the effect of post welding techniques aiming to reduce the distortions. Those processes consist in welding, on the panel plates, different combinations of seams or stiffeners (see Figure 9) in order to generate counter deformations that could reduce the residual distortions appearing during fabrication.

An analysis has been realised in order to evaluate the economic profitability of those techniques taking into account the fact that they reduce the supplementary works of straightening and adjusting.

The first step of this study was to assess the distortions reductions engendered by the techniques. We have here again exploited the data from the deformations measure campaign to evaluate, for different scantling families, the distortions occurring with and without post welding processes. The distortions reductions were directly obtained by subtracting those last values. The Figure 10a presents the deformations reductions associated to each technique; a negative value means that the post welding process engenders higher distortions than the basic case, a positive value means that the technique reduces efficiently the residual deformations.

The second step is to assess reparation (straightening and adjusting) costs avoided by the utilisation of post welding processes. We have exploited here the relations, detailed in previous chapters, between supplementary workloads and distortions. By introducing the distortions reductions in those formulas, we obtained for each process, the avoided reparation costs.

The third step is to compare the global earnings associated to the different post welding processes in order to point out the best one. The earning is defined by the avoided costs of reparation (straightening and adjusting) minus the process cost. The Figure 10b presents the earnings associated to each technique; a positive value means that the technique is profitable

Figure 9. Some post welding techniques to reduce deformations.

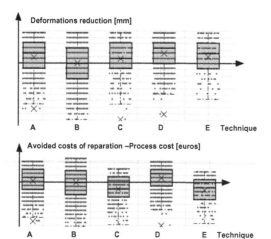

Figure 10. (a) Deformations reduction, (b) Earnings associated to techniques.

since its utilisation cost engenders less straightening and adjusting costs.

6 ESTIMATION OF SUPPLEMENTARY WORKLOADS IN FUNCTION OF ESTIMATED DISTORTIONS

If we want to evaluate the supplementary workloads (straightening and steel working) of a ship for which the residual welding distortions are not known, a solution can be to introduce the deformations given by the formula detailed in chapter 3.1 as an input of the relations relative to straightening and steel working.

7 CONCLUSIONS

The thin plates that are more and more utilised in shipbuilding are strongly subjected to welding distortions. The supplementary operations to counter the relating problems of misalignment and lack of flatness involve non negligible costs and delivery delays. It seems thus important to characterize their economic impact on the hull production.

This paper summarises several data mining analyses that exploited production data from the Saint Nazaire shipyard. A first study permitted, through a statistical analysis of 13 ships data, to establish a relation linking the straightening workload to the sections characteristics. Then, further to a distortions measurement campaign, the analyses have progressed; a formula was generated to estimate the distortions and others relations were established between the supplementary works (steel working and straightening of decks and sections edges) and the values of welding distortions. The generated formulas are useful to improve the research in the following domains: assessment of equipment profitability, production simulation, cost assessment of hull fabrication, structure optimization, design for production, etc.

Moreover an economic comparison of post welding techniques that aim to reduce residual distortions has been realised to point out the best process.

ACKNOWLEDGMENTS

The authors thank University of Liege, STX FRANCE (Saint-Nazaire) and PEPITe (Liège), for the collaboration within sub-project II.1 of InterSHIP (Project n° TIP3-CT-2004-506127 funded under the European Commission's Sixth Framework Programme) in which a part of the presented work has been realized. The authors thank also the European MARSTRUCT project (n° TNE3-CT-2003-506141).

REFERENCES

Bruce, G. & Morgan, G. 2006. Artificial Neuronal Networks-Application to freight rates, *International Conference COMPIT06*: 146–154.

Caprace, JD. et al. 2007. A Data Mining Analysis Applied to a Straightening Process Database, *International Conference COMPIT07*: 186–197.

De Souza, C. & Tostes, R. 2008. Shipbuilding Interim Product Identification and Classification System Based on Intelligent Analysis Tools, *International Conference COMPIT08*: 481–493.

Losseau, N. et al. 2008. Estimation of welding distortions and straightening workload through a data mining analysis, *International Conference DFE08, Miskolc, April 2008*.

3D numerical model of austenitic stainless steel 316L multipass butt welding and comparison with experimental results

A.P. Kyriakongonas & V.J. Papazoglou

School of Naval Architecture and Marine Engineering, Shipbuilding Technology Laboratory, National Technical University of Athens, Greece

ABSTRACT: A three-dimensional (3D) numerical simulation model of the multi-pass butt welding of AISI 316L austenitic stainless steel plates is developed with the use of finite element analysis, based on the ANSYS® software. The uncoupled thermo-mechanical analysis of the model performed aims at reliably predicting the residual stresses field and deformation due to welding. All the major physical phenomena associated with the welding process, such as heat conduction and convection, heat radiation, and convection heat losses are taken into account in the model development. The thermal and mechanical material properties are introduced as temperature dependent functions, due to the high temperatures of the weld pool and the high temperature gradients that are present during the welding process. During austenitic stainless steel welding any phase changes occurring are considered negligible and are thus ignored in the model. The model's accuracy is evaluated by comparing it with the experimental results from a multi-pass 316L austenitic stainless steel butt welded joint.

1 INTRODUCTION

Welding is an efficient and reliable metal joining method, used widely in the marine industry. The aggressiveness of the marine service environment, where welded structures are exposed, can provoke corrosion and other phenomena. As a result the mechanical capacity of the structures is degraded and can become subject to failure. Austenitic stainless steels can exhibit good corrosion resistance and mechanical properties and are thus suitable for application in the marine industry. In addition, austenitic stainless steels are considered weldable materials, if proper precautions are followed.

Due to the intense concentration of heat, during welding, the regions near the weld line undergo severe thermal cycles. The thermal cycles cause non-uniform heating and cooling in the material, thus generating inhomogeneous plastic deformation and residual stresses in the welded joint. With austenitic stainless steels the above phenomena are more intense, since they exhibit low thermal conductivity than carbon steels and high thermal expansion coefficients (Lippold & Kotecki, 2005). It is well known that the presence of residual stresses can be detrimental to the performance of a welded structure. Tensile residual stresses, which are usually present in the weld area region, are generally detrimental, increasing the susceptibility of a weld to fatigue damage, stress corrosion cracking and fracture (Leggatt, 2008). When assessing the risk for growth of defects, such as surface flaws, the welding residual stresses may give a

larger contribution to the total stress field than stresses caused by design loads (Radaj, 1992). The distribution of welding residual stresses depends on several factors such as structural dimensions, material properties, phase transformations, restraint conditions, heat input, number of weld passes and welding sequence.

The introduction of finite element analysis in the welding design process and the practice of computation welding mechanics (CMW) provide the ability to construct predictive models for the welded structures. Over the last decade, a number of finite element models have been proposed to predict the mechanical phenomena due to welding. Brickstad & Josefson (1998) employed two-dimensional (2D) axisymmetric models to numerically simulate a series of multi-pass circumferential butt-welds of stainless steel pipe up to 40 mm thick in a non-linear thermo-mechanical finite element analysis. Deng & Murakawa (2006), constructed three-dimensional (3D) and two-dimensional (2D) simulation models for a two-pass butt weld of an austenitic stainless steel pipe and compared them with experimental results. They showed that, with use of 3D models, detailed information of the welding temperature and residual stress fields can be captured, while with 2D axisymmetric models a large amount of computational time can be saved without any significant loss of accuracy in the results. Based on that conclusion, Deng et al. (2008) performed an uncoupled 2D thermo-mechanical simulation of the 14-pass butt-welding of an austenitic stainless steel pipe, obtaining good agreement with experimental results and saving computational time. Many other researchers adopted this assumption and performed 3D simulations only in single-pass welding processes. Deng et al. (2007) constructed a 3D numerical simulation model for a two-sided fillet welded joint and performed an analysis to predict the angular change that occurs in this type of joint. Moraitis & Labeas (2008) conducted an uncoupled thermo-mechanical analysis of a lap joint, in order to predict the distortion and residual stresses, resulting from laser beam welding (LBW). The capacity of powerful processors, which are very common in the present days, allowed Malik et al. (2008) not to worry about the computational time and to perform a detailed transient thermo-mechanical analysis of arc welded thin-walled cylinders to investigate the residual stress field. The existence of powerful computers allowed the performance of 3D simulations of multi-pass welding without any considerable loss in computational time. One of the most recent multi-pass welding simulations was that of Mousari & Miresmaeili (2008), who performed a 3D numerical simulation of a 304L stainless steel plate 3-pass gas tungsten arc welding (GTAW) process.

In the present study a 3D numerical simulation of multi-pass butt-welded austenitic stainless steel 316L plates is attempted, in order to predict the resulting residual stresses and deformation. The finite element model is constructed based on the experimental multi-pass welding of 316L plates. Measurements obtained during carefully designed experiments were used to evaluate the model's accuracy.

2 EXPERIMENTAL PROCEDURE

The austenitic stainless steels plates were welded with the use of a welding robotic arm, shown in Figure 1. The plate dimensions were 700×300 mm^2 and 8 mm thick. The flux cored arc welding (FCAW) process was used in the experiments, with a mixture of 82% Ar—18% CO_2 gas providing the shielding atmosphere. The flux cored wire had similar chemical composition to that of the austenitic stainless steel plates and its diameter was 1.2 mm. The welding conditions of the experiment are presented in Table 1. A bevel treatment of 30° was given to the plates and during

welding a ceramic backing-strip was used (Fig. 2) in order to obtain a well-shaped weld metal profile.

During the welding process, the thermal cycles and vertical deformation were measured with the use of thermocouples and LVDTs (Linear Variable Differential Transformers), as shown in Figure 3.

The thermocouples were positioned in the mid-length and on the top surface of the plate at various distances from the weld line. The LVDTs were positioned on the unconstrained stainless steel plate in order to capture the vertical deformation during welding and upon cooling.

In the as-welded condition strain gauge rosettes (Fig. 4) were applied on the plate's top surface in order to measure the residual stresses with the use of the hole-drilling method.

Table 1. Welding conditions.

Voltage (V)	Welding current (A)	Welding speed (mm/min)	Gas flow (l/min)	Passes
24	160	300	18	3

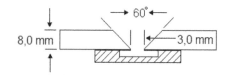

Figure 2. The groove beveled shape and the ceramic backing-strip position.

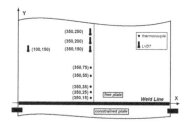

Figure 3. Positions of thermocouples and LVDTs.

Figure 1. The welding robotic arm used in the experiments.

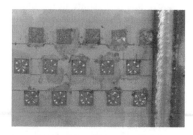

Figure 4. Position of rosette strain gauges for the hole-drilling method.

3   3D FINITE ELEMENT ANALYSIS

The temperature fields and the evolution of residual stresses and deformation were investigated by means of the finite element method. In order to capture the spatial temperature distribution during the welding process and predict the residual stress field and the deformation, a 3D finite element model, based on the ANSYS® code, was developed. The thermo-mechanical behavior of the welded joint during welding and upon cooling was simulated using the uncoupled formulation. The heat conduction problem was first solved independently from the stress problem in order to obtain the temperature history of the welding process. Thus, the temperature distribution and its history were computed during the thermal analysis. Then, the obtained temperature history was employed as a thermal load in the subsequent mechanical analysis.

The thermal and mechanical properties of austenitic stainless steel 316L (Dong, 2001) were inputted in the model as temperature dependent functions, due to the high temperatures and temperature gradients, which are present during welding. The filler and base metal were attributed the same temperature dependent properties, presented in Figures 5 and 6.

The meshed model is presented in Figure 7. The simulation model has the same dimensions as the experimental weld joint. In the weld area and its vicinity the mesh is quite fine, in order to capture the high temperature gradients and high temperature temperatures that occur during welding. The meshed model consisted of 35000 elements and 40000 nodes.

Figure 7.   Simulation model and mesh division.

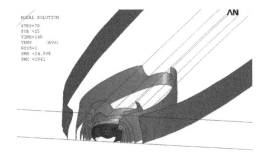

Figure 8.   Volumetric heat source model.

3.1   Thermal analysis

The element type employed for the thermal analysis was the ANSYS® SOLID70 3D solid element, which has a 3D thermal conduction capability. SOLID70 consists of eight nodes, with temperature being the only degree of freedom at each node and can be applied to a 3D steady-state or transient analysis.

The moving welding arc, along with the weld metal deposition, was modeled as a moving volumetric heat flux (Fig. 8) during the thermal analysis.

The calculation of the volumetric heat flux derives from the heat input equation of the welding arc process, Q_w (J/m), presented in Equation 1.

$$\dot{Q}_w = \eta \times \frac{V \times I}{u} \tag{1}$$

The arc voltage (V), the welding current (I) and the welding speed (u) are the welding conditions of the experiment, derived from Table 1, while the arc efficiency coefficient (η) varies for each welding pass from 0.6 to 0.8.

In order to capture the phenomenon of the filler metal deposition during the welding process, the module of the "birth and death of elements" was utilized. The "death" of elements implies the deactivation of their stiffness matrix, by multiplying it with a stiffness

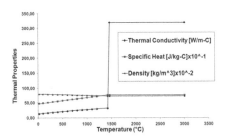

Figure 5.   Thermal properties of AISI 316L.

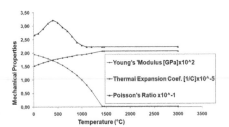

Figure 6.   Mechanical properties of AISI 316L.

373

Figure 9. "Killed" elements in the weld groove.

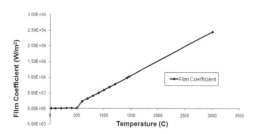

Figure 10. Combined filmcoefficient of AISI 316L.

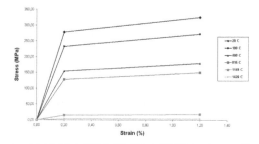

Figure 11. Stress-strain curves for AISI 316L.

coefficient, with a default value of 10^{-6}. Before the onset of the welding moving heat source model, all elements in the groove were deactivated except from the elements representing the tack welds, as shown in Figure 9. The "killed" elements were reactivated when the volumetric heat source reached their coordinates.

The thermal analysis was completed in 235 time steps, from which 70 steps represented each welding pass, 5 steps represented the inter-pass cooling while cooling to ambient temperature was achieved in 15 time steps. In order to simulate the convection and radiation to the environment, a combined temperature dependent film coefficient was employed on every free surface of the simulation model. The combined film coefficient is plotted in Figure 10.

3.2 *Mechanical analysis*

The same finite element model that was used in the thermal analysis was employed in the mechanical analysis as well, except from the element type, which was changed to ANSYS® SOLID185 3D solid element, used for 3D modeling of solid structures. It is defined by eight nodes having three degrees of freedom at each node, namely translations in the nodal x, y and z directions. The element has plasticity, hyper-plasticity, stress stiffening, creep, large deflection and large strain capabilities. It also has mixed formulation capability for simulating deformations of nearly incompressible elastoplastic materials and fully incompressible hyperelastic materials.

The input of temperature dependent stress-strain curves (Fig. 11) in the material model, before the onset of the mechanical analysis, is required.

During the welding process, because solid-state phase transformations do not occur, or are considered negligible, in stainless steels the total strain rate can be decomposed into three components: a) the elastic strain, b) the plastic strain, and c) the thermal strain. The elastic strain was modeled with the using the isotropic Hook's law with temperature dependent Young's modulus and Poisson's ratio, as shown in Figure 6. The thermal strain was computed using the temperature dependent coefficient of thermal expansion, also presented in Figure 6. For the plastic strain, a rate-independent plastic model was employed with the following features: the Von Misses yield surface, linear kinematic hardening model (Fig. 11) and the temperature dependent mechanical properties. Kinematic hardening was taken into account as an important feature because material points typically undergo both loading and unloading during the welding process (Radaj, 1992).

4 RESULTS AND DISCUSSION

4.1 *Thermal analysis results*

The experimental welding process was completed in three welding passes. Each welding pass commenced when the reading of the thermocouple closer to the weld line dropped to approximately 100°C. The numerical simulation of the thermal analysis followed the exact same sequence. The measured and predicted thermal cycles, namely the reading of the thermocouple nearest to the weld line and the nodal temperature of the node which is positioned at the same coordinates as the thermocouple, are presented in Figure 12.

A good agreement between the numerical and the experimental results is observed. During the cooling of the first and second pass a small variation between the experiment and the model is notable. This variation is attributed to the "birth and death" module, which is not able to effectively simulate the absence of subsequent weld deposits. This happens because the

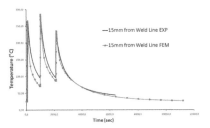

Figure 12. Numerical and experimental thermal cycle results.

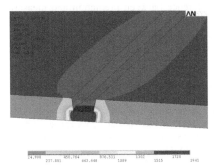

Figure 13. Moving heat source at the end of 1st pass.

elements, representing the subsequent deposits, are an obstacle to the interaction of the just-deposited weld metal with the environment.

The maximum calculated temperature was observed in the weld pool (Fig. 13). The temperature value reached approximately 1800°C, which is considered to be accurate, since austenitic stainless steels melt at approximately 1450°C and exhibit relatively low thermal conductivity.

4.2 Mechanical analysis results

The residual stress field of the austenitic stainless steel joint is presented in Figure 14. It can be readily noted that high tensile stresses are present in the area of the weld metal and the adjacent area, while in larger distances the stresses become compressive. The above observation can be proved also by the numerical results presented in Figure 15, along with the experimental measurements obtained by the hole-drilling method.

High tensile stresses are present in the weld metal and the heat affected zone, revealing a wide stress field. These tensile stresses turn to zero at a larger distance and then turn into compressive stresses of moderate magnitude. The compressive stress field extends until the end of the plate, where the stresses are practically zero. The numerical model is in good agreement with the experimental results, except from a small variation at a region between 80 and 100 mm from the weld line. The experimental values show, in that

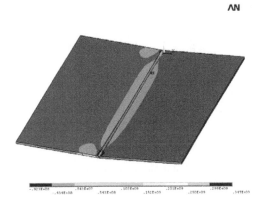

Figure 14. Residual stress field of the welded joint after cooling to ambient temperature.

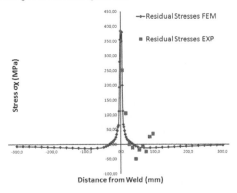

Figure 15. Stress field in the direction of the weld line.

Figure 16. Vertical deformation of the unconstrained plate.

area, the existence of tensile stresses, which could be residual stresses from the austenitic stainless steel plate manufacturing process (e.g. hot rolling). During welding a redistribution of the stress field is feasible at smaller distances. Hence, the tensile residual stresses that appear over 80 mm are stresses prior to welding. The numerical model is unable to predict this kind of stresses, because it starts with a stress-free material.

In Figure 16 the vertical deformation of the unconstrained plate is presented.

The vertical deformation results from the angular change, which is one of the basic types of deformation

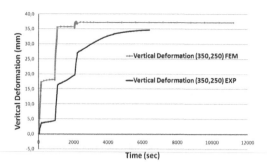

Figure 17. Vertical deformation evolution of the unconstrained plate.

in welded joints. The measurements recorded by the LVDT, which had the larger distance from the weld line, and the corresponding nodal results from the numerical model are presented in Figure 17.

The numerical and experimental values of the final deformation are in good agreement, showing a variation of about 5%. The forms of the curves are similar, exhibiting high deformation rates every time a weld pass is deposited. The curves do not match, however, a fact that can be attributed to the geometry of the weld metal assumed in the simulation model. In particular, the weld metal was assumed to have a plain trapezoidal shape, in order to match the V-shaped weld metal profile. The actual weld metal geometry, though, is in fact far more complicated and cannot be plotted accurately. In addition, the endeavor of a more complicated weld metal shape would cost in CPU-time.

5 CONCLUSIONS

A 3D simulation model was constructed in order to predict the residual stress field and vertical deformation of the austenitic stainless steel 316L multi-pass butt welding process. The prediction capability of the numerical model was evaluated through a comparison with experimental results. The austenitic stainless steel plates were 700x300 mm^2 and 8 mm thick. The numerical model had the exact same dimensions. According to the simulated and experimental results, the following conclusions can be drawn.

Based on the simulation results of the 3D model, the temperature distribution around the heat source is very steady when the welding torch moves from one end of the weld line to the other. The simulated results of the thermal cycles are in good agreement with the experimental results, as measured by the thermocouple. A small variation was observed during cooling. This variation can be attributed to the module, which was used to model the filler metal addition.

High tensile residual stresses are present in the weld metal and its adjacent area, designating a wide stress

field. These stresses turn into zero at a larger distance and become compressive stresses, of moderate magnitude. The residual stress field predicted by the model is in good agreement with the experimental results. The experimental results revealed an area with tensile stresses far from the weld line, which are assumed to be residual stresses prior to welding. The numerical model is unable to predict such stresses, since it considers a stress-free plate at the beginning of the analysis.

The final vertical deformation predicted is in good agreement with the experimental measurements, with the difference between prediction and measurement varying up to 5%. The forms of the deformation curves are different due to the geometrical features of the weld metal.

REFERENCES

Brickstad, B. & Josefson, B.L. 1998. A parametric study of residual stresses in multi-pass butt-welded stainless steel pipes, *International Journal of Pressure Vessels and Piping 75*: 11–25

Deng, D. & Murakawa, H. 2006. Numerical simulation of temperature field and residual stress in multi-pass welds in stainless steel pipe and comparison with experimental measurements, *Computational Materials Science 37*: 269–277

Deng, D., Liang, W. & Murakawa, H. 2007. Determination of welding deformation in fillet-welded joint by means of numerical simulation and comparison with experimental measurements, *Computational Materials Science 183*: 219–225

Deng, D. & Murakawa, H. 2008. Numerical and experimental investigations on welding residual stress in multi-pass butt-welded austenitic stainless steel pipe, *Computational Materials Science 42*: 234–244

Dong, P. 2001. Residual stresses analyses of a multi-pass girth weld: 3-D special shell versus axisymmetric models, *Journal of Pressure Vessels Technology 123*: 207–213

Leggatt, H. 2008. Residual stresses in welded structures, *International Journal of Pressure Vessels and Piping 85*: 144–151

Lippold, J.C. & Kotecki, D.J. 2005. *Welding metallurgy and weldability of stainless steels*, New Jersey, John Wiley & Sons, Inc.

Malik, A.M., Qureshi, E.M., Dar, N.U. & Khan, I. 2008. Analysis of circumferential arc welded thin-walled cylinders to investigate the residual stress field, *Thin-Walled Structures 46*: 1391–1401

Moraitis, G.A. & Labeas, G.N. 2008. Residual stress and distortion calculation of laser beam welding for aluminum plates, *Journal of Materials Processing Technology 198*: 260–269

Mousavi Akbari, S.A.A. & Misesmaeili, R. 2008. Experimntal and numerical analyses of residual stress distributions in TIG welding process for 304L stainless steel, *Journal of Materials Processing Technology 208*: 383–394

Radaj, D. 1992. *Heat effects of welding—Temperature field, Residual Stress, Distortion*, Berlin, Springer-Verlag.

Adhesive joints

Analysis and Design of Marine Structures – Guedes Soares & Das (eds)
© 2009 Taylor & Francis Group, London, ISBN 978-0-415-54934-9

Fabrication, testing and analysis of steel/composite DLS adhesive joints

S. Hashim & J. Nisar
University of Glasgow, Glasgow, UK

N. Tsouvalis & K. Anyfantis
National Technical University of Athens, Athens, Greece

P. Moore
TWI, Cambridge, UK

I. Chirica
University 'Dunarea de Jos' of Galati, Romania

C. Berggreen, A. Orsolini & A. Quispitupa
Technical University of Denmark, Denmark

D. McGeorge & B. Hayman
Det Norske Veritas AS, Norway

S. Boyd
University of Southampton, Southampton, UK

K. Misirlis, J. Downes & R. Dow
Newcastle University, Newcastle, UK

E. Juin
Centre of Maritime Technologies e.V, Hamburg, Germany

ABSTRACT: This paper aims to provide a guide on the design and fabrication of thick adherend double lap shear joints (DLS), often referred to as butt connections/joints in ship structures including patch repair. The specimens consist of 10 mm steel inner adherend and various outer adherend materials including 0/90 WR GFRP and 0/90 UD CFRP laminates and steel. The focus here is on CFRP/steel joint due to availability of test data. The thickness of the outer adherend varies from 3 mm to 6 mm. Shear overlaps of 25–200 mm were considered. The overall objectives are (i) to assess the quality of the standard fabrication method, (ii) to determine joint strength and overlap plateau for various specimens with a range of material combinations and (iii) to understand aspects of failure and design of joint under quasi-static loading. The paper presents experimental and numerical details with key conclusions.

1 INTRODUCTION

Structural adhesives are gaining wide recognition by industry as they offer engineers greater flexibility to achieve economic and technical advantages. In the marine industry there is potential for adhesives in various types of constructions, for example, bonding hybrid thick steel and composite joint, typically 5–10 mm. Advantages of composite materials in the marine industry are their corrosive resistant properties, their lightweight characteristics and their potential for fire resistant design (Hashim et al. 2004). Applications included superstructures for ships and offshore platforms as well as their suitability as a repair method for cracks and corroded areas (Echtermeyer et al. 2005).

The lack of a universally applicable criterion for predicting the static load carrying capacity of adhesively bonded joints means that analytical design optimisation of bonded structures is not possible (Richardson et al. 1995). Also, the use of composite materials within bonded joints further complicates the analysis of these joints due to their inherent orthotropic material properties including weakness of matrix resin (Potter et al. 2001). One of the most widely used

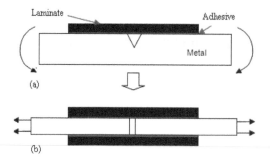

Figure 1. Overlap similarity between (a) patch repair and (b) DLS joint.

connections for adhesive bonding is the DLS joint. Although many investigations have been conducted on double strap joints in the configuration of steel as the outer adherend and composite as the inner adherend, the same cannot be said about the reverse situation. These joints could be considerably longer than standard bonded joints which can benefit from having thicker adherends with the effects of extending the shear of the adhesive more effectively within joint than with thinner adherend due to stiffness limitation. In patch repair for example, the DLJ seems to give good resemblance to a patch repaired crack. This technology is currently used for repair work onboard floating production storage and offloading units (FPSO's). Figure 1 shows the resemblance between the two cases where both designs rely on determining the shear stress level in the joint.

Toughened epoxy adhesives are often needed for these joints but the adhesive and its interface remain the weak link in this case. Therefore, it is essential to understand structural behaviour and failure, assuming a good fabrication process is followed. Besides the adhesive and joint geometry, the type of the composite adherend in terms of its stiffness and strength plays an important part in achieving a high bearing load capacity. The matrix resin is often the main weakness within bonded composites.

This paper is a joint effort of various partners within MARSTRUCT Research Network of Excellence on marine structures. The study involves fabrication, testing and numerical analysis undertaken at various institutions (see above) for benchmarking purposes.

2 EXPERIMENTAL PROGRAMME

2.1 Fabrication

The bonding process would typically require several operations including surface roughening, degreasing, marking, adhesive application, positioning and clamping, curing and removal from clamps. The bonding

process was set out taking long term bond performance into consideration. The bonding surfaces of the steel components were prepared by grit blasting. Teflon sheeting was applied to one end of each steel bar to ensure that all the loading was through shear along the straps rather than a tensile load between the steel bars. The bonding surfaces of the straps were prepared by light abrasion using 120 grit silicon carbide paper. All bonding surfaces were cleaned using a cleaning agent.

Markings were put on the steel bars and straps to ensure correct fit-up of the joint when being bonded. The adhesive used was Araldite 2015, a two-part toughened epoxy adhesive which was mixed and applied by spatula on two stages. The first stage is to prime the surfaces with a thin layer of adhesive. The second stage is to applying more adhesive and spread the excess amount. Finally to close the joint such that the entire bondline is filled with adhesive as shown in Figure 2. The clamping jigs were sprayed with Teflon spray to prevent the specimens sticking. The clamps were tightened evenly to give a uniform thickness of the adhesive layer—typically 0.5 mm. The curing of the adhesive joints was done for 1 hour

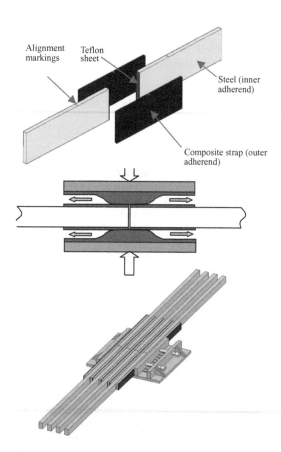

Figure 2. Aspects of adhesive bonding processes.

in an oven at 85°C. Finally, any excess adhesive from the joint was mechanically removed so that there were no effective adhesive fillets within the double lap joint. It was assumed that this would reduce variability and make modelling of the specimen configurations easier.

2.2 Mechanical testing

The bonded specimens incorporated various design and materials parameters including overlaps ranging from 25 mm to 200 mm as well as various materials combinations.

Table 1 shows the designations of the 38 specimens tested in this programme. These specify "overlap length/strap thickness & material/ fabrication site". In the designation C, P, S and H stand for CFRP, GFRP, low strength steel and high strength steel, respectively. The composites details are 0/90 WR GFRP/polyester (P) and 0/90 UD CFRP/epoxy (E). The bonded specimens were then tested to destruction under monotonic loading on universal tensile testing machines at ambient temperature. Besides failure loads, both strain and displacement values were recorded.

Figure 3 shows the test arrangement including position of the strain gauges and details of the DLS joints. Again all test results are shown in Table 1 which shows the superiority of steel/steel and CFRP/steel joint strength in comparison to the GFRP equivalent. The GFRP are considerably weaker due to resin and stiffness limitations. Therefore the focus of this paper will be on the CFRP. Figure 4 shows the trend of failure

Table 1. Test results of DLS specimens.

Specimen ID	Max load (kN)	Displ. (mm)	Max strain SG1 (με)	Max strain SG2 (με)
25/3C/G1	25	0.237	2136	422
25/3C/G2	26	0.239	1910	419
50/3C/G2	52.53		15900	4480
50/3C/G3	47.6			
50/3C/G4	45		4076	861
50/3C/G1	41.6			
75/3C/G1	70.75	1.28	7313	1257
75/3C/G2	68	1.13	7001	1456
100/3C/G1	62.9	1.297	7307	1133
100/3C/G2	72.2	1.447	8043	1611
25/6C/G1	21.25			
25/6C/G2	21.12		1140	550
50/6C/G2	54	0.6	1921	965
50/6C/G1	45.6			
75/6C/G1	80.7	1.067	3425	1651
75/6C/G2	61.2	0.751	2797	1100
100/6C/G1	94.85			
100/6C/S1	72.4	1.19		
200/6C/D1	74.7			
200/6C/G1	87.2			
200/6E/G1	47.58	3.23	8977	840
200/5P/D1	45.1			
200/5P/G1	22.87	0.676	2981	324
200/6H/G2	96.3	3.39	1350	13808
50/3S/G3	59			
25/6S/G1	37.82			
25/6S/G2	28.8			
50/6S/G2	58.1	0.59	761	1268
50/6S/D1	44.7			
50/6S/D2	54.7			
50/6S/D3	48.4			
50/6S/G1	49.6			
75/6S/G1	71.1	0.745	1253	1405
75/6S/G2	78.7	0.787	1343	1541
100/6S/S1	81.3		7190	31080
100/6S/G2	73.32			
200/6S/D1	100.3			
200/6S/D2	106.9			

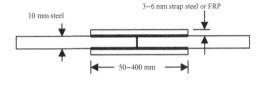

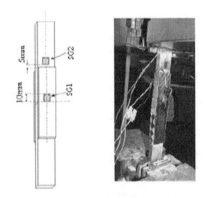

Figure 3. Test set-up and specimen details.

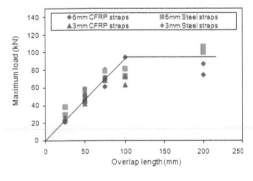

Figure 4. Failure loads of steel/steel and steel/CFRP joints with different overlaps and outer adherends thickness.

load versus overlap length with reference to steel/steel and steel/CFRP specimens only. The GFRP results are not included. From the figure (and table) the following remarks may be made:

- The failure load seems to be proportional to the overlap length up to 100 mm overlap where a plateau is reached
- Steel/steel are slightly stronger than steel/CFRP joints
- Thinner CFRP straps in longer joints exhibit a slightly lower strength than equivalent thicker ones
- Steel/steel joint with high strength tensile steel shows little difference in strength in comparison with the equivalent joints with low strength steel.

Investigation of the fractured surfaces and joints suggests that steel/CFRP is failing nearer the interface between the surface ply (0-direction) and the adhesive.

This is perhaps due to resin or adhesive failure starting at the middle of the joint. It is also possible that tensile failure of the laminate occurred. The steel/steel joint is failing at the adhesive but after yielding of the steel itself. Figure 5 shows typical failed long overlap joints.

2.3 Imaging systems

In connection with the mechanical testing, both standard low rate (ARAMIS 4M) and high speed (ARAMIS HHS with two Photon APX-RS high speed cameras) digital image correlation (DIC) measurements were carried out at the Technical University of Denmark (DTU) to measure full surface 3-D displacement and 2-D strain fields at the joint while undergoing testing.

(a)

(b)

Figure 5. Failure surface of (a) 75/6C/G1 joint and (b) deformed steel for 200/6H/G2 joint (see Table 1).

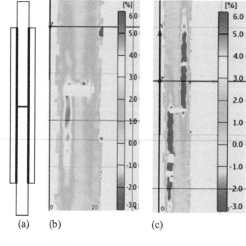

(a)   (b)   (c)

Figure 6. DIC major principal strain measurements for (a) long specimen 200/6C/G1, showing (b) failure initiating and (c) damage propagation along the adhesive bondlines.

The standard DIC system operating at 1 frame per second with 4 mega-pixel image resolutions was primarily used to capture the deformation of the joint specimen throughout the entire loading history, while the high speed DIC system operating at 3000 frames per second and 1 mega-pixel resolution captured only the failure incident.

Figure 6 shows a typical contour of the first principle strain which is plotted for the specimen 200/6C/G1 just prior (b) to and after (c) crack propagation has initiated. The DIC measurements clearly indicate for this specimen that propagation failure is starting at the centre of the specimen propagating towards the strap/joint ends.

3 NUMERICAL MODELLING

The finite element analysis (FEA) in this section is focused on CFRP/steel specimens (models) designated 25/6C/G1, 50/6C/G2, 100/6C/S1 and 200/6C/G1 (see Table 1). 2-D non-linear models were constructed in ABAQUS (as well as other software) using eight-noded solid quadrilateral plane strain elements (as well as others). The boundary conditions and loading are shown in Figure 7. The 0.5 mm adhesive bondline was divided into five layers through thickness with a standardised fine mesh towards the join ends.

This allowed stress and strain data to be taken along paths created at the upper interface of the adhesive with the outer adherend, and at the lower interface of the adhesive with the inner adherend. The nodes on the free surface were ignored due to stress singularity problems which normally occur here. The first point on each path was taken at 0.05 mm in from the free edge. In a similar fashion the 0.1 mm matrix resin adjacent to the adhesive bondline was modelled into two layers to account for stress details within the resin. The material properties used for the analyses are based on the values from Table 2. These were obtained from various sources including materials manufacturers and laboratory tests. The steel and adhesive adherends were modelled as elasto-plastic. The composite was modelled as elastic isotropic layer.

The 6 mm CFRP adherend is modelled into layers representing plies at 0/90. The plies were separated by 0.1 mm matrix resin. In some cases the whole composite was treated as one isotropic material with average Young's modulus. The two cases seem to give little difference in the overall behaviour of the joints within the elastic limits. However, in order to study the localised stresses it was necessary to have something in between. Therefore it was decided that only the first two plies at 0/90 sequence were modelled into layers while the

rest of the composite (outer adherend) was modelled as a bulk isotropic material.

The validation of the numerical models showed good correlation between the experimental and numerical strain values at the corresponding locations (see Figure 3). Figure 8 compares the results for models 50/6C/G2 and 200/6C/G1. The former was modelled as multiple composite layers while the latter assumed average property for the CFRP. In fact both gave reasonable agreements, especially within the elastic limits of the adherends. The property of the steel in the long overlap seems to be more difficult to correlate (SG2).

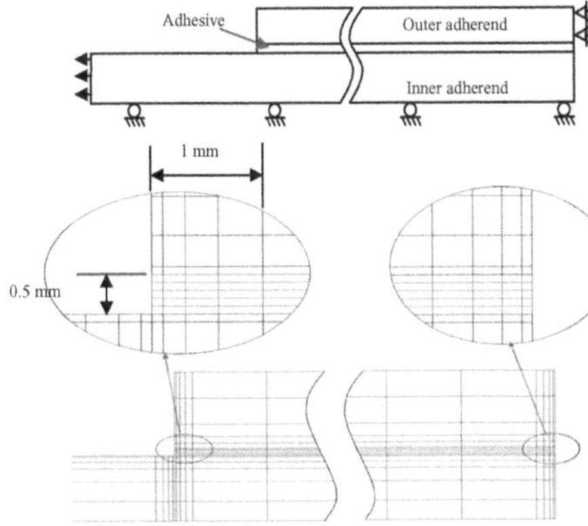

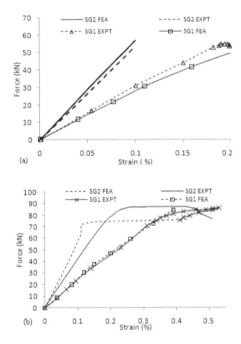

Figure 8. Force/strain curves for (a) 50/6C/G2 and (b) 200/6C/G1 until joint failure.

Figure 7. FEA model for DLS joint showing standardised mesh.

Table 2. Materials properties.

Property		GFRP WR	Polyester resin	CFRP	Epoxy resin	Adhesive	Steel
Young's modulus [GPa]	E11	25.5	3.6	126.3	4	18	210
	E22	5.8	–	5.5	–	–	–
		3.1	–	3.3	–	–	–
Shear modulus [GPa]	G12	0.15	0.35	0.3	0.38	0.36	0.3
Poisson's ratio	v12 v21	0.15	–	0.0131	–	–	
Tensile strength [MPa]		300	26	1400	65	40	395
Shear modulus [GPa]		62	–	137.2	–	26	–

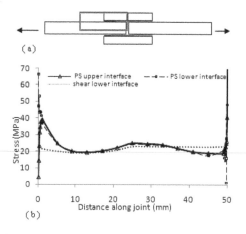

(a)

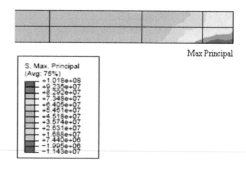

Max Principal

Figure 9. (a) Joint location and (b) stresses distributions in the adhesive at failure for 50/6C/G2-PS is max principal stress.

Figure 10. Contour of maximum principal stress in the first resin layer for 100/6C/S1 at failure load.

Therfore a more detailed steel property is needed. In addition the experimental curve for the long overlap suggests that the CFRP underwent some plasticity, evident from the knee after 70 kN load for the 200/6C/G1 specimen/model.

Figure 9 shows adhesive stresses distributions along the joint at interface with the inner and outer adherends, for the 50 mm overlap. The maximum principal (PS) stresses are as expected high nearer the edge of the joints and tend to be higher at the right hand side of the joint (upper interface). Also the adhesive seems to reach full plasticity in shear.

Besides these critical stresses within the adhesive, Figure 10 shows another critical location which is the resin at the interface with the adhesive to the right hand side/centre of the joint. The stress contour shows a very high level of principal stress (100 MPa) for the brittle epoxy resin which could be the mean source of failure initiation and propagation. The strength of the epoxy resin is 65 MPa (see Table 3). The level of resin stress appears to be the same for all the models, at corresponding failure loads.

Figure 11 shows shear stress distributions along the upper interface of the adhesive with the CFRP laminate (outer adherend). All the four models exhibited similar peel and principal stresses at the ends at failure, especially at the right hand side of the joint. Ideally the 25 mm overlap should exhibit complete plasticity in shear but this does not seem to have been reached, presumably due to peel either within the adhesive or resin. The short/thick CFRP strap exhibited high local stresses at both ends of the joint.

The experimental results with the 3 mm CFRP strap gave a higher failure load for the 25 mm overlaps (see Table 1). The 50 mm seemed to produce full plastic behaviours and interestingly both the 100 mm

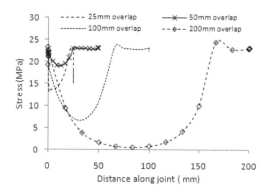

Figure 11. Shear stress distributions along the upper interface of 25/6C/G1, 50/6C/G2, 100/6C/S1 and 200/6C/G1 at failure load.

and 200 mm developed a similar size plastic zone, i.e. 60 mm or so. A possible failure criterion here is that following the plasticity in shear, a brittle fracture can develop leading to steady crack propagation. This requires further work and the main challenge is that cracks are more likely to initiate and propagate through the interface rather than cohesively in the middle of the adhesive bondline. The thickness of the adhesive is also an important factor. An added complication is the nature of the composite adherend including surface resin.

4 DISCUSSION

The standard fabrication method used in this study can be adopted for practical application and the test results from fabricated specimens suggest the bonding process is robust. However, some joints, especially long ones, showed some scatter and this applies to both steel/steel and steel/CFRP. The steel/steel displayed highest strength. The test results for long overlaps (200 mm) suggest a limited static strength advantage over 100 mm overlap. Results from an intermediate overlap length of 150 mm overlap would be useful to confirm this. The shorter overlap joints were able to carry loads up to the level where the entire bondline yielded in shear causing joint fractures. The results from the high strength steel seem to have no advantage in this case. On the contrary some of the low tensile strength steel joints produced higher strength! Therfore, further mechanical testing is required to assess the fabrication process and to understand joint behaviour and failure.

The FEA strategy of modelling the first two plies of the CFRP and their resin layers seems to yield encouraging results while accounting for maximum stress and stress values at critical spots, including the matrix resin. The results from both experiments including imaging system and FEA confirm that failure tends to start at the centre of the DLS joint at the interface between the adhesive and CFRP laminate for the configurations where the inner adherend is stiffer than the two outer adherends. The exception is the long steel/steel specimens where the combined stiffness of the outer adherends is larger than that of the inner adherend such that the largest load transfer through the bondline occurs at the free ends of the outer adherends to the left hand side of the joint. In those cases, the fracture started from the free ends of the overlaps (i.e. from the left).

For the shortest overlaps, there was a tendency to premature failure presumably due to high peel stresses at either ends of the joint as can be seen from the FEA results. The longer overlap specimens were shown to accommodate large shear plastic zones of about 60 mm length before fracture occurred. The failure suggests that the fracture was stable initially and turned brittle when insufficient undamaged bondline remained. The failure initiation may also start by rupture/fracture of the resin.

If the bond is very strong, as could occur when combining long tapered overlaps, strong adhesive and stiff straps, its capacity could be sufficient to reach loads where the steel adherend yields. Yielding of the steel adherend should tend to occur outside the overlap, where all the loading is carried by the steel adherend. That would tend to create large local deformations that would have to be accommodated by the bondline at its most highly loaded spot. One would expect this to cause local rupture of the bondline where it adheres to the yielding steel. But this local fracture would have the effect of extending the part of the inner adherend that is fully loaded into the bondline thus allowing the yielding of the adherend to propagate and overstrain another part of bondline. Hence one would expect a progressive development

of this mechanism (zipper effect) leading eventually to fracture of the entire bonded joint. This will require a damage model such as bi-linear traction-separation law for a cohesive zone model (CZM) (Crocombe et al. 2008) or similar damage models for bonded joints.

Key conclusions from this study include; (i) the initial work showed a limited advantage in using high strength steel for long overlaps although this may change with a stronger adhesive; (ii) CFRP composite provides double the joint strength of the equivalent GFRP; (iii) the overlap plateau for the tested steel and CFRP joints appears to be limited to 100 mm, under quasi-static loading; (iv) a considerable length of the bondline was loaded into the inelastic range before fracture, typically up to 60 mm for the stronger joints, carrying much of the applied loading, thus showing that attempts at predicting failure of such joints would have to account for nonlinear inelastic adhesive behaviour; (v) various failure criteria can be used for long overlaps including fracture of the bondline after it is partially plasticised.

ACKNOWLEDGEMENT

This work has been performed within the context of the Network of Excellence on Marine Structures (MARSTRUCT) partially funded by the European Union through the Growth Programme under contract TNE3-CT-2003-506141.

REFERENCES

Crocombe, A.D., Graner Solana, A., Abdel Wahab, M.M., Ashcroft, I.J. (2008). Fatigue behaviour in adhesively-bonded single lap joints, Proc. Science and Technology of Adhesion and Adhesives, Oxford, pp. 187–190.

Echtermeyer, A.T., McGeorge, D., Sund, O.E., Anderson, H.W., Fischer, K.P. (2005). Repair of FSPO with Bonded Composite Patches. Fourth international conference on composite materials for offshore applications. Houston, TX. CEAC.

Hashim, S.A., Knox, E.M. (2004). Aspect of Joint evaluation in thick-adherend applications, The J. Adhesion, 80, 569–583.

Potter, K.D., Guild, F.J., Harvey, H.J., Wisnom, M.R., Adams, R.D. (2001). Understanding and control of adhesive crack propagation in bonded joints between carbon fibre composite adherends I. Experimental. International Journal of Adhesion and Adhesives. pp. 435–443.

Richardson, G., Crocombe, A.D., Smith, P.A. (1995). Failure prediction in adhesives joints by various techniques including the modelling of crack development. Proc SAEIV, Bristol, pp. 44–50.

Analysis and Design of Marine Structures – Guedes Soares & Das (eds)
© 2009 Taylor & Francis Group, London, ISBN 978-0-415-54934-9

The effect of surface preparation on the behaviour of double strap adhesive joints with thick steel adherents

K.N. Anyfantis & N.G. Tsouvalis

National Technical University of Athens, School of Naval Architecture and Marine Engineering,
Shipbuilding Technology Laboratory, Athens, Greece

ABSTRACT: One of the major factors determining the integrity of an adhesive bond is the preparation of the bonding surfaces. The present study is an experimental investigation of the effect of the surface preparation procedure on the response of a steel-to-steel double strap adhesive joint. Two procedures for preparing the bonding surfaces are investigated, namely grit blasting (GB) and simple sandpaper (SP). The behaviour of the joints, in terms of the force-displacement and strains-displacement responses was monitored and compared for both cases. The joints with SP surface preparation exhibited slightly lower stiffness and lower strength than the joints with GB surface preparation, while the latter failed at a lower displacement. In both cases, failure initiated at the free edges of the joints and the dominating failure mode was interfacial. In addition to the above experimental measurements, results are also presented from the application of a properly modified analytical model.

1 INTRODUCTION

The significance of adhesive bonding of thick steel adherents is increasing in marine applications due to its numerous advantages compared with other joining methods. In this bonding technology, surface preparation or surface treatment of the joining adherents is a very important issue for securing a good adhesion, since surface roughness is one of the major parameters that control the state of the adhesion (Adams 1997). One basic consequence of the surface treatment is the enhancement of the degree of mechanical interlocking between adhesive polymers and substrates, which, according to the literature, seems to increase the strength of the joint (Baker & Chester 1992, Bishop 2005, Chang 1981, Chester & Roberts 1989, Katona & Batterman 1983). Roughening the surfaces of the adherents can be achieved either by mechanical or by chemical methods. Since when using a chemical treatment method, immersion of the substrates into a chemical solution is inevitable, it is impractical to apply such a method to the large scale structures incorporated in the marine industry. Therefore, this study is focused on the application of mechanical surface treatment methods on the joining of metal adherents.

According to Harris & Beevers (1999), grit blasting is a very promising technique because it introduces chemical changes on the surfaces of the adherents, which in turn affect the strength and the structural integrity of the joint. Nevertheless, there is a value of the surface roughness which is optimum for the strength of the adhesion. The value of the surface roughness does not only affect the strength of the adhesive-adherent interface, but also has a profound and critical influence on the stresses near this interface (Uehara & Sakurai 2002, Morais et al. 2007). The process of adhesion is in itself a subject of considerable research, as are joint failure modes (Shalid & Hashim 2002, Sargent 1994).

The present study examines the influence of the surface preparation on the global response of double strap joints (DSJs) loaded in tension along the adherents. This is done experimentally by testing two sets of bonded specimens of thick steel adherents. As for the surface preparation, two procedures are taken under consideration, namely grit blasting (GB) and the use of a simple sandpaper (SP), in order to produce different surface roughness. A simple analytical model is also developed by modifying a corresponding model provided by Xiao et al. (2004) and its validity against the produced experimental results is investigated.

2 ANALYTICAL MODEL

The lap shear joint is the first adhesive joint studied and its structural analysis has been the subject of numerous research efforts. Volkersen (1938) and Goland & Reissner (1944) are the pioneer researchers of the deformation and stress analysis of the lap shear joint. Both references are based on one dimensional elasticity solutions for peel and shear stress distributions in the adhesive layer. The equations obtained from the theory of elasticity do not give a direct relation between the load and deflection (i.e. the global stiffness response) of a lap joint. The analytical model presented by Xiao et al (2004) is focused on the stiffness prediction of a double lap shear joint (DLS).

In the present study, Xiao's assumptions are adopted, aiming at modifying his model to account for DSJs. Figure 1 is a schematic view of a DSJ in its undeformed (a) and deformed (b) state. Half of the joint is shown in this figure, to the right of its axis of symmetry y. In addition, the thickness of the adhesive layer is shown exaggerated. The joint is loaded by a tensile axial load F in x-direction. The part of the inner adherent outside the strap joint is denoted as segment 1, whereas segment 2 signifies half of the length of each strap, from its axis of symmetry to its edge. L_1 and L_2 are the lengths of segments 1 and 2, respectively, whereas u_1 and u_2 are their corresponding axial deformations and u_3 is the axial deformation of the adhesive layer.

The realistic assumption that load F applied to the joint is uniform over the length of segment 1, results in the development of a constant tensile stress σ over the length L_1, which is given by the following simple equation:

$$\sigma_1 = \frac{F}{b \cdot t_1} \tag{1}$$

where b is the width of the joint and t_1 the thickness of the inner adherent. Furthermore, Equations 2 and 3 below describe the well known linear elastic stress-strain and strain-displacement relations, respectively, in segment 1,

$$\sigma_1 = \varepsilon_1 \cdot E_S \tag{2}$$

$$\varepsilon_1 = \frac{u_1}{L_1} \tag{3}$$

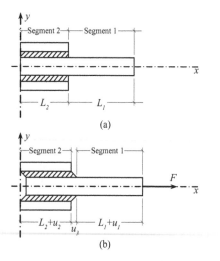

(a)

(b)

Figure 1. Schematic view of a DSJ in its undeformed (a) and deformed (b) state (half the joint is shown).

where ε denotes strain and E_S is the modulus of elasticity of the inner adherent (equal to that of steel in the present case).

Substitution of Equations 2 and 3 into Equation 1 yields the following expression for the total axial deformation u_{1T} of both segments 1 (the one shown in Figure 1 and its symmetric), as a function of the applied load F.

$$u_{1T} = 2 \cdot u_1 = \frac{2 \cdot L_1 \cdot F}{b \cdot t_1 \cdot E_S} \tag{4}$$

Generating a free body diagram of segment 2, shear traction forces develop at the interface of the strap and the adhesive layer. Applying symmetry conditions to segment 2 at y-axis, the eccentricity of these shear traction forces introduces a bending effect on the straps. This effect is very small since the lever arm (half of the strap thickness) is very small. The above fact has been verified from finite element analyses of the current problem, where it was shown that y-displacement of the free edge of the strap was negligible compared to the other displacements. Moreover, strap bending affects more the peel stresses in the adhesive and not the overall axial stiffness of the joint which is the objective of the present analytical approach. Thus, by neglecting bending of the straps, it can be considered that segment 2 is deforming only in the axial (x) direction. Following the same procedure as that followed for segment 1 and taking into account that each strap transfers a load $F/2$, the following expression is derived for the total axial deformation of each strap:

$$u_{2T} = 2 \cdot u_2 = \frac{L_2 \cdot F}{b \cdot t_2 \cdot E_S} \tag{5}$$

where t_2 is the thickness of each strap and E_S is the strap modulus of elasticity (equal again to that of steel in the present case).

With regard to the behaviour of the adhesive, it is assumed that it deforms under a uniform shear stress state and shear forces are uniformly distributed over its interfaces with either the strap or the inner adherent. Thus the constant shear stress developed in the adhesive, τ_{av}, is given by:

$$\tau_{av} = \frac{F}{2 \cdot b \cdot L_2} \tag{6}$$

whereas the axial deformation of the adhesive is

$$u_3 = t_a \cdot \gamma \tag{7}$$

where, t_α is the thickness of each adhesive layer and γ the assumed uniform shear strain of the adhesive. Moreover, the shear stress of the adhesive can be expressed as

$$\tau_{av} = G \cdot \gamma \qquad (8)$$

where G is the shear modulus of the adhesive material.

Substituting Equations 7 and 8 into Equation 6, the total axial deformation of the adhesive layers of the joint is given by the following expression:

$$u_{3T} = 2 \cdot u_3 = \frac{F \cdot t_a}{b \cdot L_2 \cdot G} \qquad (9)$$

According to the initially made assumption for a linear response of the DSJ, its load-axial deformation behaviour is described by the following simple formula:

$$u_T = K \cdot F \qquad (10)$$

where u_T is the total axial deformation of the DSJ and K its axial stiffness. Thus, taking into account that

$$u_T = u_{1T} + u_{2T} + u_{3T} \qquad (11)$$

and substituting Equations 4, 5 and 9 into Equation 11 and the result into Equation 10, the following analytical expression for the axial stiffness K of the DSJ as a function of the material properties of its constituents and its geometric characteristics is obtained:

$$K = \frac{1}{b} \left[\frac{2 \cdot L_1}{t_1 \cdot E_S} + \frac{t_a}{L_2 \cdot G} + \frac{L_2}{t_2 \cdot E_S} \right] \qquad (12)$$

This definition of the axial stiffness will be compared to the corresponding experimental results in the next section.

3 EXPERIMENTAL PROCEDURE

3.1 Materials

All substrates (inner adherents and strap adherents) were made of normal marine grade steel. It is well known that the properties of metal adherents can be very influential on the strength of the joint (Bishopp, 2005). Since all substrates are very thick with respect to the adhesive layer thickness in the present study, it can be stated with confidence that there will be no yielding of the metal substrates prior to failure of the adhesive and the consequent total failure of the joint. Thus no plasticity properties of steel are needed.

The adhesive used to bond the specimens was Araldite 2015, a relatively stiff two-component epoxy adhesive manufactured by Huntsman Container Corporation Ltd. The material properties of both the steel and the adhesive are listed in Table 1.

3.2 Geometry of specimens

The geometry of the specimens is given in Figure 2, showing a side and a top view of them. The thickness of the adhesive layer was 0.5 mm and provisions had been taken during manufacturing to keep this value constant for all specimens. Before bonding the straps, the two inner 10 mm adherents were placed in contact, without any adhesive in-between them.

Two strain gage sensors (SG-1 and SG-2) were placed on each specimen, the first on the free surface of one of the straps and the second on the free surface of the inner adherent, at the positions shown in Figure 2. They both had a gage length of 10 mm. Their aim was to monitor strains at these two parts in order to facilitate the evaluation of the performance of the joint after the test. Six specimens were manufactured in total, three for each method of surface preparation.

3.3 Preparation and fabrication of specimens

The primary scope of the experimental program was to study the influence of the surface preparation method on the strength of DSJs. Hence, two texturing techniques were used for the bonding surface preparation of the substrates in order to generate different levels of surface roughness, namely grit blasting (GB) and use of simple sandpaper (SP).

All adherents were first degreased with acetone before the application of the specific surface preparation procedure. SP treatment was performed using first a coarse sandpaper (100), followed by a finer one (200). With regard to the GB specimens, the common in the shipbuilding industry Sa2½ near-white grit blast cleaning was applied (approx. 2.5 mm grit size), according to the Swedish standards. At the end of both surface preparation procedures, the specimens were cleaned with solvent and then let to dry. The sur-

Table 1. Material properties.

	Steel	Adhesive (Araldite 2015)
E (GPa)	210	2
ν	0.3	0.36

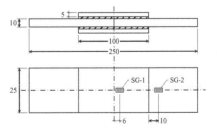

Figure 2. Geometry of the tested specimens (dimensions in mm).

face preparation methods were applied to the bonding surfaces of both the inner adherents and the straps.

After the above pre-treatment, the average surface roughness, Ra, of the bonding surfaces of the adherents was measured using a portable roughness measurement instrument with a 1 mm diameter stylus tip. The sampling length was 8 mm and the cutoff limit was 0.8 mm for each measurement. Measured values of Ra are listed in Table 2 for every one of the six specimens, indicating, besides a very good repeatability of measurements, that the average roughness attained by using the GB procedure was clearly larger than that attained by the SP method (approximately double).

During the bonding procedure of the straps onto the inner adherents, some very small 0.5 mm diameter Cu particles were spread onto the bonding surfaces. These particles functioned as spacers for the bond and kept the adhesive layer thickness as close to the value of 0.5 mm as possible, ensuring at the same time a constant and uniform adhesive layer thickness among all specimens. Thus, as subsequent corresponding measurements indicated, the thickness of the cured adhesive layer was 0.5 ± 0.01 mm.

After the assembly, the specimens were cured in an oven under a uniform pressure loading. According to the adhesive material manufacturer, curing procedure consisted of heating the specimens in 60°C for 4 hours, followed by a slow cooling to ambient temperature. The specimens were left in ambient temperature for 48 hours before performing the tests.

3.4 Test parameters

Specimens were loaded by a uniaxial static tensile displacement, applied with a speed of 0.1 mm/min by an MTS hydraulic testing machine, equipped with a 100 kN load cell.

During each test, a Canon EOS 350D series high resolution camera with a macro lens was capturing close images of the adhesive layer of the joints, every 5 seconds.

4 RESULTS AND DISCUSSION

The global response of the tested DSJ specimens in terms of force-axial displacement curves is depicted

in Figure 3. All six experimental curves of this figure denote an initially linear response, followed by increasing non-linearities, as the adhesive layer begins to be substantially deformed and partially damaged and the joint begins to progressively debond. Moreover, it seems that the sandpaper treated specimens exhibit larger non-linearities, since their failure is more progressive than that of their corresponding GB specimens and two out of the three fail at larger axial displacements. On the contrary, grit blasted specimens exhibit a slightly greater stiffness than that of the sandpaper treated ones, leading also to slightly higher failure load values. These failure loads are listed in Table 3 for all specimens. The experimental measurements of Figure 3 exhibit a very good repeatability, apart from the response of one of the SP treated specimens, which will be explained later.

The results from applying the analytical Equation 10 in the present problem are also presented in Figure 3 (dotted line). It is evident that the analytical model predicts very well the initial part of the force-displacement response, being bounded by its own basic assumptions of linear elastic behaviour.

Figures 4 and 5 present the variation of strains measured by SG-1 on the strap and SG-2 on the inner adherent, respectively, as a function of the applied loading. Once more, strain measurements exhibit a good repeatability, except the same one case which will be discussed later. The behaviour of strains from both sensors is linear, as expected, reaching maximum

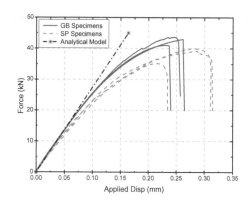

Figure 3. Reaction forces vs. applied displacement for the tested DSJ specimens.

Table 2. Average roughness of GB and SP prepared bonding surfaces.

Specimen	Average roughness (μm)
GB-1	4.07
GB-2	4.23
GB-3	4.16
SP-1	2.38
SP-2	2.29
SP-3	2.17

Table 3. Failure loads of specimens.

Specimen	Failure load (kN)
GB-1	41.0
GB-2	43.0
GB-3	43.7
SP-1	35.2
SP-2	39.2
SP-3	40.0

values just before the failure of the joint. These maximum values are well below the yield point of steel ($<<2000\ \mu\varepsilon$). This behaviour is justified by the fact that the steel adherents are quite stiff and, therefore, the weak point of the joint is the adhesive layer which fails first. Thus the adherents are not heavily loaded, resulting in low maximum strains.

An exception to the above normal strains behaviour is that exhibited by the strain measured at location SG-2 of specimen SP-1 (grey curve that diverges from linearity in Fig. 5). Such a behaviour, where, after a certain load level, strain gradually stops to increase and begins to decrease, denotes a bending behaviour of the DSJ, with the compressive side being that of the strain gage (the upper one in Fig. 1). Taking into account the perfectly symmetric geometry of the joint with respect to x-axis (see Fig. 1), the only cause that can result in such a bending behaviour is the unsymmetric way of gradual failure of the adhesive layers, below and above the inner adherent. Thus, it seems that in specimen SP-1, the lower adhesive layers started to fail before and more rapidly than the upper ones, thus forcing the specimen to bend upwards, creating a compressive strain component on its upper surface, which, when superimposed to the global tensile strain, is responsible for the particular shape of the corresponding curve in Figure 5. This bending behaviour

had a negative effect on specimen SP-1, resulting in a decreased failure load compared to that of the other specimens which, more or less, exhibited a pure tensile behaviour up to their failure. As Figure 3 shows and Table 3 indicates, this decrease is significant, being approximately equal to 12%. Therefore, the above example makes clear that the aim should always be a completely symmetric DSJ with identical and uniform adhesive layers on both sides of the inner adherent.

The analytical model of section 2 was also applied to calculate the strains in the strap (segment 2) and the inner adherent (segment 1) and the results are also plotted in Figures 4 and 5 (dotted lines). The very good prediction of the analytical model is evident, since, as discussed before, strains at these two locations behave almost linearly up to the joint failure.

Figures 6, 7 and 8 in the sequence are three images taken by the high resolution camera and showing a detail of one of the GB specimens, near one of the edges of the straps. Figure 6 shows the initial intact condition before load application, whereas Figure 7 shows the crack (small white line) initiating at the free edge of the adhesive layer. Finally, Figure 8 presents the final debonding of the specimen.

Figure 6.   Initial state—no trace of crack.

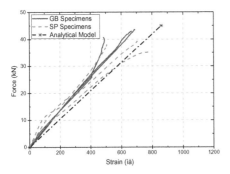

Figure 4.   Variation of strains from SG-1 on the strap.

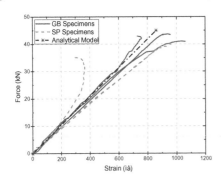

Figure 5.   Variation of strains from SG-2 on the inner adherent.

Figure 7.   Crack initiation at the adhesive—adherent interface.

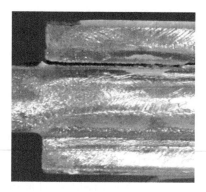

Figure 8. Final joint debonding.

In all specimens tested, failure crack initiated at the free edge of the adhesive layer, at the interface between the strap and the adhesive layer. In the sequence, as the crack propagated further, it traversed the adhesive layer and continued its propagation at the interface between the adhesive layer and the inner adherents.

5 CONCLUSIONS

The short experimental study performed investigated the effect of the bonding surfaces preparation method on the response of a double strap joint under tensile loading. The differences obtained in the stiffness and the failure loads of the joints between the two surface preparation methods investigated are in the order of 10%, being small but definitely not negligible. Although there is no proved quantified connection between the surface roughness of the bonding surfaces and the failure load of the joint, it is noteworthy that the present experimental results showed that doubling the surface roughness resulted in an approximately 10% increase of the joints' failure load. The specimens whose bonding surfaces had been grit blasted exhibited a slight advantageous response.

The particular strain behaviour of one of the specimens emphasized the importance of attaining a good quality for all the bonds of a DSJ, otherwise unsymmetry conditions are introduced during the gradual adhesive failure, resulting in a bending behaviour and consequently in an earlier total failure of the joint.

The simple analytical model presented which calculates the displacement, strain and stress response of a DSJ in its linear response range proved to be very efficient, since it predicted the experimentally measured corresponding magnitudes with great accuracy. Moreover, the available experimental data can be very effectively used for the calibration of corresponding finite element models, since they come from a very well defined case, with no uncertainties in the material properties.

ACKNOWLEDGMENTS

This work has been performed within the context of the Network of Excellence on Marine Structures (MARSTRUCT) partially funded by the European Union through the Growth Programme under contract TNE3-CT-2003–506141. The authors gratefully acknowledge Huntsman Container Corporation Ltd for supplying the adhesive material.

REFERENCES

Adams, R.D., Comyn, J. & Wake, W.C. 1997. *Structural adhesive joints in engineering*, London, Chapman & Hall.

Adams, R.D. & Peppiatt, N.A. 1974. Stress analysis of adhesively bonded lap joints, *Journal of Strain Analysis*, 9: 185–196.

Baker, A.A. & Chester, J.R. 1992. Minimum surface treatments for adhesively bonded repairs, *International Journal of Adhesion and Adhesives*, 12: 73–78.

Bishopp, J. 2005. *Handbook of Adhesives and Sealants*, McGRAW Hill Inc.

Chang, V.W. 1981. Enhancement of adhesive joint strength by surface texturing, *Journal of Applied Polymer Science*, 26: 1759–1776.

Chester, J.R. & Roberts, D.J. 1989. Void minimization in adhesive joints, *International Journal of Adhesion and Adhesives*, 9: 129–138.

Goland, M. & Reissner, E. 1944. The stresses in cemented joints, *Journal of Applied Mechanics*, 11: 17–27.

Harris, F.A. & Beevers, A. 1999. The effects of grit-blasting on surface properties for adhesion, *International Journal of Adhesion & Adhesives* 19: 445–452.

Kalnins, M., Sirmacs, A. & Malers, L. 1997. On the importance of some surface and interface characteristics in the formation of the properties of adhesive joints, *International Journal of Adhesion and Adhesives*, 17: 365–372.

Katona, R.T. & Batterman, C.H. 1983. Surface roughness effects on the stress analysis of adhesive joints, *International Journal of Adhesion and Adhesives*, 32: 85–91.

Morais, B.A., Pereira, B.A., Teixeira, P.J. & Cavaleiro, N.C. 2007. Strength of epoxy adhesive-bonded stainless-steel joints, *International Journal of Adhesion and Adhesives*, 27: 679–686.

Sargent, P.J. 1994. Adherend surface morphology and its influence on the peel strength of adhesive joints bonded with modified phenolic and epoxy structural adhesives, *International Journal of Adhesion and Adhesives*, 14: 21–30.

Shahid, M. & Hashim, A.S. 2002. Effect of surface roughness on the strength of cleavage joints, *International Journal of Adhesion and Adhesives*, 22: 235–244.

Uehara, K. & Sakurai, M. 2002. Bonding strength of adhesives and surface roughness of joined parts, *Journal of Materials Processing Technology*, 27: 178–181.

Volkersen, O. 1938. Die Niektraftverteilung in Zugbeanspruchtenmit Konstanten Laschenquerschriften, *Luftfahrtforschung*, 15: 41–47.

Xiao, X., Foss, P.H. & Schroeder, J.A. 2004. Stiffness prediction of the double lap shear joint. Part1:Analytical solution, *International Journal of Adhesion & Adhesives*, 24: 229–237.

Analysis and Design of Marine Structures – Guedes Soares & Das (eds)
© 2009 Taylor & Francis Group, London, ISBN 978-0-415-54934-9

Pultrusion characterisation for adhesive joints

J.A. Nisar & S.A. Hashim
Department of Mechanical Engineering, University of Glasgow, Glasgow, UK

P.K. Das
Department of Naval Architecture and Marine Engineering, Universities of Glasgow and Strathclyde, Glasgow, UK

ABSTRACT: This paper describes the experimental and numerical techniques used to characterise the adhesion of GFRP pultrusions for double lap shear (DLS) adhesive joints, often referred to in ship application as butt joints. The overall objectives of the study are (i) to devise laboratory techniques which mimic pultrusion moulding using UD glass rovings and vinyl ester resin, (ii) to maximise the level of adhesion of the pultrusion and hence to improve structural efficiency of joints and (iii) to determine location and level of critical stresses and failure initiation through thickness of the composite laminate. The experiments include moulding, coating, bonding and testing of pultruded laminates and joints. Both macro and meso scale specimens and models were considered, supported by finite element analysis (FEA).

1 INTRODUCTION

This study seeks to assess moulding and bonding of pultruded composites in relation to adhesive butt joints for decks and similar structures. To minimise cost in the fabrication of aluminium alloy ship structures, it has long been common practice to make use of extruded aluminium planks for the deck and bulkhead components of lightweight, high-speed catamarans and the topsides of passenger cruise ships (Freeman et al. 2000). Although these materials are more expensive to manufacture, per tonne, they are considerably easier and cheaper to assemble and can be optimised for material distribution across the section. In principle, it would appear that pultruded polymer composite planks could offer similar benefits to the designers and fabricators of composite catamarans, offering the possibility of panel fabrication similar to steel and aluminium materials, rather than complex, labour intensive GFRP mould and hand lay-up techniques (Boyd et al. 2005). Pultrusion is one of the few continuous processes for composite manufacture, which could potentially make it one of the cheapest. A schematic design concept of a plank section is illustrated in Figure 1, which may have dimensions of 500 mm wide and 100 mm high with a wall thickness of 5 to 4 mm and is suitable for decking structures.

A pultruded section may often contain uni-directional (UD) rovings (reinforcements) at the centre, with the skin of random (continuous stand) mats as shown in Figure 2. Narrow strips may be based entirely on UD reinforcements. Another replacement for the random fabric in wide sections and channel sections could be a stitched or woven (0/90) mat.

These could enhance stiffness and strength better than a random mat as the surface layer. While pultrusion design allows good side-by-side joints, it is often more difficult to achieve this for end-to-end connection and therefore, bolted joints are often considered. The introduction of holes in the composite leads to high stress concentrations and to counteract this, thicker (and heavier) walls will be needed.

The structural efficiency of the composite lap-shear joint depends on the properties of the adherends including resin and fabric and its arrangement, adhesive type, joint geometry and surface preparation. Earlier research conducted at Glasgow University (Boyd et al. 2004) and elsewhere demonstrated that it is currently unreasonable to expect better than 38% structural efficiency for basic design DLS joints based on commercial pultrusion. This has been improved to over 50% by introducing low viscosity resin coating

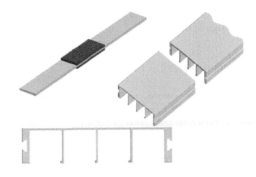

Figure 1. Concept of pultruded plank and connection.

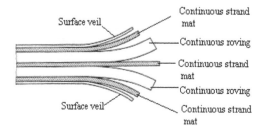

Figure 2. Layers detail of GFRP pultruded section.

of the pultrusions prior to bonding (Hashim 2009). The properties of the pultrusion materials and their moulding process limit the degree of the adhesion due to the inherent weaknesses of polyester or vinyl ester resins and use of an internal mould release agent in the moulding process. Besides the possibility of making a limited improvement by controlling the fabric constructions, its volume fraction and type of fillers, other materials and design measures are necessary to improve this further (Fozzard et al. 2005).

Previous experimental research on composite DLS joints including pultrusions and other composites has shown that failure initiates in the adhesive spew fillet or in the outer mat layers of the pultruded laminate at the joint edge (Keller et al. 2005). Cracks then easily propagate between the mat layers where there is little through thickness reinforcement and an inherent stress concentration. Final failure occurs when the surface layer delaminates from the adherend in the overlap region through a combination of through thickness tensile (peeling) and shear stresses including transverse shear in the composite. A pultruded material made entirely from glass UD rovings' fabric has no surface ply as such, so the delamination mechanism detailed above does not take place. The failure occurs a few filaments deep into the composite and the strength of this type of material in a DLS joint is superior to the more traditional "sandwich" lay-up with surface mats. Therefore further improvement for pultruded sections in general is needed by combination of measures including arrangement of top reinforcement layer, epoxy resin coating to improve surface adhesion and design of the joint.

2 EXPERIMENTS

Two phases were considered for the experimental programme including moulding and bonding laminates using commercial pultrusion materials.

2.1 Phase 1—strip laminates & DLS joints

"Pseudo-pultrusion" process was developed to produce 120 × 120 mm, 3 mm thick flat composite laminates based on the glass rovings, vinyl ester resin, internal mould release agent and calcium based filler powder. The process was carried out in accordance with the manufacturer's recommendations include mixing of resin and its hardening and initiating components and organisation of the glass roving (type 4800 from Formax). Figure 3 shows aspects of the moulding process of the small laminate and the detail of bonded specimen made of this to asses the bondability of the moulded laminates. The moulding includes organising the glass roving, resin mixing, impregnation and squeeze-out and curing while the fabric is under tension. The laminate was then placed in a hot press which represents pultrusion die conditions. Several trial and error attempts were made to optimise the nominal volume fraction. It was found that the higher the weight fraction the poorer the quality of the laminate in term of adhesion Fibre fraction of about 70% by weight (volume fraction $V_f = 50\%$) was found to be more effective than 80%. The laminates were then cut into the required adherend dimensions using parallel mounted diamond impregnated circular saw blades mounted in a horizontal axis-milling machine. The adherends were then prepared for bonding of the DLS specimens, by abrasion and solvent wiping. Some of the surfaces were coated the with 0.1 mm low viscosity epoxy resin prior to bonding. This produces intimate contact to the composite surface masking micro surface defects and hence better wetting and adhesion by the thixotropic adhesive Araldite

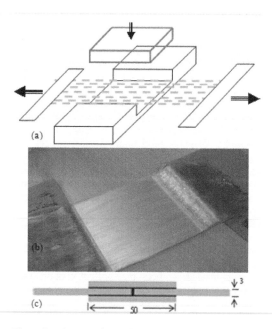

Figure 3. Aspects of 'pseudo pultrusion', showing (a) concept of die/mould arrangement, (b) moulded laminate and (c) bonded DLS specimen.

Table 1. Moulding conditions and shear test result from DLS specimens.

Fibre content (%wt)	Strap mat	Coating	Inter mould realease	Filler (phr*)	Failure load (KN)	Standard deviatior	Efficient (%)
68.4	GFRP	No	No	10	17.6	–	39.1
70.1	GFRP	No	No	0	17.4	–	38.7
72	GFRP	No	Yes	0	16.3	2.4	36.2
70	Steel	No	Yes	10	18.2	1.5	40.4
70	Steel	Yes	Yes	10	22.6	1.1	50.2

*Parts per hundred (phr).

2015 used here. Both inner and outer adherends were based on the pultruded GFRP materials (3 mm thick). However some of the specimens were based hybrid steel/GFRP where the outer adherend is steel (3 mm). It is also possible to considered other good adhesion materials e.g. CFRP/epoxy.

Having the outer adherend based on steel instead of the pultruded GFRP meant that delamination is limited to the inner adherend. The two-part toughened epoxy adhesive Araldite 2015 (Huntsman UK) was applied through a mixing nozzle to the adherends followed by clamping and curing at 85 C for 1 hr. The bonded specimens were tested under a monotonic tensile loading on a Zwick/Roell tensile testing machine at cross head speed of 0.5 mm/min, at ambient laboratory conditions. A 5 mm extensometer of 50 mm gauge length was used to measure displacement within the joints, including the straps and adhesives. The results of the testing are shown in Table 1, which include joint efficiency and failure initiation. The efficiency is based on the ratio of the failure load of the joint to the adherend tensile load capacity (45 kN max). The following remarks can be made from the table:

- Using filler slightly improves the strength
- The combined effect of using internal mould release, filler and steel as the outer adherend improves the overall joint strength by about 10% but it's difficult to quantify individual effects
- Combining steel with resin coating resulted in improving joint strength by 38% - significant.

Following each test, the fracture surfaces of the joints were examined to assess the locus of failure. All failures initiated near the edge of the joint and then propagated towards the interfaces of the adhesive or resin with adherends, leading to filament delamination deep within the surface rovings. In the case of coated specimens the failure of rovings is deeper beneath the composite surface.

To study the improvement of the joint structural efficiency, further DLS joints testing was carried out on steel/GFRP joints with different overlaps. Strap lengths of 50 mm, 75 mm, 100 mm and 150 mm were

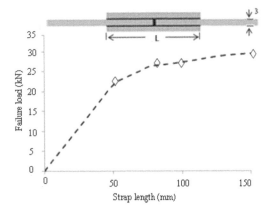

Figure 4. Failure loads of DLS joints at different strap length.

considered. The GFRP in this case is based on equivalent UD pultrusion strips from manufacturer (Excel Composites UK) due to difficulties of moulding longer strips in the laboratory. These have the same materials and their properties of the in-house produced composites. The supplied pultruded strips were machined by milling to remove the surfacing veil and cut to the required dimensions. The steel/GFRP DLS joints were prepared, coated, bonded and tested in the same way before. Figure 4 shows the average test results plotted against various strap lengths. The results demonstrate further improvement in the load carrying capacity of the joints with longer overlaps. The joint with the 150 mm long strap, i.e. 75 mm overlap, produced failure load of about 30 kN. The structural efficiency of this is about 67%—the ratio of joint failure load to that of the 3 mm GFRP inner adherend.

2.2 *Phase 2—meso-scale laminates and joints*

To understand failure of DLS joints and the architecture of surface layer fabric, it was necessary to produce more laminates more effectively. The moulding of macro laminates is very useful however, this could be time consuming and difficult to control considering the diversity of moulding materials used.

Therefore small laminates (meso-scale) were developed for moulding and bonding. These were produced to represent local shear and peel stresses that are expected in DLS joints. Figure 5 shows the details of two meso-laminates which are 2 × 20 mm, 1 mm thick with surface coating.

Model LL represents pultrusions with UD top fabric and LT represents 0/90 inlaid (stitched) top layer of pultrusions. These laminates are intended for bonding in tensile and shear joints as shown in Figure 6. To produce the laminates, a special jig was designed and manufactured to represent pultrusion close mould conditions (see Figure 6). The tension was applied by turning screw threads which are in turn aligning the fibres and reducing the possibility of the fibres between each layer twisting around one another. The copper plate

(mould) was then placed on top of its female counterpart and a weight was applied on top. The composite was cured, cut to size and coated with the epoxy resin.

Then laminates were bonded to aluminium blocks to form the meso-scal shear and tensile specimens. The bond area of the laminate of interest was limited to 10 × 10 mm using PTFE sheeting, to eliminate edge delamination effects. This was done to ensure that failure initiation starts at specific location of the joint namely at the interface between the adhesive bond line and the laminate. Finally, the specimens were tested under monotonic tensile loading using a small load cell on the Zwick/Roell tensile testing machine. Table 2 gives average failure load from three test results for the shear and tensile specimens.

The following remarks can be made from Table 2:

• Failure load for shear LL is about 10% higher than shear LT
• Failure load of tensile LL and TL are quite similar. In both cases, the load is perpendicular to the roving (filaments).

Examinations of fractured specimens and their surfaces showed that failure is taking place within the filaments of the roving at about 0.2 mm below the epoxy coating layer. Figure 7 shows a microscopic image of such a failure in a shear LT specimen where

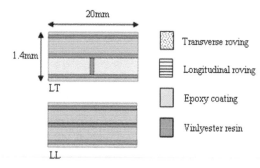

Figure 5. Meso-laminates with epoxy coating.

Table 2. Test result of meso scale specimens.

Load/Type	Load (kN)	Standard deviation (kN)	Internal mould release	Filler	Fibre content (%wt)
Shear LL	2.4	0.28	Yes	No	70
Shear LT	2.2	0.34	Yes	No	70
Tension LT and LL	1.4	–	Yes	No	70

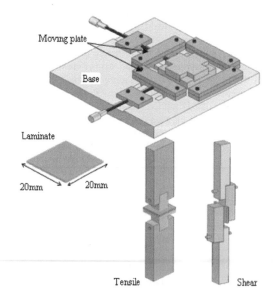

Figure 6. Moulding and testing of meso-scale laminate.

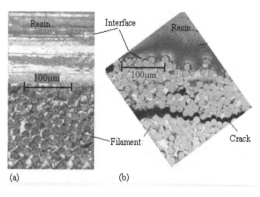

Figure 7. Moulded roving images showing glass roving/filaments (a) before failure and (b) after failure, in shear LT specimen.

crack/delamination took place within the top roving (filaments), below the resin coating.

3 MODELLING

This aims to understand failure and behaviours of macro (DLS) and meso-scale models and to see the correlation between the two. Therfore, 2-D models were constructed in ABAQUS to represent these. The models used eight-noded quadrilateral plane stress elements. Failure load was applied to all models including the long overlap joints. The macro model was based on steel/GFRP joints and meso models were based on shear LL and tensile LT. The material properties used in this are shown in Table 3. These are based on various sources including materials manufacturers, lab tests and use of "rules of mixture" for composites and the Halpin—Tsai equation for transverse modulus of elasticity. The adherends and coating resin were modelled as linear elastic. The composite was modelled with orthotropic properties. Elastic-full plastic properties were considered for the adhesive. The adhesive bondline thickness was modelled at 0.2 mm and 0.1 mm was considered for coating resin.

The adhesive was divided into four layers and the resin into two layers to allow for determination of through thickness stresses. The critical stresses in the adhesive are shear and peel/cleavage. However, the focus was on the composite top layer due to peel/transverse stresses pulling of the rovings just below the epoxy resin coating. The stress distribution in the 25 mm overlap DLS model is shown in Figure 8. The shear stress distribution along the lower interface between adhesive and inner adherend (GFRP) is showing some plasticity for half the overlap. This could cause fracture failure in shear within the elastic plastic zone. Perhaps more critically is the peel/transverse stresses at the surface of the composite which concentrated at the joint's end. Though all joints have failed at the composite just below the surface, at the localised interfaces, the composite stresses at the surface may be taken as a good indicator of the sub surface stresses

in the composite. The level of this stress is about 18 MPa.

In addition shear stress distributions in DLS joints with different overlaps/strap lengths were also analysed. The stress distributions for the four tested specimens (Figure 4) are shown in Figure 9 against normalised overlaps. The figure shows that the plastic stress proportion within individual joints decreases with increasing the overlaps. The peel stresses within the composite for all the joints were very similar at the corresponding failure loads, regardless of the overlap length.

In a similar way, the meso models were constructed for the meso shear and tensile models. These include modelling multi-through thickness materials i.e. adhesive, coating resin, matrix resin, and impregnated rovings. Figure 10 shows the shear LL model and relevant stress distributions. The focus here is on the peel stresses within the interfaces of the adhesive and composite which are maximum at the right-hand side of the joint. The composite stress is about 16 MPa. The behaviour and failure stresses of the tensile LT model (same as LL in this case) are shown in Figure 11. Again

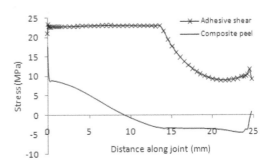

Figure 8.   Stress distributions along bondline and composite surface, in DLS model.

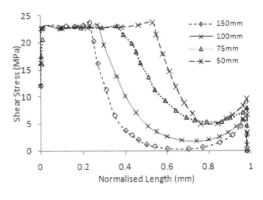

Figure 9.   Adhesive shear stresses in DLS models with various overlaps, at failure load.

Table 3.   Materials properties.

Property	GFRP(UD)	Steel	Aluminum	Adhesive	Coating resin
Young's modulus [GPa]	39.6 (Long) 7.2 (Trans)	210	73	2	3
Poisson's ratio	0.26	0.3	0.3	0.36	0.37
Tensile strength [MPa]	600 (long) 20 (Trans)	300	180	40	50

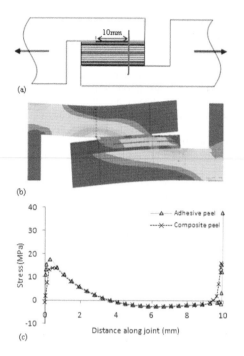

(a)

(b)

(c)

Figure 10. (a) Shear LL model detail, (b) principal stress (PS) and (c) stresses distributions along adhesive and laminate, at failure load.

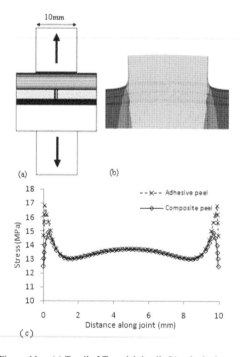

(a)   (b)

(c)

Figure 11. (a) Tensile LT model detail, (b) principal stress (PS) contour and (c) stresses distributions along adhesive and laminate, at failure load.

the critical peel stresses are at the edge of the joint with maximum peel in the composite surface of 15 MPa.

4 DISCUSSION

The initial results of the experimental programme have confirmed that there is a weakness in the polyester resin based pultruded material. The surface glass rovings of the UD pultruded adherend were delaminated within filament level resulting in joint failure at a relatively low load in comparison with the tensile capacity of the loaded adherend. Coating the adherend with low viscosity epoxy resin has been very effective and this may well be feasible in an industrial context. A key conclusion of this study is that combining a high adhesion outer adherend such as the steel, together with the coating and increasing the overlap can result in up to 67% strength improvement—the ratio of failure load of joint to the tensile strength of loaded/inner adherend.

This represents a marked improvement of pultruded joint capacity, considering the high tensile strength of the UD GFRP material in this study (600 MPa). This can only be effective materials and design measures are combined. The two "pseudo-pultrusion" moulding methods were found to be very effective tools to verify the bondability of pultrusions. However, the production of the large number of laminates for DLS joints is very time consuming. Therefore, "pseudo pultrusion" of meso laminate is necessary to study more moulding conditions and type of fabrics. Further type of fabrics are under consideration include combination mat and various sizes of rovings.

Good agreement was found between the macro and meso FEA models, with respect to composite transverse/peel stresses just below the edge of the adhesive bondline. The work demonstrated the importance of having detailed of through thickness layers in the modelling. This will enable a better understanding of failure and behaviour of adhesively bonded pultruded composites. The numerical results are encouraging in terms of failure stress correlation in relation to peel stresses of the composite. A maximum scatter of 20% in level of failure stress among the meso and macro models is vey good, considering the complexity of the composite material and their delamination stresses. These stress values are also within the range of transverse strength of the composite (see table 3). Furthermore, the level of scatter in the experimental results (standard deviation) is quite large for the meso-specimens and this was used as the average failure load, contributing further to the scatter.

Finally, because of the complex nature of the composite and materials the assessment and analysis require a constitutive modelling approach where-by multiscale models and modelling are important. Therefore this study considers both macro (DLS) and meso-models. Future work will include micro-scale models to understand failure at filament levels.

ACKNOWLEDGEMENT

This work was supported the ORSAS research scheme, Network of Excellence on Marine Structures (MARSTRUCT), The Carnegie Trust and Excel Composite UK. Such support is gratefully acknowledged.

REFERENCES

Boyd, S.W., Day, A.H. & Winkle, I.E., 2005, Geometric considerations for the design of production—Friendly high speed ship hull forms, Proc. IMechE Part M: Journal of Engineering for the Maritime Environment, 219, Vol. 2 , 65–76.

Boyd, S.W., Winkle, I.E. & Day, A.H., 2004, Bonded butt joints in pultruded GRP panels—an experimental study, Int. J. Adhesion and Adhesives, Vol. 24, 263–275.

Freeman, S. & Green, J., 2000, Aluminium in marine environment: an update, Proc OCEANS 2000, RI, USA 1591–1595.

Fozzard, O., Hashim, S.A. & Winkle, I.E., 2005, Design, construction and bonding of preformed pultruded planks for ship structures, MARSTRUCT Report No. MAR-D4-3-1-UGS-01.

Hashim, S.A., 2009, Strength of resin coated adhesive bonded double lap-shear pultrusion joints at ambient temperature, In. J. Adhesion and Adhesives, Vol 29, 294–301.

Keller, T. & Vallee, T., 2005, Adhesively bonded lap joints from pultruded GFRP profiles. Part 1: stress-strain analysis and failure modes, Composites: Part B, Vol. 36, 331–339.

Buckling of composite plates

Analysis and Design of Marine Structures – Guedes Soares & Das (eds)
© 2009 Taylor & Francis Group, London, ISBN 978-0-415-54934-9

Studies of the buckling of composite plates in compression

B. Hayman
Det Norske Veritas AS, Høvik and University of Oslo, Oslo, Norway

C. Berggreen, C. Lundsgaard-Larsen, A. Delarche & H.L. Toftegaard
Technical University of Denmark, Kongens Lyngby, Denmark

R.S. Dow, J. Downes & K. Misirlis
University of Newcastle, Newcastle, UK

N. Tsouvalis & C. Douka
National Technical University of Athens, Athens, Greece

ABSTRACT: As part of the MARSTRUCT Network of Excellence on Marine Structures, a series of studies has been carried out into the buckling of glass fibre reinforced polymer plates with in-plane compression loading. The studies have included fabrication and testing of square, laminated panels with various thicknesses and initial geometrical imperfections, material testing, advanced FE modelling studies and finally parametric studies covering a range of slendernesses and imperfection amplitudes. The paper provides an overview of the studies, which involved several participants in the Network.

1 INTRODUCTION

1.1 Background

Composite structures consisting of plates or plate-like elements are used widely in wind turbine blades and in some ships, particularly naval ships. For plates that are made of steel or aluminium, design strength curves have been established that specify the strength under in-plane compressive loading as a function of a slenderness parameter. Such curves take account of geometric imperfections, and have been established on the basis of extensive numerical, analytical and experimental studies. In contrast, design of fibre reinforced (FRP) structures against buckling is almost invariably treated in terms of the elastic critical load of the ideal structure, at best modified by a knock-down factor based on limited test data. A separate check for local material failure is performed but usually this does not consider the interaction with buckling nor does it take account of imperfections in a systematic way. This is probably due to the fact that relatively few test results are available for buckling of FRP structures, and there is little published information on the manufacturing imperfections that need to be taken into account.

1.2 MARSTRUCT studies

As part of the MARSTRUCT Network of Excellence on marine structures, a wide-ranging series of studies has been carried out into the buckling of glass fibre reinforced polymer plates with in-plane compression loading. The studies have consisted of the following:

- A series of square laminated panels was fabricated at two locations using similar layups but two different production methods. The panels were of three different thicknesses. Some were made as flat as possible while others were fabricated with intentional out-of-flatness imperfections.
- Extensively instrumented tests were performed on these panels with in-plane compressive loading. It was intended that the test rig should provide effectively clamped conditions at the panel edges but some movement was observed. A digital image correlation (DIC) system was used to measure the deformation of each panel, including the movements at the edges.
- Material testing was performed in a round-robin study to establish the mechanical properties of the laminates (both elastic properties and strength).
- Advanced FE modelling studies were performed at several institutions and the results compared with the panel tests to validate the methods and establish the best practices for predicting the behaviour up to ultimate failure. Various degradation models were used in the range of non-linear material behaviour.

- Parametric studies of plates with idealised edge conditions were performed, covering a range of slenderness ratios and imperfection amplitudes.

The paper provides an overview of the work done, together with the main conclusions. Full details of each phase of the work will be reported separately in future publications.

2 PANEL TESTS

2.1 Objectives

Square panel specimens were tested with in-plane compression loading. The objectives were to observe the behaviour up to failure and to obtain comprehensive test data for use in validation of modelling approaches.

2.2 Plate specimens

Plate specimens for three panel series were produced at NTUA (Series 1 and 2) and by Vestas Wind Systems A/S on behalf of DTU (Series 3). The composite systems used by the two fabricators were similar but not identical, and the production methods followed were also different.

The DTU material was a pre-preg E-glass fibre/epoxy system. The fibre reinforcement comprised two types of E-glass fabrics: a 1200 g/m^2 UD fabric and a 600 $g/m^2 \pm 45°$ biaxial, knitted, non-crimp fabric. The NTUA material was a wet lay-up E-glass/epoxy system. The fibre reinforcement comprised a 623 g/m^2 UD glass fabric (with 50 g/m^2 in weft direction) together with a 306 $g/m^2 \pm 45°$ biaxial non-crimp fabric, and an epoxy resin with low viscosity (600–750 mPa. s at 25°C).

With UD representing the unidirectional layers and by BIAX the biaxial layers, the lay-ups of the three series are the following:

Series 1: [BIAX/4xUD/BIAX/3xUD]$_S$
Series 2: [BIAX/4xUD/BIAX/4xUD/BIAX/3xUD]$_S$
Series 3: [BIAX/4xUD/BIAX/4xUD/BIAX/2xUD]$_S$

All lay-ups are symmetrical, and the weight of the UD layers is 88% for Series 1 and 2 and 87% for Series 3. The thicknesses of most specimens were measured using a 3-D contact digitiser, and the nominal thicknesses of Series 1–3 were approximately 9, 15 and 20 mm respectively.

Each plate had a nominal total length, L, of 400 mm (parallel to the load direction) and width, B, of 380 mm (Fig. 1). When the plate was mounted in the text fixture, its unsupported length a and width b were both 320 mm (Fig. 1). Each of the three plate series comprised 9 plates. Three of these plates were perfectly plane, three had a small maximum imperfection and the remaining three had a large maximum imperfection.

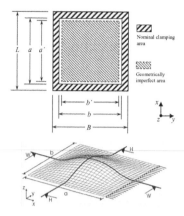

Figure 1. Geometry of the test plates (top) and shape of geometric imperfections (bottom).

Figure 2. Vacuum bagging technique followed by NTUA.

The values of the imperfection amplitudes, δ_o, were defined as a function of the unsupported width of the plate, as follows:

- small max. imperf.: $\delta_o = 0.01 \times 320 = 3.2$ mm
- large max. imperf.: $\delta_o = 0.03 \times 320 \times 9.6$ mm

The shape of the geometric imperfection was a scaled first buckling mode-shape of a corresponding fully clamped plate with dimensions $a' = b' = 300$ mm (Fig. 1). This left a 10 mm wide plane strip near the plate's clamped edges, in order to ensure easy fitting into the test fixture.

All plates were produced using manual lay-up with vacuum bagging. For the NTUA plates with imperfections a convex mould, numerically machined from aluminium plate, was placed under the composite layup. The glass fabrics were laid up and impregnated with resin by hand (Fig. 2, top). Then the whole arrangement was covered by the various vacuum materials and the air and excessive resin were pumped out (Fig. 2, bottom). The vacuum pressure used was 0.6 bar. The DTU panels (Fig. 3) were made with machined moulds against both faces. The laminate was under vacuum of at least 0.9 bar and curing was performed for 2 hours at 120°C, ensuring at least 98% cure and removing excessive resin from the specimen.

2.3 Test setup

All panel tests were carried out at DTU in an Instron 8508 5 MN servo-hydraulic test machine using a special test rig. Figure 4 shows a schematic 2-D view of the test-rig. The test plates are inserted between the side flanges (No. 4 in Figure 4) of the two towers and

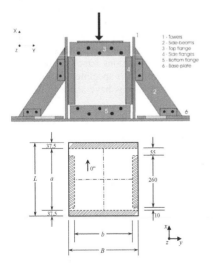

Figure 3. Hand lay-up in a double sided mould used for DTU panels (top, upper mould not positioned yet). Peel ply being removed prior to positioning of upper mould (bottom).

are bolted between the top and bottom flanges (No. 3 and No. 5 in Figure 4, respectively). In this way, along the two vertical towers, the test plate is free to move in x-and y-directions, whereas out-of-plane movement and rotation are prohibited. Along its horizontal edges (top and bottom flange), the test plate is considered rotationally fixed.

Figure 4 (bottom) presents a schematic view of the plate clamping devices, with detailed dimensions as measured from the test-rig. Thus, as indicated, the effective dimensions of the plate after being installed into the test-rig are $a' = 325$ mm and $b' = 320$ mm.

An advanced, non-contact digital image correlation (DIC) system (ARAMIS 4M) was applied to monitor deformations during testing. Two 4 megapixel digital cameras monitor the same specimen area, but with a relative angle to each other. A suitable speckle pattern is sprayed onto the specimen surface so that the digital images can be analysed digitally and a surface 3D displacement field and a 2D strain field can be generated for each time frame. A frame rate of 1 frame per second was used for all specimens. The digital cameras and the specimen speckle pattern can be seen in Figure 5 (top).

The DIC system and its cameras were placed on one side of the specimen (the convex side for plates with imperfections), while LVDTs and strain gauges were placed on the opposite (concave) side. Figure 5 shows a plate mounted in the test-rig, ready to be tested. Figure 5 (bottom) shows the concave side of the plate with positions of LVDTs and strain gauges. In all cases the compressive loading was applied in the form of linearly increasing compressive displacement with a speed of 1 mm/min.

Figure 5. The two sides of a plate specimen installed in the test-rig, ready for testing.

Figure 4. Test-rig (top) and actual plate boundary conditions (bottom). Dimensions in mm.

2.4 Test results

Table 1 represents the ultimate failure loads for the individual tested panel specimens. Generally, all specimens, irrespective of imperfection magnitude, experienced out-of-plane deformations prior to ultimate failure, see the representative DIC out-of-plane measurements in Figure 6. They were thus influenced by a mixture of bending deformations due to specimen imperfections, bending deformations due to test rig deformations and instability/buckling behaviour of the plate specimens.

For the specimens in Series 1, which have a high slenderness ratio, the results in Table 2 indicate a decrease in compressive strength when an initial imperfection is introduced. However, approximately the same compressive strength is observed for specimens with both small and large imperfections, suggesting a rather small imperfection sensitivity. It should be noted that these conclusions are based on only two specimens for each of the cases with no imperfection and with large imperfection, and must therefore be taken with reservations.

For the specimens in Series 2 with a medium slenderness ratio, the compressive strength is seen to decrease consistently with increasing imperfection

Table 2. Comparison of average material properties (in MPa).

| Property | NTUA material | | DTU material | |
	NTUA test	DTU test	NTUA test	DTU test
E_{1t}	29658	33170	48634	56235
E_{1c}	38671	37238	50619	56209
E_{2t}	6563	9338	18535	20422
E_{2c}	8501	9536	12325	15729
G_{12}	2034	2169	4800	4264
v_{12}	0.290	0.268	0.274	0.284
X_t	559	698	968	1141
X_c	253	191	915	952
Y_t	60	43	24	22
Y_c	59	69	118	127
S	31	30	65	64

magnitude, and there is a higher imperfection sensitivity than for the specimens in Series 1.

Finally, for Series 3 with a small slenderness ratio, close to that at which the critical stress for elastic buckling is close to the compressive strength of the laminate material, a significant drop in compressive strength can be observed when the maximum imperfection magnitude is increased from 3.2 to 9.6 mm. However, comparing the compressive strength of the specimens having small imperfections with those having no imperfection, a slight increase can be observed. Even though the imperfection sensitivity can be expected to be maximum at a relative slenderness ratio close to unity, the compression strength of the specimens with no imperfection should in theory be expected higher. However, as all specimens, and especially the Series 3 specimens, are also influenced by test rig deformations introducing displacements and rotations of the panel boundaries, and thus acting as gradually increasing geometrical imperfections, no general conclusion can be drawn regarding the compressive strength of the specimens without initial geometrical imperfections based on the test results. Furthermore, only two specimens have been tested, so that any conclusions have again to be taken with caution.

Table 1. Failure loads.

Imperf. mm	Series 1 kN	Series 2 kN	Series 3 kN
0	N/A	1218	2250
	415	1092	2070
	390	1170	N/A
3.2	294	906	2380
	213	882	2303
	309	930	2327
9.6	294	750	1543
	320	780	1934
	N/A	792	1892

3 MATERIAL TESTING

3.1 Objectives

Material testing was performed in order to provide reliable ply properties for use in the modelling studies described in Section 4.

3.2 Materials and fabrication methods

In order to establish in-plane ply properties for the plates to be tested, a round-robin material characterisation test programme was performed by DTU and

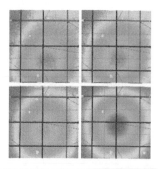

Figure 6. Out-of-plane displacement contour plot (measured with DIC) for a Serices 3 panel without imperfection for increasing applied in-plane loads. The last contour plot (bottom right) is taken immediately prior to ultimate failure.

NTUA. Each partner manufactured two identical sets of UD specimens from its own material, using exactly the same UD materials and fabrication method as for the panel test plates (see Section 2.2). The first set was tested by the partner who produced it and the second was tested by the other partner.

The DTU material was a pre-preg E-glass fibre/epoxy system. The plates from which the material specimens were cut were made from 1200 g/m² UD fabric by hand lay-up and vacuum bagging in a double sided mould. The specimens produced by DTU had, on average, a fibre volume fraction of 0.62 and a ply thickness of 0.8 mm.

The NTUA material test specimens were made with a 623 g/m² UD glass fabric (with 50 g/m² in weft direction) and the same low-viscosity epoxy resin as for the panel specimens. The vacuum bag moulding method was used, with a single sided mould. The specimens produced by NTUA had, on average, a fibre volume fraction of 0.43 and a ply thickness of 0.55 mm.

The testing program included measurement of the tensile, compressive and shear behaviour (moduli and strengths) of both DTU and NTUA materials. Tensile properties at 0° and 90° were measured in accordance with the ASTM D3039M standard, compressive properties at 0° and 90° in accordance with ISO 14126, and shear properties in accordance with ASTM D5379 (Iosipescu shear test). Three material characterisation plates were fabricated by each partner, one having a thickness of approximately 1 mm for the 0° tensile test specimens, one having a thickness of approx. 2 mm for the 90° tensile and 0° compressive test specimens, and one having a thickness of approx. 4 mm for the shear and 90° compressive test specimens. The DTU tests were carried out at Risø DTU (Risø National Laboratory for Sustainable Energy), while the NTUA tests were carried out at the Shipbuilding Technology Laboratory of NTUA.

3.3 Material property test results

The respective averaged results of the tests performed at DTU and NTUA are given in Table 2, where E is the Young's modulus, G is the shear modulus, v is the Poisson's ratio, X is the strength in the fibre direction, Y is the strength normal to the fibres and S is the shear strength. Subscripts 1 and 2 denote the directions parallel and normal to the fibres, respectively, while subscripts t and c denote tension and compression, respectively.

Table 2 reveals large differences between the properties measured for the DTU and NTUA materials, as well as significant discrepancies between the measurements by the two laboratories of the same property for the same material. The measured properties of the DTU material are generally higher than those measured for the NTUA composite. This fact can be mainly

attributed to the higher quality of the raw materials used by DTU (pre-pregs against conventional composites) and the more advanced manufacturing method. These two factors led generally to composite materials of much better quality, considerably higher fibre volume fraction and smaller thickness variations. An additional reason may also be some problems encountered with the curing of the NTUA epoxy resin, which were discovered after the specimen fabrication and testing.

Although the differences between the measured E_{1c} values at the two laboratories were acceptable, the discrepancies between the measurements of the same mechanical property for the same material were significant for the rest of the tests. In general, the NTUA measurements gave lower values than those of DTU for the same material, with the exception of shear modulus G_{12}. The greatest discrepancies were noticed for the compressive tests normal to the fibres, which usually present the greatest sensitivity to test conditions. A possible contributory factor could be that the quality of the bonding of the tabs to the specimens tested at NTUA was not as good as that achieved at DTU.

4 VALIDATION OF FINITE ELEMENT MODELS

4.1 Introduction

In order to establish a valid modelling approach to be implemented in the parametric studies, Newcastle University (UNEW) and the Technical University of Denmark (DTU) performed a series of finite element analyses (FEA) to compare with the panel test results. All analyses have been performed using Abaqus/CAE and involve both geometric non-linearity and non-linear material behaviour using a progressive failure model. The outcome of these analyses is described in the following paragraphs. Preliminary studies that focused especially on the modelling of the boundary conditions achieved in the test rig are reported by Berggreen et al. (2007).

4.2 Element selection

Due to the anisotropic behaviour of fibre reinforced laminates, with a fibre modulus much greater than that for the matrix, the effects from transverse shear deformation are more significant than in isotropic materials. To predict these effects accurately in the large range of plate slenderness ratios covered in the parametric study the conventional 8-node quadratic quadrilateral thick shell element with reduced integration (S8R type in Abaqus) and six degrees of freedom was chosen.

The S8R type element represents thick shells that allow for large rotations and transverse shear deformation. First order shear deformation theory is used which, together with a Simpson's integration scheme for the constitutive relationships through the shell thickness, allows for interlaminar stresses to develop (Abaqus 2007).

In the reduced formulation of the S8R element a penalty method is applied in the integration scheme for the calculation of the transverse shear strains that prevents shear locking behaviour. The shear stiffness takes the form:

$$K = kGtAf_p \qquad (1)$$

where k = correction factor based on experimental results; G = transverse shear modulus; t = section thickness; A = reference area assigned to an integration point; f_p = dimensionless factor defined as:

$$f_p = \left(1 + 0.25 \times 10^{-4} \frac{A}{t^2}\right)^{-1} \qquad (2)$$

4.3 Progressive failure

Prior to damage initiation the material behaviour is linearly elastic. Damage is initiated according to the Hashin quadratic failure criteria (Hashin & Rotem 1973) the general forms of which are:

Fibre tension: $\quad F_f^t = \left(\dfrac{\sigma_{11}}{X_t}\right)^2 \qquad (3a)$

Fibre compression: $\quad F_f^c = \left(\dfrac{\sigma_{11}}{X_c}\right)^2 \qquad (3b)$

Matrix tension: $\quad F_m^t = \left(\dfrac{\sigma_{22}}{Y_t}\right)^2 + \left(\dfrac{\tau_{12}}{S}\right)^2 \qquad (3c)$

Matrix compression: $\quad F_m^c = \left(\dfrac{\sigma_{22}}{Y_c}\right)^2 + \left(\dfrac{\tau_{12}}{S}\right)^2 \quad (3d)$

where X_t, X_c, Y_t, Y_c and S are as defined in Section 3.3; σ_{ii}, τ_{ij} = effective stress tensors; $i,j = 1, 2$ represent local co-ordinates.

At the onset of damage the material response is computed (Abaqus 2007) from:

$$\sigma = \frac{\varepsilon}{D} \begin{pmatrix} (1-d_f)E_1 & (1-d_f) & 0 \\ sym. & (1-d_m)v_{21}E_1 & 0 \\ sym. & (1-d_m)E_2 & (1-d_s) \\ & sym. & GD \end{pmatrix} \qquad (4)$$

where ε = strain; d_f, d_m = damage variables corresponding to failure in the fibres and matrix; d_s = additional damage variable for shear; v_{21} = minor Poisson ratio; $D = 1 - (1 - d_f)(1 - d_m) v_{12} v_{21}$.

The material properties are then degraded linearly up to failure, which occurs at a maximum strain of twice the strain at the initiation of failure (Matzenmiller et al. 1995).

4.4 Finite element model

Both sets of material property definitions described in Section 3.3 were employed in the FEA models to predict the collapse of the tested panels. The elasticity moduli from the tensile and compressive tests were averaged for the definition of single properties in the longitudinal and transverse directions. For the remaining material property definitions transverse isotropy has been assumed with the transverse shear modulus in the 23 plane equal to the in-plane shear modulus to ensure stability in the analysis. Due to limitations on space only results obtained with the NTUA data set are presented in this paper. These are the material data adopted in the parametric studies described in Section 5.

The initial geometric imperfections imposed in the FEA models are the same in magnitude and shape as those built into the test panels during fabrication. To nucleate buckling in the case of the panels with no initial imperfections, an imperfection with a magnitude (δ_o) of 5% of the panel's thickness has been imposed. The effective panel dimensions of 320 × 320 mm have been used in the models with simply supported and clamped boundary conditions. The area included in the model was later reduced to 300 × 300 mm and the displacements and rotations measured at the edges of this region by the DIC system were imposed as boundary conditions. This case is referred to as having "nonlinear boundary conditions" in the results. The rotations imposed at the boundaries induce out of plane displacements in the panel. Thus, for the panels without imperfections, it was unnecessary to introduce an initial imperfection to nucleate buckling with these boundary conditions.

A mesh refinement study was performed to determine the number of elements required to achieve convergence of the FEA model. From this study it was concluded that a mesh of 46 × 46 S8R elements would be used in the 320 × 320 mm models. For the case of the 300 × 300 mm models the same size of elements has been considered and the number of elements in the mesh was reduced accordingly.

The boundary conditions applied in the 320× 320 mm models are defined in Tables 3 and 4 with the axis directions defined in Figure 1. In order to define the nonlinear boundary conditions in the smaller models the nodes at the edges were divided into sets of

Table 3. Clamped boundary condition (CC).

Loaded edge		Reaction edge		Unloaded edges	
X	displacement	X	0	X	free
Y	0	Y	0	Y	0
Z	0	Z	0	Z	0
RX	0	RX	0	RX	0
RY	0	RY	0	RY	0
RZ	0	RZ	0	RZ	0

master and slave nodes. The master nodes are the ones where the displacements as measured during the tests are applied. These are located 30 mm apart. The nodes located between two master nodes are grouped into a set of slave nodes with displacements linearly interpolated from their neighbouring master nodes.

4.5 Numerical results

Eight cases were considered in total for the validation of the numerical models: six from the NTUA lay-up (thin and medium thickness panels) and two from the DTU lay-up (thick panels). These are presented in terms of average stress (MPa) vs. end shortening (mm) in Figures 7–14 and the values of average stress at failure are compared in Table 5.

Table 4. Simply supported boundary condition (SS).

Loaded edge		Reaction edge		Unloaded edges	
X	displacement	X	0	X	free
Y	0	Y	0	Y	0
Z	0	Z	0	Z	0
RX	0	RX	0	RX	free
RY	free	RY	free	RY	0
RZ	0	RZ	0	RZ	0

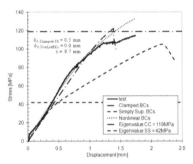

Figure 7. End shortening of Series 1 panel S1-0-2 with no initial imperfections.

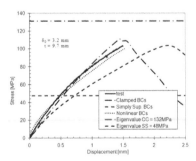

Figure 8. End shortening of Series 1 panel S1-32-2 with 3.2 mm initial imperfection.

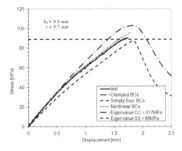

Figure 9. End shortening of Series 1 panel S1-96-1 with 9.6 mm initial imperfection.

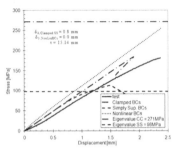

Figure 10. End shortening of Series 2 panel S2-0-2 with no initial imperfections.

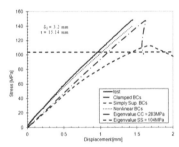

Figure 11. End shortening of Series 2 panel S2-32-2 with 3.2 mm initial imperfection.

The solutions with the simply supported and clamped boundary conditions represent limiting cases between which the tested panels are expected to perform. This was observed in most cases, the main exceptions being the cases where no initial geometric imperfections were present (see Table 5). More representative modelling of the actual boundary conditions can improve the solutions (as in Figure 14 for the case when the panel is modelled with simple supports at the loaded edges with the other edges clamped), but these need further investigation before they can be generalised for the full range of panels tested. With the displacement measurements of the boundaries applied in the FEA models, more accurate solutions were

409

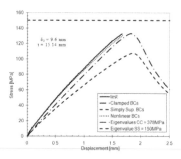

Figure 12. End shortening of Series 2 panel S2-96-2 with 9.6 mm initial imperfection.

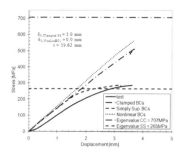

Figure 13. End shortening of Series 3 panel S3-0-2 with no initial imperfection.

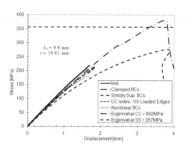

Figure 14. End shortening of Series 3 panel S3-96-1 with 9.6 mm initial imperfection.

obtained in most cases. Clear exceptions were the thick and medium thickness panels with no initial imperfections (Fig. 10, 13). For these panels the initial part of the measured end shortening response shows a stiffening behaviour that is not seen for the other panels and is probably not well represented in the modelling.

The buckling loads as predicted from the linear eigenvalue analysis for the first buckling mode are also presented. The solutions for the panels with no initial geometric imperfections demonstrate the influence of the deformations of the test rig at the boundaries. The failure load of the thin panel (Fig. 7) is very close to

Table 5. FEA correlation with panel tests when various boundary conditions are assumed (failure stresses in MPa, deviations from test results in %).

Panel	Test MPa	FEA-CC		FEA-SS		FEA-NonL	
		MPa	Dev%	MPa	Dev%	MPa	Dev%
S1-0-2	115	123	6.96	105	−8.70	133	15.65
S1-32-2	106	111	4.72	104	−1.89	100	−5.66
S1-96-1	91	103	13.19	88	−3.30	96	5.49
S2-0-2	183	184	0.55	113	−38.25	256	39.89
S2-32-2	149	148	−0.67	113	−24.16	140	−6.04
S2-96-2	133	133	0.00	107	−19.55	130	−2.26
S3-0-2	287	517	80.14	289	0.70	564	96.52
S3-96-1	218	382	75.23	277	27.06	181	−16.97

the eigenvalue solution with clamped boundary conditions. As the panel thick-ness is increased, and the slenderness is reduced, the supports exhibit significant deformations. The medium thickness panel (Fig. 10) fails between the critical loads from the simply supported and clamped solutions. In the case of the thick panel (Fig. 13), the rigidity of the supports is further reduced, relative to the plate, and the failure of the panel is very close to the simply supported critical load. This is also demonstrated by the good correlation between the test results and the nonlinear solution with simply supported boundary conditions.

The validation studies are still in progress as a number of issues require more detailed investigation.

5 PARAMETRIC STUDIES

5.1 *Definition of parameters*

The parametric study, which is still in progress, is confined to the idealised case of simply supported boundary conditions. It incorporates three lay-up configurations for two plate geometries: an aspect ratio of 1 (square plates) and an aspect ratio of 4 (long plates). The width, b, of the panels is set to 500 mm in all cases. Here attention will be confined to the square plates, for which the imposed imperfections have a half sine-wave form with amplitudes of 0.1%, 1%, 2% and 3% of the panel width.

The three lay-up configurations are presented in Table 6 In the triaxial lay-up (case A) the thicknesses for each layer are scaled up to achieve the desired plate thickness and slenderness. In the other two lay-up configurations (quadriaxial and woven roving) increased thickness is achieved by increasing the number of plies.

The material property definitions for the first two cases are the same as the NTUA data set for the DTU material (see Table 2, average of tensile and compressive properties). For the woven roving material (WR),

Table 6. Definition of lay-up configurations.

Case A:	Triaxial Lay-up	$[-45/+45/0_4/+45/-45/0_4/$
		$-45/+45/0_3]_S$
Case B:	Quadriaxial Lay-up	$[0/+45/90/-45]_{X,S}$
Case C:	Woven Roving	$[0]_X$

Table 7. Assumed WR material properties (MPa).

E_1	17180	ν_{12}	0.17	G_{12}	3520
E_2	17180	ν_{13}	0.27	G_{13}	5150
E_3	10800	ν_{23}	0.27	G_{23}	5150
X_t	238.6	Y_t	238.6	S_{12}	80.9
X_c	324.5	Y_c	324.5	S_{23}	60.7

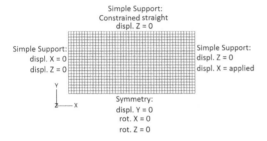

Figure 15. Applied boundary conditions for parametric studies.

properties have been adapted from an earlier study referred to by Hayman & Echtermeyer (1999). These are provided in Table 7

For the purposes of the parametric study, the FEA model had some of its features altered. The mesh consists of the same size of elements as in the validation study which increases significantly the number of elements and the analysis time. By examining the first buckling mode shapes for all configurations it was concluded that half symmetry would be applied about an axis parallel to the panel's length (x-direction; see Figure 15).

To simulate the effect of surrounding structure (stiffeners and adjacent panels) the unloaded edge was constrained to remain straight but free to displace in the y-direction.

5.2 Parametric study results

The results presented from the parametric study are confined to square panels ($a/b = 1$). An initial investigation of the post-buckling strength of the long plates ($a/b = 4$) showed that applying a half sine-wave imperfection can increase significantly the ultimate strength of the panel and also lead to a complex behaviour as the plate deformation changes to the

preferred buckling mode with roughly square buckles. Further investigations will take place before the completion of this study.

In Figures 16–18 the ultimate strength (MPa) for a range of b/t values between 10 and 50 is presented. In these results it is observed that in the triaxial lay-up initial imperfections have very little effect on the panel strength for b/t values greater than 30. In the

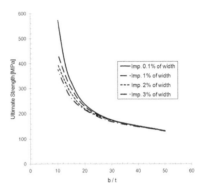

Figure 16. Case A: Triaxial lay-up; aspect ratio $= 1$.

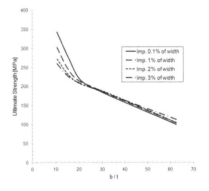

Figure 17. Case B: Quadriaxial lay-up; aspect ratio $= 1$.

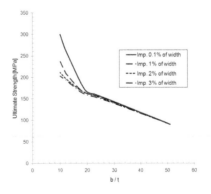

Figure 18. Case C: WR lay-up; aspect ratio $= 1$.

quadriaxial and WR lay-up configurations this effect is further extended down to b/t values of 25 or less.

Note that the triaxial lay-up considered here has about 87% of its reinforcement in the loading direction. This contrasts with the quadriaxial lay-up, which is quasi-isotropic, and the WR lay-up, which is balanced, having equal amounts of reinforcement in the $0°$ and $90°$ directions.

6 CONCLUSIONS

A comprehensive, collaborative study of the buckling and postbuckling behaviour of composite plates under compressive loading, with and without built-in geometric imperfections, has been carried out. This has included instrumented tests that provide a substantial database for validating modelling approaches. Corresponding material properties for use in modelling work have also been established.

In laboratory testing of plates subjected to buckling phenomena, it is extremely difficult or impossible to obtain clearly defined boundary conditions. However, using a DIC measurement system it is possible to monitor the actual displacements and rotations at the plate boundaries during a test. These can be imposed in a FEA in place of idealised boundary conditions for the purposes of validating the modelling approach. Once validated, the modelling approach can be used to predict the behaviour of plates with other, idealised boundary conditions. It is then possible to perform parametric studies for plates with given boundary conditions.

An advanced modelling approach, based on FEA, has been described. With a few exceptions, this modelling approach appears to be able to reproduce the main aspects of the behaviour of the panels tested with in-plane compression, provided the boundary conditions are modelled correctly. This includes estimates of the ultimate collapse loads. There is some uncertainty about the boundary conditions in some of the tests and some issues require further investigation. However, sufficient confidence in the modelling approach was established to permit the conduct of a limited parametric study.

Parametric studies performed on simply supported, square plates with three different composite lay-ups indicated very little sensitivity to geometric imperfections in the shape of the first buckling mode provided the b/t ratio is greater than a value in the range 20 to 25, depending on the lay-up. For thicker plates, i.e. with lower b/t, however, the failure load is more appreciably reduced by the presence of such imperfections.

Only a brief outline of the work has been presented here. More detailed descriptions of the experimental and modelling studies will be presented in future publications.

ACKNOWLEDGEMENTS

This work has been performed within the context of the Network of Excellence on Marine Structures (MARSTRUCT) partially funded by the European Union through the Growth Programme under contract TNE3-CT-2003-506141. The provision of test specimens by Vestas Wind Systems A/S is highly appreciated.

REFERENCES

Abaqus v6.7 Documentation 2007.

Berggreen, C., Jensen, C. & Hayman, B. 2007. Buckling Strength of Square Composite Plates with Geometrical Imperfections—Preliminary Results. MARSTRUCT International Conference on *Advancements in Marine Structures*, Glasgow, UK, March 12–14.

Hashin, Z. & Rotem, A. 1973. A Fatigue Failure Criterion for Fiber Reinforced Materials. *Journal of Composite Materials*, v 7, pp. 448–464.

Hayman, B. & Echtermeyer, A.T. 1999. European Research on Composites in High Speed Vessels. Fifth International Conference on Fast Sea Transportation (FAST' 99), Seattle, USA.

Matzenmiller, A., Lubliner, J. & Taylor, R.L. 1995. A Constitutive Model for Anisotropic Damage in Fiber-Composites. *Mechanics of Materials*. V 20, pp. 1011–1022.

Analysis and Design of Marine Structures – Guedes Soares & Das (eds)
© *2009 Taylor & Francis Group, London, ISBN 978-0-415-54934-9*

Buckling strength parametric study of composite laminated plates with delaminations

N.G. Tsouvalis & G.S. Garganidis
National Technical University of Athens, School of Naval Architecture and Marine Engineering,
Shipbuilding Technology Laboratory, Athens, Greece

ABSTRACT: The purpose of this work is to investigate the effect of delaminations on the buckling behaviour of a marine composite hull. This is done by using the Finite Element Method to model delaminations and calculate the buckling strength of a typical marine composite panel. The parametric study is based on a marine panel clamped along all its edges and loaded in compression. The delamination was assumed to have an elliptic shape and the parameters investigated in the analysis were its shape, magnitude and location. The total number of cases investigated is 45. The eigenvalue buckling analyses led in many cases in inadmissible buckling modeshapes. A procedure for eliminating these inadmissible modeshapes in the nonlinear analyses is described in the paper. The final results indicated that the greatest effect comes from delaminations which are closer to the laminate surface, are closer to the circular shape and have the largest magnitude.

1 INTRODUCTION

Delaminations are among the most frequently encountered defects in a composite marine structure. They may either be created during the manufacturing procedure of the hull or at a later stage, during the operational life of the vessel, due for example to impact loading. The presence of these defects definitely implies unfavourable effects in bending and especially in compressive loading.

The major cause for the development of delamination away from the laminate edges where interlaminar stresses are negligible, are the various matrix structural defects. The most common defect are the air voids in-between the layers, created during manufacturing. These defects cause the initiation of the laminate failure, since application of fatigue or impact loading leads to an increase of their magnitude, to their unification in larger defects and, finally, to local delamination. In marine composite laminates like those at the vessel's bottom, deck or bulkheads, fatigue loading from the waves and impact loading from slamming, collisions or groundings, may cause delaminations.

Delaminations take place at a part of the interface between two consecutive layers, forming two separate laminate parts (sublaminates) on either side of the interface. They may have various shapes and magnitudes and be in various locations along the laminate length, width and thickness. Moreover, more than one delamination may exist in the same laminate. In any case, the two or more sublaminates consist of a smaller number of layers than the initial laminate and can deform freely, being in contact between each other or not.

In-plane compressive loading of composite marine laminates is a very often loading condition that may apply to delaminated laminates. This loading may result in partial buckling of the sublaminates for quite low loads, due to the small number of layers of the sublaminates and their consequent lower stiffness. This partial buckling may be local when one sublaminate buckles whereas the other remains plane (Fig. 1a), be global when both sublaminates buckle together towards the same side (Fig. 1b) or be of a mixed mode when each sublaminate buckles independently towards opposite sides (Fig. 1c, Short et al. 2001).

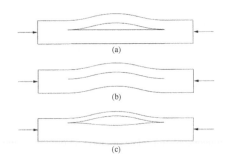

Figure 1. Buckling of a delaminated laminate: (a) local mode, (b) global mode and (c) mixed mode.

Buckling of intact composite laminated plates has been the subject of extensive literature in the past (see for example Papazoglou et al. 1992, Tsouvalis & Papazoglou 2004). Several investigators have also studied in recent years the effect of delaminations on the buckling behaviour of composite laminated plates. Kim & Kedward (1999) developed an analytical model of a laminated plate with an elliptical delamination, resulting in the prediction of the buckling mode (global or local) and in the calculation of the respective critical loads. A simple analytical methodology for the calculation of the critical load for both local and global buckling has been also developed by Shan & Pelegri (2003) for long clamped laminates under uniaxial compression.

The finite element (FE) method has been also used successively in order to study buckling of delaminated composite plates. Hwang & Liu (2001) developed a 2-D FE model of a plate with multiple delaminations of different magnitudes, concluding that the longest delamination affects most the critical buckling load, especially when this delamination is close to one of the laminate's surfaces. 3-D FE modeling makes possible the study of laminates with delaminations of various shapes and magnitudes. Kyoung et al. (1999) developed such a model to study the effect of multiple strip or circular delaminations on the buckling response of a cross-ply laminate. A 3-D FE model of a composite plate with a delamination at its mid-thickness was developed by Zor (2003) and Pekbey (2006), for the calculation of the buckling response of a simply supported and a clamped laminate, respectively. A common conclusion was that the reduction of the laminate's aspect ratio or the increase of the delamination length, result in a reduction of the buckling strength. A similar study for strip and square delaminations was performed by Lee & Park (2007), where the 3-D solid elements of the two sublaminates are connected in the areas outside the delamination. This work concluded that an accepted delamination size is that corresponding to a delamination length over laminate length ratio equal to 0.2. A corresponding recent work is that by Chirica et al. (2008), where the effect of an elliptical delamination on the buckling response of a clamped laminated plate is studied. The FE model developed is similar to that of Kyoung et al. (1999), where sublaminates are modeled using plane elements.

The scope of the present study is to investigate how the presence of delaminations affects the buckling response of a marine composite laminated plate. This goal is achieved by studying the effect of some specific parameters of the delamination on both the critical buckling load of the laminated plate and the associated modeshape (local, global or mixed mode). The parameters investigated are the shape, orientation, magnitude and through thickness location of the delamination.

2 FE MODEL DESCRIPTION AND VALIDATION

2.1 Geometry and type of elements

The present parametric study has been carried out for a typical marine composite laminate, having dimensions 1000×1000 mm and being 10 mm thick (Fig. 2). The laminate has a centrally located elliptic delamination, with axes a and b in x-and y-directions, respectively, and is loaded by a uniform compressive displacement u_x. Half of the laminate was modeled (shaded area in Figure 2), due to possible antisymmetric buckling modeshapes with respect to the y' axis of symmetry. Symmetry with respect to x' axis is assured for the cases examined.

The present parametric study was carried out using the FE code ANSYS 11.0. The FE models were developed using 3-D solid elements, since they provide more accurate results than those coming from corresponding models which incorporate 2-D elements in the x-y plane, although the latter are less computer resources demanding than the former. The model consists of two layers of solid elements in the x-y plane, one for each sublaminate on either side of the delamination. These two layers of elements are connected between each other in the intact area outside the delamination, whereas they are independent and deform separately from each other inside the delamination area (Short et al. 2001). Figure 3 provides a characteristic view of this modeling concept, showing a buckled delamination. The elements used are 8-node 3-D solids (called SOLID45 in ANSYS), they have 3 degrees of freedom per node (the three translations) and they support large displacements non-linear analyses.

In order to generate the FE mesh, a proper 2-D plane mesh is first generated in the x-y plane, which

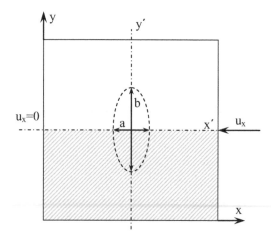

Figure 2. Geometry of the parametric study laminate.

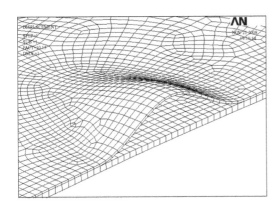

Figure 3. View of a buckled sublaminate (half laminate shown).

Figure 4. View of a typical mesh of the half-laminate.

in the sequence is extruded in the thickness direction. Thus the final 3-D solid mesh is obtained and at the same time, geometrical compatibility between the nodes of the elements of the two sublaminates is assured. The generated mesh is dense enough near the delamination's elliptical boundary where maximum stresses should normally develop and could be coarser in the rest of the areas, in order to combine adequate accuracy with acceptable solution times. However, in order to obtain 3-D solid elements with an acceptable maximum value of aspect ratio which was set equal to 20, and due to the fact that the laminate and its sublaminates were quite thin, the element size was approximately equal all over the laminate area. The mesh is dense enough for the purposes of the present study which is the prediction of the global buckling response of the plate. In case we were interested in accurate stress calculations in order, for example, to apply a failure criterion, a finer mesh would probably be necessary at the boundaries of the delamination. With regard to the number of elements in the thickness (z) direction, a comparative study indicated very small differences when using 1 or 4 elements for each one of the two sublaminates, thus the option of 1 element in the thickness direction of each sublaminate was finally adopted. Figure 4 presents a typical FE mesh of the half-laminate generated for the parametric study and consisting of approximately 7600 elements.

In the case of modeling an intact reference laminate without delamination, the same procedure was followed, the only difference being that the two "sublaminates" were connected between each other over the whole laminate area.

The typical marine composite panel taken into account for the performance of the parametric study was considered clamped along all its edges. Thus all edges had their z-translation fixed, whereas edges parallel to x-axis had also their y-translation fixed, modeling in this way the existence of stiff ship hull

reinforcing stiffeners. Symmetry conditions were applied along the laminate's x' axis of symmetry (see Fig. 2). Compressive loading was applied in x-direction on all nodes of one of its y-sides in the form of uniform displacement, thus forcing the corresponding side to remain straight and normal to the load throughout the whole loading procedure. The opposite y-side was restrained against x-displacement.

As it will be shown in section 3, the parametric study is based on a typical marine composite material consisting of many similar woven roving or woven fabric layers, thus exhibiting homogeneous orthotropic material properties. Therefore, the FE model incorporates the linear homogeneous orthotropic material model of ANSYS. In addition to this, the layered material model was also used during the verification procedure of the model, in order to compare results with other corresponding FE models found in the literature.

2.2 Type of analysis and updated FE model

The FE models developed did not account for any type of material failure, thus the delamination area remains constant throughout the analysis and cannot grow.

The first analysis type performed was the linear eigenvalue analysis, leading to the determination of the critical buckling load and the associated modeshape. However, in many cases the eigenvalue analysis resulted in inadmissible modeshapes, where the sublaminates in the area of delamination were interpenetrating between each other, as Figure 5 characteristically shows for two cases. This is happening because in the linear eigenvalue analysis there is not any restraint against these penetrations.

Such restraints can be applied with the aid of special contact pair elements, which can be incorporated only in association with a nonlinear, large displacement analysis. Thus, an updated version of the FE model was developed, which includes 4-node contact and target elements (called CONTA173 and TARG170 in ANSYS), generated on the internal surfaces of the two sublaminates, inside the delamination area. These elements match the corresponding internal surfaces

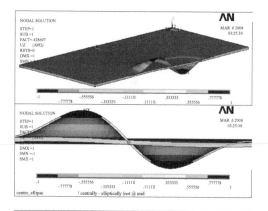

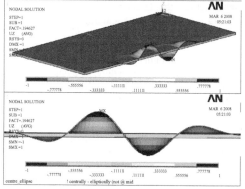

Figure 5. Two characteristic inadmissible modeshapes with interpenetrating sublaminates (half laminate shown).

of the 3-D solid elements of the sublaminates and do not allow penetration, forcing the two sublaminates to deform together when they come into contact.

When performing a non-linear analysis, an initial geometric imperfection is needed in order to trigger the buckling response. As it is normally done in such cases, a scaled form of the first modeshape has been taken into account, exhibiting a very small maximum out-of-plane initial imperfection equal to 1/100 of the laminate thickness. In the case where the first modeshape had an inadmissible geometric form, its geometry is altered by the presence of the contact pair elements which do not allow interpenetration of the sublaminates.

An important disadvantage of using a non-linear analysis for the determination of the buckling response is that it does not result directly in a specific critical buckling load, but it gives the general buckling and post-buckling response of the structure. In cases like this, one must incorporate special post-processing techniques in order to arrive to a specific buckling load value. The Southwell plot technique applied on the out-of-plane displacements is such a method and it has currently being followed in this parametric study

too; the determination of the point where the slope of the imposed axial displacement versus reaction forces curve changes is another method, also being followed in this study. Both the above methods require considerable post-processing efforts, their results being also questionable in some cases, like for example when large initial geometric imperfections exist.

A way to overcome all the above difficulties is to perform a combination of a non-linear and an eigenvalue analysis, that is to perform some steps of the non-linear analysis up to a certain fraction of the expected buckling load and then, based on the deformed geometry and the updated stiffness matrix of the model, to perform an eigenvalue analysis which directly gives the critical buckling load. This procedure has been also applied on the current parametric analysis and its results, compared to those of the two other aforementioned methods, will be presented in section 3. The fraction of the expected buckling load up to which the non-linear analysis is performed was set approximately equal to 10% after a corresponding convergence study.

2.3 Validation of the FE model

The validity of the developed FE modeling procedure was examined by comparing its results to others available in the literature. The comparison was done with the results of Zor (2003), which refer to a square simply supported laminated plate having a central strip delamination and loaded in compression. The delamination is considered to be in mid-thickness and the author has examined several different symmetric and unsymmetric lay-ups and several delamination magnitudes. FE code ANSYS was also used in this study, incorporating 8-node 3-D solid elements and eigenvalue analyses.

Table 1 presents indicative results of this comparison, referring to the case where delamination length is equal to half the laminate length. Within the framework of the present study, the buckling loads were calculated using either an eigenvalue analysis (EIG) or a non-linear analysis (N-L) or a combination of the two (N-L & EIG), as described in the previous section. Buckling load values were derived from the non-linear analyses results with the aid of the Southwell plot technique. It is evident from Table 1 that

Table 1. Comparison of critical buckling loads (in N/mm).

Lay-up	Zor (2003)	Current		
		EIG	N-L	N-L & EIG
$[0°]_4$	7.99	7.99	8.07	8.00
$[\pm30°]_2$	8.96	8.97	9.04	9.00
$[30°/-30°]_S$	9.13	9.13	9.17	9.13

the differences between the currently calculated buckling loads and those by Zor are negligible, for all types of analysis. The same is also happening for the rest of the results available in this reference, for other delamination length over laminate length ratios.

The three types of analysis result in almost identical buckling loads, regardless of whether the mode-shape of the eigenvalue analysis is admissible or not. Of course this conclusion cannot be generalized and further investigation is needed for other delamination geometries and laminate lay-ups.

Quite good comparison was also attained between the results of the present modeling and those presented by Short at al. (2001) for a $[0°/±45°/0°]_s$ square laminate with a central square delamination under compressive loading. The delamination was considered to be placed in three different through the thickness positions. Short et al. used the FE code ABAQUS in association with 20-node layered 3-D solid elements and performed only non-linear analyses. The present FE model was adapted accordingly, to incorporate the same type of elements (called SOLID186 in ANSYS). Compatibility with the contact pair elements was obtained by using the corresponding 8-node ANSYS version of these elements (called CONTA174). The comparison between the maximum out-of-plane displacements versus applied loading was quite good between the two models, especially for the case where the delamination was near the laminate surface.

3 PARAMETRIC STUDY

3.1 Definition of parameters

The effect of three delamination parameters was investigated in the present study, namely the orientation of the delamination with respect to the applied loading direction, its magnitude and its location in the thickness direction, z, of the laminate. The basic shape of the delamination was selected to be the elliptical one with axis a of the ellipsis parallel to loading and axis b normal to it (see Fig. 2). This shape is more realistic than other orthogonal delaminations taken into account in the literature.

Thus, five cases were examined with respect to the shape and orientation of the delamination, expressed by five different values of the ellipsis axes a/b ratio, equal to 0.25, 0.5, 1.0 (circle), 2.0 and 4.0. Figure 6 shows schematically the five different shape cases examined, ranging from long delaminations normal to the loading direction to long delaminations parallel to it.

Three cases were studied with respect to the delamination magnitude, expressed by different values of the ratio of the delamination area over the area of the whole laminate, A_d/A_p. The three values assigned to this ratio are 0.04 (small delamination), 0.08 (medium) and 0.12

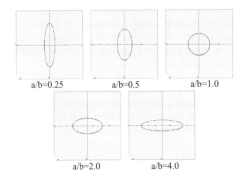

Figure 6. The different delamination shapes and orientations considered in the parametric study.

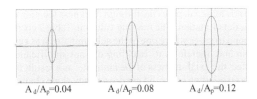

Figure 7. The different delamination magnitudes considered in the parametric study.

(large). Figure 7 gives a schematic view of these 3 magnitudes, for a/b = 0.25. Finally, if t is the laminate thickness, the delamination was considered to be located in three places through the thickness, namely at z = t/2 (mid-thickness), z = t/4 and z = t/8, with the top sublaminate being the thinnest one in the last two cases. Hence, the total number of cases investigated in this parametric study is 45. It must be noted at this point that, since the laminate thickness is constant, parameter A_d/A_p is directly connected to another parameter, a/t, for the same a/b ratio. As A_d/A_p increases, parameter a/t increases too. Therefore, parameter a/t has not been taken explicitly into account in the present parametric study.

The laminate studied is typical for marine applications, considered to be made of similar balanced fabric layers in 0° direction, thus resulting in a homogeneous orthotropic material. The material properties taken into account are typical for carbon/epoxy biaxial fabrics and are $E_x = E_y = 53100$ MPa, $E_z = 8000$ MPa, $G_{xy} = 3790$ MPa, $G_{xz} = G_{yz} = 3700$ MPa, $v_{xy} = 0.055$ and $v_{xz} = v_{yz} = 0.074$. The laminate was considered to be clamped, with the exact boundary conditions described in section 2.1.

3.2 Results and discussion

Two analyses were carried out for each case of the parametric study, namely a non-linear one and the combination of a non-linear and an eigenvalue analysis

(N-L & EIG). The critical buckling load was directly calculated from the latter, whereas three distinct post-processing procedures were applied to extract the critical buckling load from the former (see for example Parlapalli et al. 2007).

According to the first procedure, the critical buckling load was calculated from the graph of the compressive axial displacement u_x versus the applied load. Typical such results for the case of a mid-thickness large delamination ($z = t/2$ and $A_d/A_p = 0.12$) can be seen in Figure 8 for three out of the five delamination aspect ratios for reasons of simplicity. As Figure 9 in the sequence shows, the critical buckling load corresponds to the intersection of two straight lines, which are drawn tangent to the first and the last linear parts of the corresponding curve. The so extracted buckling loads are presented in Table 2 for all cases examined. In this table z denotes the through thickness location of the delamination. The missing values in Table 2 correspond to cases where one of the sublaminates was very thin and, therefore, its buckling did not significantly affect the axial stiffness of the laminate; as a result, the corresponding curve did not present any slope change.

According to the second procedure, the critical buckling load was calculated by applying the well known Southwell plot technique to the out-of-plane displacements of the bottom sublaminate. A typical variation of such displacements versus the applied load can be seen in Figure 10 for the same case as that of Figure 8 ($z = t/2$ and $A_d/A_p = 0.12$). According to the Southwell plot technique, the critical buckling load is obtained by plotting the out-of-plane displacement u_z against the out-of-plane displacement divided by the applied load (u_z/P). The slope of this line in its initial linear part gives the buckling load, as it can be characteristically seen in Figure 11 for the same case as that of Figure 9. The so extracted buckling loads are presented in Table 3 for all cases examined. The missing value in Table 3 is justified by the fact that, for this specific case, the bottom sublaminate does not buckle at all; buckling in this case is dominated by the very thin top sublaminate.

The third post-processing procedure is exactly the same to the second one, though applied to the out-of-plane displacements of the top sublaminate. The corresponding critical buckling loads are presented in Table 4, with no values missing. Comparing the

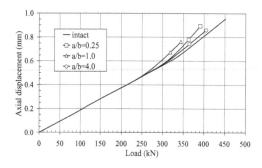

Figure 8. Axial compressive displacement versus applied load for the case $z = t/2$, $A_d/A_p = 0.12$.

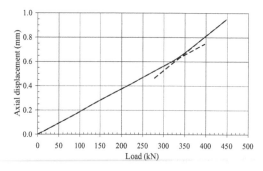

Figure 9. Determination of the critical buckling load from the axial compressive displacement versus applied load curve (case $z = t/4$, $A_d/A_p = 0.04$, $a/b = 4.0$).

Table 2. Critical buckling loads extracted from the u_x-P non-linear response (in kN).

		P_{cr}				
z	A_d/A_p	$a/b = 0.25$	$a/b = 0.5$	$a/b = 1.0$	$a/b = 2.0$	$a/b = 4.0$
t/2	0.04	337.5	337.7	337.2	336.3	337.3
	0.08	334.9	320.0	315.8	326.1	326.1
	0.12	315.7	269.1	268.0	300.9	307.9
t/4	0.04	337.6	337.2	337.9	336.8	336.3
	0.08	334.7	320.6	326.0	330.3	327.4
	0.12	328.2	279.7	301.2	322.3	315.8
t/8	0.04	327.0	318.2	322.5	326.6	337.4
	0.08	301.0	291.0	306.1	–	326.1
	0.12	277.9	277.7	289.8	–	–

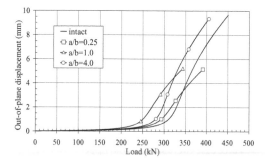

Figure 10. Out-of-plane lateral displacement of bottom sublaminate versus applied load for the case $z = t/2$, $A_d/A_p = 0.12$.

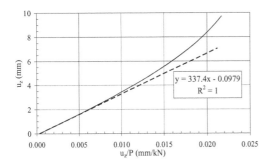

Figure 11. Determination of the critical buckling load with the Southwell plot technique (case $z = t/4$, $A_d/A_p = 0.04$, $a/b = 4.0$, bottom sublaminate).

Table 3. Critical buckling loads extracted using the Southwell plot technique at the bottom sublaminate (in kN).

		P_{cr}			
z	A_d/A_p	a/b = 0.25	a/b = 0.5	a/b = 1.0	a/b = 2.0	a/b = 4.0
t/2	0.04	339.2	339.1	338.4	337.5	336.6
	0.08	334.5	324.4	317.8	325.4	323.6
	0.12	318.3	273.0	271.5	300.3	305.7
t/4	0.04	338.9	338.1	338.3	338.0	337.4
	0.08	333.7	281.9	278.2	333.4	329.9
	0.12	328.6	208.9	195.0	325.3	315.3
t/8	0.04	298.0	246.6	255.5	330.6	339.1
	0.08	214.7	162.1	208.2	259.7	327.9
	0.12	168.4	136.3	175.0	205.3	–

Table 4. Critical buckling loads extracted using the Southwell plot technique at the top sublaminate (in kN).

		P_{cr}			
z	A_d/A_p	a/b = 0.25	a/b = 0.5	a/b = 1.0	a/b = 2.0	a/b = 4.0
t/2	0.04	339.4	339.3	338.9	337.9	337.0
	0.08	334.8	324.9	318.3	325.8	324.1
	0.12	318.5	273.3	271.7	300.6	306.1
t/4	0.04	337.3	339.9	339.9	338.5	337.9
	0.08	335.7	253.6	227.5	337.1	328.5
	0.12	295.9	173.8	154.2	295.0	316.5
t/8	0.04	260.7	142.2	130.0	293.9	340.6
	0.08	125.7	67.4	62.1	145.4	282.9
	0.12	86.2	46.1	41.3	96.3	182.0

buckling loads of Table 4 to those of Table 3, the following two special characteristics can be spotted. For a specific range of geometries, namely when delamination is at mid-thickness or when delamination is at $z = t/4$ but its magnitude is small and parallel

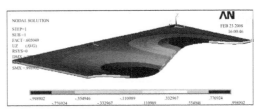

Figure 12. Buckled laminate where both sublaminates buckle together (half laminate shown).

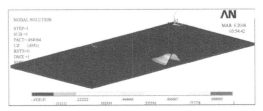

Figure 13. Buckled laminate where top sublaminate buckle before the bottom one (half laminate shown).

to the loading, the buckling loads determined from the bottom and the top sublaminate are equal, denoting that the two sublaminates buckle together in a global pattern. Figure 12 presents an example of such a global buckling shape for the case $z = t/2$, $A_d/A_p = 0.08$ and $a/b = 0.5$. For all the other cases, that is when the delamination is at $z = t/4$ and its magnitude is large or normal to the loading, or it is at $z = t/8$, the buckling loads determined from the top thin sublaminate are lower than those determined from the bottom one, denoting that the top sublaminate buckles locally before the bottom one, as expected. Figure 13 presents a corresponding local buckling shape for the case $z = t/8$, $A_d/A_p = 0.04$ and $a/b = 0.25$. Therefore, it can be concluded that, when using the Southwell plot technique, results from the thinnest sublaminate must always be processed and, therefore, the values of Table 3 from the bottom thick sublaminate are not any more taken into account.

The performance of the combination of the nonlinear and the eigenvalue analyses resulted directly in the calculation of the critical buckling loads, which are shown in Table 5. Comparing Tables 5 and 2 it is evident that the buckling loads which are calculated by these two methods are almost identical for all cases of the mid-thickness delaminations ($z = t/2$) and for some of the cases where the delamination is at $z = t/4$; the latter concern both all small delaminations, as well as the larger ones which are parallel to the load direction. All these cases correspond to a buckling shape where both sublaminates deform together globally. In all other cases (delaminations normal to

Table 5. Critical buckling loads calculated from the N-L & EIG procedure (in kN).

z	A_d/A_p	P_{cr} a/b = 0.25	a/b = 0.5	a/b = 1.0	a/b = 2.0	a/b = 4.0
t/2	0.04	338.4	338.6	338.1	337.2	336.3
	0.08	333.8	323.8	317.2	325.1	323.4
	0.12	317.4	272.4	270.7	300.0	305.4
t/4	0.04	338.4	337.8	336.1	337.5	337.2
	0.08	333.1	248.7	222.8	328.5	328.1
	0.12	290.7	168.1	148.9	311.4	314.5
t/8	0.04	247.1	135.0	124.4	295.4	338.0
	0.08	119.9	66.1	60.8	144.2	296.5
	0.12	81.3	44.0	39.6	97.0	197.1

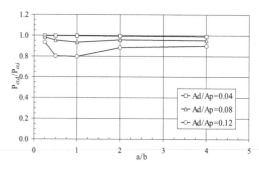

Figure 14. Reduction of buckling load versus aspect ratio of delamination, for delaminations at z = t/2.

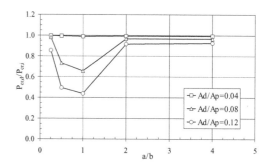

Figure 15. Reduction of buckling load versus aspect ratio of delamination, for delaminations at z = t/4.

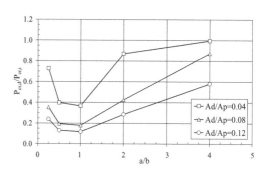

Figure 16. Reduction of buckling load versus aspect ratio of delamination, for delaminations at z = t/8.

the load or close to the surface), the thin (top) sublaminate buckles locally before the thick one without significantly affecting the axial stiffness of the whole laminate. Thus the results of Table 2 are not correct for these cases, since they correspond to the buckling of the thick sublaminate, which comes after. As a conclusion, the method of calculating the critical buckling load from the change of slope of the u_x-P curve applies with accuracy only when both sublaminates buckle together in a global buckling pattern and, therefore, it will be no further considered.

Comparing Tables 4 and 5 it can be concluded that their values are quite similar, especially for the cases where the delamination is either at z = t/2 or at z = t/4. Some greater differences occur for the case z = t/8 (reaching a maximum value of approximately 7%), but these can be justified by the small inaccuracies which are inevitable during the extraction of the critical buckling load from the out-of-plane displacement using the Southwell plot technique. Therefore, since the latter method demands large post-processing efforts in order to result in the critical buckling load, the procedure which combines a non-linear analysis with a subsequent eigenvalue analysis is finally selected for the evaluation of the parametric study results (see Table 5).

Figures 14–16 present the critical buckling load of the parametric study laminate as a function of the shape of the delamination, expressed by ratio a/b of its two axes (see Fig. 6). The critical buckling load is expressed by ratio $P_{cr,d}/P_{cr,i}$, where $P_{cr,d}$ is the critical buckling load of the laminate with the delamination and $P_{cr,i}$ is the critical buckling load of the corresponding intact laminate. This latter magnitude has been calculated to be equal to 339.5 kN by the combined N-L & EIG method. A very small imperfection similar to the first buckling modeshape was used in the non-linear part of the analysis to trigger buckling. Each figure corresponds to a specific location of the delamination through the laminate thickness and contains three curves, one for each delamination magnitude (expressed through ratio A_d/A_p, see Fig. 7).

There are some general and basic conclusions that can be drawn out from the comparative study of Figures 14–16, which are both qualitative and quantitative and provide trends of behaviour. As expected, these curves indicate that the effect of delamination becomes more important, both as the delamination magnitude increases (ratio A_d/A_p increases), and as

420

the delamination is moving closer to the laminate surface (moving from $z = t/2$ to $z = t/8$). Within the framework of the case scenarios and the parameters examined here, this effect ranges from a negligible level to an 85% decrease of the laminate's buckling load with respect to the undamaged case.

Another very important general conclusion that comes out from these curves concerns the effect of the shape and the orientation of the delamination. Thus it is first concluded that circular delaminations are the worst, whereas the longer an elliptical delamination is, the less it affects the buckling behaviour of the laminate. Moreover, and compared to the undamaged case, elliptical delaminations with their long axis normal to the loading direction decrease much more the buckling load than equivalent delaminations which are oriented parallel to the load.

An additional conclusion coming out from these curves is that the effect of small delaminations with area less than 4% of the laminate area, is negligible, if the delamination is located either at $z/2$ or at $z = t/4$ (or wherever in-between). This is happening regardless of the shape and orientation of the delamination (that is for all a/b ratios). This conclusion is in accordance with similar remarks made by Lee & Park (2007). Moreover, a delamination having an area equal to 12% of the whole laminate area has always a detrimental effect on the buckling behaviour of the laminate, effect which ranges from being just significant if the delamination is placed at $z = t/2$ (20% reduction) to being catastrophic if it is close to the surface (85% reduction).

Figure 14 for delaminations at mid-thickness indicates that the worst effect a large circular delamination can cause is a 20% reduction of the buckling load. This maximum reduction gets larger as we move towards the surface of the laminate, becoming approximately equal to 55% for delaminations at $z = t/4$ and approximately equal to 85% for delaminations at $z = t/8$, as Figures 15 and 16 indicate.

At this point, emphasis should be given to the following remark. The buckling loads shown in Table 5 and graphically represented in Figures 14–16 correspond to either a global buckling of the two sublaminates together (see Fig. 12) or to a local buckling of the thinnest sublaminate (see Fig. 13), whichever is less for each case. Thus the above annotation of the results of the present study has been made given the assumption that any buckling mode (global or local) constitutes a failure mode that should be avoided. However, as it was already shown when comparing Tables 2 and 5, there are cases where buckling of the thin sublaminate, even if it leads to its own total failure, does not significantly affect the axial stiffness of the whole laminate, which can still withstand some additional compressive load. A complete study of the whole phenomenon by the FE method can be made only by performing a post-buckling analysis of the delaminate plate, incorporating also a progressive failure model that takes into account any possible delamination growth.

4 CONCLUSIONS

The present study investigated the buckling behaviour of a delaminated typical composite marine laminate under in-plane compressive loading. The impotence of the linear eigenvalue analysis to provide accurate results for this problem was first revealed, due to the inadmissible modeshapes produced in several cases. Various techniques were applied and compared in the sequence, for the extraction of a single critical buckling load value from the results of a non-linear analysis. It was found that the combination of an initial non-linear analysis, which clears and corrects any possible modeshape inadequacies, followed by an eigenvalue analysis, result in accurate estimations of the buckling load, being at the same time the less computer resources and post-processing efforts demanding.

The parametric study performed investigated the shape, orientation, magnitude and through thickness location of the delamination. It initially verified the expected conclusions that large delaminations and delaminations close to the laminate surface have the worst effect on the buckling load. Regarding shape and orientation, it was concluded that circular delaminations are the worst, closely followed by slightly long elliptic delaminations normal to the load direction. Long delaminations parallel to the load direction have the smallest effect.

The parametric study provided also some quantitative results for the typical marine composite laminate examined. According to them, a small delamination having area equal to 4% of the total laminate area has practically no effect if it is not very close to the surface, irrespectively of its shape and orientation. However, if it's close to the surface, it causes a 20% reduction of the critical buckling load. In the case of a relatively large delamination having area equal to 12% of the total laminate area, the effects are quite worse. Such a circular delamination at the mid-thickness causes a more than 60% reduction of the buckling load, whereas, when it moves close to the surface, this reduction reaches the value of 85%.

The above remarks and conclusions, together with others that should follow in order to obtain a global view of the problem, provide valuable information towards understanding better and identifying trends with respect to the buckling behaviour of delaminated composite plates. The quantitative results can be used as data to provide guidance recommendations about whether a specific delamination found in a marine composite hull should be repaired and in what extent.

ACKNOWLEDGMENTS

This work has been performed within the context of the Network of Excellence on Marine Structures (MARSTRUCT) partially funded by the European Union through the Growth Programme under contract TNE3-CT-2003-506141.

REFERENCES

Chirica, I., Beznea, E.F., Chirica, R., Boazu, D. & Chirica, A. 2008. Buckling behavior of the delaminated ship hull panels. In Guedes Soares & Kolev (eds), *Maritime Industry, Ocean Engineering and Coastal Resources*: 161–166. London: Taylor and Francis.

Hwang, S.T. & Liu, G.H. 2001. Buckling behaviour of composite laminates with multiple delaminations under uniaxial compression. *Composite Structures* 53(2): 235–243.

Kim, H. & Kedward, K.T. 1999. A method for modelling the local and global buckling of delaminated composite plates. *Composite Structures* 44(1): 43–53.

Kyoung, W.M., Kim, C.G. & Hong, C.S. 1999. Buckling and postbuckling behavior of composite cross-ply laminates with multiple delaminations. *Composite Structures* 43(4): 257–274.

Lee, S.Y. & Park, D.Y. 2007. Buckling analysis of laminated composite plates containing delaminations using the enhanced assumed strain method. *International Journal of Solids and Structures* 44(24): 8006–8027.

Papazoglou, V.J., Tsouvalis, N.G. & Kyriakopoulos, G.D. 1992. Buckling of unsymmetric laminates under linearly varying biaxial in-plane loads combined with shear. *Composite Structures* 20(2): 155–163.

Parlapalli, M.R., Soh, K.C., Dong, W.S. & Ma, G. 2007. Experimental investigation of delamination buckling of stitched composite laminates. *Composites: Part A* 38(9): 2024–2033.

Pekbey, Y. 2006. A numerical and experimental investigation of critical buckling load of rectangular laminated composite plates with strip delamination. *Journal of Reinforced Plastics and Composites* 25(7): 685–697.

Shan, B. & Pelegri, A.A. 2003. Approximate analysis of the buckling behavior of composites with delaminations. *Journal of Composite Materials* 37(8): 673–685.

Short, G.J., Guild, F.J. & Pavier, M.J. 2001. The effect of delamination geometry on the compressive failure of composite laminates. *Composite Science and Technology* 61(14): 2075–2086.

Tsouvalis, N.G. & Papazoglou, V.J. 2004. Design buckling curves for clamped orthotropic laminated plates. *Advanced Composites Letters* 13(5): 227–235.

Zor, M. 2003. Delamination width effect on buckling loads of simply supported woven-fabric laminated composite plates made of carbon/epoxy. *Journal of Reinforced Plastics and Composites* 22(17): 1535–1546.

Analysis and Design of Marine Structures – Guedes Soares & Das (eds)
© 2009 Taylor & Francis Group, London, ISBN 978-0-415-54934-9

Buckling behaviour of the ship deck composite plates with cut-outs

I. Chirica, E.F. Beznea & R. Chirica
University "Dunarea de Jos" of Galati, Romania

ABSTRACT: Mechanical-buckling analyses were performed on square plates, made of composite materials, with central cut-outs. The cutouts were either circular holes or elliptical holes. The finite-element structural analysis method was used to study the effects of plate aspect ratio, hole geometry, and hole size on the mechanical-buckling strengths of the perforated plates. The compressive-buckling strengths of the plates could be increased considerably only under aspect ratios. The plate-buckling mode can be symmetrical or anti-symmetrical, depending on the plate boundary conditions, aspect ratio, and the hole size. In this paper, the analysis has been performed only for the plate clamped on sides. The results and illustrations provide important information for the efficient design of ship structural panels made of composite materials, having cut-outs.. The aim of the work presented in this paper is to analyze the influence of cut-out on the changes in the buckling behaviour of ship deck plates made of composite materials. For each diameters ratio there are plotted variation of the transversal displacement of the point placed in the middle of the plate, according to the pressure that has been applied. Buckling load determination for the general buckling of the plate has been made by graphical method. The post-buckling calculus has been performed to explain the complete behaviour of the plate.

1 INTRODUCTION

In ship structures, cutouts are commonly used as access ports for mechanical and electrical systems, or simply to reduce weight. Structural panels with cut-outs often experience compressive loads that are induced mechanically can result in panel buckling. Thus, the buckling behavior of the structural panels with cutouts must be fully understood in the structural design. Also, laminated polymer composites are being used in many advanced structural applications.

For an unperforated rectangular plate of finite extent (i.e., with finite length and finite width) under uniform compression, the closed-form buckling solutions are easily obtained because the prebuckling stress field is uniform everywhere in the plate. When a finite rectangular plate is perforated with a central cutout (e.g., a circular or a square hole), however, the buckling analysis becomes extremely cumbersome because the cutout introduces a load-free boundary that causes the stress field in the perforated plate to be non-uniform. Hence, the closed-form buckling solutions are practically unobtainable, and various approximate methods had to be developed to analyze such perforated plates.

The buckling of flat square plates with central circular holes under in-plane edge compression has been studied both theoretically and experimentally by various authors. The methods of theoretical analysis used by most of the past investigators were the Rayleigh-Ritz minimum energy method and the Timoshenko method. However, except for Schlack

(1964) and Kawai and Ohtsubo (1968), the theoretical analysis methods used do not allow the boundary and loading conditions to be precisely defined for larger hole sizes because the stress distributions of the infinite perforated plate are used as the pre-buckling stress solution for the finite perforated plate. Thus, most of the earlier buckling solutions are limited to small hole sizes, and are not fit for studying the effects of different plate boundary conditions on the buckling strengths of the finite plates with arbitrarily sized holes using those approximate solutions.

In the paper, the analysis of the buckling behaviour of the composite plates with central cut-outs, used in ship structures, is presented.

2 PLATES CHARACTERISTICS

The square plates (320 × 320 mm), clamped on the sides, are made of E-glass/epoxy having the material characteristics:

- unidirectional layers, UD: $t_1 = 0.59$ [mm]
- biaxial layers: $t_2 = 0.39$ [mm].

Due to the grouping of the layers with the same characteristics we consider the following macro-layers with the thicknesses (see figure 1):

$$t' = t_2/2 = 0.195[mm],$$
$$t'' = 4 \cdot t_1 = 2.36[mm],$$
$$t''' = 6 \cdot t_1 = 3.54[mm].$$

+45°	macro-layer 11	$t' = 0.195$ mm	
−45°	macro-layer 10	$t' = 0.195$ mm	
0°	macro-layer 9	$t'' = 2.36$ mm	
+45°	macro-layer 8	$t' = 0.195$ mm	
−45°	macro-layer 7	$t' = 0.195$ mm	
0°	macro-layer 6	$t'' = 3.54$ mm	
−45°	macro-layer 5	$t' = 0.195$ mm	
+45°	macro-layer 4	$t' = 0.195$ mm	
0°	macro-layer 3	$t'' = 2.36$ mm	
−45°	macro-layer 2	$t' = 0.195$ mm	
+45°	macro-layer 1	$t' = 0.195$ mm	

Figure 1. Plate lay-up.

The characteristics of the material are:

$E_x = 46[\text{GPa}], E_y = 13[\text{GPa}], E_z = 13[\text{GPa}],$
$G_{xy} = 5[\text{GPa}], G_{xz} = 5[\text{GPa}], G_{yz} = 4.6[\text{GPa}].$
$\mu_{xy} = 0.3, \mu_{yz} = 0.42, \mu_{xz} = 0.3.$
$R_x^T = 1.062\,[\text{GPa}],$
$R_x^C = 0.61\,[\text{GPa}],$
$R_y^T = 0.031\,[\text{GPa}],$
$R_y^C = 0.118\,[\text{GPa}],$
$R_{xy} = 0.72\,[\text{GPa}].$

For the material behaviour model two cases have been considered:

– linear behaviour;
– nonlinear behaviour (Tsai-Wu failure criterion).

The cut-outs are placed in the center of the plate, having an elliptical shape. Dimensions of the cut-out are:

– transverse diameter is considered as unchangeable: $dy = 100[\text{mm}]$;
– longitudinal diameter dx is considered as changeable.

The in-plane loading was applied as a uniform compressive pressure in the x direction.

In the parametric calculus, the following diameter ratios were considered: $dx/dy = \{0.5; 0.75; 1; 1.25; 1.5; 1.75; 2\}$.

3 SENSITIVITY ANALYSIS OF BUCKLING LOAD

Due to the fact, the fabrication of the plate is hand made one and the material characteristics are offen statistically determined, the sensitivity analysis of the buckling load has been performed.

So, the influences of the variation of the mechanical characteristics and layers thicknesses on the linear buckling load were analysed.

Certain sets of the values of mechanical characteristics and layers thicknesses were chosen, as are presented in tables 1 and 2.

In figures 5 and 6 the variations of the buckling loads function of diameters ratio for each set of characteristics and layers thickness are presented.

The study made possible to determine the most decisive material parameter and the most decisive layer thickness for the buckling behaviour of the plate with cut-outs.

The results presented in figure 3 show that the buckling load decreases since the diameter ratio increases. For the same diameter ratio, the variation of elastic

Table 1. Sets of characteristics.

Set	Ex [GPa)	Ey [GPa]
1	43	10
2	44.25	11.25
3	45.5	12.5
4	46.75	13.75
5	48	15

Table 2. Sets of macro-layer thicknesses.

Set	t'[mm]	t''[mm]	t'''[mm]
1	0.19	2.2	3.4
2	0.192	2.25	3.45
3	0.195	2.3	3.5
4	0.197	2.35	3.55
5	0.2	2.4	3.6

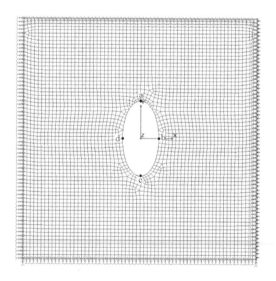

Figure 2. Mesh model of the plate.

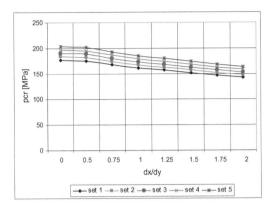

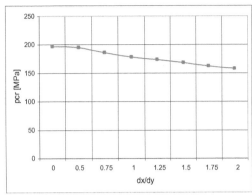

Figure 3. Variation of the buckling load function of diameters ratio for each set of mechanical characteristics.

Figure 6. Variation of buckling load (linear calculus) versus diameters ratio dx/dy.

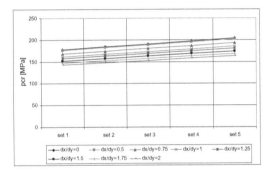

Figure 4. Variation of the buckling load function of diameters ratio for each set of layers thicknesses.

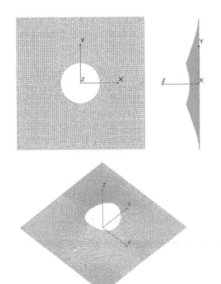

Figure 5. Deformed plate after buckling.

modulus produces a variation of buckling load. The sensitivity of the buckling load is decreasing since the diameter ratio is increasing.

In figure 4, the sensitivity analysis shows that the buckling load increases since the diameter ratio increases. For the same diameter ratio, the variation of layer thickness with 10% produces a variation with 15%–20% of buckling load. The sensitivity of the buckling load is increasing since the diameter ratio is increasing.

4 F.E. BUCKLING ANALYSIS

The analysis was carried out using COSMOS/M finite element software. For the present study, a 3-D model with 4-node SHELL4L layered composite element of COSMOS/M is used.

For unidirectional compression buckling analysis, an uniform pressure in longitudinal direction was incremental applied.

4.1 FE linear buckling analysis

In Table 3, the results of linear buckling calculus for each diameters ration are presented. In figure 7, the deformed plate after buckling is presented.

The results of the linear analysis are presented in figure 6, where the variation of the buckling load function of the ratio dx/dy is plotted.

As it is seen, the buckling load is decreasing since the ratio dx/dy is increasing.

4.2 FE buckling nonlinear calculus

To solving of geometrically and material nonlinear problems, the load is applied as a sequence of sufficiently small increments so that the structure can be

425

Table 3. Buckling load (linear calculus).

dx/dy	p_{cr} [MPa]
0	196.71
0.5	194.77
0.75	186.25
1	178.51
1.25	174.01
1.5	168.13
1.75	162.50
2	157.66

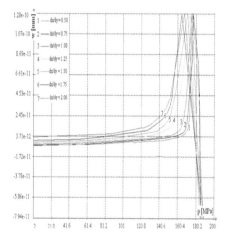

Figure 7. Variation of maximum transversal displacement versus in-pane load for each diameters ratio.

assumed to respond linearly during each increment. For each increment of load, increments of displacements and corresponding increments of stress and strain are computed. These incremental quantities are used to compute various corrective stiffness matrices (variously termed geometric, initial stress, and initial strain matrices) which serve to take into account the deformed geometry of the structure. A subsequent increment of load is applied and the process is continued until the desired number of load increments has been applied. The net effect is to solve a sequence of linear problems wherein the stiffness properties are recomputed based on the current geometry prior to each load increment. The solution procedure takes the following mathematical form

$$(K + K_I)_{i-1} \Delta d_i = \Delta Q \tag{1}$$

where

K is the linear stiffness matrix,
K_I is an incremental stiffness matrix based upon displacements at load step i-1,

Δd_i is the increment of displacement due to the i-th load increment,
ΔQ is the increment of load applied.

The correct form of the incremental stiffness matrix has been a point of some controversy. The incremental approach is quite popular (this is the procedure applied in this study). This is due to the ease with which the procedure may be applied and the almost guaranteed convergence if small enough load increments are used.

Buckling and post-buckling analysis has been performed for all types of panels.

The buckling load determination may use the Tsai-Wu failure criterion in the case of the general buckling does not occurred till the first-ply failure occurring. In this case, the buckling load is considered as the in-plane load corresponding to the first-ply failure occurring.

The Tsai-Wu failure criterion provides the mathematical relation for strength under combined stresses. Unlike the conventional isotropic materials where one constant will suffice for failure stress level and location, laminated composite materials require more elaborate methods to establish failure stresses. The strength of the laminated composite can be based on the strength of individual plies within the laminate. In addition, the failure of plies can be successive as the applied load increases. There may be a first ply failure followed by other ply failures until the last ply fails, denoting the ultimate failure of the laminate. Progressive failure description is therefore quite complex for laminated composite structures. A simpler approach for establishing failure consists of determining the structural integrity which depends on the definition of an allowable stress field. This stress field is usually characterized by a set of allowable stresses in the material principal directions.

The failure criterion is used to calculate a failure index (F.I.) from the computed stresses and user-supplied material strengths. A failure index of 1 denotes the onset of failure, and a value less than 1 denotes no failure. The failure indices are computed for all layers in each element of your model. During postprocessing, it is possible to plot failure indices of the mesh for any layer.

The Tsai-Wu failure criterion (also known as the Tsai-Wu tensor polynomial theory) is commonly used for orthotropic materials with unequal tensile and compressive strengths. The failure index according to this theory is computed using the following equation, (Altenbach et al., 2004)

$$F.I. = F_1 \cdot \sigma_1 + F_2 \cdot \sigma_2 + F_{11} \cdot \sigma_1^2$$
$$+ F_{22} \cdot \sigma_2^2 + F_{66} \cdot \sigma_6^2 + 2F_{12} \cdot \sigma_1 \cdot \sigma_2 \tag{2}$$

where

$$F_1 = \frac{1}{R_1^T} - \frac{1}{R_1^C}; \quad F_{11} = \frac{1}{R_1^T \cdot R_1^C};$$

$$F_2 = \frac{1}{R_2^T} - \frac{1}{R_2^C}; \quad F_{22} = \frac{1}{R_2^T \cdot R_2^C}; \quad F_{66} = \frac{1}{R_{12}^2}. \quad (3)$$

The coefficient F_{12} which represents the parameter of interaction between σ_1 and σ_2 is to be obtained by a mechanical biaxial test. In the equations (3), the parameters R_i^C, R_i^T are the compressive strength and tensile strength in the material in longitudinal direction (i = 1) and trasversal direction (i = 2). The parameter R12 is in-plane shear strength in the material 1–2 plane.

According to the Tsai-Wu failure criterion, the failure of a lamina occurs if

$$F.I.>1. \quad (4)$$

In COSMOS, nonlinear material is considered as a material with nonlinear behaviour (the nonlinear material curve) or case of introducing the material strength components for Failure criteria using for composites. This latest case is the case analysed in the paper.

The failure index in calculated in each ply of each element. In the ply where failure index is greater than 1, the first-ply failure occurs, according to the Tsai-Wu criterion. In the next steps, the tensile and compressive properties of this element are reduced by the failure index. If the buckling did not appeared until the moment of the first-ply failure occurring, the in-plane load corresponding to this moment is considered as the buckling load.

In the nonlinear calculus, for the buckling load, the graphical method and Tsai-Wu failure criterion were used. The values obtained for buckling load were placed in the range specified in each case in Figure 8.

In figure 8, the variations of the buckling load corresponding to the fails in the tension cases (Fail 1) and compressive cases (Fail 2), versus diameters ratio are presented.

For the panels with elliptical central cut-out the values of the buckling load is placed in the range

$$125,36 \text{ [MPa]} < \text{pcr} < 163,307 \text{ [MPa]}$$

In figure 7, the variations of maximum transversal displacement versus in-plane load for each diameters ratio are presented. Using graphical method the buckling load may be estimated by drawn an asymptote to the curve.

The postbuckling behaviour of the plate may be explained according to the curves in the figure 7, from region drawn after buckling occurring.

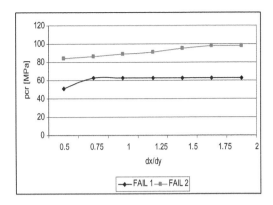

Figure 8. Variation of buckling load, corresponding to fail 1 and fail 2, versus diameters ratio.

5 CONCLUSIONS

A FEM based methodology was successfully developed for the investigation of buckling problems of composite plates with central elliptical cut-out. Two hypotheses regarding the type of material modeling were used (linear and nonlinear).

The buckling behavior of plates with central holes as presented in figure 7, is quite peculiar because, under certain boundary conditions (clamped edges) and plate aspect ratios, the mechanical-buckling strengths of the perforated plates, contrary to expectation, increase rather than decrease as the hole sizes grow larger. The conventional wisdom is that, as the hole sizes increase, the plates lose more materials and become weaker. Therefore, the buckling strengths were expected to decrease as the hole sizes increase. This was not the case. Such peculiar buckling phenomenon of the perforated plates may be explained as follows.

When the hole size becomes considerably large relative to the plate width, most of the compressive load is carried by the narrow side strips of material along the plate boundaries. As it is well known, a stronger plate boundary condition (e.g., clamped rather than simply-supported boundaries) increases the buckling strength, while the higher stress concentration decreases the buckling strength. Thus, which effects become dominant will determine the increase or decrease of the buckling strengths of the perforated plates.

For the circular-hole cases, the narrow compressed side strips are under stress concentration, which reduces the buckling strengths.

The unusual buckling characteristics of the perforated plates offer important applications in ship structural panel design. Namely, by opening holes of proper sizes in ship structural panels for weight saving, their buckling strengths can be boosted simultaneously.

The buckling load determination is too difficult without applying a graphical method. In certain cases, the using of Tsai-Wu criterion may predict so named "buckling load", if general buckling of the plate does not occurred before first-ply failure occurring.

The first failure occurring in an element is based on the Tsai-Wu failure criterion, which provides the mathematical relation for strength under combined stresses may be used.

The lack of the criterion is referring to the anticipation of the real mode to occurring the cracking.

Taking into account the mathematical formulation, the Tsai-Wu failure criterion is easy to be applied. Additionally, this criterion offers advantages concerning the real prediction of the strength at variable loadings. It is to remark that by applying linear terms, it is possible to take into account the differences between the tension and compression strengths of the material.

ACKNOWLEDGMENTS

The work has been performed in the scope of the projects:

— Project MARSTRUCT, Network of Excellence on Marine Structures, (2004-2009), which has been financed by the EU through the GROWTH Programme under contract TNE3-CT-2003-506141 (Task4.3), www.mar.ist.utl.pt/marstruct/
— Romanian Project PN2-IDEI, Code 512.

REFERENCES

Altenbach, H., Altenbach, J. and Kissing, W. 2004. *Mechanics of Composite Structural Elements*, Ed. Springer, Berlin.

Ambarcumyan, S.A. 1991. *Theory of Anisotropic Plates: Strength, Stability, and Vibrations*, Hemispher Publishing, Washington.

Adams, D.F., Carlsson, L.A. and Pipes, R.B. 2003. *Experimental Characterization of Advanced Composite materials*, Ed. Taylor & Francis Group.

Chirica, I., Beznea, E.F. and Chirica, R., 2006. *Placi compozite* (in Romanian). Edit. Fund. Univ. Dunarea de Jos, Galati, ISBN (10) 973-627-337-7; ISBN (13) 978-973-627-337-7.

Engelstad, S.P., Reddy, J.N. and Knight, N.F., Jr. 1992. *Postbuckling Response and Failure Prediction of Graphite-Epoxy Plates Loaded in Compression*, AIAA Journal, 30(8), 2106–2113.

Hilburger, M.F. 2001. *Nonlinear and Buckling Behavior of Compression-loaded Composite Shells*, Proceedings of the 6th Annual Technical Conference of the American Society for Composites, Virginia.

Kawai, T. and Ohtsubo, H. 1968. *A Method of Solution for the Complicated Buckling Problems of ElasticPlates With Combined Use of Rayleigh-Ritz's Procedure in the Finite Element Method*, AFFDLTR-68-150.

Schlack, A. L., Jr. 1964. *Elastic Stability of Pierced Square Plates*. Experimental Mechanics, June 1964: 167–172.

SRAC. 2001. Cosmos/M FEM program user guide. Structural Research & Analysis Corporation. (www.cosmosm.com)

Thurley, G.J. and Marshall, I.H. 1995. *Buckling and Postbuckling of Composite Plates*, Chapman & Hall, London.

Analysis and Design of Marine Structures – Guedes Soares & Das (eds)
© 2009 Taylor & Francis Group, London, ISBN 978-0-415-54934-9

Buckling behaviour of plates with central elliptical delamination

E.F. Beznea, I. Chirica & R. Chirica
University "Dunarea de Jos" of Galati, Romania

ABSTRACT: The aim of the work presented in this paper is to analyze the influence of delamination on the changes in the buckling behaviour of ship deck plates made of composite materials. This problem has been solved by using the finite element method. The damaged part of the structures and the undamaged part have been represented by layered shell elements. The influence of the position and the ellipse's diameters ratio of delaminated zone on the critical buckling force was investigated. Taking into account the thickness symmetry of the plates, are presented only cases of position of delamination on one side of symmetry axis. For each position of delamination there are plotted variation of the transversal displacement of the point placed in the middle of the plate, according to the pressure that has been applied. Buckling load determination for the general buckling of the plate has been done by graphical method. The post-buckling calculus has been performed to explain the complete behaviour of the plate.

1 INTRODUCTION

Laminated polymer composites are being used in many advanced structural applications. In shipbuilding, many of these structures are situated such that they are susceptible to foreign object impacts which can result in barely visible impact damage. Often, in the form of a complicated array of matrix cracks and interlaminar delaminations, these barely visible impact damages can be quite extensive and can significantly reduce a structure's load bearing capability. Since such damage is in general difficult to detect, structures must be able to function safely with delamination present. Thus there clearly exists the need to be able to predict the tolerance of structures to damage forms which are not readily detectable (Chirica & al (2006)).

Ambarcumyan (1991), Adams & al (2003) have analysed experimental characterization of advanced composite materials. When a laminate is subjected to in-plane compression, the effects of delamination on the stiffness and strength may be characterized by three sets of results, (Finn & al (1993)):

a. Buckling load;
b. Postbuckling solutions under increased load;
c. Results concerning the onset of delamination growth and its subsequent development.

Many of the analytical treatments deal with a thin near surface delamination. Such approaches are known as "thin-film" analysis in the literature (Kim & al (1999), Thurley & al (1995)). The thin-film analytical approach may involve significant errors in the post-buckling solutions.

Naganarayana and Atluri (1995) have analysed the buckling behaviour of laminated composite plates with elliptical delaminations at the centre of the plates using finite element method. They propose a multi-plate model using 3-noded quasi-conforming shell element, and use J-integral technique for computing pointwise energy release rate along the delamination crack front.

Pietropaoli & al (2008) studied delamination growth phenomena in composite plates under compression by taking into account also the matrix and fibres breakages until the structural collapse condition is reached.

The aim of the work presented in this paper is to analyze the influence of delamination on the changes in the buckling behaviour of ship deck plates made of composite materials. This problem has been solved by using the finite element method, in Beznea (2008). An orthotropic delamination model, describing mixed mode delaminating, by using COSMOS/M soft package, was applied. So, the damaged part of the structures and the undamaged part have been represented by well-known finite elements (layered shell elements). The influence of the position and the ellipse's diameters ratio of delaminated zone on the critical buckling force was investigated.

If an initial delamination exists, this delamination may close under the applied load. To prevent the two adjacent plies from penetrating, a simple numerical contact model is used.

2 BUCKLING ANALYSIS

The finite element delamination analysis was carried out using COSMOS/M finite element software. There are several ways in which the panel can be modeled for the delamination analysis. For the present study, a 3-D model with 4-node SHELL4L composite element of COSMOS/M is used. The panel is divided into two sub-laminates by a hypothetical plane containing the delamination. For this reason, the present finite element model would be referred to as two sub-laminate model. The two sub-laminates are modeled separately using 4-node SHELL4L composite element, and then joined face to face with appropriate interfacial constraint conditions for the corresponding nodes on the sub-laminates, depending on whether the nodes lie in the delaminated or undelaminated region.

The delamination model has been developed by using the surface-to-surface contact option. In case of surface-to-surface contact, the FE meshes of adjacent plies do no need to be identically. The contact algorithm of COSMOS/M has possibility to determine which node of the so-called master surface is in contact with a given node on the slave surface. Hence, the user can define the interaction between the two surfaces.

In the analysis the different layers are intentionally not connected to each other in ellipse regions. The condition is that the delaminated region does not grow. In COSMOS/M these regions were modeled by two layers of elements with coincident but separate nodes and section definitions to model offsets from the common reference plane. Thus their deformations are independent. At the boundary of the delamination zones the nodes of one row are connected to the corresponding nodes of the regular region by master slave node system.

Typically, a node in the underlaminated region of bottom sub-laminate and a corresponding node on the top sub-laminate are declared to be coupled nodes using master—slave nodes facility of COSMOS/M. The nodes in the delaminated region, whether in the top or bottom laminate, are connected by contact element. This would mean that the two sublaminates are free to move away from each other in the delaminated region, and constrained to move as a single laminate in the undelaminated region.

The square plates (320 × 320 mm), clamped on the sides, are made of E-glass/epoxy having the material characteristics:

- unidirectional layers, UD: $t_1 = 0.59$ [mm]
- biaxial layers: $t_2 = 0.39$ [mm].

These layers are grouped in macro-layers like are presented in figure 1.

$t' = t_2/2 = 0.195$ [mm],
$t'' = 4 \cdot t_1 = 2.36$ [mm],
$t''' = 6 \cdot t_1 = 3.54$ [mm].

The characteristics of the material are:

$Ex = 46$ [GPa], $Ey = 13$ [GPa], $Ez = 13$ [GPa],
$Gxy = 5$ [GPa], $Gxz = 5$ [GPa], $Gyz = 4.6$ [GPa].

$\mu_{xy} = 0.3$, $\mu_{yz} = 0.42$, $\mu_{xz} = 0.3$
$R_x^T = 1.062$[GPa], $R_x^C = 0.61$[GPa],
$R_Y^T = 0.031$[GPa], $R_Y^C = 0.118$[GPa]
$R_{xy} = 0.72$[GPa]

For the material behaviour model two cases has been considered:

- linear behaviour;
- nonlinear behaviour (Tsai-Wu failure criterion).

For perfect plate (without delaminations), the value of the buckling load is:

- linear calculus: $p_{cr} = 59.6$ [MPa];
- nonlinear calculus: $p_{cr} = 75$ [MPa].

Dimensions of the delamination area are:

- transverse diameter, dy, is considered as unchangeable: dy = 100[mm];
- longitudinal diameter, dx, is considered as changeable.

The in-plane loading was applied as a uniform compressive pressure in the x direction.

In the parametric calculus, the following diameter ratios were considered: dx/dy = {0.5; 0.75; 1; 1.25; 1.5; 1.75; 2}.

The position of the delamination along the thickness it's been considered between two neighbors layers i and $i + 1$, (i = 1,10). Acording to the table 1, we considered all the cases. Taking into account the thickness symmetry of the plates, will be presented only cases of position of delamination on one side of symmetry axis.

For each position of delamination there are plotted variation of the transversal displacement of the point placed in the middle of the plate, according to the in-plane pressure that has been applied. Buckling load determination has been done trough graphical method.

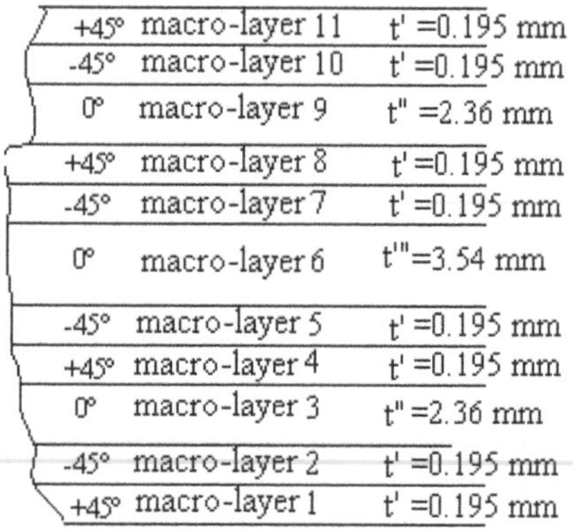

+45°	macro-layer 11	t' =0.195 mm
-45°	macro-layer 10	t' =0.195 mm
0°	macro-layer 9	t" =2.36 mm
+45°	macro-layer 8	t' =0.195 mm
-45°	macro-layer 7	t' =0.195 mm
0°	macro-layer 6	t'"=3.54 mm
-45°	macro-layer 5	t' =0.195 mm
+45°	macro-layer 4	t' =0.195 mm
0°	macro-layer 3	t" =2.36 mm
-45°	macro-layer 2	t' =0.195 mm
+45°	macro-layer 1	t' =0.195 mm

Figure 1.   Plate lay-up.

On each curve has been drawn an asymptota to part of the curve after the bifurcation point. For each position of delamination, for buckling load have been obtained values placed in a specified range.

2.1 Linearized behaviour of the plate

In the figures 2, 3, 4, 5, 6 the variation of the transversal displacement of the point placed in the middle of the plate, versus in-plane load, for each diameters ratio are presented. Each figure is made for each position of delamination.

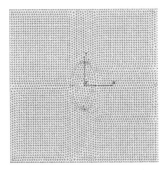

Figure 2. Mesh model of the delaminated plate.

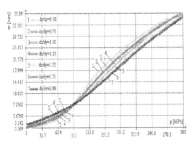

Figure 3. Variation of maximum transversal displacement versus in-plane load. Case of the delamination placed between layers 1 and 2.

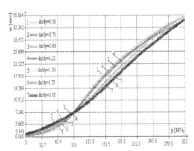

Figure 4. Variation of maximum transversal displacement versus in-plane load. Case of the delamination placed between layers 2 and 3.

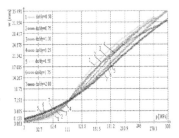

Figure 5. Variation of maximum transversal displacement versus in-plane load. Case of the delamination placed between layers 3 and 4.

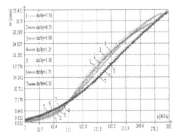

Figure 6. Variation of maximum transversal displacement versus in-plane load. Case of the delamination placed between layers 4 and 5.

As it is seen, the buckling load is increasing since the diameters ratio is increasing.

2.2 Nonlinear behaviour of the plate

Buckling problem to be solved is a nonlinear one:

– geometrically nonlinearity (existing a delamination);
– material nonlinearity. The buckling does not occur in the elastic domain. In this case, the Tsai-Wu failure criterion is used.

In COSMOS, nonlinear material is considered as a material with nonlinear behaviour (the nonlinear material curve) or case of introducing the material strength components for Failure criteria using for composites. This latest case is the case analysed in the paper.

To solving of material and geometrically nonlinear problems, the load is applied as a sequence of sufficiently small increments so that the structure can be assumed to respond linearly during each increment. For each increment of load, increments of displacements and corresponding increments of stress and strain are computed. These incremental quantities are used to compute various corrective stiffness matrices (variously termed geometric, initial stress, and initial strain matrices) which serve to take into account the deformed geometry of the structure. A subsequent increment of load is applied and the process is continued until the desired number of load increments has been applied.

The net effect is to solve a sequence of linear problems wherein the stiffness properties are recomputed based on the current geometry prior to each load increment. The solution procedure takes the following mathematical form

$$(K + K_I)_{i-1} \Delta d_i = \Delta Q \tag{1}$$

where K is the linear stiffness matrix, K_I is an incremental stiffness matrix based upon displacements at load step $i - 1$, Δd_i is the increment of displacement due to the i-th load increment, and ΔQ is the increment of load applied. The correct form of the incremental stiffness matrix has been a point of some controversy. The incremental approach is quite popular (this is the procedure applied in this study). This is due to the ease with which the procedure may be applied and the almost guaranteed convergence if small enough load increments are used.

Buckling and postbuckling analysis has been performed for all types of panels. Since the plate has a delamination, the increasing of the transversal deformation is starting from the beginning, that is so name "buckling" is starting since the inplane load is starting to increase from 0.

The buckling load determination may use the Tsai-Wu failure criterion in the case of the general buckling does not occurred till the first-ply failure occurring. In this case, the buckling load is considered as the in-plane load corresponding to the first-ply failure occurring.

The Tsai-Wu failure criterion provides the mathematical relation for strength under combined stresses. Unlike the conventional isotropic materials where one constant will suffice for failure stress level and location, laminated composite materials require more elaborate methods to establish failure stresses. The strength of the laminated composite can be based on the strength of individual plies within the laminate. In addition, the failure of plies can be successive as the applied load increases. There may be a first ply failure followed by other ply failures until the last ply fails, denoting the ultimate failure of the laminate. Progressive failure description is therefore quite complex for laminated composite structures. A simpler approach for establishing failure consists of determining the structural integrity which depends on the definition of an allowable stress field. This stress field is usually characterized by a set of allowable stresses in the material principal directions.

The failure criterion is used to calculate a failure index (F.I.) from the computed stresses and user-supplied material strengths. A failure index of 1 denotes the onset of failure, and a value less than 1 denotes no failure. The failure indices are computed for all layers in each element of your model. During post-processing, it is possible to plot failure indices of the mesh for any layer.

The Tsai-Wu failure criterion (also known as the Tsai-Wu tensor polynomial theory) is commonly used for orthotropic materials with unequal tensile and compressive strengths. The failure index according to this theory is computed using the following equation, (Altenbach & al, 2004)

$$\text{F.I.} = F_1 \cdot \sigma_1 + F_2 \cdot \sigma_2 + F_{11} \cdot \sigma_1^2$$
$$+ F_{22} \cdot \sigma_2^2 + F_{66} \cdot \sigma_6^2 + 2F_{12} \cdot \sigma_1 \cdot \sigma_2 \tag{2}$$

where

$$F_1 = \frac{1}{R_1^T} - \frac{1}{R_1^C}; \quad F_{11} = \frac{1}{R_1^T \cdot R_1^C};$$

$$F_2 = \frac{1}{R_2^T} - \frac{1}{R_2^C}; \quad F_{22} = \frac{1}{R_2^T \cdot R_2^C}; \quad F_{66} = \frac{1}{R_{12}^2}. \tag{3}$$

The coefficient F_{12} which represents the parameter of interaction between σ_1 and σ_2 is to be obtained by a mechanical biaxial test. In the equations (3), the parameters R_i^C, R_i^T are the compressive strength and tensile strength in the material in longitudinal direction ($i = 1$) and trasversal direction ($i = 2$). The parameter R_{12} is in-plane shear strength in the material 1–2 plane.

According to the Tsai-Wu failure criterion, the failure of a lamina occurs if

$$\text{F.I.} > 1. \tag{4}$$

Buckling load determination for the general buckling of the plate has been performed by graphical method. The post-buckling calculus was done to explain the complete behaviour of the plate.

Table 2. Buckling load range.

| Position of delamination | Tsai-Wu criterion | | | | | |
| | Graphical Method [MPa] | | Fail 1 Tension [MPa] | | Fail 1 Compersion [MPa] | |
	min p_{cr}	max p_{cr}	min p_{cr}	max p_{cr}	min p_{cr}	max p_{cr}
Layer 1 Layer 2	159	210	60	125	150	207
Layer 2 Layer 3	210	234	60	125	150	209
Layer 3 Layer 4	204	246	66	135	171	234
Layer 4 Layer 5	204	252	69	135	171	234
Layer 5 Layer 6	204	255	69	135	174	237

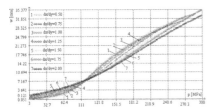

Figure 7. Variation of maximum transversal displacement versus in-plane load. Case of the delamination placed between layers 5 and 6.

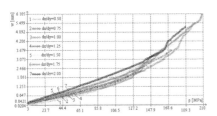

Figure 8. Variation of maximum transversal displacement versus in-plane load. Case of the delamination placed between layers 1 and 2.

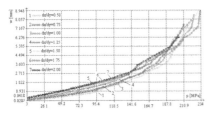

Figure 9. Variation of maximum transversal displacement versus in-plane load. Case of the delamination placed between layers 2 and 3.

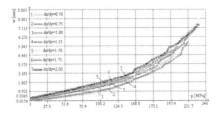

Figure 10. Variation of maximum transversal displacement versus in-plane load. Case of the delamination placed between layers 3 and 4.

Details on the buckling load values for each diameters ratio and position of delamination are presented in table 3.

Buckling load determination has been done trough graphical method, by drawing an asymptote to each

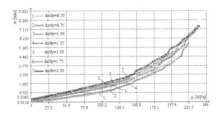

Figure 11. Variation of maximum transversal displacement versus in-plane load. Case of the delamination placed between layers 4 and 5.

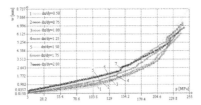

Figure 12. Variation of maximum transversal displacement versus in-plane load. Case of the delamination placed between layers 5 and 6.

Table 3. Buckling load values [MPa] for each diameters ratio and position of delamination.

Delam. pos.	Fail no.	dx/dy						
		0.5	0.75	1.0	1.25	1.50	1.75	2.0
1–2	1	60	80	90	105	111	117	125
	2	150	162	174	195	198	204	207
2–3	1	60	81	90	105	111	118	125
	2	150	168	174	201	204	206	209
3–4	1	66	87	99	111	117	119	135
	2	171	177	186	207	209	216	234
4–5	1	69	87	99	111	117	120	135
	2	171	180	186	207	210	213	234
5–6	1	69	90	99	111	117	120	135
	2	174	180	189	207	210	219	237

curve corresponding to a plate with a delamination type, under in-plane load (Table 2). The values obtained for buckling load have been placed in the range specified in the first two columns in each case in table 2.

3 CONCLUSIONS

A FEM based methodology was successfully developed for the investigation of buckling problems of composite plates with central delamination. Two hypotheses regarding the type of material modeling were used (linear and nonlinear). The FEM model is robust

433

in that it can be used to predict the global buckling loads of composite plates either on one side or both sides. Finite-elements analysis was carried out to assess the reliability of the methodology. The two-sublaminate model developed in this work provides a convenient method to model delaminated composite panels.

For the model with the linearized behaviour of material, in all cases of diameters ratio and position of delamination, for a loading of about 114 [MPa] the value of maximum transversal displacement is the same, that is about 7.28 [mm]. This situation corresponds to the same stiffness for each plate type.

For the values of in-plane loads lower than 114 [MPa], the displacement values are increasing since the diameters ratio is increasing. This trend is due to presence of the delamination area which is increasing since the diameters ratio is increasing. Bigger delamination area means loss of shear stiffness.

In the case of the in-plane loading values bigger than 114 [MPa], the displacement values are decreasing since the diameters ratio is increasing. This trend is due to the contact pressure between the layers in contact in the delamination, which is increasing since the loading force is increasing.

In the case of nonlinear material model (Figures 8, 9, 10, 11 and 12) the trend of the curves is: the transversal displacement is increasing since the diameters ratio is increasing for the same in-plane loading. At an in-plane loading of about 140–160 [MPa] for each case, a small instant jumping of transversal displacement is observed. If the calculus should be linear, the trend of the curves should be such as in the Figures 3, 4, 5, 6, 7. But, for these inplane loading values, fail of a lamina is occurring in the material.

This means that what is recover in plate stiffness after the increasing of contact pression in the delamination area, is lost due to the lamina damage occurring.

The buckling load determination is too difficult without applying a graphical method, or applying the Tsai-Wu failure criterion in the case of the general buckling not occurred till the first-ply failure occurring.

The first failure occurring in an element is based on the Tsai-Wu failure criterion, which provides the mathematical relation for strength under combined stresses was used.

The failure index is calculated in each ply of each element. In the ply where failure index is greater than 1, the first-ply failure occurs, according to the Tsai-Wu criterion. In the next steps, the tensile and compressive properties of this element are reduced by the failure index. If the buckling did not appeared until the moment of the first-ply failure occurring, the in-plane load corresponding to this moment is considered as the buckling load.

The lack of the criterion is referring to the anticipation of the real mode to occurring the cracking.

Taking into account the mathematical formulation, the Tsai-Wu failure criterion is easy to be applied. Additionally, this criterion offers advantages concerning the real prediction of the strength at variable loadings.

It is to remark that by applying linear terms, it is possible to take into account the differences between the tension and compression strengths of the material.

ACKNOWLEDGMENTS

The work has been performed in the scope of the projects:

– Project MARSTRUCT, Network of Excellence on Marine Structures, (2004–2009), which has been financed by the EU through the GROWTH Programme under contract TNE3-CT-2003-506141 (Task4.3), www.mar.ist.utl.pt/marstruct/.
– Romanian Project PN2-IDEI, Code 512.

REFERENCES

Altenbach, H., Altenbach, J., Kissing, W. 2004. *Mechanics of Composite Structural Elements*, Ed. Springer, Berlin.

Ambarcumyan, S.A. 1991. *Theory of Anisotropic Plates: Strength, Stability, and Vibrations*, Hemispher Publishing, Washington.

Adams, D.F., Carlsson, L.A., Pipes, R.B. 2003. *Experimental Characterization of Advanced Composite materials*, Ed. Taylor & Francis Group.

Beznea, E.F. 2008. *Studies and researches on the buckling behaviour of the composite panels*, Doctoral Thesis, University Dunarea de Jos of Galati.

Chirica, I., Beznea, E.F., Chirica, R. 2006. *Placi composite* (in Romanian). Edit. Fund. Univ. Dunarea de Jos, Galati, ISBN (10) 973-627-337-7; ISBN (13) 978-973-627-337-7.

Finn, S.C., Springer, G.S. 1993. *Delamination in composites plates under transverse static or impact loads—a model*, Composite Structures, vol. 23.

Kim, H, Kedward, K.T., 1999. *A Method for Modeling the Local and Global Buckling of Delaminated Composite Plates*. Composite Structures 44 (1999): 43–53.

Naganarayana, B.P., Atluri, S.N. 1995. *Strength reduction and delamination growth in thin and thick composite plates under compressive loading*, Computational Mechanics, 16 (1999): 170–189.

Pietropaoli, E., Riccio, A., Zarrelli, M. 2008. *Delamination Growth and Fibre/Matrix Progresive Damage in Composite Plates under Compression*. ECCM13, The 13-th European Conference on Composite Materials, June 2–5, 2008, Stockholm, Sweden.

Thurley, G.J., Marshall, I.H. 1995. *Buckling and Postbuckling of Composite Plates*, Chapman & Hall, London.

Methods and tools for structural design and optimization

Analysis and Design of Marine Structures – Guedes Soares & Das (eds)
© 2009 Taylor & Francis Group, London, ISBN 978-0-415-54934-9

Structural design of a medium size passenger vessel with low wake wash

Dario Boote & Donatella Mascia
Dipartimento di Ingegneria Navale e Tecnologie Marine,
University of Genova, Italy

ABSTRACT: The main features are presented related to the development of an innovative passenger ship, starting from the concept design up to the final realization of the real scale prototype. The vessel herein enlightened is represented by a very unconventional solution for the employment in the short range passenger traffic with a low environmental impact. The proposed solution has been inspired by both hydrofoil and SWATH technologies with the aim of matching relatively high transfer speeds, low environmental impact and reduced wave washing phenomena.

The Department of Naval Architecture of the University of Genova cooperated with Rodriquez Cantieri Navali to develop the complete design of this new vessel. In this paper the structural design is described, starting from a first, preliminary approach by HSC Rules. Preliminary studies, on simplified numerical models, have also been performed to evaluate the mutual interaction between the hull and the submerged structure.

The main information have been collected in order to setup a finite element model suitable to simulate the behaviour of the structure as a whole. Starting from this model subsequent implementations have been carried out improving the investigation of stress and strain distributions.

1 INTRODUCTION

The Mediterranean area around the Italian coast with its famous islands and tourist sites represents a very interesting business for medium size passenger ships. Rodriquez Cantieri Navali S.p.A of Messina and the University of Genova has been established a research program in order to perform the project of a fast passenger ship with a very low wake wash to be used in a short range transport close to the shore.

The wake wash represents the biggest limitation to the commercial development of fast vessels. This phenomenon is emphasised by increasing vessel dimensions and speed and by limited sea depth. It becomes particularly harmful close to the shore for the environment (coast erosion, sea bottom modifications) and for commercial activities (tourism, fishing and sea cultures). A possible solution, already adopted by some Governments in the North of Europe and in the Unites States, consists in promulgating dedicated laws to limit transit speed of fast ships. A more suitable approach is to develop technical solutions able to reduce the wake wash without damaging maritime traffics. This is the main aim of the team involved in this study.

Discarded then the hypothesis of a conventional monohull since the first phase of the investigation, the possible examined solutions were represented by hydrofoil, catamaran and SWATH vessels. Each of these typologies is characterised by merits and disadvantages. Hydrofoil is provided with sufficiently high speed and good manoeuvrability, easiness of construction and low management costs. On the other side the transport capacity is limited owing to the reduced deck space.

Catamaran has comparable hydrodynamic performances with very large platform area. Nevertheless it shows unsatisfactory wake wash performances and seakeeping features in rough sea.

SWATH has very good manoeuvrability and seakeeping characteristics together with a high payload vs waterline area ratio. Otherwise SWATH configuration is subject to deck slamming and green water phenomena and requires very high installed power for high speed.

From this analysis the idea took place of a new typology, able to match hydrofoil and SWATH principles, in order to achieve low environmental impact and reduced wave washing phenomena, meanwhile keeping an acceptable transfer speed.

The design of a new special vessel was then developed able to provide high performances, manoeuvrability and controllability typical of hydrofoils and, at the same time, good sea keeping qualities and low installed power, typical of SWATH solution. To synthesise all of the described characteristics, the new project was called ENVIROALISWATH.

The vessel consists of the hull, with spaces devoted to cargo and passengers, and the submerged body, holding the main propulsion system, and four foils providing the dynamic lift. Hull and submerged body are connected together by means of two column structures: a larger one in the aft part and a very thin one at fore.

The hull has a trimaran type layout with two hard chine lateral bodies and a central hull protecting the cross deck from wave impacts while supporting the fore and aft column structures. By the length of 63 m and the breadth of 15.5 m the vessel has a transport capacity of about 450 passengers and 50 cars (see fig. 1).

The submerged body has a length of 50 m, a breadth of 4.10 m and a depth of 2.6 m and it provides the 80% of the hydrostatic buoyancy. The remaining 20% is assured:

– at zero speed and in the preplaning phase by two lateral hull bodies;
– at cruise speed by four foils lifting force.

This way a cruise speed of 27 knots is guaranteed with moderate wave making, in force of the small waterplane area of the column structures.

The first phase of the design regarded the structural concept and the geometry lay-out; afterwards the preliminary scantling by several HSC Rules was laid down and then it was set-up with the aid of simplified direct calculations. A further refinement of the structure scantling has been then carried out by a finite element analysis of the hull and of the submerged body

separately; this activity is described in detail in Boote and alii (2005 and 2006a).

The investigation herein presented refers to the FEM analysis of the whole structure performed with the aim of individuating possible critical zones rising from the interaction of hull and submerged body. Detailed models of these connections have been considered and the structural behaviour analyzed in terms of stress, strain and local buckling.

2 STRUCTURE MAIN CHARACTERISTICS

The hull is composed by two lateral hulls and a central one, each other connected trough a cross deck (garage deck). Two superstructure levels, the passengers deck and the wheelhouse, are fitted inside. The height of the complete vessel results to be 10.30 m. A longitudinal structure has been chosen for the unit. The reinforced frame spacing is 1250 mm while the longitudinal stiffener spacing is about 300 mm. Bottom plating keeps a constant thickness, except in the connection to the central hull zone, where it is increased. Decks are fitted with two girders positioned at 3150 mm on both sides of the symmetry plane; girders span is partitioned by circular section steel pillars. The hull is composed by 51 frames; the after leg is positioned between the 10th and the 22th frame; the fore one is positioned between the 29th and the 30th frame. A typical hull cross section is presented in fig. 2.

Hull and submerged body structures are made of AlMg 5083 light alloy, foils supports and pillars are made of Fe510 steel; the main characteristics of the employed materials are resumed in table 1.

Lifting foils, connected to the submerged body, are composed by solid elements with variable NACA profile cross section. They are further joined to both lateral hulls by means of steel legs, NACA profile as well.

The submerged body maintains the longitudinal structure typical of the hull, with reinforced frames aligned to those of the hull. Bottom floors are longitudinally connected by two fore and aft lateral keelsons, plus a central one in the engine room. Top side shell is reinforced by beams and two fore and aft lateral girders, aligned and connected with bottom keelsons through tubular pillars at each frame. In correspondence of the main leg side walls two additional girders are arranged.

To prevent the occurrence of buckling phenomena, particularly harmful for aluminium alloy structures, longitudinal and transversal beams, as well as ordinary stiffeners, are "T" shaped elements. Plate thickness is constant over the whole structure, except for the zones around the main leg where it has been increased.

The aft leg (2750 mm high, and 1200 mm wide), is extended along twelve frames. It consists of a box

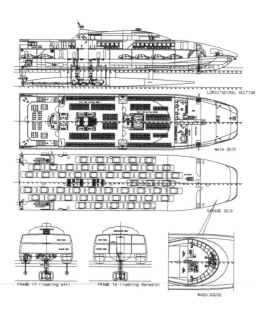

Figure 1. ENVIROALISWATH general arrangements: longitudinal section, decks and main section.

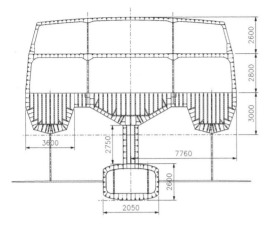

Figure 2. ENVIROALISWATH main section.

Table 1. Material mechanical characteristics.

Material	σ_{ult} (N/mm^2)	$\sigma_{y0.2}$ (N/mm^2)	τ_{ult} (N/mm^2)	$\tau_{y0.2}$ (N/mm^2)
5083-H321	309	217	183	113
5083-H111	281	168	155	77
5083-H321 Weld.	281	168	–	–
Fe510	510	355	295	205

structure with transversal diaphragms stiffened by T longitudinals, 425 mm spanned.

The chosen legs height allows proper heave movements to the vessel when sailing in rough sea. To minimise the water resistance the fore and aft edges present smooth surfaces.

3 FINITE ELEMENT MODEL

Starting from the structure obtained by HSC Rules scantling and preliminary direct calculations, a detailed structural model has been developed by the finite element code MAESTRO (Hughes, 1995).

As already mentioned the hull and the submerged body carry out different operative functions and this gives rise to significant structural and geometric peculiarities and differences. The complete model of the vessel has then been obtained by assembling the numerical models of hull and submerged body, separately schematized.

The hull has been divided into nine slices characterized by smooth geometric variations. Each slice has been modeled by means of MAESTRO library elements:

– "STRAKE" elements for orthotropic stiffened shells of decks, sides and bottoms;

– "GIRDER" elements for reinforced longitudinal beams;
– "COMPOUND" elements for transverse bulkheads.

Keelsons and other large beams have been modelled by "STRAKE" elements, in order to take into account secondary stiffeners as well. The resulting model is composed by about 9300 nodes and 14000 elements.

The same procedure has been applied to the submerged body which has been subdivided into ten modules to obtain the same detail level achieved for the hull. The numerical model is characterised by about 2800 nodes and 4500 elements.

The two parts have been then assembled by modelling the aft and fore connecting structures, represented by the long aft leg and the fore thin one. The complete FE model, shown in fig. 3 is then approximately composed by 13000 nodes and 20000 elements.

3.1 Operating conditions

In a previous study the two parts of the vessel have been separately investigated by simulating each of the missing part through proper boundary conditions. This made it possible to deal with finite element models always satisfying equilibrium conditions. The description of the work and the achieved results have been presented in Boote et alii (2006b).

In this second phase of the study attention has been focused on the behaviour of the whole vessel, in order to take into account mutual interactions between the hull and the submerged body. The same loading conditions assumed for the distinct models, previously investigated, have been considered:

– floating unit in still water at zero speed ("Hull Borne Condition");
– "flying" unit during navigation in calm sea ("Foil Borne Condition");
– "flying" unit during navigation in rough sea ("Rough Sea Condition").

In the "Hull Borne" condition the ship is sustained by the hydrostatic buoyancy provided by the submerged body and by lateral hulls. Two distinct cases have been individuated, full load and ballast condition, each one giving rise to different weight distribution in the submerged body. The maximum draft corresponds to 5.50 meters approximately and this is the condition

Figure 3. ALISWATH complete model.

which the results refer to, being the most severe in the displacement configuration.

In this condition the equilibrium pattern is pursued by a specific option of MAESTRO code ("balance" command), able to find the trim corresponding to the actual displacement and centre of gravity of the vessel.

Equilibrium conditions are obtained by a hydrostatic pressure distribution automatically applied by MAESTRO to the plate elements of the wetted surface. Nevertheless, to run FEM calculations, fictitious constraints should be provided to avoid numerical lability. The number and position of such constraints must be found by an iterative procedure in order to have zero reactions.

In the "Foil Borne" condition the ship is sustained by the hydrostatic buoyancy of the submerged body and by the hydrodynamic lift provided by the foils. The hull is completely out of the water and the draft is about 4.3 meters. No dynamic effect is applied in this phase. In this case the "balance" option of MAESTRO must be integrated by the foil lift simulated through a pressure distribution on the foil surface. The proper equilibrium condition must be individuated by an iterative procedure starting from a static pattern.

The "Rough Sea" condition is obtained from the previous one by introducing acceleration effects due to sea waves.

The additional dynamic forces are counterbalanced by a stronger lift action generated by a proper angle of attack of the foils. The values of the lift in those two conditions have been determined by CFD calculations, confirmed by seakeeping experiments in towing tank.

3.2 *Design loads and weights*

The design loads have been individuated by analysing combinations of ship speeds and sea states occurring during the ship operative life in the Mediterranean area. Adopting an exceeding probability of 1%, the Raleigh probability distribution gives a value for the "α" coefficient equal to g 1.517; the corresponding design wave $H_{1/3}$ results to be 3 meters high.

Experimental investigations on ship motions have been carried out in the "Krilov Shipbuilding Research Institute" model basin (Krilov, 2006). The tests have been performed on an ALISWATH 1:6 scale model, adopting the parameters synthesised in the following table 2. For seakeeping tests a wave height of 2 meters has been adopted. The vertical acceleration has been measured at three meaningful sections along the hull: at centre of gravity and on the fore and aft perpendicular.

Assuming a linear relationship between vertical accelerations and wave height, the design accelerations to be adopted in the numerical calculation can be derived from the experimental values. They are resumed in table 3.

Table 2. Parameters for ALISWATH model basin tests.

Sea Spectrum	JONSWAP
Heading	0° and 180°
Ship speed	27 kN
Wave significant height $H_{1/3}$	2.0 m
Wave modal period	6.0 s

Table 3. Measured and calculated values of vertical accelerations $a_v(x)$.

	Aft perp.	Centre of gravity	Fore perp.
Experimental	0.0848 g	0.0298 g	0.1397 g
Calculated	0.1286 g	0.045 g	0.212 g

By means of a separate investigation, proper dynamic amplification coefficients (D.A.C.) have been individuated capable of taking into account dynamic effects by a static analysis. These coefficients are related to:

- diagram of external impulsive force versus time;
- ratio between impulsive force duration t_0 and the first vibration mode of the structure T.

The following amplification coefficients are generally assumed:

- for the main hull D.A.C. = 1.50 (which is the most conservative value);
- for reinforced beams D.A.C. = 1.25;
- for shell and stiffeners of bottom, sides and decks D.A.C. = 1.00.

A conservative load condition has been applied by multiplying calculated acceleration by D.A.C. equal to 1.50 for all structures (see fig. 4). The considered acceleration distributions do not simultaneously act on the whole structure; each acceleration intensity represents the value having the 1% probability of being exceeded at a specific longitudinal position. The application of the diagram of fig. 4 is therefore an extremely conservative hypothesis for the study of the hull as a beam.

In "Hull Borne" and "Foil Borne" conditions the loads due to weights and payloads have been applied as follows:

- local main loads (engines, reducers, generators etc.) of hull and submerged body: on the nodes corresponding at points in which such loads are actually applied;
- structural loads of hull and submerged body: at every module as a distributed load equal to the value of the module weight smeared along its length;
- cars, passengers and consumables: at every module as pressures on the surface where they are acting.

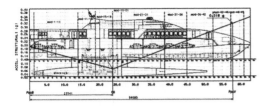

Figure 4. Distribution of the amplified (D.A.C. = 1.5) vertical accelerations $a_v(x)$.

Table 4. ALISWATH weights and loads.

ITEM	HULL [t]	TORPEDO [t]	TOTAL [t]
Structures	130	28	158
Outfitting	80	10	90
Machinery	18	35	53
Pay Load	150	–	150
TOTAL	378	73	451

In table 4 all weights and loads are resumed. The dynamic loads have been obtained as follows:

- static concentrate loads multiplied by the acceleration values relative to the longitudinal position of application points;
- static distributed loads multiplied by average acceleration calculated on the corresponding area.

4 ANALYSIS OF RESULTS

FEM calculations have been carried out on the numerical model loaded by the previously described loading conditions. Stress and strain results have been analysed on hull plates and longitudinal and transverse reinforcements. In order to assure the actual strength requirement without affecting the structure with an excessive weight, an iterative optimisation procedure has been carried out. This led to set up the final suitable structure step by step, complying the light weight requirements with the structure reliability.

The results herein presented refer then to the final structure solution on the base of which the real scale prototype has been realised.

From the analysis of results it comes out that the Hull Borne condition is characterised by very small strains and stresses. Highest stress intensity takes place at the fore leg connection to the hull bottom. The equivalent stress reaches an intensity of about 12 N/mm², far below the admissible value for welded light alloy (see table 1).

Under Foil Borne e Rough Sea conditions the equilibrium patterns correspond to different pressure distributions on fore and aft foils. In both conditions

fore foils provide a significantly higher support, giving rise to higher stress level on the fore connecting structure and in the neighbour part of the torpedo. Obtained results are examined making reference to average stress levels. Local higher values, being ascribable to coarse model refinement, need to be investigated with a more detailed mesh. The maximum stress level all over the vessel (both hull and torpedo) does not exceed 20 N/mm².

As predictable the Rough Sea condition resulted to be the most severe one being characterised by the highest stress level. For this case obtained results are presented in detail by means of stress contour plots.

As an example in fig. 5 and 6 the longitudinal and transverse stress distributions over the external shell surface are represented. In fig. 7 the equivalent Von Mises stress distributions are show on the outside and inside surfaces.

By the observation of plotted results it is possible to ascertain that the average stress is far below the maximum allowable stress for welded light alloy. Generally

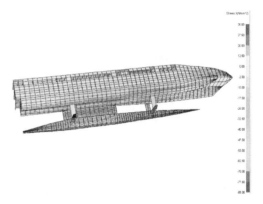

Figure 5. Longitudinal σ_x stress distribution on outer surface (N/mm²).

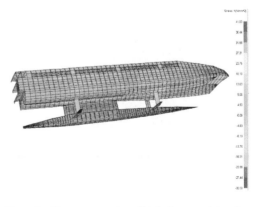

Figure 6. Transverse σ_y stress distribution on outer surface (N/mm²).

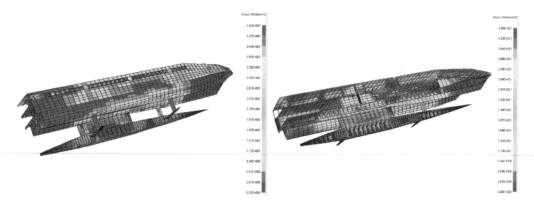

Figure 7.   Equivalent Von Mises stress σ_{VM} distribution on outer and inner surface (N/mm^2).

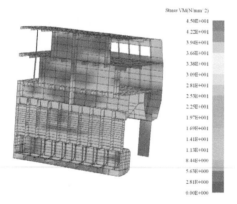

Figure 8.   Equivalent Von Mises stress σ_{VM} distribution on the aft connecting structure (N/mm^2).

Table 5.   Hull and torpedo average and maximum stresses in "Rough Sea" condition.

Hull	Shell [N/mm^2]	Beams [N/mm^2]
Average stress	20	20
Maximum stress	53	41
Submerged body		
Average stress	10	25
Maximum stress	50	65

the stress intensity does not exceed 10 N/mm^2 on the torpedo shell and 20 N/mm^2 on the hull shell.

Nevertheless some zones come out where higher stress intensities take place; these points are located mainly at the connection between hull and torpedo and on the central hull keelson (see fig. 8).

These values are ascribable partly to some stress concentration and partly to a rough schematisation at detail level because of the plate elements size utilised for the modelling of the complete vessel. The highest stresses, concentrated in the fore leg attachment to the hull, may be accepted being the leg steel made.

Similar plots can be produced for describing beam element behaviour as well. Also in this case higher stresses take place in the zone of fore and aft connections of foil legs to the bottom structures. In fig. 9 an example is shown of stress distribution on deck and side beams.

In order to synthesise achieved results for the whole structure, in the following table 5 the average and maximum stress intensities on hull and submerged body structures are presented.

For what displacements are concerned; a maximum vertical displacement of 77 mm at the fore end of the

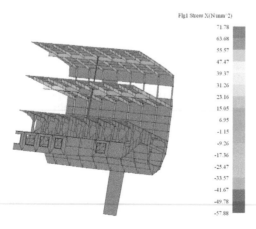

Figure 9.   Distribution of longitudinal stress σ_x on beams for a fore module in correspondence of a foil leg (N/mm^2).

torpedo has been detected. A further check on the displacements has been performed in order to assure the compatibility of the deformed shape of the aft structure with shaft alignment.

5 ANALYSIS OF DETAILS

The structural analysis of ALISWATH vessel has been completed by a detail investigation on most critical areas individuated by previous calculations. Some portions of the complete model have been schematised by the "refine mesh" command available in the MAE-STRO menu. The first one to be investigated is the fore leg connection to the submerged body (fig. 10). Particular attention has been devoted to the torpedo internal structures corresponding to the fore leg attachment. In this model the NACA profile of the fore leg has been simplified by constant thickness shell elements. This approximation didn't affect the results, being the leg structure a steel thick section with high stiffness. As an example, some results relative to strain (fig. 11) and Von Mises stress distribution (fig. 12) are presented.

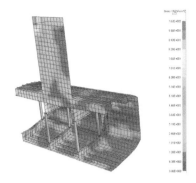

Figure 12. Von Mises equivalent stress distribution of the fore leg model.

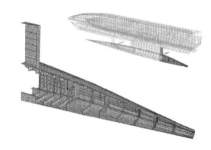

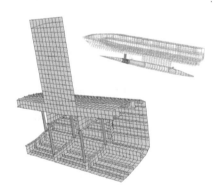

Figure 13. Refined mesh of the aft part of torpedo.

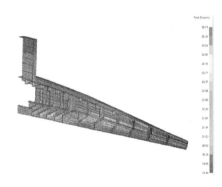

Figure 10. Refined mesh of the fore leg connection to the torpedo.

Figure 14. Deformed shape of the aft part of torpedo.

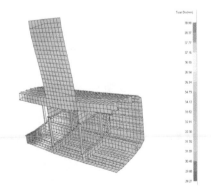

Maximum stress takes place in the steel leg where an intensity of about 100 N/mm^2 is reached at the attachment to the hull. In the light alloy structures the highest stress keeps below 70 N/mm^2.

A second examined detail regards the aft part of the submerged body (fig. 13), where displacements should be carefully verified in order to check the axis alignment. Attained results confirmed those coming out from the global model analysis: the maximum measured differential displacements between the aft

Figure 11. Deformed shape of the fore leg model.

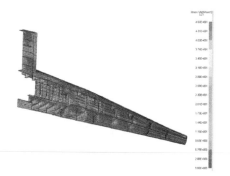

Figure 15. Von Mises equivalent stress distribution on the aft part of torpedo.

Figure 16. Refined mesh of the hull central keelson.

Figure 17. Deformed shape of the hull central keelson.

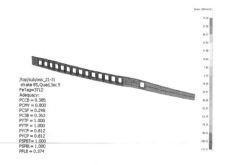

/top/sub/oes_21-31
strake 05,Quad,Sec 5
FeTag=3712
Adequacy:
PCCB = 0.385
PCMY = 0.800
PCSF = 0.248
PCSB = 0.363
PYTF = 1.000
PYTP = 1.000
PYCF = 0.812
PYCP = 0.812
PSPBT = 1.000
PSPBL = 1.000
PFLB = 0.374

Figure 18. Longitudinal σ_x stress distribution on the hull central eelson (N/mm^2) and "adequacy parameter" table.

engine flange and the propeller are maintained within 5 mm. Displaced pattern is shown in fig. 14. For what the stress level is concerned maximum Von Mises stress intensity keeps below 50 N/mm^2, as shown in fig. 15.

Owing to the very thin thickness of hull and submerged body shell and of many beam webs a buckling verification should be performed. A third analysis has then been carried out on the central hull keelson where buckling phenomena can be induced by the presence of lightening holes. This verification is based on DNV (1995) buckling strength approach implemented in MAESTRO options.

For each kind of collapse mode the code compares the elements stresses with the correspondent failure stress, obtaining ratios called "safety factors" (Paik, 2005). The "adequacy parameter" is represented by the ratio between the actual stress value and the admissible ones deriving from DNV HSC Rules. This is a non dimensional value ranging between −1 e +1: the 0.0 value indicates that the required safety factor is just met, the positive value means that the level of stress in the structure is within admissible limits while negative values mean that the structure is in critical situations. For the detail under investigation general tables and colour coded plots are provided showing the safety factors with regards to the possible collapse modes.

The refined mesh of the keelson is presented in fig. 16; deformed shape and longitudinal stress distributions are shown in fig. 17 and 18 respectively. In this last figure the table with "adequacy parameter" relative to possible collapse modes is included. Each symbol refers to a particular failure mode: the one relative to shell buckling is named "PFLB" (panel failure, local buckling).

6 ALISWATH CONSTRUCTION

The construction of the ALISWATH has been carried out in the RODRIQUEZ shipyards of Messina (Italy). The shipyard gained a very long experience in aluminium light alloy constructions since the 60's, when they started up the production of hydrofoil vessels. In 1990 a new fast ferry class, named Aquastrada, made the first appearance; since then, more than 20 ships of

Figure 19. A complete module of the hull.

Figure 20. The assembling of the modules on the building slip.

Figure 21. A view of the hull during completion.

Figure 22. Outer view of submerged body with aft connecting structure.

this kind from a length of 90 m up to 130 m, have been realised.

The construction of ALISWATH has been carried out separately for the hull and the torpedo. The hull has been divided into 6 blocks and they have been assembled in the building slip of the shipyard. Some significant stages of the construction are shown in fig. 19, 20 and 21.

The torpedo has been built in 5 blocks and assembled in the same shipyard; the complete structure is shown in fig. 22. A inner view in correspondence of the machinery room is shown in fig. 23. The hull and the

Figure 23. Inner view of submerged body in machinery zone.

submerged body, fitted with a small part of outfitting, are going to be connected each other and the whole vessel structure will be completed at the beginning of the new year.

7 CONCLUDING REMARKS

The design of the new hybrid vessel ALISWATH is the result of the cooperation between Rodriquez Shipyards and the Department of Naval Architecture of Genova University. The unit features a very low wake wash at relatively high speed. Through an iterative optimization procedure, developed by FEM computations, the structure has been modified up to reach a final version suitable for prototype construction.

The last version has been obtained by studying separately the hull and the submerged body. In the second phase of the research, herein presented, the complete numerical model has been extensively examined under the same loading conditions previously considered.

The results obtained by the two procedures showed a good fitting so confirming the reliability of all the adopted assumptions. At this stage the attention has been devoted to further investigate some structural details on which highest stress level was expected.

While the FEM model was improved the prototype construction proceeded and the necessary modifications realized, up to the completion of hull and submerged body, separately built in two distinct Rodriquez shipyards.

In the next future the hull will be launched and transferred in Messina yard, where the two parts will be connected in a dry dock.

An experimental campaign will then be performed on the real scale prototype to compare measured stress and strain values with those obtained by numerical investigation.

ACKNOWLEDGEMENTS

The authors wish to acknowledge the late lamented Mr. A. Sculati, the designer of ALISWATH, for his support to the development of this paper.

REFERENCES

Boote D., Colaianni T., Mascia D., 2005 "ENVIROAL-ISWATH: Analisi FEM delle strutture del siluro con il codice MAESTRO.", Rapporto interno DINAV ROD-STR 007, Genova (Italy).

Boote D., Colaianni T., Mascia D., 2006a, "ENVIROAL-ISWATH: Analisi FEM delle strutture dello scafo con il codice MAESTRO.", Rapporto interno DINAV ROD-STR 010, Genova (Italy).

Boote D., Colaianni T., Mascia D., Sculati A., 2006b, "ENVI-ROALISWATH: Structural Design of an Advanced Passenger Vessel", Proceedings of the International Conference on Ship and Shipping Research, NAV 2006, Genova (Italy).

Det Norske Veritas, 1995 "Buckling Strength Analysis", Classification Notes n.30.1, Hovik, Norway.

Hughes O., 1995, "MAESTRO User and Application Manuals", Maryland USA.

Krilov S.R.I., 2006, "ENVIROALISWATH: Rough Sea Investigation on 1:6 Scale Model", Project Report, St. Petersbourgh.

Paik J.K., Hughes O., Hess P.E. Renaus C., 2005, "Ultimate Limit State Design Technology for Aluminium Multi-Hull Ship Structures", Transaction SNAME, Vol. 113, 2005.

Analysis and Design of Marine Structures – Guedes Soares & Das (eds)
© 2009 Taylor & Francis Group, London, ISBN 978-0-415-54934-9

Multi-objective optimization of ship structures: Using guided search vs. conventional concurrent optimization

J. Jelovica & A. Klanac

Helsinki University of Technology, Department of Applied Mechanics, Marine Technology, Finland

ABSTRACT: Structural optimization regularly involves conflicting objectives, where beside the eligible weight reduction, increase in e.g. safety or reliability is imperative. For large structures, such as ships, to obtain a well-developed Pareto frontier can be difficult and time-demanding. Non-linear constraints, involving typical failure criteria, result in complex design space that is difficult to investigate. Evolutionary algorithms can cope with such problems. However they are not a fast optimization method. Here we aim to improve their performance by guiding the search to a particular part of Pareto frontier. For this purpose we use a genetic algorithm called VOP, and use it for optimization of the 40 000 DWT chemical tanker midship section. Beside weight minimization, increase in safety is investigated through stress reduction in deck structure. Proposed approach suggests that in the first stage one of the objectives is optimized alone, preferably more complicated one. After obtaining satisfactory results the other objective is added to optimization in the second stage. The results of the introduced approach are compared with the conventional concurrent optimization of all objectives utilizing widespread genetic algorithm NSGA-II. Results show that the guided search brings benefits particularly with respect to structural weight, which was a more demanding objective to optimize. Salient optimized alternatives are presented and discussed.

1 INTRODUCTION

Design of modern ships introduces new complex structural solutions that must follow the increasing demand for more reliable and safe products. Innovation has become necessity which ensures survival in the market, and it requires improvements of multiple conflicting ship attributes. However, available time does not follow the increasing complexity of design procedure, thus more advanced support systems are required that can assist designers. This is conveniently performed through the optimization process, but with obstacles on the way.

Early design stage lacks precise information on e.g. loading or structural details, while the bounds of some requirements, such as e.g. weight, vertical centre of gravity, nominal stress levels or length of weld meters are not precisely defined. In general it is then useful to venture into analysing correlation between them and investigate their sensitivity for the considered structural arrangement.

Complex ship structures involve large number of variables and even larger number of constraints. Variables are in structural optimization regularly discrete, whether they represent element size, material type, stiffener spacing etc. Constraints are non-linear and non-convex typically involving yielding and buckling of structural elements. These reasons confine the choice of possible optimization algorithms to those that do not require gradient calculation of constraints

and objective functions. Evolutionary algorithms have shown capability to handle such problems and provide sufficient benefits for the structure. Their prominent representative, genetic algorithm (GA), is used in this study. Several applications have shown that GA is a successful tool for practical problems in ship structural design and optimization, see e.g. Nobukawa & Zhou (1996), Klanac (2004), Romanoff and Klanac (2007), Ehlers et al. (2007), Klanac and Jelovica (2008).

Genetic algorithm operates in the space of design objectives, by having multiple design alternatives at hand when deciding where to continue the search from generation to generation. This number of available solutions is known as a population size and should grow with the number of considered variables. Literature suggests using population size in range from 50 to 500; see Osyczka (2002), Deb (2001). This lengthens the optimization process even for a simple engineering problem, so that the number of generated and evaluated designs before reaching the optima can be more than several thousands. Clearly, this can be rather costly when optimizing large ship structures, especially if *Finite Element Method* is applied for structural assessment. In any case, optimization should be short, and if it is time-consuming, it is often, for convenience, stopped prematurely, immediately after noticing some improvements in objective values, and without attaining their optimal values. Making relevant conclusions based on such results can be misleading and costly in the later stages.

Several conflicting objectives that are typically interesting for ship structural optimization, e.g. weight and safety, form a distinctive Pareto frontier which gains in size with problem. Conventional and widespread multi-objective GAs, e.g. NSGA-II (Deb et al. 2002) or SPEA2 (Zitzler et al. 2002), attempt to attain whole Pareto frontier in one run, so that in the very end all Pareto optimal designs are attained. Designer then has a possibility to consider many alternatives that possess different objective values which are then selected based on some preference. But as shown in this paper limited population size restricts such algorithms to fully reach extreme parts of Pareto frontier. Those extreme locations (for which some objective is at minimum/maximum, regardless of the others) contain possibly innovative design principles that can provide new knowledge on possibilities of the structure. Recent study by Klanac et al. (2008) showed that increased crashworthiness of a ship side can be accomplished without significant sacrifice in weight, contrary to traditional belief. Such conclusion was possible by comparing the edges of the Pareto frontier.

In this study we consider a way to avoid unnecessary increase in number of evaluated designs to reach desired parts of a Pareto frontier. Simply said, sometimes the whole frontier is not required to be contained in the final population. Optimization can consist of several parts, each exploring a different part of the frontier. User knows his preference toward objectives included in the optimization. Progress can then be monitored and re-directed if considered appropriate, for example in the case of non-satisfactory results or simply different aspect of the structure wants to be known. Alternatives are then moving along the frontier towards the instructed direction. This is based on the assumption that Pareto optimal solutions predominantly share common variable values, see Deb & Shrinivasan (2006), so that the transition along the Pareto frontier should not require significant changes in the design and should be quick. To allow this manipulation, optimization progress must be monitored in order to conclude on the proper moment for changing the direction of the search.

To show the benefits of this approach we use a simple GA called VOP, and compare it with NSGA-II, a recognized algorithm that possesses several advanced features. VOP optimizes both constraints and objectives by using the *vectorization* principle. NSGA-II concurrently optimizes all the objectives with equal importance and works by utilising existing solutions in the front. It has difficulties to operate with single-objective optimization case to sufficient extent as there is simply no frontier then and it recombines the dominated alternatives in the population for the same purpose.

To demonstrate this comparison, a structure of a 40 000 DWT and 180 m long chemical tanker will be optimized for two objectives: minima of weight and maximum of adequacy of deck structure.

In the continuation, we will use the term 'non-dominated frontier' instead of the Pareto frontier for the results we obtain, since the evolutionary algorithms strive to it for real-world engineering problems, but reach only certain designs which, when filtered, form non-dominated front.

In the following chapter we revisit theory behind the proposed approach, provide arguments for validity and show how re-formulated optimization statement can be utilized. Chapter 3 describes the VOP algorithm. Chapter 4 compares the optimization of the tanker problem by VOP and NSGA-II. Last two chapters discuss the findings and conclude the paper.

2 GUIDING MULTI-OBJECTIVE OPTIMIZATION

The original multi-objective structural optimization problem of M objectives and J constraints can be formalized with:

$$\min_{x \in X}\{f_1(x),\ldots,f_M(x)|g_j(x) \geq 0, \quad j \in [1,J]\} \qquad (1)$$

where we search for design alternatives x in the total design space X confined within variable bounds. Goal is to find such x that minimizes the objectives while satisfying all the imposed constraints. If constraints are satisfied, design is called feasible and belongs to a feasible set $\hat{\Omega}$

$$\Omega = \{x \in X|g_j(x) \geq 0, \quad j \in [1,J]\}. \qquad (2)$$

The solution of Equation 1 is a Pareto optimal alternative x^* which is non-dominated by other feasible alternatives, i.e. there is no alternative better than x^* in the objective space Y (whose feasible part is denoted with $Y^{\hat{\Omega}}$). Such alternative represents then a rational choice and it belongs to a set of Pareto optima $\hat{\Omega}$, called also the Pareto frontier, defined as:

$$\hat{\Omega} = \{x \in \Omega| \not\exists x^k, \quad f(x^k) < f(x), \quad \forall x^k \in X \backslash x\}. \qquad (3)$$

2.1 Concurrent-search multi-objective optimization

Standard concurrent-search multi-objective GAs seek the whole Pareto frontier in a single optimization run, requiring large population size to store all the encountered non-dominated solutions. Their working principle is based on recombination of such designs to yield new and better ones, thus their convergence is threatened when population size is not adequate. Progress direction of such optimizers in objective space can be

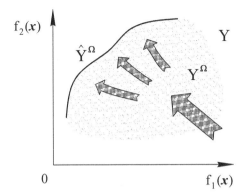

Figure 1. Standard concurrent approach to GA-based multi-objective optimization, shown in the bi-objective case (first objective to be minimized and second to be maximized).

seen in Figure 1 for the case of two objectives. Initial population spreads into multiple 'streams' where each progresses to different parts of the non-dominated frontier. Direction of advancement is pre-defined by non-domination concept, and the algorithm is intended to discover even the outermost designs in the frontier. If the results are not satisfactory enough, one can only tediously run the algorithm further, hoping to reach improvements in particular objective without a possibility to affect on the process.

2.2 Guided search approach

In this approach only a part of the Pareto frontier is searched based on the instructed direction. To allow control over this direction, specific manipulative weighting factors are applied. Before such manipulation can take place, problem statement in Equation 1 is re-defined into vectorized form following Klanac & Jelovica (2008):

$$\min_{x \in X}\{f_1(x), \dots, f_M(x), f_{M+1}(x), \dots, f_{M+J}(x)\}. \quad (4)$$

where constraints $g_j(x)$ are now treated as additional objectives $\{f_{M+1}(x), \dots, f_{M+J}(x)\}$, after being converted with the Heaviside function, given as

$$f_{M+j}(x) = \begin{cases} -g_j(x), & \text{if } g_j(x) < 0 \\ 0, & \text{otherwise} \end{cases}, \quad \forall j \in [1, J] \quad (5)$$

Control over a particular objective is gained by multiplication of its normalized value within one GA population with the weighting factor w_k:

$$\min_{x \in X}\{w_k \bar{f}_k(x)\}, \quad \forall k \in [1, M+J]$$
$$\text{s.t.} \, 0 \le w_k \le 1, \quad \sum_k w_k = 1 \quad (6)$$

where normalization of objective f_k for design i is linearly performed using

$$\bar{f}_k(x, i) = \frac{f_k - \min_{\forall x \in X^i} f_k}{\max_{\forall x \in X^i} f_k - \min_{\forall x \in X^i} f_k}. \quad (7)$$

Increased weighting factor leads to stronger minimization of corresponding objective and vice versa. For convenience, constraints can be set to share the same weighting factor. Figure 2a and b show the principle of guided search for two different directions of search. Guided search is divided in two stages, first to reach the frontier, and second to generate non-dominated designs, exploring thus the possibilities of the optimized structural arrangement. Weighting factors are altered freely by the user, in any manner and whenever desired. But the manipulation should be based on heuristics and not on random choice.

The basic idea of weight factor manipulation is to direct, or guide the 'cloud of alternatives' in search during the optimization process which would 'leave behind' a trail of 'good' alternatives.

Deb and Srinivasan (2006) indicate that the Pareto optimal design alternatives of one system possess many commonalities. If this is considered to apply for our problem, Pareto optima obtained in the first phase would share then most of the variable values

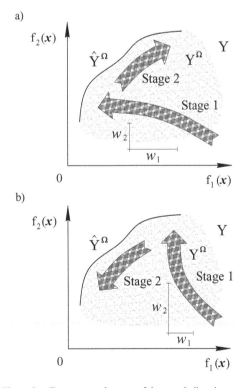

Figure 2. Two proposed routes of the search direction.

449

with the Pareto optima obtained in the later phase(s) of optimization. Decomposing, therefore, the overall optimization problem as proposed should not negatively affect on the possibility to generate Pareto optima. On the contrary, it should affect positively, since a multi-objective GA would perform then a lesser amount of variable changes to generate Pareto frontier than if optimization would have been done concurrently.

3 VOP—A GA FOR GUIDED SEARCH

To guide the search in multi-objective optimization of ship structures, we employ a GA called VOP (Klanac and Jelovica 2007, 2008, Klanac et al. 2008a, 2008b, Jelovica 2008). VOP is a binary coded algorithm consisting of: a) a fitness calculator, b) the weighted roulette wheel selector operating on the basis of computed fitness values, and c) a subroutine executing the single-point cross-over with a probability of p_C and the bit-wise mutation with a probability of p_M. These are standard operators and are, except for the fitness calculator, elaborated in Jelovica (2008), Klanac and Jelovica (2008c).

VOP's fitness function is defined as:

$$\varphi_1(\mathbf{x}, i) = \left(\max_{\mathbf{x} \in X^i} [d(\mathbf{x}, i)] - d(\mathbf{x}, i) \right)^{\bar{d}(\mathbf{x}, i)} \tag{8}$$

where the design's distance $d(\mathbf{x}, i)$ from the origin of normalized objective space is obtained as:

$$d(\mathbf{x}, i) = \left\{ \sum_k [w_k \bar{f}_k(\mathbf{x}, i)]^2 \right\}^{1/2}, \quad \forall k \in [1, M + J] \tag{9}$$

Minimization of the distance $d(\mathbf{x}, i)$ replaces the problem from Equation 6. Definition of the weighting factors w is elaborated in the actual example that follows.

4 OPTIMIZATION OF A TANKER STRUCTURE

Guided search for optimal solutions is examined on the 40 000 DWT and 180 m long chemical tanker's midship section. Its main frame longitudinal elements are optimized for the smallest allowed scantlings, while keeping the topology of the structure unchanged. The tanker's arrangement, as seen in Figure 3, is characterized with two internal longitudinal cofferdams bounded with double sides and double bottom structure. The tank's plating is built from duplex steel to resist the aggressive chemicals which are transported and is the only part of the structure with yield strength of 460 MPa. For the remaining structure, 355 MPa steel is used.

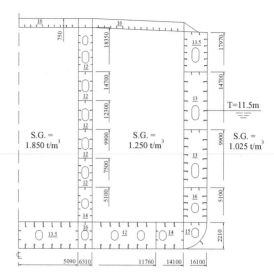

Figure 3. Half of the main frame section of a considered tanker and scantlings of the transverse structures (underlined).

4.1 Variables

The structure of one half of the ship's main frame is divided into 47 longitudinal strakes characterized by plate thickness, stiffener size and number, material type and, additionally, size of transversal elements. Former two are varied in this study while the later three are kept constant. Thus two variables are considered per each strake, so the optimization problem consists in total of 94 variables. Number of stiffeners per each strake is depicted in Figure 3 together with scantlings of the transversal structure. Considered variable values are discrete, having the step for the plates of 1 mm, a value in general appropriate for early design stage. Plate thicknesses in double bottom and double side structure are assumed to be available from 8 to 23 mm, except of stringers whose range is the same as for longitudinal bulkhead and the deck, 5 to 20 mm. Stiffener sizes are taken as standard *holland* profiles; see Rukki (2008).

4.2 Structural model and constraints

The midship section is assumed to stretch between L/4 and 3L/4 cross-sections, without the change in scantlings. It is subjected to the normal service loads, those being the hull girder loads, the cargo loads and the lateral hydrostatic loads, while ballast tanks are assumed to be empty. Pressure loads are calculated from liquid density indicated in Figure 3, while global loads are tabulated in Table 1.

The response under the hull girder loads is calculated applying the numerical Coupled Beam method of Naar et al. (2005). On top of that is added the response

Table 1. Wave loads acting on a ship.

Loading condition	Magnitude	Location
Sagging		
Vertical bending moment	2 452 000 kNm	L/4
Vertical shear force	74 880 kN	L/2
Hogging		
Vertical bending moment	2 932 000 kNm	L/4
Vertical shear force	72 960 kN	L/2

of the panel under the cargo and hydrostatic loads, calculated with uniformly loaded simple beam.

Each strake is checked for eight failure constraints concerning plate yield and buckling, stiffener yield, lateral and torsional buckling, stiffener's web and flange buckling and crossing-over. These criteria are taken from DNV (2005), Hughes (1988) and Hughes et al. (2004). The last constraint is used to ensure controlled panel collapse due to extensive in-plane loading, where plating between stiffeners should fail first; see Hughes et al. (2004). Physically this means that the panel is not allowed to consist of thick plate and weak stiffening, and the stiffener size has to rise with plate thickness. However, in this study the cross-over constraint is activated only when stresses in stiffeners and plates exceed 2/3 of their buckling or yield strength, since it is pointless to consider controlled collapse if the collapse is unlikely to occur. Altogether 376 failure criteria are calculated for each loading condition, which raises their total number to 1504. They are transformed into adequacies, effectively describing an optimization constraint. Adequacy is considered as a non-linear normalization function between the structural capacity of some structural element j, $a_j(x)$, and a loading demand acting on it, $b_j(x)$, as proposed in Hughes et al. (1980):

$$g_j(x) = \frac{a_j(x) - |b_j(x)|}{a_j(x) + |b_j(x)|} \qquad (10)$$

4.3 Objectives

Two objectives are considered in this study: minimize the total weight of hull steel (abbreviated as HULL) and maximize the adequacy of deck strakes (abbreviated as ADEQUACY). Minimizing the weight would increase the payload capacity and to certain extent provide cost savings. By introducing the latter objective, the goal is to explore the needed trade-offs when increasing the safety of some part of the structure. In this case this is the safety of deck structures which are according to experiences prone to failures, e.g. buckling or fatigue. To simplify the process, all the adequacies of the deck panels can be summed into one function which is in the end treated as the objective.

The validity of such an approach is shown in Koski & Silvenoinen (1987).

The total weight of the hull is calculated by extending the obtained cross-sectional weight for the whole length of the ship, on top of which the weight of web frames of 21.4 t each 3.56 m is added.

Maximization of deck adequacies also positively influences on the feasibility of design alternatives. Although treated as the objective, they are also the constraints which, in case of negative value, receive double penalization: first they deteriorate the distance function in Equation 9, coming from their transformation using Equation 5 and second, their negative values decrease the sum that is meant to be maximized. This will then lead to the strong penalization of the alternatives with large infeasibility, while those with smaller infeasibility will become preferred, again leading to the increase in safety. If on the other hand, the adequacy of some alternative is positive, its value as vectorized constraint will now be zero, while as objective it will remain positive, and the alternative can be freely maximized.

If HULL is presented on abscise and ADEQUACY on ordinate in the objective space on Figure 2, one can expect similar location and shape of the Pareto frontier in the tanker case.

4.4 Optimization using VOP: A guided search

The optimization is carried out with a population of 60 design alternatives. This is significantly smaller than recommended in the literature, but accounting for overall optimization time it is considered sufficient, based on some preliminary results of weight minimization for the same case reported in Klanac et al. (2008). GA parameters are kept constant during the optimization: crossover probability is set to 0.8, while the mutation probability is 0.003. Both values are set based on the literature (Deb 2001) and previous experiences of the case (Klanac et al. 2008).

Two optimization runs are performed, each following a different search direction. The intention is to see the influence of search path on the non-dominated frontier obtained in the objective space. Initially we decide to start the first run with only HULL minimization and optimize for ADEQUACY afterwards. We name this search direction 'Strategy H-A' accounting for the sequence of objective consideration. The second run takes opposite path between the two objectives and is abbreviated as 'Strategy A-H'. In fact, these two strategies correspond to the two example routes in Chapter 2, first one seen in Figure 2a and the later in Figure 2b.

To prevent any bias towards particular objective, both optimization runs are initiated with the same randomly generated population of design alternatives. Each strategy in the Stage 1 performs single-objective optimization to reach the different edge of non-

451

dominated frontier between HULL and ADEQUACY. Their weighting factors are accordingly set to emphasize improvements only in desired objective, as seen in Table 2. Reference value of the weighting factor is $2.646 \cdot 10^{-3}$, obtained from the fraction $1/(M + J)$, where M is taken as 2 and J equals to 376. Other weighting factors in the continuation are scaled relative to this value.

Stage 1 of the Strategy H-A is run until the point where the improvement rate becomes small, being the 1028th generation in this case, as seen in Figure 4. Assuming now that the further mass reductions will not be significant and that the 'light' alternatives have attained predominantly optimal variable values, Stage 2 is initiated by adding the ADEQUACY maximization to the minimization of HULL in order to generate the non-dominated frontier between them. To accomplish this, relative weighting factor of ADEQUACY is changed from 0 to 1, being now the same as for HULL. Thus the interest has moved to the 'middle' of the frontier, and the algorithm responds by starting to improve the ADEQUACY at the expense of the HULL; compare Figure 4 and Figure 5. Progress is monitored, and in the 1293th generation the interest is additionally moved towards the second objective to explore the frontier further. Optimization is stopped after 1500 generations, when it become obvious that current HULL values, nearing 8500 t, are too high to even be considered as possible solutions in reality. ADEQUACY was increased from 12.3 to 18.8 which was declared sufficient. Note that the second objective can theoretically take value from 0 to 32, consisting from four strakes having 8 constraints. But in order to have constraint value equal to one, stress in corresponding member should be zero, so obviously such a case cannot exist.

Computing time was approximately 4 days on normal PC, arising from 4 minutes that each generation requires.

Strategy A-H follows the same logic when reaching and exploring the non-dominated frontier, but it starts from the opposite direction, firstly searching for the maximum of ADEQUACY and in Stage 2 attaining the other edge of the frontier. The adjustments in the weighting factors during this optimization are presented in Table 2, with each leading the solutions to another part of the frontier. Initial increase in ADEQUACY is continued until generation 262 where certain 'plateau' is visible in Figure 5. Assuming no significant improvement in ADEQUACY is possible after that point, HULL is included in the search to attain the needed trade-offs. This leads to significant decrease of hull steel weight while retaining relatively sound values of ADEQUACY, see Figure 4 and Figure 5. What differs this strategy from the previous one is much higher number of variables that must be altered in order to come from maximal ADEQUACY design to minimal HULL solution. Also, we are in this optimization case more interested in reaching low hull steel weight, thus the algorithm is for Strategy A-H faced with demand to gain both extremes of the frontier. Therefore, the optimization is not stopped as in the Strategy H-A, but has to rapidly come from one end to the other. Optimization was terminated when meeting the stopping criterion of 1500 evaluated generations.

Table 2. Relative value of the weighting factors during the optimization for the two search directions (given values are obtained by fraction $w_{OBJECTIVE}/w_{CONSTRAINT}$).

	Stage	Generation	w_{HULL}	$w_{ADEQUACY}$	w_{CONSTR}
Strategy	1	1	1	0	1
H-A	2	1028	1	1	1
		1293	0.5	1	1
Strategy	1	1	0	1	1
A-H	2	262	1	1	1
		941	1	0.5	1
		1215	1	0.3	1

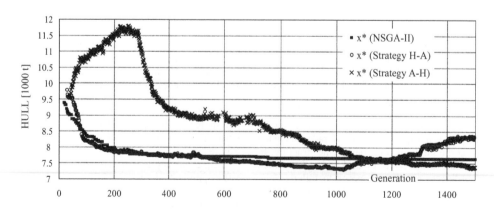

Figure 4. Progress of HULL optimization for VOP and NSGA-II, showing generation's best design.

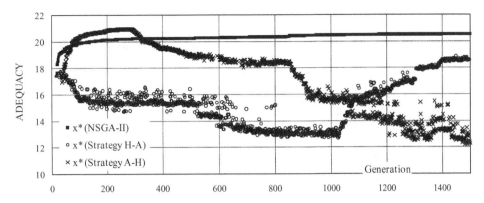

Figure 5.   Progress of ADEQUACY optimization for VOP and NSGA-II, showing generation's best design.

4.5   *Comparison with NSGA-II*

Binary coded version of NSGA-II (Deb et al. 2002) is applied to compare the results from the 'guided search' optimization described previously.

NSGA-II is characterized with:

- an elitism concept that ensures preservation of non-dominated solutions encountered from the beginning of the optimization run,
- ranking the solutions in the population according to non-dominated frontier in which it belongs to,
- giving advantage to solutions that belong to less crowded part of the objective space,
- constraint-domination which prefers any feasible solutions over the infeasible, or between two infeasible designs selects the one with less sum of violated constraints.

Optimization with NSGA-II is initiated using the same random population as VOP in Stage 1. The same crossover and mutation probabilities apply also. In difference to VOP, both HULL and ADEQUACY are considered from the beginning, and optimization is run conventionally for the same amount of generations. Optimization history for each objective is shown in Figure 4 and Figure 5.

5   DISCUSSION

Figure 6 depicts the essential comparison between the results of optimization performed with the 'guided search' approach, utilizing VOP, and with the standard multi-objective approach, utilizing NSGA-II. Minimizing solely HULL in the first stage of Strategy H-A, VOP managed to attain design with the hull weight of 7312 t, 330 t lower than with NSGA-II; see Table 3 for the alternatives located in the edges of the frontier, marked with '**' to simplify the notation. Continuing the optimization with the second stage, valuable

trade-offs between the two objectives are created up to the point where the further increase in HULL is inadequate.

It can be clearly seen in Figure 6 that the Strategy A-H performed worse. This confirms the suspicion that changing too many variables can be difficult for the optimizer. All 94 variables had to be altered here in the Stage 2, while in the previous strategy effectively only 8 lead to initial increase of ADEQUACY.

NSGA-II starts the optimization with relatively high ADEQUACY value since the initial designs have large scantlings of the structural elements and therefore the stresses in the deck are low. NSGA-II proceeds by spreading the frontier towards both ends, but the HULL as the more difficult objective, stalls the progress in that direction. Although the second objective reached much better values, their hull weight is too high for practical use; see Figure 6 and Table 3.

VOP's non-dominated frontier for the Strategy H-A nicely covers the one from NSGA-II for the HULL value in range 7600–8600 t. There are totally 191 non-dominated designs in the 'trail' left by guided search when maximizing for the second objective. Naturally, all of them are not interesting and their number is too large for this problem case, but can prove valuable when considering many objectives, or it can be filtered. NSGA-II on the other hand is limited by the population size that defines the maximal available non-dominated solutions in the end of optimization, which is 60 in this case.

5.1   *Trade-off design alternatives*

Although neither of design alternatives obtained in this study cannot be proved to be globally optimal due to complexity of the problem, we present two design alternatives in order to see how the structure resembles the imposed loading condition and the objective values. We select the design of minimal hull steel weight from the VOP's non-dominated frontier and one design

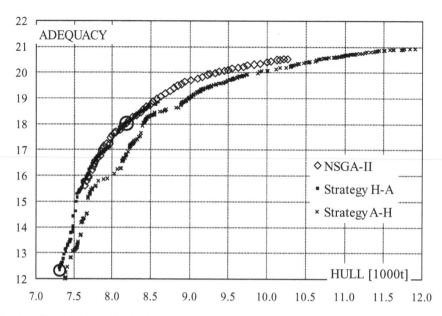

Figure 6. Overall non-dominated frontier from each of the three runs, showing also the two selected designs (O).

Table 3. Designs from the edges of non-dominated frontier (HULL values in tones).

	Guided VOP		NSGA-II	
	$x^{**}_{\mathrm{HULL(VOP)}}$	$x^{**}_{\mathrm{ADEQ(VOP)}}$	$x^{**}_{\mathrm{HULL(NSGA-II)}}$	$x^{**}_{\mathrm{ADEQ(NSGA-II)}}$
HULL	7312	8595	7641	10 262
ADEQ.	12.34	18.86	15.62	20.54

with increased ADEQUACY value for which the structural scantlings are still relatively acceptable. Both designs are shown in Figure 7 and their characteristics given in Table 4. The same table shows the differences in the adequacy values for the deck strakes between the two alternatives. Structure was not standardized before presented here nor was it given any corrosion addition.

Structural elements in double bottom, side shell and deck structure from the lowest hull steel weight design follow vertically the beam distribution of weight, as seen in Figure 7, in order to satisfy the area moment. Side has the lowest plate thicknesses and stiffener sizes, while the bottom elements are additionally increased to resist the water pressure. Inner bottom elements in a side cargo tank are larger than in the bottom because of the increased liquid density of 1.25 t/m^3, and the same is valid for the inner side. For the same tank, reduction in scantlings can be seen in the longitudinal bulkhead when going upwards in the direction of decreasing cargo pressure. The same

Table 4. Characteristics of two selected design alternatives, including the adequacy values of stiffener yield (Sy), plate yield (Py) and plate buckling (Pb) per each deck strake.

	$x^{**}_{\mathrm{HULL(VOP)}}$			$x_{\mathrm{TRADE-OFF}}$		
HULL [t]	7312			8177		
ADEQ.	12.34			18.04		
Act. con.	63			49		
Constraints	Sy	Py	Pb	Sy	Py	Pb
Deck-CL	0.462	0.15	0.349	0.599	0.33	0.564
Deck-2	0.356	0.097	0.239	0.512	0.282	0.462
Deck-3	0.432	0.131	0.226	0.575	0.317	0.564
Deck-side	0.352	0.092	0.177	0.507	0.276	0.473

happens in a central tank but with generally higher plate thicknesses and stiffener sizes due to higher density. The deck and inner bottom strakes are in that tank also larger than in neighboring strakes.

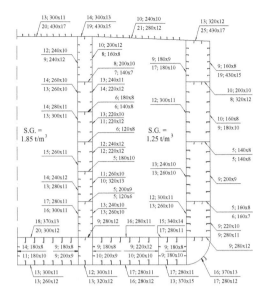

Figure 7. Scantlings of the main-frame members for the design $x^{**}_{HULL(VOP)}$ (shown above the dimension lines) and design $x_{TRADE-OFF}$ (shown below). One set of scantlings is shown for the strakes that are the same in both alternatives.

When the design of the lowest hull steel weight is compared to the one selected from the non-dominated frontier with increased ADEQUACY value, most salient differences can be observed in deck and surrounding strakes. As expected, plates are thickened and stiffeners are enlarged in the deck, but also in a sheer strake since it contributes to stress reduction in deck. The highest strakes in the bulkhead, next to the deck, posses decreased scantlings in order to yield weight savings.

Other minor differences between the two design alternatives can be assigned to the working principle of the GA which makes certain variations between possible variable values when optimizing the structure.

Reduction of stresses in the deck leads also to a decrease in number of active constraints for such solutions, as can be seen for the two previously described alternatives in the Table 4, where constraint is considered to be active if stress exceeds 3/4 of its critical value.

6 CONCLUSION

To enable more qualitative decision-making in the beginning of design process, it is helpful to posses different trade-offs between crucial objectives. We have shown on example of the main frame of 40 000 DWT chemical tanker that this can be done using *vectorized genetic algorithm*—VOP in a way adoptable to the designers' needs. Relying only on the "black-box" approach one cannot expect to gain desirable results. Certain parts of the non-dominated frontier might be undiscovered or their attainment would be stipulated with quite long optimization run. Even then the possibilities to improve the structure in certain sense would not be known.

In this study a different approach is taken: firstly to optimize for the best possible design according to specific objective, and secondly, explore the non-dominated frontier by including the other objective to certain extent. This was achieved by weighting the importance of particular objective to steer the cloud of design alternatives in desired direction. This 'guided search' resembles then the desires of the user, who can particularly benefit from it when understanding the problem at hand. In that sense, it is the best to start the optimization with the 'most difficult' objective that requires manipulation of the highest number of variables. After reaching the optima in the difficult objective, it is relatively easy to attain its other parts that depend only on several variables, the point at which one changes weight factors can be different: either satisfactory results are obtained, or improvements, in terms of objectives, became rather poor, as was in our case.

In the future, the 'guided' search methodology should be tried on more practical examples involving ship structures, to allow for further testing of the concept. Nevertheless, some fundamental issues remain to be studied further:

– Finding best strategy for three and more conflicted objectives,
– Analysis of the heuristics of weight factors,
– Application of the 'guided' search approach to a widespread algorithm, e.g. NSGA-II.

ACKNOWLEDGMENTS

The authors gratefully acknowledge the support of IMPROVE project, funded by European Union (Contract nr. 031382- FP6 2005 Transport-4), and the Technology Development Centre of Finland—TEKES, including Finnish shipbuilding industry, through the project CONSTRUCT.

REFERENCES

Deb, K. 2001. *Multi-Objective Optimization Using Evolutionary Algorithms*. Chichester: John Wiley & Sons.
Deb, K., Pratap, A., Agarwal, S., Meyarivan, T. 2002. A Fast and Elitist Multiobjective Genetic Algorithm—NSGA-II, *IEEE Transactions on Evolutionary Computation*. 6/1: 182–197.
Deb, K., Srinivasan, A. 2006. Innovization: innovating design principles thru optimization. Proc. of the 8th

annual conference on Genetic and evolutionary computation. Seattle: 1629–1636.

Det Norske Veritas 2005. *Rules for the classification of steel ships*. Høvik.

Ehlers, S., Klanac, A., Tabri, K. 2007. Increased safety of a tanker and a RO-PAX vessel by implementing a novel sandwich structure. 4th Int. Conference on Collision and Grounding of Ships. Hamburg:109–115.

Hughes, O.F. (1988), Ship Structural Design. Society of Naval Architects and Marine Engineers. New York:Wiley.

Hughes, O.F., Ghosh, B., Chen, Y. 2004. Improved prediction of Simultaneous local and overall buckling of stiffened panels. *Thinn-Walled Structures*, 42: 827–856.

Hughes, O.F., Mistree, F., Zanic, V. 1980. A Practical Method for the Rational Design of Ship Structures. *J. Ship Research* 24(2): 101–113.

Jelovica, J. 2008. *Vectorization of mathematical programming for ship structures using genetic algorithm*. Master's thesis, Univesity of Rijeka, Rijeka.Klanac, A., Jelovica, J. 2007. *Vectorization in the structural optimization of a fast ferry*, Brodogradnja (Shipbuilding), 58: 11–17.

Klanac, A., Jelovica, J. 2008. Vectorization and Constraint Grouping to Enhance Optimization of Marine Structures. *Marine Structures* doi:10.1016/j.marstruc.2008.07.001 (in press)

Klanac, A., Ehlers, S., Jelovica, J. 2008a. Rational Increase of Safety of Tankers in Collision: Structural Optimization for Crashworthiness. Submitted to *Marine Structures*. Available at: http://www.tkk.fi/Units/Ship/Personnel/Klanac/index.html

Klanac, A., Jelovica, J., Niemeläinen, M., Damagallo, S., Remes, H., Romanoff, J. 2008b. Structural Omni-Optimization of a Tanker. 7th International Conference on Computer Applications and Information Technology in the Maritime Industries—COMPIT '08, Liege.

Koski, J., Silvennoinen, R. 1987. Norm Methods and Partial Weighting in Multicriterion Optimization of Structures. *International Journal for Numerical Methods in Engineering* 24: 1101–1121.

Naar, H., Varsta, P., Kujala, P. 2004. A theory of coupled beams for strength assessment of passenger ships, *Marine Structures* 17(8): 590–611.

Nobukawa, H., Zhou, G. 1996. Discrete optimization of ship structures with genetic algorithm. *Journal of The Society of Naval Architects of Japan* 179: 293–301.

Osyczka, A. 2002. *Evolutionary Algorithms for Single and Multicriteria Design Optimization*. New York: Physica-Verlag.

Osyczka, A., Krenich, S., Tamura, H., Goldberg, D.E. 2000. A Bicriterion Approach to Constrained Optimization Problems Using Genetic Algorithms. *Evolutionary Optimization*—An International Journal on the Internet 2(1): 43–54.

Romanoff, J., Klanac, A. 2007. Design Optimization of a Steel Sandwich Hoistable Car-Decks Applying Homogenized Plate Theory. 10th International Symposium on Practical Design of Ships and Other Floating Structures—PRADS, Houston.

Rukki, 2008. *Hot Rolled Shipbuilding Profiles*.

Zitzler, E., Laumanns, M., Thiele, L. 2002. SPEA2: Improving the Strength Pareto Evolutionary Algorithm for Multiobjective Optimization. In K. Giannakoghu, et al. (eds.), *Evolutionary Methods for Design, Proc. Intern. Symp. Optimisation and Control—CIMNE, Barcelona:* 1–6.

Digital prototyping of hull structures in basic design

J.M. Varela, Manuel Ventura & C. Guedes Soares
Centre of Marine Technology and Engineering (CENTEC), Technical University of Lisbon,
Instituto Superior Técnico, Lisboa, Portugal

ABSTRACT: This work describes a computer system developed for the fast parametric generation of a 3D model of the ship hull structure. The system is intended to be used as a tool for generating an initial product model of the hull structures at the basic design stage. The model provides not only the geometry of the hull and the main structural systems but also the data describing the arrangement of plates and stiffeners of the component panels, including scantlings, spacings and materials. The objective of the system is to generate and help to evaluate alternative structural configurations, producing information on total or partial weights and centres of gravity at different levels of detail. For the intended purpose, the easiness and speed of generation of a model is much more relevant than a high geometric accuracy, and so it was adopted a simplified geometric modelling, representing curved shapes by polygonal approximations. The modelling concepts and methodology are discussed and the architecture of the system is presented, describing the main components. Finally, some results are shown from the partial modelling of a real ship that was used for testing and validation of the system.

1 INTRODUCTION

Existing CAD systems present some limitations concerning the generation of product data models. Generic CAD systems can produce a geometric model at the cost of extensive interactive work but without all the associated product data. Another category of systems such as IntelliShip and CATIA Ship Structural Detail, result from the development of tailor made customisations of generic mechanical CAD systems applied to shipbuilding. These systems although in principle able to be used on the basic design, in practice are generally used on the detailed design and are reported to originate large data files due to the conversion process used. On the other hand, specific PDM systems used in shipbuilding such as TRIBON Hull (AVEVA Group), FORAN (SENER Engineering), NUPAS-CADMATIC (Cadmatic Oy) and NAPA Steel (NAPA Oy), are mainly oriented to production and must provide high accuracy and therefore demand a lot of detailed and extensive input. These systems are not compatible neither to the time constrains nor to the degree of design development for their application during the basic design stage. More recently a new type of systems for hull design is being developed by some classification societies such as American Bureau of Shipping (SafeHull Express), Bureau Veritas (VeriSTAR Hull), Det Norske Veritas (Nauticus Hull), Germanischer Lloyd (Poseidon), RINA (Leonardo Hull). Typically these systems integrate a 3D modeller that is mainly a pre-processor for FE analysis and an rules compliance checking module. Some of these systems apply only to a limited set of hull types (for example, bulk-carriers and tankers) or to a limited part of the hull (cargo area). In addition, although oriented for the basic design stage, these are generally closed systems, i.e., they provide limited or inexistent data exchange capabilities and therefore they do not provide easy access or re-use of the developed 3D model or of the associated product data.

A system targeting the initial stages of ship design should be able to produce a 3D product data model of the hull with reduced input and should also be flexible enough to allow minor alterations of the hull form maintaining the same structural configuration or to evaluate alternative structural arrangements for the same hull form. Such a system could provide much better estimates of the hull weight by comparison to the usual statistics based empiric formulas from bibliographic references or to the accumulated historic data of the designer. Better estimates than the statistical ones due to two main reasons. First, because they could take into consideration the actual configuration of the hull (number of bulkheads, number of decks, frame spacing, etc.) instead of only a set of main dimensions and a few additional parameters used in the correlation formulas. Second, because they would not be dated, i.e., dependent of the age and the configuration of the ships' sample on which the statistics were based. In addition, it would provide the capability to divide the weights in several categories (by structural system, by plates and profiles, by planar and curved, by mild steel and high tensile steel, etc.). This type of analysis allows much improved production cost estimates of both material and labour.

Research on the development of 3D models of hull structures during the basic design started more then 20 years ago with systems such as the BRITSHIPS2 developed by BSRA and Swan Hunter shipyards for the structural design (Forrest and Parker, 1982). Since then many other systems have been developed for different applications such as evaluation of labour during the production stages (Bong et al, 1990), design and analysis of the structures (Jingen and Jensen, 1982; Na et al, 1994), generation of models for FE analysis (Kawamura et al, 1997), data exchange for class approval (Hwanga et al, 2004); design and assembling for shipbuilding (Aoyama and Nomoto, 1999) and block division and process planning (Roh and Lee, 2006).

This works describes a computational system developed for the fast generation of a 3D product data model of a part or the totality of the hull structure of a new or an existing ship. The system works with a minimal set of input data, and the resulting model has the level of detail required for the intended purpose. It is based on the object-oriented concept and was entirely implemented in C++ programming language. Due to the academic nature of the work, open source components were used as much as possible, without compromising the functionalities and performance of the system.

Standards were used where possible, namely for the database (SQL), for the data model (STEP) and data exchange (XML, DXF) and for graphical representation (OpenGL).

The system is intended to work either interactively or in batch mode, reading a command and data files and producing some output files. This latter behaviour is required to be able to be used as a part of some optimization processes.

2   SYSTEM ARCHITECTURE

The system has a modular architecture providing a set of functionalities identified in the Use-Case diagram of Fig 1.

The identified functionalities can be grouped and assigned to independent system modules with a specific role. As shown in Figure 1, these modules may be divided into two main groups according to their role: system base modules and client modules.

Base modules are the core of the system in which all the fundamental information is stored and functions to generate the model are defined. These are the *Simplified Geometry Kernel*, the *Structural Modelling Functions* and the *Ship Database*. The remaining modules work as clients of the basic modules using their functions to get information or to perform specific calculations. The application user accesses the functions and data defined and stored in the system through the client modules.

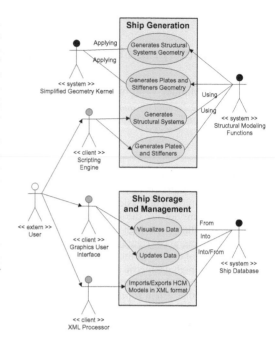

Figure 1.   Use case diagram and associated system modules.

The diagram also emphasizes the three fundamental features that must exist in the process: the *Ship Structure Generation*, and the *Ship Storage and Management*. The following sections describe the main characteristics of the system modules.

3   SIMPLIFIED GEOMETRIC KERNEL

The external hull shape is initially composed by a set of cross-section curves as defined in the traditional body plan. Next, these curves are processed into a single triangular mesh, complete with topological data, which, together with the main deck, is used both as the geometric support of the outer envelope structures and boundary of the inner structural components.

Since the required geometric accuracy of the model for the intended purpose is not high, a simplified geometry concept was adopted that consists in using linear approximations of the shapes, replacing curves by polygonal lines and surfaces by polygonal grids designated by panel-sets. A panel-set is a topological association of panels, each defined by a plane and a closed polygonal contour.

Basically, the geometric model of the ship is composed by a set of panel-sets, each being composed by a set of panels, each being defined by a closed polygonal contour as shown in the diagram of Figure 2.

An example of the subdivision into panel sets, panels and contours is also presented in Figure 2 for the

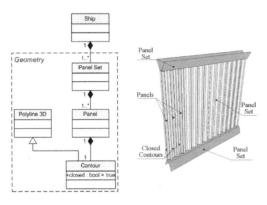

Figure 2. The geometric model of the ship is composed by a set of panel-sets, panels and closed contours.

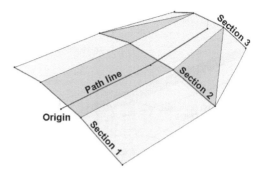

Figure 3. Main Deck generation applying the sweep method with a multi-segment path line and three cross sections.

case of a corrugated bulkhead. The bulkhead is composed by three panel sets: the upper stool, the lower stool and the corrugate. The corrugate is composed by several adjacent panels from portside to starboard (which do not correspond to the plates) and each panel has a closed contour. As a measure of the level of parameterization, the definition of a corrugated bulkhead is composed by 32 parameters. This includes 15 parameters to define the corrugate, 4 parameters to define the shape of each stool and 13 parameters for the assembly of the bulkhead.

In order to improve the performance and robustness of the geometric operations required to generate the model (which are described ahead), panel sets are implemented with topology in the form of half-edge data structures described in Kettner (1998).

The kernel was developed to provide only the elementary modelling operations for the generation of panels-sets.

After an extensive survey of the typical hull structure configuration and of the existent methods for parametric generation of surfaces and polygonal meshes used in 3D Modelling, three of those were identified as fundamental and sufficient for building the simplified geometric model of the ship:

- Sweeping
- Intersection
- Trimming

The sweeping method was selected as the more appropriated to generate the geometry of the structural systems parametrically. In fact, this method is very flexible because it uses free 3D lines for the generation of the meshes which are not restricted to any rules. By sweeping one or more cross-sections along a path line, any existent geometry in the ship can be created as long as the shapes of the cross-sections and the path are defined correctly.

As an example, the forward zone of the main deck generation is considered using this method. The user defines the sheer line along the ship which is used as the path line of the sweeping method, and a set of transversal lines are defined and positioned along the sheer to define the deck camber. These are used as the cross sections along the path as shown in Figure 3.

Grey tonalities correspond to different panels generated by the method. As can be seen between Section 2 and Section 3, the sweep method between corresponding segments of two adjacent cross sections may generate more than one panel since the segments may not be coplanar.

Simplified versions of the method, using straight lines as paths and only one cross section are normally used to generate other structural systems such as transversal bulkheads or plain decks.

Intersections between panels and panel-sets are mainly used to define path lines for the sweeping method. Simple path lines or lines which are defined by design parameters such as the sheer or the deck camber are not generated by intersections. However, trace lines of stiffeners or deck beams can only be calculated by intersections with the shell and the main deck respectively.

Finally, the trimming feature is fundamental to remove the non-existing geometry adapting the shape to the defined boundaries as shown in Figure 4.

As a rule to avoid incomplete geometry and to simplify the generation process form the user point of view, the sweeping method always generates geometries that go beyond their boundaries. The trimming process is then applied using the boundaries as the cutting objects in order to achieve the final geometry.

This feature is applied for almost all the panel-sets since their parametric definition normally includes a set of boundaries. The shell may also (and often is) be a boundary and hence a cutting object as presented in Figure 4 for the case of a longitudinal bulkhead.

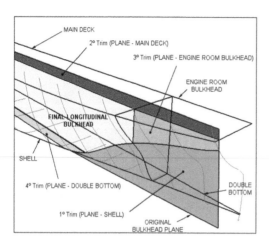

Figure 4. The trimming feature is applied to remove the non-existent geometry of the transversal bulkhead that goes beyond the shell and the main deck.

4 STRUCTURAL MODELLING KERNEL

A group of functions able to model the base geometry of each of the structural systems from a set of design parameters was developed on the top of the simplified geometric kernel.

The main five types of structural systems considered are the shell, deck, bulkhead, web frame and girder. The modelling of these systems is a two-step process. First, a generic structural system definition is specified and next, instantiations of that system can be generated according to the definition and location (s) provided.

As presented in Figure 5, Structural System definitions have two main different layers: the geometric definition and the stiffened panel (s) definition.

Common data of geometric definitions for all structural systems' definitions are the list of the boundary elements, a list of opening definitions and the required parameters to apply the sweeping method on their generation, namely the path and section (s). An additional base structure may also be provided if the path defined by the user is to be projected into this structure generating then the real path to be used by the sweeping. Paths and sections are defined in a local 2D coordinate system and then reoriented and translated to the final location depending of the type of structural system that is being generated. The remaining data provided by the geometric definition depends of the structural system being defined. Therefore, each type of structural system has its own definition template which is an extension of the generic structural system definition.

The simplest cases of the specific geometry definitions are the deck and shell definitions. In fact no

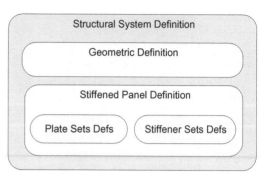

Figure 5. The structural system definition has two main layers: the geometric definition and the stiffened panel(s) definition.

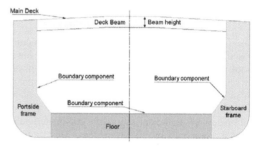

Figure 6. Web frames are composed by four panel sets: the portside and starboard panel sets, the Deck beam and the Floor.

additional information is required to the generic structural system definition.

The girder definition requires an additional parameter that indicates the orientation. Every stiffened panel that doesn't fit in the bulkhead, deck, shell or web-frame definition, is modelled as a girder.

Bulkhead definitions also provide an orientation which may be longitudinal or transversal. Additionally three types of bulkheads' definitions are considered: plane, corrugated and sandwich. The corrugated bulkhead definition provides a corrugate definition and two optional stool definitions whose parameters are according to the Hull Condition Model (HCM) described in Jaramillo and Cabos (2007).

From the geometric point of view, web-frames are subdivided into four transversal panels (Figure 6): portside and starboard panels, floor and deck beam. Each of these has its own definition which is also an extension of the generic structural system definition.

Where there are no physical boundaries defined by a limiting structure member, a boundary line is explicitly defined. The exception is the shell for which the base moulded surface is a polygonal mesh generated from a set on input cross-sections, each defined by one or more polygonal lines.

Regarding the specification of the stiffened panels, for each one, the direction of the layout of the plates and the direction of the stiffening are defined. Along each of these directions are defined sets of cross-sections, each associated to the respective plate-set or stiffener-set.

Plate-set definitions define groups of adjacent plates with the following same attributes:

- Breadth
- Thickness
- Material

Each plate-set definition may have several different groups whose length is the same for all and is given by the cross section length.

Stiffener-set definitions define groups of adjacent stiffeners with the following same attributes:

- Section type (flat bar, bulb, T bar, L bar, etc.)
- Dimensions (which depend of the type of stiffener)
- Spacing (between stiffeners)
- Web direction
- Flange direction (when applicable)
- Material

Similarly to plate-set, stiffener-set definitions may have several groups of stiffeners with the same characteristics. Also the length of each stiffener-set is given by its cross section. Figure 7 shows the most intuitive case of applying this concept for the case of the plates, which is in the cargo zone of the side shell structure. Normally in this zone each cross section represents a column of plates to which is applied a plate-set definition with different groups of equal plates.

The four plate sets have the same length which is defined by the cross section. The plate breadths of the first set are smaller than the second one. The plates of

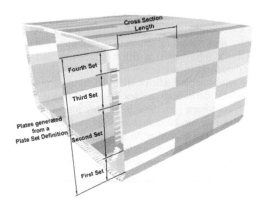

Figure 7. In the cargo zone, the concept of cross section and plate-set definition is visualized as a column of a specific length with different adjacent groups of plates with the same characteristics.

the second and third set have the same breadth however their thicknesses are different. Finally the fourth set which is composed by a single plate has different breadth and thickness.

Naturally this concept is also applied to the hull inner structure.

5 SHIP DATABASE

The structure of a ship, even if simplified, has a large number of components and different uses may require many types of queries beyond those to be made available by the system. For the proposed system, a relational database (*www.mysql.com*) that supports the standard SQL language (Date and Darwen, 1996) was adopted.

In order to make the most wide usage of the hull data generated, a data model was developed that takes into consideration both the STEP AP218 standard (ISO, 2004) and the HCM (Jaramillo and Cabos, 2007), oriented to hull maintenance data exchange.

The database stores all the specified entities in the PDM including the structures' definitions and instances data, the hull cross-sections defining the hull form and general data such as the ship main dimensions and the frame spacing table.

Due to the Constructive Solid Geometry (CSG) approach of the system, geometry and structures must become available to be re-used on subsequent modelling operations. Therefore, this data, which is often too large to be kept in memory, is stored in the database immediately after its generation.

When generated by the Structural Modelling Functions, geometry and structures are immediately sent to the database by using specific functions of each structure class to store the data in the corresponding module of the database. This data, which is often too large to be kept in memory, remains available in the database to be acceded and used on the generation of subsequent structures that depend of the previous generated ones.

Therefore, during the generation of the ship, input functions and queries to database are constantly being called by the Structure Generation Functions.

The query of a ship instance also provides some parameters necessary to evaluate the cost of different alternatives under study. The parameters considered are the following:

- Weight and coordinates of the centres of gravity, separately for plates and stiffeners
- Painting areas
- Lengths of welding seams

The results are computed and displayed for the regions of the ship selected in the tree view.

6 SCRIPTING ENGINE

The development of a ship hull product model can be taken to different levels of detail, depending on the intended use. A complete modelling work of a ship requires a lot of repetitive tasks (for example, when generating web frames with similar arrangement and scantlings, but adapted to the local hull shape) that may be performed by programmed loop sequences with the appropriate parameters. Nevertheless, there are always exceptions that must be dealt with in an interactive way. Thus, order to improve the efficiency of the modelling work it is important for the user to be able to automate some of these repetitive tasks, avoiding human inter-action. Moreover, concerning the modelling approach described above, the sequence of tasks is relevant to model the ship correctly. Thus, a script stores not only the information required for modelling but also the sequence of function calls. For this purpose the system is provided with a scripting capacity.

The modelling process requires high-level cus-tomized commands, which can access both the system modelling functions to generate the structures and the database to store and retrieve data. For this purpose the Python scripting language was adopted (www. python. org) and extended with a set of new commands. These new commands were developed to perform specific tasks in the modelling process and can be classified by their functionality in four main groups:

- Definition of global ship data
- Definition of typical structural systems shape con-figuration and scantlings
- Geometric modelling operations to obtain the shape from the topological boundaries specification
- Creation of the actual instances of the structural systems

The structure of a typical script is presented in Figure 8.

Although an interpreted scripting language such a Python can be extremely useful for this type of mod-elling tasks, if not used correctly, it may become a severe bottleneck of the application and the user may experiment many difficulties before even modelling a small part o the ship. This happens mainly due to the absence of efficient parsers to detect syntax errors or debuggers to monitor the tasks being performed on run-time.

In order to minimize the drawbacks of using the scripting language, composed scripts are used to model the ship. This means that the user is able to divide the full script that models the entire ship into smaller scripts that may be called maintaining the same seq-uence. This approach also allows the user to test the smaller scripts and easily detect eventual errors.

Some rough script debug functionalities were also developed by writing messages to a log file with the

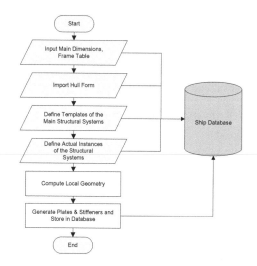

Figure 8. Logical structure of a generating script.

tasks being performed by the application while run-ning each script.

Extended scripting functions are used as wrappers for the structural modelling functions that are writ-ten in the application development language, which is C++. Thus, parameters provided by the user into the Python scripting functions are very similar to the ones provided internally to the structure generation functions.

7 DATA EXCHANGE PROCESSOR

Hull data can be exchanged in XML format in accor-dance to HCM, developed to provide support for the hull condition assessment (Jaramillo and Cabos, 2007), providing support for the planning, viewing and analysing the results from thickness measurement campaigns (Jaramillo et al, 2005).

Therefore, although the user is able to export the entire ship, including structure definitions, using the GUI, the information is filtered internally and only the data that can be mapped into the HCM is exported. The same happens when a HCM file is imported. The structural definitions are not known (the model may have even been generated in another modelling sys-tem) and the structure members are mapped into the available ones according to a mapping table defined by the designer. If no correspondence exists in the mapping table, structure members are mapped into the *Other Members* table of the database.

8 GRAPHICS USER INTERFACE

Due to the eventual large amount of the components, the visualization of a complete 3D model can be not only very time consuming but also very difficult to percept by the user, even with the shading capabilities. Therefore, the capability to display only the selected items and to control the type of representation is relevant when browsing the model.

Plates and stiffeners allow different types of graphical representation, as polygonal lines (wireframe representations) or as their base surfaces (surface representation), without thickness. Plates can be represented either by their boundary contour or as polygonal surface meshes. Stiffeners can be represented either by their trace lines or by a polygonal mesh. Another type of representation is available, in which the plates and stiffeners are represented with the respective thicknesses as meshed solids. Currently this option is only implemented as a file export capability based on the DXF industrial standard format (Autodesk, 2007). The 3D solid model can be imported in most of the existing CAD systems and provide the basis for the initial study of the layout of equipment and of piping systems.

Similarly to the other modules or sub-systems of the application, the GUI is naturally centred in the database where all the data is stored. Briefly, it allows the user to access, visualize, change and store the information contained in the database.

Three main types of visualization and management of the database content are provided:

- 3D viewer
- 2D viewer
- Hierarchical tree of the structural systems

The 3D window is a typical fly-through viewer in which objects with 3D graphical representation are displayed. The following types of objects that are displayed in this viewer:

- Hull form sections
- Structure members geometry
- Structures (plates and stiffeners)
- Measured points

Figure 9 presents screen captures of the five types of objects displayed by the 3D viewer.

As can be seen in Figure 9 as well as in Figure 11 and Figure 12, different tonalities are applied to each plate and stiffener to allow an easier perception of the structures that really interest for this type of application.

Selecting structures (plates or stiffeners) in the viewer through the mouse cursor allows visualizing in an appropriate way (through tables, pictures, forms, etc.) its corresponding information, which is stored in the database.

A typical drawing which is widely used in Naval Architecture to identify plates, stiffeners trace lines is the hull expansion. Therefore, the system provides the 2D viewer that represents the plates and stiffeners of the ship outer shell. For this purpose, the shell shape is expanded in the ship transverse direction, using

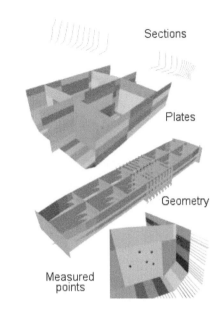

Figure 9. The 3D viewer displays four types of objects with graphical representation.

Figure 10. The corrugated bulkhead is one of the Structure Members that requires specific modelling functions incorporated in the scripting language.

the concept commonly used in the naval architecture drawings.

The hierarchical tree allows the user to visualize and manage the all the objects in the database. Even objects that don't have graphical representation such as the structural system definitions are visualized and accessed from the tree.

OpenGL (Shreiner et al, 2005) was selected as the base engine for the graphic representation of the system.

The user interface is of a typical windows based application where commands can be given from hierarchical menus. Data associated with elements selected

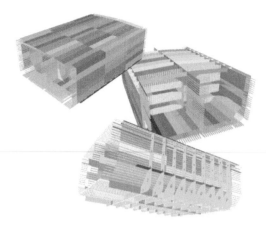

Figure 11. 3D model of the tank generated for validation purposes in a measurement campaign simulation.

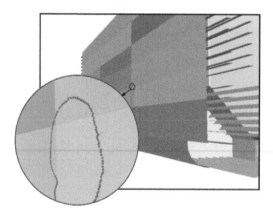

Figure 12. The large number of points measured by the crawler in the shell are imported to the system through an XML file and represented in the 3D model.

interactively on the graphic display can be viewed and edited through dialog-boxes.

The main window of the application shows a tree view, which displays all the contents of the database organized by their types. The selection of entities on the tree is the beginning of most operations carried out by the system.

The selection of an entity on the tree implies the selection of all the child entities.

9   VALIDATION AND TESTING

For the validation of the developed system a set of test cases were prepared, specifying typical structural arrangements of the cargo area of two merchant ships,

namely an oil tanker and a bulk carrier. Each of these ship types were selected due to some specific aspects of the inner structure shape and arrangements.

The oil tanker modelled, a SuezMax with a double-hull, allows checking the correctness of the generation of the web frames, provided with openings. The longitudinal bulkheads can have horizontal or vertical knuckles.

A typical bulk-carrier was also modelled in order to test the modelling functionalities with some specific aspects such as the existence of corrugated bulkheads with upper and lower stools (Figure 10), as well as of the hopper and wing tanks composed by panels inclined both in the transversal and longitudinal directions.

Finally, the system modelling, visualization and data exchange capabilities were tested in the context hull maintenance support which required the representation of an existing hull. This type of application sets a higher level of requirements since it is expected to reproduce some particularities of the structural arrangement in order to guarantee the correct mapping between the model and the plates and stiffeners existing on the ship.

Analyzing the structural drawings and writing the Python scripts to conclude this task took about three days of work. However, modelling the complete cargo zone would take only a few hours more due to the similarity of the cargo tanks. On a PC computer the script runs in about five minutes and the final result is presented in Figure 11.

In the scope of the project "CAS—Condition Assessment of Aging Ships for Realtime Structural Maintenance Decision" (Cabos et al, 2008), a partial model of an existing tanker was developed to support the recording and visualization of data from a thickness measurement campaign that was carried out in the real ship. Figure 12 presents the visualization of the measured points.

In such scenario, the product data model becomes crucial for the success of the tests. In fact, the location of the measured points in the hull must be mapped with a level of accuracy enough to identify the matching part (plate or stiffener). The identification of the part provides not only access to its properties (material, thickness, scantlings) but also to the attributes of the parent structural member which determine material renewal criteria.

10   CONCLUSIONS

A system for the generation of simplified 3D models of ship hull structures was developed. The system, intended to be used as a basic design tool, was designed with a limited level of detail and accuracy in order to gain in terms of speed of visualization and actualization of the model.

The prototype system demonstrated that for the level of detail required in this type of application, the geometry kernel may be very simple, composed by small sub-set of the features that normally are used in commercial 3D modelling applications.

The hull model can be entirely generated automatically, by running a script file that populates the ship's database. The resulting structures can be viewed either in wireframe or surface representations or exported as a solid model.

The scope of database query operations can vary from the full hull, to a set of structural members or down to individual parts. According to the developed product data model, not only is it stored the shape, material and scantlings of every plate and stiffener of the hull but also the relation to the associated structural system, type, functionality, location and some additional attributes such as symmetry and watertightness.

Future developments of the system will increase the detail of the model by adding new entities (e.g. flanges, brackets, compartments, etc.) and by improving the description of existing ones (e.g. adding attributes to the information on welding seams).

Exporting capabilities of the system will also be improved in order to exchange as much data as possible with other Ship Design Systems.

ACKNOWLEDGEMENTS

The present paper has been partially funded by the project "Condition Assessment of aging ships for Real-Time Structural Maintenance Decision (CAS)", financed by the European Union through the Sustainable Surface Transport program of the 6th framework programme under contract TST4-CT-2005-516561. The first author has been funded by the Portuguese Science and Technology Foundation (FCT), through grant nnumber SFRH/BD/39312/2007.

REFERENCES

Aoyama, K., Nomoto, T. (1997), "Information models and functions for CIM in Shipbuilding", Journal of Maritime Science and Technology, No. 2, pp. 148–162.

Autocad (2008). DXF Reference, Autodesk, January 2007.

Bong, H., Hills, W. e Caldwell, J. (1990), "Methods of Incorporating Design-for-Production Considerations into Concept Design Investigations", Journal of Ship Production, Vol. 6, No. 2, pp. 69–80.

Cabos, C., Jaramillo, D., Stadie-Frohbös, G., Renard, P., Ventura, M., Dumas, B. (2008), "Condition Assessment Scheme for Ship Hull Maintenance", Proceedings of the Computer Applications and Information Technology in the Maritime Industries, COMPIT'2008, Liege, April 2008, pp. 222–243.

Date, C.J., Darwen, Hugh. "A Guide to the SQL Standard", Addison-Wesley Professional, 4th Edition, 1996.

Forrest, P. e Parker, M. (1982), "Steelwork Design Using Computer Graphics", Transactions of the Royal Institution of Naval Architects, pp. 1–25.

Hwanga, Ho-Jin, Hana, Soonhung and Kim, Yong-Dae (2004), "Mapping 2D Midship Drawings into a 3D Ship Hull Model Based on STEP AP218", Computer-Aided Design Vol. 36, pp. 537–547.

ISO 10303–218 (2004). "Industrial automation systems and integration—Product data representation and exchange—Part 218: Application Protocol: Ship Structures".

Jaramillo, David, Cabos, Christian and Renard, Philippe (2005), "Efficient Data Management for Hull Condition Assessment", Proceedings of the International Conference on Computer Applications in Shipbuilding, ICCAS'2005, Pusan, Korea, Sept. 2005.

Jaramillo, David, Cabos, Christian, "Specification of HCM (Hull Condition Data Model)", deliverable report D1.3.1 from CAS Project, 2007.

Jingen, G. e Jensen, J. (1982), "A Rational Approach to Automatic Design of Ship Sections", Proceedings of Computer Applications in the Automation of Shipyard and Ship Design (ICCAS IV), pp. 117–124.

Kawamura, Yasumi, Ohtsubo, Hideomi and Suzuki, Katsuyuki (1997), Development of a Finite Element Modeling System for Ship Structures, Journal of Marine Science Technology (1997) Vol. 2, No. 1, pp. 35–51.

Kettner, L. (1998). Designing a Data Structure for Polyhedral Surfaces, in Proceedings of 14th Annual ACM Symposium on Comput. Geometry, pp. 146–154.

MySQL 5.0 Reference Manual (www.mysql.com).

Na, S.-S., Kim, Y.-D., e Lee, K.-Y. (1994), "Development of an Interactive Structural Design System for the Midship Part of Ship Structures", Proceedings of the 8th International Conference on Computer Applications for Shipbuilding (ICCAS'94), Bremen, Germany.

Python Language (www.python.org).

Roh, Myung-Il and Lee, Kyu-Yeul (2006), "An Initial Hull Structural Modeling System for Computer-Aided Process Planning in Shipbuilding", Advances in Engineering Software, Vol. 37, pp. 457–476.

Shreiner, Dave, Woo, Mason, Neider, Jackie and Davis, Tom, "The OpenGL Programming Guide: The Official Guide to Learning OpenGL, Version 2", Addison-Wesley Professional, 5th Edition, 2005.

Structural reliability safety and environmental protection

Still water loads

Analysis and Design of Marine Structures – Guedes Soares & Das (eds)
© 2009 Taylor & Francis Group, London, ISBN 978-0-415-54934-9

Probabilistic presentation of the total bending moments of FPSO's

L.D. Ivanov
American Bureau of Shipping, Houston, TX, USA

A. Ku & B.Q. Huang
Energo Engineering, Houston, TX, USA

V.C.S. Krzonkala
ABS Consulting, Brazil

ABSTRACT: The wide spread practice for presentation of the still water loads acting on the hull structure of FPSO's is to use the individual amplitude statistics. When the probability density function of the loads is built, its integration provides the probability of exceeding any given level of loads. Also, once it is available, one can apply the principles of extreme value statistics to obtain the highest value in any number of cycles. The individual amplitude statistics is used in fatigue strength calculations while the extreme value statistics can be used in ultimate strength calculations. An example is given for the probability density functions of a sample FPSO when the two approaches are applied for assessment of the total hull girder load that can be applied for calculation of its ultimate strength.

1 INTRODUCTION

Traditionally, in shipbuilding, the total shear forces and bending moments are calculated as a sum of the still water and wave-induced loads (see, e.g., Hughes, 1983). This practice has been in existence for more than a century. It has been introduced for convenience in the calculations because the variability of these two processes is different. In reality, the load acting on the hull is the total load but its calculation is difficult because of the uncertainties in predicting the sea environment and loading patterns the ship will sustain over its life. The uncertainty related to wave-induced loads is greater than that for still water loads. As to the still water loads, the presumption is that they can be controlled and more accurately predicted, especially now with the availability of onboard computers.

The situation with FPSOs is different from that for ships. In most cases, they are operating in benign sea environment. In addition, even if they are located in harsh sea environment, they work as stationary structures. Thus, prediction of the sea loads for FPSOs is more accurate due to the existence of comprehensive data for the world oceans and seas (see, e.g., OCEANOR Oceanographic Company). Again, the final goal is to predict the total load acting on the FPSO hull structure. It requires statistical analysis of the total shear forces and bending moments for each specific FPSO type and sea environment, which includes statistical analysis of its two components. In

this sense, collection of statistical data for still water loads and their analysis is still a worthy activity.

With this in mind, the paper presents the results of statistical analysis of still water and wave-induced bending moments of an FPSO using individual amplitude statistics (i.e., statistical analysis of all individual amplitudes) and extreme value statistics, which can be used in predicting the probabilistic distribution of the total bending moments.

2 RAW STATISTICAL DATA FOR THE STILL WATER BENDING MOMENTS

Records of the still water bending moments (M_{SW}) within the parallel middle body of a FPSO for dozens of months from 2003 till 2007 have been analyzed and shown in Figs. 1–7. In general, the records were made once a day (in the afternoon) although some days no records were made (this explains the non-equidistant points in the graphs). The sign change of M_{SW} is due to the change of the loading pattern at the time of recording.

One should note that the recorded M_{SW} were not at the same sections although all of them were within the region of 0.4 L_{BP}. However, for the statistical analysis, the data were mixed on the ground that within the parallel middle body (or within the region of 0.4 L_{BP}) the hull structure scantlings are the same. For slow-going ships such as tankers (the FPSO was refurbished from a tanker), this region may be even longer. Therefore,

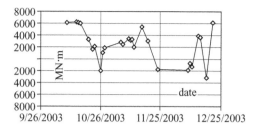

Figure 1. M_{sw} distribution for Oct.–Dec. 2003.

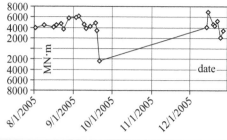

Figure 3. M_{sw} distribution for **Aug. 2005–Dec. 2005.**

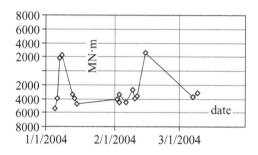

Figure 2. M_{sw} distri bution for **Jan. 2004–March 2004.**

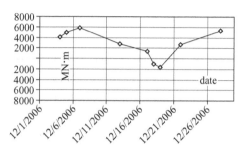

Figure 4. M_{sw} distribution for **Dec. 2006.**

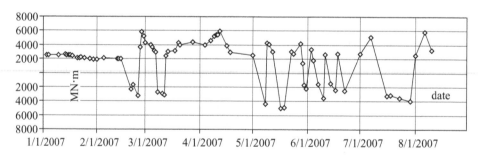

Figure 5. M_{sw} distribution for **Jan. 2007–Aug. 2007.**

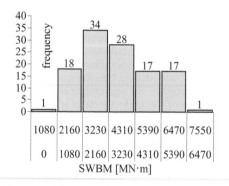

Figure 6. Histogram of the hogging still water bending moments (116 records).

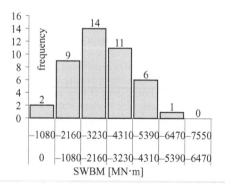

Figure 7. Histogram of the sagging still water bending moments (43 records).

470

the location of the recorded M_{SW} within this region is not very important because the calculated bending stresses within the region will be the same.

3 PROBABILISTIC DISTRIBUTION THE BENDING MOMENTS DERIVED BY THE INDIVIDUAL AMPLITUDE STATISTICS

3.1 Still water loads

The history of probabilistic presentation of still water loads is given in the paper of Ivanov and Wang (2008). In this paper, goodness of fit has been performed by the computer program EasyFit (see Mathwave Data Analysis & Simulation Company) to analyze the raw statistical data. Two criteria have been used—those of Kolmogorov-Smirnov and Andersen-Darling. The results do not always coincide but, in general, they are close to each other. Following the usual practice in shipbuilding, the hogging and sagging M_{SW} are analyzed separately. Preference is given to probabilistic distributions that are most commonly used, such as Gaussian, Rayleigh, Weibull, Log-normal (see Fig. 8 and Fig. 9). The notation "R-i" denotes the ranking of the corresponding theoretical probabilistic distribution when Kolmogorov-Smirnov criterion is used.

The shape parameter of Weibull distribution in Fig. 8 is ≈1 and this is the reason for the Weibull distribution to almost coincide with the exponential distribution. One can notice in Fig. 8 and Fig. 9 that probabilistic distributions that fit hogging M_{SW} may not fit the sagging M_{SW}. For this reason only the Gaussian distribution is shown in Fig. 9 because the other distributions shown in Fig. 8 are rejected by the Kolmogorov-Smirnov criterion. The derived probabilistic distributions should be truncated because the M_{SW} cannot decrease/increase indefinitely. The lower boundary should not be zero. It is possible that at a section within the parallel middle body (or the region of 0.4 L_{BP}) the M_{SW} is equal to zero. However, we are interested in the maximum M_{SW} within the region. It is not possible that at each section the M_{SW} will be equal to zero (it would be zero only in the case when the distribution laws of the buoyancy and gravity forces within this region are exactly the same, which cannot happen in real ships). In the paper, the numerical value of the lower boundary is determined based on the available records for the M_{SW} (it is equal to 6% of the design M_{SW} which is ±10780 MN · m). The upper boundary is the permissible M_{SW} in the Rules (on the premise that the operators strictly follow the loading manual for operation of the FPSO). Mark the boundaries of the M_{SW} as follows:

b_u = maximum possible value of the M_{SW}
b_l = minimum possible value of the M_{SW}

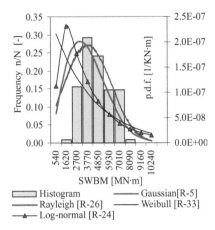

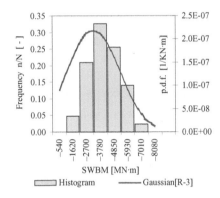

Figure 8. Histogram of hogging M_{SW} and several p.d.f. that fit the statistical data.

Figure 9. Histogram of sagging M_{SW} and Gaussian p.d.f. that fit the statistical data.

In the paper, Gaussian distribution of the M_{SW} is recommended. For this case, one can find the equation of the truncated normal distribution based on the premise that the area below the truncated normal distribution with boundaries b_u, b_l should be equal to unity (as for a normal distribution with boundaries $-\infty$, $+\infty$). The difference between the ordinates of the two probability density functions (p.d.f.) will be a constant C_a, which can be determined in the following way:

$$\int_{b_l}^{b_u} f_c(y)\, dy = C_a\left[F(b_u) - F(b_l)\right] = 1 \quad (1)$$

$$C_a = \frac{1}{F(b_u) - F(b_l)} \quad (2)$$

471

$F(b_u)$, $F(b_l)$ = cumulative distribution function (c.d.f.), correspondingly, for $y = b_u$ and $y = b_l$, i.e.

$$F(b_u) = \Phi\left(\frac{b_u - \bar{y}}{\sigma_y}\right) \qquad F(b_l) = \Phi\left(\frac{b_l - \bar{y}}{\sigma_y}\right)$$

(3)

Thus, the truncated normal distribution function of the M_{SW}, $f_c(y)$, will be:

$$f_c(y) = \frac{1}{\Phi\left(\frac{b_u - \bar{y}}{\sigma_y}\right) - \Phi\left(\frac{b_l - \bar{y}}{\sigma_y}\right)} f(y)$$

(4)

$$f(y) = \frac{1}{\sigma_y\sqrt{2\pi}} \exp\left[-\left(\frac{y - \bar{y}}{\sigma_y\sqrt{2}}\right)^2\right]$$

(5)

where $y = M_{SW}$, $f(y)$ = Gaussian p.d.f. of M_{SW} $\bar{y}$ = mean value of "y", σ_y = standard deviation of "y", Φ = Laplace integral of probability.

$$\Phi\left(\frac{y - \bar{y}}{\sigma_y}\right) = \frac{2}{\sqrt{2\pi}} \int_0^{\frac{y-\bar{y}}{\sigma_y}} \exp\left[-\left(\frac{y - \bar{y}}{\sigma_y\sqrt{2}}\right)^2\right] dy$$

(6)

The truncated Gaussian distributions treated separately for hogging and sagging M_{SW} are shown in Fig. 10. In principle, there is only one probabilistic distribution for the M_{SW}. Its separate treatment for hogging and sagging is performed for convenience in the calculations. The area below each truncated Gaussian p.d.f. (for hogging and sagging) in Fig. 10

is equal to unity because the two cases are treated separately.

However, the phenomenon (i.e., M_{SW}) is one and should have one probabilistic distribution but with two modes. Therefore, the following method is used to build the bi-modal p.d.f. for hogging and sagging. The major assumptions are:

- The sum of the areas below the two p.d.f. is equal to unity
- The ratio between the two areas below the p.d.f. is equal to the ratio between the number of cases with hogging M_{SW} and sagging M_{SW}.

In mathematical terms, these assumptions can be presented in the following way:

$$\left| \begin{array}{l} K_s \int_{b_l}^{b_u} f_s(y)\,dy + K_h \int_{b_l}^{b_u} f_h(y)\,dy = 1 \\[2mm] K_h \int_{b_l}^{b_u} f_h(y)\,dy / K_s \int_{b_l}^{b_u} f_s(y)\,dy = \alpha \end{array} \right.$$

(7)

where:

K_s = unknown coefficient to multiply each ordinate of the p.d.f. for sagging M_{SW}
K_h = unknown coefficient to multiply each ordinate of the p.d.f. for hogging M_{SW}
α = coefficient equal to the ratio between the number of cases with hogging M_{SW}, N_h, and the number of cases with sagging M_{SW}, N_s, i.e.

$$\alpha = N_h / N_s$$

(8)

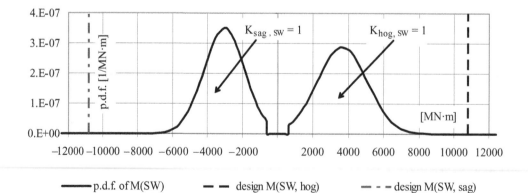

Figure 10. Truncated Gaussian distribution of M_{SW} when hogging and sagging are treated separately.

However, the area below each p.d.f. is equal to unity, i.e.

$$\int_{b_l}^{b_u} f_s(y)\, dy = 1 \qquad \int_{b_l}^{b_u} f_h(y)\, dy = 1 \qquad (9)$$

Thus, Eq. (7) is reduced to the following system of equations:

$$\left|\begin{array}{l} K_h + K_s = 1 \\ K_h/K_s = \alpha \end{array}\right. \qquad (10)$$

The solution of Eq. (10) is simple, i.e.:

$$K_s = 1/(1+\alpha) \qquad K_h = 1 - K_s \qquad (11)$$

Naturally, if the total number of records of sagging and hogging M_{SW} and the separate number of records for sagging (or hogging) M_{SW} are known, the unknown coefficient will be only one. Once the coefficients K_s and K_h are known, the bi-modal p.d.f. can be calculated.

Fig. 11 and Fig. 12 show the histogram and the corresponding probability density function for hogging and sagging M_{SW} when they are considered as two sides of one phenomenon (i.e., the p.d.f. of M_{SW} is a bi-modal one). It is worth noting that the area below the two parts of the bi-modal p.d.f. is equal to unity.

As an example, the result for the FPSO under consideration is shown in Fig. 13 (for other FPSO, the upper and lower boundaries may be different but the shape of the bi-modal p.d.f. will be very similar to that on Fig. 13). Fig. 14 presents the p.o.e. in logarithmic scale, which is a convenient way to illustrate the behavior of the functions in the asymptotic tails.

It never happens that the maximal M_{SW} equals zero. This is the reason that within the range of $M_{SW} = -633400$ and $633400\ \mathrm{KN \cdot m}$, the ordinates of the p.d.f. are zero (for other FPSO this range might be different but a range with zero ordinates of the p.d.f. of M_{SW} will still exist).

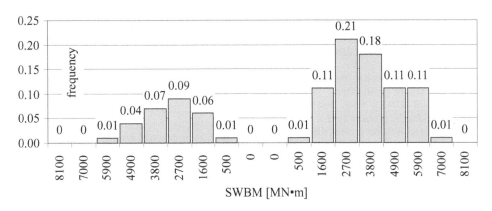

Figure 11.   Histogram of sagging and hogging M_{SW} considered as two sides of one phenomenon.

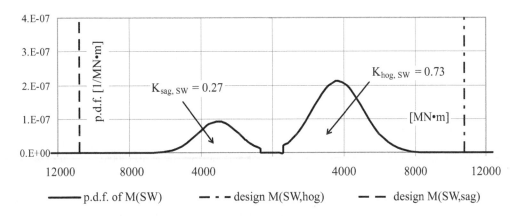

Figure 12.   Truncated Gaussian distribution of M_{SW} when hogging and sagging are treated as two sides of one phenomenon.

473

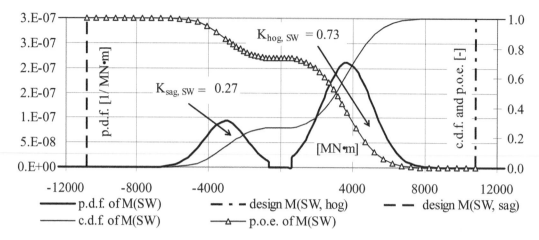

Figure 13. Truncated Gaussian distribution for hogging and sagging M_{SW} derived by individual amplitude statistics while treating the data for hogging and sagging as two sides of one phenomenon (design $M_{SW} = \pm 10780390$ KN · m).

One should be careful when using the c.d.f. in Fig. 13. By definition, each ordinate of the c.d.f. is equal to the area below the p.d.f., i.e. it represents the probability that "y" is smaller than the selected value of "y". In this case, this definition is not valid. An example of how to use the so derived probabilistic distributions is given below.

If one wants to calculate, e.g., the probability that the sagging M_{SW} will be greater than -1835 MN · m, the probability will be 0.23. Note, that the ordinate of the c.d.f. provides the probability that (in physical terms) sagging M_{SW} will be greater than the given value (not smaller, as by the traditional mathematical definition of the c.d.f.). If one wants to calculate the probability that the hogging M_{SW} will be greater than $+1880$ MN · m, the probability will be $1-0.33 = 0.67$ (0.33 is the ordinate of the c.d.f. for hogging M_{SW} at $M_{SW} = 1880$ MN · m). Following the two examples, one can calculate the probability that the M_{SW} will be greater than any given positive or negative (i.e., hogging or sagging) M_{SW} using the c.d.f. for negative M_{SW} and the p.o.e.—for positive M_{SW}.

3.2 Wave induced bending moment

The wave-induced bending moment (M_W) is calculated following the ABS Rules, 2008, for hogging, M_{wh}, and sagging, M_{ws}, bending moment:

$$M_{wh} = 190 \beta C_1 L^2 B C_B \cdot 10^{-3} \quad [\text{KN} \cdot \text{m}] \quad (12)$$

$$M_{ws} = -110 \beta C_1 L^2 B$$
$$\times (C_B + 0.7).10^{-3} \quad [\text{KN} \cdot \text{m}] \quad (13)$$

where:

$$C_1 = 10.75 - \left(\frac{300 - L}{100}\right)^{1.5}$$
$$\text{for} \quad 90 \leq L \leq 300 \text{ m}$$
$$C_1 = 10.75 \quad \text{for} \quad 300 \leq L \leq 350 \text{ m} \quad \Biggr\} \quad (14)$$
$$C_1 = 10.75 - \left(\frac{350 - L}{150}\right)^{1.5}$$
$$\text{for} \quad 350 \leq L \leq 500 \text{ m}$$

$\beta =$ environmental severity factor (in the example, it is equal to 0.67).

The probability of exceedence (p.o.e.) of the design wave-induced bending moment in the Classification Societies Rules is assumed to be 10^{-8} (it means one exceedence in one hundred million load cycles, which means approximately within around ship's service life $T_o = 20$ years). It is also known that the probabilistic distribution is assumed to be two-parameter Weibull distribution (Ochi, 1989), i.e.

$$f(y) = \frac{\lambda}{\alpha} \left(\frac{y}{\lambda}\right)^{\lambda - 1} \exp\left[-\frac{1}{2}\left(\frac{y}{\alpha}\right)^{\lambda}\right]$$
$$F(y) = 1 - \exp\left[-\frac{1}{2}\left(\frac{y}{\alpha}\right)^{\lambda}\right] \quad (15)$$

where $f(y) =$ p.d.f., $F(y) =$ c.d.f., $\alpha =$ scale parameter, $\lambda =$ shape parameter. The parameters λ and α are obtained following the method developed by Kamenov-Toshkov et al. (2006), $\lambda = 0.8258$, $\alpha = -196090$ KN·m for sagging and $\alpha = 183740$ KN · m for hogging.

The values of the M_W however cannot reach plus/minus infinity because the Ocean wave energy is not infinite, indicating that its probabilistic distributions

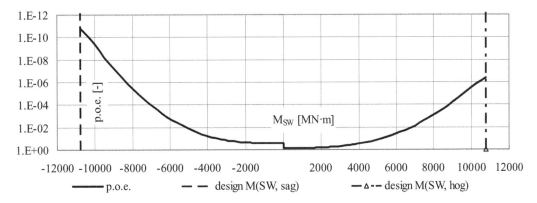

Figure 14. P.o.e. of M_{SW} in logarithmic scale (individual amplitude statistics are used).

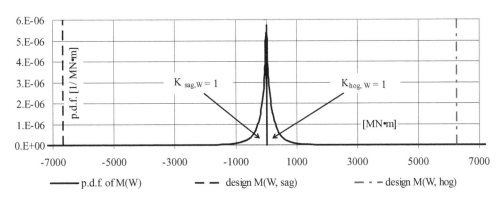

Figure 15. Weibull p.d.f. of M_W when hogging and sagging are treated as separately (individual amplitude statistics are applied).

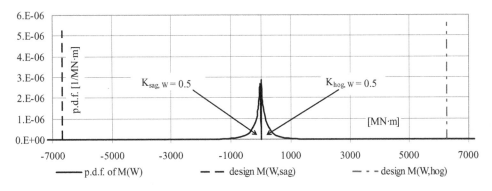

Figure 16. Weibull p.d.f. of M_W when hogging and sagging are treated as separately as two sides of one phenomenon (individual amplitude statistics are applied).

should be truncated. The ordinates of the truncated probabilistic distributions are obtained in the same way as are for the truncated probabilistic distributions already done for the still water loads. For hogging and sagging M_W, the lower boundary is zero (it is possible to have calm sea) while the upper boundary could be either the design M_W or above it to consider extraordinary large wave load. In the paper, it is assumed 10% above the design M_W (the value can easily be replaced when more accurate data is available). The

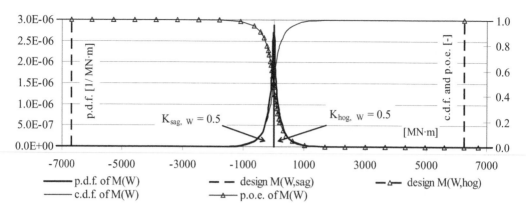

Figure 17. Distribution functions of M_W when hogging and sagging are treated as two sides of one phenomenon (individual amplitude statistics are applied). The design M_W for hogging is 6258260 [KN · m] and −6678920 [KN · m] for sagging.

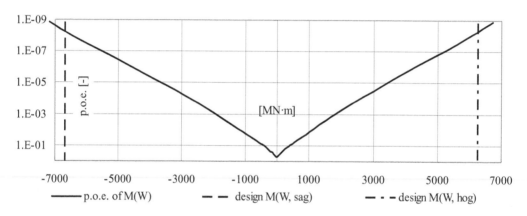

Figure 18. P.o.e. of M_W when hogging and sagging are treated as two sides of one phenomenon in log-scale (individual amplitude statistics are applied).

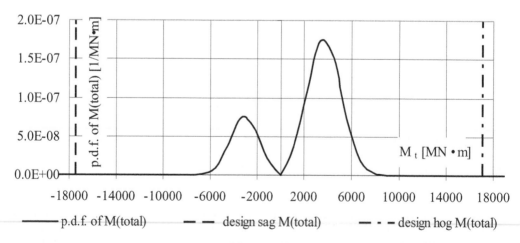

Figure 19. P.d.f. of M_t obtained by differentiating Eq. (16); individual amplitude statistics are used; design M (total, sag) = −17459310 [KN · m]; design M (total, hog) = 17038660 [KN · m]).

476

bi-modal p.d.f. of M_W when hogging and sagging are treated separately is shown in Fig. 15. As for still water loads, it is not possible for the ship to be exposed simultaneously to hogging and sagging wave-induced bending moment, M_W. Therefore, if the wave-induced load cycles within a given ship's life-span are taken as 100%, as a first approximation, 50% hogging and 50% sagging within the given life-span can be assumed (i.e., $K_{sag,W} = 0.5$ and $K_{hog,W} = 0.5$). The bi-modal p.d.f. s are calculated and shown in Fig. 16 and Fig. 17 together with the design M_W calculated with the formulae in ABS Rules, 2008. The p.o.e. of M_W is shown in logarithmic scale in Fig. 18.

3.3 Probabilistic distribution of the total bending moment

There are several proposals in which it is possible to combine M_{SW} and M_W (e.g., Ferry Borges, Castanheta, 1968, 1971; Guedes Soares, 1992; Söding, 1979;

Turkstra, 1970). Any of them can be used provided the final results are calibrated against real experience from ship operation. The difficulty in calculating the p.d.f. of the total bending moment, M_t, is in the fact that there is insufficient data for the total loads acting on the ship structure during relatively long periods (e.g., two or three years). The attention of the researchers was concentrated on deriving data for the probabilistic distribution of the wave-induced loads as the key contributor to the uncertainties in loads' assessment. Less attention was paid to the combination of the still water and wave-induced loads. The general thinking was that the still water loads are controllable and the uncertainties in their calculations are smaller than those for the wave loads.

Under these circumstances, the authors calculated the probabilistic distributions of the total bending moments by the rules for the composition of the distribution laws of the constituent variables (Guedes Soares, 1992, Söding, 1979, Suhir, 1997]):

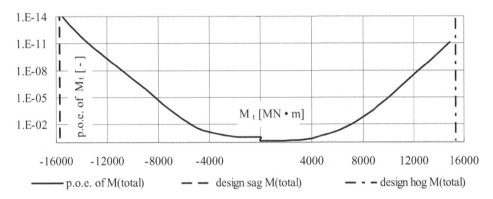

Figure 20. P.o.e. of M_t in log scale and the design values of M_t (individual amplitude statistics are applied).

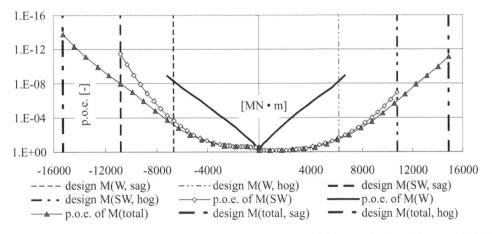

Figure 21. P.o.e. of M_{SW}, M_W and M_t in log scale and their design values (individual amplitude statistics are applied). The coefficient ϕ in Eq. (20) is 0.875.

$$F_{M_T}(M_t) = \int_0^\infty F_{M_{SW}}(M_t - M_W) f_{M_W}(M_W) \, d(M_W) \quad (16)$$

where

$f_{M_t}(M_t)$ = p.d.f. of the total bending moment M_t
$F_{M_t}(M_t)$ = c.d.f. of M_t,
$f_{M_W}(M_W)$ = p.d.f. of M_W
$F_{M_{SW}}(M_{SW} = M_t - M_W)$ = c.d.f. of M_{SW}
$f_{M_{SW}}(M_{SW})$ = p.d.f. of M_{SW}
$F_{M_W}(M_W = M_t - M_{SW})$ = c.d.f. of M_W.

The distribution functions of M_t derived by Eq. (16) are shown in Fig. 19, the p.o.e. of M_t is shown in Fig. 20. For comparison, all p.o.e. of M_t are shown in Fig. 21.

4 ANALYSIS BY THE STATISTICS OF EXTREMES

When the ship's hull ultimate strength is to be checked, the extreme value statistics (Kotz, Nadarajah, 2000) for still water and wave-induced bending moments are used. In this case, this approach is more reasonable than the use of individual amplitude statistics. Type one of the probabilistic distributions of extremes is

derived for hogging and sagging M_W and M_{SW} by the formula:

$$F_e(y_e) = \exp\{-\exp[-\alpha(y_e - \mu)]\} \quad (17)$$

where α and μ are parameters of the extreme value distribution type one and y_e is the corresponding random variant (i.e., M_{SW} or M_W). The parameters α and μ are calculated following the procedure described by Ochi, (1989).

The parameter μ is the probable extreme value expected to occur in "n" observations. It can be evaluated from the initial c.d.f. of M_{SW} or M_W for which the probability of exceeding this value is $1/n$, i.e.:

$$F(\mu) = 1 - 1/n \quad (18)$$

where F is the initial c.d.f. of M_W or M_{SW} calculated for μ.

For the wave-induced bending moment, this is the design wave-induced bending moment (calculated by the formulae in ABS Rules, 2008, for sagging and hogging with p.o.e. 10^{-8}).

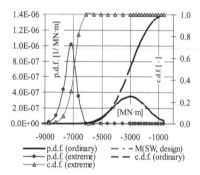

Figure 22. P.d.f. and c.d.f. of sag M_{SW} obtained by extreme value statistics and individual amplitude statistics.

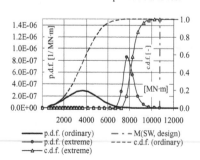

Figure 23. P.d.f. and c.d.f. of hog M_{SW} obtained by extreme value statistics and individual amplitude statistics.

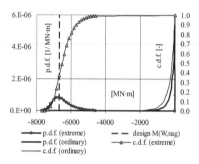

Figure 24. P.d.f. and c.d.f. of sag M_W obtained by extreme value statistics and individual amplitude statistics.

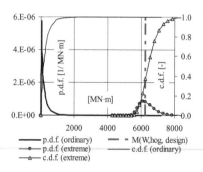

Figure 25. P.d.f. and c.d.f. of hog M_W obtained by extreme value statistics and individual amplitude statistics.

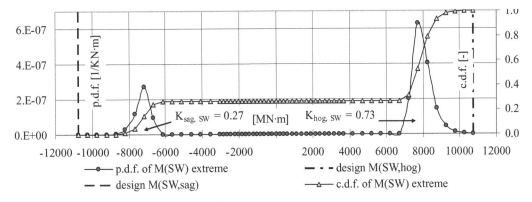

Figure 26. Probabilistic distributions of M_{SW} when sagging and hogging are treated as two sides of one phenomenon (extreme value statistics are applied).

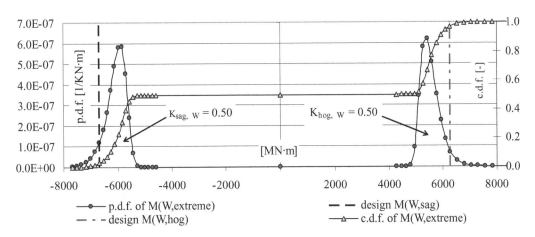

Figure 27. Probabilistic distributions of the extreme value of M_W when sagging and hogging are treated as two sides of one phenomenon.

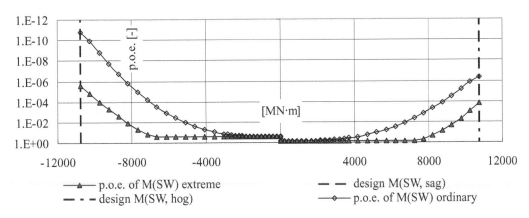

Figure 28. P.o.e. in log-scale of M_{SW} when hogging and sagging are treated as two sides of one phenomenon (extreme and individual amplitude statistics are applied).

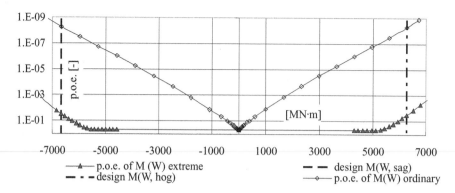

Figure 29.   P.o.e. in log-scale of M_W when hogging and sagging are treated as two sides of one phenomenon (extreme and individual amplitude statistics are applied).

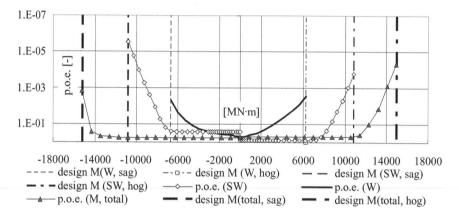

Figure 30.   P.o.e. of M_{SW}, M_W and M_t in logarithmic scale when sagging and hogging are treated as two sides of one phenomenon (extreme value statistics are applied).

The other parameter α is calculated by the formulae Ochi (1989):

$$\alpha = \frac{f(\mu)}{1 - F(\mu)} \qquad (19)$$

where f is the initial p.d.f. of M_W or M_{SW} calculated for μ.

For the wave-induced bending moments the following values for the parameters α and μ were obtained:

For sagging: $\mu = 6678920\,[\text{KN} \cdot \text{m}]$, $\alpha = 2.28 * 10^{-6}$
For hogging: $\mu = 6258260\,[\text{KN} \cdot \text{m}]$, $\alpha = 2.43 * 10^{-6}$

For the still water bending moments the following values for the parameters α and μ are obtained:

For sagging: $\mu = -6825100\,[\text{KN} \cdot \text{m}]$,
$\qquad \alpha = 3.05 * 10^{-6}$
For hogging: $\mu = 7345000\,[\text{KN} \cdot \text{m}]$, $\alpha = 2.05 * 10^{-6}$

The results of the calculations with Eq. (17) for M_{SW} and M_W are shown in Figs. 22–25 together with the results obtained when individual amplitude statistics are used.

All distribution functions for M_{SW} and M_W obtained by the extreme value statistics are shown in Fig. 26 and Fig. 27. The p.o.e. of the design M_{SW} and M_W, when extreme statistics and ordinary statistics are used, is given in Fig. 28 and Fig. 29 in logarithmic scale while the p.o.e. of M_{SW}, M_W and M_t are shown in Fig. 30 in logarithmic scale as well.

After all necessary parameters were calculated, a parametric study was performed to find the relationship between the probability of exceeding any given limit either presented in absolute numbers (i.e., the value of the permissible M_t for sagging and hogging) or as a function of the coefficient ϕ in the following formula:

permissible $M(\text{total}) = \phi(\text{design } M_{SW} + \text{design} M_W)$
$$\qquad\qquad (20)$$

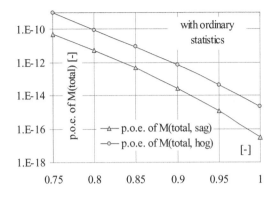

Figure 31. P.o.e. of any given limit value of M (total, sag) presented as a portion of $(M_{SW} + M_W)$ by the coefficient ϕ (see Eq. (20)).

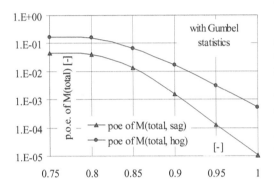

Figure 32. P.o.e. of any given limit value of M (total, hog) presented as a portion of $(M_{SW} + M_W)$ by the coefficient ϕ (see Eq. (20)).

The results of the parametric study are graphically illustrated in Fig. 31 and Fig. 32.

5 COMMENTS ON THE NUMERICAL RESULTS

When individual amplitude statistics are used, the p.o.e. of the design M_{SW} and M_W is negligible. The p.o.e. of the design wave-induced bending moment (hogging and sagging) is 10^{-8}. The p.o.e. of the design still water bending moment is around 2.2×10^{-7} for hogging and 1.3×10^{-11} for sagging.

When the design total bending moment M_t is calculated with coefficient $\phi = 0.875$ (see Eq. (20)), the p.o.e. of the design total bending moment is given in Fig. 21. One should emphasize that the absolute value of the total design bending moment is given here only as an example. The procedure, however, is

a general one and can be applied to any other value of the permissible total bending moment.

The calculations showed that, when extreme value statistics are applied, the p.o.e. of the design wave-induced bending moment and the design total bending moment in class rules becomes substantial.

It is a well established practice to apply the extreme value statistics for control of the hull girder ultimate strength, which depends on M_t. However, even with the application of extreme value statistics, the hull girder collapse is a very unlikely event. One of the reasons is that this FPSO operates in benign environment. Another reason is the practical experience and theoretical knowledge incorporated in classification societies' Rules, which ensure high reliability level of the hull girder. One should also bear in mind that the permissible stresses are method dependent. Hence, when extreme value statistics are applied, one should recalibrate the permissible stresses in the rules to reflect the application of new methodology.

All calculations in the paper refer to the so-called "life time" approach, i.e. all probabilistic distributions refer to specific service life of the FPSO (in the example, 20 years service life). This approach does not contradict the so-called "annual" approach when the loads, the corresponding geometric properties and the probability of failure are also calculated on the annual basis. In the former approach, an "average" reliability of the hull structure is calculated while the latter approach provides information for the year when structural failure may occur. Both approaches are needed for concise assessment of the hull structure reliability at the design and operation stage.

6 CONCLUSION

A method is proposed and tested to present the probabilistic distributions of hogging and sagging still water and wave-induced loads by bi-modal p.d.f. The sum of the areas below the two parts of the p.d.f.'s (for hogging and sagging) should be equal to unity. The area below each p.d.f can be determined from statistical data or assumed based on experience.

The parameters of the probabilistic distributions of the wave-induced loads are calculated by the method proposed by Kamenov-Toshkov et al (2006), which allows for determining these parameters for any duration of the service life. The probabilistic distribution of the total loads is calculated by the rules of the composition of the distribution laws of the constituent variables. This allows for determining the p.o.e. of any given or assumed total load.

The application of extreme value statistics for calculation of the probability that the total hull girder bending moment will exceed the design total bending moment confirmed the very high reliability of the hull girder.

REFERENCES

American Bureau of Shipping (2008) *Rules for Building and Classing Steel Vessels*.

Ferry Borges J. and Castanheta M. (1968, 1971) *Structural Safety*, Laboratorio Nacional de Engenharia Civil, Lisboa, Portugal.

Guedes Soares C. Combination of primary load effects in ship structures, *Probabilistic Engineering Mechanics*, vol. 7, 1992, pp. 103–111.

Guedes Soares C., Dogliani, M., Ostergaard, C., Parmentier, G., and Pedersen, P.T. Reliability Based Ship Structural Design, *Transactions of the Society of Naval Architects and Marine Engineers* (SNAME), 1996; vol. 104, pp. 357–389.

Hughes O.F., (1983) *Ship Structural Design. A Rationally-Based, Computer-Aided, Optimization Approach*, John Wiley and Sons, New York, Chichester, Brisbane, Toronto, Singapore.

Ivanov L.D. & Wang Ge (2008) Probabilistic Presentation of the Still Water Loads. Which Way Ahead?, *Proceedings of the 27th International Conference on Offshore Mechanics and Arctic Engineering, OMAE2008*, June 15–20, 2008, Estoril, Portugal, paper 2008–57011.

Kamenov-Toshkov L., Ivanov L.D. & Garbatov Y.I., (2006) Wave-induced design bending moment assessment for any given ship's operational life, *Ship and Offshore Structures*, vol. 1, No. 3, pp. 221–227.

Kotz S. & Nadarajah S. (2000) *Extreme Value Distributions: Theory and Applications*, Imperial College Press.

Mathwave Data Analysis & Simulation—EasyFit computer program

OCEANOR. *World Wave Atlas*, Oceanographic Company of Norway ASA, N-7462 Trondheim, Norway.

Ochi M.K. (1989) *Applied Probability and Stochastic Processes in Engineering and Physical Sciences*, John Wiley & Sons, New York, Chichester, Brisbane, Toronto, Singapore.

Söding H. (1979) The prediction of still-water wave bending moments in Containerships, *Schiffstechnik*; vol. 26, pp. 24–48.

Suhir E., *Applied Probability for Engineers and Scientists*, McGraw-Hill, 1997

Turkstra C.J., (1970) *Theory of Structural Design Decisions*, Study No. 2, Solid Mechanics Division, University of Waterloo, Canada.

Analysis and Design of Marine Structures – Guedes Soares & Das (eds)
© *2009 Taylor & Francis Group, London, ISBN 978-0-415-54934-9*

Stochastic model of the still water bending moment of oil tankers

L. Garrè & E. Rizzuto

University of Genoa- DINAV, Department of Naval Architecture, Genoa, Italy

ABSTRACT: In this paper a predictive model is adopted for quantifying the stochastic variability of the global static bending load acting on a tanker ship in sagging. The purpose of the investigation is to provide quantitative information on the uncertainties affecting the prediction of still water load effect to be possibly utilised in a reliability evaluation of the hull girder strength. The various sources of variability in bending loads are reviewed, then the attention is focussed on the influence of the uncertainties affecting the cargo weight distribution on board. Loading modes corresponding to uniform filling levels are analysed in particular, based on the note that such conditions are both the most frequent and those inducing the highest sagging moments. The loading process is simulated by a stochastic variation of the filling levels in the various tanks around the nominal values. The characteristics of the probability distributions for filling levels are derived from a specific analysis carried out on the detailed reports of two sister ships, compiled at the end of the loading process. These data reflect also geometrical and operational constraints of the specific ships. The fluctuation of filling levels represents a disturbance in input to the loading process that is partially controlled by means of a feed-back based on trim checks. The corresponding variations in the still water bending moment (SWBM) are identified, with retro-actions triggered by different levels of deviation from the predicted trim conditions. The advantages of the predictive procedure are seen in a direct modelling of physical quantities subjected to uncertainties and of the control process put in place by the crew, allowing a better understanding of the reasons for the variability of the bending loads. Further, the model allows to focus on a specific loading condition that can be adopted as reference situation for the structural checks. The results in terms of mean values and coefficients of variation are compared with those provided by other procedures proposed in literature and considerations are made on a possible improvement in the stochastic model of SWBM in reliability evaluations.

1 INTRODUCTION

In all reliability analyses it is essential that the stochastics model of the involved variables reflects as much precisely as possible the physics of the problem considered: any deviation from reality on this aspect of the analysis implies an unrealistic re-allocation of the probability of failure, with differences in the total value and/or in its distribution over the space of the state variables (which, according to structural reliability theory, implies also a re-location of the design point).

In the verification of the longitudinal strength of ships, as in many other structural problems involving beams in bending conditions, three main random variables are considered: bending capacity, static and dynamic loads. The latter two correspond, respectively, to stillwater bending moment (occurring in static conditions) and wave induced bending moments (associated to the action of waves and the motion of ships in a seaway). Since the earlier applications of the reliability analysis to this problem, many efforts have been dedicated to the systematic investigation of the factors affecting the statistical description of the bending strength of the hull girder and of the

wave-induced global bending loads, in particular the vertical ones. Such investigations have been based on (and at the same time have stimulated the development of) progressively updated predictive models. Less systematic has been the study of the static loads, for which the investigation effort has been mostly confined to analyses of the frequency of occurrence of bending loads derived indirectly from operational data. Only a few works on container ships have been dedicated to the development of predictive models for these kind of loads.

The drawbacks of frequency analyses as respect to predictive models are, as always, that high value load events are difficult to be traced in records because they are rare and that the real situations at the basis of a record are not known with precision, which can make difficult to identify different effects in results. A specific problem is that the analysed load data are based on computations performed by the crew by means of the loading instrument present on board and not directly measured: therefore, such data may differ from the actual value because of uncertain input data in the prediction.

The present work moves once more from the analysis of a limited set of operational data, but such data are intensively examined to calibrate a predictive model able to identify the major effects influencing the shape and the characteristics of the 'actual' probability distribution of static loads.

2 CRITICAL REVIEW OF EXISTING MODELS FOR STILLWATER BENDING LOADS

2.1 Rule values

A typical design load for the verification of the longitudinal strength of ships is contained in the Unified Requirement (UR) S7 issued by the international Association of Classification Societies (IACS 1989). The formulation actually regards the minimum modulus for the midship section, but in association with an allowable stress value, provides a quantification of the total design bending load for the hull girder. Later, the UR S11 (IACS 2001) was issued, containing a design value for the wave induced bending load. Since then, the difference between the total bending load (UR S7) and the wave bending load (UR S11) has been assumed as a reference value for the allowable stillwater bending moment (in the following: SWBM Rule value: RV). This requirement has been since then included in all the Rules issued by single members of IACS and in the draft formulations derived by joint projects: the Joint Tanker Project (JTP) and the Joint Bulker Project (JBP). Presently, the same requirement is included in the Common Structural Rules (CSR, IACS 2008). All the mentioned design values (total bending load, static and dynamic components) are formulated with reference to a first yielding limit state.

While the UR S11 was, to some extent, checked against direct computations (see e.g. Nitta et al 1992), UR S7 and the SWBM RV seem not to have a direct validation in terms of first principles. In any case, these formulae contain only macro-geometrical characteristics of the hull (length, breadth, block coefficient) and cannot reflect the internal subdivision of spaces which originates the weight distribution and, as a consequence, the stillwater bending load.

What above does not constitute a criticism of empirical formulations with a key role in a prescriptive requirements, but indicates that the SWBM RV is not likely to be a value statistically representative of actual static bending loads.

2.2 Other models for normative purposes

An interesting analysis on the statistical description of the hull girder loads is contained in the background document regarding the development of the Common Structural Rules for Tankers, IACS (2006). A stochastic model for the variable corresponding to the SWBM (sagging) was derived from the analysis of a set of double hull tankers and of their loading manuals. Such model is made, for each individual ship, of a normal distribution featuring a mean value of 70% of the maximum sagging SWBM value of the loading manual and a standard deviation of 20% of the same value (coefficient of variation: 28.6%). This variable was used in conjunction with an additional normal uncertainty factor with unit mean and 0.10 standard deviation. It is noted that this stochastic model is entirely based on the loading manual and it represents the first departure at a normative level from the RV. It is also noted that all the loading conditions (in sagging) of the manual were considered implicitly as equi-probable.

The model above described was fed into a reliability evaluation of the sample set of ships and the following calibration process gave as output a partial safety factor of 0.85 to be applied to the characteristic value (chosen equal to the RV). Despite this, the final outcome of the study was a formulation with a partial safety factor of 1 for the SWBM in the check of the longitudinal strength of tankers (which implies to 'force' one of the coordinates of the design point in the reliability assessment to correspond to the RV). This represents the present situation in the check of the vertical hull girder ultimate bending capacity in the CSR for Tankers, IACS (2008).

2.3 Studies based on frequencies analysis

Pioneer studies dating back to the beginning of the 70ies can be found in the literature. Among those, Ivanov & Madjarov (1975) dealt with maximum values of SWBM along cargo ships in fully loaded conditions at departure and arrival. They performed SWBM computations on the basis of realized cargo loading plans of the considered fleet and normalized them with respect to reference values of SWBM based on minimum hull section modulus. They found that variability of these non-dimensional maximum SWBM values, whose occurrence was assumed at mid-ship section, can be modelled by means of the normal distribution, which may be truncated in the case of on board control systems.

A different approach has been followed in Guedes Soares & Moan (1982) and Guedes Soares & Moan (1988), who analysed an extensive collection of records of bending moments and shear forces computed by on board loading instruments for different ships' type. The ships considered included oil tankers, OBO, bulk and chemical carriers. They applied descriptive statistics to the collected data in terms of dependence of the load effects on type and size of ship and amount of carried cargo in each trip. They employed a statistical model for data analysis expressing the uncertainty of the load effects of a ship as the sum of a basic level of uncertainty common to the whole population, plus specific contributions associated to the particular type, size and voyage of a particular ship. Results of this analysis were indications of the variability of the load effects in terms of mean values and standard deviations for each ship type considered in the analysis. Furthermore, they realised a statistical regression using ship's length, deadweight, mean length of the tanks, ship types and deadweight as governing variables. They found high influence of ship's length, deadweight on the load effects' regression. In addition, they found also a remarkable importance of the ship's type as regression variable. This latter result underlines how certain typologies show clear differences with respect to all the others classes considered. The dependency of the load effects on the longitudinal coordinate was also studied.

As a general comment to all the works moving from a statistical analysis of records, it can be said that the larger the sample that has been used, the better the quality of the statistics, but, on the other hand, this brings to a loss of the references to the single ship characteristics (internal subdivision of spaces, definition of loading conditions, etc.). Unavoidably, loading conditions that are formally similar but in practice very different are considered together, leading to very high variances and comparatively low mean values in the loads. A final comment regards the fact that records reflect anyway an estimate of loads as predicted by the officer in charge of the loading/unloading process. All those uncertainties that affect the model, the procedure adopted or the input data in the loading instruments are not reflected in the statistics.

2.4 Studies based on predictive models

The effect of a truncation of the load distribution at maximum allowable values as a consequence of human control has been investigated by Guedes Soares (1990) by means of a predictive model. Starting from statistical data showing truncation of the distribution of measured SWBM in tankers and containerships, the author defined a Monte Carlo simulation model to account for possible re-distribution of cargo on board in case of high exceedance of the maximum allowable SWBM predicted by the loading instrument. This effect was modelled by assuming that acceptance level of the exceedance of the load factor is higher for small exceedances and vice versa. In case of rejection of the simulated loading case leading to high SWBM, a new loading case was re-simulated accordingly to a different normal distribution. The effect of the human control on the maximum SWBM distribution was then estimated by comparison of the two distributions of maximum SWBM, the one derived on the basis of the simulated scenarios with no correction and the one derived on the set of the re-distributed loadings.

Söding (1979) dealt with prediction of still water bending moment in containerships. He applied a stochastic model for the number of empty and filled containers in each bay. In addition, he considered also the contribution given by ballast operations,

employing simulations of possible loading and ballast conditions which were aimed at a minimization of either the still water bending moment or the total amount of ballast. The optimization was performed under some constraints regarding the deadweight, the longitudinal trim and the metacentric height in the loaded condition. Results of the simulations were cumulative distribution of the still water bending moment acting at the midship section and corresponding design values. A further different approach to containerships, completely theoretical, has been applied by Ditlevsen & Friis-Hansen (2002). The authors developed a stochastic model for the loading on board, assuming the load field as Gaussian distributed. They moreover considered ballasting to correct for heel due to asymmetrical loadings and trim. Both fields were conditional on a given draft and trim of the ship. This model has been applied in order to derive probabilistic descriptions of the sectional forces in the hull girder of a fully loaded containership in terms of second moment statistics.

As a general comment to the predictive models developed in the literature and above recalled, it can be said that a key point is in the representation of the decision making process of the officer in charge of the loading plan. In the work by Guedes Soares (1990), the acceptance criteria was set directly on the load effect, relaxing however the criterion to account for some cases exceeding the threshold value. In the other works, the values of still water bending moments were derived conditioning on other operative quantities (trim, deadweight, etc.) that are under the direct control of the crew. In other words, the probability distribution of the load effect is derived imposing a control on other operative variables: a truncation on such variables does not imply a truncation in the derived load, but a coherent representation of the 'tails' of the distribution is obtained.

3 CASE STUDY AND ANALYSIS OF RECORDS

3.1 Characteristics of the ships

As mentioned, operational data have been used to set up the model. They regard two sister tanker ships, which are shown in Figure 1 and whose main characteristics are listed in Table 1.

The ships ('Suez Max' size) feature eight side-to-side tanks and two twin slop tanks placed next

Figure 1. The sister tankers considered in the analysis (out of scale).

Table 1. Main characteristics of the sister tankers.

LBP	264 m
B	45 m
D	24 m
DWT	150000 t
N° tanks	10

to the engine room. Absence of longitudinal central bulkhead is rather atypical for this class of ships. This case corresponds to a twin tanks configuration with a unitary coefficient of correlation between the fillings in the tanks. With respect to independently loaded twin tanks this leads to a higher variability of the cargo weight distribution. As usual in oil tankers, the maximum nominal capacity of the ship corresponds to 98% of the total volume of the tanks, because an allowance of 2% is foreseen to prevent over flows in the loading process. The enforcement of this margin is facilitated through alarm sensors set in the proximity of the maximum allowable level. Cargo densities between 0.902 and 0.95 t/m³ are contained in the loading instructions. The first value corresponds to the saturation of the nominal volume capacity of the ship and of the nominal maximum deadweight (corresponding to freeboard limitations with all the tanks containing consumable items (fuel, lubricating oil, fresh water etc.) completely full.

3.2 Analysis of loading records

The records that have been analysed are contained in so-called Ullage Reports, filled in by the crew just after completing the loading process, before leaving the terminal. Such reports provide in particular, for each trip: type and density of the cargo and filling levels in the different tanks, as they are surveyed by the sensors on board. A collection of these reports for a set of nearly 70 voyages was available and it has been used to identify the typical scenarios of the tankers under analysis. A statistical analysis based upon the available sample, strictly speaking, was not possible due to the limited number of observations. Nevertheless, it has been assumed that data at hand were enough to derive information about the typical loading situations at departure and to identify those more recurrent. Such situations are characterised by density and total quantity of the oil and by the loading mode (i.e. the way the cargo weight is distributed longitudinally).

3.3 Density of the cargo

The actual density of crude oil can range from 0.8 to 1.0 t/m³, but these can be considered quite extreme

values. As it can be seen by Figure 2, the modal value of the distribution as emerged from the reports is approximately 0.87 t/m³. The saturation density of 0.902 t/m³ has been exceeded only three times, but two of these occurrences were associated to the same voyage, as the associated oils were loaded at the same time. Generally, only a single type of oil is loaded at a time, but it may happen that two different lots of oil can be carried on a single voyage. When two different kinds of oil are boarded at the same time, a strong correlation between their densities has been found. The density of cargo for tankers has relatively small variations in comparison with dry bulk cargoes which range between 0.7 to 3 t/m³ (and more, for specific types of ore).

The cargo density in case of uniform filling levels of the tanks along the ship influences the total payload (in tons), the weight distribution and, accordingly, the trim of the ship and the static bending load.

3.4 Total cargo volume and filling distribution

As can be seen from Figure 3, the most common condition is represented by a full load cargo volume around the saturation of the nominal volume capacity of the ship (98% of the total volume), with uniform filling levels of about 98% realised in the various tanks.

This is indeed the predominant loading mode, corresponding to nearly 50% of the observations. Another

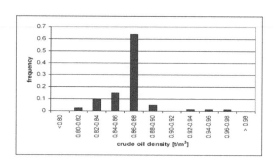

Figure 2. Oil densities.

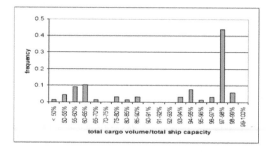

Figure 3. Histogram of cargo volume (% of total ship capacity).

15% of the cases is associated to cargo quantities corresponding to 93% to 96% of the total volume. A uniform distribution of filling levels is still realised in these cases. In the records they mainly correspond to slightly smaller lots of cargo, but in a couple of cases the lower filling ratio was also justified by high oil densities. The remaining observations correspond to smaller netweights for the specific travel: the total filling ratios are spread between 50% and 70% with a small group of realisations placed around 80%. In all these cases, the strategy put into practice was to keep empty some tanks and to adopt a relatively high level (depending on the various situations) in the others.

More in detail, grouping the realisations according to the number of tanks used, it is possible to identify the following situations:

– uniform loading in all tanks (for total cargo volumes above 93% of the total tank volume);
– tanks 2 and 8 plus either tank 5 or 6 empty. These combinations correspond to a total cargo volume in the range between 52% and 60%, depending mainly on the level in the slop tanks;
– tanks 3, 6 and slop tanks empty corresponding to a range of total cargo volume between 60% and 70%, depending on the level in tank 8. It is noticeable also for this condition the use of tank 5 as trim corrector;

A further small group of records were placed between 80% and 90% of the total capacity, not corresponding to a definite loading mode.

The above dispositions are reported in Table 2, listing also the typical filling levels realized in each tank. It is noted that, as regards the definition of loading modes, tankers have a quite different situation than bulk carriers, for which the alternate holds mode is commonly adopted also for high values of the netweight in the presence of high cargo densities.

It should also be remarked that the above loading modes are not univocally related to the cargo volume, because in the middle-low range of filling ratios, for specific ranges of cargo volumes, there is an overlapping of the different possible modes. As it will be underlined later, the existence of different possible combinations complicates considerably the analysis of records performed on the SWBM data alone, since similar quantities of cargo can correspond to different loading modes and to substantial differences in the SWBM response.

An important note regards the shape of the histogram reporting the filling ratios of all tanks in all the realisation of the uniform loading mode corresponding to a total cargo volume of 98%. The histogram shows a clear asymmetrical shape, with an upper limit of the filling ratios around 98.5%, slightly higher than the formal limit of 98%, but lower than the volume saturation (100%). This was justified by the presence of alarms set in order to prevent significant exceeding of the allowable limit (and possible damages to the pumping system, the pressure release valve and/or the inert gas system because of overflows). The lower tail has a less defined trend.

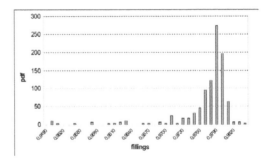

Figure 4. Histogram of fillings corresponding to a total cargo volume of 98%.

Table 2. Identified recurrent loading modes.

Load configuration		Filling levels		
Tank no.	Uniform (%)	Partial 1 (%)	Partial 2 (%)	Partial 3 (%)
Tank 1	95–98	95–98	96	90–98
Tank 2	95–98	0	0	90–98
Tank 3	95–98	95–98	96	0
Tank 4	95–98	87–98	96	90–98
Tank 5	95–98	25–40	0	90–98
Tank 6	95–98	0	87–98	0
Tank 7	95–98	87–98	96	90–98
Tank 8	95–98	0	0	70–98
Slop Tank Sb.	95–98	18–50	60–80	0
Slop Tank Pb.	95–98	18–50	60–80	0
Cargo Volume	95–98	52–60	52–60	60–70

It is interesting to note that, in the loading mode corresponding to all tanks filled with uniform levels, the variance in the fillings of the single tanks around the average value increases when the average level decreases. This effect is induced by the above mentioned constraint represented by the maximum allowed level, which reduces the variance in the levels when the target is close to the threshold. In other words, there is only a single way of saturating the total capacity of a tank, while the number of possible realizations of lower filling ratios increases with the amount of void spaces present in the tank. The same applies to the total volume capacity of the ship.

3.5 SWBM dependence on oil density and cargo volume

The dependency of SWBM on oil density and total cargo volume on board has been studied. Figure 5 and Figure 6 report respectively curves of minimum and maximum SWBM occurring in the central part of the ship (0.4L centered in the midship section) as a function of the cargo volume; the parameter is the cargo density and the groups of curves relate to the loading modes identified in Table 2. In computing

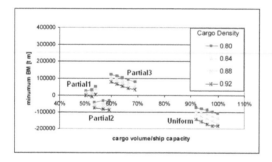

Figure 5. SWBM dependence on payload distribution (minimum values).

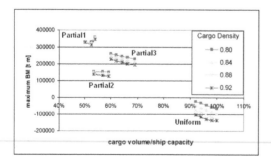

Figure 6. SWBM dependence on payload distribution (maximum values).

these static load values, additional checks have been performed in order to avoid exceeding the maximum allowable values for SWBM (-329200 *tm* for sagging and 361500 *tm* for hogging) and for the free board draft. Only in the case of the highest density considered, negative trim conditions occurred. This is due to the forward position of the cargo centre of gravity. These cases have been corrected through ballasting according to a minimum-quantity-of-ballast principle.

A first note on these preliminary results regards the dependence of SWBM on the loading modes. There are modes in which the tanker is completely in sagging or completely in hogging conditions in the central region, but in other ones (e.g. partial2) both sagging and hogging moments are present.

It is apparent that in these situations a relatively small variation of the filling levels can bring the maximum SWBM in modulus from a negative to a positive value (or vice-versa). In addition to that, as already mentioned, there are ranges of the filling ratios (e.g. the one between 50% and 60%) where more loading modes can be adopted. In these cases, the different loading strategies chosen by the crew lead to completely different load distributions.

4 UNCERTAINTY IN SWBM PREDICTION

In common practice, computations of static bending loads on the hull girder are based on the application of the beam column model, represented by a mono-dimensional beam in pure bending conditions under the action of purely transverse forces (normal to the axis). Such forces are represented by weight and buoyancy distributed along the hull girder. This model implies approximations with respect to the actual distributions of the external forces, which generate uncertainties in the evaluation of bending loads (see Rizzuto 2006). In the following, however, the focus is on the uncertainties on the data in input to the model rather than on those connected with the model itself.

4.1 Uncertainties arising from input data

The buoyancy distribution for a ship in a given loading condition is easily derived from geometrical computations on the hull shape. Possible deviations between the geometrical surface as defined in the hull body and the physical external surface of the ship are considered negligible and not considered hereafter.

As known, the weight distribution on board is made up of a time invariant component (light ship weight), and a time-variant part, due to deadweight (cargo, consumables and ballast). For tanker and bulk carrier ships, the former component is of the order of 10 to 15% of the total full load displacement (depending on the size, the type and the specific characteristics of the unit). This weight is to some extent exposed to uncertainties, as the actual weight and position on board of the structure steel, of machinery and of the outfitting may not be known exactly even after construction. However, from trim and stability tests after construction, discrepancies of more than 1% of the lightship weight are rarely surveyed.

Within the deadweight, consumables can represent a few percents: even though they deserve some attention, because of the location of the tanks that can induce trim and bending effects, the most important source of variability in the weight distribution and, accordingly, in the bending moment distribution is by far represented by the payload (or by ballast, if applicable), which constitute the largest part of the displacement for a commercial ship and change significantly from trip to trip.

The ballast weight ranges around 30% to 40% of the design (full load) deadweight, but the distribution is such that quite high hogging SWBM arise, often. On the other hand, wave loads are generally lower in ballast (due to a lower displacement). In addition, ballast conditions are much more repetitive than cargo load conditions and tend to include either empty or completely filled tanks. These two conditions are easy to realise (no problems with overflows) and verify and, accordingly, do not imply a high level of uncertainty.

Based on what above, in the following, the attention will be focussed in particular on the payload weight distribution in full load, as major source of dispersion and uncertainty in the weight distribution and in the sagging bending load evaluation.

4.2 Factors affecting the realization of the foreseen loading plan

The degree of similarity between the realised weight distribution and the planned one quantifies the accuracy with which the loading process is carried out and controlled. Such accuracy is very much affected by the possibility of monitoring the level of cargo in the various compartments during the process.

In the case of tankers, level sensors are fitted in each tank, featuring a high precision (less than 1 cm, from specifications of radar-type transducers). This should allow to monitor the loading process with an excellent accuracy and, in principle, to 'tune' the actual filling level to the foreseen one. This 'tuning' does not occur in reality (as shown by the ullage survey records) with the accuracy claimed by the instruments. Further, the variability in the realised filling levels around the target values is not constant (as it would be if the driving factor were the sensitivity of the sensors) but, as mentioned above, is strongly dependent on the target value. The reason for this mismatch between the accuracy of the control instrument and the accuracy of the loading process can be due to the fact that sensors'

readings are performed in transient conditions during the pumping process. In addition, there can be timing imprecision by the operator when switching the pumps from a tank to another after reaching the desired level. In any case, the final result in terms of variance of the levels distribution on board appears to be much more driven by the presence of physical limitations (tank close to full) than by an actual control capability of the loading process.

When the loading process is complete, in static conditions, a precise picture of the actual final distribution is however available thanks to the presence of the mentioned level sensors. In principle, these precise information can be used to update the predictions of static bending loads and if, for any reason, the final prediction of the bending loads overcomes a predetermined threshold, a back up is provided by the possibility of a re-distribution of the content of tanks or of the use of ballast to lower the structural loads. On the other hand, it is to be noted that, most times, the final checks on the levels concentrate on the aspect of the total cargo quantity more than on the weight distribution on board for structural purposes. Therefore, the majority of operational data on still water loads come from the previous phase (loading plan formulation) and do not reflect the further uncertainty connected with the actual realisation of the loading plan.

5 SIMULATIONS

5.1 Focus of the study

The present study concentrates on the aspects that make the realisation of a load plan different from its foreseen characteristics, with the aim of quantifying the uncertainties on the bending load in the final cargo distribution. The numerical simulations that are described below regard the specific type of ship and of loading condition here considered, but the general framework can be applied to any kind of ship and condition. Data from the final level surveys are believed to be of a better quality than the bending load estimate based on the loading plan: they have been used to calibrate a probability distribution for the levels in input to the SWBM evaluation. The resulting predictive model can also be run with a very high number of realisations, thus allowing a much better description of the tails of the load effect distribution with respect to a direct statistical analysis.

A uniform loading mode has been selected as the target condition for the analysis of the statistical variation of the SWBM distribution. With this aim, sets of simulations reproducing the variability of the filling levels around this reference case, as resulted by the analysis of the records, have been realized.

489

5.2 Range of parameters analysed in the study

As regards the target values of the filling levels in each tank, the preliminary classification of the uniform loading mode indicated the most representative reference cases corresponding to a total volume of 98% of the saturation volume. A further value has been assumed equal to 95%, as it also constitutes a representative realization in the uniform loading class (Table 2). Together with the variability of fillings, also the density of carried oil has been considered. Records show that the modal value for oil density is in the range between 0.86 t/m^3 and 0.88 t/m^3 (Figure 2). The latter has been considered in the analysis together with a further value equal to 0.902 t/m^3, as it corresponds to nominal saturation of all the tanks. The entire class of families of simulations, as originated by the possible combinations of oil density, mean and standard deviation of the fillings, counts up to 8 different sets.

Finally, consumption of consumables implies that still water loading effects depend on time and change from departure to arrival. This effect may additionally imply the need for trim corrections through ballasting, thus increasing the variability of the SWBM distribution during voyage. These joint effects have not been accounted for in the present analysis, as no changes of consumables have been considered. It is however remarked that the considered levels of consumables were derived as means of from the records. In this respect, the assumed values represent a mean reference configuration around which the variations due to consumption during an 'average voyage' can be assumed to occur. In the simulations, ballast tanks are assumed void.

5.3 Probability distribution function of filling levels

In order to proceed with simulations, a realistic description of the statistical behaviour of the filling levels in the single tanks was sought. The rather limited number of data did not allow to derive directly a sufficiently accurate statistical characterisation. Accordingly, the shape of the probability distribution for filling levels has been inferred on the basis of the tendencies emerged by the previous preliminary analysis. The conclusions of that part of the study on the histograms of filling ratios in the single tanks can be summarised as:

– operative maximum value (bound) of about 98.5%;
– no definite operative lower bound (physical limitation: filling level = 0%);
– mean value corresponding to the ratio between the total cargo volume and the total ship capacity (also corresponding to the target value for the filling ratio of all tanks: uniform loading mode);
– standard deviations inversely proportional to the mean;

To represent the probability distributions of the filling levels Beta distributions were selected. They are limited distributions characterised by four parameters, that can be related to the upper and lower bounds and to the first and second moment of the distribution. For details about the analytical formulation, see e. g. Hahn & Shapiro (1967). The upper and lower bounds were set to be respectively 98.5% and 0, both corresponding to physical constraints (respectively the maximum level allowed by the control system and the empty tank). The values assumed for the mean value μ_y and the standard deviation σ_y are reported in Table 3.

In Figure 7 a comparison between the histogram of filling levels and the Beta pdfs associated to the assumed stochastic description is presented; the case shown corresponds to $\mu_y = 98\%$. As can be seen, the frequency histogram of the fillings is bounded between the assumed distributions which, accordingly, appear to reproduce well the variability of the levels in the tanks as emerged from the analysis of the ullage reports. Furthermore, given the limited number of available records, a precise computation of covariance terms among the different tanks was not possible. Simulations are therefore carried assuming independent fillings among the tanks.

5.4 Simulation rejection criteria

A proper model of the loading conditions at departure must include also a suitable treatment of the unrealistic realizations potentially originated by the simulation

Table 3.   Assumed second moment representation.

Moments of beta distribution	Lower mean		Higher mean	
μ_y	95%	95%	98%	98%
σ_y	0.7%	1.4%	0.1%	0.2%
CoV	0.71%	1.5%	0.1%	0.2%

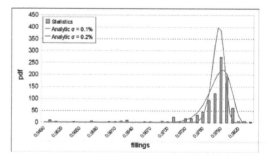

Figure 7.   Histogram of realized fillings and assumed distributions.

490

process. Such realizations could for instance correspond to abnormal trim conditions, that would be immediately detected and corrected by the crew still during the loading process. A proper model of this aspect would imply to account for the decision process, whose final outcome is, in practice, an alteration of the joint probability distribution of the filling levels. As some combinations are inherently 'forbidden', their probability is transferred on other 'allowed' combinations.

An approximate model of the above described effect is represented by a rejection policy: in this cases the combinations that are outside the envelope of possible realisations are simply deleted from the statistics. This procedure is not completely equivalent to the actual situation, because the probability corresponding to the 'deleted' cases is transferred on all the other combinations instead of on selected ones. This anyway produces an effect on the probabilistic description of the filling levels which is in the same direction as the 'real' one, including a modification of the marginal distributions of the filling in the single tanks as well as of the correlation between them. In the present case, a rejection criteria based on a certain number of operational limitations has been adopted. These constraints (that are assumed to be fulfilled at departure) have been selected as:

- Heel angles less than 0.076 degrees, (corresponding to a difference between port and starboard draft of approximately 6 cm);
- Drafts (forward and aft) diverging from the expected ones (computed with all the tanks filled at the mean value) by less than ±3 cm;
- Draft amidship less than the maximum allowable one (given by freeboard regulations);
- SWBM less than the maximum allowable value;

The first condition was necessary because an heel angle was generated by asymmetrical fillings in the slop tanks, while all other tanks feature an inherently centred barycentre (and do not contribute to heel). The third and fourth conditions were never exceeded, while

the highest number of rejections were due to the second requirement. A sensitivity analysis was performed on the forward/aft draft tolerance.

5.5 Results

The minimum target population for statistical analysis has been assumed equal to 650 valid conditions, i.e. loading cases respecting the previously mentioned conditions. Oil weight distributions realised according to the mentioned Beta probability distribution of filling levels were generated. For each realisation, the loading instrument program adopted on board was used to compute the trim and bending moment in that condition. The rejection criteria above mentioned were applied and statistics were generated with the values complying with the requirements.

Simulations have been run until the size of the sampled valid cases was large enough according to the criterion above. Being the reference loading mode characterized by full loaded tanks, all the simulations resulted in predominant sagging and negligible hogging moments. Moreover, the longitudinal positions where the maximum sagging SWBM occurred did not change considerably among the different simulations. In Table 4 are reported the values of mean μ_{BM}, standard deviation σ_{BM} and coefficient of variation CoV of maximum sagging SWBM for the oil density of 0.88 t/m^3 and different characterizations of the pdf of filling levels. The left column is associated to the entire set of simulations, the right one to the selected population satisfying the above limitations.

As can be seen, increasing the standard deviation of the fillings leads to a corresponding increment of the variability of the loading effects, according to a tendency somewhat linear between σ_y and σ_{BM} (if the trim control is not activated: 'all simulation' values). This can be seen as an increased noise in the response when increasing the amplitude of the perturbation in input. When the control on trim is active, the effect is negligible at low values of σ_y, whereas it becomes more effective for high variability of the filling levels

Table 4. Moments of simulated SWBM distributions.

$\rho = 0.88$ t/m^3	All simulations			Selected simulations		
	μ_{BM} 10^3tm	σ_{BM} tm	CoV	μ_{BM} 10^3tm	σ_{BM} tm	CoV
$\mu_y = 98\%$						
$\sigma_y = 0.1\%$	−146.5	715	0.5%	−146.5	715	0.5%
$\sigma_y = 0.2\%$	−146.5	1478	1.0%	−146.5	1420	1.0%
$\mu_y = 95\%$						
$\sigma_y = 0.7\%$	−127.5	5044	4.0%	−127.2	4333	3.4%
$\sigma_y = 1.4\%$	−127.9	9807	7.7%	−126.9	8585	6.7%

in input, with visible reductions in the coefficient of variations of nearly 0.5%–1%. A small change in the mean values is also observed.

The same results have been analysed by means of normal probability plots. As known, such plots can be prepared reporting the absolute value of the samples of a population on the x axis and so-called beta index β on the y axis. Beta index is a measure of the exceedance probability of a given value y_i of the population, according to the well known relation of structural analysis $P[Y > y_i] = \Phi(-\beta)$ in which Φ is the normal standard cumulative distribution (see for instance Ditlevsen & Madsen (1996)). A well-known property of this type of plots is that Gaussian-distributed variables are placed along straight lines. Plotting the simulated populations of maximum SWBM on it is therefore possible to assess the 'level of Gaussianity' of this variable obtained for each set. Normal plots for the sets associated to the 0.88 t/m^3 density are reported in Figure 8 and Figure 9. Similar behaviours have been obtained for the second oil density considered in the analysis (0.902 t/m^3).

Each plot pictures the entire and selected populations together with a straight red line representing a Gaussian distribution fitting the first and third quartiles of the selected population. The plots show how the obtained populations fit quite well into straight lines, thus suggesting that Gaussian distribution are suitable in practice to model this variable both for the selected and unselected sets. In the present case, the fluctuation in the total bending load can be seen as the linear combination of contributions coming from the fluctuations of levels in the single tanks (at least for small deviations from the nominal value). The above described Gaussian-type distribution for the SWBM could be related to the so-called central limit theorem, which states that a linear combination of a high number of identically independent distributed variables approaches a Gaussian distribution. Additionally, it is noted that the originating beta distributions for the filling levels are not very dissimilar in shape from a Gaussian distribution, in the range close to the mean value, given also the relatively small standard deviations.

When the situations corresponding to the tails of the filling level distributions are considered, the effects of the shape of the beta distribution (in particular the asymmetry generated by the upper bound) tend to be stronger (in the input and, accordingly, in the output SWBM). Additionally, the hypothesis of a linear combination of the various levels on the SWBM becomes less and less realistic. One or both of these facts can explain the deviations of the tails from the straight lines for the unselected sets. In addition to that, for the selected set, also correlations between the filling levels in the tanks tend to arise.

This analysis confirms that the application of a rejection policy changes noticeably the distribution in case of high variability of the fillings, as can be seen from Figure 7 and Figure 8. This effect, evidenced by the change in the standard deviations of Table 4, is here represented by the different slopes featured by the curves of the entire and selected populations in the two plots. A further effect of the rejection policy based on draft control is also detectable on the tails of the distributions: the tails are reduced, as the selected set is considerably less extended than the entire set. Finally, slight non-Gaussianity is somewhat observed in correspondence of the tails of the distributions. This effect can be associated to the arise of nonlinearities in the SWBM computation as the filling levels deviate significantly from the mean values. In other words, whereas for small perturbations of the levels around their means the loading effect can be considered linear with respect to these variables, this approximation is no longer appropriate increasing the magnitude of the deviations.

As already outlined by Table 5, the importance of the draft control is dependent on the standard deviation of the input filling levels. For less dispersed variables, changes in trim are more contained and therefore less prone to be rejected (as also shown in Figure 8) with a resulting lower impact of the checks. The above effects depend also upon the applied tolerances on the draft. Broadening such tolerances leads to a reduced impact of the rejection policy with the resulting distribution

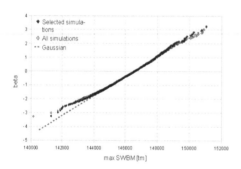

Figure 8. $\rho = 0.88$; $\mu_y = 98\%$; $\sigma_y = 0.2\%$.

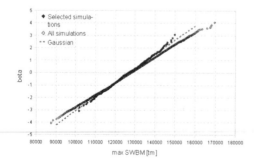

Figure 9. $\rho = 0.88$; $\mu_y = 95\%$; $\sigma_y = 1.4\%$.

Table 5. Effect of different tolerances on draft checks.

Applied tolerances	±3 cm	±6 cm	±10 cm	no limitations
μ_{BM} tm	−126922	−126363	−126899	−127916
σ_{BM} tm	8585	8570	8902	9807
CoV	6.7%	6.8%	7.0%	7.7%
BM_{max} tm	−101886	−101017	−93239	−87416
BM_{min} tm	−149865	−154980	−154980	−169395
ξ_{max}	20%	20%	27%	32%
ξ_{min}	−18%	−22%	−22%	−32%
No of valid simulations	650	1515	3060	28750

approaching that of the unrestricted population. It is noteworthy that the tolerance initially assumed for the analysis (3 cm) is particularly narrow and can be actually considered as the higher precision level realisable in a draft survey. Standard practice is likely to be less precise. In Table 5 the influence of tolerances is studied more in details again for the case ρ 0.88 t/m^3, μ_y 95% and σ_y 1.4%. Each column collects results of different applied tolerances, corresponding respectively to ±3 cm, ±6 cm, ±10 cm and no limitations.

In the Table, the effect is shown also in terms of extreme values BM_{max} and BM_{min} of the simulations, also reported in the last rows in the form of percent deviations (in modulus) from the mean SWBM. As shown, the percent deviation for the extremes of the whole population correspond to ±32%. Application of the rejection policy in this case reduces the same deviations down to ξ_{max} = 20% and ξ_{min} = −18%.

It is here remarked that each of the above statistical representations must be considered with reference to the number of simulations upon which it has been derived, which is reported in the last row. Accordingly, all the values listed must be considered as estimates of the real ones, necessarily obtained from large samples. In this respect, the column associated to the whole population of simulations (last column) gives a more accurate estimate of both the central moments and the tail values, as compared with the other columns, because of the much broader sample involved.

The simulations generating the upper 5%-percentile of the bending load values have been considered separately with the aim of identifying the characteristics of the originating filling levels distributions (they correspond to the highest absolute values of sagging). In this subset of cases, a tendency is observed corresponding to the tanks placed at midship filled slightly more than the reference mean value and, conversely, those aft and fore slightly less. Such cases result in almost null draft changes with respect to the reference values, but in a noticeable increase of the sagging bending moment (up to about 20%).

6 REMARKS

What above reported has been derived on a specific ship, but the general trends seem reasonably extensible to a vast majority of oil tanker ships.

From this viewpoint, the present results can be used in the probabilistic modelisation of the static bending effects for the verification of the longitudinal strength of the hull girder in sagging. As it has been outlined, the uniform condition results to be the most frequently realized during operation of the considered tankers and, additionally, induces the highest magnitudes of sagging SWBM. For these reasons, the uniform loading mode is considered as a preferred reference for the development of reliability-based analyses involving sagging-induced effects. Assuming as reference the magnitude of the bending moment originated for filling levels at 98% (M_{ref} = −146469 tm), the non-dimensional values reported in Table 6 are computed for the obtained SWBM distributions (similar values have been obtained for the other oil density considered in the analysis). It is worthwhile remembering that the Rule value for sagging is −329200 tm, i.e. more than two times the above reference.

As it can be seen, even considering the most dispersed results, the coefficient of variation of the SWBM is less than 8%. The uncertainty in the SWBM with respect to the nominal full load condition results therefore to be characterised by mean values comparatively close to unity and very small standard deviations, if compared with analogous values derived from other analyses. This is due to a different interpretation of the origin of the uncertainty, which is here identified only in the inaccuracy of the loading process. In the model proposed in IACS (2006) also the effects of different cargo quantities and loading modes were included, thus resulting in a lower mean value and a larger standard deviation (respectively 70% and 20% of the worst sagging bending moment from the loading instructions). In other works based on the statistical analysis of records coming from different ships this effect is even more pronounced.

Table 6. Non dimensional moments of SWBM distributions.

$\rho 0.88t/m^3$	All simulations		Selected simulations	
	μ_{BM}/M_{ref}	CoV	μ_{BM}/M_{ref}	CoV
$\mu_y = 98\%$				
$\sigma_y = 0.1\%$	1	0.5%	1	0.5%
$\sigma_y = 0.2\%$	1	1.0%	1	1.0%
$\mu_y = 95\%$				
$\sigma_y = 0.7\%$	0.87	4.0%	0.87	3.4%
$\sigma_y = 1.4\%$	0.87	7.7%	0.87	6.7%

493

The key point to be evaluated in setting the stochastic model for static loads should be, from the authors' viewpoint, the reference scenario for the hull girder. As regards the wave induced sagging bending moment, such reference corresponds generally to the life time extreme for environmental conditions corresponding to North Atlantic: the stochastic prediction of such extreme load is usually conditional on a full load uniform condition. As regards the static load, it seems reasonable to model the variability in a corresponding loading condition (which is also the most probable one), instead of modelling the variability corresponding to all loading conditions (in some cases without a proper weighting of the frequency of occurrence). This can be seen as a way to improve the consistency of the analysis and also to account for a possible influence of the correlation between the two types of load. Another reason for confining the model to a single loading condition, is, from a different perspective, that the conservative approach implicit in the North Atlantic description can be better balanced in static loads by considering the loading mode corresponding to the highest (sagging) moment.

7 CONCLUSIONS

The paper presents the results of a predictive model for the variability of SWBM acting on a tanker as originated by fluctuations of cargo levels in the various tanks. The definition of the model has been based on a preliminary statistical analysis of loading reports on two sister tankers. Such analysis allowed the identification of a few typical loading modes, among which the most frequent one is the uniform load condition, featuring in total more than 60% of the whole population of available occurrences. This loading mode is characterized by high filling levels (ranging between 90% and 98%) realised with uniform distributions of oil levels in all the tanks; it also corresponds to the highest values of sagging bending moment. A further outcome of the data analysis regards the accuracy in the realisation of the loading plan and variability around the reference levels are small. It is moreover noted that the standard deviations of the filling values are inversely proportional to the quantity stored in each tank. This result is justified by the fact that approaching the maximum capacity of the tanks, 'ullages' (void space in the tanks above the oil level) are smaller and allow a smaller variability of the way the cargo is distributed on board.

On the basis of these trends, sets of simulations of the identified uniform loadings have been performed obtaining that variability in SWBM is mainly dependent on the accuracy with which the loading process is carried out; as mentioned this is in turn dependent on the amount of ullage. In those cases in which the variance in SWBM is higher, the application of

draft checks on the realized trims have a beneficial influence, reducing the variance of the bending effect and the tails of the probability distribution. However, even employing rather accurate tolerances in the draft checks, marked deviations of SWBM from the reference mean value can be found.

Even though with some limitations, Gaussian distributions seem to be appropriate to model the SWBM for this class of loadings.

In any case, the standard deviations and mean values obtained analysing a single loading mode appear to be respectively higher an lower than those obtained by less selective analyses reported in literature.

The present quantitative evaluation of the variability of still water load effects, specifically related to the full load uniform loading mode, suggests, in respect to previous applications, a possible re-consideration of the stochastic model for static loads within the reliability assessment of the hull girder strength.

REFERENCES

Ditlevsen, O. & Madsen, H.O. 1996. Structural Reliability Methods, John Wiley & Sons, New York, ISBN: 0471960861.

Ditlevsen, O. & Friis-Hansen, P. 2002. A Stochastic Stillwater Response Model. Journal of Ship Research 46: 16–30.

Guedes Soares, C. & Moan, T. 1982. Statistical Analysis of Still-Water Bending Moments and Shear Forces on Tankers, Ore and Bulk Carriers. Norwegian Maritime Research 10: 33–47.

Guedes Soares, C. & Moan, T. 1988. Statistical Analysis of Stillwater Load Effects in Ship Structures. Transaction SNAME 96: 129–156.

Guedes Soares, C. 1990. Influence of Human Control on the Probability Distribution of Maximum Still-water Load Effects in Ships. Marine Structures 3: 319–339.

Hahn, G.J. & Shapiro, S.S. 1967. Statistical Models in Engineering, John Wiley & Sons, New York, ISBN: 0471040657.

IACS 1989 Unified Requirement S7: Minimum longitudinal strength standards http://www.iacs.org.uk/publications/

IACS 2001 Unified Requirement S11: Longitudinal strength standard http://www.iacs.org.uk/publications/

IACS 2006. Background documentation of the Common Structural Rules for Double Hull Oil Tankers—section 9/1.

IACS 2008. Common Structural Rules for Double Hull Oil Tankers—section 9/1.

Ivanov, L.D. & Madjarov, H. 1975. The Statistical Estimation of Still-Water Bending Moments for Cargo Ships. Shipping World and Shipbuilder 168: 759–762.

Nitta, A., Arai, H. & Magaino, A. 1992. Basis of IACS Unified Longitudinal Strength Standard. Marine Structures 5: 1–21.

Rizzuto, E. 2006. Uncertainties in Still Water Loads of Tankers and Bulkers. Proc. Int. Conference on Ship and Shipping Research (NAV), Genoa, Italy.

Söding, H. 1979. The Prediction of Still-Water Bending Moments in Containerships. Schiffstechnik 26: 24–48.

Analysis and Design of Marine Structures – Guedes Soares & Das (eds)
© 2009 Taylor & Francis Group, London, ISBN 978-0-415-54934-9

Statistics of still water bending moments on double hull tankers

J. Parunov & M. Ćorak
University of Zagreb, Faculty of Mechanical Engineering and Naval Architecture, Zagreb, Croatia

C. Guedes Soares
*Centre for Marine Technology and Engineering (CENTEC), Technical University of Lisbon,
Instituto Superior Técnico, Lisboa, Portugal*

ABSTRACT: The paper deals with statistical description of the still water bending moments, which is important step in ship structural reliability assessments. Mean values and standard deviations of still water bending moments are calculated by analysing data from loading manuals of double hull tankers of different sizes. Loading conditions are grouped according to three characteristic modes of operation of tankers: full load, partial load and ballast. Partial loading condition is further divided into two sub modes: one giving hogging and the other one giving sagging bending moment at midship. Ballast condition is also divided in two separate sub modes: normal ballast and emergency ballast condition. Statistical properties for these modes and sub modes are compared and their dependency on ship size is investigated. Loading conditions giving maximum bending moments in hogging and sagging are identified and these values are compared to minimum design requirements of still water bending moments from Common Structural Rules. Statistics of differences between still water bending moments at departure and arrival are also analysed and presented in tabular and graphical forms. The paper ends with conclusions that may be useful for researchers dealing with ship structural reliability and also for ship designers applying Common Structural Rules.

1 INTRODUCTION

Recently published MSC 81/INF.6 (IMO 2006) guidance for ship structural reliability analysis, also adopted in Common Structural Rules (CSR) for Double Hull Oil Tankers (ABS & al. 2006), represents first official and generally accepted proposal for ship structural reliability assessment by authorities responsible for ship safety. IMO guidance proposes probabilistic models of random variables relevant for hull girder reliability of double hull tankers, procedure for structural safety assessment as well as target reliability levels that should be respected.

The present study is motivated by the fact that much more space and relevance in MSC 81/INF.6 is paid to probabilistic presentation of wave induces load effects, compared to simplified models adopted for still water bending moments (SWBMs). Thus, the statistical model of SWBMs in MSC 81/INF.6 is simply assumed to be normal distribution with mean value equal to 0.7 of the maximum value from the loading manual and with the standard deviation of 0.2 times the maximum value. That model is obtained based on statistical analysis of data from loading manuals of eight test double hull tankers of different sizes. Mean values and standard deviations represent approximately average values resulting from that statistical analysis. It is interesting to mention that only full load conditions

are included in the MSC 81/INF.6 statistical analysis while the emergency ballast and segregation loading conditions are omitted. Authors of the MSC 81/INF.6 elaborate that full load conditions are giving sagging bending moments and as such are relevant for sagging failure mode, which is critical failure mode regarding ultimate bending moment failure of oil tankers.

The intention of the present study is to shed some more light on statistical properties of SWBMs presented in MSC 81/INF.6. More precisely, this study has intention to find out following answers:

– what are statistical properties of other modes of operation of oil tankers as partial load and ballast;
– if there is a correlation between ship size and statistical parameters of the SWBMs;
– which of loading conditions from loading manuals are actually giving largest still water sagging bending moments;
– if the minimum design SWBM proposed in CSR is conservative with respect to the maximum values from loading manuals;
– what is the relevance of the emergency ballast condition with respect to design SWBMs;
– how important are differences between SWBMs at departure and arrival for different modes of operation.

To find out answers to these questions, authors studied loading manuals of six double hull oil/chemical/ product tankers of different sizes ranging from 170 m to 270 m in length. Results of the study could be of interest not only for researchers performing structural reliability studies, but also for practical ship designers to better understand and estimate SWBMs already in early design stage.

2 LITERATURE REVIEW

Studies of SWBMs have been performed by researchers in two different ways: either to study actual loading conditions and corresponding SWBMs that ship experienced during operation, either to study ship loading manuals with limited number of design loading conditions included. The former approach is generally considered to be more accurate but it has disadvantage that it is very difficult to collect satisfactory number of actual operational data from ships in service. Furthermore, ship designs evaluate through years and therefore data collected in past may for different reasons become of limited relevance for future designs. Typical example is appearance of double hull tankers that have substantially different subdivision than their single hull predecessors. Therefore, statistical data about SWBMs collected on single hull tankers can't be used on double hull structures. The other approach, i.e. analysis of SWBMs from design loading conditions described in loading booklets is much more easily accessible, but it has serious drawback that it is not possible to know which of design loading conditions and how often is actually used in operation.

The classical contribution to the probabilistic treatment of the still water loads is the paper by Guedes Soares & Moan (1988) who carried out statistical analysis of large number of data registered onboard ships at departures. They found out that Gaussian distribution is suitable for modelling of still water effects for different ships types. Parameters of distribution (means and standard deviations) are presented as regression equations. These data are collected for single hull oil tankers and as such are hardly applicable to double hull structures. Furthermore, it was assumed that SWBMs at departure and arrival are similar, which is an assumption that is not always correct.

Guedes Soares & Dogliani (2000) proposed to model still water loads at random time instant as a normal distribution with mean and standard deviation equal to their average values for departure and arrival conditions respectively. Such model was confirmed by simulation and has subsequently been employed in number of reliability studies of double hull oil tankers.

The uncertainties in SWBMs of oil tankers and bulk carriers are studied by Rizzuto (2006), He studied different sources of uncertainties, where the most important one was found to be the existence of different and often alternative loading modes. It was suggested to separate loading condition in those giving sagging and hogging bending moments rather than to consider all loading conditions together. This approach of separating partial loading conditions on those inducing hog and sag at midship is adopted also in the present paper.

Ivanov & Wang (2008) studied loading manuals of 22 double hull and 12 single hull tankers to define probability distribution functions of still water load effects. Tankers are grouped according to their deadweight in three categories and statistical parameters are calculated for each of them. In addition, different distribution types are fitted for different load effects and the best distributions are finally proposed. It was found that Weibull two-parameters distribution is the best fit to still water bending moments in most cases.

3 SHIPS

Six double hull tankers of different sizes are analysed in the present paper. Their particulars are specified in the Table 1.

Ships nos. 3 & 4 are chemical and oil product tanker respectively. Ships nos. 1 & 6 are Aframax tankers, while ships 2 & 5 are Suezmax oil tankers. The chemical tanker (ship no. 3) used in the study is double hull tanker intended for carriage of different chemicals and oil, with structural configuration in compliance to the CSR. Differences in operational scenarios between chemical and oil tankers are ignored in the present paper.

4 STATISTICS OF STILL WATER BENDING MOMENTS

Statistics of SWBMs is presented separately for full load conditions, partial load conditions inducing hogging at midship and partial load conditions giving sagging at midship. For ballast and emergency ballast loading conditions, there were not enough data to calculate standard deviations, so only mean values are calculated. It should be mentioned that arrival and departure loading conditions are considered together.

Results of statistical analysis are presented in Tables 2–6 in the form similar as in MSC 81/INF.6, i.e.

Table 1. Particulars of analysed ships.

Ship no.	Length, m	Breadth, m	Height, m
1	235	42	21
2	258	46	22.5
3	176	30	16
4	175	40	18
5	268	48	23
6	236	42	21

Table 2. Statistics of SWBM for full load.

Ship no.	Mean/max.	St. dev/max	(Max-mean)/st. dev	N
1	42%	24%	2.38	18
2	80%	9%	2.33	4
3	58%	57%	0.75	14
4	51%	26%	1.92	8
5	69%	20%	1.53	6
6	65%	24%	1.46	4
Average	61%	27%	1.73	–

Table 3. Statistics of SWBM for partial load giving sagging.

Ship no.	Mean/max.	St. dev/max	(Max-mean)/st.dev	N
1	83%	12%	1.43	8
2	63%	10%	3.87	3
3	–	–	–	–
4	37%	10%	6.07	7
5	83%	13%	1.36	9
6	72%	16%	1.75	4
Average	67%	12%	2.90	–

Table 4. Statistics of SWBM for partial load giving hogging.

Ship no.	Mean/max.	St. dev/max	(Max-mean)/st. dev	N
1	69%	12%	2.58	38
2	57%	16%	2.67	12
3	81%	16%	1.14	4
4	33%	5%	12.93	9
5	61%	19%	2.08	30
6	60%	14%	2.77	21
Average	60%	14%	4.03	–

Table 5. Mean values of SWBM for ballast.

Ship no.	Mean/max.
1	84%
2	75%
3	89%
4	87%
5	98%
6	99%
Average	89%

Table 6. Mean values of SWBM for emergency ballast.

Ship No.	Mean/max.
1	84% (sag)
2	91% (sag)
3	38% (hog)
4	37% (hog)
5	95% (hog)
6	32% (sag)
Average	63%

mean values and standard deviations of each mode of operation are normalized by the maximum value from the loading manual. Loading conditions giving hogging are normalized by maximum hogging bending moment, while loading conditions giving sagging are normalized by maximum sagging bending moments. In addition, number of points N used for calculation of statistical properties is also shown in Tables 2–4.

Statistics of SWBMs for full load conditions are fairly close to those reported in MSC 81/INF.6, which reads 63%, 23% and 1.62 for mean/max, st. dev/max and (max-mean)/st.dev respectively. Mean values for partial loading conditions in hog and sag are quite similar to mean values for full load. However, the variability of normalized SWBM is noticeably lower for partial loading conditions compared to full load condition. Maximum values and mean values for those two loading conditions are separated 2.9 and 4.03 standard deviations for hog and sag respectively, so the probability that maximum SWBM will be exceeded in the partial loading condition is much lower than in the full load condition. That could explain why partial loading conditions are omitted in MSC 81/INF.6.

SWBMs in ballast conditions are for all ships very close to maximum hogging bending moment, since the ballast condition for double hull tanker is mostly unique condition, with only variation during the voyage. Therefore, there were not enough data to calculate standard deviation of SWBMs for ballast and emergency ballast conditions.

Emergency ballast condition is sometimes giving large sagging while in some cases hogging SWBMs. It seems that when sagging bending moment is produced in emergency ballast condition, then the SWBM is very high and close to the maximum for all loading conditions. When emergency ballast results in hogging bending moment, then the SWBM is moderate. However, number of data is limited and these conclusions are only provisory.

Trends of normalized SWBMs with respect to ship length are presented in Figures 1 & 2. Figures 1 & 2 show points from Tables 2–6 together with linear trend lines adopted to them.

One may notice that mean/max. values of SWBMs for all groups of loading conditions, except ballast and emergency ballast are increasing with increasing ship length. For ballast condition, SWBM is almost constant and at very high level for different ship lengths. Data for emergency ballast condition are sparse and rather scattered, so although general trend appears to be insensitive to ship length this conclusion should be taken with care.

Normalized standard deviations (st.dev/max) of SWBMs in full load conditions have clear tendency to reduce by increasing ship length. The same trend is reported in MSC 81/INF.6. Normalized standard

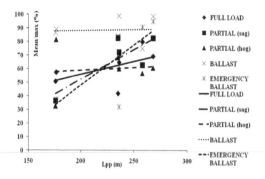

Figure 1. Mean/max (%) for different groups of loading conditions as a function of ship length.

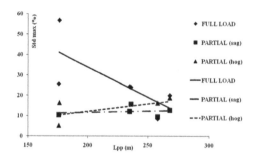

Figure 2. St.dev/max(%) for different groups of loading conditions as a function of ship length.

deviations of SWBMs in partial loading conditions are slightly increasing by increasing ship length.

5 LOADING CONDITIONS GIVING MAXIMUM SWBM AND COMPARISON WITH MINIMUM CSR DESIGN SWBM

It is of interest to analyse in which of loading conditions maximum still water bending moments are induced and how these values compare to minimum required design still water bending moments from CSR. These data are presented in Table 7.

Concerning loading conditions giving largest bending moments, one may notice from Table 7 that ballast and partial loading conditions are the most important conditions for maximum hogging SWBM, while the emergency ballast and full load are critical loading conditions for maximum sagging SWBM. However, it should be noted that full load is dominant sagging condition for smaller tankers (ships nos. 3 & 4) while emergency ballast and partial load conditions are inducing maximum sagging SWBMs for large tankers. It also appear that in all cases maximum hogging bending moments are induced for ships at departure, while the maximum sagging bending moments are induced for arrival loading conditions.

Table 7. Maximum SWBMs and corresponding loading conditions.

Ship no.	Max SWBM (hog), MNm			Max SWBM (sag), MNm		
	CSR	Loading manual	Load cond.	CSR	Loading manual	Load cond.
1	2543	**2975**	PL(d)	−1920	−1707	EB(a)
2	3399	**4032**	PL(d)	−2572	**−3061**	EB(a)
3	919	822	BL(d)	−694	−131	FL(a)
4	1159	**1780**	BL(d)	−893	**−1209**	FL(a)
5	3749	**3891**	BL(d)	−2890	−2836	PL(a)
6	2425	**2587**	BL(d)	−1872	**−1895**	EB(a)

*PL – partial loading condition; BL – ballast loading condition; FL – full load condition; EB – emergency ballast condition; (d)–departure; (a) arrival.

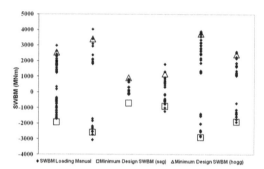

Figure 3. SWBMs from loading manual and minimum design SWBMs from CSR.

It may be seen from Table 7 that in majority of cases, minimum design SWBMs from CSR are exceeded by maximum SWBMs from loading manuals. This is unexpected, as in all cases reported in MSC 81/INF.6, rule minimum design requirements are higher than maximum values from loading manuals. This could be the valuable information for designers, as they should be aware that preliminary design SWBMs could be exceeded in a later phase when the final loading booklet is available.

Results of this analysis are summarized in Figure 3.

One may notice from Figure 3 that sagging an hogging bending moments are well separated for large tankers (ship nos. 1, 2 ,5 & 6) and it would indeed be reasonable to use separate distributions to describe them. For smaller tankers (ship nos. 3 & 4), however, this separation is not so clear and the usage of one single distribution may be more appropriate.

5.1 The emergency ballast loading condition

As may be seen from Table 7, the emergency ballast loading condition is governing condition for sagging SWBM for two Aframax and one Suezmax tanker. Furthermore, in two cases the sag SWBM in the emergency ballast condition is larger than minimum rule SWBM given in CSR.

Considerations of the importance of the emergency ballast condition require analysis of ship's operational aspects and measures that shipmaster takes in heavy weather. One of measures that shipmaster can undertake to improve ship performance in heavy weather is to replace the normal ballast by the emergency ballast condition, which is the condition when one of cargo tanks is filled with seawater. This is done for several reasons—to reduce bow slamming, propeller emerging and to improve ship maneuverability. Dynamic ship loads caused by slamming may be so dangerous to even cause hull girder collapse. One example of such hull girder collapse in normal ballast condition in heavy weather is the sinking of bulk carrier "Flare" in 1998 (Flare, 2008). The forensic analysis has shown that ship probably would not be lost if normal ballast condition have been replaced by the emergency ballast condition. Therefore, question is imposed weather the emergency ballast condition should be considered in ship structural reliability studies instead of normal ballast, especially if maximum still water bending moment in sagging are often produced in that loading condition.

6 DIFFERENCES BETWEEN SWBM AT DEPARTURE AND ARRIVAL

The next aspect of the problem which deserves attention is how much SWBMs change during one voyage. The differences in the load effects during the voyages result from the gradual consumption of fuel and eventually from any redistribution of fuel that may occur. In the MSC 81/INF.6 all seagoing departure and arrival loading conditions are considered together and as equally probable. However, according to Guedes Soares and Dogliani (2000) significant differences could sometimes occur between departure and arrival and separate statistical models should be applied to describe them.

The statistical analysis of differences between SWBMs at departure and arrival is presented in Tables 8–12. The analysis is performed in a way that the absolute values of differences between SWBMs

Table 8. Statistics of arrival/departure differences of SWBM for full load.

Ship no.	Mean/max.	St. dev/max	N
1	12%	11%	9
2	6%	0%	2
3	21%	26%	7
4	37%	0%	4
5	0%	0%	3
6	12%	14%	2
Average	15%	9%	–

Table 9. Statistics of arrival/departure differences of SWBM for partial load giving sagging.

Ship no.	Mean/max.	St. dev/max	N
1	10%	11%	4
2	–	–	–
3	–	–	–
4	–	–	–
5	8%	7%	3
6	21%	–	1
Average	13%	9%	–

Table 10. Statistics of arrival/departure differences of SWBM for partial load giving hogging.

Ship no.	Mean/max.	St. dev/max	N
1	7%	4%	18
2	2%	9%	6
3	44%	36%	2
4	20%	–	1
5	21%	7%	11
6	10%	8%	9
Average	17%	13%	–

Table 11. Mean values of arrival/departure differences of SWBM for ballast.

Ship no.	Mean/max.
1	32%
2	4%
3	22%
4	27%
5	3%
6	2%
Average	15%

Table 12. Mean values of arrival/departure differences of SWBM for emergency ballast.

Ship no.	Mean/max.
1	33%
2	18%
3	1%
4	22%
5	–
6	18%
Average	18%

at arrival and departure for each loading condition are calculated firstly. Then, mean values and standard deviations of these differences are calculated for each ship and each mode of operation. Finally, means and standard deviations are normalized by maximum values from loading manual. If SWBM in particular loading condition is sagging, then maximum value is taken as maximum sagging SWBM. The same consideration is applied for hogging conditions, where differences in SWBMs are divided by maximum hogging SWBM for all loading conditions. It should

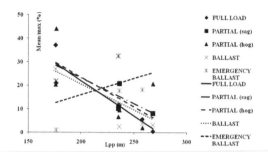

Figure 4. Trend lines of normalized mean differences of SWBM at arrival and departure as functions of ship length.

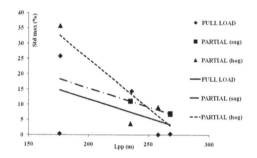

Figure 5. Trend lines of normalized standard deviations of differences of SWBM at arrival and departure as functions of ship length.

be noted that in some of loading conditions separate data for departure and arrival were not available in loading manuals in which case empty cell appears in the tables. Number of points N used for calculation of statistical properties is also shown in Tables 8–10.

It may be concluded that the normalized mean values of differences between SWBMs at departure and arrival are approximately the same in average sense for all modes of operation. Averaged normalized difference of SWBM is in the range of 15–18% of maximum SWBM. This is certainly not a negligible difference, what supports approach of Guedes Soares and Dogliani (2000) to consider separately SWBMs at departure and arrival for their probabilistic characterization.

Trend lines of differences in SWBM between arrival and departure with ship length, are presented in Figures 4 & 5.

One may observe that there is clear tendency that both means and standard deviations of normalized differences are decreasing by increasing ship length. The only exception is the emergency ballast condition with increasing mean value with ship length, but this result should be taken with care because of the small data sample available.

Decrease of means and standard deviations with ship length could be explained by the fact that the relative influence of consumables and their disposition on SWBMs is decreasing because of predominant influence of ship and cargo weight.

Reduction of standard deviation with ship length means that differences between SWBM at arrival and departure are becoming smaller. In other words, SWBMs at departure and arrival are correlated since the knowledge of departure values enables immediately determination of SWBM at arrival.

7 CONCLUSIONS

Study of statistical characteristics of SWBMs on double hull tankers based on data from loading manuals is presented in the paper. The main purpose of the study is improvement of statistical models to be used in ship structural reliability studies.

Findings that could have consequences on the ship structural reliability analysis are as follows:

− there is a strong dependency between ship length and mean value and standard deviation of SWBM in full load condition. Therefore, the model proposed in MSC 81/INF.6 could be refined;
− differences between SWBMs at departure and arrival are significant and they should be considered separately, e.g. following approach proposed by Guedes Soares & Dogliani (2000).

Furthermore, it was found that minimum design still water bending moment given in CSR is exceeded by the maximum values from the loading manual for most of the studied ships. Finally, the question of the appropriate role of the emergency ballast loading condition in ship structural reliability assessment is opened in the paper.

REFERENCES

ABS, DNV, LLOYD'S REGISTER. 2006. *Common Structural Rules for Double Hull Oil Tankers.*

FLARE, 2008, https://www.tc.gc.ca/tcss/TSB-SS/Marine/2000/m98n0001/M98N0001_p5.htm

Guedes Soares, C. & Dogliani, M. 2000. Probabilistic Modeling of Time-varying Still-water Load Effects in Tankers. *Marine Structures.*, 13:129–143.

Guedes Soares, C. & Moan, T. 1988. Statistical Analysis of Stillwater Load Effects in Ship Structures, *Transactions SNAME*, 96:129–156.

International Maritime Organization (IMO). 2006. *Goal-based New Ship Construction Standards*, MSC 81/INF.6.

Ivanov, L.D. & Wang, G. 2008. Probabilistic Presentation of the Still-water Loads. Which Way Ahead?., *Proceedings of the 27th International Conference on Offshore Mechanics and Arctic Engineering*, Estoril, Portugal, ASME Paper no.OMAE08-57011.

Rizzuto, E. 2006. Uncertainties in Still-water Loads of Tankers and Bulkers., *Proceedings NAV International Conference on Ship and Shipping Research*, Genoa, Italy.

Ship structural reliability

Analysis and Design of Marine Structures – Guedes Soares & Das (eds)
© 2009 Taylor & Francis Group, London, ISBN 978-0-415-54934-9

Structural reliability of the ultimate hull girder strength of a PANAMAX container ship

J. Peschmann, C. Schiff & V. Wolf
Germanischer Lloyd AG, Hamburg, Germany

ABSTRACT: The presented paper is related to the reliability regarding the hull girder of container vessels subjected to vertical bending. From a classification society's point of view the inherent reliability of existing rules is of interest. For a wide acceptance and practical use of structural reliability analyses (SRA) the comparability of the results is of great importance. The paper focuses on these topics. It identifies issues related to the reproducibility and the comparability of the results of SRA and will encourage further discussion. In particular aspects of the determination of the still-water bending moment and the ultimate bending strength of the hull cross section have been investigated and the results are discussed.

1 MOTIVATION

In the last few years the structural reliability of ships in connection with further developing the standards and rules (Goal Based Standards of IMO und Common Structural Rules (CSR) of IACS) has been comprehensively discussed. Spectacular accidents like the breaking apart of the Prestige and the MSC Napoli were in part responsible for these discussions and developments. In the CSR and also in the rules of the GL (Germanischer Lloyd, 2008) an examination of the ultimate capacity of the hull girder by means of an analytical method (Smith's method) has been introduced, which considers a progressive collapse of the structure mathematically. The result of this calculation is the ultimate capacity of the calculated hull girder cross section which is compared to the load of the still-water bending moment and the wave bending moment, taking into consideration the partial safety factors.

To calibrate the partial safety factors a sufficient knowledge of the uncertainties of the parameter and the method is necessary.

An extensive comparative analysis of the reliability of the hull girder strength compared to the applied vertical bending moment was published by Moan et al (2006). This analysis was motivation to carry out further research regarding the reliability of container ships.

2 DESCRIPTION OF THE MOAN METHOD

One important idea in the analysis of Moan (Moan et al, 2006) is to evaluate the reliability of the various tested types of vessels (tanker, bulk carrier and container ships) on the basis of the existing rules.

Generally, to evaluate the ultimate load of the hull girder the following limit state function is defined:

$$G_{(x)} = M_U - (M_{SW} + M_{VW}) \quad (1)$$

The equation describes the comparison between the maximal load carrying capacity M_U and the carrying load (for the different loading cases) which in this case is to be combined from the still-water bending moment M_{SW} and the wave bending moment M_{VW}. Taking into consideration the various factors of influence χi, the following limit state function used in (Moan et al. 2006) arises.

$$G_{(x)} = M_{Uc} \cdot \chi_{Rm1} \cdot \chi_{Rm2} - (\Psi_{SW} \cdot \chi_{SW} \cdot M_{SWc})$$
$$+ (\chi_{annual} \cdot \chi_{WC} \cdot \chi_{nl} \cdot \chi_{envir} \cdot \chi_{hwa}$$
$$\times \chi_{IACS} \cdot M_{VWc}) \quad (2)$$

The relevance and the derivation of the factors of influence can be found in Moan et al (2006).

The factor ψ_{SW} takes into consideration that generally the maximum value for still-water bending moment and wave bending moment do not occur at the same time.

The characteristic value for the structural ultimate strength M_{Uc} is determined in Moan et al (2006) by the characteristic values of still-water and wave bending moment using partial safety factors. The limit state function is then standardised by the vertical wave bending moment. The characteristic value of the still-water moment is expressed by its average plus k-times its standard deviation.

$$M_{swc} = \mu + k\sigma = M_{sw,rule} (\overline{\mu} + k\overline{\sigma}) \quad (3)$$

$\bar{\mu}$ and $\bar{\sigma}$ are the average and the standard deviation related to the still-water bending moment $M_{SW,rule}$ as defined by backward calculation of the IACS Unified Requirements (IACS, 2007).

The result is the following limit state function:

$$g_{HG} = \hat{\chi}_{Rm1}\,\hat{\chi}_{Rm2}\,\chi_{c,r}\,\gamma_R$$

$$\times \left(\gamma_{sw} \frac{M_{sw,rule}}{M_{vwc}} (\bar{\mu} + k\bar{\sigma}) + \gamma_{vw} \right)$$

$$- \psi_{sw}\hat{\chi}_{sw} \frac{M_{sw,rule}}{M_{vwc}} (\bar{\mu} + k\bar{\sigma})$$

$$- \chi_{annual}\,\hat{\chi}_{vw}\,\hat{\chi}_{nl}\,\chi_{envir}\,\chi_{hwa}\,\chi_{IACS} \qquad (4)$$

3 RECALCULATION—USED DATA AND DERIVED RESULTS

It was then attempted to comprehend the results from Moan et al (2006) by carrying out calculations using the programme COMREL. However, the distribution function used for the factor χ_{SW} (see equation 8) was not supported by the software COMREL. Therefore it was attempted to approximate this function by using a Gumbel distribution. The data for the other parameters χ_i were taken from Moan et al (2006) and are listed in table 1.

The results of the recalculation are compared to the results from Moan et al (2006) in table 2. Although the calculated results show the same tendency as shown in figure 1, the absolute result of the reliability index and the failure probability differ considerably. The cause for these differences lies in the deviation of the distribution function according to Moan et al (2006) and the Gumbel distribution which mainly shows considerable differences in the absolute values within the "tails" of the function.

4 EXPERIENCE CONCERNING RELIABILITY CALCULATION OF CONTAINER SHIPS

4.1 Calculation of the ultimate strength

The ultimate strength moment of a ship is determined in Moan et al (2006) by addition of the still water and wave bending moment which were calculated back according to IACS UR S7 and S11, taking into consideration the partial safety factors.

If the reliability method is applied to a certain ship, the ultimate strength and the distribution of the ultimate strength should be determined by using the direct method of calculation based on the actual geometry of the hull girder cross section, used materials

Table 1. Parameters and data of the distribution functions used for the recalculation and the PANAMAX vessel.

Parameter	Recalculation of (Moan et al. 2006)			PANAMAX – Ship		
	Distribution	Mean value	Standard deviation	Distribution	Mean value	Standard deviation
$\hat{\chi}_{Rm1}$	normal	1.05	0.1	normal	1.05	0.1
$\hat{\chi}_{Rm2}$	normal	1.1	0.1	normal	1.1	0.1
$\hat{\chi}_{sw}$	Gumbel	1.09 (1.35) *	0.05 (0.08) *	Gumbel	1.093	0.051
$\hat{\chi}_{vw}$	normal	0.9	0.1	normal	0.9	0.1
$\hat{\chi}_{nl}$	normal	0.9 (1.4) **	0.02 (0.07) **	normal	0.896 (1.289) **	0.026 (0.077) **
χ_{annual}	fixed	0.83	–	normal	0.816	0.009
$\chi_{c,r}$	fixed	1	–	fixed	1	–
χ_{envir}	fixed	1	–	fixed	1	–
χ_{hwa}	fixed	0.85	–	fixed	0.85	–
χ_{IACS}	fixed	1	–	fixed	1	–
sw	fixed	0.85	–	fixed	0.85	–
γ_R	fixed	1.1	–	fixed	1.1	–
γ_{sw}	fixed	1	–	fixed	1	–
γ_{vw}	fixed	1.2	–	fixed	1.2	–
$\bar{\mu}$	fixed	0.6	–	fixed	0.709	–
$\bar{\sigma}$	fixed	0.15 (0.25)*	–	fixed	0.127	–
k_{1_year}	fixed		–	fixed	1.732	–
k_{20_years}	fixed		–	fixed	2.865	–
$M_{sw,rule}$ [kNm]	fixed	–	–	fixed	2975609	–
M_{vwc} [kNm]	fixed	–	–	fixed	5357000 (3533536)***	–

* value in brackets valid for the generic vessel; ** value in brackets valid for sagging; *** value in brackets used for a second calculation.

504

Table 2. Parameters and data of the distribution functions used for the recalculation and the PANAMAX vessel.

$\dfrac{M_{sw,rule}}{M_{vwc}}$			Sagging		Hogging		
			0.6	0.6*	1.0	1.0*	1.1*
k = 1.5	Moan et. al., 2006	P_f	4.8E-03	1.0E-02	6.7E-05	3.7E-04	5.4E-04
		β	2.59	2.33	3.82	3.38	3.27
	GL	P_f	2.5E-03	1.0E-02	3.2E-05	1.0E-03	1.5E-03
		β	2.81	2.32	4.00	3.09	2.97
k = 2.24	Moan et. al., 2006	P_f	1.9E-03	2.5E-03	2.0E-05	6.7E-05	9.3E-05
		β	2.90	2.80	4.11	3.82	3.74
	GL	P_f	2.6E-03	1.3E-02	4.8E-05	2.0E-03	2.9E-03
		β	2.80	2.23	3.90	2.88	2.76

* generic vessel.

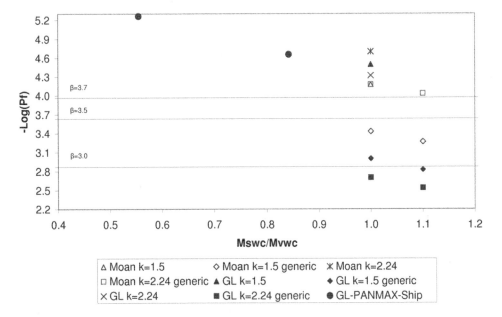

Figure 1. Results of the recalculation of the failure probability acc. to the paper of Moan et al (2006) and the example PANAMAX ship-only results for the hogging case are shown.

etc. The distribution of the particular parameters and their influence such as geometry (density and pre-deformation), material properties (yield stress), etc. need to be considered. To carry this out, a Monte-Carlo simulation or a surface response procedure should be applied.

The present analysis is restricted to the direct calculation of the ultimate strength of the observed PANAMAX container ship concentrating on selected extreme points in the design space to show the influence of certain parameters on the ultimate strength of the hull girder cross section.

For the calculation the analytical Smith's method was used. Various program codes were applied which

implement the procedure. In table 3 the applied codes and their properties are listed.

The calculations were performed based on the data of the example PANAMAX ship. The main data of the ship are listed in table 4 while the cross section of the hull is shown in figure 2.

Figure 3 shows a comparison of the results. In spite of the same calculation procedure there are significant differences in the results. The causes for these differences can be found mainly in the definition and calculation of the load-end shortening curves especially in the description of the so called "Hard Corner" elements. This explanation is supported by the following: In the sagging case (negative values of the vertical

Table 3. Properties of used program codes (Smith's method).

Code	Definition of elements	Load-end-shortening-curves	assumed stress condition*
GL-Excel	by hand	by program using buckling code of GL-Rules	uniaxial
GL-CSR	by hand	by hand using Common Structural Rules for Bulk Carrier	Free definition of load-end-shortening curves
HULLCOL (Gordo)	by hand	by program described in Gordo & Guedes Soares, 1993	uniaxial
HULLCOL (Gordo) modified	by hand	by program described in Gordo & Guedes Soares, 1993	plane stress condition

* assumed stress condition for the determination of the load-end-shortening curves.

Table 4. Main data of the example container ship.

Scantling length L	278.35 m
Length between perp.	283.20 m
Breadth	32.20 m
Draught	13.55 m
Block coefficient	0.70
Speed	24.00 kn

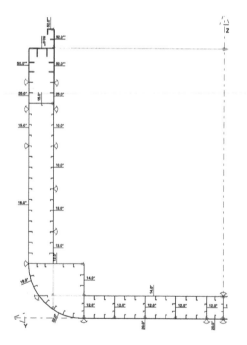

Figure 2. Cross section of the example PANAMAX container ship.

bending moment) where the deck-structure is under compression, larger deviation in the results occurs, whereas in the hogging case (positive values of the vertical bending moment) where the double bottom is under compression the results are relatively close together. The deck structure is in the models described by using hard corner elements. In some codes these elements fails only by yielding when the stress in the element reaches the nominal yielding stress. In other codes the reduction of the effective width under compression is considered, which reduces the strength of these elements considerably. In the hogging case this difference is not of such importance caused by the buckling failure of the stiffened panels in the double bottom structure, which is defined in a similar way in all codes.

Figure 4 shows the results for the ultimate moment of the hull cross section where respectively 50% and 100% of the corrosion addition are deducted for the geometry of the hull cross section in line with the building regulation of the Germanischer Lloyd (Germanischer Lloyd 2008). The ultimate moment is reduced by 10% for the sagging case and by about 15% in the hogging case if the entire (100%) corrosion addition is deducted.

Figure 5 shows the results for the ultimate bending moment of the examined hull cross section taking into consideration the different yield stress of the materials (see table 5). The results for the nominal value of the yield stress and the characteristic value only differ slightly (around 1%) because the characteristic value (the value above which 95% of the expected values fall) only differs slightly from the nominal value. This applies to the normal distribution of the yield stress as well as to the log-normal distribution. The third curve shows the results based on the mean values of the material yield stress. For this the values for the ultimate stress are about 10% higher. The calculations were done taking into consideration a reduction of 50% of the corrosion addition in line with the GL rules. The assumed distribution of the yield stress was the normal distribution (the corresponding characteristic values of table 5 were used).

Table 6 shows the calculated ultimate strength and the comparison to the ultimate strength for the example PANAMAX container ship which were compiled acc. to Moan et al (2006) (last two lines of table 6). It

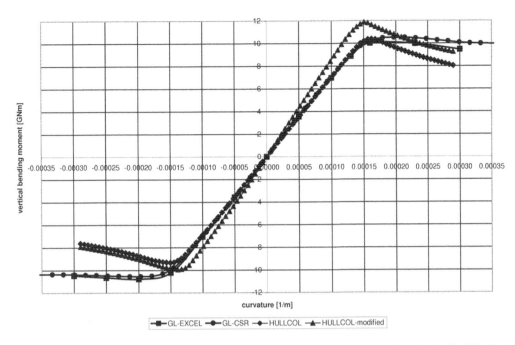

Figure 3. Moment-curvature curves calculated by using different codes (squares for GL-Excel, bullets for GL-CSR, rhombi's for HULLCOL and triangles for the modified version of HULLCOL).

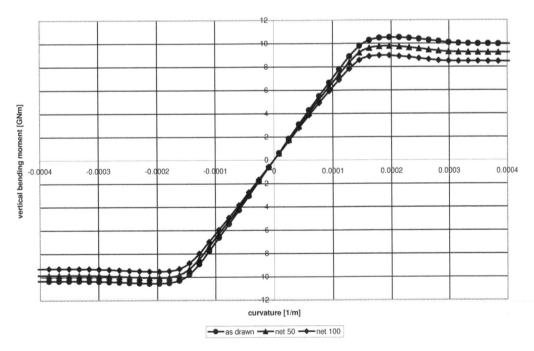

Figure 4. Moment-curvature curves considering thickness reduction of plates of 50% (triangles) and 100% (rhombi's) of the corrosion allowance in comparison with the as drawn thickness (bullets).

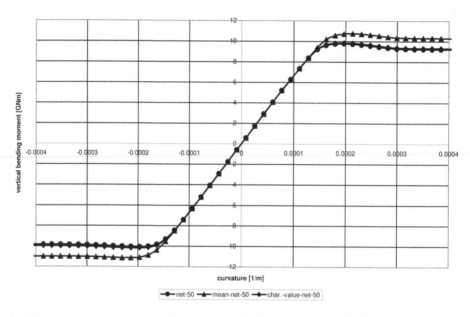

Figure 5. Moment-curvature curve calculated by using nominal (bullets), mean- (triangles) and characteristic value (rhombi's) for the yield stress of the material.

Table 5. Nominal, mean and characteristic values of yield stress of the used steel.

			Characteristic values (95%)		
Nominal value	Mean value [MPa]	COV [%]	Normal [MPa]	Lognormal [MPa]	Gumbel [MPa]
235	286	7,1	252,5	253,4	259,4
315	353	6,0	318,0	318,9	325,2
355	398	5,7	361,0	362,0	368,7

Table 6. Comparison of ultimate bending moments as calculated based on geometry of hull girder and on backward calculation of IACS UR S11 and S7 (IACS 2007).

	GL-Excel			GL-CSR			HULLCOL		
Calculation	as drawn	−50% tc	−100% tc	as drawn	−50% tc	−100% tc	Char. Value	Mean value	as drawn
Hogging [GNm]	10.039	9.313	8.568	10.527	9.781	8.978	9.867	10.764	10.390
Sagging [GNm]	−10.775	−10.149	−9.644	−10.546	−10.059	−9.529	−10.175	−11.145	−9.319
Mu/Mu IACS Hogging	1.26	1.17	1.08	1.33	1.23	1.13	1.24	1.36	1.31
Mu/Mu IACS Sagging	1.34	1.26	1.20	1.31	1.25	1.18	1.26	1.38	1.16

can be observed that the ultimate strength which was calculated on the basis of the actual structure is considerably higher than the values estimated acc. to Moan et al (2006).

These findings imply an increase of the reliability index and a decline in the failure probability be

cause the calculated ultimate strength The calculated ultimate moment capacities considering the deduction of the corrosion addition or using the nominal value of the yield stress of the material and a deduction of the plate thicknesses of 50% of the corrosion addition at the same time describes values at the lower allowable limit for the examined cross section. It is therefore expected, that a subsequent reliability analysis will result in an increase of the reliability index and a decline in the failure probability respectively.

The section modulus of the hull cross section can be reduced by around 10% in comparison to the new construction according to IACS UR S7 before a substitution of the structural elements has to be carried out. The reduction of the scantlings by around 50% of the corrosion addition t_c according to the GL rules means for the here examined ship a reduction of the section modulus by about 4%.

4.2 Derivation of the still water bending moment

For the estimation of vessel's still-water bending moment two alternatives exist. The set up of a statistic from actual loading conditions based on data from load master computers or based on data from loading manuals.

The computation of the characteristic value of the still-water bending moment is done by using the following equation.

$$M_{SW,c} = \mu_M + k\sigma_M \tag{5}$$

Equation 5 in Moan (et al 2006).

The factor k can be calculated using the data of the loading manual and $M_{SW,C}$ is the maximum still-water bending moment taken from the loading manual. Another method is based on the expected maximum still-water bending moment of the voyages within a certain reference period. The latter is completely independent of the actual values of the still-water bending moment, only the number of voyages within the reference period is of importance.

$$M_{SW,mp} = \mu_M - \sigma_M \Phi^{-1}(1/N) \tag{6}$$

Equation 6 in Moan et al (2006).

Here Φ is the standard normal distribution, which gives the following description for the factor k.

$$k = -\Phi^{-1}(1/N) \tag{7}$$

N is the number of voyages within the reference period.

For the extreme value distribution function of the minimum and maximum of the still-water bending moment the following formulation has been assumed in Moan et al (2006).

$$f_{x_{\max}^N} = N f_x(x) (F_x(x))^N \tag{8}$$

Here a normal distribution of the still water bending moments is assumed. In the equation are $f(x)$ the density function and $F(x)$ the cumulative density or distribution function of the still-water bending

moments respectively (a normal distribution in this case). N is the number of voyages.

IACS UR S11 offers the possibility to estimate a minimum value of the still-water bending moment by using the definition of the minimum hull section modulus given in IACS UR S7 according to the following equation.

$$M_{SW} = M_{tot} - M_{WV} = 0.175 \cdot C_0 L^2 B$$
$$\times (CB + 0.7) - M_{WV} \tag{9}$$

M_{tot} is the sum of the still water- and vertical wave bending moment, C_0 is the wave coefficient as a function of the ship length L, B is the breadth of the ship and CB the block coefficient. The computation of M_{WV} has to be done separately for the hogging and sagging case. This leads to one hogging and one sagging case for the still-water bending moment.

This value is a minimum value which can be used as a start value during the design process.

For container vessels GL proposes the use of a different start value which is dependent of the amount of containers. For most ships this value is about 10% higher than the Stillwater bending moment according to the backward calculation from IACS UR S7 and S11.

In the real design process the used still-water bending moment depends on the actual loading conditions and so predefined by the ship-owner. From the structural drawing plans the maximum permissible total moment can be calculated for every frame section using the permissible stresses according to IACS UR S11. Subtracting the vertical wave bending moment according to IACS UR S11, the permissible still-water bending moment can be derived. The value for the mid-ship section is normally given in the relevant structural drawings. This two still-water bending moments (permissible and predefined by owner) may differ significantly from the minimum value of the still-water bending moment calculated by backward calculation of IACS UR S7 and S11.

Therefore data from loading manuals or data from the loading computer system of similar ships should be used for reliability analyses.

The data evaluation from data of Hansen (1990), which were the basis for many other published analyses like Guedes Soares & Dias (1996) and Östergaard et al (1996), by use of equation 7 and 8 according to Moan et al (2006) lead to mean values and standard deviation of the still-water bending moment as presented in table 7.

The data have been taken from 9 PANAMAX container ships 5 of them have been operating between Europe and East-Asia and 4 of them between Europe and East Coast of North America. The evaluation was performed for all ships together and for the 4 and 5 ships separately fort he different operation areas. The computed values are in good accordance with values

Table 7. Results of data analysis for PANAMAX container ships based on data from (Hansen 1990) and comparison with data from Moan (Moan et al. 2006) for the generic vessel.

	4 Ships (Far-East)	5 Ships (North-Atlantic)	PANAMAX (bothroutes)	Values from (Moan et al. 2006)
Mean value	74.5%	69.5%	70.9%	62.0%
Standard deviation	12.9%	12.3%	12.7%	14.0%
k	1.978	2.474	2.297	1.750
k (24 voyages 1 year)	1.732	1.732	1.732	1.732
k (480 voyages 20 years)	2.865	2.865	2.865	2.865
k loading manual	1.402	1.402	1.402	
Mmax (1 year)	96.8%	90.9%	92.8%	99.6%
Mmax (20 years)	111.4%	104.8%	107.2%	124.1%

from Moan et al (2006) but they do not scatter so much. All values are based on still-water bending moments according IACS Rules (100%).

The factor k in table 7 has been calculated as follows:

The values in line "k" are based on the evaluation of the data from Hansen (1990) as described above. The values of line "k (24 voyages 1 year)" and "k (480 voyages 20 years)" are based on equation 6 and 7 assuming 24 voyages per year (480 voyages within 20 years). The number of voyages is based on Hansen (1990). He showed that ships cross the North Atlantic about two time per month. The line "k loading manual" is based on the evaluation of the loading manual of the sample ship. The values for the maximum Stillwater bending moment to be expected for 1 and 20 years have been computed on the basis of 24 voyages per year.

4.3 Example calculation of the reliability of a PANAMAX container ship

The reliability of the sample PANAMAX container vessel, as described in the previous sections of the paper, was estimated with the limit state function from equation 4 using COMREL software. The characteristic value of the ultimate strength (ultimate bending moment) was calculated by adding the still-water and wave bending moment according IACS under consideration of partial safety factors as described in Moan et al (2006). It was not based on the existing steel structure as described in section 4.1 of this paper. In a first calculation the wave bending moment according IACS is used yielding to a ratio of still-water bending moment to wave bending moment of 0.842. In a second calculation wave bending moments taken from direct calculations were used and the above mentioned ratio decreases to 0.555. The factor k describing the characteristic value of the still-water bending moment is based on the assumption of 24 voyages per year as described in the section before. Table 1 show the parameters used as well as the associated distribution

Table 8. Probability of failure and reliability index for the PANAMAX vessel.

		Msw/Mwv = 0.555		Msw/Mwv = 0.842	
		Sagging	Hogging	Sagging	Hogging
1 Year	P_f	8.0E-4	5.5E-6	1.16E-3	2.21E-5
	β	3.156	4.397	3.047	4.085
20 Years	P_f	8.6E-3	1.08E-4	8.89E-3	2.52E-4
(χ_{annual} = 1)	β	2.381	3.700	2.370	3.479

functions. The values in grey fields are identical with those from Moan et al (2006) and all other values were taken from own investigations. The calculated probabilities of failure and the reliability indices are listed in table 8. In figure 1 the values for the hogging case are indicated by circles.

Relatively high reliability values have been found but the ratio Stillwater to wave bending moment differ significantly to the ratios given in Moan et al (2006). As the limit state function used is dependent on this moment ratio but not from the load carrying capacity of the structure the result can be taken as an indicator only. The actual reliability of a ship should be examined using the ultimate bending moment capacity of the real steel structure.

5 SUMMARY AND CONCLUSION

The present paper deals with the estimation of the reliability of hull girders with respect to vertical bending moment. It is based on the procedure presented by Moan et al. (2006). The following topics were investigated and discussed:

1. Still-water bending moment

For reliability investigation of a ship realistic data from operating ships should be used and evaluated statistically. The authors evaluated data from late 80ths of the last century (Hansen, 1990). For further investigations larger and younger ships should be used for collecting still-water bending data. It must be expected that changed trade routes and streams of goods will influence the still-water loads of container vessels. For this it is of interest to evaluate current data. In this content the definition of "voyage" is important as it is used with different meanings in different analyses. The question is whether one voyage is from one harbour to the next or a North-Atlantic passage for ships operating in that area. In the second case trades between European harbours or trades between East American harbours are neglected. It must be expected that the distribution of the still-water bending moment over the ships length differ for trips along the coast from that for trips crossing the Atlantic. This question is important for the statistic evaluation of the data.

The recalculation from data of Moan et al (2006) lead to differences which can be explained by the use of different distribution functions of the maxima of still-water bending moments. This shows that the distribution function of parameters or data has to be analysed carefully. The problem is pronounced by the fact that data are normally available in the vicinity of the mean value only. The missing information for the "tails" of the distribution function is essential for reliability analyses. The question is how to ensure that distribution functions are realistic or confidential, also in cases where the tails are described insufficient.

2. *Ultimate bending moment*

For reliability analyses the ultimate bending moment should be based on real structural designs in any case. Also for classes of ships, evaluations of direct calculated ultimate load carrying capacities of the structures should be used as they differ significantly from backwards calculated values according to IACS Unified Requirements in many cases.

Reason is that values derived from IACS are minimum requirements only. In the contrary to that, the allowable or maximum permissible still-water bending moment under consideration of the real hull section modulus can be calculated with the permissible stress as well as the maximum wave hogging and sagging moments according to IACS UR S11.

ACKNOWLEDGEMENT

The present paper has been partly prepared within the "MARSTRUCT—Network of Excellence on Marine Structures" http://www.mar.ist.utl.pt/marstruct/, which has been funded by the European Union Through the Growth program under contract TNE3-CT-2003-506141.

REFERENCES

Germanischer Lloyd, 2008. Rules for Classification and Construction, I Ship Technology, Part 1 Seagoing Ships, Chapter 1 Hull Structures. Hamburg: Germanischer Lloyd AG.

Gordo, J.M. & Guedes Soares, C., 1993. Approximate Load Shortening Curves For Stiffened Plates Under Uniaxial Compression. In: D. Faulkner, M.J. Cowling, A. Incecik, P.K. Das (eds), *Integrity of Offshore structures—5*: 189–211. Glasgow: University of Glasgow.

Guedes Soares, C. & Dias, S., 1996. Probabilistic Models of Still-Water Load Effects in Containers. *Marine Structures* 9: 287–312.

Hansen, H.-J. 1990. Erfassung und Standardisierung der Häufigkeitsverteilungen von Glattwasserbelastungen der Längs- und Querverbände. Abschlussbericht MTK 440 7: Hamburg: Germanischer Lloyd.

IACS 2006. Common Structural Rules for Bulk Carriers. Ch05, Appendix 1: International Association of Classification Societies. April 2006, http://www.iacs.org.uk/publications/default.aspx

IACS 2007. (International Association of Classification Societies), UR S 2007 (Unified Requirements concerning Strength of Ships, Version 2007), www.iacs.org.uk

Moan, T. Shu, Z. Drummen, I. Amlashi, H. 2006. Comparative Reliability Analysis of Ships—considering different Ship Types and the Effect of Ship Operations on Loads. *Transactions Society of Naval Architects and Marine Engineers SNAME*.

Östergaard, C., Otto, S., Teixeira, A. & Guedes Soares, C., 1996. A Reliability Based Proposal for Modern Structural Design Rules of the Ultimate Vertical Bending Moment of Containerships. *Jahrbuch der Schiffbautechnischen Gesellschaft* 90: 515ff. Berlin: Springer.

Analysis and Design of Marine Structures – Guedes Soares & Das (eds)
© 2009 Taylor & Francis Group, London, ISBN 978-0-415-54934-9

Sensitivity analysis of the ultimate limit state variables for a tanker and a bulk carrier

A.W. Hussein & C. Guedes Soares
*Centre for Marine Technology and Engineering (CENTEC), Technical University of Lisbon,
Instituto Superior Técnico, Lisboa, Portugal*

ABSTRACT: Reliability analyses are performed for a SUEZMAX tanker ship and a HANDYMAX bulk carrier. Three loading conditions are considered for the bulk carrier, namely: homogenous loading condition, alternative hold loading condition and ballast condition. For the tanker ship only the full load condition is considered. After performing the reliability analyses, the importance of variables included in the ultimate limit state function for both ships is calculated and compared. Since the ultimate sagging bending moment is the most critical capacity variable, the effect of modifying the structural design by increasing the deck plating by a certain factor while keeping the scantlings of the sides, bulkheads and bottom structure constant is studied. The influence of increasing the scantlings on the reliability and the sensitivity of the variables is studied.

1 INTRODUCTION

The design, analysis, and planning of any engineering system require the basic concept that the capacity should be greater or at least satisfy the demand. Structural elements must have adequate strength to permit proper functioning during their intended service life. As this must be accomplished under conditions of uncertainty, probabilistic analyses are needed in the development of such reliability-based design of panels and fatigue details of ship structures.

In recent years, ship structural design has been moving towards a more rational and reliability-based design procedure related to limit states design. The reliability based design approach takes into account more information than deterministic methods. This information includes uncertainties in the strength of various structural elements, in loads and load combinations, and modeling errors in analysis procedures. Reliability-based design formats are more flexible and rational than working stress formats because they provide consistent levels of safety over various types of structure.

The aim of structural reliability analysis is to represent the problem as realistically as possible, reflecting the uncertainties involved, and hence avoiding undue conservatism due to generalizations. The main outcome is a prediction of a nominal annual probability of failure together with the sensitivity of this result to the various parameters involved.

The limit state function used in the reliability analysis represents a state associated with collapse and includes all the design variables. In case of ultimate capacity limit state, it includes the ultimate bending moment, the wave induced loads and the still water bending moment. Changing the values of these variables affects the results of the reliability assessment and for this reason, a sensitivity analysis is required. Sensitivity analysis is used to determine how much the model output is changed by each of the variables of the model.

Guedes Soares and Teixeira (2000) studied the changes in notional reliability levels that result from redesigning a traditional single-hull bulk carrier to become a double-hull structure. The sensitivity analysis was presented for the two dominating load conditions, full load for sagging and ballast load for hogging. The authors compared the sensitivity parameters for the two bulk carriers and a tanker ship. It was concluded that the importance of the uncertainty on the ultimate strength remains constant in the two load conditions for the tanker. However, the overall importance of the wave-induced load variables decreases from full to ballast load condition. For the bulk carriers, the importance of the vertical still-water bending moment increases for both loading conditions.

Guedes Soares and Parunov (2008) redesigned an existing SUEZMAX tanker to comply with the new Common Structural Rules (CSR) requirement for ultimate vertical bending moment capacity. The reliability assessment was performed for as-built and corroded states of the existing ship and a reinforced design configuration complying with CSR. Sensitivity analysis and a parametric study were performed and it was concluded that although the overall sensitivity to uncertainty in wave loads is the largest, each of its component uncertainties has smaller impact on results than the uncertainty in ultimate strength calculation.

Parunov and Guedes Soares (2008) studied the reliability of an AFRAMAX ship redesigned according to the new Common Structural Rules. The sensitivity analyses showed that the most important random variable is the modeling uncertainty in ultimate strength calculation.

Parunov et al. (2007) calculated the changes in hull-girder reliability resulting from the new generation oil tankers which differ from traditional oil tankers by their unusual form. The sensitivity analyses for full load in sagging and ballast load in hogging for "corroded" hulls were assessed. The sensitivity factors were calculated and by analyzing the results it was obvious that the uncertainty of ultimate longitudinal strength is the most important random variable. In ballast condition, the importance of the vertical stillwater bending moment overcomes that of the ultimate longitudinal strength. The importance of the wave-induced load and the corresponding uncertainty decreases from full load to ballast.

Khan and Das (2008) made a reliability analysis for a double skin tanker and a bulk carrier considering the combined effect of vertical and horizontal loads. It was concluded that the wave-induced vertical bending moment and vertical ultimate strength are the most important parameters. The ultimate strength in vertical bending is the most important factor in the structural design, but in case of the bulk carrier, it is noticed that the model uncertainty factor for horizontal strength showed significant sensitivity.

In the present paper reliability assessments are performed for a bulk carrier and a tanker. Since the new Common Structural Rules CSR (IACS, 2006) defined the sagging condition in full load for tanker to be the most critical limit state, the reliability is calculated for the tanker only for that condition, while the reliability indices are calculated for three loading condition for the bulk carrier. Sensitivity analyses are made to compare the importance of the variables included in the limit state function for both ships.

To modify the structural design, the deck plating is increased by a certain factor while keeping the scantlings of the sides, bulkheads and bottom structure constant and assuming that the loads remain unchanged after applying this factor. The influence of increasing the scantling on the reliability and the sensitivity of the variables is also studied to know whether the importance's of the variables change with different designs or not.

Table 1. Ships principal dimensions.

	Tanker	Bulk carrier
L_{BP}(m)	264	176
B(m)	45.1	30
D(m)	23.8	16.2
C_B	0.83	0.821

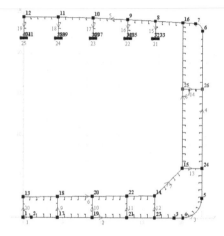

Figure 1. Tanker 264 m length.

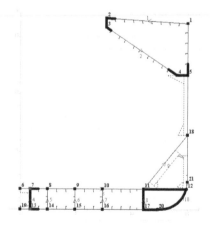

Figure 2. Bulk Carrier 177 m length.

2 SHIPS PARTICULARS

Two ships are under consideration in this study; a SUEZMAX tanker ship and a HANDYMAX Bulk carrier. The particulars of the two ships are presented in Table 1 and the cross sections are shown in Figures 1 and 2.

3 ULTIMATE CAPACITY

The longitudinal strength is the most fundamental strength parameters of a ship structure. The hull girder ultimate capacity is probably the most critical failure mode for ships, as the capacity in hogging is usually significantly higher than in sagging.

The ultimate strength is calculated based on the net scantlings + 50% of the corrosion addition as defined in the IACS New Common Structural Rules. The progressive collapse method is used to calculate the strength and the stress strain curves are as defined in the CSR. The method enables to perform progressive collapse analysis on the cross-section of a hull girder subjected to longitudinal bending.

The cross-section is divided into small elements composed of stiffeners and attached plating. At the beginning, the average stress–average strain relationships of individual elements are derived under the axial load considering the influences of yielding and buckling according to the equations defined in the new CSR. Then, a progressive collapse analysis is performed assuming that plane cross-sections remain plane and each element behaves according to its average stress–average strain relationships.

The ultimate hull girder bending moment capacity M_U is defined as the peak value of the curve of the vertical bending moment M versus the curvature R of the ship cross section. The curve M-R is obtained by means of an incremental-iterative approach. The bending moment M_i which acts on the hull girder transverse section due to the imposed curvature R_i is calculated for each step of the incremental procedure. This imposed curvature corresponds to an angle of rotation of the hull girder transverse section about its effective horizontal neutral axis, which induces an axial strain ε in each hull structural element. In the sagging condition, the structural elements below the neutral axis are lengthened, whilst elements above the neutral axis are shortened.

The stress σ induced in each structural element by the strain ε is obtained from the stress-strain curve $\sigma - \varepsilon$ of the element, which takes into account the behavior of the structural element in the non-linear elasto-plastic domain. The force in each structural element is obtained from its area times the stress and these forces are summed to derive the total axial force on the transverse section. The element area is taken as the total net area of the structural element. This total force may not be zero as the effective neutral axis may have moved due to the non linear response. Hence it is necessary to adjust the neutral axis position, recalculate the element strains, forces and total sectional force and iterate until the total force is zero.

Once the position of the new neutral axis is known, then the correct stress distribution in the structural elements is obtained. The bending moment M_i about the new neutral axis due to the imposed curvature R_i is then obtained by summing the moment contribution given by the force in each structural element.

$$M_i = \sum \sigma_i \cdot A_i \cdot (z_i - Z_{NA\_i}) \tag{1}$$

A computer program was developed to calculate the ultimate strength according to the progressive collapse method and considering the failure modes defined by the New IACS CSR. The program and the compatibility with other programs are well explained in Hussein et al. (2007). The bending capacity was modeled as lognormal distribution with COV equal to 0.08. Table 2 shows the results. The uncertainties for each case are presented according to the comparison of the program with others.

4 STOCHASTIC MODEL OF STILL WATER BENDING MOMENT

4.1 SWBM for the tanker ship

The stochastic model of still water bending moment is considered as normally distributed with a mean value of 0.7 times the maximum value in the loading manual, with a standard deviation of 0.2 times the maximum value, as proposed in Horte et al. (2007).

For full load condition the weight distribution for the tanker is as shown in Figure 3. The maximum still water bending moment is 2226.7 MN·m. The stochastic values are presented in Table 3. The model uncertainty is taken as normal distribution with mean value equal to 1 and COV equal to 0.1.

4.2 SWBM for the bulk carrier

For the bulk carrier three different load conditions are defined, namely: homogeneous hold loading condition, alternate hold loading condition and ballast. In each loading condition, a percentage of ship life can be

Table 2. Stochastic model of UBM.

		μ	σ
Bulk	Sag BM	3142	251
	Hog BM	4311	345
	Uncertainty	0.85	0.05
Tanker	Sag BM	11062	885
	Hog BM	11999	960
	Uncertainty	1.05	0.1

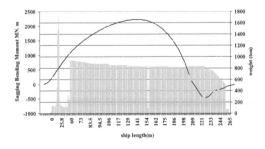

Figure 3. Still water bending moment and weight distribution in full loading condition.

515

Table 3. Stochastic model of SWBM tanker 2.

	SWBM	Distribution	μ	σ
Tanker 2	2226.7	Normal	1558	312

Table 4. Operational profile adopted for bulk carriers.

Condition	Fraction of ship life
LC1—Alternate Loading Conditions	20%
LC2—Homogeneous Loading Condition	20%
LC3—Ballast Loading Condition	40%
Harbour	20%

Table 5. Maximum SWBM at each loading condition.

Load condition	Max SWBM (MN · m)
LC1: Alternate	1101
LC2: Homogenous	117
LC3: Ballast	1160

Table 6. Stochastic model of still water BM in Mn · m.

Condition	Max	Distribution	Mean	Std.
Alternate	1101	Normal	771	220
Homogeneous	117		234	67
Ballast	1160		812	232

value. The stochastic model of the SWBM is presented in Table 6. The same model uncertainty will be assumed for bulk carrier.

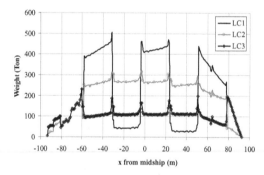

Figure 4. Weight distributions of the three loading conditions.

identified according to an estimate of the operational profile for the ship type. Table 4 shows the operational profile used in reliability calculations as found in the ship manual.

A homogeneous hold loading condition refers to the carriage of cargo, evenly distributed in all cargo holds. This condition is usually adopted for low density cargoes such as coal and grain, but may also be permitted for high-density cargoes under certain conditions. However, heavy cargo, such as iron ore, is often carried in alternate holds. This type of cargo distribution raises the ship's centre of gravity, which eases the ship's rolling motion. Figure 4 illustrates the weight distributions of the three loading conditions of the bulk carrier according to the loading manual. Table 5 shows the maximum still water bending moment for each case.

The still water bending moment will be assumed normally distributed with mean value equal to 0.7 of the maximum allowable value defined in the load manual. The standard deviation is taken as 0.2 of that

5 STOCHASTIC MODE OF WAVE BENDING MOMENT

Due to the random nature of the ocean, wave loads are stochastic processes both in the short-term and long-term. The short term vertical wave induced bending moment corresponds to a steady sea state, which is considered stationary within several hours. Within one sea state the amplitude of the wave induced load effects follows a Rayleigh distribution. The response is modeled as a Gaussian zero mean stationary stochastic process described by its variance R:

$$R = \int_0^\infty S_R(\omega) \cdot d\omega \qquad (2)$$

where $S_R(\omega)$ is the response spectrum given by the product of the non linear transfer function $H(\omega)$ for a specified relative heading and significant wave height and the seaway spectrum $S_H(\omega)$.

$$S_R(\omega) = S_H(\omega) \cdot H^2(\omega) \qquad (3)$$

The amplitude of a Gaussian zero mean stationary stochastic process with a narrow band assumption, follow a Rayleigh distribution such that the probability of exceeding the amplitude x is given by Longuett-Higgins, (1952):

$$Q(x|R) = e^{\left(-\frac{x^2}{2R}\right)} \qquad (4)$$

The probability distribution of the wave induced load effects that occur during long term operation of the ship in the seaway is obtained by the weighting the conditional Rayleigh distribution by the probability of occurrence of the various sea states in the ship route,

Guedes Soares and Viana (1988), Guedes Soares and Trovão, (1991).

$$Q_L(x) = \int Q_s(x|R) \cdot f_R(r) \cdot dr \qquad (5)$$

where $f_R(r)$ is the probability density of the response variation in the considered sea states which depends on the wave height, wave period, ship heading and speed.

The sea waves can not be adequately described by only a frequency spectrum. In general, the patterns observed in the ocean shows the existence of many components travelling in various directions. The directional spectrum represents the distribution of the wave energy both in frequency of the wave components and also in direction θ. The spreading function G is a function of both direction and frequency. The directional spectrum $S(\omega, \theta)$ can be presented as:

$$S(\omega, \theta) = S(\omega) \cdot G(\omega, \theta)$$

The directional function fulfills the requirement:

$$\int_{\theta_{min}}^{\theta_{max}} G(\theta, \omega)d\theta = 1 \qquad (6)$$

The functional form of $G(\theta, \omega)$ has no universal shape and several proposed formulas are available. It is customary to consider G to be independent of frequency such that we have:

$$G(\theta) = \frac{2}{\pi} \cos^2(\theta - \psi) \quad \text{if } |\theta| \leq \pi/2$$
$$G(\theta) = 0.0 \qquad |\theta| \geq \pi/2 \qquad (7)$$

where ψ is the principal direction of spectrum and θ is a counter clockwise measured angle from the principal wave direction.

The resultant probability distribution is fitted to the Weibull distribution given by:

$$F_x(x) = 1 - e^{-(\frac{x}{w})^k} \qquad (8)$$

where w and k are the scale and shape parameter to be estimated from a Weibull fit of $F_{VBM}(s)$ to

$$1 - Q_L(x) = P(VBM \leq x) \qquad (9)$$

The Weibull model fitted to the long term distribution describes the distribution of the peaks at random point in time. However, the main target is to get the probability distribution of the maximum amplitude of wave induced effects in n cycles where n corresponds to the mean number of load cycles expected during a ship's lifetime. Gumbel, (1958) has shown that whenever the initial distribution of a variable has an exponential tail, the distribution of the largest value in n observations follow an extreme distribution. Thus, the distribution of extreme values of the wave induced

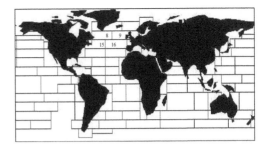

Figure 5. Global wave statistics ocean areas.

bending moment over the time period T is obtained as Gumbel law.

$$F_e(x_e) = \exp\left[-\exp\left(-\frac{x_e - x_n}{\sigma}\right)\right] \qquad (10)$$

where x_n and σ are parameters of the Gumbel distribution that can be estimated from the Weibull distribution using the following equation (Guedes Soares, 1984).

$$x_n = w \cdot [\ln(n)]^{1/k} \qquad (11)$$

$$\sigma = \frac{w}{k}[\ln(n)]^{\frac{1-k}{k}} \qquad (12)$$

where w and k are the Weibull parameter and n is the return period associated with one year of operation. The uncertainty associated with these calculations is taken as normal distribution with mean value equal to 1 and COV equal to 0.1.

5.1 Wave induced bending moment for tanker ship

The short term structural response due to waves in terms of midship bending moment is obtained by linear hydrodynamic analysis and assuming the Pierson Moskowitz (PM) spectrum. The long term response is then computed using IACS North Atlantic scatter diagram (IACS, 2000), covering areas 8, 9, 15 and 16, Figure 5.

Table 7 shows the stochastic modeling of wave bending moment of the ship assuming 42.5% of the year in full load condition. Figure 6 shows the Weibull fit for the wave load for the tanker. The resultant probability distribution is fitted by a Weibull model, which describes the distribution of the peaks at a random point in time.

However, one is normally interested in having the probability distribution of the maximum amplitude of wave induced effects in n cycles where n corresponds to the mean number of load cycles expected during a ship's lifetime.

The Gumbel parameters x_n and σ can be estimated from the initial Weibull distribution using the following equations 11 and 12. Table 7 shows the stochastic

517

Table 7. Stochastic model of extreme wave induced moment for the tanker.

Load condition		Ballast	Full load
Fraction of ship life		42.5%	42.5%
Weibull	w	282	292
Parameters	k	0.92	0.95
Gumbel parameters	n	1.91E+06	1.91E+06
	σ_e	387	358
	x_n	5144	4903
	Mean	5367	5110
	Std	496	459

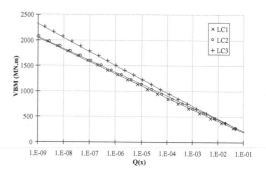

Figure 7. Long Term distributions for the three loading condition of the bulk carrier.

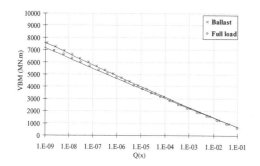

Figure 6. Wave load long term distribution for the tanker.

Table 8. Stochastic model of extreme wave induced moment for the bulk carrier.

Load cond.	Weibull param.		Gumbel moments (One Year)		
	w	k	n	Mean V.	Std. dev.
LC 1	84.2	0.950	9.0E+05	1384.	130.6
LC 2	97.4	0.993	9.0E+05	1419.	128.3
LC 3	90.1	0.933	1.8E+06	1640.	150.1

modeling of wave bending moment of the ship assuming 42.5% of the year in full load condition. Figure 6 shows the Weibull fit for the wave load for the tanker.

5.2 Wave induced bending moment for the bulk carrier

The same steps are repeated for the bulk carrier for the three loading conditions. Table 8 shows the long term distribution for the three loading conditions. The Gumbel parameters are calculated for one year. Figure 7 shows the Weibull fit for the wave load for the bulk carrier.

Table 8 shows the stochastic model of extreme wave induced bending moment for the adopted operational profile. For each loading condition, the distribution of the extreme wave load is calculated over the period

of time that the ship spends in this condition T_C. The average of wave period is assumed to be 7 seconds in North Atlantic.

6 LIMIT STATE FUNCTION AND RELIABILITY ASSESSMENT

The limit state can be defined as a condition beyond which the structure no longer satisfies the design performance requirements. The ultimate limit state is a state associated with collapse and denotes inability to sustain increased load. In the present reliability assessment a time independent first order reliability formulation corresponding to one year operation is considered. The reliability calculations were carried out using the computer program COMREL Gollwitzer (1988). The limit state equation considered corresponds to the hull girder failure under vertical bending (Guedes Soares et al., 1996, IACS, 2006):

$$g(x) = M_u \cdot X_R - [M_{wn} \cdot X_{st} \cdot X_{nl} + M_{sw} \cdot X_{sw}] \quad (13)$$

where, M_U is the ultimate capacity with a model uncertainty factor X_R. M_{WV} is the wave bending moment with model uncertainty factors; X_{st} for the linear response calculation and X_{nl} for nonlinear effects. M_{SW} is the random still water bending moment with a model uncertainty factor X_{SW}.

The reliability calculations were carried out using the computer program COMREL Gollwitzer, (1988). For the tanker in sagging condition, the reliability index was 2.95. While for the bulk carrier, the reliability index in sagging and hogging for the three loading conditions are presented in Table 9. One can notice that the reliability index of bulk carrier in sagging is very low in alternative and ballast conditions. While in the homogenous loading condition, the reliability level is acceptable.

Table 9. Reliability index of bulk carrier.

	Sag	Hog
LC1: Alternate	1.429	2.562
LC2: Homogeneous	2.867	3.885
LC3: Ballast	1.182	2.32

Table 10. Sensitivity Factors for the tanker design.

UBM	WBM	SWBM	X_u	X_{sw}	X_{nl}	X_{st}
0.40	−0.48	−0.20	0.57	−0.11	−0.34	−0.34

Table 11. Sensitivity analyses for the bulk carrier.

Sagging	Alternative	Homogenous	Ballast
UBM	0.46	0.41	0.46
WBM	−0.34	−0.61	−0.32
SWBM	−0.53	−0.16	−0.55
xsw	−0.22	−0.06	−0.22
xu	0.35	0.32	0.35
xnl	−0.33	−0.41	−0.33
xst	−0.33	−0.41	−0.33

Hogging	Alternative	Homogenous	Ballast
UBM	0.35	0.31	0.36
WBM	−0.27	−0.42	−0.26
SWBM	−0.40	−0.13	−0.41
xsw	−0.17	−0.05	−0.17
xu	0.70	0.72	0.69
xnl	−0.25	−0.31	−0.25
xst	−0.25	−0.31	−0.25

7 SENSITIVITY ANALYSES

A sensitivity analyses is done to know the importance of all variables and their influence on the ship reliability. Positive sensitivity indicates that an increase in this variable leads to an increase in the reliability and vice versa. The sensitivity factor for a give limit state function is given by:

$$\alpha_i = \frac{1}{\sqrt{\sum_i^n \left(\frac{\partial g(x)}{\partial x_i}\right)^2}} \cdot \frac{\partial g(x)}{\partial x_i} \quad (14)$$

Table 10 shows the sensitivity factors for the tanker ship. One can notice that the still water bending moment and the corresponding uncertainty have the least importance of the variables. The wave bending moment has high importance. The ultimate strength and the corresponding uncertainty have very high importance.

Table 11 shows the sensitivity factors for sagging and hogging conditions for the three loading conditions for the bulk carrier. The ultimate capacity importance is the highest and it does not change too much from one loading condition to the other, while the wave bending moment importance increases in the homogenous loading condition. The importance of the still water bending moment decreases in the homogenous loading condition.

One can also notice that the sensitivities of the variables in the alternative loading condition are always close to the ones on the ballast condition, while they are different from the homogenous loading condition. From Figures 8 and 9, one can notice that the SWBM, the WBM and the SWBM uncertainty differs from one loading condition to the other both in sagging and hogging, otherwise there is no big difference between the other variables.

7.1 Comparison between sensitivity factors of the tanker and the bulk carrier

If one wants to compare the sensitivity factor for the tanker and the bulk carrier, the full loading homogenous condition is to be compared with the full loading condition of the tanker, Figure 10. Figure 11 also shows a comparison between the three loading conditions for the bulk carrier and the tanker. The comparison indicates that:

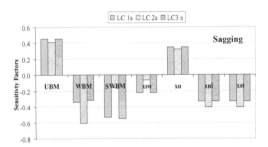

Figure 8. Sensitivity factors in sagging

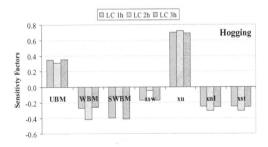

Figure 9. Sensitivity factors in hogging.

• The importance of the UBM does not vary too much between the two cases, while the importance of the

519

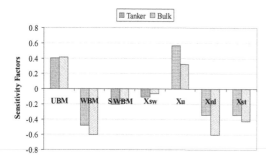

Figure 10. Sensitivity factors in the bulk carrier and tanker. (Homogenous loading).

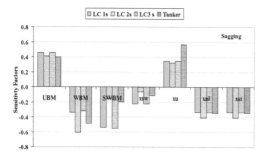

Figure 11. Sensitivity factors in sagging for the tanker and bulk carrier.

corresponding uncertainty is almost doubled for the tanker compared with the bulk carriers.

- The importance of the wave induced loads and the corresponding uncertainties are higher in the bulk carrier.
- The importance of the still water bending moment in the tanker is higher than that in the bulk carrier. The importance of the corresponding uncertainty is small for both ships although in the tanker it is as twice as it for the bulk carrier.

Figure 11 shows a comparison between the sensitivities of the variables in all loading conditions for the bulk carrier and the full loading condition in tanker. One can notice from the figure that the importance of the WBM, SWBM and the SWBM uncertainty differs between all the cases.

8 DESIGN MODIFICATION FACTOR

As seen in Table 9, the reliability index in sagging condition is very small for the bulk carrier. Besides the sagging capacity is the only limit state considered for tankers according to the IACS CSR. If one wants to modify the structural design, eventually to change

the reliability level the efficient way is to increase the thickness of the deck plating by a certain factor while keeping the scantlings of the sides, bulkheads and bottom structure constant.

The increase of deck plating thickness can be defined by a multiplicative design modification factor DMF. The ultimate strength is calculated using the developed program for each case of the DMF. Figure 12 shows the relation between the UBM and the DMF which seems to be linear for the tanker and the bulk carrier. The slope of the tanker curve appears to be steeper than the bulk carrier, but this might be attributed to the difference of size of the two ships and consequently the different in scantlings which lead to more increase in tanker UBM.

8.1 Reliability index versus DMF

The reliability is calculated for each design for the tanker and the bulk carrier in sagging condition. The increase in reliability index is linear for all cases as shown in Figure 13.

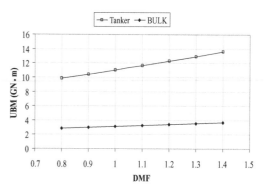

Figure 12. UBM versus DMF.

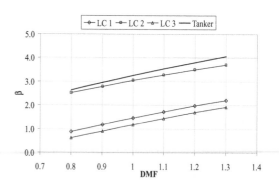

Figure 13. Reliability index versus DMF.

520

8.2 Sensitivity factors versus DMF

Sensitivity analyses are done for all cases. The objective behind this type of analyses is to know whether the importance of the variables change with different designs or not. Figure 14 to Figure 17 show how the sensitivity factors change with the design modification factor for the bulk carrier and the tanker designs.

From Figures 14 to 17, one can conclude the following:

• The importance of the UBM increased for alternative load condition and ballast condition. While decreased almost with the same percentage for the

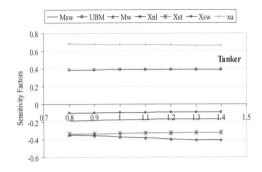

Figure 17. Sensitivity Factors for the tanker.

tanker and homogenous load condition of the bulk carrier.

• The importance of the wave induced bending moment increases with the DMF with almost the same percentage for the tanker and the three conditions of the bulk carrier.
• The SWBM importance decreased for all the cases although the decrease was steeper for the tanker and homogenous load condition in bulk carrier.
• The importance of the uncertainties associated with the still water bending moment increased slightly for the alternative loading condition and the ballast condition for bulk carrier, while decreased sharply for the other two cases taking into account that the importance of the variable was already low.
• Slight decrease in the importance associated with the WBM for all the cases took place.
• And finally the importance of the uncertainty associated with the UBM increased slightly with the DMF for all the cases except for the homogenous loading condition of bulk carrier it s value decreased slightly.

One can conclude generally that the homogenous loading condition for the bulk carrier and the full load condition of the tanker shared the same trend of increase or decrease the variables importance, while the alternative loading condition and the ballast loading condition of the bulk carrier shared another trend.

9 CONCLUSIONS

A reliability assessment has been performed for a SUEZMAX tanker ship and a HANDYMAX bulk carrier designed according to the new IACS CSR, considering the ultimate limit state. Three loading conditions are considered for the bulk carrier while only the full load condition is considered for the tanker.

The reliability index for the bulk carrier was very low in the sagging condition. The tanker gave an acceptable reliability level. The sensitivity factors for

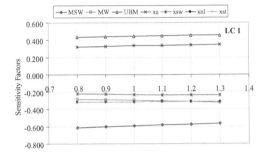

Figure 14. Sensitivity Factors. Bulk Carrier LC 1.

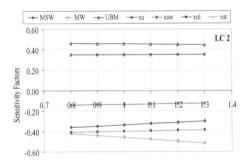

Figure 15. Sensitivity Factors. Bulk Carrier LC 2.

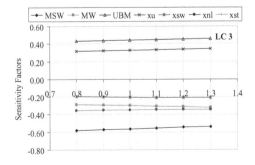

Figure 16. Sensitivity Factors. Bulk Carrier LC 3.

the design variables were calculated to indicate the important variables which need consideration during the ship design process.

The general conclusions are:

- The UBM and the WBM have almost the same importance in all the cases. The importance of the ultimate capacity uncertainty in the tankers is almost the double as that for the bulk carrier.
- The importance of the still water bending moment in the tanker is higher than that in the bulk carrier. The importance of the corresponding uncertainty is small for both ships although in the tanker, it is about twice as it for the bulk carrier.

To improve the reliability in sagging, a design modification factor is applied to the deck plating and the sensitivity of the variables was reassessed. One can conclude that:

- The importance of the still water bending moment decreased for all the cases although the decrease was steeper for the homogenous loading condition of the bulk carrier and the tanker.
- The importance of the wave induced bending moment increases with the increase of the DMF for all the cases.

ACKNOWLEDGEMENTS

The present paper has been prepared within the project "MARSTRUCT—Network of Excellence on Marine Structures" http://www.mar.ist.utl.pt/marstruct which has been funded by the European Union through the Growth program under contract TNE3-CT-2003-506141.

REFERENCES

Gollwitzer, S., Abdo, T. & Rackwitz, R. (1988) "FORM—Program Manual, Munich".

Guedes Soares, C. (1984) "Probabilistic Models for Load Effects in Ship Structures, Division of Marine Structures", *The Norwegian Institute of Technology*, Report UR-84-38.

Guedes Soares, C. & Viana, P.C. (1988) "Sensitivity of the response of marine structures to wave climatology". In: B.A. Schreffler and O.C. Zienkiewicz, Editors, *Computer Modelling in Ocean Engineering*. Rotterdam. pp. 487–492.

Guedes Soares, C. & Trovão, M.F.S. (1991) "Influence of wave climate modelling on the long-term prediction of wave induced responses of ship structures". In: W.G. Price, P. Temarel and A.J. Keane, Editors, *Dynamics of Vehicles and Structures in Waves*, Elsevier Science Publishers. pp. 1–10.8.

Guedes Soares, C. & Teixeira, A.P. (2000) "Structural reliability of two bulk carriers designs". *Marine Structures*. 13(2):107–128.

Guedes Soares, C. & Parunov, J. (2008) "Structural Reliability of a SUEZMAX Oil Tanker Designed According to New Common Structural Rules", *Journal of Offshore Mechanics and Arctic Engineering*, Vol.130, 021003-1/021003-10.

Guedes Soares, C., Dogliani, M., Ostergaard, C., Parmentier, G., & Pedersen, P.T. (1996) "Reliability Based Ship Structural Design". *Transactions of the Society of Naval Architects and Marine Engineers* (SNAME). Vol.104: 357–389.

Gumbel, E.J. (1958) "*Statistics of Extremes*", Columbia University Press, New York.

Horte, T., Wang, G. & White, N. (2007) "Calibration of the Hull Girder Ultimate Capacity Criterion for Double Hull Tankers", *Proceedings of the Conference on Practical Design of Ships and Offshore Structures*, ABS Houston, USA.

Hussein, A.W., Teixeira, A.P. & Guedes Soares, C. (2007) "Assessment of the IACS Common Structural Hull Girder Check Applied to Double Hull Tankers", Maritime Industry, *Ocean Engineering and Coastal Resources*, C. Guedes Soares and P. Kolev (Eds), Taylor and Francis, pp. 175–183.

IACS. (2000) Recommendation No. 34, "*Standard Wave Data*", 1992, Rev. 1 June 2000, Corr. Nov. 2001.

IACS (2006) "Common *Structural Rules for Double Hull Oil Tankers*", International Association of Classification Societies, London, http://www.iacs.org

Khan, I.A. & Das, P.K. (2008) "Sensitivity analysis of the random design variables applicable to ship structures", *Maritime Industry, Ocean Engineering and Coastal Resources*, C. Guedes Soares & P. Kolev (Eds) Taylor and Francis, pp. 185–192.

Longuet-Higgins, M.S. (1952) "On the statistical distribution of the height of sea waves" *Journal of Marine Research*, 11, pp. 245–266.

LR, ABS & DNV. (2005) "Background Documents: Common Structural Rules for Double Hull Oil Tankers". 2nd Draft.

Parunov, J., & Guedes Soares, C. (2008) "Effects of Common Structural Rules on hull-girder reliability of an AFRAMAX oil tanker", *Reliability Engineering and System Safety*, Vol. 93, pp. 1317–1327.

Parunov, J., Senjanovic, I., & Guedes Soares, C. (2007). "Hull-girder reliability of New Generation Oil Tankers". *Marine Structures*, Vol. 20, pp. 49–70.

Analysis and Design of Marine Structures – Guedes Soares & Das (eds)
© *2009 Taylor & Francis Group, London, ISBN 978-0-415-54934-9*

Ultimate strength and reliability assessment of laminated composite plates under axial compression

N. Yang & P.K. Das
Universities of Glasgow & Strathclyde, UK

X.L. Yao
Harbin Engineering University, China

ABSTRACT: The objective of this work is to investigate the ultimate strength and reliability of laminated composite plates under axial compression. A non-linear finite element technique including the multi-frame restart analysis is developed to perform the progressive failure analysis. The Tsai-Wu criterion is adopted to identify the material failure of structures. The numerical accuracy is validated with the results published in the literatures and test result performed by DTU and NTUA. Finite Element method coupled with response surface method is used for reliability assessment.

1 INTRODUCTION

The use of composite materials in all types of engineering structures such as aerospace, automotive, underwater structures, and sports equipments has been increased over the last few decades due to inherently high strength-to-weight ratio, low thermal expansion, resistance to corrosion, better durability and excellent damage tolerance. It is well known that composite plate can sustain a much higher load after the first failure occurrence of damage. Therefore, the ability to predict the ultimate failure is essential for predicting the performance of composite structures and developing reliable, safe designs which exploit the advantages offered by composite materials.

The objectives of the present work it to investigate the process of progressive failure analysis of laminated plates under the action of uni-axial compression. A non-linear finite element technique including the multi-frame restart analysis is developed to perform the progressive failure analysis. The Tsai-Wu criterion is adopted to identify the material failure of structures. The results are compared with those obtained from experiments. Smith (1990) have shown that marine composite material have large statistical variation in their mechanical properties, which means that large variability may exist in the prediction of the ultimate strength in composite materials. Thus the reliability assessment for the ultimate strength of plate in composite materials is necessary.

2 THE PROGRESSIVE FAILURE ANALYSIS

2.1 *The procedure of progressive failure analysis*

Generally, the analytical methods, Luo & Wang (1992) such as double Fourier methods, Rayleigh-Ritz method can be used for prediction of composite structure with simple geometries and boundary condition. For structures of more general geometry, finite element methods are likely to be the most effective means to get the accurate results. A progressive failure analysis is used to predict the ultimate strength of composite panel under axial compressive load. The structural analysis is performed by using finite element software ANSYS 11.0 (2007). To implement and perform the progressive failure methodology, the ANSYS-APDL (ANSYS Parametric Design Language) is used as a subroutine. The detailed flow chart of the solution is shown in Figure 1. The procedure used for simulating the progressive failure is explained in the following.

Generally speaking, in the first step, a linear static analysis is performed to find out the bifurcation point. The next step is to perform a linear stability analysis in order to form the initial geometric imperfection which serves as a trigger for the geometric nonlinear analysis. The geometry of the finite element model can be update according to the displacement results of the linear stability analysis by scale factor and creates a revised geometry at the deformed configuration directly. Then the initial small pressure value P_0 (or displacement, e.g. 10% of the expected ultimate load)

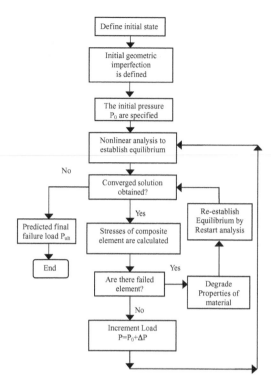

Figure 1. Flow chart of progressive analysis methodology.

is specified to ensure no element failure. Based on this load level, nonlinear finite element analysis is run to establish equilibrium equation. A displacement (or force) controlled convergence criterion is employed to control the analysis. Then the stresses on the on-axis stresses of composite elements are determined from the nonlinear analysis solution. These stresses are then used to determine failure according to adopted failure criteria. If failure is detected somewhere by a failure criterion, certain material properties of that element shall be reduced in locations where the failure is detected, the engineering material constants corresponding to that particular mode of failure are updated by multiplying an extremely small value such as 1×10^{-10} instead of zero and the layer number and element number of the failed element are recorded. After the element material properties have been degraded, the historical database for the element material properties is updated for the current load step in the nonlinear analysis. The static equilibrium is re-established after the material properties have been degraded by restarting analysis at the current load step. When a failure mechanism has occurred somewhere in a ply, this ply is not checked for that failure mechanism in the same region (e.g. element) any more. If no failure is observed, the material properties in the layer are not changed and an incremental pressure

ΔP (or displacement) is added until some fails. In what follows, the loading procedure is cycled until the solution is done.

2.2 Failure criterion and material degradation

A failure criterion is required to access whether the laminated plate has failed or not under a load system. Among existing failure criteria, the most commonly used is the Tsai-Wu criteria, Tsai & Wu (1971) and it is adopted in this analysis.

As indicated previously, the principal stress associated with each layer of element of the laminate plate can be computed and the failure criterion is based on these principle stresses. According to the Tsai-Wu criterion, the failure occurs when the following criterion is satisfied in any one of the lamina.

$$F = \sum_{i=1}^{6} F_i \sigma_i + \sum_{i=1}^{6} \sum_{j=1}^{6} F_{ij} \sigma_i \sigma_j \geq 1 \tag{1}$$

where, σ_i are stresses in material directions and F_{ij}, F_i are the tensor strength factors which are expressed as:

$$F_1 = \frac{1}{X_t} - \frac{1}{X_c} \quad F_2 = \frac{1}{Y_t} - \frac{1}{Y_c} \quad F_3 = \frac{1}{Z_t} - \frac{1}{Z_c}$$

$$F_{11} = \frac{1}{X_t X_c} \quad F_{22} = \frac{1}{Y_t Y_c} \quad F_{33} = \frac{1}{Z_t Z_c}$$

$$F_{44} = \frac{1}{R^2} \quad F_{55} = \frac{1}{S^2} \quad F_{66} = \frac{1}{T^2}$$

$$F_{12} = -\frac{1}{2} / \sqrt{X_t X_c Y_t Y_c}$$

$$F_{13} = -\frac{1}{2} / \sqrt{X_t X_c Z_t Z_c}$$

$$F_{12} = -\frac{1}{2} / \sqrt{Y_t Y_c Z_t Z_c} \tag{2}$$

where X_t, Y_t and X_c, Y_c are the tensile and compressive strength of lamina in the x, y direction, respectively. S is shear strength in the principal material plane.

If failure is detected in a particular layer of element of the laminated plate, a reduction in the corresponding lamina modulus is introduced, but only in the corresponding element. This causes the corresponding changes in the corresponding element's laminate stiffness according to a material property degradation model. The Tsai-Wu failure criterion identifies an element failure, but it cannot identify the modes of failure, if failure occurred, then the following terms were used to determine the failure mode, Engelstad et al. (1992).

$$H_1 = F_1 \sigma_1 + F_{11} \sigma_1^2; \quad H_2 = F_2 \sigma_2 + F_{22} \sigma_2^2;$$

$$H_4 = F_{44} \sigma_4^2; \quad H_5 = F_{55} \sigma_5^2; \quad H_6 = F_{66} \sigma_6^2 \tag{3}$$

Table 1. Material degradation.

Primary failure mode	Fibre failure H_1	Matrix failure H_2	Shear failure		
			H_3	H_4	H_5
Degraded properties	E_1	E_2	G_{23}	G_{13}	G_{12}

The largest H_i term was selected to be the dominant failure mode and the corresponding material properties is reduced. The material degradation for this validation analysis is summarized in Table 1.

3 VALIDATION STUDY

3.1 Example 1

3.1.1 Problem statement

The first example used in this validation study is a 24-ply orthotropic flat rectangular plate loaded in axial compression reported by Starnes & Rouse (1981). The panel is 508-mm-long by 178-mm-wide with a stacking sequence $[45/-45/0_2/45/-45/0_2/45/-45/0/90]_s$. The boundaries are clamped for loaded edges (along width) and simple supported along the unloaded sides (along length) shown in Figure 2. The panel is discretized into 40 elements and 14 elements along the length and the width, respectively. The mechanical parameters of materials are listed in Table 2.

The first buckling mode from the linear stability analysis has two longitudinal half-waves with a buckling mode line at the panel mid-length and one transverse half-wave along the panel width. An initial geometric imperfection is formed by using the first buckling mode shape with amplitude 5% of the panel thickness in order to past the critical buckling point.

3.1.2 Results comparison

Similar progressive failure analyses have also been performed by Engelstad et al. (1992), Sleight (1999) and Chen (2007). Table 2 summarizes analytical first ply failure and final failure load of this study, test results and other analytical results. Figure 3 shows the comparison of experimental and analytical end-shortening results as a function of the applied load. Figure 4 shows the comparison of experimental and analytical out-of-plane deflection near a point of maximum deflection as a function of the applied load.

The results obtained from progressive failure analysis agree reasonably well with the experimental results. The final failure load predicted by Tsai-Wu failure criterion is slightly higher than results from experimental results. The load at which the first ply failure occurs predicted by Tsai-Wu failure criterion is 10.22% higher than Christensen's criterion and lower than 5.04% Hashin's criterion.

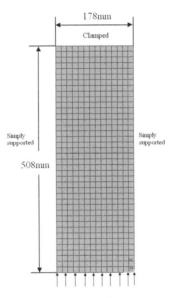

Figure 2. Geometry, loading and boundary conditions of panel.

Table 2. Material properties and Strength properties of graphite-epoxy composite material, Starnes et al. (1981).

Mechanical properties	Values	Strength properties	Values
E_1	131.0 GPa	X_t	1400 MPa
E_2	13.0 GPa	X_c	1138 MPa
G_{12}	6.4 GPa	$Y_t = Z_t$	80.9 MPa
G_{13}	6.4 GPa	$Y_c = Z_c$	189 MPa
G_{23}	1.7 GPa	R	62.0 MPa
V_{12}	0.38	$S = T$	69.0 MPa

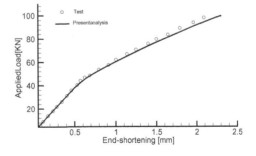

Figure 3. End-shortening vs. Applied Load.

3.2 Example 2

3.2.1 Problem statement

These examples are part of the validation studies of a selected number of compressive tests made by DTU &

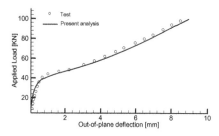

Figure 4. Out-of-plane deflection vs. Applied Load.

NTUA (2007) on three series of composite plates in order to investigate the influence of various sizes of geometrical imperfections on the compressive strength of composite plates with respect to their relative slenderness ratio. A schematic 2-D view of the test rig shows in Figure 5. The plate is 400-mm-long (along the loaded axis) by 380-mm-wide. The unsupported length and width of the plate after it is fixed in the testing fixture was 320×320 mm. The imperfection shape is formed with artificial geometrical imperfections by using the first buckling mode shape of a corresponding fully clamped plate with dimensions L = B = 300 mm with following material data, DTU & NTUA (2007):

$E_1 = 46000$ MPa $v_{12} = 0.30$ $G_{12} = 5000$ MPa
$E_2 = 13000$ MPa $v_{23} = 0.42$ $G_{23} = 4600$ MPa
$E_3 = 13000$ MPa $v_{13} = 0.30$ $G_{13} = 5000$ MPa

In finite element analysis, the model of active area is modelled by 30×30 finite element mesh shown in Figure 6. The geometry imperfection can be update according the displacement results of the linear stability analysis by scale factor and creates a revised geometry at the deformed configuration directly. The boundary conditions are given by the experimental data extracted from NTUA and DTU test. The displacement and rotation values are set as table parameters which are on a function of time. All amplitudes values are input into ANSYS as table parameters from an external file. Then the boundary condition is applied by Tabular boundary condition. The displacements values of nodes between these measurement points are linearly interpolated directly. The ply thickness, stacking sequence and material properties of all these panels are summarized in Table 4, DTU & NTUA (2007).

3.2.2 Results comparison

Two different types of boundary condition (BC with rotation and BC without rotation) were considered in this study. The stresses for the experimental results are taken as the total force on the cross-sectional area of the specimen's width. For the finite element analysis, the

Table 3. The experimental results and the corresponding estimated results reported in the literatures (unit: KN).

	First ply Failure load	Final Failure Load (KN)	End shortening at failure (mm)	Failure criteria
Present analysis	92.26	99.66	2.30	Tsai-Wu
Chen et al.	–	101.3	2.4	Tsai-Wu
Sleight	82.83	99.8	–	Christensen
et al. (1997)	96.91	104.6	–	Hashin
Engelstad				Maximum
et al.	–	111.4	–	stress
Test results	–	98.0	2.1	–

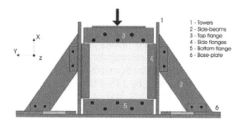

Figure 5. Test-rig.

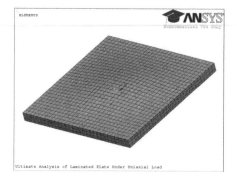

Figure 6. Finite element model with imperfection shape.

stress is taken on the active area's width of plate. Here, only the values of maximum imperfection assumed with amplitude 3% of the unsupported width of the panel were given in Figures 7–9.

For the boundary condition with rotations, only medium panel (Figure 8) can get to the final load due to the convergence problem. The comparison showed good agreement on stress/displacement graph. The maximum stress and out of plate displacement predicted by progressive failure analysis are 15.08% and 24.09% lower than the results from the test results.

Table 4. Material properties and dimension properties, DTU & NTUA (2007).

	Thin plate	Medium plate	Thick plate
Thickness of UD	0.59 mm	0.59 mm	0.93 mm
Thickness of BIAX	0.36 mm	0.36 mm	0.48 mm
Stacking Sequence	[BIAX/4× UD/BIAX/ 3×UD]S	[BIAX/4× UD/BIAX/ 4 × UD/ BIAX/3× UD]S	[BIAX/4× UD(0°)/BIAX/ 4 × UD(0°)/BIA X/2 × UD(0°]S

Mechanical properties	Values	Values
E_1	35204 MPa	56210 MPa
E_2	9437 MPa	18075 MPa
G_{12}	2169 MPa	4264 MPa
v_{12}	0.268	0.284
X_t	698 MPa	1141 MPa
X_c	191 MPa	952 MPa
$Y_t = Z_t$	43 MPa	22 MPa
$Y_c = Z_c$	69 MPa	127 MPa
S	30 MPa	64 MPa

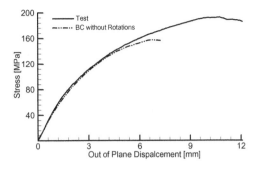

Figure 9. Out-of-plane deflection vs. Applied Load of thick plate.

For the boundary condition without rotations, all the calculations were able to continue until the final load. Figure 7–9 show the analytical results correlate reasonably well to the experimental results, however, the progressive failure analysis under-predicted the final failure loads from the test, 17.73% from the thin panel, 29.75% from the medium panel and 15.71% from the thick panel, respectively. The influence of the rotations seems great for thin and medium panels. For thick panel, the progressive failure analysis prediction is in good agreement without rotations.

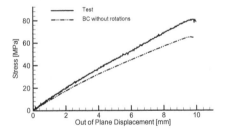

Figure 7. Out-of-plane deflection vs. Applied Load of Thin plate.

4 RELIABILITY

4.1 Introduction

The structural reliability theory deals mainly with the assessment of these uncertainties and the methods of quantifying and rationally including them in the design process. The limit state equation g for current problem can be defined by using resistance R and load S as follows

$$G = R - S \tag{5}$$

The probability of failure of the structural member is defined when $G \leq 0$ by

$$P_f = P(R - S \leq 0) = \int_{G(X) \leq 0} f_{RS}(r, s) dr ds \tag{6}$$

where $f_{RS}(r, s)$ is the joint probability density function of R (resistance) and S (load). If R and S are statistically independent normally distributed random variables, the probability of failure is given from

$$P_f = P(R - S \leq 0) = \Phi(-\beta) \tag{7}$$

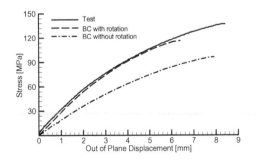

Figure 8. Out-of-plane deflection vs. Applied Load of medium plate.

$$\beta = \frac{\mu_R - \mu_S}{\sqrt{\sigma_R^2 + \sigma_S^2}} \qquad (8)$$

where $\Phi(\cdot)$ is the cumulative distribution function for a standard normal variables. In general, First Order Reliability method (FORM) and Second Order Reliability method (SORM) are powerful tools to obtain the probability integral. If the limit state function has a non-closed-form expression, FORM and SORM are not directly applicable. Therefore, Monte Carlo simulation (MCS) and Response Surface Method (RSM) seem feasible. Generally, the accuracy of MCS increases as the sample size increases. It may be the most widely used, but it is not the most efficient, especially in complex systems.

The response surface method is one of the latest developments in the field of structural reliability analysis. The response surface method is based on the fundamental assumption that the influence of the random input variables on the random output parameters can be approximated by mathematical function. Such approximations are referred to as Response Surfaces (RS). In the original conceptual form of the Response Surface technique, polynomials are used to approximate real limit state functions. The second-order model is widely used in response surface methodology because it is very flexible and easy to estimate. The second-order response surface is

$$G(x) = A + X^T B + X^T C X \qquad (9)$$

wheree X are the vector of basic variables and A, B, C are the coefficients of constant term, linear terms and quadratic term respectively. Once the response surface is obtained, the reliability analysis can be made with this response surface as limit-state surface in place of the complicated original limit-state surface of the model. In this paper, the determination of response surface is based on the method given by LeiYu et al. (2002).

4.2 Model

A symmetric laminated plate of 500 mm × 500 mm with simple supported boundaries has been studied in the present investigation. The laminate is consisting of 34 layers with the lamination sequence of $[-45/45/0_4/-45/45/0_4/-45/45/0_3]_s$ fibre orientations The lamina thickness of UD layers and the BIAX are 1.39 mm and 0.12 mm, respectively. Due to structural symmetry, only an half of the plate is modelled. The accuracy of this spatial discretization is established by considering a finer finite element mesh of 36 elements along the length and 18 elements along the width. After analysis, the mesh size had a small effect on the prediction of the ultimate strength, but

the computational time of the model with coarse mesh size is only 20% of that with finer mesh.

4.3 Limit state function

The reliability of a structure is defined as the probability that the structure will perform its intended function without failing. In the present context, the reliability is the probability that the panel will carry a given load without ultimate failure.

$$G = X_u \sigma_u - \sigma \qquad (10)$$

where X_u = Ultimate strength modelling uncertainty factor; σ_u = Estimated ultimate panel strength; σ = Stress at panel.

In general, the reliability assessment of structure requires information on the probability distribution of variables, however, the probability distribution of these variables are generally indeterminate. In this paper, the applied stress on the panel, the material properties, initial geometrical imperfection and the material strength are considered as independent random variables and they are randomly generated according to their assumed probability distribution. The coefficients of variation (COV) are defined by DNV offshore standard, DNV-OS-C501 (2003). All basic random variables and the coefficients of variation used in this analysis are shown in Table 5. The general purpose structural reliability analysis program CALREL, Liu et al. (1989) developed by the University of California is used to perform these analyses. The FORM, SORM and MCS are used to predict the reliability and probabilistic failure on the fitted response surface and results are shown in Table 6.

4.4 Sensitivity analysis

Sensitivity analysis is an important part of structural reliability assessment. It can provide guidance to designers regarding the relative importance among

Table 5. Probabilistic variables.

Basic variables	Mean value	Coefficient of variation (COV)	Distribution type
X_u	1.0	0.1	Normal
σ	60 MPA	0.1	Normal
δ	0.5 mm	0.05	Normal
E_1	49627 MPA	0.05	Lognormal
E_2	15430 MPA	0.1	Lognormal
G_{12}	4800 MPA	0.1	Lognormal
X_T	968 MPA	0.05	Lognormal
X_C	915 MPA	0.05	Lognormal
Y_T	24 MPA	0.1	Lognormal
Y_C	118 MPA	0.15	Lognormal
S	65 MPA	0.1	Lognormal

Table 6. Probabilities of failure and Reliability indices.

	Reliability index β	Probability of failure P_f
FORM	3.5437	1.973×10^{-4}
SORM	3.7019	1.070×10^{-4}
MCS	3.6995	1.080×10^{-4}

Table 7. Sensitivity factors.

Variables	Sensitivity factor α_i	Variables	Sensitivity factor α_i
X_u	0.8297	X_T	0.2408
σ	0.5012	X_C	0.0006
δ	0.0008	Y_T	0.0232
E_1	0.0332	Y_C	0.0266
E_2	0.0091	S	0.0001
G_{12}	0.0019		

the basic random variables. Furthermore, variables having a small sensitivity factor might be assumed to be deterministic rather than random variable in subsequent analysis, this reduces the dimensionality of the space of variables.

The importance factor α_i is considered as a measure of the sensitivity of β with respect to the basic variables. The importance of the contribution of the each basic variable towards the uncertainty of the limit state function can be obtained from the sensitivity factor given by

$$\alpha_i = -\frac{1}{\sqrt{\sum_{i=1}^{\infty}(\partial g(x)/\partial x_i)^2}} \frac{\partial g(x)}{\partial x_i} \quad (11)$$

The sensitivity factors for the all variables are listed in Table 7.

4.5 Discussion

It can be seen from the Table 6, the reliability index and probability of failure is 3.7019 and 1.070×10^{-4}, respectively, which indicates that the ultimate strength of this panel might be enough against the load by DNV standard, DNV-OS-C501 (2003), in which the target reliability of low safety class is $P_f = 10^{-4}$ for Brittle failure type. For the results obtained from different methods, assuming that the results obtained from MCS are the most accurate, the reliability index β calculated by SORM is very close to the exact result with accuracy of less than 1%. Table 7 shows that sensitivity factors of the modelling uncertainty factor X_u, applied compressive stress σ, fibre tensile strength X_T are generally significant values, which indicates that

these variables are more sensitive and fibre compressive strength X_c, small initial displacement δ and shear strength S are negligible compared to other variables.

5 CONCLUSION

In this paper, a non-linear finite element technique including the multi-frame restart analysis is developed to predict the first failure load and final collapse load by progressive failure analysis. Tsai-Wu failure criteria is used to predict the failure mechanism and a constant degradation method is adopted to degrade the material properties after failure. The numerical accuracy is evaluated with the results published in the literatures and test result performed by DTU and NTUA. Reliability analysis is performed on finite element method coupled with response surface method. Through this investigation, the following key points can be concluded:

1. The procedure of progressive failure analysis is used to predict the ultimate strength of composite panels under axial compressive load using the multi-frame restart analysis with Tsai-Wu failure criterion is feasible. However, only three failure mode i.e. matrix cracking, fibre breakage and fibre/matrix interface failure are considered in present model. Delamination which has a large influence on the compression strength is not included.

2. The modelling uncertainty factor X_u, applied compressive stress σ and fibre tensile strength X_T are sensitive to other variables and fibre compressive strength X_c, small initial displacement δ and shear strength S are negligible compared to other variables.

ACKNOWLEDGEMENTS

This paper has been prepared within the project "MARSTRUCT-Network of Excellence on Marine Structures", (www.mar.ist.utl.pt/marstruct/), which has been funded by the European Union through the Growth Programme under contract TNE3-CT-2003-506141.

REFERENCES

ANSYS User Manuals version 11.0, ANSYS Inc, USA.
Christian, B. et al. 2007 Buckling of Imperfect Composite plates: Testing and Validation of Numerical Models. *DTU report.*
Composite components 2003. Offshore Standard *DNV-OS-C501.*
Engelstad, S.P., Reddy, J.N., Knight Jr., N.F., 1992. Post buckling response and failure prediction of graphite-epoxy

529

plates loaded in compression. *AIAA Journal 30*, *2106–2113*.

Liu, P.-L., Lin, H.-Z. & Der Kiureghian, A.D. 1989. CALREL User Manual. Department of Civil Engineering, University of California at Berkeley.

LeiYu, P.K. Das & YunlongZheng, 2002. "Stepwise response surface method and its application in reliability analysis of ship hull structure" *Journal of Offshore Mechanics and Arctic Engineering, vol. 124, Issue 4, pp. 226–230.*

Luo, Z.D. & Wang, Z.M. 1992. Advances in mechanics of composite materials. *Peking University Publisher*, Beijing, P.R. China.

Nian-Zhong Chen, C. Guedes Soares, 2007. Reliability assessment for ultimate longitudinal strength of ship hulls in composite materials. *Probability Engineering Mechanics. vol. 22, Issue 4, pp. 330–342.*

Nicholas, T. et al. 2007. Buckling of Imperfect Composite plates: Testing and Validation of Numerical Models. *NTUA report.*

Sleight David, W. Progressive Failure Analysis Methodology for Laminated Composite Structures. 1999. *Technical Report: NASA-99-tp209107.*

Smith, C.S. 1990. Design of marine structures in composite materials. *Elsevier Applied Science*, London.

Starnes, J.H. & Jr., Rouse, M., 1981. Postbuckling and failure characteristics of selected flat rectangular graphite-epoxy plates loaded in compression. *AIAA Paper 81–0543.*

Tsai, S.W. & Wu, E.M., 1971. A general theory for strength for anisotropic materials. *Journal of Composite Materials 5, 58–80.*

Environmental impact

Analysis and Design of Marine Structures – Guedes Soares & Das (eds)
© 2009 Taylor & Francis Group, London, ISBN 978-0-415-54934-9

Modelling of environmental impacts of ship dismantling

I.S. Carvalho, P. Antão & C. Guedes Soares
Centre for Marine Technology and Engineering (CENTEC), Technical University of Lisbon,
Instituto Superior Técnico, Lisboa, Portugal

ABSTRACT: Ships at the end of their economic life, and intended to dismantling, are considered as residues, according to the international and EU laws on residues. Furthermore, they are considered as dangerous residues if holding important amounts of dangerous substances, or when tanks have not been emptied of the respective dangerous substance load. This study presents the results of modelling the environmental impact of ships dismantling activity. The modelling was made for different ship types and for different scenarios. The data used for the modelling was collected at a Portuguese dismantling shipyard. For the modelling a commercial software tool (SimaPro LCA) was used. From the different environmental impact assessment methods the following were used: CML1992, 2 CML Baseline 2000, Eco-indicator 99 (H) and IMPACT 2002+. The work indicated that using different environmental assessment methods leads to similar conclusions about the relative environmental impact of the various ships. Furthermore it was shown that the environmental impact cannot be only related to ship type and size as its composition is a major factor.

1 INTRODUCTION

The ship dismantling process is typically considered as an "unclean" activity due to environmental implications that usually result from it. In fact, when one considers the maritime activity globally one may see that ship dismantling is responsible for a number of operational fatalities and considerable negative visual impacts in the areas of transformation. This seems to be a common issue in every dismantling site but with more emphasis in countries of southeast Asia, like India, Bangladesh or China. Several studies over the years highlight the effects of ship breaking in the local environment in these countries (Islam & Hossain, 1986; Reddy *et al.*, 2003, Reddy *et al.*, 2005, Reddy *et al.*, 2006).

Although the number of European shipyards involved in this activity and their contribution to the total scrapped ships is small in the global context, since it represents only 3% of the GRT of the ships dismantling (Knapp *et al.*, 2008), the political concerns surrounding the activity is high. These concerns lead the European Commission to launch a Green Paper on better ship dismantling (EU, 2007). In this document it is highlighted that an estimated 5.5 million tonnes of materials of potential environmental concern will end up in dismantling yards (in particular oil sludge, oils, paints, PVC and asbestos) and that most of the present shipyards or dismantling areas do not have adequate facilities to prevent environmental pollution. Also the fact that the number of expected dismantled ships in the upcoming decades is to be increased substantially (particularly due to single hull tankers) only emphasizes the problem.

In Portugal, this problem is not significant mainly due to the low number of shipyards involved in ship's demolition activities. In fact, in 2001 ship dismantling activity only represented 0.1% of the 564 companies related to maritime activities in Portugal (Carvalho *et al.*, 2008), generating 0.1% of the total trade volume. Despite its low importance to the Portuguese economy, the dismantling activity in Portugal is diversified as it covers different types of ships. Having information on dismantling of different types of ships it is possible to determine how the environmental impact of the dismantling activity is dependent on ship type and size. This is in fact the objective of this paper which is based on data collected in one of the dismantling yards in the Lisbon area.

The present work is aimed at:

- Evaluating the environmental impacts of the ships dismantling activity;
- Understanding the environmental advantages of recycling the materials and equipment as compared with the sinking or ship abandonment in fluvial waters;
- Understanding the environmental advantage of dismantling as a function of the type or class of the ship.

2 SHIP DISMANTLING CHARACTERIZATION

2.1 *Ship dismantling process*

The process of ship' dismantling starts with the acquisition of the ship by a purchaser (specialized broker or the operator of the demolition shipyard) when the ship-owner decides to put term to the economic life of the ship. Usually, the decision to end the ship' life is taken when the maintenance costs start to exceed the forecasted incomes or when the ship stops to be interesting for the second hand market. Typically, the ship sails unloaded towards the dismantling shipyard by its own propulsion means.

The present characterization of the dismantling processes is based on a Portuguese dismantling shipyard, environmentally certified for the activity, according to norm NP EN ISO 14001:2004. It is located south of Lisbon, on the river side. The areas of reception of residues are well confined but they are not covered. The facilities include several docks and quays. Currently, the shipyard undergoes a process of physical and environment improvement, in an attempt of being adapted to its social and environmental surroundings, as well as to the European and international legal framework.

Conceptually, the dismantling process can be divided in five distinct phases: inspection and attainment of the license for ship dismantling, dismantling of the ship, transformation of residues and delivery of the transformed residues to the respective final destination. The majority of the processes involved are common to other industrial sectors, e.g., metallurgic and wood transformation industries. The maritime authorities attend the beginning and the end of process of dismantling. The process is complete when a certificate is issued by the maritime authorities attesting the ships end of life.

2.2 Produced and recycled residues by the dismantling process

Although ships at the end of life are to be considered as residues according to the European and international law, also the activity of dismantling generates residues, particularly in the transformation phase.

Figure 1 is a schematic representation of materials generally removed from ships. This representation is based on the collected information in the dismantling yard. Being the ship an assembly of materials, after its inspection and attainment of the license its dismantling starts by cleaning and decontaminating the ship from dangerous solid and liquid substances such as mineral wool (e.g., asbestos) and sludge water and oil, and undifferentiated materials. After this, the ship is cut into big blocks, which are then grinding or cut into small pieces. From these activities, results what can be considered as the subassemblies or building blocks of the ship. It comprises structural and non structural steel (that is reused or recycled in proper foundries), non-ferrous metallic materials, glass and wood (that are recycled in proper foundries), polymeric and composite materials, essentially from furniture that are land filled as well as dangerous solid substances such as mineral wool like asbestos. To the cleaning, transformation and transportation phases of ship dismantling both energy and fuel are required. Among the subassemblies there are the residues from the cut and grinding processes.

The amounts of materials removed from dismantled ships are dependent on the ship types and respective gross tonnages. For this study the information collected on 23 ships was used. From these 23 ships, 18 are merchant vessels and 5 are military ships. Of the merchant ships 9 are transportation ships (1 reefer ship, 6 bulk carriers, 1 oil tanker, and 1 container ship), 3 are fishing vessels and 6 are auxiliary ships (2 barges, 1 floating dock, 1 is a workshop ship, and 2 tugs).

The average age of the dismantled ships was of 25 years, although some of the military ships would have up to 40 years and the reefer ship would have only 10 years.

Figures 2–5 show the material proportions (with respect to its gross tonnages) removed from those ships. In these figures, the characterization of the removed materials is made as a function of their respective final destination. Thus, the materials are divided into seven types of materials: ferrous materials

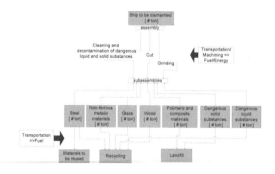

Figure 1. Schematic representation of the processes and produced residues from ship dismantling and final destination of the removed materials.

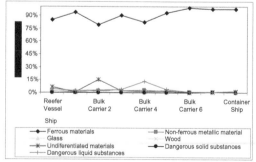

Figure 2. Characterization of the removed materials from the merchant vessels.

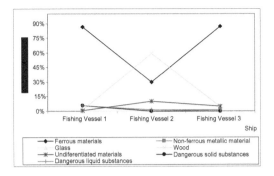

Figure 3. Characterization of the removed materials from the fishing vessels.

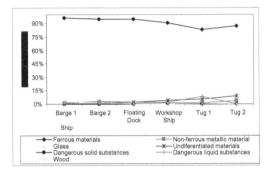

Figure 4. Characterization of the removed materials from the auxiliary vessels.

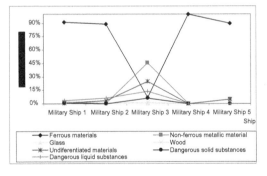

Figure 5. Characterization of the removed materials from the military vessels.

(structural and not structural steel), non-ferrous metallic materials, glass, wood, undifferentiated, dangerous solid substances and dangerous liquid substances.

To the majority of the ships the ferrous materials are the predominant material constituent. The importance of the remaining materials varies according to the type of the ship.

Figure 2 characterizes the composition of the merchant vessels of the collected sample of ships. For these vessels the gross tonnages vary from 1100 ton to 13188 ton, and steel is the major constituent material (being between 79%–89% of the gross tonnage of the ship). Also, glass and dangerous solid substances are almost present in negligible percentages or are even inexistent. The proportion of each type of material, such as non-ferrous metallic materials, wood, undifferentiated and dangerous liquid substances will vary with ship type.

For merchant ships steel is the principal constituent material, lying between 79%–98%. The second most important materials are the undifferentiated materials (e.g., from furniture) and dangerous liquid substances (e.g., from sludge water and oil). Undifferentiated materials are present in a considerably amount (approximately 15%) in the bulk carrier 2 and in the reefer vessel (in approximately 7%) but have a rather negligible expression in the other vessels of this class. Dangerous liquid substances are present in a considerably amount in bulk carrier 4 and are between 1% and 3% for the other vessels. Materials such as glass, wood, non-ferrous materials and dangerous solid substances are negligible or inexistent in the composition of vessels of the figure 2.

Figure 3 shows the proportion of materials in fishing vessels of the collected sample. For these vessels, the gross tonnages vary from 101 ton to 1730 ton. Alike what happened in the merchant vessels, in the fishing vessels the steel is not always the main material and wood plays always an important role in the sense that if it is not the major material it is the second most important one. Also, undifferentiated materials from furniture, solid and liquid dangerous materials from isolation and water and oil sludge are also materials present as second major materials.

It is worth to note that in fishing vessel 2 about 30% of the removed materials were steel. Wood accounted for 60% of the gross tonnage of this ship. This was a traditional fishing vessel, a small one with about 101 ton. Undifferentiated materials and wood accounted for about 5% each and are the second major percentages of the constituent materials. Non-ferrous metallic, undifferentiated materials and glass have a rather negligible expression or are inexistent, accounting globally for 3% of the gross tonnage of the ship.

Figure 4 shows the material proportions of the auxiliary ships in the sample of ships. For this type of vessel steel is always the main constituent material. Wood among undifferentiated materials is an important constituent material in barges. Solid and liquid dangerous materials and glasses are have a rather negligible expression or are inexistent.

In figure 5 the material proportions composition of the military ships is shown. Also, for this class of ships a pattern of the constituent materials can de attributed but with some deviations. The pattern of

535

constituent materials follows the ones observed in the ones described above, being the steel the most important material with the exception of one of the ships in which steel account for only 7% approximately, being the non-metallic material the major constituent materials of that ship. The importance of the other materials for the composition of the ship differs in importance.

Figures 6–9 show the importance of the recycling of the removed materials from ships. In these figures the characterization of the removed materials is made as a function of their respective final destination, according to figure 1 above. The ships were ordered according to its gross tonnage, from the largest to the smallest one.

Figure 6 presents the allocation of the final destination of the materials removed from merchant vessels. It shows the importance of recycling. At least 85% of the materials removed have recycling as the final destination. The second most common final destination is the landfill. Nevertheless this has a relatively low expression. Bulk carrier 2 had about 15% of undifferentiated materials in its composition showing the biggest percentage of materials that goes to landfill (about 15%).

Figure 7 presents the allocation of the final destination of the materials removed from fishing vessels. Recycling is also the most important final destination

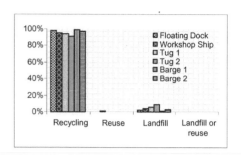

Figure 8. Characterization of the final destination of the materials removed from the auxiliary ships.

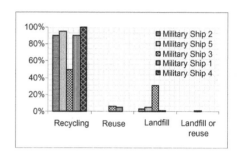

Figure 9. Characterization of the final destination of the materials removed from military ships.

of the materials removed from these vessels. At least 90% of the removed materials are recycled. Landfill has a low expression (less than 6%) due to the low proportions of polymeric and composite materials in the compositions of these ships.

In figure 8 are presented the allocation of the final destination of the materials removed from auxiliary vessels. It is seen the importance of the recycling. At least 90% of the materials removed have recycling as the final destination. The presence of undifferentiated materials as the second most important material of its composition explains the importance of the landfill as the final destination of materials removed from tugs, despite it has a rather small importance (less than 10%).

Figure 9 presents the allocation of the final destination of the materials removed from military ships. The most important feature is that in military ship 3 recycling and landfill has almost the same importance as final destinations of the materials removed from this ship. On the contrary to the other military ships from the collected sample recycling is the most important final destination. As shown in figure 5, military ship 3 had about 24% of undifferentiated materials, which was the second most constituent material, with a rather small gross tonnage, explaining the pattern observed in figure 9.

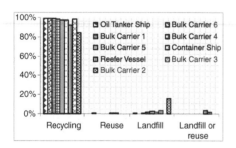

Figure 6. Characterization of the final destination of the materials removed from the merchant ships.

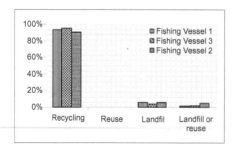

Figure 7. Characterization of the final destination of the materials removed from the fishing vessels.

3 MODELING

The dismantling activity can be thought of as a part of vessels environmental impact along its life cycle. On the other hand, the vessels environmental impact is composed of several contributions along their life cycle. According to Figure 10, assuming the life cycle of a ship is composed of three phases: ship production phase, operational phase and disposal phase, all materials/fuels and energy expenditures in each phase must be accounted for to assess environmental impacts of a vessel.

On every phase direct environmental impacts are generated: direct emission into air, soil and water mainly due to energy consumption; indirect environmental impacts are generated by materials consumption and output wastes.

In the present work, the model includes the materials/ fuels and energy expenditures needed for ship dismantling.

The environmental impact was assessed using the commercial software SimaPro v7.1.8. As impact assessment methods the following standard methods were chosen CML1992, CML 2 Baseline 2000, Ecoindicator 99 (H) and IMPACT 2002+.

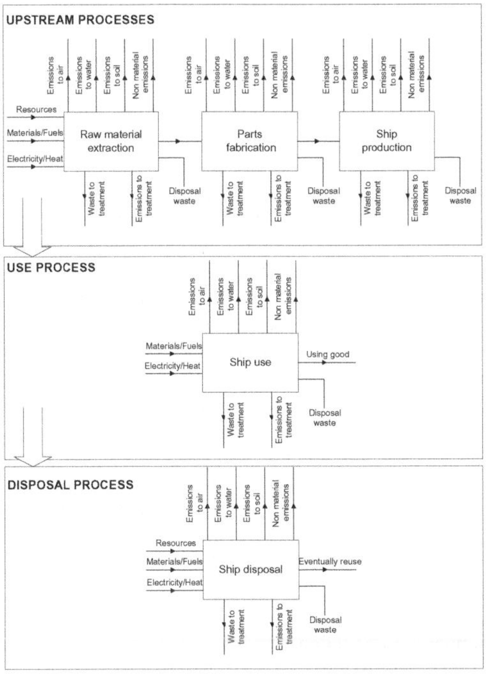

Referring to figure 11, the ISO standard allows the use of impact category indicators that are between the inventory result (i.e., the emissions) and the endpoint (i.e., major impact assessment categories). Indicators that are chosen between the inventory result and the endpoints are sometimes called as midpoint indicators. The environmental models for each impact category are extended up to endpoint level, as the impact category indicators, which relate to the same endpoint, have a common unit and as such they can be added.

CML 1992, CML 2 Baseline 2000 and IMPACT 2002+ are typical examples of midpoint methods while Eco-indicator 99 is a typical example of an endpoint method.

Eco-indicator 99 is based on its predecessor Ecoindicator 95 both of which developed under the authority of Dutch Ministry of Housing, Spatial Planning and the environment. The procedures and intermediate results are shown in figure 12. A clear distinction is made between intermediate results (the lower boxes) and the procedures to go from one intermediate result to the other.

This methodology has a top-down approach meaning that it starts by defining the required result of assessment.

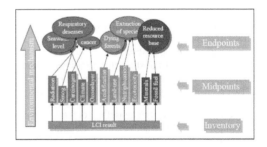

Figure 11. General overview of the structure of an impact assessment method. LCI results are characterized to produce a number of impact category indicators (Goedkoop *et al.*, 2008).

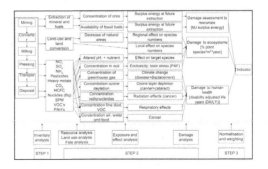

Figure 12. General overview of Eco-indicator 99 methodology (Goedkoop *et al.*, 2008).

Figure 10. Diagram of life cycle of a ship: in every phase direct emissions and environmental impacts are generated. All input materials and output wastes lead to additional indirect environmental impact.

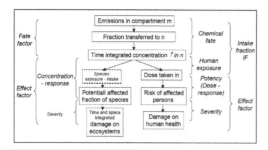

Figure 13. General overview of IMPACT 2002+ Methodology (Goedkoop *et al.*, 2008).

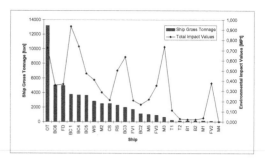

Figure 14. Total environmental impact, normalized and weighted, for the dismantling scenario and all ships, assessed by Eco-indicator 99 (H) method.

The IMPACT 2002+ (IMPact Assessment of Chemical Toxicants) estimates the cumulative toxicological risk and potential impacts associated with environmental emissions (organics and metals) to human health and ecosystem. It consists of a fate model, meaning a combined midpoint and damage approach (figure 13). All midpoint scores are expressed in units of a reference substance and related to four damage categories: human health, ecosystem quality, climate change, and resources.

For the modelling four scenarios have been considered:

- Dismantling of the ship (the real scenario), where each removed and transformed material follows for the respective final destination;
- Total recycling of the ship; this fictitious scenario allows the validation of the real scenario;
- Two scenarios for ships abandonment or sinking.

The scenarios have been modelled for the 23 considered ships, as described above. It was assumed that the composition of each material is identical for all ship types and it is identical for the different parts of ship' structure (for example, it is assumed that the structural and not structural steel compositions are identical).

To model the dismantling, transformation and transport processes, whenever possible the real consumptions were considered, but when this was not possible the database was used to obtain representation values. It was assumed, also, that the process is continuous, i.e., the effect of residues storage was not considered in the open sky during a period of time.

4 RESULTS AND DISCUSSION

As the result of the modelling, it was observed that from the four impact assessment methods used, both methods Eco-indicator 99 (H) and IMPACT 2002+ are actually sufficient and convenient to assess the environmental impacts of the proposed model scenarios and ships, since in essence they both contain all

the information provided by the other two methods, and together the information given by each other is complementary, both in qualitative terms (through the different individual and grouped categories of environmental impacts) and in quantitative terms (through different assessment approaches, as discussed above). One pitfall is that, when grouping individual impact categories into general ones, the relative importance of the generated environmental impacts is different, meaning that in the Ecoindicator 99 (H) the majority of the impacts of ship' dismantling are related to ecosystem quality while in the IMPACT 2002+ they are related to human health. Therefore, henceforward, the results are presented and discussed taking these two methods as reference.

Figure 14 shows the obtained results for the total environmental impacts, generated for the dismantling scenario, evaluated by the Eco-indicator 99 (H). In this figure the environmental impacts are values normalised and weighted; the normalization reference value is the average environmental load due to one average European inhabitant over one year; the weighting values adopted were the default values of the software. The units are reported in *points* (the basic unit of the evaluation methods, or in multiples of it). Also, henceforward, in the figures, M stands for military ship, T stands for a tug vessel, B stands for a barge, FV stands for a fishing vessel, BC stands for a bulk carrier, WS stands for a workshop vessel, CS stands for a container ship, RS stands for a refrigerant ship, OT stands for a oil tanker ship and FD stands for a floating dock.

Despite the uncertainties, both in data and in the modelling, some conclusions about the relative importance of the positive and negative environmental impacts can be taken.

It is worth noting that a positive environmental impact generates a positive environmental load so it is damaging to the environment while a negative environmental impact generates a negative environmental load so it is does not harm the environment. According to this assessment method, the respective dismantling of all 23 ships has positive environmental load. It can be concluded that the environmental impacts cannot be directly correlated to the gross tonnage of a ship, or its class since it is observed that two ships with the same gross tonnage display a non-correlated environmental load, meaning that a ship with a gross tonnage may display a lower environmental load. Instead, the relative proportion and types of the respective constituent materials are more important variables.

Consider for example the ships OT, BC6, WS, M3 and FV2, whose gross tonnages (and main constituent materials) are, respectively, and as presented above: 13188 ton ($\approx$100% steel), 5000 ton ($\approx$100% steel), 2820 ton ($\approx$90% steel), 642 ton ($\approx$50% nonferrous metals, $\approx$50% polymeric, composite and dangerous liquid substances), and 101 ton ($\approx$60% wood, $\approx$30% non-ferrous metals). Among these ships, OT has the larger gross tonnage, about two and a half times of BC6 but both have nearly 100% of steel as its major constituent material. According to both methods, the environmental impact generated by both scenarios is about the double for ship OT. Nevertheless, this linear like behaviour is not generally observed. Considering now ships BC6 and WS, the former has about twice the gross tonnage of ship WS, the same percentage of steel, and the environmental impact generated by both scenarios is about the same but having ship WS a slightly higher environmental impact.

On the other hand, comparing ship WS with ship FV2, despite the gross tonnage (differing by a factor of about 30) and composition being quite different, the environmental impacts are about the same. Within this sample, the ship M3 is the second ship with less gross tonnage but also the second with higher environmental load, for both scenarios, for this contributing the non-metallic and polymeric materials contribution as the two major classes of constituting materials.

Figure 15 shows the obtained results for the total environmental impacts, generated for the dismantling scenario, evaluated by the IMPACT 2002+.

There is a huge difference in the scales of the calculated values for both methods: environmental loads calculated by eco-indicator 99 (H) are one thousand times those calculated by IMPACT 2002+. Therefore, one can say that both methods set the upper and lower border lines, respectively, for the environmental impacts. Nevertheless, the pattern of high non-linearity between the environmental impacts and tonnage of the ship or its class it is also verified.

Comparing now the four modelled scenarios, it was found out that the recycling scenario presents a similar behaviour to the dismantling scenario. This is shown in figure 16, where the total environmental impacts from recycling and dismantling are presented for merchant ships. Merchant ships were chosen because it is the class of ships that comprises more ships, from the col-

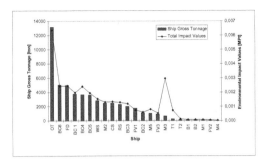

Figure 15. Total environmental impact, normalized and weighted, for the dismantling scenario and all ships, assessed by IMPACT 2002+ method.

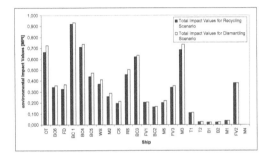

Figure 16. Total environmental impact, normalized and weighted, for the dismantling and recycling scenarios, and all ships, assessed by Eco-indicator 99 (H) method.

lected sample. In this figure, the impact values were assessed by Eco-indicator 99 (H).

The similarity of the behaviour between recycling and dismantling scenarios is not surprising since about 60%–100% of the materials/equipments removed from ships are recycled which by itself is not surprising also since the ferrous metals are the major constituent material, laying between 79,05%–98,24% for merchant vessels, 83,58%–96,25% for the auxiliary ships, 29,63%–87,48% for fishing vessels and 7,17%–100% for military ships.

For these different classes of ships, the maximum values of the calculated environmental load are linked to merchant vessels and military vessels, which are about twice the ones calculated for the fishing vessels or auxiliary ones. Also, it was observed that the abandonment and sinking scenarios have a similar behaviour. This is shown in figure 17, where the total environmental impacts from abandonment and sinking scenarios are represented for merchant ships, as well, assessed by Eco-indicator 99 (H). It is observed that the environmental impacts generated by recycling and dismantling scenarios are smaller when compares to those generated by the scenarios of abandonment and sinking.

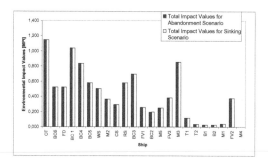

Figure 17. Total environmental impact, normalized and weighted, for the abandonment and sinking scenarios, and all ships, assessed by Eco-indicator 99 (H) method.

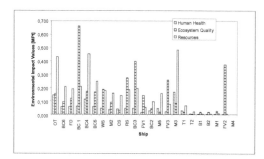

Figure 18. Environmental impact, per category of environmental impact, normalized and weighted, for the dismantling scenarios, and all ships, assessed by Eco-indicator 99 (H) method.

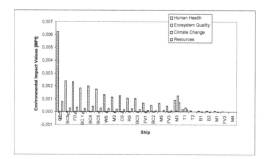

Figure 19. Environmental impact, per category of environmental impact, normalized and weighted, for the dismantling scenarios, and all ships, assessed by IMPACT 2002+.

Figures 18 and 19 show the behaviour of the environmental impacts generated for the dismantling scenario splitting the total environmental impact into general categories of impacts, assessed by Ecoindicator 99 (H) and IMPACT 2002+, respectively.

Eco-indicator 99 (H) accounts for three general categories: human health, ecosystem quality and resources. According to this method, two patterns can be defined: 1) the ecosystem quality impact category is dominant over the other two or 2) vice-versa, in each of these cases both human health and resources present similar values. As in the previous results it should be expected to find a linear like behaviour according to the tonnage of the ships. It should be expected that the scenarios for ships with either smallest and highest gross tonnage should have the biggest impacts onto resources and human health since the expected environmental effects should be quite unbalanced to dismantling small ships or quite proportional to dismantle a large vessel. But this, once more, is not verified. To both patterns contribute ships among smallest, intermediate and highest gross tonnage, which seems to reveal that also the type and proportions of materials play a role, with special importance in those ships with smallest gross tonnage.

To exemplify it consider the division of the sample of 23 ships into classes of weight such as: in class I will include all the ships with $0 \leq ton < 1000$ (i.e., ships M1, M3, M4, R1, R2, B1, B2, PS2 and PS3), class II will include all the ships with $1000 \leq ton \geq 3000$ (i.e., ships M2, M5, G2, G3, PS1, O, PC and F) and class III includes the ships with $ton \geq 3000$ (i.e., G1, G4, G5, G6, PT and DF). Taking the fishing vessel as example both smallest and interme intermediate classes (in gross tonnage basis) obey to the first pattern defined above while in the case of the merchant vessels, for instance the big oil tanker obey to the second one.

Those patterns are quite different from those of IMPACT 2002+. This method accounts for four general impact categories: human health, ecosystem quality, climate change and resources and it is observed that all classes of ships impact more onto human health impact category, with exception of some military ships, to which climate change has the biggest contribution.

Despite this, the fact is that ship dismantling activity, while not being an activity free from environmental impacts, has a contribution of minimizing those impacts through valorisation by reusing, recycling or treatment of the materials and equipments removed from those ships. A particularly interesting result is that it does not contribute for climate change when impacts are assessed by IMPACT 2002+ method as it has negative environmental load, as it is shown in figure 19. It was found out that the main individual impact category contributing for this was the negative environmental load on ozone layer's depletion, meaning that ship dismantling does not contribute for ozone layer's depletion.

5 CONCLUSIONS

The present work analyzed and discussed the environmental impacts from ship dismantling. The modelling was performed for a sample of 23 different ships, from different classes, and taking into account four different scenarios.

According to this work, the ship dismantling activity is not an activity free from environmental impacts but those impacts can be minimized through valorisation by reusing, recycling or treatment of the materials/equipments removed from those ships. In particular, the scenario of ship dismantling does not contribute for ozone layer's depletion, meaning that it has negative magnitude in this impact category. Furthermore, it was found out that the negative impacts of this scenario were smaller than the positive impacts generated by the scenarios that modelled ship abandonment or sinking.

Furthermore, it was verified in this work that there is no direct correlation between the magnitude of the environmental impacts and the gross tonnage of a ship or its class. Instead, the proportions of the constituent materials and their types play an important role in this relationship.

The model contains uncertainties, both in the data and in the modelling as well, due to the need of simplification of the problem, that were not modelled in this study.

ACKNOWLEDGMENTS

The work presented was performed within the project MARSTRUCT, funded partially by the European Commission, through the contract n° TNE3-CT-2003-506141. The second author acknowledges the financial support of the Portuguese Foundation for Science and Technology under the contract BD/31272/2006.

REFERENCES

Carvalho, I.S, Antão, P. Guedes Soares, C., 2008, "Modelling of environmental impacts of ship dismantling". *O Sector Marítimo Português*, in press (in Portuguese).

European Commission, 2007, Green paper on better ship dismantling. COM (2007) 269 final, Brussels; 2007. 38.

Islam, K.L., Hossain, M.M., 1986, Effect of ship scrapping activities on the soil and sea environment in the coastal area of Chittagong, Bangladesh, Marine *Pollution Bulletin*, Volume 17, Issue 10, October 1986, pp. 462–463.

Knapp, S., Kumar, S.N., Remijn, A.B., 2008, Econometric analysis of the ship demolition market, Marine Policy, Volume 32, Issue 6, November 2008, pp. 1023–1036.

Neser, G., Ünsalan, D., Tekoðul, N., Stuer-Lauridsen, F., 2008, The shipbreaking industry in Turkey: environmental, safety and health issues, Journal of *Cleaner Production*, Volume 16, Issue 3, February 2008, pp. 350–358.

Reddy, S.M., Basha, S., Joshi, H.V., Ramachandraiah, G., 2005, Seasonal distribution and contamination levels of total PHCs, PAHs and heavy metals in coastal waters of the Alang–Sosiya ship scrapping yard, Gulf of Cambay India, *Chemosphere*, Volume 61, Issue 11, December 2005, pp. 1587–1593.

Reddy, S.M., Basha, S., Adimurthy, S., Ramachandraiah, G., 2006, Description of the small plastics fragments in marine sediments along the Alang-Sosiya ship-breaking yard India Estuarine, Coastal and Shelf Science, Volume 68, Issues 3–4, July 2006, pp. 656–660.

Reddy, S.M., Basha, S., Kumar, V.G.S., Joshi, H.V., Ghosh, P.K., 2003, Quantification and classification of ship scraping waste at Alang–Sosiya India, Marine Pollution Bulletin, Volume 46, Issue 12, December 2003, pp. 1609–1614.

Goedkop, M., Schryver, A., Oele, M., 2008, Introduction to LCA with SimaPro 7, Pré-Consultants.

Analysis and Design of Marine Structures – Guedes Soares & Das (eds)
© *2009 Taylor & Francis Group, London, ISBN 978-0-415-54934-9*

Fuel consumption and exhaust emissions reduction by dynamic propeller pitch control

Massimo Figari
Naval Architecture & Marine Engineering Department (DINAV), University of Genoa, Italy

C. Guedes Soares
Centre for Marine Technology and Engineering (CENTEC), Technical University of Lisbon, Instituto Superior Técnico, Lisboa, Portugal

ABSTRACT: The aim of this paper is to focus attention on how a proper control strategy can save both: fuel consumption and exhaust emissions. A ship with a Controllable Pitch Propeller, driven by a diesel engine or a gas turbine, is traditionally controlled via the "combinator" that set a proper combination of pitch and shaft speed depending on the bridge telegraph (lever) position. The propulsion control based on a 'static' combinator curve does not assure the best use of the propulsion system in terms of power, consumption and emissions. The idea proposed here is to shift from the classic paradigm of 'static' combinator curve (or curves) to a 'dynamic set point' of the propulsion plant. A 'ship performance code' has been developed and used to evaluate the potential benefit of the adoption of the 'dynamic set point' control scheme with respect to the traditional 'combinator' control scheme. Some simulations have been performed and results compared with full scale data measured during normal ship service. The technological implementation seems straightforward, at the present state of the art, for what the automation (propulsion controllers) is concerned, instead, it will probably require some improvements for what the pitch control system is concerned.

1   ENVIRONMENTAL ISSUES IN SHIP DESIGN

The effectiveness of an environmental protection strategy depends on the integration of the environmental considerations in the decision-making process.

The application of this concept to maritime transport leads to the introduction of environmental issues into the ship design process (design spiral) at the same level of importance as other 'traditional' issues like stability, propulsion, strength, and so on (Benvenuto et al., 1996, 1998, Guedes Soares et al. 2007). The concept can be visualized in Figure 1.

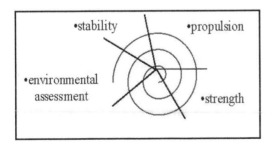

Figure 1.   Ship design spiral.

As far as sustainability concepts have to be implemented for ships and shipping activities, emissions reduction represents an important goal to be achieved.

Emission reduction can be obtained by adopting concurrent strategies: lowering emission factors (by adopting better engines and/or better fuels) and lowering the energy required by the ship (propulsion energy, electric energy, auxiliary energy).

Emission reduction produces concurrent benefits: environmental benefits (pollution reduction) and economic benefit (ship running costs reduction). For these reasons it is pursued by all the parties of the shipping community.

To share and balance the reduction strategies and to maximize the benefits, a numerical evaluation of the key parameters of the process is necessary for a given ship design.

Energy required by the ship and related exhaust emissions are considered fundamental parameters for trade-off considerations at design level. These parameters are heavily affected by ships' operational conditions, leading to the necessity of 'smart' ship control strategies to maximize the benefits in all conditions.

The aim of this paper is to focus attention on how a proper control strategy can save both: fuel consumption and exhaust emissions. Details of the adopted methods and simulation results are reported in the next paragraphs.

2 SHIP ENERGETIC BALANCE

2.1 General

One of the major aspects that influences the environmental and economical performance of a ship during her entire lifetime is the required energy because it is directly related to the burned fuel.

A common ship designers' goal is the increase of propulsion efficiency, instead, rarely they try to reduce the global amount of energy required by the ship. One of the reasons is probably because no straightforward calculation methods are available.

A scheme of ship energetic balance is reported in Figures 13 and 14 (end of the paper). The energy required by ship is the sum of the energy requirements of the different ship services and it is directly related to the fuel consumption, as shown in:

$$E = \sum_i (\dot{m}_{\text{fuel}} \cdot LHV)_i \cdot h \quad \text{ENERGY} \tag{1}$$

where:

LHV = lower heating value of the fuel
h = voyage time
$\dot{m}_{\text{fuel}}$ = fuel flow rate

To compare different ships or different transport means or different technological solutions some specific energy indexes, as the one defined as (Ben-venuto et al. 1996, 1998):

$$E_S = \frac{\text{Ship energy}}{\text{Cargo capacity} \cdot \text{Travelled distance}} \left[\frac{MJ}{t \cdot km} \right] \tag{2}$$

The precise evaluation of the energy required by the ship involves the knowledge of many parameters. The parameters related to the propulsion are well known and normally evaluated during ship design. For most ship types the propulsion represents the great part of the energy, but for passenger ships and ferries, the ship energetic balance is highly influenced by the electric and auxiliary services. Most of the parameters related to the electric energy requirement and auxiliary services requirement are, at the moment, of difficult evaluation. They were traditionally considered of scarce importance in the past and not evaluated, but nowadays their evaluation becomes of the same importance as the propulsion system. In the following a brief discussion concerning the development of a code able to evaluate propulsion and electric energy requirement is reported.

2.2 Propulsion energy

The precise evaluation of the energy required by the propulsion involves the knowledge of the ship resistance, the propeller efficiency and the prime mover specific fuel consumption. The calculation method is shown in Figure 14 (end of the paper).

Since all the above quantities are not constant, but greatly depend on the ship speed and on the enviromental conditions, a code has been developed to perform the calculations for all the interesting ship operational conditions (Altosole et al., 2007).

The results of the 'service performance code' have been tested by the service data collected onboard two ships: the Ro-Ro Pax vessel 'La Superba' and the new Italian Navy Aircraft Carrier 'Cavour'.

A number of voyages have been simulated by the code and results have been compared with the measured data (Fasce, 2008, Borlenghi, 2008). The differences between measured and calculated data are reported in Table 1 and Figure 2.

2.3 Electric energy

The precise evaluation of the electric energy required by the ship, the associated fuel consumption and exhaust emissions have become a matter of concern very recently.

Very few data are available in literature and also ship operators are not yet very familiar to record consumption data related to electric energy. The method used in the ship performance code to assess fuel consumption and exhaust emissions of electric generators

Table 1. Average differences between simulated and measured quantities.

	Propulsion power [%]	Fuel consumption [%]
RO/RO-PAX	4.1	4
Aircraft Carrier	3	3

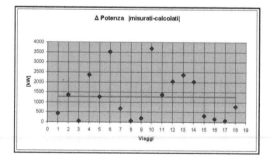

Figure 2. Difference between simulated and measured propulsion power for 18 voyages.

is very similar to the one adopted for the propulsion engines. The input data are the number of generators working at a given ship speed and the load factor of each generator. Using the engine fuel consumption maps it is straightforward to compute the fuel flow rate. As already stated above, a number of voyages have been simulated by the code and results have been compared with the measured data (Borlenghi, 2008). In particular some direct measurements have been performed to assess the in service energy requirement of the RO-RO pax La Superba in several operating conditions. At the same time the fuel flow rate related to electric generation has been measured. The average difference, among 18 voyages, between simulated and measured generators fuel consumption is 6.6%, slightly more than for propulsion.

2.4 Auxiliary services energy

The importance of this category of energy depends on ship type and also on owner's choices. For most ships the most important contribution to this category comes from fuel heating. In fact, for ship using HFO, it is necessary to maintain the fuel in the storage tanks at a temperature of about 45°C and to warm up the fuel fed to the users to a temperature around 120°C. For the heating service a low pressure steam plant is usually used. The energy required to produce the steam comes from the exhaust gas heat recovery (in this case no associated fuel consumption exists) and/or from an auxiliary boiler. Despite the importance of the heating service, very few data are available in literature and also ship operators are not yet very familiar to record consumption data related to this service.

At the moment this portion of ship energy balance is not implemented in the code, even if work is in progress in order to assess also this aspect.

3 EXHAUST EMISSIONS

The exhaust emissions are directly related to the fuel consumption and to the energy conversion systems. The emission evaluation can be assessed, in a simplified way, by using the fuel consumption and the specific emission indexes [g/kWh] that characterize the emissions of the conversion system (the engine).

The specific emissions for different exhaust components have been evaluated from the works published by Cooper (2001). Cooper measured the exhaust emissions of main engines and auxiliary engines of three ferries during normal operation. From the analysis of published data the following specific emission for diesel engines have been extracted and used for the assessment (Table 2). Cooper's data covers a wide working range of the engines, giving the opportunity to assess the exhaust emissions for different ship speeds.

Table 2. Marine diesel engines specific exhaust emissions.

	Specific emissions g/kWh	
	Engine load 53%	Engine load 90%
CO_2	688	671
NOx	14.6	11.7
CO	0.72	0.44
VOC	0.14	0.23
PM	0.29	0.10
SO_2	0.39	0.38

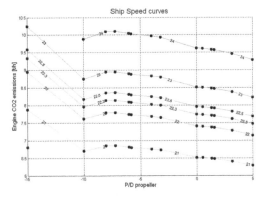

Figure 3. CO_2 exhaust emissions versus pitch.

SO_2 is mainly related to the sulfur content of the bunker fuel. To be able to assess SO_2 emission for different fuels the following approach has been adopted.

$$SO_2 = m \cdot S \cdot (1 - SO_4) \cdot \frac{64}{32} \quad (3)$$

where:

$SO_2 = SO_2$ exhaust emissions [t]
$SO_4 = 0.02247$
m = fuel mass [t]
S = sulfur content (generally $0.015 - 0.045$)

Carbon dioxide CO_2 is related to the quantity of burnt fuel, and IMO suggests the use of the following relationship:

$$CO_2 = m \cdot C \cdot 3.664 \quad (4)$$

where:

$CO_2 = CO_2$ exhaust emissions [t]
m = fuel mass [t]
$C = 0.85$ HFO, 0.875 MDO

In Figure 3 an example of code output is presented. The figure shows the CO_2 flow rate [t/h] with respect to propeller pitch setting [P/D], each curve represent a ship speed (from 21 to 24 knots).

The exhaust emission assessment, available as an output of the 'service performance code', is not validated yet, due to the difficulties of the measurements of the exhaust emission onboard ships in service.

The study presented in this section describes the usefulness of this kind of assessment from the point of view of minimization of fuel consumption and exhaust emissions of the ships. In fact from the figure above it is possible to identify the pitch setting (P/D) that minimize the CO_2 exhaust emissions in air.

4 PROPULSION CONTROL

A ship with a Controllable Pitch Propeller, driven by a diesel engine or a gas turbine, is traditionally controlled via the "combinator" that set a proper combination of propeller pitch and shaft speed depending on the bridge telegraph (lever) position. The scheme is reported in Figure 4.

The 'proper' combination of pitch and revolution is generally evaluated at design condition ("ideal" condition) with no waves and other added resistance conditions.

Generally the combinator is optimized at the design speed (i.e. 'full ahead' position of the telegraph) and the "optimum" is influenced by fuel consumption considerations and also by propeller noise and cavitation issues. For ship speed different from design speed the 'proper' combination very often is not calculated but it is guessed from experience; normally the 'combinator' curve is represented by a linear relationship between the design point and zero speed.

In contrast to ideal conditions, ships at sea normally operate at various draft, at different operational speeds, with different hull and propeller cleanness, with all environmental conditions. Vessels with multiple engines can also operate in different modes, depending on the number of engines simultaneously in operation.

It is straightforward to realize that, since each 'real' condition is associated to a particular resistance curve, a particular propeller performance curve, a particular number of engines operating simultaneously, the 'optimum' working point of the propulsion system is different for each experienced condition.

The propulsion control based on a 'static' combinator curve does not assure the best use of the propulsion system in terms of power, consumption and emissions.

The idea proposed here is to shift from the classic paradigm of 'static' combinator curve (or curves) to a 'dynamic set point' of the propulsion plant. To implement the idea some technological improvements in the pitch control systems will be probably necessary (Godjevac, 2008) while it seems straight forward, at the present state of the art, for what concerns the propulsion controllers due to the high performances of the available industrial PLC (Figari et al., 2008).

4.1 Dynamic set point

The evaluation of the optimal set point of the propulsion system requires a code that is able to calculate, in real time, the shaft speed and the propeller pitch that satisfy a minimum of an objective function. To reduce air pollution from ships valuable objective functions may be: fuel consumption flow rate, total fuel consumption over a voyage, emission components (CO_2, SO_2, etc) flow rate, total emissions over a voyage, or a combination of above.

Figure 5 shows a particular output of the code that allows to identify the minimum fuel consumption for each ship speed. The propeller pitch is represented on the x-axis, the fuel flow rate is represented on the y-axis, each curve represents the fuel consumption at a particular ship speed. The circles identify the optimum pitch (with respect to fuel consumption only) for all the considered ship speeds. Optimum pitch is related to shaft revolutions via the relationship presented in Figure 6.

To be adopted as a set point, the values 'optimum pitch' and 'related shaft revolution must produce a 'required power' that lay inside the engine load diagram. To check this condition Figure 7 can be used. The figure is a different way to represent the system working points: the engine load diagram and the propeller power, at different pitch settings and at different ship speeds, are represented in terms of fuel consumption and shaft speed.

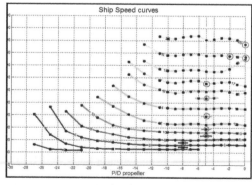

Figure 5. Fuel consumption versus pitch at constant speed, optimum pitch identification.

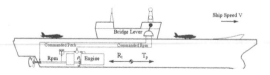

Figure 4. Propulsion control scheme.

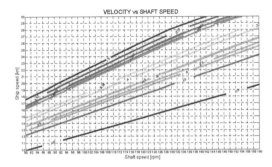

Figure 6.   Ship speed versus shaft speed at constant pitch.

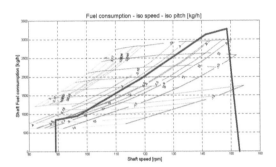

Figure 7.   Fuel flow rate versus shaft speed at constant pitch and speed.

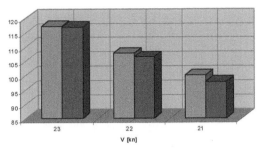

Figure 8.   Fuel Consumption: 'optimum set point' (magenta) and 'combinator' (blue).

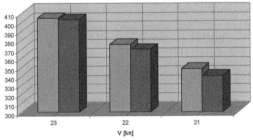

Figure 9. CO_2 emissions: 'optimum set point' (magenta) and 'combinator' (blue).

5   SIMULATION RESULTS

The 'ship performance code' has been used to evaluate the potential benefit of the adoption of the 'dynamic set point' control scheme with respect to the traditional 'combinator' control scheme. To achieve the objective some simulations have been performed and results compared with full scale data measured during normal ship service.

5.1   Genova-Palermo regular service

The regular service between Genova and Palermo is operated by GNV (Grandi Navi Veloci) with M/V La Superba. The records of the voyages in the years 2005-2007 were made available by the operator; in the same period the DINAV research group made 4 voyages to directly measure machinery working data.

Among all, a number of 18 voyages have been selected due to their similarity in operating conditions (weather, displacement, speed).

First, each voyage has been simulated by the 'ship performance code' in order to quantify fuel consumption and exhaust emissions. The calculated fuel consumption (using the 'combinator' law installed onboard) has been compared with the measured one,

giving an average difference of 4% for the main engines and 6.6% of the diesel generators (as previously explained in paragraphs 2.2 and 2.3).

For each voyage, simulations have been performed to identify the 'optimum' combination of pitch and shaft speed. Figure 8 shows a comparison between the fuel consumption using the ship 'combinator' and using the 'optimum set point', the average fuel saving is about 1%.

The same trend can be found in CO_2 emissions, as shown in Figure 9.

5.2   Moderate weather conditions

To verify the benefit of the 'dynamic set point' some weather conditions have been simulated by increasing the ship resistance. For each weather condition and for each speed the 'optimum set point' has been identified, the related fuel consumption has been compared with the fuel consumption calculated with the 'combinator'.

The results in terms of fuel consumption are reported in Figure 10. The figure shows, the fuel saved per voyage between the 'optimum set point' and the 'combinator' for different ship speed and for different sea conditions (calm sea, added resistance 15% and 20%). Similar results have been obtained for the exhaust emissions.

For this ship the benefit is maintained also with moderate weather conditions, and the benefit increases

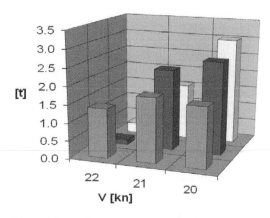

Figure 10. Fuel save between 'optmimum set point' and 'combinator', calm sea(blu), 15% added resistance (magenta), 20% added resistance (white).

Table 3. Considered added resistance versus sea state.

$H_{1/3}$ [m]	Proba-bility	Sea state	Added resistance	Number voyages
0 m < $H_{1/3}$ < 1	0.345	0–3	0%	99
1 m < $H_{1/3}$ < 2	0.357	4–5	15%	103
$H_{1/3}$ > 2	0.298	>5	Not considered	86

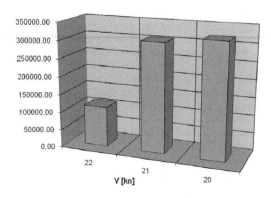

Figure 11. Estimated fuel saving over 1 year between 'optimum set point' and 'combinator'.

at the lower speed where the 'shape' of the engine load diagram is wider.

5.3 Benefit over one year of operation

To quantify the benefits in terms of operating costs, the fuel saving figure referred to 1 voyage has been shared over a reference operating period, in this case 1 year.

Through the analysis of the weather statistics of the Western Mediterranean area, the probability of

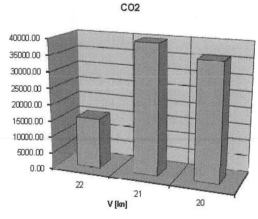

Figure 12. Estimated exhaust emissions cost saving between 'optimum set point' and 'combinator'.

occurrence of the significant wave heights have been inferred. The wave height has been correlated with a sea state and the sea state has been correlated to an added resistance, the latter step being somewhat arbitrary.

Considering a fuel price of 500 [$/t], the estimate fuel saving adopting the 'optimum set point' for each voyage is reported in Figure 11.

The cost reduction is not only related to the fuel cost, but it is also related to the reduction of exhaust emissions. In fact there are States around the world where the ship exhaust emissions are subjected to a taxation fee. Moreover, IMO is considering the possibility to adopt a taxation scheme on CO_2 emissions.

In order to quantify the benefit of the 'optimum set point' control scheme due to reduction of exhaust emissions, the following hypotheses have been adopted: CO_2 emission fee 23 [€/t], NOx emission fee 2 [€/t]. Results have been reported in Figure 12.

6 CONCLUSIONS

The aim of the society and of the shipping community is to have new ships with reduced fuel consumption and reduced exhaust emissions as compared to the past. To accommodate this aim design procedures are necessary but not yet completely available, in particular the evaluation of the energy balance and the prediction of exhaust emissions are becoming important issues in ship design.

A proper control strategy can assure that the benefits related to fuel saving and exhaust emissions reduction would be effective over the entire operational profile and would not be confined to the 'design' condition.

In this paper the idea of the 'dynamic set point', to control the propulsion, is presented to give matter of discussion to the interested parties.

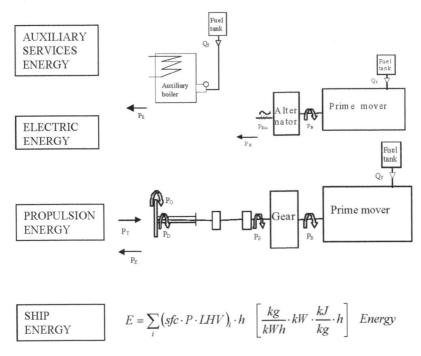

$$E = \sum_{i}\left(sfc \cdot P \cdot LHV\right)_{i} \cdot h \quad \left[\frac{kg}{kWh} \cdot kW \cdot \frac{kJ}{kg} \cdot h\right] \quad Energy$$

Figure 13.   Ship energetic balance.

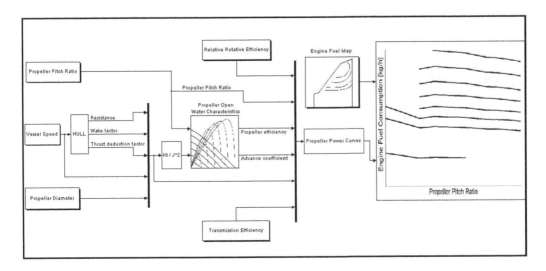

Figure 14.   Propulsion energy evaluation.

The practical implementation of the idea is absolutely feasible from the 'controller' point of view, instead, it is not evaluated yet from the pitch hydraulic system point of view. Some problems to be evaluated may arise from the precision required in the pitch setting.

ACKNOWLEDGMENTS

This research is co-sponsored by European Union through Six Framework Programme, MARSTRUCT Project, Task 6.5, and contract number TNE3–CT–2003-506141.

REFERENCES

Altosole, M., Borlenghi, M., Capasso, M. & Figari, M. 2007. "Computer-Based design tool for a fuel efficient—low emissions marine propulsion plant", *Proceedings of ICMRT 2007*, Ischia, Italy.

Benvenuto, G., Figari, M., Migliaro, C. & Rossi, E. 1996. "Environmental impacts of land and maritime transports in urban areas". *2nd International Conference on Urban Transport & the Environment in the 21stCentury*, Barcelona, Spain.

Benvenuto, G. & Figari, M. 1998. "Environmental Impact Assessment of Short Sea Shipping", *Transactions SNAME*, Vol. 105.

Borlenghi, M., Figari, M., Carvalho, I.S. & Guedes Soares, C. 2007. "Modelling and assessment of ferries' environmental impacts", Maritime Industry, Ocean Engineering and Coastal Resources, pp. 1135–1144.

Borlenghi, M. 2008. "*Environmental impact assessment and mitigation studies for RO/RO & passenger vessels*", PhD Thesis (in Italian), Supervisor: M. Figari.

Cooper, D.A. 2001. "Exhaust emissions from high speed passenger ferries", *Atmospheric Environment*, n°. 35, Pergamon.

Fasce, L. 2008. "*Computer code for the minimisation of fuel consumption and exhaust emissions of ships*", MSc Thesis in Naval Architecture and Marine Engineering (in Italian), Supervisor: M. Figari.

Figari, M., Altosol, A., Benvenuto, G., Campora, U., Bagnasco, A., D'Arco, S., Giuliano, M., Giuffra, V., Spadoni, A., Zanichelli, A., Michetti, S., Ratto, M. 2008. "Real Time simulation of the propulsion plant dynamic behavior of the aircraft carrier Cavour", *Proceedings of 9th International Naval Engineering Conference (INEC 2008)*, Hamburg, Germany.

Godjevac, M. 2008. "Towards high performance pitch control", *Wartsila Technical Journal*, 1.

Analysis and Design of Marine Structures – Guedes Soares & Das (eds)
© 2009 Taylor & Francis Group, London, ISBN 978-0-415-54934-9

Author index

Alan Klanac, 447
Albert Ku, 469
Aleksi Laakso, 133
Alexander N. Minaev, 207
Alexandru Ioan, 53
Amdahl, J. 305, 345
Anders Ulfvarson, 199
Antão, P. 533
Anyfantis, K. 379
Anyfantis, K.N. 387
Apurv Bansal, 113

Bart de Leeuw, 261
Beiqing Huang, 469
Benson, S. 121
Berggreen, C. 379, 403
Boyd, S. 379
Brizzolara, S. 13

Cabos, C. 93
Carvalho, I.S. 533
Cesare Mario Rizzo, 333
Chen, Y. 13
Chun, M.J. 315
Chun, S.E. 301
Claude Daley, 113, 139
Clemens Schiff, 503
Couty, N. 13
Czujko, J. 315

Dario Boote, 437, 67
Das, P.K. 323, 393, 523
Delarche, A. 403
Diebold, L. 13
Donatella Mascia, 437
Douka, C. 403
Dow, R. 379
Dow, R.S. 121, 403
Downes, J. 121, 379, 403
Drake, K.R. 27

Elena-Felicia Beznea, 423, 429
Enrico Rizzuto, 483

Feargal P. Brennan, 261
Feltz, O. 247
Fernandez Francisco Aracil, 365
Francesco Cecchini, 67
Fricke, W. 247

Garbatov, Y. 267
Garganidis, G.S. 413
Garrè, L. 483
Gaspar, B. 267
Giovanni Carrera, 333
Guedes Soares, C. 145, 193, 215,
    231, 267, 293, 457, 495, 513,
    533, 543

Ha, M.K. 301
Hashim, S. 379
Hashim, S.A. 393
Hayman, B. 379, 403
Hoflack, S. 13
Hubertus von Selle, 255
Hussein, A.W. 513

Igor Skalski, 193
Ingmar Pill, 103
Ionel Chirica, 53, 379, 423, 429

Jacob Abraham, 139
Jang, Y.S. 315
Jani Romanoff, 133
Jasmin Jelovica, 447
Jean David Caprace, 365
Jelena Vidic-Perunovic, 37
Jeong, J.S. 315
Jing Tang Xing, 83
Jörg Peschmann, 503
Jørgen Amdahl, 163
Jørgen Juncher Jensen, 37
José Varela, 457
Joško Parunov, 495
Juin, E. 379

Kim, B.J. 301, 315
Kim, D.K. 181
Kim, G.S. 315
Kim, M.S. 181
Kim, S.H. 315
Kim, T.H. 301
Kim, Y.S. 315
Klas Vikgren, 199
Kristjan Tabri, 103
Kyriakongonas, A.P. 371

Lars Børsheim, 345
Leira, B.J. 345

Leonard Domnisoru, 53
Lundsgaard-Larsen, C. 403
Lyuben D. Ivanov, 469

Maciej Taczala, 155
Malenica, Š. 45
Manfred Scharrer, 255
Manuel Ventura, 457
Marc Wilken, 93
Maro Ćorak, 495
Masanobu Nishimoto, 279
Massimo Figari, 543
Matteo Paci, 333
McGeorge, D. 379
Md. Mobesher Ahmmad,
    223
Mika Bäckström, 239
Mingyi Tan, 83
Misirlis, K. 379, 403
Moirod, N. 13
Moore, P. 379

Natalie A. Gladkova, 207
Nguyen, T.-H. 305
Nicolas Losseau, 365
Nisar, J. 379
Nisar, J.A. 393

Of, G. 93
Øivind Espeland, 345
Olaf Doerk, 255
Orsolini, A. 379

Pahos, S.J. 323
Paik, J.K. 181, 301, 315
Papazoglou, V.J. 371
Pawel Domzalicki, 193
Petri Varsta, 133
Philippa Moore, 357
Philippe Rigo, 365

Qing-hai Du, 77
Quispitupa, A. 379

Raluca Chirica, 423, 429

Saad-Eldeen, S. 231
Samuelides, M.S. 305

Savio, L. 13
Schipperen, I. 45
Seppo Kivimaa, 239
Sergey V. Gnedenkov, 207
Shin, Y.S. 315
Sireta, F.X. 45
Smith, T.W.P. 27
Souto Iglesias, A. 13
Steinback, O. 93
Suh, Y.S. 301
Sutherland, L.S. 293

Temarel, P. 13
Toftegaard, H.L. 403
Tomašević, S. 45

Torgeir Moan, 3, 173
Toulios, M. 305
Tsouvalis, N. 379, 403
Tsouvalis, N.G. 387, 413
Tuitman, J.T. 45

Viktor Wolf, 503
Viviane C.S. Krzonkala, 469
Viviani, M. 13
Vladimir V. Goriaynov, 207

Wei-cheng Cui, 77
Witkowska, M. 145
Woo, J.H. 315
Wrobel, P. 27

Xiong Liang Yao, 523

Yang, N. 523
Yasumi Kawamura, 279
Ye Ping Xiong, 83
Yoichi Sumi, 223, 279
Yordan Garbatov, 193, 215

Zheng-quan Wan, 77
Zhenhui Liu, 163
Zhi Shu, 3, 173
Zilakos, I. 305